Categorical Data Analysis

Categorical Data Analysis

Second Edition

ALAN AGRESTI

University of Florida
Gainesville, Florida

WILEY-
INTERSCIENCE

A JOHN WILEY & SONS, INC., PUBLICATION

For ordering and customer service, call 1-800-CALL-WILEY.

Library of Congress Cataloging-in-Publication Data Is Available

ISBN 0-471-36093-7

Printed in the United States of America

10 9 8 7 6 5 4 3 2

To Jacki

Contents

*Sections marked with an asterisk are less important for an overview.

Preface

The explosion in the development of methods for analyzing categorical data that began in the 1960s has continued apace in recent years. This book provides an overview of these methods, as well as older, now standard, methods. It gives special emphasis to generalized linear modeling techniques, which extend linear model methods for continuous variables, and their extensions for multivariate responses.

Today, because of this development and the ubiquity of categorical data in applications, most statistics and biostatistics departments offer courses on categorical data analysis. This book can be used as a text for such courses. The material in Chapters 1–7 forms the heart of most courses. Chapters 1–3 cover distributions for categorical responses and traditional methods for two-way contingency tables. Chapters 4–7 introduce logistic regression and related logit models for binary and multicategory response variables. Chapters 8 and 9 cover loglinear models for contingency tables. Over time, this model class seems to have lost importance, and this edition reduces somewhat its discussion of them and expands its focus on logistic regression.

In the past decade, the major area of new research has been the development of methods for repeated measurement and other forms of clustered categorical data. Chapters 10–13 present these methods, including marginal models and generalized linear mixed models with random effects. Chapters 14 and 15 present theoretical foundations as well as alternatives to the maximum likelihood paradigm that this text adopts. Chapter 16 is devoted to a historical overview of the development of the methods. It examines contributions of noted statisticians, such as Pearson and Fisher, whose pioneering efforts—and sometimes vocal debates—broke the ground for this evolution.

Every chapter of the first edition has been extensively rewritten, and some substantial additions and changes have occurred. The major differences are:

- A new Chapter 1 that introduces distributions and methods of inference for categorical data.
- A unified presentation of models as special cases of generalized linear models, starting in Chapter 4 and then throughout the text.

xiii

- Greater emphasis on logistic regression for binary response variables and extensions for multicategory responses, with Chapters 4–7 introducing models and Chapters 10–13 extending them for clustered data.
- Three new chapters on methods for clustered, correlated categorical data, increasingly important in applications.
- A new chapter on the historical development of the methods.
- More discussion of "exact" small-sample procedures and of conditional logistic regression.

In this text, I interpret *categorical data analysis* to refer to methods for categorical response variables. For most methods, explanatory variables can be qualitative or quantitative, as in ordinary regression. Thus, the focus is intended to be more general than contingency table analysis, although for simplicity of data presentation, most examples use contingency tables. These examples are often simplistic, but should help readers focus on understanding the methods themselves and make it easier for them to replicate results with their favorite software.

Special features of the text include:

- More than 100 analyses of "real" data sets.
- More than 600 exercises at the end of the chapters, some directed towards theory and methods and some towards applications and data analysis.
- An appendix that shows, by chapter, the use of SAS for performing analyses presented in this book.
- Notes at the end of each chapter that provide references for recent research and many topics not covered in the text.

Appendix A summarizes statistical software needed to use the methods described in this text. It shows how to use SAS for analyses included in the text and refers to a web site (www.stat.ufl.edu/ ~ aa/cda/cda.html) that contains (1) information on the use of other software (such as R, S-plus, SPSS, and Stata), (2) data sets for examples in the form of complete SAS programs for conducting the analyses, (3) short answers for many of the odd-numbered exercises, (4) corrections of errors in early printings of the book, and (5) extra exercises. I recommend that readers refer to this appendix or specialized manuals while reading the text, as an aid to implementing the methods.

I intend this book to be accessible to the diverse mix of students who take graduate-level courses in categorical data analysis. But I have also written it with practicing statisticians and biostatisticians in mind. I hope it enables them to catch up with recent advances and learn about methods that sometimes receive inadequate attention in the traditional statistics curriculum.

The development of new methods has influenced—and been influenced by—the increasing availability of data sets with categorical responses in the social, behavioral, and biomedical sciences, as well as in public health, human genetics, ecology, education, marketing, and industrial quality control. And so, although this book is directed mainly to statisticians and biostatisticians, I also aim for it to be helpful to methodologists in these fields.

Readers should possess a background that includes regression and analysis of variance models, as well as maximum likelihood methods of statistical theory. Those not having much theory background should be able to follow most methodological discussions. Sections and subsections marked with an asterisk are less important for an overview. Readers with mainly applied interests can skip most of Chapter 4 on the theory of generalized linear models and proceed to other chapters. However, the book has distinctly higher technical level and is more thorough and complete than my lower-level text, *An Introduction to Categorical Data Analysis* (Wiley, 1996).

I thank those who commented on parts of the manuscript or provided help of some type. Special thanks to Bernhard Klingenberg, who read several chapters carefully and made many helpful suggestions, Yongyi Min, who constructed many of the figures and helped with some software, and Brian Caffo, who helped with some examples. Many thanks to Rosyln Stone and Brian Marx for each reviewing half the manuscript and Brian Caffo, I-Ming Liu, and Yongyi Min for giving insightful comments on several chapters. Thanks to Constantine Gatsonis and his students for using a draft in a course at Brown University and providing suggestions. Others who provided comments on chapters or help of some type include Patricia Altham, Wicher Bergsma, Jane Brockmann, Brent Coull, Al DeMaris, Regina Dittrich, Jianping Dong, Herwig Friedl, Ralitza Gueorguieva, James Hobert, Walter Katzenbeisser, Harry Khamis, Svend Kreiner, Joseph Lang, Jason Liao, Mojtaba Ganjali, Jane Pendergast, Michael Radelet, Kenneth Small, Maura Stokes, Tom Ten Have, and Rongling Wu. I thank my co-authors on various projects, especially Brent Coull, Joseph Lang, James Booth, James Hobert, Brian Caffo, and Ranjini Natarajan, for permission to use material from those articles. Thanks to the many who reviewed material or suggested examples for the first edition, mentioned in the Preface of that edition. Thanks also to Wiley Executive Editor Steve Quigley for his steadfast encouragement and facilitation of this project. Finally, thanks to my wife Jacki Levine for continuing support of all kinds, despite the many days this work has taken from our time together.

ALAN AGRESTI

Gainesville, Florida
November 2001

CHAPTER 1

Introduction: Distributions and Inference for Categorical Data

From helping to assess the value of new medical treatments to evaluating the factors that affect our opinions and behaviors, analysts today are finding myriad uses for categorical data methods. In this book we introduce these methods and the theory behind them.

Statistical methods for categorical responses were late in gaining the level of sophistication achieved early in the twentieth century by methods for continuous responses. Despite influential work around 1900 by the British statistician Karl Pearson, relatively little development of models for categorical responses occurred until the 1960s. In this book we describe the early fundamental work that still has importance today but place primary emphasis on more recent modeling approaches. Before outlining the topics covered, we describe the major types of categorical data.

1.1 CATEGORICAL RESPONSE DATA

A *categorical variable* has a measurement scale consisting of a set of categories. For instance, political philosophy is often measured as liberal, moderate, or conservative. Diagnoses regarding breast cancer based on a mammogram use the categories normal, benign, probably benign, suspicious, and malignant.

The development of methods for categorical variables was stimulated by research studies in the social and biomedical sciences. Categorical scales are pervasive in the social sciences for measuring attitudes and opinions. Categorical scales in biomedical sciences measure outcomes such as whether a medical treatment is successful.

Although categorical data are common in the social and biomedical sciences, they are by no means restricted to those areas. They frequently occur in the behavioral sciences (e.g., type of mental illness, with the categories schizophrenia, depression, neurosis), epidemiology and public health (e.g., contraceptive method at last intercourse, with the categories none, condom, pill, IUD, other), genetics (type of allele inherited by an offspring), zoology (e.g., alligators' primary food preference, with the categories fish, invertebrate, reptile), education (e.g., student responses to an exam question, with the categories correct and incorrect), and marketing (e.g., consumer preference among leading brands of a product, with the categories brand A, brand B, and brand C). They even occur in highly quantitative fields such as engineering sciences and industrial quality control. Examples are the classification of items according to whether they conform to certain standards, and subjective evaluation of some characteristic: how soft to the touch a certain fabric is, how good a particular food product tastes, or how easy to perform a worker finds a certain task to be.

Categorical variables are of many types. In this section we provide ways of classifying them and other variables.

1.1.1   Response–Explanatory Variable Distinction

Most statistical analyses distinguish between *response* (or *dependent*) *variables* and *explanatory* (or *independent*) *variables*. For instance, regression models describe how the mean of a response variable, such as the selling price of a house, changes according to the values of explanatory variables, such as square footage and location. In this book we focus on methods for categorical response variables. As in ordinary regression, explanatory variables can be of any type.

1.1.2   Nominal–Ordinal Scale Distinction

Categorical variables have two primary types of scales. Variables having categories without a natural ordering are called *nominal*. Examples are religious affiliation (with the categories Catholic, Protestant, Jewish, Muslim, other), mode of transportation to work (automobile, bicycle, bus, subway, walk), favorite type of music (classical, country, folk, jazz, rock), and choice of residence (apartment, condominium, house, other). For nominal variables, the order of listing the categories is irrelevant. The statistical analysis does not depend on that ordering.

Many categorical variables *do* have ordered categories. Such variables are called *ordinal*. Examples are size of automobile (subcompact, compact, midsize, large), social class (upper, middle, lower), political philosophy (liberal, moderate, conservative), and patient condition (good, fair, serious, critical). Ordinal variables have ordered categories, but distances between categories are unknown. Although a person categorized as moderate is more liberal than a person categorized as conservative, no numerical value describes *how much more* liberal that person is. Methods for ordinal variables utilize the category ordering.

An *interval variable* is one that *does* have numerical distances between any two values. For example, blood pressure level, functional life length of television set, length of prison term, and annual income are interval variables. (An internal variable is sometimes called a *ratio variable* if ratios of values are also valid.)

The way that a variable is measured determines its classification. For example, "education" is only nominal when measured as public school or private school; it is ordinal when measured by highest degree attained, using the categories none, high school, bachelor's, master's, and doctorate; it is interval when measured by number of years of education, using the integers $0, 1, 2, \ldots$.

A variable's measurement scale determines which statistical methods are appropriate. In the measurement hierarchy, interval variables are highest, ordinal variables are next, and nominal variables are lowest. Statistical methods for variables of one type can also be used with variables at higher levels but not at lower levels. For instance, statistical methods for nominal variables can be used with ordinal variables by ignoring the ordering of categories. Methods for ordinal variables cannot, however, be used with nominal variables, since their categories have no meaningful ordering. It is usually best to apply methods appropriate for the actual scale.

Since this book deals with categorical responses, we discuss the analysis of nominal and ordinal variables. The methods also apply to interval variables having a small number of distinct values (e.g., number of times married) or for which the values are grouped into ordered categories (e.g., education measured as < 10 years, $10-12$ years, > 12 years).

1.1.3 Continuous–Discrete Variable Distinction

Variables are classified as *continuous* or *discrete*, according to the number of values they can take. Actual measurement of all variables occurs in a discrete manner, due to precision limitations in measuring instruments. The continuous–discrete classification, in practice, distinguishes between variables that take lots of values and variables that take few values. For instance, statisticians often treat discrete interval variables having a large number of values (such as test scores) as continuous, using them in methods for continuous responses.

This book deals with certain types of discretely measured responses: (1) nominal variables, (2) ordinal variables, (3) discrete interval variables having relatively few values, and (4) continuous variables grouped into a small number of categories.

1.1.4 Quantitative–Qualitative Variable Distinction

Nominal variables are *qualitative*—distinct categories differ in quality, not in quantity. Interval variables are *quantitative*—distinct levels have differing amounts of the characteristic of interest. The position of ordinal variables in the quantitative–qualitative classification is fuzzy. Analysts often treat them as qualitative, using methods for nominal variables. But in many respects, ordinal variables more closely resemble interval variables than they resemble nominal variables. They possess important quantitative features: Each category has a *greater* or *smaller* magnitude of the characteristic than another category; and although not possible to measure, an underlying continuous variable is usually present. The political philosophy classification (liberal, moderate, conservative) crudely measures an inherently continuous characteristic.

Analysts often utilize the quantitative nature of ordinal variables by assigning numerical scores to categories or assuming an underlying continuous distribution. This requires good judgment and guidance from researchers who use the scale, but it provides benefits in the variety of methods available for data analysis.

1.1.5 Organization of This Book

The models for categorical response variables discussed in this book resemble regression models for continuous response variables; however, they assume binomial, multinomial, or Poisson response distributions instead of normality. Two types of models receive special attention, logistic regression and loglinear models. Ordinary *logistic regression models*, also called *logit models*, apply with *binary* (i.e., two-category) responses and assume a binomial distribution. Generalizations of logistic regression apply with multicategory responses and assume a multinomial distribution. *Loglinear models* apply with count data and assume a Poisson distribution. Certain equivalences exist between logistic regression and loglinear models.

The book has four main units. In the first, Chapters 1 through 3, we summarize descriptive and inferential methods for univariate and bivariate categorical data. These chapters cover discrete distributions, methods of inference, and analyses for measures of association. They summarize the non-model-based methods developed prior to about 1960.

In the second and primary unit, Chapters 4 through 9, we introduce models for categorical responses. In Chapter 4 we describe a class of *generalized linear models* having models of this text as special cases. We focus on models for binary and count response variables. Chapters 5 and 6 cover the most important model for binary responses, logistic regression. In Chapter 7 we present generalizations of that model for nominal and ordinal multicategory response variables. In Chapter 8 we introduce the modeling of multivariate categorical response data and show how to represent association and interaction patterns by loglinear models for counts in the table that cross-classifies those responses. In Chapter 9 we discuss model building with loglinear and related logistic models and present some related models.

In the third unit, Chapters 10 through 13, we discuss models for handling repeated measurement and other forms of clustering. In Chapter 10 we present models for a categorical response with matched pairs; these apply, for instance, with a categorical response measured for the same subjects at two times. Chapter 11 covers models for more general types of repeated categorical data, such as longitudinal data from several times with explanatory variables. In Chapter 12 we present a broad class of models, *generalized linear mixed models*, that use random effects to account for dependence with such data. In Chapter 13 further extensions and applications of the models from Chapters 10 through 12 are described.

The fourth and final unit is more theoretical. In Chapter 14 we develop asymptotic theory for categorical data models. This theory is the basis for large-sample behavior of model parameter estimators and goodness-of-fit statistics. Maximum likelihood estimation receives primary attention here and throughout the book, but Chapter 15 covers alternative methods of estimation, such as the Bayesian paradigm. Chapter 16 stands alone from the others, being a historical overview of the development of categorical data methods.

Most categorical data methods require extensive computations, and statistical software is necessary for their effective use. In Appendix A we discuss software that can perform the analyses in this book and show the use of SAS for text examples. See the Web site *www.stat.ufl.edu/~aa/cda/cda.html* to download sample programs and data sets and find information about other software.

Chapter 1 provides background material. In Section 1.2 we review the key distributions for categorical data: the binomial, multinomial, and Poisson. In Section 1.3 we review the primary mechanisms for statistical inference, using maximum likelihood. In Sections 1.4 and 1.5 we illustrate these by presenting significance tests and confidence intervals for binomial and multinomial parameters.

1.2 DISTRIBUTIONS FOR CATEGORICAL DATA

Inferential data analyses require assumptions about the random mechanism that generated the data. For regression models with continuous responses, the normal distribution plays the central role. In this section we review the three key distributions for categorical responses: *binomial, multinomial,* and *Poisson.*

1.2.1 Binomial Distribution

Many applications refer to a fixed number n of binary observations. Let $y_1, y_2, \ldots, y_n$ denote responses for n independent and identical trials such that $P(Y_i = 1) = \pi$ and $P(Y_i = 0) = 1 - \pi$. We use the generic labels "success" and "failure" for outcomes 1 and 0. *Identical trials* means that the probability of success π is the same for each trial. *Independent trials* means that the $\{Y_i\}$ are independent random variables. These are often called *Bernoulli trials*. The total number of successes, $Y = \sum_{i=1}^{n} Y_i$, has the *binomial distribution* with index n and parameter π, denoted by bin(n,π).

The probability mass function for the possible outcomes y for Y is

$$p(y) = \binom{n}{y}\pi^y(1 - \pi)^{n-y}, \qquad y = 0,1,2,\ldots,n, \tag{1.1}$$

where the binomial coefficient $\binom{n}{y} = n!/[y!\,(n-y)!]$. Since $E(Y_i) = E(Y_i^2)$ $= 1 \times \pi + 0 \times (1 - \pi) = \pi$,

$$E(Y_i) = \pi \quad \text{and} \quad \text{var}(Y_i) = \pi(1 - \pi).$$

The binomial distribution for $Y = \sum_i Y_i$ has mean and variance

$$\mu = E(Y) = n\pi \quad \text{and} \quad \sigma^2 = \text{var}(Y) = n\pi(1 - \pi).$$

The skewness is described by $E(Y - \mu)^3/\sigma^3 = (1 - 2\pi)/\sqrt{n\pi(1 - \pi)}$. The distribution converges to normality as n increases, for fixed π.

There is no guarantee that successive binary observations are independent or identical. Thus, occasionally, we will utilize other distributions. One such case is sampling binary outcomes without replacement from a finite population, such as observations on gender for 10 students sampled from a class of size 20. The *hypergeometric distribution*, studied in Section 3.5.1, is then relevant. In Section 1.2.4 we mention another case that violates these binomial assumptions.

1.2.2 Multinomial Distribution

Some trials have more than two possible outcomes. Suppose that each of n independent, identical trials can have outcome in any of c categories. Let $y_{ij} = 1$ if trial i has outcome in category j and $y_{ij} = 0$ otherwise. Then $y_i = (y_{i1}, y_{i2}, \ldots, y_{ic})$ represents a multinomial trial, with $\sum_j y_{ij} = 1$; for instance, $(0,0,1,0)$ denotes outcome in category 3 of four possible categories. Note that y_{ic} is redundant, being linearly dependent on the others. Let $n_j = \sum_i y_{ij}$ denote the number of trials having outcome in category j. The counts $(n_1, n_2, \ldots, n_c)$ have the *multinomial distribution*.

Let $\pi_j = P(Y_{ij} = 1)$ denote the probability of outcome in category j for each trial. The multinomial probability mass function is

$$p(n_1, n_2, \ldots, n_{c-1}) = \left(\frac{n!}{n_1!\,n_2!\cdots n_c!}\right)\pi_1^{n_1}\pi_2^{n_2}\cdots\pi_c^{n_c}. \tag{1.2}$$

Since $\sum_j n_j = n$, this is $(c-1)$-dimensional, with $n_c = n - (n_1 + \cdots + n_{c-1})$. The binomial distribution is the special case with $c = 2$.

For the multinomial distribution,

$$E(n_j) = n\pi_j, \qquad \text{var}(n_j) = n\pi_j(1 - \pi_j), \qquad \text{cov}(n_j, n_k) = -n\pi_j\pi_k.$$

$$(1.3)$$

We derive the covariance in Section 14.1.4. The marginal distribution of each n_j is binomial.

1.2.3 Poisson Distribution

Sometimes, count data do not result from a fixed number of trials. For instance, if y = number of deaths due to automobile accidents on motorways in Italy during this coming week, there is no fixed upper limit n for y (as you are aware if you have driven in Italy). Since y must be a nonnegative integer, its distribution should place its mass on that range. The simplest such distribution is the *Poisson*. Its probabilities depend on a single parameter, the mean μ. The Poisson probability mass function (Poisson 1837, p. 206) is

$$p(y) = \frac{e^{-\mu}\mu^y}{y!}, \qquad y = 0, 1, 2, \ldots . \tag{1.4}$$

It satisfies $E(Y) = \text{var}(Y) = \mu$. It is unimodal with mode equal to the integer part of μ. Its skewness is described by $E(Y - \mu)^3/\sigma^3 = 1/\sqrt{\mu}$. The distribution approaches normality as μ increases.

The Poisson distribution is used for counts of events that occur randomly over time or space, when outcomes in disjoint periods or regions are independent. It also applies as an approximation for the binomial when n is large and π is small, with $\mu = n\pi$. So if each of the 50 million people driving in Italy next week is an independent trial with probability 0.000002 of dying in a fatal accident that week, the number of deaths Y is a bin(50000000, 0.000002) variate, or approximately Poisson with $\mu = n\pi = 50,000,000(0.000002) = 100$.

A key feature of the Poisson distribution is that its variance equals its mean. Sample counts vary more when their mean is higher. When the mean number of weekly fatal accidents equals 100, greater variability occurs in the weekly counts than when the mean equals 10.

1.2.4 Overdispersion

In practice, count observations often exhibit variability exceeding that predicted by the binomial or Poisson. This phenomenon is called *overdispersion*. We assumed above that each person has the same probability of dying in a fatal accident in the next week. More realistically, these probabilities vary, due to factors such as amount of time spent driving, whether the person wears a seat belt, and geographical location. Such variation causes fatality counts to display more variation than predicted by the Poisson model.

Suppose that Y is a random variable with variance $\text{var}(Y \mid \mu)$ for given μ, but μ itself varies because of unmeasured factors such as those just described. Let $\theta = E(\mu)$. Then unconditionally,

$$E(Y) = E[E(Y \mid \mu)], \qquad \text{var}(Y) = E[\text{var}(Y \mid \mu)] + \text{var}[E(Y \mid \mu)].$$

When Y is conditionally Poisson (given μ), for instance, then $E(Y) = E(\mu) = \theta$ and $\text{var}(Y) = E(\mu) + \text{var}(\mu) = \theta + \text{var}(\mu) > \theta$.

Assuming a Poisson distribution for a count variable is often too simplistic, because of factors that cause overdispersion. The *negative binomial* is a related distribution for count data that permits the variance to exceed the mean. We introduce it in Section 4.3.4.

Analyses assuming binomial (or multinomial) distributions are also sometimes invalid because of overdispersion. This might happen because the true distribution is a mixture of different binomial distributions, with the parameter varying because of unmeasured variables. To illustrate, suppose that an experiment exposes pregnant mice to a toxin and then after a week observes the number of fetuses in each mouse's litter that show signs of malformation. Let n_i denote the number of fetuses in the litter for mouse i. The mice also vary according to other factors that may not be measured, such as their weight, overall health, and genetic makeup. Extra variation then occurs because of the variability from litter to litter in the probability π of malformation. The distribution of the number of fetuses per litter showing malformations might cluster near 0 and near n_i, showing more dispersion than expected for binomial sampling with a single value of π. Overdispersion could also occur when π varies among fetuses in a litter according to some distribution (Problem 1.12). In Chapters 4, 12, and 13 we introduce methods for data that are overdispersed relative to binomial and Poisson assumptions.

1.2.5   Connection between Poisson and Multinomial Distributions

In Italy this next week, let y_1 = number of people who die in automobile accidents, y_2 = number who die in airplane accidents, and y_3 = number who die in railway accidents. A Poisson model for (Y_1, Y_2, Y_3) treats these as independent Poisson random variables, with parameters (μ_1, μ_2, μ_3). The joint probability mass function for $\{Y_i\}$ is the product of the three mass functions of form (1.4). The total $n = \Sigma Y_i$ also has a Poisson distribution, with parameter $\Sigma \mu_i$.

With Poisson sampling the total count n is random rather than fixed. If we assume a Poisson model but condition on n, $\{Y_i\}$ no longer have Poisson distributions, since each Y_i cannot exceed n. Given n, $\{Y_i\}$ are also no longer independent, since the value of one affects the possible range for the others.

For c independent Poisson variates, with $E(Y_i) = \mu_i$, let's derive their conditional distribution given that $\Sigma Y_i = n$. The conditional probability of a set of counts $\{n_i\}$ satisfying this condition is

$$P\big[(Y_1 = n_1, Y_2 = n_2, \ldots, Y_c = n_c) \mid \textstyle\sum Y_j = n\big]$$

$$= \frac{P(Y_1 = n_1, Y_2 = n_2, \ldots, Y_c = n_c)}{P(\textstyle\sum Y_j = n)}$$

$$= \frac{\Pi_i[\exp(-\mu_i)\mu_i^{n_i}/n_i!]}{\exp(-\textstyle\sum\mu_j)(\textstyle\sum\mu_j)^n/n!} = \frac{n!}{\Pi_i n_i!} \prod_i \pi_i^{n_i}, \qquad (1.5)$$

where $\{\pi_i = \mu_i/(\Sigma\mu_j)\}$. This is the multinomial $(n, \{\pi_i\})$ distribution, characterized by the sample size n and the probabilities $\{\pi_i\}$.

Many categorical data analyses assume a multinomial distribution. Such analyses usually have the same parameter estimates as those of analyses assuming a Poisson distribution, because of the similarity in the likelihood functions.

1.3 STATISTICAL INFERENCE FOR CATEGORICAL DATA

The choice of distribution for the response variable is but one step of data analysis. In practice, that distribution has unknown parameter values. In this section we review methods of using sample data to make inferences about the parameters. Sections 1.4 and 1.5 cover binomial and multinomial parameters.

1.3.1 Likelihood Functions and Maximum Likelihood Estimation

In this book we use *maximum likelihood* for parameter estimation. Under weak regularity conditions, such as the parameter space having fixed dimension with true value falling in its interior, maximum likelihood estimators have desirable properties: They have large-sample normal distributions; they are asymptotically consistent, converging to the parameter as n increases; and they are asymptotically efficient, producing large-sample standard errors no greater than those from other estimation methods.

Given the data, for a chosen probability distribution the *likelihood function* is the probability of those data, treated as a function of the unknown parameter. The maximum likelihood (ML) estimate is the parameter value that maximizes this function. This is the parameter value under which the data observed have the highest probability of occurrence. The parameter value that maximizes the likelihood function also maximizes the log of that function. It is simpler to maximize the log likelihood since it is a sum rather than a product of terms.

We denote a parameter for a generic problem by β and its ML estimate by $\hat{\beta}$. The likelihood function is $\ell(\beta)$ and the log-likelihood function is $L(\beta) = \log[\ell(\beta)]$. For many models, $L(\beta)$ has concave shape and $\hat{\beta}$ is the point at which the derivative equals 0. The ML estimate is then the solution of the likelihood equation, $\partial L(\beta)/\partial\beta = 0$. Often, β is multidimensional, denoted by $\boldsymbol{\beta}$, and $\hat{\boldsymbol{\beta}}$ is the solution of a set of likelihood equations.

Let SE denote the standard error of $\hat{\beta}$, and let $\text{cov}(\hat{\boldsymbol{\beta}})$ denote the asymptotic covariance matrix of $\hat{\boldsymbol{\beta}}$. Under regularity conditions (Rao 1973, p. 364), $\text{cov}(\hat{\boldsymbol{\beta}})$ is the inverse of the *information matrix*. The (j, k) element of the information matrix is

$$-E\left(\frac{\partial^2 L(\boldsymbol{\beta})}{\partial\beta_j\partial\beta_k}\right). \qquad (1.6)$$

The standard errors are the square roots of the diagonal elements for the inverse information matrix. The greater the curvature of the log likelihood, the smaller the standard errors. This is reasonable, since large curvature implies that the log likelihood drops quickly as $\boldsymbol{\beta}$ moves away from $\hat{\boldsymbol{\beta}}$; hence, the data would have been much more likely to occur if $\boldsymbol{\beta}$ took a value near $\hat{\boldsymbol{\beta}}$ rather than a value far from $\hat{\boldsymbol{\beta}}$.

1.3.2 Likelihood Function and ML Estimate for Binomial Parameter

The part of a likelihood function involving the parameters is called the *kernel*. Since the maximization of the likelihood is with respect to the parameters, the rest is irrelevant.

To illustrate, consider the binomial distribution (1.1). The binomial coefficient $\binom{n}{y}$ has no influence on where the maximum occurs with respect to π. Thus, we ignore it and treat the kernel as the likelihood function. The binomial log likelihood is then

$$L(\pi) = \log\left[\pi^y(1-\pi)^{n-y}\right] = y\log(\pi) + (n-y)\log(1-\pi). \quad (1.7)$$

Differentiating with respect to π yields

$$\partial L(\pi)/\partial\pi = y/\pi - (n-y)/(1-\pi) = (y-n\pi)/\pi(1-\pi). \quad (1.8)$$

Equating this to 0 gives the likelihood equation, which has solution $\hat{\pi} = y/n$, the sample proportion of successes for the n trials.

Calculating $\partial^2 L(\pi)/\partial\pi^2$, taking the expectation, and combining terms, we get

$$-E\left[\partial^2 L(\pi)/\partial\pi^2\right] = E\left[y/\pi^2 + (n-y)/(1-\pi)^2\right] = n/[\pi(1-\pi)].$$
$$(1.9)$$

Thus, the asymptotic variance of $\hat{\pi}$ is $\pi(1 - \pi)/n$. This is no surprise. Since $E(Y) = n\pi$ and $\text{var}(Y) = n\pi(1 - \pi)$, the distribution of $\hat{\pi} = Y/n$ has mean and standard error

$$E(\hat{\pi}) = \pi, \qquad \sigma(\hat{\pi}) = \sqrt{\frac{\pi(1 - \pi)}{n}}.$$

1.3.3 Wald–Likelihood Ratio–Score Test Triad

Three standard ways exist to use the likelihood function to perform large-sample inference. We introduce these for a significance test of a null hypothesis $H_0: \beta = \beta_0$ and then discuss their relation to interval estimation. They all exploit the large-sample normality of ML estimators.

With nonnull standard error SE of $\hat{\beta}$, the test statistic

$$z = (\hat{\beta} - \beta_0)/\text{SE}$$

has an approximate standard normal distribution when $\beta = \beta_0$. One refers z to the standard normal table to obtain one- or two-sided P-values. Equivalently, for the two-sided alternative, z^2 has a chi-squared null distribution with 1 degree of freedom (df); the P-value is then the right-tailed chi-squared probability above the observed value. This type of statistic, using the nonnull standard error, is called a *Wald statistic* (Wald 1943).

The multivariate extension for the Wald test of $H_0: \boldsymbol{\beta} = \boldsymbol{\beta}_0$ has test statistic

$$W = (\hat{\boldsymbol{\beta}} - \boldsymbol{\beta}_0)'\big[\text{cov}(\hat{\boldsymbol{\beta}})\big]^{-1}(\hat{\boldsymbol{\beta}} - \boldsymbol{\beta}_0).$$

(The prime on a vector or matrix denotes the transpose.) The nonnull covariance is based on the curvature (1.6) of the log likelihood at $\hat{\boldsymbol{\beta}}$. The asymptotic multivariate normal distribution for $\hat{\boldsymbol{\beta}}$ implies an asymptotic chi-squared distribution for W. The df equal the rank of $\text{cov}(\hat{\boldsymbol{\beta}})$, which is the number of nonredundant parameters in $\boldsymbol{\beta}$.

A second general-purpose method uses the likelihood function through the ratio of two maximizations: (1) the maximum over the possible parameter values under H_0, and (2) the maximum over the larger set of parameter values permitting H_0 or an alternative H_a to be true. Let ℓ_0 denote the maximized value of the likelihood function under H_0, and let ℓ_1 denote the maximized value generally (i.e., under $H_0 \cup H_a$). For instance, for parameter vector $\boldsymbol{\beta} = (\beta_0, \beta_1)'$ and $H_0: \beta_0 = 0$, ℓ_1 is the likelihood function calculated at the $\boldsymbol{\beta}$ value for which the data would have been most likely; ℓ_0 is the likelihood function calculated at the β_1 value for which the data would have been most likely, when $\beta_0 = 0$. Then ℓ_1 is always at least as large as ℓ_0, since ℓ_0 results from maximizing over a restricted set of the parameter values.

The ratio $\Lambda = \ell_0/\ell_1$ of the maximized likelihoods cannot exceed 1. Wilks (1935, 1938) showed that $-2\log\Lambda$ has a limiting null chi-squared distribution, as $n \rightarrow \infty$. The df equal the difference in the dimensions of the parameter spaces under $H_0 \cup H_a$ and under H_0. The *likelihood-ratio test statistic* equals

$$-2\log\Lambda = -2\log(\ell_0/\ell_1) = -2(L_0 - L_1),$$

where L_0 and L_1 denote the maximized log-likelihood functions.

The third method uses the *score statistic*, due to R. A. Fisher and C. R. Rao. The score test is based on the slope and expected curvature of the log-likelihood function $L(\beta)$ at the null value β_0. It utilizes the size of the *score function*

$$u(\beta) = \partial L(\beta)/\partial \beta,$$

evaluated at β_0. The value $u(\beta_0)$ tends to be larger in absolute value when $\hat{\beta}$ is farther from β_0. Denote $-E[\partial^2 L(\beta)/\partial\beta^2]$ (i.e., the information) evaluated at β_0 by $\iota(\beta_0)$. The score statistic is the ratio of $u(\beta_0)$ to its null SE, which is $[\iota(\beta_0)]^{1/2}$. This has an approximate standard normal null distribution. The chi-squared form of the score statistic is

$$\frac{[u(\beta_0)]^2}{\iota(\beta_0)} = \frac{[\partial L(\beta)/\partial\beta_0]^2}{-E[\partial^2 L(\beta)/\partial\beta_0^2]},$$

where the partial derivative notation reflects derivatives with respect to β that are evaluated at β_0. In the multiparameter case, the score statistic is a quadratic form based on the vector of partial derivatives of the log likelihood with respect to $\boldsymbol{\beta}$ and the inverse information matrix, both evaluated at the H_0 estimates (i.e., assuming that $\boldsymbol{\beta} = \boldsymbol{\beta}_0$).

Figure 1.1 is a generic plot of a log-likelihood $L(\beta)$ for the univariate case. It illustrates the three tests of H_0: $\beta = 0$. The Wald test uses the behavior of $L(\beta)$ at the ML estimate $\hat{\beta}$, having chi-squared form $(\hat{\beta}/SE)^2$. The SE of $\hat{\beta}$ depends on the curvature of $L(\beta)$ at $\hat{\beta}$. The score test is based on the slope and curvature of $L(\beta)$ at $\beta = 0$. The likelihood-ratio test combines information about $L(\beta)$ at both $\hat{\beta}$ and $\beta_0 = 0$. It compares the log-likelihood values L_1 at $\hat{\beta}$ and L_0 at $\beta_0 = 0$ using the chi-squared statistic $-2(L_0 - L_1)$. In Figure 1.1, this statistic is twice the vertical distance between values of $L(\beta)$ at $\hat{\beta}$ and at 0. In a sense, this statistic uses the most information of the three types of test statistic and is the most versatile.

As $n \rightarrow \infty$, the Wald, likelihood-ratio, and score tests have certain asymptotic equivalences (Cox and Hinkley 1974, Sec. 9.3). For small to moderate sample sizes, the likelihood-ratio test is usually more reliable than the Wald test.

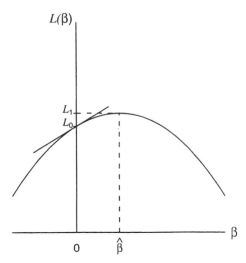

FIGURE 1.1   Log-likelihood function and information used in three tests of H_0: $\beta = 0$.

1.3.4   Constructing Confidence Intervals

In practice, it is more informative to construct confidence intervals for parameters than to test hypotheses about their values. For any of the three test methods, a confidence interval results from inverting the test. For instance, a 95% confidence interval for β is the set of β_0 for which the test of H_0: $\beta = \beta_0$ has a P-value exceeding 0.05.

Let z_a denote the z-score from the standard normal distribution having right-tailed probability a; this is the $100(1 - a)$ percentile of that distribution. Let $\chi^2_{df}(a)$ denote the $100(1 - a)$ percentile of the chi-squared distribution with degrees of freedom df. $100(1 - \alpha)\%$ confidence intervals based on asymptotic normality use $z_{\alpha/2}$, for instance $z_{0.025} = 1.96$ for 95% confidence. The Wald confidence interval is the set of β_0 for which $|\hat{\beta} - \beta_0|/\text{SE} < z_{\alpha/2}$. This gives the interval $\hat{\beta} \pm z_{\alpha/2}(\text{SE})$. The likelihood-ratio-based confidence interval is the set of β_0 for which $-2[L(\beta_0) - L(\hat{\beta})] < \chi^2_1(\alpha)$. [Recall that $\chi^2_1(\alpha) = z^2_{\alpha/2}$.]

When $\hat{\beta}$ has a normal distribution, the log-likelihood function has a parabolic shape (i.e., a second-degree polynomial). For small samples with categorical data, $\hat{\beta}$ may be far from normality and the log-likelihood function can be far from a symmetric, parabolic-shaped curve. This can also happen with moderate to large samples when a model contains many parameters. In such cases, inference based on asymptotic normality of $\hat{\beta}$ may have inadequate performance. A marked divergence in results of Wald and likelihood-ratio inference indicates that the distribution of $\hat{\beta}$ may not be close to normality. The example in Section 1.4.3 illustrates this with quite different confidence intervals for different methods. In many such cases, inference can instead utilize an exact small-sample distribution or "higher-order" asymptotic methods that improve on simple normality (e.g., Pierce and Peters 1992).

The Wald confidence interval is most common in practice because it is simple to construct using ML estimates and standard errors reported by statistical software. The likelihood-ratio-based interval is becoming more widely available in software and is preferable for categorical data with small to moderate n. For the best known statistical model, regression for a normal response, the three types of inference necessarily provide identical results.

1.4   STATISTICAL INFERENCE FOR BINOMIAL PARAMETERS

In this section we illustrate inference methods for categorical data by presenting tests and confidence intervals for the binomial parameter π, based on y successes in n independent trials. In Section 1.3.2 we obtained the likelihood function and ML estimator $\hat{\pi} = y/n$ of π.

1.4.1   Tests about a Binomial Parameter

Consider H_0: $\pi = \pi_0$. Since H_0 has a single parameter, we use the normal rather than chi-squared forms of Wald and score test statistics. They permit tests against one-sided as well as two-sided alternatives. The Wald statistic is

$$z_W = \frac{\hat{\pi} - \pi_0}{\text{SE}} = \frac{\hat{\pi} - \pi_0}{\sqrt{\hat{\pi}(1 - \hat{\pi})/n}} . \qquad (1.10)$$

Evaluating the binomial score (1.8) and information (1.9) at π_0 yields

$$u(\pi_0) = \frac{y}{\pi_0} - \frac{n - y}{1 - \pi_0}, \qquad \iota(\pi_0) = \frac{n}{\pi_0(1 - \pi_0)} .$$

The normal form of the score statistic simplifies to

$$z_S = \frac{u(\pi_0)}{[\iota(\pi_0)]^{1/2}} = \frac{y - n\pi_0}{\sqrt{n\pi_0(1 - \pi_0)}} = \frac{\hat{\pi} - \pi_0}{\sqrt{\pi_0(1 - \pi_0)/n}} . \qquad (1.11)$$

Whereas the Wald statistic z_W uses the standard error evaluated at $\hat{\pi}$, the score statistic z_S uses it evaluated at π_0. The score statistic is preferable, as it uses the actual null SE rather than an estimate. Its null sampling distribution is closer to standard normal than that of the Wald statistic.

The binomial log-likelihood function (1.7) equals $L_0 = y\log\pi_0 + (n - y)\log(1 - \pi_0)$ under H_0 and $L_1 = y \log \hat{\pi} + (n - y)\log(1 - \hat{\pi})$ more generally. The likelihood-ratio test statistic simplifies to

$$-2(L_0 - L_1) = 2\left(y \log \frac{\hat{\pi}}{\pi_0} + (n - y) \log \frac{1 - \hat{\pi}}{1 - \pi_0}\right).$$

Expressed as

$$-2(L_0 - L_1) = 2\left(y \log \frac{y}{n\pi_0} + (n - y) \log \frac{n - y}{n - n\pi_0}\right),$$

it compares observed success and failure counts to fitted (i.e., null) counts by

$$2\sum \text{observed} \, \log \frac{\text{observed}}{\text{fitted}}. \tag{1.12}$$

We'll see that this formula also holds for tests about Poisson and multinomial parameters. Since no unknown parameters occur under H_0 and one occurs under H_a, (1.12) has an asymptotic chi-squared distribution with df $= 1$.

1.4.2 Confidence Intervals for a Binomial Parameter

A significance test merely indicates whether a particular π value (such as $\pi = 0.5$) is plausible. We learn more by using a confidence interval to determine the range of plausible values.

Inverting the Wald test statistic gives the interval of π_0 values for which $|z_W| < z_{\alpha/2}$, or

$$\hat{\pi} \pm z_{\alpha/2} \sqrt{\frac{\hat{\pi}(1 - \hat{\pi})}{n}}. \tag{1.13}$$

Historically, this was one of the first confidence intervals used for any parameter (Laplace 1812, p. 283). Unfortunately, it performs poorly unless n is very large (e.g., Brown et al. 2001). The actual coverage probability usually falls below the nominal confidence coefficient, much below when π is near 0 or 1. A simple adjustment that adds $\frac{1}{2}z_{\alpha/2}^2$ observations of each type to the sample before using this formula performs much better (Problem 1.24).

The score confidence interval contains π_0 values for which $|z_S| < z_{\alpha/2}$. Its endpoints are the π_0 solutions to the equations

$$(\hat{\pi} - \pi_0)/\sqrt{\pi_0(1 - \pi_0)/n} = \pm z_{\alpha/2}.$$

These are quadratic in π_0. First discussed by E. B. Wilson (1927), this interval is

$$\hat{\pi}\left(\frac{n}{n + z_{\alpha/2}^2}\right) + \frac{1}{2}\left(\frac{z_{\alpha/2}^2}{n + z_{\alpha/2}^2}\right)$$

$$\pm z_{\alpha/2}\sqrt{\frac{1}{n + z_{\alpha/2}^2}\left[\hat{\pi}(1 - \hat{\pi})\left(\frac{n}{n + z_{\alpha/2}^2}\right) + \left(\frac{1}{2}\right)\left(\frac{1}{2}\right)\left(\frac{z_{\alpha/2}^2}{n + z_{\alpha/2}^2}\right)\right]}.$$

The midpoint $\tilde{\pi}$ of the interval is a weighted average of $\hat{\pi}$ and $\frac{1}{2}$, where the weight $n/(n + z_{\alpha/2}^2)$ given $\hat{\pi}$ increases as n increases. Combining terms, this midpoint equals $\tilde{\pi} = (y + z_{\alpha/2}^2/2)/(n + z_{\alpha/2}^2)$. This is the sample proportion for an adjusted sample that adds $z_{\alpha/2}^2$ observations, half of each type. The square of the coefficient of $z_{\alpha/2}$ in this formula is a weighted average of the variance of a sample proportion when $\pi = \hat{\pi}$ and the variance of a sample proportion when $\pi = \frac{1}{2}$, using the adjusted sample size $n + z_{\alpha/2}^2$ in place of n. This interval has much better performance than the Wald interval.

The likelihood-ratio-based confidence interval is more complex computationally, but simple in principle. It is the set of π_0 for which the likelihood-ratio test has a P-value exceeding α. Equivalently, it is the set of π_0 for which double the log likelihood drops by less than $\chi_1^2(\alpha)$ from its value at the ML estimate $\hat{\pi} = y/n$.

1.4.3 Proportion of Vegetarians Example

To collect data in an introductory statistics course, recently I gave the students a questionnaire. One question asked each student whether he or she was a vegetarian. Of $n = 25$ students, $y = 0$ answered "yes." They were not a random sample of a particular population, but we use these data to illustrate 95% confidence intervals for a binomial parameter π.

Since $y = 0$, $\hat{\pi} = 0/25 = 0$. Using the Wald approach, the 95% confidence interval for π is

$$0 \pm 1.96\sqrt{(0.0 \times 1.0)/25}, \quad \text{or} \quad (0,0).$$

When the observation falls at the boundary of the sample space, often Wald methods do not provide sensible answers.

By contrast, the 95% score interval equals (0.0, 0.133). This is a more believable inference. For H_0: $\pi = 0.5$, for instance, the score test statistic is $z_S = (0 - 0.5)/\sqrt{(0.5 \times 0.5)/25} = -5.0$, so 0.5 does not fall in the interval. By contrast, for H_0: $\pi = 0.10$, $z_S = (0 - 0.10)/\sqrt{(0.10 \times 0.90)/25} = -1.67$, so 0.10 falls in the interval.

When $y = 0$ and $n = 25$, the kernel of the likelihood function is $l(\pi) = \pi^0(1 - \pi)^{25} = (1 - \pi)^{25}$. The log likelihood (1.7) is $L(\pi) = 25\log(1 - \pi)$. Note that $L(\hat{\pi}) = L(0) = 0$. The 95% likelihood-ratio confidence interval is the set of π_0 for which the likelihood-ratio statistic

$$-2(L_0 - L_1) = -2[L(\pi_0) - L(\hat{\pi})]$$

$$= -50\log(1 - \pi_0) \leq \chi_1^2(0.05) = 3.84.$$

The upper bound is $1 - \exp(-3.84/50) = 0.074$, and the confidence interval equals $(0.0, 0.074)$. [In this book, we use the natural logarithm throughout, so its inverse is the exponential function $\exp(x) = e^x$.] Figure 1.2 shows the likelihood and log-likelihood functions and the corresponding confidence region for π.

The three large-sample methods yield quite different results. When π is near 0, the sampling distribution of $\hat{\pi}$ is highly skewed to the right for small n. It is worth considering alternative methods not requiring asymptotic approximations.

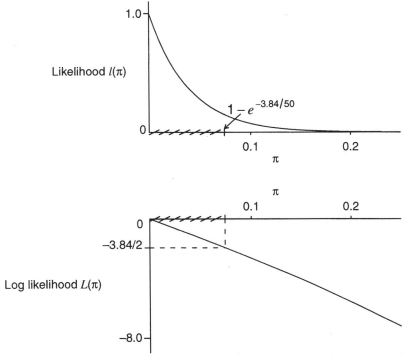

FIGURE 1.2 Binomial likelihood and log likelihood when $y = 0$ in $n = 25$ trials, and confidence interval for π.

1.4.4 Exact Small-Sample Inference*[1]

With modern computational power, it is not necessary to rely on large-sample approximations for the distribution of statistics such as $\hat{\pi}$. Tests and confidence intervals can use the binomial distribution directly rather than its normal approximation. Such inferences occur naturally for small samples, but apply for any n.

We illustrate by testing $H_0: \pi = 0.5$ against $H_a: \pi \neq 0.5$ for the survey results on vegetarianism, $y = 0$ with $n = 25$. We noted that the score statistic equals $z = -5.0$. The exact P-value for this statistic, based on the null bin(25, 0.5) distribution, is

$$P(|z| \geq 5.0) = P(Y = 0 \text{ or } Y = 25) = 0.5^{25} + 0.5^{25} = 0.00000006.$$

$100(1 - \alpha)\%$ confidence intervals consist of all π_0 for which P-values exceed α in exact binomial tests. The best known interval (Clopper and Pearson 1934) uses the *tail method* for forming confidence intervals. It requires each one-sided P-value to exceed $\alpha/2$. The lower and upper endpoints are the solutions in π_0 to the equations

$$\sum_{k=y}^{n} \binom{n}{k} \pi_0^k (1 - \pi_0)^{n-k} = \alpha/2 \quad \text{and} \quad \sum_{k=0}^{y} \binom{n}{k} \pi_0^k (1 - \pi_0)^{n-k} = \alpha/2,$$

except that the lower bound is 0 when $y = 0$ and the upper bound is 1 when $y = n$. When $y = 1, 2, \ldots, n - 1$, from connections between binomial sums and the incomplete beta function and related cumulative distribution functions (cdf's) of beta and F distributions, the confidence interval equals

$$\left[1 + \frac{n - y + 1}{y F_{2y, 2(n-y+1)}(1 - \alpha/2)} \right]^{-1} < \pi < \left[1 + \frac{n - y}{(y + 1) F_{2(y+1), 2(n-y)}(\alpha/2)} \right]^{-1},$$

where $F_{a, b}(c)$ denotes the $1 - c$ quantile from the F distribution with degrees of freedom a and b. When $y = 0$ with $n = 25$, the Clopper–Pearson 95% confidence interval for π is $(0.0, 0.137)$.

In principle this approach seems ideal. However, there is a serious complication. Because of discreteness, the actual coverage probability for any π is at least as large as the nominal confidence level (Casella and Berger 2001, p. 434; Neyman 1935) and it can be much greater. Similarly, for a test of $H_0: \pi = \pi_0$ at a fixed desired size α such as 0.05, it is not usually possible to achieve that size. There is a finite number of possible samples, and hence a finite number of possible P-values, of which 0.05 may not be one. In testing H_0 with fixed π_0, one can pick a particular α that can occur as a P-value.

---

[1]Sections marked with an asterisk are less important for an overview.

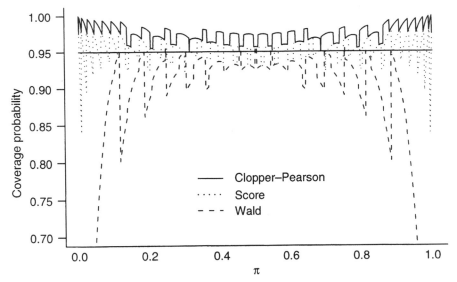

FIGURE 1.3 Plot of coverage probabilities for nominal 95% confidence intervals for binomial parameter π when $n = 25$.

For interval estimation, however, this is not an option. This is because constructing the interval corresponds to inverting an entire range of π_0 values in H_0: $\pi = \pi_0$, and each distinct π_0 value can have its own set of possible P-values; that is, there is not a single null parameter value π_0 as in one test.

For any fixed parameter value, the actual coverage probability can be much larger than the nominal confidence level. When $n = 25$, Figure 1.3 plots the coverage probabilities as a function of π for the Clopper–Pearson method, the score method, and the Wald method. At a fixed π value with a given method, the coverage probability is the sum of the binomial probabilities of all those samples for which the resulting interval contains that π. There are 26 possible samples and 26 corresponding confidence intervals, so the coverage probability is a sum of somewhere between 0 and 26 binomial probabilities. As π moves from 0 to 1, this coverage probability jumps up or down whenever π moves into or out of one of these intervals. Figure 1.3 shows that coverage probabilities are too low for the Wald method, whereas the Clopper–Pearson method errs in the opposite direction. The score method behaves well, except for some π values close to 0 or 1. Its coverage probabilities tend to be near the nominal level, not being consistently conservative or liberal. This is a good method unless π is very close to 0 or 1 (Problem 1.23).

In discrete problems using small-sample distributions, shorter confidence intervals usually result from inverting a single two-sided test rather than two one-sided tests. The interval is then the set of parameter values for which the P-value of a two-sided test exceeds α. For the binomial parameter, see Blaker (2000), Blyth and Still (1983), and Sterne (1954) for methods. For observed outcome y_o, with Blaker's approach the P-value is the minimum of the two one-tailed binomial probabilities $P(Y \geq y_o)$ and $P(Y \leq y_o)$ plus an attainable probability in the other tail that is as close as possible to, but not greater than, that one-tailed probability. The interval is computationally more complex, although available in software (Blaker gave S-Plus functions). The result is still conservative, but less so than the Clopper–Pearson interval. For the vegetarianism example, the 95% confidence interval using the Blaker exact method is $(0.0, 0.128)$ compared to the Clopper–Pearson interval of $(0.0, 0.137)$.

1.4.5   Inference Based on the Mid-P-Value*

To adjust for discreteness in small-sample distributions, one can base inference on the *mid-P-value* (Lancaster 1961). For a test statistic T with observed value t_o and one-sided H_a such that large T contradicts H_0,

$$\text{mid-}P\text{-value} = \tfrac{1}{2}P(T = t_o) + P(T > t_o),$$

with probabilities calculated from the null distribution. Thus, the mid-P-value is less than the ordinary P-value by half the probability of the observed result. Compared to the ordinary P-value, the mid-P-value behaves more like the P-value for a test statistic having a continuous distribution. The sum of its two one-sided P-values equals 1.0. Although discrete, under H_0 its null distribution is more like the uniform distribution that occurs in the continuous case. For instance, it has a null expected value of 0.5, whereas this expected value exceeds 0.5 for the ordinary P-value for a discrete test statistic.

Unlike an exact test with ordinary P-value, a test using the mid-P-value does not guarantee that the probability of type I error is no greater than a nominal value (Problem 1.19). However, it usually performs well, typically being a bit conservative. It is less conservative than the ordinary exact test. Similarly, one can form less conservative confidence intervals by inverting tests using the exact distribution with the mid-P-value (e.g., the 95% confidence interval is the set of parameter values for which the mid-P-value exceeds 0.05).

For testing $H_0: \pi = 0.5$ against $H_a: \pi \neq 0.5$ in the example about the proportion of vegetarians, with $y = 0$ for $n = 25$, the result observed is the most extreme possible. Thus the mid-P-value is half the ordinary P-value, or 0.00000003. Using the Clopper–Pearson inversion of the exact binomial test but with the mid-P-value yields a 95% confidence interval of $(0.000, 0.113)$ for π, compared to $(0.000, 0.137)$ for the ordinary Clopper–Pearson interval.

The mid-P-value seems a sensible compromise between having overly conservative inference and using irrelevant randomization to eliminate problems from discreteness. We recommend it both for tests and confidence intervals with highly discrete distributions.

1.5 STATISTICAL INFERENCE FOR MULTINOMIAL PARAMETERS

We now present inference for multinomial parameters $\{\pi_j\}$. Of n observations, n_j occur in category j, $j = 1, \ldots, c$.

1.5.1 Estimation of Multinomial Parameters

First, we obtain ML estimates of $\{\pi_j\}$. As a function of $\{\pi_j\}$, the multinomial probability mass function (1.2) is proportional to the kernel

$$\prod_j \pi_j^{n_j} \quad \text{where} \quad \text{all } \pi_j \geq 0 \quad \text{and} \quad \sum_j \pi_j = 1. \qquad (1.14)$$

The ML estimates are the $\{\pi_j\}$ that maximize (1.14).

The multinomial log-likelihood function is

$$L(\boldsymbol{\pi}) = \sum_j n_j \log \pi_j.$$

To eliminate redundancies, we treat L as a function of $(\pi_1, \ldots, \pi_{c-1})$, since $\pi_c = 1 - (\pi_1 + \cdots + \pi_{c-1})$. Thus, $\partial \pi_c / \partial \pi_j = -1$, $j = 1, \ldots, c - 1$.

Since

$$\frac{\partial \log \pi_c}{\partial \pi_j} = \frac{1}{\pi_c} \frac{\partial \pi_c}{\partial \pi_j} = -\frac{1}{\pi_c},$$

differentiating $L(\boldsymbol{\pi})$ with respect to π_j gives the likelihood equation

$$\frac{\partial L(\boldsymbol{\pi})}{\partial \pi_j} = \frac{n_j}{\pi_j} - \frac{n_c}{\pi_c} = 0.$$

The ML solution satisfies $\hat{\pi}_j / \hat{\pi}_c = n_j / n_c$. Now

$$\sum_j \hat{\pi}_j = 1 = \frac{\hat{\pi}_c \left(\sum_j n_j \right)}{n_c} = \frac{\hat{\pi}_c n}{n_c},$$

so $\hat{\pi}_c = n_c / n$ and then $\hat{\pi}_j = n_j / n$. From general results presented later in the book (Section 8.6), this solution does maximize the likelihood. Thus, the ML estimates of $\{\pi_j\}$ are the sample proportions.

1.5.2 Pearson Statistic for Testing a Specified Multinomial

In 1900 the eminent British statistician Karl Pearson introduced a hypothesis test that was one of the first inferential methods. It had a revolutionary impact on categorical data analysis, which had focused on describing associations. Pearson's test evaluates whether multinomial parameters equal certain specified values. His original motivation in developing this test was to analyze whether possible outcomes on a particular Monte Carlo roulette wheel were equally likely (Stigler 1986).

Consider $H_0: \pi_j = \pi_{j0}$, $j = 1, \ldots, c$, where $\Sigma_j \pi_{j0} = 1$. When H_0 is true, the expected values of $\{n_j\}$, called *expected frequencies*, are $\mu_j = n\pi_{j0}$, $j = 1, \ldots, c$. Pearson proposed the test statistic

$$X^2 = \sum_j \frac{(n_j - \mu_j)^2}{\mu_j}. \tag{1.15}$$

Greater differences $\{n_j - \mu_j\}$ produce greater X^2 values, for fixed n. Let X_o^2 denote the observed value of X^2. The *P*-value is the null value of $P(X^2 \geq X_o^2)$. This equals the sum of the null multinomial probabilities of all count arrays (having a sum of n) with $X^2 \geq X_o^2$.

For large samples, X^2 has approximately a chi-squared distribution with df $= c - 1$. The *P*-value is approximated by $P(\chi_{c-1}^2 \geq X_o^2)$, where χ_{c-1}^2 denotes a chi-squared random variable with df $= c - 1$. Statistic (1.15) is called the *Pearson chi-squared statistic*.

1.5.3 Example: Testing Mendel's Theories

Among its many applications, Pearson's test was used in genetics to test Mendel's theories of natural inheritance. Mendel crossed pea plants of pure yellow strain with plants of pure green strain. He predicted that second-generation hybrid seeds would be 75% yellow and 25% green, yellow being the dominant strain. One experiment produced $n = 8023$ seeds, of which $n_1 = 6022$ were yellow and $n_2 = 2001$ were green. The expected frequencies for $H_0: \pi_{10} = 0.75$, $\pi_{20} = 0.25$ are $\mu_1 = 8023(0.75) = 6017.25$ and $\mu_2 = 2005.75$. The Pearson statistic $X^2 = 0.015$ (df $= 1$) has a *P*-value of $P = 0.90$. This does not contradict Mendel's hypothesis.

Mendel performed several experiments of this type. In 1936, R. A. Fisher summarized Mendel's results. He used the reproductive property of chi-squared: If $X_1^2, \ldots, X_k^2$ are independent chi-squared statistics with degrees of freedom $\nu_1, \ldots, \nu_k$, then $\Sigma_i X_i^2$ has a chi-squared distribution with df $= \Sigma_i \nu_i$. Fisher obtained a summary chi-squared statistic equal to 42, with df $= 84$. A chi-squared distribution with df $= 84$ has mean 84 and standard deviation $(2 \times 84)^{1/2} = 13.0$, and the right-tailed probability above 42 is $P = 0.99996$. In other words, the chi-squared statistic was so small that the fit seemed *too* good.

Fisher commented: "The general level of agreement between Mendel's expectations and his reported results shows that it is closer than would be expected in the best of several thousand repetitions.... I have no doubt that Mendel was deceived by a gardening assistant, who knew only too well what his principal expected from each trial made." In a letter written at the time (see Box 1978, p. 297), he stated: "Now, when data have been faked, I know very well how generally people underestimate the frequency of wide chance deviations, so that the tendency is always to make them agree too well with expectations." In summary, goodness-of-fit tests can reveal not only when a fit is inadequate, but also when it is better than random fluctuations would have us expect. [R. A. Fisher's daughter, Joan Fisher Box (1978, pp. 295–300), and Freedman et al. (1978, pp. 420–428, 478) discussed Fisher's analysis of Mendel's data and the accompanying controversy. Despite possible difficulties with Mendel's data, subsequent work led to general acceptance of his theories.]

1.5.4 Chi-Squared Theoretical Justification*

We now outline why Pearson's statistic has a limiting chi-squared distribution. For a multinomial sample $(n_1, \ldots, n_c)$ of size n, the marginal distribution of n_j is the $\text{bin}(n, \pi_j)$ distribution. For large n, by the normal approximation to the binomial, n_j (and $\hat{\pi}_j = n_j/n$) have approximate normal distributions. More generally, by the central limit theorem, the sample proportions $\hat{\boldsymbol{\pi}} = (n_1/n, \ldots, n_{c-1}/n)'$ have an approximate multivariate normal distribution (Section 14.1.4). Let $\boldsymbol{\Sigma}_0$ denote the null covariance matrix of $\sqrt{n}\,\hat{\boldsymbol{\pi}}$, and let $\boldsymbol{\pi}_0 = (\pi_{10}, \ldots, \pi_{c-1,0})'$. Under H_0, since $\sqrt{n}\,(\hat{\boldsymbol{\pi}} - \boldsymbol{\pi}_0)$ converges to a $N(\mathbf{0}, \boldsymbol{\Sigma}_0)$ distribution, the quadratic form

$$n(\hat{\boldsymbol{\pi}} - \boldsymbol{\pi}_0)' \boldsymbol{\Sigma}_0^{-1} (\hat{\boldsymbol{\pi}} - \boldsymbol{\pi}_0) \qquad (1.16)$$

has distribution converging to chi-squared with df $= c - 1$.

In Section 14.1.4 we show that the covariance matrix of $\sqrt{n}\,\hat{\boldsymbol{\pi}}$ has elements

$$\sigma_{jk} = \begin{cases} -\pi_j \pi_k & \text{if } j \neq k \\ \pi_j(1 - \pi_j) & \text{if } j = k \end{cases}.$$

The matrix $\boldsymbol{\Sigma}_0^{-1}$ has (j, k)th element $1/\pi_{c0}$ when $j \neq k$ and $(1/\pi_{j0} + 1/\pi_{c0})$ when $j = k$. (You can verify this by showing that $\boldsymbol{\Sigma}_0 \boldsymbol{\Sigma}_0^{-1}$ equals the identity matrix.) With this substitution, direct calculation (with appropriate combining of terms) shows that (1.16) simplifies to X^2. In Section 14.3 we provide a formal proof in a more general setting.

This argument is similar to Pearson's in 1900. R. A. Fisher (1922) gave a simpler justification, the gist of which follows: Suppose that $(n_1, \ldots, n_c)$ are independent Poisson random variables with means $(\mu_1, \ldots, \mu_c)$. For large

$\{\mu_j\}$, the standardized values $\{z_j = (n_j - \mu_j)/\sqrt{\mu_j}\}$ have approximate standard normal distributions. Thus, $\Sigma_j z_j^2 = X^2$ has an approximate chi-squared distribution with c degrees of freedom. Adding the single linear constraint $\Sigma_j(n_j - \mu_j) = 0$, thus converting the Poisson distributions to a multinomial, we lose a degree of freedom.

When $c = 2$, Pearson's X^2 simplifies to the square of the normal score statistic (1.11). For Mendel's data, $\hat{\pi}_1 = 6022/8023$, $\pi_{10} = 0.75$, $n = 8023$, and $z_S = 0.123$, for which $X^2 = (0.123)^2 = 0.015$. In fact, for general c the Pearson test is the score test about multinomial parameters.

1.5.5 Likelihood-Ratio Chi-Squared

An alternative test for multinomial parameters uses the likelihood-ratio test. The kernel of the multinomial likelihood is (1.14). Under H_0 the likelihood is maximized when $\hat{\pi}_j = \pi_{j0}$. In the general case, it is maximized when $\hat{\pi}_j = n_j/n$. The ratio of the likelihoods equals

$$\Lambda = \frac{\Pi_j(\pi_{j0})^{n_j}}{\Pi_j(n_j/n)^{n_j}}.$$

Thus, the likelihood-ratio statistic, denoted by G^2, is

$$G^2 = -2\log\Lambda = 2\sum n_j \log(n_j/n\pi_{j0}). \tag{1.17}$$

This statistic, which has form (1.12), is called the *likelihood-ratio chi-squared statistic*. The larger the value of G^2, the greater the evidence against H_0.

In the general case, the parameter space consists of $\{\pi_j\}$ subject to $\Sigma_j\pi_j = 1$, so the dimensionality is $c - 1$. Under H_0, the $\{\pi_j\}$ are specified completely, so the dimension is 0. The difference in these dimensions equals $(c - 1)$. For large n, G^2 has a chi-squared null distribution with df $= c - 1$.

When H_0 holds, the Pearson X^2 and the likelihood ratio G^2 both have asymptotic chi-squared distributions with df $= c - 1$. In fact, they are asymptotically equivalent in that case; specifically, $X^2 - G^2$ converges in probability to zero (Section 14.3.4). When H_0 is false, they tend to grow proportionally to n; they need not take similar values, however, even for very large n.

For fixed c, as n increases the distribution of X^2 usually converges to chi-squared more quickly than that of G^2. The chi-squared approximation is usually poor for G^2 when $n/c < 5$. When c is large, it can be decent for X^2 for n/c as small as 1 if the table does not contain both very small and moderately large expected frequencies. We provide further guidelines in Section 9.8.4. Alternatively, one can use the multinomial probabilities to generate exact distributions of these test statistics (Good et al. 1970).

1.5.6 Testing with Estimated Expected Frequencies

Pearson's X^2 (1.15) compares a sample distribution to a hypothetical one $\{\pi_{j0}\}$. In some applications, $\{\pi_{j0} = \pi_{j0}(\boldsymbol{\theta})\}$ are functions of a smaller set of unknown parameters $\boldsymbol{\theta}$. ML estimates $\hat{\boldsymbol{\theta}}$ of $\boldsymbol{\theta}$ determine ML estimates $\{\pi_{j0}(\hat{\boldsymbol{\theta}})\}$ of $\{\pi_{j0}\}$ and hence ML estimates $\{\hat{\mu}_j = n\pi_{j0}(\hat{\boldsymbol{\theta}})\}$ of expected frequencies in X^2. Replacing $\{\mu_j\}$ by estimates $\{\hat{\mu}_j\}$ affects the distribution of X^2. When $\dim(\boldsymbol{\theta}) = p$, the true df $= (c - 1) - p$ (Section 14.3.3). Pearson failed to realize this (Section 16.2).

We now show a goodness-to-fit test with estimated expected frequencies. A sample of 156 dairy calves born in Okeechobee County, Florida, were classified according to whether they caught pneumonia within 60 days of birth. Calves that got a pneumonia infection were also classified according to whether they got a secondary infection within 2 weeks after the first infection cleared up. Table 1.1 shows the data. Calves that did not get a primary infection could not get a secondary infection, so no observations can fall in the category for "no" primary infection and "yes" secondary infection. That combination is called a *structural zero*.

A goal of this study was to test whether the probability of primary infection was the same as the conditional probability of secondary infection, given that the calf got the primary infection. In other words, if π_{ab} denotes the probability that a calf is classified in row a and column b of this table, the null hypothesis is

$$H_0: \pi_{11} + \pi_{12} = \pi_{11}/(\pi_{11} + \pi_{12})$$

or $\pi_{11} = (\pi_{11} + \pi_{12})^2$. Let $\pi = \pi_{11} + \pi_{12}$ denote the probability of primary infection. The null hypothesis states that the probabilities satisfy the structure that Table 1.2 shows; that is, probabilities in a trinomial for the categories (yes–yes, yes–no, no–no) for primary–secondary infection equal $(\pi^2, \pi(1 - \pi), 1 - \pi)$.

Let n_{ab} denote the number of observations in category (a, b). The ML estimate of π is the value maximizing the kernel of the multinomial likelihood

$$(\pi^2)^{n_{11}}(\pi - \pi^2)^{n_{12}}(1 - \pi)^{n_{22}}.$$

TABLE 1.1 Primary and Secondary Pneumonia Infections in Calves

Primary Infection	Secondary Infection[a]	
	Yes	No
Yes	30 (38.1)	63 (39.0)
No	0 (—)	63 (78.9)

Source: Data courtesy of Thang Tran and G. A. Donovan, College of Veterinary Medicine, University of Florida.

[a] Values in parentheses are estimated expected frequencies.

TABLE 1.2 Probability Structure for Hypothesis

Primary Infection	Secondary Infection		Total
	Yes	No	
Yes	π^2	$\pi(1 - \pi)$	π
No	—	$1 - \pi$	$1 - \pi$

$$\left(\pi^2\right)^{n_{11}}\left(\pi - \pi^2\right)^{n_{12}}\left(1 - \pi\right)^{n_{22}}.$$

The log likelihood is

$$L(\pi) = n_{11} \log \pi^2 + n_{12} \log(\pi - \pi^2) + n_{22} \log(1 - \pi).$$

Differentiation with respect to π gives the likelihood equation

$$\frac{2n_{11}}{\pi} + \frac{n_{12}}{\pi} - \frac{n_{12}}{1 - \pi} - \frac{n_{22}}{1 - \pi} = 0.$$

The solution is

$$\hat{\pi} = (2n_{11} + n_{12})/(2n_{11} + 2n_{12} + n_{22}).$$

For Table 1.1, $\hat{\pi} = 0.494$. Since $n = 156$, the estimated expected frequencies are $\hat{\mu}_{11} = n\hat{\pi}^2 = 38.1$, $\hat{\mu}_{12} = n(\hat{\pi} - \hat{\pi}^2) = 39.0$, and $\hat{\mu}_{22} = n(1 - \hat{\pi}) = 78.9$. Table 1.1 shows them. Pearson's statistic is $X^2 = 19.7$. Since the $c = 3$ possible responses have $p = 1$ parameter (π) determining the expected frequencies, df $= (3 - 1) - 1 = 1$. There is strong evidence against H_0 $(P = 0.00001)$. Inspection of Table 1.1 reveals that many more calves got a primary infection but not a secondary infection than H_0 predicts. The researchers concluded that the primary infection had an immunizing effect that reduced the likelihood of a secondary infection.

NOTES

Section 1.1: Categorical Response Data

1.1. Stevens (1951) defined (nominal, ordinal, interval) scales of measurement. Other scales result from mixtures of these types. For instance, *partially ordered* scales occur when subjects respond to questions having categories ordered except for don't know or undecided categories.

Section 1.3: Statistical Inference for Categorical Data

1.2. The score method does not use $\hat{\beta}$. Thus, when β is a model parameter, one can usually compute the score statistic for testing H_0: $\beta = \beta_0$ without fitting the model. This is advantageous when fitting several models in an exploratory analysis and model fitting is computationally intensive. An advantage of the score and likelihood-ratio methods is that they apply even when $|\hat{\beta}| = \infty$. In that case, one cannot compute the Wald statistic. Another disadvantage of the Wald method is that its results depend on the parameterization; inference based on $\hat{\beta}$ and its SE is not equivalent to inference based on a nonlinear function of it, such as $\log \hat{\beta}$ and its SE.

Section 1.4: Statistical Inference for Binomial Parameters

1.3. Among others, Agresti and Coull (1998), Blyth and Still (1983), Brown et al. (2001), Ghosh (1979), and Newcombe (1998a) showed the superiority of the score interval to the Wald interval for π. Of the "exact" methods, Blaker's (2000) has particularly good properties. It is contained in the Clopper–Pearson interval and has a nestedness property whereby an interval of higher nominal confidence level necessarily contains one of lower level.

1.4. Using continuity corrections with large-sample methods provides approximations to exact small-sample methods. Thus, they tend to behave conservatively. We do not present them, since if one prefers an exact method, with modern computational power it can be used directly rather than approximated.

1.5. In theory, one can eliminate problems with discreteness in tests by performing a supplementary randomization on the boundary of a critical region (see Problem 1.19). In rejecting the null at the boundary with a certain probability, one can obtain a fixed overall type I error probability α even when it is not an achievable P-value. For such randomization, the one-sided P – value is

$$\text{randomized } P\text{-value} = U \times P(T = t_o) + P(T > t_o),$$

where U denotes a uniform $(0, 1)$ random variable (Stevens 1950). In practice, this is not used, as it is absurd to let this random number influence a decision. The mid P-value replaces the arbitrary uniform multiple $U \times P(T = t_o)$ by its expected value.

Section 1.5: Statistical Inference for Multinomial Parameters

1.6. The chi-squared distribution has mean df, variance 2 df, and skewness $(8/\text{df})^{1/2}$. It is approximately normal when df is large. Greenwood and Nikulin (1996), Kendall and Stuart (1979), and Lancaster (1969) presented other properties. Cochran (1952) presented a historical survey of chi-squared tests of fit. See also Cressie and Read (1989), Koch and Bhapkar (1982), Koehler (1998), and Moore (1986b).

PROBLEMS

Applications

1.1 Identify each variable as nominal, ordinal, or interval.

    a. UK political party preference (Labour, Conservative, Social Democrat)

    b. Anxiety rating (none, mild, moderate, severe, very severe)

    c. Patient survival (in number of months)

    d. Clinic location (London, Boston, Madison, Rochester, Montreal)

   e. Response of tumor to chemotherapy (complete elimination, partial reduction, stable, growth progression)

   f. Favorite beverage (water, juice, milk, soft drink, beer, wine)

   g. Appraisal of company's inventory level (too low, about right, too high)

1.2 Each of 100 multiple-choice questions on an exam has four possible answers, one of which is correct. For each question, a student guesses by selecting an answer randomly.

   a. Specify the distribution of the student's number of correct answers.

   b. Find the mean and standard deviation of that distribution. Would it be surprising if the student made at least 50 correct responses? Why?

   c. Specify the distribution of (n_1, n_2, n_3, n_4), where n_j is the number of times the student picked choice j.

   d. Find $E(n_j)$, $\text{var}(n_j)$, $\text{cov}(n_j, n_k)$, and $\text{corr}(n_j, n_k)$.

1.3 An experiment studies the number of insects that survive a certain dose of an insecticide, using several batches of insects of size n each. The insects are sensitive to factors that vary among batches during the experiment but were not measured, such as temperature level. Explain why the distribution of the number of insects per batch surviving the experiment might show overdispersion relative to a $\text{bin}(n, \pi)$ distribution.

1.4 In his autobiography *A Sort of Life*, British author Graham Greene described a period of severe mental depression during which he played Russian Roulette. This "game" consists of putting a bullet in one of the six chambers of a pistol, spinning the chambers to select one at random, and then firing the pistol once at one's head.

   a. Greene played this game six times and was lucky that none of them resulted in a bullet firing. Find the probability of this outcome.

   b. Suppose that he had kept playing this game until the bullet fired. Let Y denote the number of the game on which it fires. Show the probability mass function for Y, and justify.

1.5 Consider the statement, "Please tell me whether or not you think it should be possible for a pregnant woman to obtain a legal abortion if she is married and does not want any more children." For the 1996 General Social Survey, conducted by the National Opinion Research Center (NORC), 842 replied "yes" and 982 replied "no." Let π denote the population proportion who would reply "yes." Find the P-value for testing H_0: $\pi = 0.5$ using the score test, and construct a 95% confidence interval for π. Interpret the results.

1.6 Refer to the vegetarianism example in Section 1.4.3. For testing H_0: $\pi = 0.5$ against H_a: $\pi \neq 0.5$, show that:

    a. The likelihood-ratio statistic equals $2[25\log(25/12.5)] = 34.7$.

    b. The chi-squared form of the score statistic equals 25.0.

    c. The Wald z or chi-squared statistic is infinite.

1.7 In a crossover trial comparing a new drug to a standard, π denotes the probability that the new one is judged better. It is desired to estimate π and test H_0: $\pi = 0.5$ against H_a: $\pi \neq 0.5$. In 20 independent observations, the new drug is better each time.

    a. Find and sketch the likelihood function. Give the ML estimate of π.

    b. Conduct a Wald test and construct a 95% Wald confidence interval for π. Are these sensible?

    c. Conduct a score test, reporting the P-value. Construct a 95% score confidence interval. Interpret.

    d. Conduct a likelihood-ratio test and construct a likelihood-based 95% confidence interval. Interpret.

    e. Construct an exact binomial test and 95% confidence interval. Interpret.

    f. Suppose that researchers wanted a sufficiently large sample to estimate the probability of preferring the new drug to within 0.05, with confidence 0.95. If the true probability is 0.90, about how large a sample is needed?

1.8 In an experiment on chlorophyll inheritance in maize, for 1103 seedlings of self-fertilized heterozygous green plants, 854 seedlings were green and 249 were yellow. Theory predicts the ratio of green to yellow is 3:1. Test the hypothesis that 3:1 is the true ratio. Report the P-value, and interpret.

1.9 Table 1.3 contains Ladislaus von Bortkiewicz's data on deaths of soldiers in the Prussian army from kicks by army mules (Fisher 1934; Quine and Seneta 1987). The data refer to 10 army corps, each observed for 20 years. In 109 corps-years of exposure, there were no deaths, in 65 corps-years there was one death, and so on. Estimate the mean and test whether probabilities of occurrences in these five categories follow a Poisson distribution (truncated for 4 and above.)

TABLE 1.3   Data for Problem 1.9

Number of Deaths	Number of Corps-Years
0	109
1	65
2	22
3	3
4	1
≥ 5	0

1.10 A sample of 100 women suffer from dysmenorrhea. A new analgesic is claimed to provide greater relief than a standard one. After using each analgesic in a crossover experiment, 40 reported greater relief with the standard analgesic and 60 reported greater relief with the new one. Analyze these data.

Theory and Methods

1.11 Why is it easier to get a precise estimate of the binomial parameter π when it is near 0 or 1 than when it is near $\frac{1}{2}$?

1.12 Suppose that $P(Y_i = 1) = 1 - P(Y_i = 0) = \pi$, $i = 1, \ldots, n$, where $\{Y_i\}$ are independent. Let $Y = \sum_i Y_i$.

   a. What are var(Y) and the distribution of Y?

   b. When $\{Y_i\}$ instead have pairwise correlation $\rho > 0$, show that var$(Y) > n\pi(1 - \pi)$, overdispersion relative to the binomial. [Altham (1978) discussed generalizations of the binomial that allow correlated trials.]

   c. Suppose that heterogeneity exists: $P(Y_i = 1|\pi) = \pi$ for all i, but π is a random variable with density function $g(\cdot)$ on $[0, 1]$ having mean ρ and positive variance. Show that var$(Y) > n\rho(1 - \rho)$. (When π has a beta distribution, Y has the *beta-binomial distribution* of Section 13.3.)

   d. Suppose that $P(Y_i = 1|\pi_i) = \pi_i$, $i = 1, \ldots, n$, where $\{\pi_i\}$ are independent from $g(\cdot)$. Explain why Y has a bin(n, ρ) distribution unconditionally but not conditionally on $\{\pi_i\}$. (*Hint:* In each case, is Y a sum of independent, identical Bernoulli trials?)

1.13 For a sequence of independent Bernoulli trials, Y is the number of successes before the kth failure. Explain why its probability mass function is the *negative binomial*,

$$p(y) = \frac{(y + k - 1)!}{y!(k - 1)!} \pi^y (1 - \pi)^k, \qquad y = 0, 1, 2, \ldots .$$

[For it, $E(Y) = k\pi/(1 - \pi)$ and $\mathrm{var}(Y) = k\pi/(1 - \pi)^2$, so $\mathrm{var}(Y) > E(Y)$; the Poisson is the limit as $k \to \infty$ and $\pi \to 0$ with $k\pi = \mu$ fixed.]

1.14 For the multinomial distribution, show that

$$\mathrm{corr}(n_j, n_k) = -\pi_j \pi_k / \sqrt{\pi_j(1 - \pi_j)\pi_k(1 - \pi_k)} .$$

Show that $\mathrm{corr}(n_1, n_2) = -1$ when $c = 2$.

1.15 Show that the moment generating function (mgf) for the binomial distribution is $m(t) = (1 - \pi + \pi e^t)^n$, and use it to obtain the first two moments. Show that the mgf for the Poisson distribution is $m(t) = \exp(\mu[\exp(t) - 1])$, and use it to obtain the first two moments.

1.16 A likelihood-ratio statistic equals t_o. At the ML estimates, show that the data are $\exp(t_o/2)$ times more likely under H_a than under H_0.

1.17 Assume that $y_1, y_2, \ldots, y_n$ are independent from a Poisson distribution.
   a. Obtain the likelihood function. Show that the ML estimator $\hat{\mu} = \bar{y}$.
   b. Construct a large-sample test statistic for $H_0: \mu = \mu_0$ using (i) the Wald method, (ii) the score method, and (iii) the likelihood-ratio method.
   c. Construct a large-sample confidence interval for μ using (i) the Wald method, (ii) the score method, and (iii) the likelihood-ratio method.

1.18 Inference for Poisson parameters can often be based on connections with binomial and multinomial distributions. Show how to test $H_0: \mu_1 = \mu_2$ for two populations based on independent Poisson counts (y_1, y_2), using a corresponding test about a binomial parameter π. [*Hint:* Condition on $n = y_1 + y_2$ and identify $\pi = \mu_1/(\mu_1 + \mu_2)$.] How can one construct a confidence interval for μ_1/μ_2 based on one for π?

1.19 A researcher routinely tests using a nominal $P(\text{type I error}) = 0.05$, rejecting H_0 if the P-value ≤ 0.05. An exact test using test statistic T

has null distribution $P(T = 0) = 0.30$, $P(T = 1) = 0.62$, and $P(T = 2) = 0.08$, where a higher T provides more evidence against the null.

a. With the usual P-value, show that the actual P(type I error) = 0.

b. With the mid-P-value, show that the actual P(type I error) = 0.08.

c. Find P(type I error) in parts (a) and (b) when $P(T = 0) = 0.30$, $P(T = 1) = 0.66$, $P(T = 2) = 0.04$. Note that the test with mid-P-value can be conservative or liberal. The exact test with ordinary P-value cannot be liberal.

d. In part (a), a randomized-decision test generates a uniform random variable U from $[0, 1]$ and rejects H_0 when $T = 2$ and $U \leq \frac{5}{8}$. Show the actual P(type I error) = 0.05. Is this a sensible test?

1.20 For a binomial parameter π, show how the inversion process for constructing a confidence interval works with (**a**) the Wald test, and (**b**) the score test.

1.21 For a flip of a coin, let π denote the probability of a head. An experiment tests H_0: $\pi = 0.5$ against H_a: $\pi \neq 0.5$, using $n = 5$ independent flips.

a. Show that the true null probability of rejecting H_0 at the 0.05 significance level is 0.0 for the exact binomial test and $\frac{1}{16}$ using the large-sample score test.

b. Suppose that truly $\pi = 0.5$. Explain why the probability that the 95% Clopper–Pearson confidence interval contains π equals 1.0. (*Hint:* Is there any possible y for which both one-sided tests of H_0: $\pi = 0.5$ have P-value ≤ 0.025?)

1.22 Consider the Wald confidence interval for a binomial parameter π. Since it is degenerate when $\hat{\pi} = 0$ or 1, argue that for $0 < \pi < 1$ the probability the interval covers π cannot exceed $[1 - \pi^n - (1 - \pi)^n]$; hence, the infimum of the coverage probability over $0 < \pi < 1$ equals 0, regardless of n.

1.23 Consider the 95% binomial score confidence interval for π. When $y = 1$, show that the lower limit is approximately $0.18/n$; in fact, $0 < \pi < 0.18/n$ then falls in an interval only when $y = 0$. Argue that for large n and π just barely below $0.18/n$ or just barely above $1 - 0.18/n$, the actual coverage probability is about $e^{-0.18} = 0.84$. Hence, even as $n \to \infty$, this method is not guaranteed to have coverage probability ≥ 0.95 (Agresti and Coull 1998; Blyth and Still 1983).

1.24 From Section 1.4.2 the midpoint $\tilde{\pi}$ of the score confidence interval for π is the sample proportion for an adjusted data set that adds $z_{\alpha/2}^2/2$

observations of each type to the sample. This motivates an adjusted Wald interval,

$$\tilde{\pi} \pm z_{\alpha/2} \sqrt{\tilde{\pi}(1 - \tilde{\pi})/n^*}, \qquad \text{where } n^* = n + z_{\alpha/2}^2.$$

Show that the variance $\tilde{\pi}(1 - \tilde{\pi})/n^*$ at the weighted average is at least as large as the weighted average of the variances that appears under the square root sign in the score interval (*Hint:* Use Jensen's inequality). Thus, this interval contains the score interval. [Agresti and Coull (1998) and Brown et al. (2001) showed that it performs much better than the Wald interval. It does not have the score interval's disadvantage (Problem 1.23) of poor coverage near 0 and 1.]

1.25 A binomial sample of size n has $y = 0$ successes.

  a. Show that the confidence interval for π based on the likelihood function is $[0.0, 1 - \exp(-z_{\alpha/2}^2/2n)]$. For $\alpha = 0.05$, use the expansion of an exponential function to show that this is approximately $[0, 2/n]$.

  b. For the score method, show that the confidence interval is $[0, z_{\alpha/2}^2/(n + z_{\alpha/2}^2)]$, or approximately $[0, 4/(n + 4)]$ when $\alpha = 0.05$.

  c. For the Clopper–Pearson approach, show that the upper bound is $1 - (\alpha/2)^{1/n}$, or approximately $-\log(0.025)/n = 3.69/n$ when $\alpha = 0.05$.

  d. For the adaptation of the Clopper–Pearson approach using the mid-P-value, show that the upper bound is $1 - \alpha^{1/n}$, or approximately $-\log(0.05)/n = 3/n$ when $\alpha = 0.05$.

1.26 For the geometric distribution $p(y) = \pi^y(1 - \pi)$, $y = 0, 1, 2, \ldots$, show that the tail method for constructing a confidence interval [i.e., equating $P(Y \geq y)$ and $P(Y \leq y)$ to $\alpha/2$] yields $[(\alpha/2)^{1/y}, (1 - \alpha/2)^{1/(y+1)}]$. Show that all π between 0 and $1 - \alpha/2$ *never* fall above a confidence interval, and hence the actual coverage probability exceeds $1 - \alpha/2$ over this region.

1.27 A statistic T has discrete distribution with cdf $F(t)$. Show that $F(T)$ is *stochastically larger* than uniform over $[0, 1]$; that is, its cdf is everywhere no greater than that of the uniform (Casella and Berger 2001, pp. 77, 434). Explain why an implication is that a P-value based on T has null distribution that is stochastically larger than uniform.

1.28 Suppose that $P(T = t_j) = \pi_j$, $j = 1, \ldots$. Show that $E(\text{mid-}P\text{-value}) = 0.5$. [*Hint:* Show that $\sum_j \pi_j(\pi_j/2 + \pi_{j+1} + \cdots) = (\sum_j \pi_j)^2/2$.]

1.29 For a statistic T with cdf $F(t)$ and $p(t) = P(T = t)$, the *mid-distribution function* is $F_{\text{mid}}(t) = F(t) - 0.5p(t)$ (Parzen 1997). Given $T = t_o$, show that the mid-P-value equals $1 - F(t_o)$. (It also satisfies $E[F_{\text{mid}}(T)] = 0.5$ and $\text{var}[F_{\text{mid}}(T)] = (1/12)\{1 - E[p^2(T)]\}$.)

1.30 Genotypes AA, Aa, and aa occur with probabilities $[\theta^2, 2\theta(1 - \theta), (1 - \theta)^2]$. A multinomial sample of size n has frequencies (n_1, n_2, n_3) of these three genotypes.

　a. Form the log likelihood. Show that $\hat{\theta} = (2n_1 + n_2)/(2n_1 + 2n_2 + 2n_3)$.

　b. Show that $-\partial^2 L(\theta)/\partial\theta^2 = [(2n_1 + n_2)/\theta^2] + [(n_2 + 2n_3)/(1 - \theta)^2]$ and that its expectation is $2n/\theta(1 - \theta)$. Use this to obtain an asymptotic standard error of $\hat{\theta}$.

　c. Explain how to test whether the probabilities truly have this pattern.

1.31 Refer to Section 1.5.6. Using the likelihood function to obtain the information, find the approximate standard error of $\hat{\pi}$.

1.32 Refer to Section 1.5.6. Let a denote the number of calves that got a primary, secondary, and tertiary infection, b the number that received a primary and secondary but not a tertiary infection, c the number that received a primary but not a secondary infection, and d the number that did not receive a primary infection. Let π be the probability of a primary infection. Consider the hypothesis that the probability of infection at time t, given infection at times $1, \ldots, t - 1$, is also π, for $t = 2, 3$. Show that $\hat{\pi} = (3a + 2b + c)/(3a + 3b + 2c + d)$.

1.33 Refer to quadratic form (1.16).

　a. Verify that the matrix quoted in the text for Σ_0^{-1} is the inverse of Σ_0.

　b. Show that (1.16) simplifies to Pearson's statistic (1.15).

　c. For the z_S statistic (1.11), show that $z_S^2 = X^2$ for $c = 2$.

1.34 For testing $H_0: \pi_j = \pi_{j0}$, $j = 1, \ldots, c$, using sample multinomial proportions $\{\hat{\pi}_j\}$, the likelihood-ratio statistic (1.17) is

$$G^2 = -2n \sum_j \hat{\pi}_j \log(\pi_{j0}/\hat{\pi}_j).$$

Show that $G^2 \geq 0$, with equality if and only if $\hat{\pi}_j = \pi_{j0}$ for all j. (*Hint:* Apply Jensen's inequality to $E(-2n \log X)$, where X equals $\pi_{j0}/\hat{\pi}_j$ with probability $\hat{\pi}_j$.)

1.35 The chi-squared mgf with df $= \nu$ is $m(t) = (1 - 2t)^{-\nu/2}$, for $|t| < \frac{1}{2}$. Use it to prove the reproductive property of the chi-squared distribution.

1.36 For the multinomial $(n, \{\pi_j\})$ distribution with $c > 2$, confidence limits for π_j are the solutions of

$$\left(\hat{\pi}_j - \pi_j\right)^2 = \left(z_{\alpha/2c}\right)^2 \pi_j(1 - \pi_j)/n, \qquad j = 1, \ldots, c.$$

a. Using the Bonferroni inequality, argue that these c intervals simultaneously contain all $\{\pi_j\}$ (for large samples) with probability at least $1 - \alpha$.

b. Show that the standard deviation of $\hat{\pi}_j - \hat{\pi}_k$ is $[\pi_j + \pi_k - (\pi_j - \pi_k)^2]/n$. For large n, explain why the probability is at least $1 - \alpha$ that the Wald confidence intervals

$$\left(\hat{\pi}_j - \hat{\pi}_k\right) \pm z_{\alpha/2a}\left\{\left[\hat{\pi}_j + \hat{\pi}_k - \left(\hat{\pi}_j - \hat{\pi}_k\right)^2\right]/n\right\}^{1/2}$$

simultaneously contain the $a = c(c - 1)/2$ differences $\{\pi_j - \pi_k\}$ (see Fitzpatrick and Scott 1987; Goodman 1965).

CHAPTER 2

Describing Contingency Tables

In this chapter we introduce tables that display relationships between categorical variables. We also define parameters that summarize their association. Parameters in Section 2.2 are used to compare groups on the proportions of responses in the outcome categories. The *odds ratio* has special importance, appearing as a parameter in models discussed later. In Section 2.3 we extend the scope by controlling for a third variable. The association can change dramatically under a control. The chapter's primary focus is binary variables, which have only two categories, but in Section 2.4 we present parameters for nominal and ordinal multicategory variables. First, in Section 2.1, we introduce basic terminology and notation.

2.1 PROBABILITY STRUCTURE FOR CONTINGENCY TABLES

The joint distribution between two categorical variables determines their relationship. This distribution also determines the marginal and conditional distributions.

2.1.1 Contingency Tables and Their Distributions

Let X and Y denote two categorical response variables, X with I categories and Y with J categories. Classifications of subjects on both variables have IJ possible combinations. The responses (X, Y) of a subject chosen randomly from some population have a probability distribution. A rectangular table having I rows for categories of X and J columns for categories of Y displays this distribution. The *cells* of the table represent the IJ possible outcomes. When the cells contain frequency counts of outcomes for a sample, the table is called a *contingency table*, a term introduced by Karl Pearson (1904). Another name is *cross-classification table*. A contingency table with I rows and J columns is called an $I \times J$ (or I-by-J) table.

TABLE 2.1   Cross-Classification of Aspirin Use and Myocardial Infarction

	Myocardial Infarction		
	Fatal Attack	Nonfatal Attack	No Attack
Placebo	18	171	10,845
Aspirin	5	99	10,933

Source: Preliminary report: Findings from the aspirin component of the ongoing Physicians' Health Study. *New Engl. J. Med.* **318**: 262–264 (1988).

Table 2.1, a 2×3 contingency table, is from a report on the relationship between aspirin use and heart attacks by the Physicians' Health Study Research Group at Harvard Medical School. The Physicians' Health Study was a 5-year randomized study of whether regular aspirin intake reduces mortality from cardiovascular disease. Every other day, physicians participating in the study took either one aspirin tablet or a placebo. The study was *blind*—those in the study did not know whether they were taking aspirin or a placebo. Of the 11,034 physicians taking a placebo, 18 suffered fatal heart attacks over the course of the study, whereas of the 11,037 taking aspirin, 5 had fatal heart attacks.

Let π_{ij} denote the probability that (X, Y) occurs in the cell in row i and column j. The probability distribution $\{\pi_{ij}\}$ is the *joint distribution* of X and Y. The *marginal distributions* are the row and column totals that result from summing the joint probabilities. We denote these by $\{\pi_{i+}\}$ for the row variable and $\{\pi_{+j}\}$ for the column variable, where the subscript "$+$" denotes the sum over that index; that is,

$$\pi_{i+} = \sum_j \pi_{ij} \quad \text{and} \quad \pi_{+j} = \sum_i \pi_{ij}.$$

These satisfy $\sum_i \pi_{i+} = \sum_j \pi_{+j} = \sum_i \sum_j \pi_{ij} = 1.0$. The marginal distributions provide single-variable information.

In most contingency tables (such as Table 2.1), one variable, say Y, is a response variable and the other (X) is an explanatory variable. When X is fixed rather than random, the notion of a joint distribution for X and Y is no longer meaningful. However, for a fixed category of X, Y has a probability distribution. It is germane to study how this distribution changes as the category of X changes. Given that a subject is classified in row i of X, $\pi_{j|i}$ denotes the probability of classification in column j of Y, $j = 1, \ldots, J$. Note that $\sum_j \pi_{j|i} = 1$. The probabilities $\{\pi_{1|i}, \ldots, \pi_{J|i}\}$ form the *conditional distribution* of Y at category i of X. A principal aim of many studies is to compare conditional distributions of Y at various levels of explanatory variables.

TABLE 2.2   Estimated Conditional Distributions for
Breast Cancer Diagnoses

| Breast | Diagnosis of Test | | |
Cancer	Positive	Negative	Total
Yes	0.82	0.18	1.0
No	0.01	0.99	1.0

Source: Data from W. Lawrence et al., *J. Natl. Cancer Inst.*
90: 1792–1800 (1998).

2.1.2   Sensitivity and Specificity

The results in Table 2.2 are from a recent article about various methods of attempting to diagnose breast cancer. Based on a literature survey, the authors reported these results for the impact of using mammography together with clinical breast examination. Let X = true disease status (i.e., whether a woman truly has breast cancer) and let Y = diagnosis (positive, negative), where a positive outcome predicts that a woman has breast cancer. The probabilities estimated in Table 2.2 are conditional probabilities of Y given X.

With diagnostic tests for a disease, the two correct diagnoses are a positive test outcome when the subject has the disease and a negative test outcome when a subject does not have it. Given that the subject has the disease, the conditional probability that the diagnostic test is positive is called the *sensitivity*; given that the subject does not have the disease, the conditional probability that the test is negative is called the *specificity* (Yerushalmy 1947). Ideally, these are both high.

For a 2×2 table with the format of Table 2.2, sensitivity is $\pi_{1|1}$ and specificity is $\pi_{2|2}$. In Table 2.2, the estimated sensitivity of combined mammography and clinical examination is 0.82. Of women with breast cancer, 82% are diagnosed correctly. The estimated specificity is 0.99. Of women not having breast cancer, 99% were diagnosed correctly.

2.1.3   Independence of Categorical Variables

When both variables are response variables, descriptions of the association can use their joint distribution, the conditional distribution of Y given X, or the conditional distribution of X given Y. The conditional distribution of Y given X relates to the joint distribution by

$$\pi_{j|i} = \pi_{ij}/\pi_{i+} \qquad \text{for all } i \text{ and } j.$$

Two categorical response variables are defined to be *independent* if all joint probabilities equal the product of their marginal probabilities,

$$\pi_{ij} = \pi_{i+}\pi_{+j} \qquad \text{for } i = 1,\ldots,I \quad \text{and} \quad j = 1,\ldots,J. \qquad (2.1)$$

TABLE 2.3   Notation for Joint, Conditional, and Marginal Probabilities

	Column		
Row	1	2	Total
1	π_{11} $(\pi_{1\|1})$	π_{12} $(\pi_{2\|1})$	π_{1+} (1.0)
2	π_{21} $(\pi_{1\|2})$	π_{22} $(\pi_{2\|2})$	π_{2+} (1.0)
Total	π_{+1}	π_{+2}	1.0

When X and Y are independent,

$$\pi_{j|i} = \pi_{ij}/\pi_{i+} = (\pi_{i+}\pi_{+j})/\pi_{i+} = \pi_{+j} \qquad \text{for } i = 1,\ldots,I.$$

Each conditional distribution of Y is identical to the marginal distribution of Y. Thus, two variables are independent when $\{\pi_{j|1} = \cdots = \pi_{j|I}$, for $j = 1,\ldots,J\}$; that is, the probability of any given column response is the same in each row. When Y is a response and X is an explanatory variable, this is a more natural way to define independence than (2.1). Independence is then often referred to as *homogeneity* of the conditional distributions.

Table 2.3 displays notation for joint, conditional, and marginal distributions for the 2×2 case. Sample distributions use similar notation, with p or $\hat{\pi}$ in place of π. For instance, $\{p_{ij}\}$ denotes the sample joint distribution. The cell frequencies are denoted $\{n_{ij}\}$, and $n = \sum_i \sum_j n_{ij}$ is the total sample size. Thus,

$$p_{ij} = n_{ij}/n.$$

The sample proportion of times that subjects in row i made response j is

$$p_{j|i} = p_{ij}/p_{i+} = n_{ij}/n_{i+},$$

where $n_{i+} = np_{i+} = \sum_j n_{ij}$.

2.1.4   Poisson, Binomial, and Multinomial Sampling

The probability distributions introduced in Section 1.2 extend to cell counts in contingency tables. For instance, a Poisson sampling model treats cell counts $\{Y_{ij}\}$ as independent Poisson random variables with parameters $\{\mu_{ij}\}$. The joint probability mass function for potential outcomes $\{n_{ij}\}$ is then the product of the Poisson probabilities $P(Y_{ij} = n_{ij})$ for the IJ cells, or

$$\prod_i \prod_j \exp(-\mu_{ij})\mu_{ij}^{n_{ij}}/n_{ij}! \,.$$

When the total sample size n is fixed but the row and column totals are not, a *multinomial sampling* model applies. The IJ cells are the possible outcomes. The probability mass function of the cell counts has the multinomial form

$$[n!/(n_{11}! \cdots n_{IJ}!)] \prod_i \prod_j \pi_{ij}^{n_{ij}} .$$

Often, observations on a response Y occur separately at each setting of an explanatory variable X. This case normally treats row totals as fixed, and for simplicity, we use the notation $n_i = n_{i+}$. Suppose that the n_i observations on Y at setting i of X are independent, each with probability distribution $\{\pi_{1|i}, \ldots, \pi_{J|i}\}$. The counts $\{n_{ij}, j = 1, \ldots, J\}$ satisfying $\sum_j n_{ij} = n_i$ then have the multinomial form

$$\frac{n_i!}{\prod_j n_{ij}!} \prod_j \pi_{j|i}^{n_{ij}}. \tag{2.2}$$

When samples at different settings of X are independent, the joint probability function for the entire data set is the product of the multinomial functions (2.2) from the various settings. This sampling scheme is *independent multinomial sampling*, also called *product multinomial sampling*.

Independent multinomial sampling also results under the following conditions: Suppose that $\{n_{ij}\}$ result from either independent Poisson sampling with means $\{\mu_{ij}\}$ or multinomial sampling over the IJ cells with probabilities $\{\pi_{ij} = \mu_{ij}/n\}$. When X is an explanatory variable, it is sensible to perform statistical inference conditional on the totals $\{n_i = \sum_j n_{ij}\}$ even when their values are not fixed by the sampling design. Conditional on $\{n_i\}$, the cell counts $\{n_{ij}, j = 1, \ldots, J\}$ have the multinomial distribution (2.2) with response probabilities $\{\pi_{j|i} = \mu_{ij}/\mu_{i+}, j = 1, \ldots, J\}$, and cell counts from different rows are independent. With this conditioning, we treat the row totals as fixed and analyze the data as if they formed separate independent samples.

Sometimes both row and column margins are naturally fixed. The appropriate sampling distribution is then the *hypergeometric*. In Section 3.5.1 we discuss this case, which is less common.

2.1.5 Seat Belt Example

Researchers in the Massachusetts Highway Department plan to study the relationship between seat-belt use (yes, no) and outcome of an automobile crash (fatality, nonfatality) for drivers involved in accidents on the Massachusetts Turnpike. They will summarize results in the format shown in Table 2.4. They plan to catalog all accidents on the turnpike for the next year, classifying each according to these variables. The total sample size is

TABLE 2.4   Seat-Belt Use and Results of Automobile Crashes

Seat-Belt Use	Result of Crash	
	Fatality	Nonfatality
Yes		
No		

then a random variable. They might treat the numbers of observations at the four combinations of seat-belt use and outcome of crash as independent Poisson random variables with unknown means $\{\mu_{11}, \mu_{12}, \mu_{21}, \mu_{22}\}$.

Suppose, instead, that the researchers randomly sample 200 police records of crashes on the turnpike in the past year and classify each according to seat-belt use and outcome of crash. For this study, the total sample size n is fixed. They might then treat the four cell counts as a multinomial random variable with $n = 200$ trials and unknown joint probabilities $\{\pi_{11}, \pi_{12}, \pi_{21}, \pi_{22}\}$.

Suppose, instead, that police records for accidents involving fatalities were filed separately from the others. The researchers might instead randomly sample 100 records of accidents with a fatality and randomly sample 100 records of accidents with no fatality. This approach fixes the column totals in Table 2.4 at 100. They might then regard each column of Table 2.4 as an independent binomial sample. Yet another approach, the traditional experimental design, takes 200 subjects and randomly assigns 100 of them to wear seat belts; the 200 then all are forced to have an accident. The recorded results would then be independent binomial samples in each row, with fixed row totals of 100 each. (Obviously, traditional designs common in some experimental science may not be ethical for humans. This is especially true in medical studies.)

2.1.6   Types of Studies

Table 2.5 comes from one of the first studies of the link between lung cancer and smoking, by Richard Doll and A. Bradford Hill. In 20 hospitals in London, England, patients admitted with lung cancer in the preceding year were queried about their smoking behavior. For each of the 709 patients admitted, researchers studied the smoking behavior of a noncancer patient at the same hospital of the same gender and within the same 5-year grouping on age. The 709 *cases* in the first column of Table 2.5 are those having lung cancer and the 709 *controls* in the second column are those not having it. A smoker was defined as a person who had smoked at least one cigarette a day for at least a year.

Normally, whether lung cancer occurs is a response variable and smoking behavior is an explanatory variable. In this study, however, the marginal

**TABLE 2.5   Cross-Classification of Smoking by
Lung Cancer**

Smoker	Lung Cancer	
	Cases	Controls
Yes	688	650
No	21	59
Total	709	709

Source: Based on data reported in Table IV, R. Doll and A. B.
Hill, *British Med. J.*, Sept. 30, 1950, pp. 739–748.

distribution of lung cancer is fixed by the sampling design, and the outcome
measured is whether the subject ever was a smoker. The study, which uses a
retrospective design to "look into the past," is called a *case–control study.* Such
studies are common in health-related applications. Often, the two samples
are matched, as in this study. Sometimes the samples of cases and controls
are independent rather than matched. For instance, another early case–control study on lung cancer and smoking sampled subjects by sending letters to
the estates of physicians who had died of some type of cancer in 1950 or
1951, and observations were cross-classified on type of cancer and the
subject's smoking behavior (see, e.g., Cornfield 1956).

One might want to compare smokers with nonsmokers in terms of the
proportion who suffered lung cancer. These proportions refer to the conditional distribution of lung cancer, given smoking behavior. Instead, case–control studies provide proportions in the reverse direction, for the conditional
distribution of smoking behavior, given lung cancer status. For those in Table
2.5 with lung cancer, the proportion who were smokers was $688/709 = 0.970$,
while it was $650/709 = 0.917$ for the controls.

When we know the proportion of the population having lung cancer, we
can use Bayes' theorem to compute sample conditional distributions in the
direction of main interest (Problem 2.21). Otherwise, using a retrospective
sample, we cannot estimate the probability of lung cancer at each category of
smoking behavior. For Table 2.5 we do not know the population prevalence
of lung cancer, and the patients suffering it were probably sampled at a rate
far in excess of their occurrence in the general population.

By contrast, imagine a study that samples subjects from the population of
teenagers and then 60 years later measures the rates of lung cancer for the
smokers and nonsmokers. Such a sampling design is *prospective.* There are
two types of prospective studies. *Clinical trials* randomly allocate subjects to
the groups who will be smokers and nonsmokers. In *cohort studies*, subjects
make their own choice about whether to smoke, and the study observes in
future time who develops lung cancer. Yet another approach, a *cross-sectional design*, samples subjects and classifies them simultaneously on both
variables.

Prospective studies usually condition on the totals $\{n_i = \sum_j n_{ij}\}$ for categories of X and regard each row of J counts as an independent multinomial sample on Y. *Retrospective studies* usually treat the totals $\{n_{+j}\}$ for Y as fixed and regard each column of I counts as a multinomial sample on X. In *cross-sectional studies*, the total sample size is fixed but not the row or column totals, and the IJ cell counts are a multinomial sample.

Case–control, cohort, and cross-sectional studies are called *observational studies*. They simply observe who chooses each group and who has the outcome of interest. By contrast, a clinical trial is an *experimental* study, the investigator having the advantage of experimental control over which subjects receive each treatment. Such studies can use the power of randomization to make the groups balance roughly on other variables that may be associated with the response. Observational studies are common but have more potential for biases of various types.

2.2 COMPARING TWO PROPORTIONS

Many studies are designed to compare groups on a binary response variable. Then Y has only two categories, such as (success, failure) for outcome of a medical treatment. With two groups, a 2×2 contingency table displays the results. The rows are the groups and the columns are the categories of Y. This section presents parameters for comparing the groups.

2.2.1 Difference of Proportions

For subjects in row i, $\pi_{1|i}$ is the probability that the response has outcome in category 1 ("success"). With only two possible outcomes, $\pi_{2|i} = 1 - \pi_{1|i}$, and we use the simpler notation π_i for $\pi_{1|i}$. The *difference of proportions* of successes, $\pi_1 - \pi_2$, is a basic comparison of the two rows. Comparison on failures is equivalent to comparison on successes, since

$$(1 - \pi_1) - (1 - \pi_2) = \pi_2 - \pi_1.$$

The difference of proportions falls between -1.0 and $+1.0$. It equals zero when the rows have identical conditional distributions. The response Y is statistically independent of the row classification when $\pi_1 - \pi_2 = 0$.

When both variables are responses, conditional distributions apply in either direction. One can also compare the two columns, such as by the difference between the proportions in row 1. This usually is not equal to the difference $\pi_1 - \pi_2$ comparing the rows.

2.2.2 Relative Risk

A value $\pi_1 - \pi_2$ of fixed size may have greater importance when both π_i are close to 0 or 1 than when they are not. For a study comparing two treatments on the proportion of subjects who die, the difference between 0.010 and 0.001 may be more noteworthy than the difference between 0.410 and 0.401, even though both are 0.009. In such cases, the ratio of proportions is also informative.

The *relative risk* is defined to be the ratio

$$\pi_1/\pi_2. \tag{2.3}$$

It can be any nonnegative real number. A relative risk of 1.0 corresponds to independence. For the proportions just given, the relative risks are $0.010/0.001 = 10.0$ and $0.410/0.401 = 1.02$. Comparing the rows on the second response category gives a different relative risk, $(1 - \pi_1)/(1 - \pi_2)$.

2.2.3 Odds Ratio

For a probability π of success, the *odds* are defined to be

$$\Omega = \pi/(1 - \pi).$$

The odds are nonnegative, with $\Omega > 1.0$ when a success is more likely than a failure. When $\pi = 0.75$, for instance, then $\Omega = 0.75/0.25 = 3.0$; a success is three times as likely as a failure, and we expect about three successes for every one failure. When $\Omega = \frac{1}{3}$, a failure is three times as likely as a success. Inversely,

$$\pi = \Omega/(\Omega + 1) .$$

For instance, when $\Omega = \frac{1}{3}$, then $\pi = 0.25$.

Refer again to a 2×2 table. Within row i, the odds of success instead of failure are $\Omega_i = \pi_i/(1 - \pi_i)$. The ratio of the odds Ω_1 and Ω_2 in the two rows,

$$\theta = \frac{\Omega_1}{\Omega_2} = \frac{\pi_1/(1 - \pi_1)}{\pi_2/(1 - \pi_2)} \tag{2.4}$$

is called the *odds ratio*.

For joint distributions with cell probabilities $\{\pi_{ij}\}$, the equivalent definition for the odds in row i is $\Omega_i = \pi_{i1}/\pi_{i2}$, $i = 1, 2$. Then the odds ratio is

$$\theta = \frac{\pi_{11}/\pi_{12}}{\pi_{21}/\pi_{22}} = \frac{\pi_{11}\pi_{22}}{\pi_{12}\pi_{21}}. \tag{2.5}$$

An alternative name for θ is the *cross-product ratio*, since it equals the ratio of the products $\pi_{11}\pi_{22}$ and $\pi_{12}\pi_{21}$ of probabilities from diagonally opposite cells (Yule 1900, 1912).

2.2.4 Properties of the Odds Ratio

The odds ratio can equal any nonnegative number. The condition $\Omega_1 = \Omega_2$ and hence (when all cell probabilities are positive) $\theta = 1$ corresponds to independence of X and Y. When $1 < \theta < \infty$, subjects in row 1 are more likely to have a success than are subjects in row 2; that is, $\pi_1 > \pi_2$. For instance, when $\theta = 4$, the odds of success in row 1 are four times the odds in row 2. This does not mean that the *probability* $\pi_1 = 4\pi_2$; that is the interpretation of a *relative risk* of 4.0. When $0 < \theta < 1$, $\pi_1 < \pi_2$. When one cell has zero probability, θ equals 0 or ∞.

Values of θ farther from 1.0 in a given direction represent stronger association. Two values represent the same association, but in opposite directions, when one is the inverse of the other. For instance, when $\theta = 0.25$, the odds of success in row 1 are 0.25 times the odds in row 2, or equivalently, the odds of success in row 2 are $1/0.25 = 4.0$ times the odds in row 1. When the order of the rows is reversed or the order of the columns is reversed, the new value for θ is the inverse of the original value.

For inference, we shall see it is convenient to use $\log \theta$. Independence corresponds to $\log \theta = 0$. The log odds ratio is symmetric about this value— reversal of rows or of columns results in a change in its sign. Two values for $\log \theta$ that are the same except for sign, such as $\log 4 = 1.39$ and $\log 0.25 = -1.39$, represent the same strength of association.

The odds ratio does not change value when the orientation of the table reverses so that the rows become the columns and the columns become the rows. This is clear from the symmetric form of (2.5). It is unnecessary to identify one classification as the response variable in order to use θ. In fact, although (2.4) defined it in terms of odds using $\pi_i = P(Y = 1 \mid X = i)$, one could just as well define it using reverse conditional probabilities. With a joint distribution, conditional distributions exist in each direction, and

$$\theta = \frac{\pi_{11}\pi_{22}}{\pi_{12}\pi_{21}} = \frac{P(Y=1 \mid X=1)/P(Y=2 \mid X=1)}{P(Y=1 \mid X=2)/P(Y=2 \mid X=2)}$$

$$= \frac{P(X=1 \mid Y=1)/P(X=2 \mid Y=1)}{P(X=1 \mid Y=2)/P(X=2 \mid Y=2)}. \tag{2.6}$$

In fact, the odds ratio is equally valid for prospective, retrospective, or cross-sectional sampling designs. The sample odds ratio estimates the same parameter in each case.

For cell counts $\{n_{ij}\}$, the sample odds ratio is

$$\hat{\theta} = n_{11}n_{22}/n_{12}n_{21}.$$

This does not change when both cell counts within any row are multiplied by a nonzero constant or when both cell counts within any column are multiplied by a nonzero constant. An implication is that the sample odds ratio estimates the same characteristic (θ) even when the sample is disproportionately large or small from marginal categories of a variable. For a retrospective study of the association between vaccination and catching a certain strain of flu, the sample odds ratio estimates the same characteristic with a random sample of (1) 100 people who got the flu and 100 people who did not, or (2) 40 people who got the flu and 160 people who did not. The sample versions of the difference of proportions and relative risk (2.3) are invariant to multiplication of counts within rows by a constant, but they change with multiplication within columns or with row–column interchange.

2.2.5 Aspirin and Heart Attacks Revisited

We illustrate the three association measures with Table 2.1 on aspirin use and heart attacks. The table differentiates between fatal and nonfatal heart attacks, but we combine these outcomes for now. Of the 11,034 physicians taking placebo, 189 suffered heart attacks, a proportion of $189/11{,}034 = 0.0171$. Of the 11,037 taking aspirin, 104 had heart attacks, a proportion of 0.0094. The sample difference of proportions is $0.0171 - 0.0094 = 0.0077$. The relative risk is $0.0171/0.0094 = 1.82$. The proportion suffering heart attacks of those taking placebo was 1.82 times the proportion suffering heart attacks of those taking aspirin. The sample odds ratio is $(189 \times 10{,}933)/(10{,}845 \times 104) = 1.83$. The odds of heart attack for those taking placebo was 1.83 times the odds for those taking aspirin.

2.2.6 Case–Control Studies and the Odds Ratio

With retrospective sampling designs, such as case–control studies, it is possible to estimate conditional probabilities of form $P(X = i \mid Y = j)$. It is usually not possible to estimate the probability $P(Y = j \mid X = i)$ of an outcome of interest or the difference of proportions or relative risk for that outcome. It is possible to estimate the odds ratio, however, since by (2.6) it is determined by conditional probabilities in *either* direction.

To illustrate, we revisit Table 2.5 on $X =$ smoking behavior and $Y =$ lung cancer. The data were two binomial samples on X at fixed levels of Y. Thus, we can estimate the probability a subject was a smoker, given the outcome on whether the subject had lung cancer; this was $688/709$ for the cases and $650/709$ for the controls. We cannot estimate the probability of lung cancer, given whether one smoked, which is more relevant. Thus, we cannot estimate differences or ratios of probabilities of lung cancer. The difference of proportions and relative risk are limited to comparisons of the probabilities of being a smoker. However, we can compute the odds ratio using the sample analog of (2.6),

$$\frac{(688/709)/(21/709)}{(650/709)/(59/709)} = \frac{688 \times 59}{650 \times 21} = 3.0.$$

Moreover, by (2.6), interpretations can use the direction of interest, even though the study was retrospective: The estimated odds of lung cancer for smokers were 3.0 times the estimated odds for nonsmokers.

2.2.7 Relationship between Odds Ratio and Relative Risk

From definitions (2.3) and (2.4),

$$\text{odds ratio} = \text{relative risk}\left(\frac{1 - \pi_2}{1 - \pi_1}\right).$$

Their magnitudes are similar whenever the probability π_i of the outcome of interest is close to zero for both groups. We saw this similarity in Section 2.2.5 for the aspirin study, where the heart attack proportion was less than 0.02 for each group. The relative risk was 1.82 and the odds ratio was 1.83.

Because of this similarity, when each π_i is small, the odds ratio provides a rough indication of the relative risk when it is not directly estimable, such as in case–control studies (Cornfield 1951). For instance, for Table 2.5, if the probability of lung cancer is small regardless of smoking behavior, 3.0 is also a rough estimate of the relative risk; that is, smokers had about 3.0 times the relative frequency of lung cancer as nonsmokers.

2.3 PARTIAL ASSOCIATION IN STRATIFIED 2 × 2 TABLES

An important part of most studies, especially observational studies, is the choice of control variables. In studying the effect of X on Y, one should control any covariate that can influence that relationship. This involves using some mechanism to hold the covariate constant. Otherwise, an observed effect of X on Y may actually reflect effects of that covariate on both X and Y. The relationship between X and Y then shows *confounding*. Experimental studies can remove effects of confounding covariates by randomly assigning subjects to different levels of X, but this is not possible with observational studies.

Suppose that a study considers effects of passive smoking, the effects on a nonsmoker of living with a smoker. To analyze whether passive smoking is associated with lung cancer, a cross-sectional study might compare lung cancer rates between nonsmokers whose spouses smoke and nonsmokers whose spouses do not smoke. The study should attempt to control for age, socioeconomic status, or other factors that might relate both to spouse smoking and to developing lung cancer. Otherwise, results will have limited usefulness. Spouses of nonsmokers may tend to be younger than spouses of smokers, and younger people are less likely to have lung cancer. Then a lower proportion of lung cancer cases among spouses of nonsmokers may merely reflect their lower average age.

In this section we discuss the analysis of the association between categorical variables X and Y while controlling for a possibly confounding variable Z. For simplicity, the examples refer to a single control variable. In later chapters we treat more general cases and discuss the use of models to perform statistical control.

2.3.1 Partial Tables

We control for Z by studying the XY relationship at fixed levels of Z. Two-way cross-sectional slices of the three-way contingency table cross classify X and Y at separate categories of Z. These cross sections are called *partial tables*. They display the XY relationship while removing the effect of Z by holding its value constant.

The two-way contingency table obtained by combining the partial tables is called the *XY marginal table*. Each cell count in the marginal table is a sum of counts from the same location in the partial tables. The marginal table, rather than controlling Z, ignores it. The marginal table contains no information about Z. It is simply a two-way table relating X and Y but may reflect the effects of Z on X and Y.

The associations in partial tables are called *conditional associations*, because they refer to the effect of X on Y conditional on fixing Z at some level. Conditional associations in partial tables can be quite different from associations in marginal tables. In fact, it can be misleading to analyze only marginal tables of a multiway contingency table. The following example illustrates.

2.3.2 Death Penalty Example

Table 2.6 is a $2 \times 2 \times 2$ contingency table–two rows, two columns, and two layers–from an article that studied effects of racial characteristics on whether persons convicted of homicide received the death penalty. The 674 subjects classified in Table 2.6 were the defendants in indictments involving cases

TABLE 2.6   Death Penalty Verdict by Defendant's Race and Victims' Race

Victims' Race	Defendant's Race	Death Penalty		Percent Yes
		Yes	No	
White	White	53	414	11.3
	Black	11	37	22.9
Black	White	0	16	0.0
	Black	4	139	2.8
Total	White	53	430	11.0
	Black	15	176	7.9

Source: M. L. Radelet and G. L. Pierce, *Florida Law Rev.* **43**: 1–34 (1991). Reprinted with permission from the *Florida Law Review.*

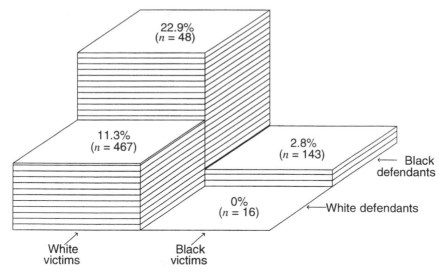

FIGURE 2.1   Percent receiving death penalty.

with multiple murders in Florida between 1976 and 1987. The variables in Table 2.6 are Y = death penalty verdict, having the categories (yes, no), X = race of defendant, and Z = race of victims, each having the categories (white, black). We study the effect of defendant's race on the death penalty verdict, treating victims' race as a control variable. Table 2.6 has a 2 × 2 partial table relating defendant's race and the death penalty verdict at each category of victims' race.

For each combination of defendant's race and victims' race, Table 2.6 lists and Figure 2.1 displays the percentage of defendants who received the death penalty. These describe the conditional associations. When the victims were white, the death penalty was imposed 22.9% − 11.3% = 11.6% more often for black defendants than for white defendants. When the victims were black, the death penalty was imposed 2.8% more often for black defendants than for white defendants. *Controlling* for victims' race by keeping it fixed, the death penalty was imposed more often on black defendants than on white defendants.

The bottom portion of Table 2.6 displays the marginal table. It results from summing the cell counts in Table 2.6 over the two categories of victims' race, thus combining the two partial tables (e.g., 11 + 4 = 15). Overall, 11.0% of white defendants and 7.9% of black defendants received the death penalty. *Ignoring* victims' race, the death penalty was imposed less often on black defendants than on white defendants. The association reverses direction compared to the partial tables.

Why does the association change so much when we ignore versus control victims' race? This relates to the nature of the association between victims' race and each of the other variables. First, the association between victims'

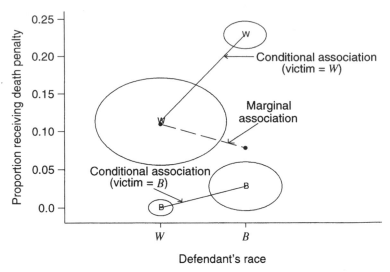

FIGURE 2.2   Proportion receiving death penalty by defendant's race, controlling and ignoring victims' race.

race and defendant's race is extremely strong. The marginal table relating these variables has odds ratio $(467 \times 143)/(48 \times 16) = 87.0$. Second, Table 2.6 shows that, regardless of defendant's race, the death penalty was much more likely when the victims were white than when the victims were black. So whites are tending to kill whites, and killing whites is more likely to result in the death penalty. This suggests that the marginal association should show a greater tendency than the conditional associations for white defendants to receive the death penalty. In fact, Table 2.6 has this pattern.

Figure 2.2 illustrates why the marginal association differs so from the conditional associations. For each defendant's race, the figure plots the proportion receiving the death penalty at each category of victims' race. Each proportion is labeled by a letter symbol giving the category of victims' race. Surrounding each observation is a circle having area proportional to the number of observations at that combination of defendant's race and victims' race. For instance, the W in the largest circle represents a proportion of 0.113 receiving the death penalty for cases with white defendants and white victims. That circle is largest because the number of cases at that combination $(53 + 414 = 467)$ is largest. The next-largest circle relates to cases in which blacks kill blacks.

We control for victims' race by comparing circles having the same victims' race letter at their centers. The line connecting the two W circles has a positive slope, as does the line connecting the two B circles. Controlling for victims' race, this reflects the death penalty being more likely for black defendants than for white defendants. When we add results across victims' race to get a summary result for the marginal effect of defendant's race on the death penalty verdict, the larger circles, having the greater number of cases, have greater influence. Thus, the summary proportions for each defendant's race, marked on the figure by periods, fall closer to the center of the larger circles than to the center of the smaller circles. A line connecting the summary marginal proportions has negative slope, indicating that overall the death penalty was more likely for white defendants than for black defendants.

The result that a marginal association can have a different direction from each conditional association is called *Simpson's paradox* (Simpson 1951, Yule 1903). It applies to quantitative as well as categorical variables. Statisticians commonly use it to caution against imputing causal effects from an association of X with Y. For instance, when doctors started to observe strong odds ratios between smoking and lung cancer, statisticians such as R. A. Fisher warned that some variable (e.g., a genetic factor) could exist such that the association would disappear under the relevant control. However, other statisticians (such as J. Cornfield) showed that with a very strong XY association, a very strong association must exist between the confounding variable Z and both X and Y in order for the effect to disappear or change under the control (Breslow and Day 1980, Sec. 3.4).

2.3.3 Conditional and Marginal Odds Ratios

Odds ratios can describe marginal and conditional associations. We illustrate for $2 \times 2 \times K$ tables, where K denotes the number of categories of a control variable, Z. Let $\{\mu_{ijk}\}$ denote cell expected frequencies for some sampling model, such as binomial, multinomial, or Poisson sampling.

Within a fixed category k of Z, the odds ratio

$$\theta_{XY(k)} = \frac{\mu_{11k} \mu_{22k}}{\mu_{12k} \mu_{21k}} \tag{2.7}$$

describes conditional XY association in partial table k. The odds ratios for the K partial tables are called XY *conditional odds ratios*. These can be quite different from marginal odds ratios. The XY marginal table has expected frequencies $\{\mu_{ij+} = \sum_k \mu_{ijk}\}$. The XY marginal odds ratio is

$$\theta_{XY} = \frac{\mu_{11+} \mu_{22+}}{\mu_{12+} \mu_{21+}} .$$

Sample values of $\theta_{XY(k)}$ and θ_{XY} use similar formulas with cell counts substituted for expected frequencies. We illustrate for the association between defendant's race and the death penalty in Table 2.6. In the first partial table, victims' race is white and

$$\hat{\theta}_{XY(1)} = \frac{53 \times 37}{414 \times 11} = 0.43.$$

The sample odds for white defendants receiving the death penalty were 43% of the sample odds for black defendants. In the second partial table, victims' race is black and the estimated odds ratio equals $\hat{\theta}_{XY(2)} = (0 \times 139)(16 \times 4) = 0.0$, since the death penalty was never given to white defendants with black victims.

Estimation of the marginal odds ratio uses the 2×2 marginal table within Table 2.6, collapsing over victims' race, or $(53 \times 176)/(430 \times 15) = 1.45$. The sample odds of the death penalty were 45% higher for white defendants than for black defendants. Yet within each victims' race category, those odds were smaller for white defendants. This reversal in the association after controlling for victims' race illustrates Simpson's paradox.

2.3.4 Marginal versus Conditional Independence

More generally, X may have I categories and Y may have J categories. An $I \times J \times K$ table describes the relationship between X and Y, controlling for Z. If X and Y are independent in partial table k, then X and Y are called *conditionally independent at level k* of Z. When Y is a response, this means that

$$P(Y = j | X = i, Z = k) = P(Y = j | Z = k), \qquad \text{for all } i, j. \quad (2.8)$$

More generally, X and Y are said to be *conditionally independent given Z* when they are conditionally independent at every level of Z, that is, when (2.8) holds for all k. Then, given Z, Y does not depend on X.

Suppose that a single multinomial applies to the entire three-way table, with joint probabilities $\{\pi_{ijk} = P(X = i, Y = j, Z = k)\}$. Then

$$\pi_{ijk} = P(X = i, Z = k) \, P(Y = j | X = i, Z = k),$$

which under conditional independence of X and Y, given Z, equals

$$= \pi_{i+k} P(Y = j | Z = k) = \pi_{i+k} P(Y = j, Z = k)/P(Z = k) \, .$$

Thus, conditional independence is then equivalent to

$$\pi_{ijk} = \pi_{i+k} \pi_{+jk} / \pi_{++k} \qquad \text{for all } i, j, \text{ and } k. \quad (2.9)$$

TABLE 2.7 Expected Frequencies Showing That Conditional Independence Does Not Imply Marginal Independence

Clinic	Treatment	Response	
		Success	Failure
1	A	18	12
	B	12	8
2	A	2	8
	B	8	32
Total	A	20	20
	B	20	40

Conditional independence does not imply marginal independence (Yule 1903). For instance, summing (2.9) over k on both sides yields

$$\pi_{ij+} = \sum_k \left(\pi_{i+k} \pi_{+jk} / \pi_{++k} \right).$$

All three terms in the summation involve k, and this does not simplify to $\pi_{ij+} = \pi_{i++} \pi_{+j+}$, marginal independence.

For $2 \times 2 \times K$ tables, X and Y are conditionally independent when the odds ratio between X and Y equals 1 at each category of Z. The expected frequencies $\{\mu_{ijk}\}$ in Table 2.7 illustrate this relation for $Y =$ response (success, failure), $X =$ drug treatment (A, B), and $Z =$ clinic (1, 2). From (2.7), the conditional XY odds ratios are

$$\theta_{XY(1)} = \frac{18 \times 8}{12 \times 12} = 1.0, \qquad \theta_{XY(2)} = \frac{2 \times 32}{8 \times 8} = 1.0.$$

Given the clinic, response and treatment are conditionally independent. The marginal table combines the tables for the two clinics. Its odds ratio is $\theta_{XY} = (20 \times 40)/(20 \times 20) = 2.0$, so the variables are not marginally independent.

Ignoring the clinic, why are the odds of a success for treatment A twice those for treatment B? The conditional XZ and YZ odds ratios give a clue. The odds ratio between Z and either X or Y, at each fixed category of the other variable, equals 6.0. For instance, the XZ odds ratio at the first category of Y equals $(18 \times 8)/(12 \times 2) = 6.0$. The conditional odds (given response) of receiving treatment A at clinic 1 are six times those at clinic 2, and the conditional odds (given treatment) of success at clinic 1 are six times those at clinic 2. Clinic 1 tends to use treatment A more often, and clinic 1 also tends to have more successes. For instance, if patients at clinic 1 tended to be younger and in better health than those at clinic 2, perhaps they had a better success rate regardless of the treatment received.

It is misleading to study only the marginal table, concluding that successes are more likely with treatment A. Subjects within a particular clinic are likely to be more homogeneous than the overall sample, and response is independent of treatment in each clinic.

2.3.5 Homogeneous Association

A $2 \times 2 \times K$ table has *homogeneous XY association* when

$$\theta_{XY(1)} = \theta_{XY(2)} = \cdots = \theta_{XY(K)}.$$

Then the effect of X on Y is the same at each category of Z. Conditional independence of X and Y is the special case in which each $\theta_{XY(k)} = 1.0$.

Under homogeneous XY association, homogeneity also holds for the other associations. For instance, the conditional odds ratio between two categories of X and two categories of Z is identical at each category of Y. For the odds ratio, homogeneous association is a symmetric property. It applies to any pair of variables viewed across the categories of the third. When it occurs, there is said to be *no interaction* between two variables in their effects on the other variable.

When interaction exists, the conditional odds ratio for any pair of variables changes across categories of the third. For X = smoking (yes, no), Y = lung cancer (yes, no), and Z = age (< 45, $45-65$, > 65), suppose that $\theta_{XY(1)} = 1.2$, $\theta_{XY(2)} = 3.9$, and $\theta_{XY(3)} = 8.8$. Then smoking has a weak effect on lung cancer for young people, but the effect strengthens considerably with age. Age is called an *effect modifier*; the effect of smoking is modified depending on its value.

For the death penalty data (Table 2.6), $\hat{\theta}_{XY(1)} = 0.43$ and $\hat{\theta}_{XY(2)} = 0.0$. The values are not close, but the second estimate is unstable because of the zero cell count. Adding $\frac{1}{2}$ to each cell count, $\hat{\theta}_{XY(2)} = 0.94$. Because $\hat{\theta}_{XY(2)}$ is unstable and because further variation occurs from sampling variability, these partial tables do not necessarily contradict homogeneous association in a population. In Section 6.3 we show how to analyze whether sample data are consistent with homogeneous association or conditional independence.

2.4 EXTENSIONS FOR $I \times J$ TABLES

For 2×2 tables, a single number such as the odds ratio can summarize the association. For $I \times J$ tables, it is rarely possible to summarize association by a single number without some loss of information. However, a set of odds ratios or another summary index can describe certain features of the association.

2.4.1 Odds Ratios in $I \times J$ Tables

Odds ratios can use each of the $\binom{I}{2} = I(I - 1)/2$ pairs of rows in combination with each of the $\binom{J}{2} = J(J - 1)/2$ pairs of columns. For rows a and b and columns c and d, the odds ratio $(\pi_{ac}\pi_{bd})/(\pi_{bc}\pi_{ad})$ uses four cells in a rectangular pattern. There are $\binom{I}{2}\binom{J}{2}$ odds ratios of this type. This set of odds ratios contains much redundant information.

Consider the subset of $(I - 1)(J - 1)$ *local odds ratios*

$$\theta_{ij} = \frac{\pi_{ij}\pi_{i+1,j+1}}{\pi_{i,j+1}\pi_{i+1,j}}, \qquad i = 1,\ldots,I - 1, \quad j = 1,\ldots,J - 1. \quad (2.10)$$

Figure 2.3 shows that local odds ratios use cells in adjacent rows and adjacent columns. These $(I - 1)(J - 1)$ odds ratios determine all odds ratios formed from pairs of rows and pairs of columns. To illustrate, in Table 2.1, the sample local odds ratio is 2.08 for the first two columns and 1.74 for the

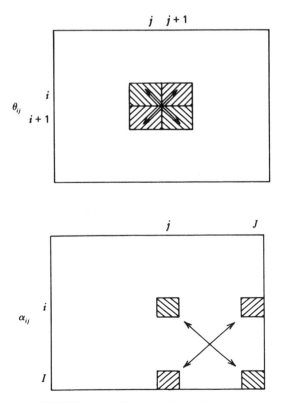

FIGURE 2.3 Odds ratios for $I \times J$ tables.

second and third columns. In each case, the more serious outcome was more prevalent for the placebo group. The product of these two odds ratios is 3.63, which is the odds ratio for the first and third columns.

Construction (2.10) for a minimal set of odds ratios is not unique. Another basic set is

$$\alpha_{ij} = \frac{\pi_{ij}\pi_{IJ}}{\pi_{Ij}\pi_{iJ}}, \qquad i = 1, \ldots, I-1, \quad j = 1, \ldots, J-1. \qquad (2.11)$$

This uses the rectangular pattern of cells determined by the cell in row i and column j and the cell in the last row and last column. Figure 2.3 illustrates.

Given the marginal distributions $\{\pi_{i+}\}$ and $\{\pi_{+j}\}$, when $\{\pi_{ij} > 0\}$, conversion of the probabilities into the set of odds ratios (2.10) or (2.11) does not discard information. The cell probabilities determine the odds ratios, and given the marginals, the odds ratios determine the cell probabilities. In this sense, $(I-1)(J-1)$ parameters can describe any association in an $I \times J$ table. Independence is equivalent to all $(I-1)(J-1)$ odds ratios equaling 1.0.

For three-way $I \times J \times K$ tables, sets of odds ratios in the partial tables describe the conditional association. Homogeneous XY association means that any conditional odds ratio formed using two categories of X and two categories of Y is the same at each category of Z.

2.4.2   Summary Measures of Association

An alternative way to describe association uses a single summary index. We discuss this first for nominal variables and then ordinal variables. The most interpretable indices for nominal variables have the same structure as R-squared for interval variables. It and the more general intraclass correlation coefficient and correlation ratio (Kendall and Stuart 1979) describe the proportional reduction in variance from the marginal distribution of the response Y to the conditional distributions of Y given an explanatory variable X.

Let $V(Y)$ denote a measure of variation for the marginal distribution $\{\pi_{+j}\}$ of Y, and let $V(Y|i)$ denote this measure computed for the conditional distribution $\{\pi_{1|i}, \ldots, \pi_{J|i}\}$ of Y at the ith setting of X. A proportional reduction in variation measure has the form

$$\frac{V(Y) - E[V(Y|X)]}{V(Y)}, \qquad (2.12)$$

where $E[V(Y|X)]$ is the expectation of the conditional variation taken with respect to the distribution of X. For the marginal distribution $\{\pi_{i+}\}$ of X, $E[V(Y|X)] = \sum_i \pi_{i+} V(Y|i)$.

For a nominal response, Theil (1970) proposed an index using the variation measure $V(Y) = \Sigma \pi_{+j} \log \pi_{+j}$, called the *entropy*. For contingency tables, the proportional reduction in entropy equals

$$U = -\frac{\Sigma_i \Sigma_j \pi_{ij} \log(\pi_{ij}/\pi_{i+}\pi_{+j})}{\Sigma_j \pi_{+j} \log \pi_{+j}}, \qquad (2.13)$$

called the *uncertainty coefficient*. This measure is well defined when more than one $\pi_{+j} > 0$. It takes value between 0 and 1: $U = 0$ is equivalent to independence of X and Y; $U = 1$ is equivalent to a lack of conditional variation, in the sense that for each i, $\pi_{j|i} = 1$ for some j.

Various measures of form (2.12) describe association in $I \times J$ tables (e.g., Problems 2.38 and 2.39). A difficulty with them is developing intuition for how large a value constitutes a strong association. What does it mean, for instance, to say that there is a 30% reduction in entropy? Summary measures seem easier to interpret and more useful when both classifications are ordinal, as discussed next.

2.4.3 Ordinal Trends: Concordant and Discordant Pairs

In Table 2.8 the variables are income and job satisfaction, measured for the black males in a national (U.S.) sample. Both classifications are ordinal, job satisfaction with the categories very dissatisfied (VD), little dissatisfied (LD), moderately satisfied (MS), and very satisfied (VS).

When X and Y are ordinal, a monotone trend association is common. As the level of X increases, responses on Y tend to increase toward higher levels, or responses on Y tend to decrease toward lower levels. For instance, perhaps job satisfaction tends to increase as income does. A single parameter can describe this trend. Measures analogous to the correlation describe the degree to which the relationship is monotone. Some measures are based on classifying each pair of subjects as concordant or discordant. A pair is *concordant* if the subject ranked higher on X also ranks higher on Y. The

TABLE 2.8   Cross-Classification of Job Satisfaction by Income

| Income (dollars) | Job Satisfaction | | | |
	Very Dissatisfied	Little Dissatisfied	Moderately Satisfied	Very Satisfied
< 15,000	1	3	10	6
15,000–25,000	2	3	10	7
25,000–40,000	1	6	14	12
> 40,000	0	1	9	11

Source: 1996 General Social Survey, National Opinion Research Center.

pair is *discordant* if the subject ranking higher on X ranks lower on Y. The pair is *tied* if the subjects have the same classification on X and/or Y.

We illustrate for Table 2.8. Consider a pair of subjects, one in the cell $(< 15, \text{VD})$ and the other in the cell $(15-25, \text{LD})$. This pair is concordant, since the second subject ranks higher than the first both on income and on job satisfaction. The subject in cell $(< 15, \text{VD})$ forms concordant pairs when matched with each of the three subjects classified $(15-25, \text{LD})$, so these two cells provide $1 \times 3 = 3$ concordant pairs. The subject in the cell $(< 15, \text{VD})$ is also part of a concordant pair when matched with each of the other $(10 + 7 + 6 + 14 + 12 + 1 + 9 + 11)$ subjects ranked higher on both variables. Similarly, the three subjects in the $(< 15, \text{LD})$ cell are part of concordant pairs when matched with the $(10 + 7 + 14 + 12 + 9 + 11)$ subjects ranked higher on both variables.

The total number of concordant pairs, denoted by C, equals

$$
\begin{aligned}
C = \; & 1(3 + 10 + 7 + 6 + 14 + 12 + 1 + 9 + 11) \\
& + 3(10 + 7 + 14 + 12 + 9 + 11) + 10(7 + 12 + 11) \\
& + 2(6 + 14 + 12 + 1 + 9 + 11) + 3(14 + 12 + 9 + 11) \\
& + 10(12 + 11) + 1(1 + 9 + 11) + 6(9 + 11) + 14(11) = 1331.
\end{aligned}
$$

The total number of discordant pairs of observations is

$$
D = 3(2 + 1 + 0) + 10(2 + 3 + 1 + 6 + 0 + 1) + \cdots + 12(0 + 1 + 9) = 849.
$$

In this example, $C > D$, suggesting a tendency for low income to occur with low job satisfaction and high income with high job satisfaction.

Consider two independent observations from a joint probability distribution $\{\pi_{ij}\}$. For that pair, the probabilities of concordance and discordance are

$$
\Pi_c = 2 \sum_i \sum_j \pi_{ij} \left(\sum_{h > i} \sum_{k > j} \pi_{hk} \right), \qquad \Pi_d = 2 \sum_i \sum_j \pi_{ij} \left(\sum_{h > i} \sum_{k < j} \pi_{hk} \right).
$$

Here i and j are fixed in the inner summations, and the factor of 2 occurs because the first observation could be in cell (i, j) and the second in cell (h, k), or vice versa. Several association measures for ordinal variables utilize the difference $\Pi_c - \Pi_d$.

2.4.4 Ordinal Measure of Association: Gamma

Given that a pair is untied on both variables, $\Pi_c / (\Pi_c + \Pi_d)$ is the probability of concordance and $\Pi_d / (\Pi_c + \Pi_d)$ is the probability of discordance. The difference between these probabilities is

$$\gamma = \frac{\Pi_c - \Pi_d}{\Pi_c + \Pi_d}, \tag{2.14}$$

called *gamma* (Goodman and Kruskal 1954). The sample version is $\hat{\gamma} = (C - D)/(C + D)$.

Like the correlation, gamma treats the variables symmetrically—it is unnecessary to identify one classification as a response variable. Also like the correlation, gamma has range $-1 \le \gamma \le 1$. A reversal in the category orderings of one variable causes a change in the sign of γ. Whereas the absolute value of the correlation is 1 when the relationship between X and Y is perfectly linear, only monotonicity is required for $|\gamma| = 1$, with $\gamma = 1$ if $\Pi_d = 0$ and $\gamma = -1$ if $\Pi_c = 0$. Independence implies that $\gamma = 0$, but the converse is not true. For instance, a U-shaped joint distribution can have $\Pi_c = \Pi_d$ and hence $\gamma = 0$.

2.4.5 Gamma for Job Satisfaction Example

For Table 2.8, $C = 1331$ and $D = 849$. Hence,

$$\hat{\gamma} = (1331 - 849)/(1331 + 849) = 0.221.$$

Only a weak tendency exists for job satisfaction to increase as income increases. Of the untied pairs, the proportion of concordant pairs is 0.221 higher than the proportion of discordant pairs.

NOTES

Section 2.2: Comparing Two Proportions

2.1. Breslow (1996) presented an interesting overview of the development of methods for case–control studies.

2.2. For 2×2 tables, Edwards (1963) showed that functions of the odds ratio are the only statistics that are invariant both to row–column interchange and to multiplication within rows or within columns by a constant. For $I \times J$ tables, Altham (1970) gave related results. Yule (1912, p. 587) had argued that multiplicative invariance is a desirable property for measures of association, especially when proportions sampled in various marginal categories are arbitrary. Goodman (2000) showed five ways of viewing association in a 2×2 table and proposed a general measure that includes all five.

Section 2.3: Partial Association in Stratified 2×2 Tables

2.3. Paik (1985) proposed circle diagrams of type Figure 2.2 to summarize three-way tables. Friendly (2000) discussed graphical presentation of categorical data. For more on Simpson's paradox and when it can happen, see Blyth (1972), Davis (1989), Dong (1998),

Samuels (1993), and Simpson (1951). Good and Mittal (1989) extended it to an *amalgamation paradox*, whereby a marginal measure is greater than the maximum or less than the minimum of the partial table measures.

Section 2.4: Extensions for $I \times J$ Tables

2.4. For continuous variables, samples can be fully ranked (i.e., no ties occur), so $C + D = \binom{n}{2}$ and $\hat{\gamma} = (C - D)/\binom{n}{2}$. This is *Kendall's tau*. Agresti (1984, Chaps. 9 and 10) and Kruskal (1958) surveyed ordinal measures of association. These also apply when one variable is ordinal and the other is binary. When Y is ordinal and X is nominal with $I > 2$, no measure presented in Section 2.4 is very helpful. Ordinal modeling approaches (Section 7.2) use a parameter for each category of X; comparing parameters compares the ordinal response for pairs of categories of X.

PROBLEMS

Applications

2.1 An article in the *New York Times* (Feb. 17, 1999) about the PSA blood test for detecting prostate cancer stated: "The test fails to detect prostate cancer in 1 in 4 men who have the disease (false-negative results), and as many as two-thirds of the men tested receive false-positive results." Let $C(\overline{C})$ denote the event of having (not having) prostate cancer, and let $+ (-)$ denote a positive (negative) test result. Which is true: $P(- \mid C) = \frac{1}{4}$ or $P(C \mid -) = \frac{1}{4}$? $P(\overline{C} \mid +) = \frac{2}{3}$ or $P(+ \mid \overline{C}) = \frac{2}{3}$? Determine the sensitivity and specificity.

2.2 A diagnostic test has sensitivity = specificity = 0.80. Find the odds ratio between true disease status and the diagnostic test result.

2.3 Table 2.9 is based on records of accidents in 1988 compiled by the Department of Highway Safety and Motor Vehicles in Florida. Identify the response variable, and find and interpret the difference of proportions, relative risk, and odds ratio. Why are the relative risk and odds ratio approximately equal?

TABLE 2.9 Data for Problem 2.3

Safety Equipment in Use	Injury	
	Fatal	Nonfatal
None	1601	162,527
Seat belt	510	412,368

Source: Florida Department of Highway Safety and Motor Vehicles.

2.4 Consider the following two studies reported in the *New York Times*.

   a. A British study reported (Dec. 3, 1998) that of smokers who get lung cancer, "women were 1.7 times more vulnerable than men to get small-cell lung cancer." Is 1.7 the odds ratio or the relative risk?

   b. A National Cancer Institute study about tamoxifen and breast cancer reported (Apr. 7, 1998) that the women taking the drug were 45% less likely to experience invasive breast cancer then were women taking placebo. Find the relative risk for (**i**) those taking the drug compared to those taking placebo, and (**ii**) those taking placebo compared to those taking the drug.

2.5 A study (E. G. Krug et al., *Internat. J. Epidemiol.*, **27**: 214–221, 1998) reported that the number of gun-related deaths per 100,000 people in 1994 was 14.24 in the United States, 4.31 in Canada, 2.65 in Australia, 1.24 in Germany, and 0.41 in England and Wales. Use the relative risk to compare the United States with the other countries. Interpret.

2.6 A newspaper article preceding the 1994 World Cup semifinal match between Italy and Bulgaria stated that "Italy is favored 10–11 to beat Bulgaria, which is rated at 10–3 to reach the final." Suppose that this means that the odds that Italy wins are $\frac{11}{10}$ and the odds that Bulgaria wins are $\frac{3}{10}$. Find the probability that each team wins, and comment.

2.7 In the United States, the estimated annual probability that a woman over the age of 35 dies of lung cancer equals 0.001304 for current smokers and 0.000121 for nonsmokers (M. Pagano and K. Gauvreau, *Principles of Biostatistics*, Duxbury Press, Pacific Grove, CA. 1993, p. 134).

   a. Find and interpret the difference of proportions and the relative risk. Which measure is more informative for these data? Why?

   b. Find and interpret the odds ratio. Explain why the relative risk and odds ratio take similar values.

2.8 For adults who sailed on the *Titanic* on its fateful voyage, the odds ratio between gender (female, male) and survival (yes, no) was 11.4. (For data, see R. J. M. Dawson, *J. Statist. Ed.* **3**, 1995.)

   a. What is wrong with the interpretation, "The probability of survival for females was 11.4 times that for males"? Give the correct interpretation. When would the quoted interpretation be approximately correct?

   b. The odds of survival for females equaled 2.9. For each gender, find the proportion who survived.

2.9   In an article about crime in the United States, *Newsweek* (Jan. 10, 1994) quoted FBI statistics for 1992 stating that of blacks slain, 94% were slain by blacks, and of whites slain, 83% were slain by whites. Let Y = race of victim and X = race of murderer. Which conditional distribution do these statistics refer to, $Y|X$, or $X|Y$? What additional information would you need to estimate the probability that the victim was white given that a murderer was white? Find and interpret the odds ratio.

2.10   A research study estimated that under a certain condition, the probability that a subject would be referred for heart catheterization was 0.906 for whites and 0.847 for blacks.

     a. A press release about the study stated that the odds of referral for cardiac catheterization for blacks are 60% of the odds for whites. Explain how they obtained 60% (more accurately, 57%).

     b. An Associated Press story later described the study and said "Doctors were only 60% as likely to order cardiac catheterization for blacks as for whites." Explain what is wrong with this interpretation. Give the correct percentage for this interpretation. (In stating results to the general public, it is better to use the relative risk than the odds ratio. It is simpler to understand and less likely to be misinterpreted. For details, see *New Engl. J. Med.* **341**: 279–283, 1999.)

2.11   A 20-year cohort study of British male physicians (R. Doll and R. Peto, *British Med. J.* **2**: 1525–1536, 1976) noted that the proportion per year who died from lung cancer was 0.00140 for cigarette smokers and 0.00010 for nonsmokers. The proportion who died from coronary heart disease was 0.00669 for smokers and 0.00413 for nonsmokers.

     a. Describe the association of smoking with each of lung cancer and heart disease, using the difference of proportions, relative risk, and odds ratio. Interpret.

     b. Which response is more strongly related to cigarette smoking, in terms of the reduction in number of deaths that would occur with elimination of cigarettes? Explain.

2.12   Table 2.10 refers to applicants to graduate school at the University of California at Berkeley, for fall 1973. It presents admissions decisions by gender of applicant for the six largest graduate departments. Denote the three variables by A = whether admitted, G = gender, and D = department. Find the sample AG conditional odds ratios and the marginal odds ratio. Interpret, and explain why they give such different indications of the AG association.

TABLE 2.10   Data for Problem 2.12

| Department | Whether Admitted | | | |
| | Male | | Female | |
	Yes	No	Yes	No
A	512	313	89	19
B	353	207	17	8
C	120	205	202	391
D	138	279	131	244
E	53	138	94	299
F	22	351	24	317
Total	1198	1493	557	1278

Source: Data from Freedman et al. (1978, p.14). See also P. Bickel et al., *Science* **187**: 398–403 (1975).

2.13 State three "real-world" variables X, Y, and Z for which you expect a marginal association between X and Y but conditional independence controlling for Z.

2.14 Based on 1987 murder rates in the United States, an Associated Press story reported that the probability that a newborn child has of eventually being a murder victim is 0.0263 for nonwhite males, 0.0049 for white males, 0.0072 for nonwhite females, and 0.0023 for white females.

   a. Find the conditional odds ratios between race and whether a murder victim, given the gender. Interpret. Do these variables exhibit homogeneous association?

   b. Half the newborns are of each gender, for each race. Find the marginal odds ratio between race and whether a murder victim.

2.15 At each age level, the death rate is higher in South Carolina than in Maine, but overall, the death rate is higher in Maine. Explain how this could be possible. (For data, see H. Wainer, *Chance* **12**: 44, 1999.)

2.16 A study of the death penalty for cases in Kentucky between 1976 and 1991 (T. Keil and G. Vito, *Amer. J. Criminal Justice* **20**: 17–36, 1995) indicated that the defendant received the death penalty in 8% of the 391 cases in which a white killed a white, in 2% of the 108 cases in which a black killed a black, in 12% of the 57 cases in which a black killed a white, and in 0% of the 18 cases in which a white killed a black. Form the three-way contingency table, obtain the conditional odds ratios between the defendant's race and the death penalty verdict, interpret those associations, study whether Simpson's paradox occurs, and explain why the marginal association is so different from the conditional associations.

2.17 An estimated odds ratio for adult females between the presence of squamous cell carcinoma (yes, no) and smoking behavior (smoker, nonsmoker) equals 11.7 when the smoker category has subjects whose smoking level s is $0 < s < 20$ cigarettes per day; it is 26.1 for smokers with $s \geq 20$ cigarettes per day (R. C. Brownson et al., *Epidemiology* **3**: 61–64, 1992). Show that the estimated odds ratio between carcinoma (yes, no) and the smoking levels ($s \geq 20$, $0 < s < 20$) equals 2.2.

2.18 Table 2.11 refers to a retrospective study of lung cancer and tobacco smoking among patients in several English hospitals. The table compares male lung cancer patients with control patients having other diseases, according to the average number of cigarettes smoked daily over a 10-year period preceding the onset of the disease.

  a. Find the sample odds of lung cancer at each smoking level and the five odds ratios that pair each level of smoking with no smoking. As smoking increases, is there a trend? Interpret.

  b. If the log odds of lung cancer is linearly related to smoking level, the log odds in row i satisfies $\log(\text{odds}_i) = \alpha + \beta i$. Show that this implies that the local odds ratios are identical.

  c. Using these data, can you estimate the probability of lung cancer at each level of smoking? Are the estimated odds ratios in part (a) meaningful? Explain.

  d. Show that the disease groups are *stochastically ordered* with respect to their distributions on smoking of cigarettes (see Problem 2.34 and Section 7.3.4). Interpret.

TABLE 2.11   Data for Problem 2.18

Daily Average Number of Cigarettes	Disease Group	
	Lung Cancer Patients	Control Patients
None	7	61
< 5	55	129
5–14	489	570
15–24	475	431
25–49	293	154
50 +	38	12

Source: Reprinted with permission from R. Doll and A. B. Hill, *British Med. J.* **2**: 1271–1286 (1952).

TABLE 2.12 Data for Problem 2.19

| | Wife's Rating of Sexual Fun | | | |
Husband's Rating	Never or Occasionally	Fairly Often	Very Often	Almost Always
Never or occasionally	7	7	2	3
Fairly often	2	8	3	7
Very often	1	5	4	9
Almost always	2	8	9	14

Source: Reprinted with permission from Hout et al. (1987).

2.19 Table 2.12 summarizes responses of 91 married couples in Arizona to a question about how often sex is fun. Find and interpret a measure of association between wife's response and husband's response.

2.20 Table 2.13 is from an early study on the death penalty in Florida. Analyze these data and show that Simpson's paradox occurs.

TABLE 2.13 Data for Problem 2.20

| Victim's Race | Defendant's Race | Death Penalty | |
		Yes	No
White	White	19	132
	Black	11	52
Black	White	0	9
	Black	6	97

Source: Reprinted with permission from M. L. Radelet, *Amer. Sociol. Rev.* **46**: 918–927 (1981)

Theory and Methods

2.21 For a diagnostic test of a certain disease, π_1 denotes the probability that the diagnosis is positive given that a subject has the disease, and π_2 denotes the probability that the diagnosis is positive given that a subject does not have it. Let ρ denote the probability that a subject does have the disease.

   a. Given that the diagnosis is positive, show that the probability that a subject does have the disease is

$$\pi_1 \rho / \left[\pi_1 \rho + \pi_2 (1 - \rho) \right].$$

b. Suppose that a diagnostic test for HIV+ status has both sensitivity and specificity equal to 0.95, and $\rho = 0.005$. Find the probability that a subject is truly HIV+ , given that the diagnostic test is positive. To better understand this answer, find the joint probabilities relating diagnosis to actual disease status, and discuss their relative sizes.

2.22 Binomial parameters for two groups are graphed, with π_1 on the horizontal axis and π_2 on the vertical axis. Plot the locus of points for a 2×2 table having **(a)** relative risk $= 0.5$, **(b)** odds ratio $= 0.5$, and **(c)** difference of proportions $= -0.5$.

2.23 Let D denote having a certain disease and E denote having exposure to a certain risk factor. The *attributable risk* (AR) is the proportion of disease cases attributable to that exposure (see Benichou 1998).
a. Let $P(\overline{E}) = 1 - P(E)$. Explain why

$$AR = \left[P(D) - P(D|\overline{E}) \right]/P(D).$$

b. Show that AR relates to the relative risk RR by

$$AR = \left[P(E)(RR - 1) \right]/\left[1 + P(E)(RR - 1) \right].$$

2.24 For a 2×2 table of counts $\{n_{ij}\}$, show that the odds ratio is invariant to **(a)** interchanging rows with columns, and **(b)** multiplication of cell counts within rows or within columns by $c \neq 0$. Show that the difference of proportions and the relative risk do not have these properties.

2.25 For given π_1 and π_2, show that the relative risk cannot be farther than the odds ratio from their independence value of 1.0.

2.26 Explain why for three events E_1, E_2, and E_3 and their complements, it is possible that $P(E_1|E_2) > P(E_1|\overline{E}_2)$ even if both $P(E_1|E_2 E_3) < P(E_1|\overline{E}_2 E_3)$ and $P(E_1|E_2 \overline{E}_3) < P(E_1|\overline{E}_2 \overline{E}_3)$. (*Hint:* Use Simpson's paradox for a three-way table.)

2.27 Let $\pi_{ij \mid k} = P(X = i, Y = j|Z = k)$. Explain why XY conditional independence is

$$\pi_{ij|k} = \pi_{i+|k} \pi_{+j|k} \quad \text{for all } i \text{ and } j \text{ and } k.$$

2.28 For a $2 \times 2 \times 2$ table, show that homogeneous association is a symmetric property, by showing that equal XY conditional odds ratios is equivalent to equal YZ conditional odds ratios.

2.29 Smith and Jones are baseball players. Smith has a higher batting average than Jones in each of K years. Is is possible that for the combined data from the K years, Jones has the higher batting average? Explain, using an example to illustrate.

2.30 When X and Y are conditionally dependent at each level of Z yet marginally independent, Z is called a *suppressor variable*. Specify joint probabilities for a $2 \times 2 \times 2$ table to show that this can happen **(a)** when there is homogeneous association, and **(b)** when the association has opposite direction in the partial tables.

2.31 Show that the $\{\alpha_{ij}\}$ in (2.11) determine **(a)** all $\binom{I}{2}\binom{J}{2}$ odds ratios formed from pairs of rows and pairs of columns, **(b)** all $\{\theta_{ij}\}$ in (2.10), and vice versa.

2.32 Refer to Problem 2.31. When all rows and columns have positive probability, show that independence is equivalent to all $\{\alpha_{ij} = 1\}$.

2.33 For $I \times J$ contingency tables, explain why the variables are independent when the $(I - 1)(J - 1)$ differences $\pi_{j|i} - \pi_{j|I} = 0$, $i = 1, \ldots,$ $I - 1$, $j = 1, \ldots, J - 1$.

2.34 A $2 \times J$ table has ordinal response. Let $F_{j|i} = \pi_{1|i} + \cdots + \pi_{j|i}$. When $F_{j|2} \leq F_{j|1}$ for $j = 1, \ldots, J$, the conditional distribution in row 2 is *stochastically higher* than the one in row 1. Consider the *cumulative odds ratios*

$$
\theta_j = \frac{F_{j|1} / (1 - F_{j|1})}{F_{j|2} / (1 - F_{j|2})}, \qquad j = 1, \ldots, J - 1.
$$

a. Show that $\log \theta_j \geq 0$ for all j is equivalent to row 2 being stochastically higher than row 1. Explain why row 2 is then more likely than row 1 to have observations at the high end of the ordinal scale.

b. If all local log odds ratios are nonnegative, $\log \theta_j \geq 0$ for $1 \leq j \leq J - 1$ (Lehmann 1966). Show by counterexample that the converse is not true.

2.35 Suppose that $\{Y_{ij}\}$ are independent Poisson variates with means $\{\mu_{ij}\}$. Show that $P(Y_{ij} = n_{ij})$ for all i, j, conditional on $\{Y_{i+} = n_i\}$, satisfy independent multinomial sampling [i.e., the product of (2.2) for all i] within the rows.

2.36 For 2×2 tables, Yule (1900, 1912) introduced

$$Q = \frac{\pi_{11}\pi_{22} - \pi_{12}\pi_{21}}{\pi_{11}\pi_{22} + \pi_{12}\pi_{21}},$$

which he labeled Q in honor of the Belgian statistician Quetelet. It is now called *Yule's Q*.

a. Show that for 2×2 tables, Goodman and Kruskal's $\gamma = Q$.

b. Show that Q falls between -1 and 1.

c. State conditions under which $Q = -1$ or $Q = 1$.

d. Show that Q relates to the odds ratio by $Q = (\theta - 1)/(\theta + 1)$, a monotone transformation of θ from the $[0, \infty]$ scale onto the $[-1, +1]$ scale.

2.37 When X and Y are ordinal with counts $\{n_{ij}\}$:

a. Explain why the $\binom{n}{2}$ pairs of observations partition into $C + D + T_X + T_Y - T_{XY}$, where $T_X = \Sigma n_{i+}(n_{i+} - 1)/2$ pairs are tied on X, T_Y pairs are tied on Y, and T_{XY} pairs are tied on X and Y.

b. For each ordered pair of observations (X_a, Y_a) and (X_b, Y_b), let $X_{ab} = \text{sign}(X_a - X_b)$ and $Y_{ab} = \text{sign}(Y_a - Y_b)$. Show that the sample correlation for the $n(n-1)$ distinct (X_{ab}, Y_{ab}) pairs is

$$\tau_b = \frac{C - D}{\left\{\left[\binom{n}{2} - T_X\right]\left[\binom{n}{2} - T_Y\right]\right\}^{1/2}}.$$

This ordinal measure, called *Kendall's tau-b* (Kendall 1945), is less sensitive than gamma to the choice of response categories.

c. Let $d = (C - D)/\left[\binom{n}{2} - T_X\right]$. Explain why d is the difference between the proportions of concordant and discordant pairs out of those pairs untied on X (Somers 1962). (For 2×2 tables, d equals the difference of proportions, and tau-*b* equals the correlation between X and Y.)

2.38 Goodman and Kruskal (1954) proposed an association measure (tau) for nominal variables based on variation measure

$$V(Y) = \Sigma \pi_{+j}(1 - \pi_{+j}) = 1 - \Sigma \pi_{+j}^2.$$

a. Show $V(Y)$ is the probability that two independent observations on Y fall in different categories (called the *Gini concentration index*).

Show that $V(Y) = 0$ when $\pi_{+j} = 1$ for some j and $V(Y)$ takes maximum value of $(J - 1)/J$ when $\pi_{+j} = 1/J$ for all j.

b. For the proportional reduction in variation, show that $E[V(Y|X)]$ $= 1 - \Sigma_i \Sigma_j \pi_{ij}^2 / \pi_{i+}$. [The resulting measure (2.12) is called the *concentration coefficient*. Like U, $\tau = 0$ is equivalent to independence. Haberman (1982) presented generalized concentration and uncertainty coefficients.]

2.39 The measure of association *lambda* for nominal variables (Goodman and Kruskal 1954) has $V(Y) = 1 - \max\{\pi_{+j}\}$ and $V(Y|i) = 1 - \max_j\{\pi_{j|i}\}$. Interpret lambda as a proportional reduction in prediction error for predictions which select the response category that is most likely. Show that independence implies $\lambda = 0$ but that the converse is not true.

CHAPTER 3

Inference for Contingency Tables

In this chapter we introduce inferential methods for contingency tables. Many of these methods also play a vital role in analyses of later chapters for which categorical data need not have contingency table form. The methods assume Poisson, multinomial, or independent binomial sampling.

In Section 3.1 we present confidence intervals for measures of association for 2×2 tables such as the odds ratio. Section 3.2 covers chi-squared tests of the hypothesis of independence between two categorical variables. Like any significance test, these have limited usefulness. In Section 3.3 we show how to follow-up the test using residuals or the partitioning property of chi-squared to extract components that describe the evidence about the association. In Section 3.4 we present more powerful inference applicable with ordered categories. The methods of Sections 3.1 through 3.4 assume large samples. In Sections 3.5 and 3.6 we introduce small-sample methods.

3.1 CONFIDENCE INTERVALS FOR ASSOCIATION PARAMETERS

The accuracy of estimators of association parameters is characterized by standard errors of their sampling distributions. In this section we present large-sample standard errors and confidence intervals.

3.1.1 Interval Estimation of Odds Ratios

The sample odds ratio $\hat{\theta} = n_{11}n_{22}/n_{12}n_{21}$ for a 2×2 table equals 0 or ∞ if any $n_{ij} = 0$, and it is undefined if both entries in a row or column are zero. Since these outcomes have positive probabilities, the expected value and variance of $\hat{\theta}$ and $\log \hat{\theta}$ do not exist. (In fact, this is also true for ML estimators of model parameters presented in later chapters.) In terms of bias and mean-squared error, Gart and Zweiful (1967) and Haldane (1956) showed that the amended estimators

$$\tilde{\theta} = \frac{(n_{11} + 0.5)(n_{22} + 0.5)}{(n_{12} + 0.5)(n_{21} + 0.5)}$$

and $\log \tilde{\theta}$ behave well (Problem 14.4).

The estimators $\hat{\theta}$ and $\tilde{\theta}$ have the same asymptotic normal distribution around θ. Unless n is quite large, however, their distributions are highly skewed. When $\theta = 1$, for instance, $\hat{\theta}$ cannot be much smaller than θ (since $\hat{\theta} \geq 0$), but it could be much larger with nonnegligible probability. The log transform, having an additive rather than multiplicative structure, converges more rapidly to normality. An estimated standard error for $\log \hat{\theta}$ is

$$\hat{\sigma}\left(\log \hat{\theta}\right) = \left(\frac{1}{n_{11}} + \frac{1}{n_{12}} + \frac{1}{n_{21}} + \frac{1}{n_{22}}\right)^{1/2}. \tag{3.1}$$

We derive this formula in Section 3.1.7.

By the large-sample normality of $\log \hat{\theta}$,

$$\log \hat{\theta} \pm z_{\alpha/2}\, \hat{\sigma}\left(\log \hat{\theta}\right) \tag{3.2}$$

is a Wald confidence interval for $\log \theta$. Exponentiating (taking antilogs of) its endpoints provides a confidence interval for θ. Woolf (1955) proposed this interval. It works quite well, usually being a bit conservative (i.e., actual coverage probability higher than the nominal level).

When $\hat{\theta} = 0$ or ∞, Woolf's interval does not exist. When $\hat{\theta} = 0$, one should take 0 as the lower limit and when $\hat{\theta} = \infty$, one should take ∞ as the upper limit. The other bound can use the Woolf formula following some adjustment, such as Gart's (1966), which replaces $\{n_{ij}\}$ by $\{n_{ij} + 0.5\}$ in the estimator and standard error. A less ad hoc approach forms the interval by inverting score tests (Cornfield 1956) or likelihood-ratio tests for θ, as we discuss in Section 3.1.8.

3.1.2 Aspirin and Myocardial Infarction Example

We illustrate inference for the odds ratio with Table 3.1 based on a Swedish study of the association between aspirin use and myocardial infarction similar to that described in Section 2.2.5. The study randomly assigned 1360 patients who had already suffered a stroke to an aspirin treatment (one low-dose tablet a day) or to a placebo treatment. Table 3.1 reports the number of deaths due to myocardial infarction during a follow-up period of about 3 years.

The sample odds ratio $\hat{\theta} = 1.56$ is close to $\tilde{\theta} = 1.55$, since no cell count is especially small. The standard error (3.1) of $\log \hat{\theta} = 0.445$ is $\hat{\sigma}(\log \hat{\theta}) = 0.307$.

**TABLE 3.1   Swedish Study on Aspirin Use and
Myocardial Infarction**

	Myocardial Infarction		
	Yes	No	Total
Placebo	28	656	684
Aspirin	18	658	676

Source: Based on results described in *Lancet* **338**: 1345–1349
(1991).

A 95% confidence interval for $\log\theta$ in the population this sample represents
is $0.445 \pm 1.96(0.307)$, or $(-0.157, 1.047)$. The corresponding interval for θ is
$[\exp(-0.157), \exp(1.047)]$, or $(0.85, 2.85)$. The estimate of the true odds ratio
is rather imprecise.

Since the confidence interval for θ contains 1.0, it is plausible that the true
odds of death due to myocardial infarction are equal for aspirin and placebo.
If there truly is a beneficial effect of aspirin but the odds ratio is not large, it
may require a large sample size to show that benefit because of the relatively
small number of myocardial infarction cases (Problem 3.21).

3.1.3   Interval Estimation of Difference of Proportions

The difference of proportions and the relative risk compare conditional
distributions of a response variable for two groups. For these measures, we
treat the samples as independent binomials. For group i, y_i has a binomial
distribution with sample size n_i and a probability π_i of a "success" response.

The sample proportion $\hat{\pi}_i = y_i/n_i$ has expectation π_i and variance
$\pi_i(1 - \pi_i)/n_i$. Since $\hat{\pi}_1$ and $\hat{\pi}_2$ are independent, their difference has

$$E(\hat{\pi}_1 - \hat{\pi}_2) = \pi_1 - \pi_2$$

and standard error

$$\sigma(\hat{\pi}_1 - \hat{\pi}_2) = \left[\frac{\pi_1(1 - \pi_1)}{n_1} + \frac{\pi_2(1 - \pi_2)}{n_2} \right]^{1/2}. \tag{3.3}$$

The estimate $\hat{\sigma}(\hat{\pi}_1 - \hat{\pi}_2)$ uses formula (3.3) with π_i replaced by $\hat{\pi}_i$. Then

$$(\hat{\pi}_1 - \hat{\pi}_2) \pm z_{\alpha/2}\, \hat{\sigma}(\hat{\pi}_1 - \hat{\pi}_2) \tag{3.4}$$

is a Wald confidence interval for $\pi_1 - \pi_2$. Like the Wald interval (1.13) for a
single proportion, it usually has true coverage probability less than the
nominal confidence coefficient, especially when π_1 and π_2 are near 0 or 1.
More complex but better methods are cited in Section 3.1.8, Note 3.2, and
Problem 3.23.

3.1.4 Interval Estimation of Relative Risk

The sample relative risk is $r = \hat{\pi}_1/\hat{\pi}_2$. Like the odds ratio, it converges to normality faster on the log scale. The asymptotic standard error of $\log r$ is

$$\sigma(\log r) = \left(\frac{1 - \pi_1}{\pi_1 n_1} + \frac{1 - \pi_2}{\pi_2 n_2} \right)^{1/2}. \tag{3.5}$$

The Wald interval exponentiates endpoints of $\log r \pm z_{\alpha/2} \hat{\sigma}(\log r)$. It works well but can be somewhat conservative. We discuss an alternative method in Section 3.1.8.

For Table 3.1, the sample proportion of myocardial infarction deaths was 0.0409 for subjects taking placebo and 0.0266 for subjects taking aspirin. The sample relative risk is $0.0409/0.0266 = 1.54$. The 95% confidence interval for the log relative risk of $\log(1.54) \pm 1.96(0.297)$ translates to $(0.86, 2.75)$ for the relative risk. We infer that the death rate for those taking placebo was between 0.86 and 2.75 times that for those taking aspirin. The Wald 95% confidence interval for $\pi_1 - \pi_2$ is $0.014 \pm 1.96(0.0098)$ or $(-0.005, 0.033)$. According to either measure, substantial public health benefits could result from taking aspirin, but no effect or a slight negative effect are also plausible. Results for the larger study described in Section 2.2.5 do show a benefit.

3.1.5 Deriving Standard Errors with the Delta Method*

A simple and useful method exists of deriving standard errors for large-sample inferences. Let T_n denote a statistic that is asymptotically normally distributed about a parameter θ, the subscript n expressing its dependence on sample size. Suppose that an estimator is a function $g(T_n)$ of T_n. Then, under mild conditions, $g(T_n)$ itself has a large-sample normal distribution. The standard error depends on how fast $g(t)$ changes for t near θ.

Specifically, for large n, suppose that T_n is normally distributed about θ with standard error $\sigma/\sqrt{n}$. That is, as $n \to \infty$, the cdf of $\sqrt{n}(T_n - \theta)$ converges to the cdf of a normal random variable with mean 0 and variance σ^2. This limiting behavior is an example of *convergence in distribution*, denoted by

$$\sqrt{n}(T_n - \theta) \overset{d}{\to} N(0, \sigma^2).$$

Let g be a function that is at least twice differentiable at θ. Using the Taylor series expansion for $g(t)$ in a neighborhood of $t = \theta$, in Section 14.1.2 we show

$$\sqrt{n}[g(T_n) - g(\theta)] \approx \sqrt{n}(T_n - \theta)g'(\theta)$$

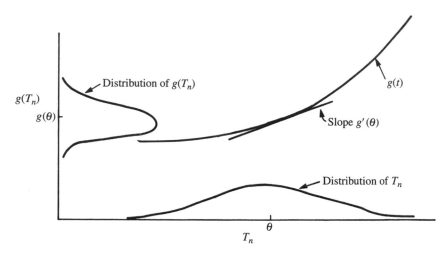

FIGURE 3.1   Depiction of delta method.

for large n, where $g'(\theta) = \partial g/\partial t$ evaluated at $t = \theta$. Recall if a variate $Y \sim N(0, \sigma^2)$, then $cY \sim N(0, c^2\sigma^2)$. Thus,

$$\sqrt{n}\,[g(T_n) - g(\theta)] \xrightarrow{d} N\big(0, [g'(\theta)]^2 \sigma^2\big). \qquad (3.6)$$

In other words, $g(T_n)$ is approximately normal around $g(\theta)$ with variance $[g'(\theta)]^2\sigma^2/n$.

Figure 3.1 portrays this result. Locally around θ, $g(t)$ is approximately linear, with slope $g'(\theta)$. Then $g(T_n)$ is approximately normal, since linear transformations of normal random variables are themselves normal. The dispersion of $g(T_n)$ values about $g(\theta)$ is about $|g'(\theta)|$ times the dispersion of T_n values about θ. If the slope of g at θ is $\frac{1}{2}$, then g maps a region of T_n values into a region of $g(T_n)$ values only about half as wide.

Result (3.6) is called the *delta method*. Since $g'(\theta)$ and $\sigma^2 = \sigma^2(\theta)$ usually depend on the unknown parameter θ, the asymptotic variance is unknown. Confidence intervals and tests substitute T_n for θ and use the result that $\sqrt{n}\,[g(T_n) - g(\theta)]/\,|g'(T_n)|\,\sigma(T_n)$ is asymptotically standard normal. For instance,

$$g(T_n) \pm 1.96|g'(T_n)|\,\sigma(T_n)/\sqrt{n}$$

is a large-sample Wald 95% confidence interval for $g(\theta)$.

3.1.6   Delta Method Applied to Sample Logit*

We illustrate the delta method for a function of the ML estimator $T_n = \hat{\pi} = y/n$ of the binomial parameter π, for y successes in n trials. Since $E(Y) = n\pi$ and $\text{var}(Y) = n\pi(1 - \pi)$, $E(\hat{\pi}) = \pi$ and $\text{var}(\hat{\pi}) = \pi(1 - \pi)/n$. Also, $\hat{\pi}$

has a large-sample normal distribution by the central limit theorem. So do many functions of $\hat{\pi}$.

The log odds function of $\hat{\pi}$,

$$g(\hat{\pi}) = \log[\hat{\pi}/(1 - \hat{\pi})],$$

is called the sample *logit*. Evaluated at π, its derivative equals $1/\pi(1 - \pi)$. By the delta method, the asymptotic variance of the sample logit is $\pi(1 - \pi)/n$ (the variance of $\hat{\pi}$) multiplied by the square of $[1/\pi(1 - \pi)]$. That is

$$\sqrt{n}\left(\log\frac{\hat{\pi}}{1 - \hat{\pi}} - \log\frac{\pi}{1 - \pi}\right) \xrightarrow{d} N\left(0, \frac{1}{\pi(1 - \pi)}\right).$$

The asymptotic normality of $\hat{\pi}$ propagates to asymptotic normality of $\log[\hat{\pi}/(1 - \hat{\pi})]$.

The asymptotic variance is the variance of the normal distribution that approximates the true distribution, for large n. It is *not* an approximation for the variance of the true distribution. For $0 < \pi < 1$, the asymptotic variance $[n\pi(1 - \pi)]^{-1}$ of the sample logit is finite. By contrast, the true variance does not exist: Since $\hat{\pi} = 0$ or 1 with positive probability, the logit can equal $-\infty$ or ∞ with positive probability. The probability of an infinite logit converges to zero rapidly as n increases. For large n, the distribution of the sample logit looks essentially normal with mean $\log[\pi/(1 - \pi)]$ and standard deviation $[n\pi(1 - \pi)]^{-1/2}$. Thus, for the logit, the asymptotic variance actually has greater use than the true variance. Incidentally, related to this, the bootstrap is not helpful for approximating standard errors for many discrete measures, because it mimics the true rather than the more relevant asymptotic standard error.

3.1.7 Delta Method for Log Odds Ratio*

Standard errors for the log odds ratio and the log relative risk result from a multiparameter version of the delta method. Suppose that $\{n_i, i = 1, \ldots, c\}$ have a multinomial $(n, \{\pi_i\})$ distribution. The sample proportion $\hat{\pi}_i = n_i/n$ has mean and variance

$$E(\hat{\pi}_i) = \pi_i \quad \text{and} \quad \text{var}(\hat{\pi}_i) = \pi_i(1 - \pi_i)/n. \tag{3.7}$$

In Section 14.1.4 we show that for $i \neq j$, $\hat{\pi}_i$ and $\hat{\pi}_j$ have covariance

$$\text{cov}(\hat{\pi}_i, \hat{\pi}_j) = -\pi_i\pi_j/n. \tag{3.8}$$

The sample proportions $(\hat{\pi}_1, \hat{\pi}_2, \ldots, \hat{\pi}_{c-1})$ have a large-sample multivariate normal distribution. For functions of them, the delta method implies the following result, proved in Section 14.1.4:

Let $g(\pi)$ denote a differentiable function of $\{\pi_i\}$, with sample value $g(\hat{\pi})$ for a multinomial sample. Let

$$\phi_i = \frac{\partial g(\pi)}{\partial \pi_i}, \qquad i = 1,\ldots,c.$$

Then as $n \to \infty$, the distribution of $\sqrt{n}[g(\hat{\pi}) - g(\pi)]/\sigma$ converges to standard normal, where

$$\sigma^2 = \sum \pi_i \phi_i^2 - \left(\sum \pi_i \phi_i \right)^2. \tag{3.9}$$

The asymptotic variance depends on $\{\pi_i\}$ and the partial derivatives of the measure with respect to $\{\pi_i\}$. In practice, replacing $\{\pi_i\}$ and $\{\phi_i\}$ in (3.9) by their sample values yields an ML estimate $\hat{\sigma}^2$ of σ^2. Then $\hat{\sigma}/\sqrt{n}$ is an estimated standard error for $g(\hat{\pi})$. A large-sample Wald confidence interval for $g(\pi)$ is

$$g(\hat{\pi}) \pm z_{\alpha/2}\hat{\sigma}/\sqrt{n}.$$

With the substitution of $\hat{\sigma}$ for σ in (3.9), the limiting distribution is still standard normal, but convergence is slower. The equivalence in the large-sample distribution is justified as follows: The sample proportions converge in probability to $\{\pi_i\}$, by the weak law of large numbers. Since $\hat{\sigma}$ is a continuous function of the sample proportions, it converges in probability to σ, and $\sigma/\hat{\sigma}$ converges in probability to 1. Now

$$\sqrt{n}\, \frac{g(\hat{\pi}) - g(\pi)}{\hat{\sigma}} = \sqrt{n}\, \frac{g(\hat{\pi}) - g(\pi)}{\sigma}\, \frac{\sigma}{\hat{\sigma}}.$$

The first term on the right-hand side converges in distribution to standard normal, by (3.9), and the second term converges in probability to 1. Thus, their product also has a limiting standard normal distribution.

We now apply the delta method to the log odds ratio, taking $g(\pi) = \log \theta = \log \pi_{11} + \log \pi_{22} - \log \pi_{12} - \log \pi_{21}$. Since

$$\phi_{11} = \partial(\log \theta)/\partial \pi_{11} = 1/\pi_{11}$$

$$\phi_{12} = -1/\pi_{12}, \qquad \phi_{21} = -1/\pi_{21}, \qquad \phi_{22} = 1/\pi_{22},$$

$\sum_i \sum_j \pi_{ij} \phi_{ij} = 0$ and $\sigma^2 = \sum_i \sum_j \pi_{ij} \phi_{ij}^2 = \sum_i \sum_j (1/\pi_{ij})$. The asymptotic standard error of $\log \hat{\theta}$ for a multinomial sample $\{n_{ij}\}$ is

$$\sigma\left(\log \hat{\theta}\right) = \sigma/\sqrt{n} = \left(\sum_i \sum_j 1/n\pi_{ij} \right)^{1/2}.$$

Since $n\hat{\pi}_{ij} = n_{ij}$, the estimated standard error is (3.1).

The delta method also applies directly with θ to obtain $\hat{\sigma}(\hat{\theta})$ and a Wald confidence interval $\hat{\theta} \pm z_{\alpha/2} \hat{\sigma}(\hat{\theta})$. This is not recommended; $\hat{\theta}$ converges more slowly than $\log \hat{\theta}$ to normality, this interval could contain negative values, and it does not give results equivalent to those obtained with the Wald interval using $1/\hat{\theta}$ and its standard error.

3.1.8 Score and Profile Likelihood Confidence Intervals*

Standard errors obtained with the delta method appear in Wald confidence intervals. However, intervals based on inverting Wald tests sometimes work poorly for small to moderate n. Alternative intervals result from inverting likelihood-ratio or score tests. Although computationally more complex, these methods often perform better.

We illustrate first with the score method for the difference of proportions. The score test (Mee 1984; Miettinen and Nurminen 1985) of H_0: $\pi_1 - \pi_2 = \Delta$ has the test statistic

$$z(\Delta) = \frac{(\hat{\pi}_1 - \hat{\pi}_2) - \Delta}{\sqrt{\hat{\pi}_1(\Delta)[1 - \hat{\pi}_1(\Delta)]/n_1 + \hat{\pi}_2(\Delta)[1 - \hat{\pi}_2(\Delta)]/n_2}}$$

where $\hat{\pi}_i(\Delta)$ denotes the ML estimate of π_i subject to the constraint $\pi_1 - \pi_2 = \Delta$. That is, $\hat{\pi}_1(\Delta)$ and $\hat{\pi}_2(\Delta)$ are the values of π_1 and π_2 satisfying $\pi_1 - \pi_2 = \Delta$ that maximize the product of the two binomial probability mass functions. These values do not have closed-form expressions and are determined using numerical methods. The score confidence interval is the set of Δ such that $|z(\Delta)| < z_{\alpha/2}$. Computations for such intervals require iteration (Nurminen 1986).

For the relative risk also, slightly better performance results with an interval using the score method (Bedrick 1987; Gart and Nam 1988; Koopman 1984, Miettinen and Nurminen 1985; Nurminen 1986). Cornfield (1956) and Miettinen and Nurminen (1985) showed the score interval for the odds ratio. We prefer not to use a continuity or finite-sampling correction with these intervals, as then performance is too conservative. The fact that the score intervals are computationally more complex than Wald intervals should not be an impediment to their use in this modern era of computing, as the principle behind them is simple. However, currently they are not available in standard software.

For a confidence interval based on the likelihood-ratio test, we illustrate with the odds ratio. The multinomial likelihood for a 2×2 table is a function of $\{\pi_{11}, \pi_{12}, \pi_{21}\}$. Equivalently, it can be expressed in terms of $\{\theta, \pi_{1+}, \pi_{+1}\}$ (recall Section 2.4.1). Thus, in inverting a likelihood-ratio test of H_0: $\theta = \theta_0$ to check whether θ_0 belongs in the confidence interval, there are two *nuisance parameters*. Their null ML estimates $\hat{\pi}_{1+}(\theta_0)$ and $\hat{\pi}_{+1}(\theta_0)$ that maximize the likelihood under the null vary as θ_0 does.

The *profile log-likelihood function* is $L(\theta_0, \hat{\pi}_{1+}(\theta_0), \hat{\pi}_{+1}(\theta_0))$, viewed as a function of θ_0. For each θ_0 this function gives the maximum of the ordinary log likelihood subject to the constraint $\theta = \theta_0$. Evaluated at $\theta_0 = \hat{\theta}$, this is the maximized log likelihood $L(\hat{\theta}, \hat{\pi}_{1+}, \hat{\pi}_{+1})$, which occurs at the sample proportions $\hat{\pi}_{1+} = n_{1+}/n$ and $\hat{\pi}_{+1} = n_{+1}/n$. The profile likelihood confidence interval for θ is the set of θ_0 for which

$$-2\left[L(\theta_0, \hat{\pi}_{1+}(\theta_0), \hat{\pi}_{+1}(\theta_0)) - L(\hat{\theta}, \hat{\pi}_{1+}, \hat{\pi}_{+1}) \right] < \chi_1^2(\alpha) .$$

This contains all θ_0 not rejected in likelihood-ratio tests of nominal size α.

The profile likelihood approach is available with some software (e.g., for SAS, see Table A.2 in Appendix A). A related approach, discussed in Section 6.7.1, uses a *conditional likelihood function* that eliminates the nuisance parameters by conditioning on their sufficient statistics. This is beneficial when there are many nuisance parameters. An advantage of score and likelihood-based intervals is that unlike the Wald, they are not adversely affected when the sample relative risk or odds ratio is 0 or ∞.

In this section we have discussed interval estimation. Significance tests normally refer to a null hypothesis value of 0.0 for the log odds ratio, log relative risk, and difference of proportions. These are special cases of independence applied to 2×2 tables. In the next section we present tests of independence for two-way contingency tables.

3.2   TESTING INDEPENDENCE IN TWO-WAY CONTINGENCY TABLES

For multinomial sampling with probabilities $\{\pi_{ij}\}$ in an $I \times J$ contingency table, the null hypothesis of statistical independence is H_0: $\pi_{ij} = \pi_{i+} \pi_{+j}$ for all i and j. For independent multinomial samples in the I rows, independence corresponds to homogeneity of each outcome probability among the rows. Our discussion refers to a single multinomial sample, but the same tests apply with independent multinomial samples.

3.2.1   Pearson and Likelihood-Ratio Chi-Squared Tests

In Section 1.5.2 we introduced the Pearson X^2 statistic (1.15) for tests about multinomial probabilities. A test of H_0: independence uses X^2 with n_{ij} in place of n_i and with $\mu_{ij} = n\pi_{i+} \pi_{+j}$ in place of μ_i. Here $\mu_{ij} = E(n_{ij})$ under H_0. Usually, $\{\pi_{i+}\}$ and $\{\pi_{+j}\}$ are unknown. Their ML estimates are the sample marginal proportions $\hat{\pi}_{i+} = n_{i+}/n$ and $\hat{\pi}_{+j} = n_{+j}/n$, so estimated expected frequencies are $\{\hat{\mu}_{ij} = n\hat{\pi}_{i+} \hat{\pi}_{+j} = n_{i+} n_{+j}/n\}$. Then X^2 equals

$$X^2 = \sum_i \sum_j \frac{\left(n_{ij} - \hat{\mu}_{ij}\right)^2}{\hat{\mu}_{ij}} . \tag{3.10}$$

Pearson (1900, 1904, 1922) claimed that replacing $\{\mu_{ij}\}$ by estimates $\{\hat{\mu}_{ij}\}$ would not affect the distribution of X^2. Since the contingency table has IJ categories, he argued that X^2 is asymptotically chi-squared with df $= IJ - 1$. On the contrary, since $\{\hat{\mu}_{ij}\}$ require estimating $\{\pi_{i+}\}$ and $\{\pi_{+j}\}$, by Section 1.5.6

$$\text{df} = (IJ - 1) - (I - 1) - (J - 1) = (I - 1)(J - 1).$$

The dimensions of $\{\pi_{i+}\}$ and $\{\pi_{+j}\}$ reflect the constraints $\Sigma_i \pi_{i+} = \Sigma_j \pi_{+j} = 1$. R. A. Fisher (1922) corrected Pearson's error (see Section 16.2). His article introduced the notion of *degrees of freedom*. (Pearson had dealt with an indexed family of chi-squared distributions but had not dealt explicitly with "degrees of freedom.")

The score test produces the X^2 statistic. The likelihood-ratio test produces a different one. For multinomial sampling, the kernel of the likelihood is

$$\prod_i \prod_j \pi_{ij}^{n_{ij}}, \quad \text{where all } \pi_{ij} \geq 0 \quad \text{and} \quad \sum_i \sum_j \pi_{ij} = 1.$$

Under H_0: independence, $\hat{\pi}_{ij} = \hat{\pi}_{i+} \hat{\pi}_{+j} = n_{i+} n_{+j}/n^2$. In the general case, $\hat{\pi}_{ij} = n_{ij}/n$. The ratio of the likelihoods equals

$$\Lambda = \frac{\prod_i \prod_j (n_{i+} n_{+j})^{n_{ij}}}{n^n \prod_i \prod_j n_{ij}^{n_{ij}}}.$$

The likelihood-ratio chi-squared statistic is $-2 \log \Lambda$. Denoted by G^2, it equals

$$G^2 = -2 \log \Lambda = 2 \sum_i \sum_j n_{ij} \log(n_{ij}/\hat{\mu}_{ij}) \tag{3.11}$$

where $\{\hat{\mu}_{ij} = n_{i+} n_{+j}/n\}$. The larger the values of G^2 and X^2, the more evidence exists against independence.

In the general case, the parameter space consists of $\{\pi_{ij}\}$ subject to the linear restriction $\Sigma_i \Sigma_j \pi_{ij} = 1$, so the dimension is $IJ - 1$. Under H_0, $\{\pi_{ij}\}$ are determined by $\{\pi_{i+}\}$ and $\{\pi_{+j}\}$, so the dimension is $(I - 1) + (J - 1)$. The difference in these dimensions equals $(I - 1)(J - 1)$. For large samples, G^2 has a chi-squared null distribution with df $= (I - 1)(J - 1)$. So G^2 and X^2 have the same limiting null chi-squared distribution. In fact, they are then asymptotically equivalent; $X^2 - G^2$ converges in probability to zero (Section 14.3.4). The limiting results for multinomial sampling also hold with other sampling schemes (Roy and Mitra 1956, Watson 1959).

These results apply as n grows, and hence $\{\mu_{ij} = n\pi_{ij}\}$ grow, for a fixed number of cells. As they grow, the multinomial distribution for $\{n_{ij}\}$ is better approximated by a multivariate normal, and X^2 and G^2 have more nearly chi-squared distributions. The convergence to chi-squared is quicker for X^2 than G^2. The approximation is usually poor for G^2 when $n/IJ < 5$. When I or J is large, it can be decent for X^2 when some expected frequencies are as small as 1 but most exceed 5. In Section 9.8.4 we provide further guidelines. Small-sample methods (Section 3.5) are available whenever it is doubtful whether n is sufficiently large.

3.2.2 Education and Religious Fundamentalism Example

Table 3.2 cross-classifies the degree of fundamentalism of subjects' religious beliefs by their highest degree of education. The table also contains the estimated expected frequencies for H_0: independence. For instance, $\hat{\mu}_{11} = n_{1+}n_{+1}/n = (424 \times 886)/2726 = 137.8$. The chi-squared statistics are $X^2 = 69.2$ and $G^2 = 69.8$, with df $= (3 - 1)(3 - 1) = 4$. The P-values are < 0.0001. These statistics provide extremely strong evidence of an association.

3.3 FOLLOWING-UP CHI-SQUARED TESTS

Like any significance test, chi-squared tests of independence have limited usefulness. A small P-value indicates strong evidence of association but provides little information about the nature or strength of the association. Statisticians have long warned about dangers of relying solely on results of chi-squared tests rather than studying the nature of the association (e.g., Berkson 1938; Cochran 1954). In this section we discuss ways to follow up the tests to learn more about the association.

TABLE 3.2   Education and Religious Beliefs

| | Religious Beliefs | | | |
Highest Degree	Fundamentalist	Moderate	Liberal	Total
Less than high school	178	138	108	424
	$(137.8)^1$	(161.5)	(124.7)	
	$(4.5)^2$	(−2.6)	(−1.9)	
High school or junior college	570	648	442	1660
	(539.5)	(632.1)	(488.4)	
	(2.6)	(1.3)	(−4.0)	
Bachelor or graduate	138	252	252	642
	(208.7)	(244.5)	(188.9)	
	(−6.8)	(0.7)	(6.3)	
Total	886	1038	802	2726

Source: 1996 General Social Survey, National Opinion Research Center.
[1]Estimated expected frequencies for testing independence; [2]standardized Pearson residuals.

3.3.1 Pearson and Standardized Residuals

A cell-by-cell comparison of observed and estimated expected frequencies helps show the nature of the dependence. Under H_0, larger differences $(n_{ij} - \hat{\mu}_{ij})$ tend to occur in cells with larger μ_{ij}. For Poisson sampling, for instance, the standard deviation of n_{ij} and hence $(n_{ij} - \mu_{ij})$ is $\sqrt{\mu_{ij}}$; the standard deviation of $(n_{ij} - \hat{\mu}_{ij})$ is less than that of $n_{ij} - \mu_{ij}$ but is proportional to $\sqrt{\mu_{ij}}$. Thus, this raw difference is insufficient. The *Pearson residual*, defined for a cell by

$$e_{ij} = \frac{n_{ij} - \hat{\mu}_{ij}}{\hat{\mu}_{ij}^{1/2}}, \tag{3.12}$$

attempts to adjust for this. Pearson residuals relate to the Pearson statistic by $\sum_i \sum_j e_{ij}^2 = X^2$.

Under H_0, $\{e_{ij}\}$ are asymptotically normal with mean 0. However, in Section 14.3.2 we show that their asymptotic variances are less than 1.0, averaging $[(I - 1)(J - 1)]/(\text{number of cells})$. Comparing Pearson residuals to standard normal percentage points provides conservative indications of cells having lack of fit.

A *standardized Pearson residual* that is asymptotically standard normal results from dividing it by its standard error (Haberman 1973a; see also Section 14.3.2). For H_0: independence, this is

$$\frac{n_{ij} - \hat{\mu}_{ij}}{\left[\hat{\mu}_{ij}(1 - p_{i+})(1 - p_{+j}) \right]^{1/2}}. \tag{3.13}$$

A standardized Pearson residual that exceeds about 2 or 3 in absolute value indicates lack of fit of H_0 in that cell. Larger values are more relevant when df is larger and it becomes more likely that at least one is large simply by chance.

3.3.2 Education and Religious Fundamentalism Revisited

Table 3.2 also shows standardized Pearson residuals for testing independence. For instance, $n_{11} = 178$ and $\hat{\mu}_{11} = 137.8$. The relevant marginal proportions equal $p_{1+} = 424/2726 = 0.156$ and $p_{+1} = 886/2726 = 0.325$. The standardized Pearson residual (3.13) for this cell equals

$$(178 - 137.8)/[(137.8)(1 - 0.156)(1 - 0.325)]^{1/2} = 4.5.$$

This cell shows a much greater discrepancy between n_{11} and $\hat{\mu}_{11}$ than expected if the variables were truly independent.

Table 3.2 shows large positive residuals for subjects with less than a high school education and fundamentalist views and for subjects with a bachelor's or graduate degree and liberal views. This means that significantly more subjects were at these combinations than H_0: independence predicts. Similarly, there were fewer subjects with high levels of education and fundamentalist views and with low levels of education and liberal views than independence predicts.

Odds ratios describe this trend. The 2×2 table constructed from the first and last rows and the first and last columns of Table 3.2 has a sample odds ratio of $(178 \times 252)/(108 \times 138) = 3.0$. For those with a bachelor's or graduate degree, the estimated odds of selecting liberal instead of fundamentalist were 3.0 times the estimated odds for those with less than a high school education.

3.3.3 Partitioning Chi-Squared

Let Z denote a standard normal random variable. Then Z^2 has a chi-squared distribution with df = 1. A chi-squared random variable with df = ν has representation $Z_1^2 + \cdots + Z_\nu^2$, where $Z_1, \ldots, Z_\nu$ are independent standard normal variables. Thus, a chi-squared statistic having df = ν has partitionings into independent chi-squared components—for example, into ν components each having df = 1. Conversely, if X_1^2 and X_2^2 are independent chi-squared random variables having degrees of freedom ν_1 and ν_2, then $X^2 = X_1^2 + X_2^2$ has a chi-squared distribution with df = $\nu_1 + \nu_2$. Another supplement to a chi-squared test partitions its test statistic so that the components represent certain aspects of the effects. A partitioning may show that an association reflects primarily differences between certain categories or groupings of categories.

We begin with a partitioning for the test of independence in $2 \times J$ tables. We partition G^2, which has df = $(J - 1)$, into $J - 1$ components. The jth component is G^2 for a 2×2 table where the first column combines columns 1 through j of the full table and the second column is column $j + 1$. That is, G^2 for testing independence in a $2 \times J$ table equals a statistic that compares the first two columns, plus a statistic that combines the first two columns and compares them to the third column, and so on, up to a statistic that combines the first $J - 1$ columns and compares them to the last column. (In Section 9.2.4 we justify this partitioning.) Each component statistic has df = 1.

It might seem more natural to compute G^2 for the $(J - 1)$ separate 2×2 tables that pair each column with a particular one, say the last. However, these component statistics are not independent and do not sum to G^2 for the full table. (This is beyond our scope at this stage but relates to the contrasts of log probabilities that form the log odds ratios for the two tables not being orthogonal.)

For an $I \times J$ table, independent chi-squared components result from comparing columns 1 and 2 and then combining them and comparing them to column 3, and so on. Each of the $J - 1$ statistics has df = $I - 1$. More refined partitions contain $(I - 1)(J - 1)$ statistics, each having df = 1. One such partition (Lancaster 1949) applies to the $(I - 1)(J - 1)$ separate 2×2 tables

$$
\begin{array}{c|c}
\displaystyle\sum_{a<i}\sum_{b<j} n_{ab} & \displaystyle\sum_{a<i} n_{aj} \\
\hline
\displaystyle\sum_{b<j} n_{ib} & n_{ij}
\end{array}
\tag{3.14}
$$

for $i = 2,\ldots,I$ and $j = 2,\ldots,J$. For others, see Gilula and Haberman (1998) and Goodman (1969a, 1971b).

3.3.4 Origin of Schizophrenia Example

Table 3.3 classifies a sample of psychiatrists by their school of psychiatric thought and by their opinion on the origin of schizophrenia. Here $G^2 = 23.04$ with df $= 4$. To understand this association better, we partition G^2 into four independent components. The partitioning (3.14) applies to the subtables shown in Table 3.4.

The first subtable compares the eclectic and medical schools of psychiatric thought on whether the origin of schizophrenia is biogenic or environmental given that the classification was in one of these two categories. For this subtable, $G^2 = 0.29$, with df $= 1$. The second subtable compares these two schools on the proportion of times the origin was ascribed to be a combination, rather than biogenic or environmental. This subtable has $G^2 = 1.36$,

TABLE 3.3   **Most Influential School of Psychiatric Thought and Ascribed Origin of Schizophrenia**

School of Psychiatric Thought	Origin of Schizophrenia		
	Biogenic	Environmental	Combination
Eclectic	90	12	78
Medical	13	1	6
Psychoanalytic	19	13	50

Source: Reprinted with permission, based on data from B. J. Gallagher III, B. J. Jones, and L. P. Barakat, *J. Clin. Psychol.* **43**: 438–443 (1987).

TABLE 3.4   **Subtables Used in Partitioning Chi-Squared for Table 3.3**[a]

	Bio	Env		Bio + Env	Com		Bio	Env		Bio + Env	Com
Ecl	90	12	Ecl	102	78	Ecl + Med	103	13	Ecl + Med	116	84
Med	13	1	Med	14	6	Psy	19	13	Psy	32	50

[a]Bio, biogenic; Com, combination; Ecl, eclectic; Env, environmental; Psy, psychoanalytic

with df = 1. The sum of these two components equals G^2 for testing independence with the first two rows of Table 3.3. There is little evidence of a difference between the eclectic and medical schools of thought on the ascribed origin of schizophrenia.

Next we combine the eclectic and medical schools and compare them to the psychoanalytic school. The third subtable in Table 3.4 compares them for the (biogenic, environmental) classification, giving $G^2 = 12.95$ with df = 1. The fourth subtable compares them for the (biogenic or environmental, combination) split, giving $G^2 = 8.43$ with df = 1.

The psychoanalytic school seems more likely than the other schools to ascribe the origins of schizophrenia as being a combination. Of those who chose either the biogenetic or environmental origin, members of the psychoanalytic school were somewhat more likely than the other schools to choose the environmental origin. The sum of these four G^2 components equals the value of 23.04 for testing independence in the full table.

3.3.5   Rules for Partitioning

Goodman (1968, 1969a, 1971b) and Lancaster (1949, 1969) gave rules for determining independent components of chi-squared. For forming subtables, among the necessary conditions are the following:

1. The df for the subtables must sum to df for the full table.
2. Each cell count in the full table must be a cell count in one and only one subtable.
3. Each marginal total of the full table must be a marginal total for one and only one subtable.

For a certain partitioning, when the subtable df values sum properly but the G^2 values do not, the components are not independent.

For the G^2 statistic, exact partitionings occur. The Pearson X^2 need not equal the sum of the X^2 values for the subtables. It is valid to use the X^2 statistics for the separate subtables; they simply need not provide an exact algebraic partitioning of X^2 for the full table. When the null hypotheses all hold, X^2 does have an asymptotic equivalence with G^2, however. In addition, when the table has small counts, in large-sample chi-squared tests it is safer to use X^2 to study the subtables.

3.3.6   Limitations of Chi-Squared Tests

Chi-squared tests of independence merely indicate the degree of evidence of association. They are rarely adequate for answering all questions about a data set. Rather than relying solely on results of these tests, investigate the nature of the association: Study residuals, decompose chi-squared into components, and estimate parameters such as odds ratios that describe the strength of association.

The chi-squared tests also have limitations in the types of data to which they apply. For instance, they require large samples. Also, the $\{\hat{\mu}_{ij} = n_{i+}n_{+j}/n\}$ used in X^2 and G^2 depend on the marginal totals but not on the order of listing the rows and columns. Thus, X^2 and G^2 do not change value with arbitrary reorderings of rows or of columns. This implies that they treat both classifications as nominal. When at least one variable is ordinal, test statistics that utilize the ordinality are usually more appropriate. We present such tests in Section 3.4.

3.3.7 Why Consider Independence?

Any idealized structure such as independence is unlikely to hold in any given practical situation. With large samples such as in Table 3.2 it is not surprising to obtain a small P-value. Given this and the limitations just mentioned, why even bother to consider independence as a possible representation for a joint distribution? One reason refers to the benefits of model parsimony. If the independence model approximates the true probabilities well, then unless n is very large, the model-based estimates $\{\hat{\pi}_{ij} = n_{i+}n_{+j}/n^2\}$ of cell probabilities tend to be better than the sample proportions $\{p_{ij} = n_{ij}/n\}$. The independence ML estimates smooth the sample counts, somewhat damping the random sampling fluctuations.

The mean-squared error (MSE) formula

$$\text{MSE} = \text{variance} + (\text{bias})^2$$

explains why the independence estimators can have smaller MSE. Although they may be biased, they have smaller variance because they are based on estimating fewer parameters ($\{\pi_{i+}\}$ and $\{\pi_{+j}\}$ instead of $\{\pi_{ij}\}$). Hence, MSE can be smaller unless n is so large that the bias term dominates the variance.

We illustrate using Table 3.5, which has $\pi_{ij} = \pi_{i+}\pi_{+j}[1 + \delta(i - 2)(j - 2)]$ for $\pi_{i+} = \pi_{+j} = \frac{1}{3}$. Here $-1 < \delta < 1$, with $\delta = 0$ equivalent to independence. Independence approximates the relationship well when δ is close to zero. The total MSE values of the two estimators are

$$\text{MSE}(\{p_{ij}\}) = \sum_i \sum_j E(p_{ij} - \pi_{ij})^2 = \sum_i \sum_j \text{var}(p_{ij})$$

$$= \sum_i \sum_j \pi_{ij}(1 - \pi_{ij})/n = \left(1 - \sum_i \sum_j \pi_{ij}^2\right)\Big/n$$

$$\text{MSE}(\{\hat{\pi}_{ij}\}) = \sum_i \sum_j E(\hat{\pi}_{ij} - \pi_{ij})^2.$$

TABLE 3.5   Cell Probabilities for Comparison of Estimators

$(1 + \delta)/9$	$1/9$	$(1 - \delta)/9$
$1/9$	$1/9$	$1/9$
$(1 - \delta)/9$	$1/9$	$(1 + \delta)/9$

TABLE 3.6 Comparison of Total MSE($\times 10,000$)for Sample Proportion and Independence Estimators

n	$\delta = 0$		$\delta = 0.1$		$\delta = 0.2$		$\delta = 0.6$		$\delta = 1.0$	
	p	$\hat{\pi}$	p	$\hat{\pi}$	p	$\hat{\pi}$	p	$\hat{\pi}$	p	$\hat{\pi}$
10	889	489	888	493	887	505	871	634	840	893
50	178	91	178	95	177	110	174	261	168	565
100	89	45	89	50	89	65	87	220	84	529
500	18	9	18	14	18	28	17	186	17	500
∞	0	0	0	5	0	20	0	178	0	494

For Table 3.5,

$$\text{MSE}(\{p_{ij}\}) = \frac{1}{n}\left(\frac{8}{9} - \frac{4\delta^2}{81}\right)$$

and rather tedious calculations yield

$$\text{MSE}(\{\hat{\pi}_{ij}\}) = \frac{1}{n}\left(\frac{4}{9} + \frac{4}{9n}\right) + \frac{4\delta^2}{81}\left(1 - \frac{2}{n} + \frac{2}{n^2} - \frac{2}{n^3}\right).$$

Table 3.6 lists the total MSE values for various δ and n. When $\delta = 0$, $\text{MSE}(\{p_{ij}\}) = 8/9n$, whereas $\text{MSE}(\{\hat{\pi}_{ij}\}) \approx 4/9n$ for large n. The independence estimator is then much better than the sample proportions. When the table is close to independence ($\delta \approx 0$) and n is not large, MSE is only about half as large for the independence estimator. When $\delta \neq 0$, the inconsistency of $\{\hat{\pi}_{ij}\}$ is reflected by $\text{MSE}(\{\hat{\pi}_{ij}\}) \to 4\delta^2/81$ [whereas $\text{MSE}(\{p_{ij}\}) \to 0$] as $n \to \infty$. When the table is close to independence, however, the independence estimator has a lower total MSE even for moderately large n (e.g., for $n = 500$ when $\delta = 0.1$).

3.4 TWO-WAY TABLES WITH ORDERED CLASSIFICATIONS

The X^2 and G^2 chi-squared tests ignore some information when used to test independence between ordinal classifications. When rows and/or columns are ordered, more powerful tests usually exist.

3.4.1 Linear Trend Alternative to Independence

When the row variable X and the column variable Y are ordinal, a positive or negative trend in the association is common. One approach to inference, described later in this section, uses an ordinal measure of monotone trend.

A more popular analysis assigns scores to categories and summarizes the *linear trend*.

A test statistic that is sensitive to positive or negative linear trends utilizes correlation information. Let $u_1 \leq u_2 \leq \cdots \leq u_I$ denote scores for the rows, and let $v_1 \leq v_2 \leq \cdots \leq v_J$ denote column scores. The scores have the same ordering as the categories. They assign distances between categories and actually treat the measurement scale as interval, with greater distances between categories that are farther apart.

The sum $\sum_i \sum_j u_i v_j n_{ij}$ weights cross-products of scores by their frequency. It relates to the covariation of X and Y. For the scores chosen, the correlation r between X and Y equals the standardization of this sum to the -1 to $+1$ scale (in fact, r equals this sum when both sets of scores are linearly transformed for the n subjects to have a mean of 0 and standard deviation of 1). The larger the correlation is in absolute value, the farther the data fall from independence in this linear dimension.

A statistic for testing independence against the two-sided alternative of nonzero true correlation is

$$M^2 = (n - 1)r^2. \qquad (3.15)$$

This statistic increases as $|r|$ or n do. For large samples, it is approximately chi-squared with df $= 1$ (Mantel 1963). Large values contradict independence, so as with X^2 and G^2, the P-value is the right-tailed probability above the value observed. A small P-value does not imply that the association is linear, merely that searching for a linear component to the association helped to build power against H_0. The test treats the variables symmetrically.

3.4.2 Job Satisfaction Example Revisited

Table 2.8 showed job satisfaction and income for 96 subjects. The ordinary chi-squared statistics for testing independence are $X^2 = 6.0$ and $G^2 = 6.8$ with df $= 9$ (P-values $= 0.74$ and 0.66). These statistics show little evidence of association, but they ignore the ordering of rows and columns. With scores $(1, 2, 3, 4)$ for job satisfaction and scores $\{7.5, 20, 32.5, 60\}$ for income that approximate midpoints of categories in thousands of dollars, the correlation is $r = 0.200$. The linear trend test statistic $M^2 = (96 - 1)(0.200)^2 = 3.81$. This shows some evidence of association ($P = 0.051$). The evidence is stronger for the one-sided (positive trend) alternative, using $M = \sqrt{n-1}\, r = 1.95$ ($P = 0.026$).

The nontrivial evidence of positive association may be surprising, since X^2 and G^2 have such unimpressive values. When a positive or negative trend exists, analyses designed to detect that trend can provide much smaller P-values than analyses that ignore it.

3.4.3   Monotone Trend Alternatives to Independence

Ordinal variables do not have a specified metric. Detecting a linear trend alternative to independence requires assigning scores to X and Y, treating them as interval variables. Alternatively, a strict ordinal analysis with the weaker alternative of monotonicity uses an ordinal measure of association, such as gamma (Section 2.4.4).

For large random samples, sample gamma has approximately a normal sampling distribution. The standard error (SE) follows from the delta method (Problem 3.27). Gamma is the basis of an ordinal test of independence using test statistic $z = \hat{\gamma}/\text{SE}$. A confidence interval describes the strength of positive or negative monotone association.

For Table 2.8 on income and job satisfaction, in Section 2.4.5 we showed that $\hat{\gamma} = 0.221$. The sample has a weak tendency for job satisfaction to be higher at higher income levels. Software (e.g., PROC FREQ in SAS) reports a standard error of 0.117 for gamma. There is some evidence that $\gamma > 0$, since $z = 0.221/0.117 = 1.89$ ($P = 0.03$ for the one-sided alternative). An approximate 95% confidence interval for γ is $0.221 \pm 1.96(0.117)$, or $(-0.01, 0.45)$. The true association between income and job satisfaction is at best moderately positive.

3.4.4   Extra Power with Ordinal Tests

For testing independence, X^2 and G^2 refer to the most general alternative, whereby cell probabilities exhibit *any* type of statistical dependence. Their df value of $(I - 1)(J - 1)$ reflects an alternative hypothesis that has $(I - 1)(J - 1)$ more parameters than the null hypothesis—the nonredundant odds ratios that describe the association [such as (2.10)]. These statistics are designed to detect any pattern for these parameters. In achieving this generality, they sacrifice sensitivity for detecting particular patterns.

By contrast, the analyses for ordinal row and column variables attempt to describe association using a single parameter. For instance, M^2 uses the correlation. When a chi-squared test statistic refers to a single parameter [such as M^2 or $(\hat{\gamma}/\text{SE})^2$ do], it has df $= 1$. When the association truly has a positive or negative trend, an ordinal test has a power advantage over the tests using X^2 or G^2. Since df equals the mean of the chi-squared distribution, a relatively large M^2 value with df $= 1$ falls farther out in its right-hand tail than a comparable value of X^2 or G^2 with df $= (I - 1)(J - 1)$; falling farther out in the tail produces a smaller P-value. The potential discrepancy in power increases as I and J increase. In Section 6.4 we present the theory behind such a power comparison.

3.4.5   Choice of Scores

Often, it is unclear how to assign scores to statistics that require them, such as M^2. Cochran (1954) noted that "any set of scores gives a *valid* test, provided that they are constructed without consulting the results of the experiment. If the set of scores is poor, in that it badly distorts a numerical scale that really does underlie the ordered classification, the test will not be sensitive. The scores should therefore embody the best insight available about the way in which the classification was constructed and used." Ideally, the scale is chosen by a consensus of experts, and subsequent interpretations use that same scale.

How sensitive are analyses to the choice or scores? There is no simple answer, but different scoring systems can give quite different results (e.g., Graubard and Korn 1987). For most data sets, different choices of monotone scores give similar results. Scores that are linear transforms of each other, such as (1, 2, 3, 4) and (0, 2, 4, 6), have the same absolute correlation and hence the same M^2. Results *may* depend on the scores, however, when the data are highly unbalanced, with some categories having many more observations than others.

Table 3.7 illustrates the potential dependence. It refers to a prospective study of maternal drinking and congenital malformations. After the first three months of pregnancy, the women in the sample completed a questionnaire about alcohol consumption. Following childbirth, observations were recorded on the presence or absence of congenital sex organ malformations. When a variable is nominal but has only two categories, statistics that treat it as ordinal are still valid. For instance, we can artificially regard malformation as ordinal, treating "present" as "high" and "absent" as "low." With only two rows, any set of distinct row scores is a linear transformation of any other set and gives the same M^2 value. Alcohol consumption, measured as the average number of drinks per day, is an ordinal explanatory variable. This groups a naturally continuous variable, and we first use the scores $\{v_1 = 0, v_2 = 0.5, v_3 = 1.5, v_4 = 4.0, v_5 = 7.0\}$, the last score being somewhat arbitrary. For this choice, $M^2 = 6.57$, for which the P-value is 0.010. By contrast, for the equally spaced row scores (1, 2, 3, 4, 5), $M^2 = 1.83$, giving a much weaker conclusion ($P = 0.18$).

An alternative approach uses the data to form the scores automatically, by using ranks as the category scores. All subjects in a category receive the average of the ranks that would apply for a complete ranking of the sample from 1 to n. These are called *midranks*. The 17,114 subjects at level 0 for

TABLE 3.7 **Example for which Results Depend on Choice of Scores**

Malformation	Alcohol Consumption (average number of drinks per day)				
	0	< 1	1–2	3–5	≥ 6
Absent	17,066	14,464	788	126	37
Present	48	38	5	1	1

Source: Reprinted with permission from the Biometric Society (Graubard and Korn 1987).

alcohol consumption share ranks 1 through 17,114. Each receives the average of these ranks, which is the midrank $(1 + 17,114)/2 = 8557.5$. Similarly, the midranks for the last four categories are 24,365.5, 32,013, 32,473, and 32,555.5. These scores yield $M^2 = 0.35$ and a weaker conclusion yet $(P = 0.55)$.

Why does this happen? Adjacent categories having relatively few observations necessarily have similar midranks. The midranks are similar for the final three categories, since those categories have few observations compared with the first two categories. This scoring scheme treats alcohol consumption level 1–2 drinks (category 3) as much closer to consumption level ≥ 6 drinks (category 5) than to consumption level 0 drinks (category 1). This seems inappropriate. It is usually better to select scores that reflect distances between categories. When uncertain about this choice, a sensitivity analysis should be performed, selecting two or three sensible choices and checking whether results are similar. Equally spaced scores often provide a reasonable compromise when the category labels do not suggest obvious choices, such as the categories (liberal, moderate, conservative) for political philosophy.

When X and Y are both ordinal and M^2 uses midrank scores, the correlation on which M^2 is based is called *Spearman's rho*.

3.4.6 Trend Tests for $I \times 2$ and $2 \times J$ Tables

When I or J equal 2, the tests based on linear or monotonic trend simplify to well-established procedures. With binary X, $2 \times J$ tables occur in comparisons of two groups, such as when the rows represent two treatments. Using scores $\{u_1 = 0, u_2 = 1\}$ for levels of X, the covariation measure $\Sigma_i \Sigma_j u_i v_j n_{ij}$ in M^2 simplifies to $\Sigma_j v_j n_{2j}$. This term sums the scores on Y for all subjects in row 2. Divided by the number of subjects in row 2, it gives the mean score for that row. In fact, M^2 is then directed toward detecting differences between the two row means of the scores on Y.

With midrank scores for Y, the test using M^2 for $2 \times J$ tables is sensitive to differences in mean ranks for the two rows. This test is called the *Wilcoxon* or *Mann–Whitney test*. Most nonparametric statistics textbooks present this test for fully ranked response data, whereas the $2 \times J$ table is an extended case in which sets of subjects in the same category of Y are tied and use midranks. The large-sample version of that nonparametric test uses a standard normal z statistic. The square of the statistic is equivalent to M^2, using arbitrary row scores and midranks for the columns. It is also asymptotically equivalent to test statistics based on the numbers of concordant and discordant pairs, such as the one using gamma.

When Y has two levels, the table has size $I \times 2$. The linear trend statistic then refers to a linear trend in the probability of either response category, such as the probability of malformation as a function of alcohol consumption. The test in that case, often called the *Cochran–Armitage trend test*, is presented in Section 5.3.5.

3.4.7 Nominal–Ordinal Tables

The tests using the correlation or gamma are appropriate when both classifications are ordinal. When one is nominal with more than two categories, other statistics are needed. One is based on summarizing the variation among means on the ordinal variable in the various categories of the nominal variable. We defer discussion of this case to Section 7.5.3, Note 3.6, and Problem 3.28.

3.5 SMALL-SAMPLE TESTS OF INDEPENDENCE

The inferential methods of the preceding four sections are large-sample methods. When n is small, alternative methods use *exact* small-sample distributions rather than large-sample approximations. In this section we describe small-sample tests of independence, starting with one that R. A. Fisher proposed for 2×2 tables.

3.5.1 Fisher's Exact Test for 2×2 Tables

In Section 3.5.7 we show that a distribution not depending on unknown parameters results from conditioning on the marginal totals of the contingency table. These are usually not naturally fixed. For Poisson sampling nothing is fixed, for multinomial sampling only n is fixed, and for independent binomial sampling in the two rows only the row marginal totals are fixed. In any of these cases, under H_0: independence, conditioning on both sets of marginal totals yields the hypergeometric distribution

$$p(t) = P(n_{11} = t) = \frac{\binom{n_{1+}}{t}\binom{n_{2+}}{n_{+1} - t}}{\binom{n}{n_{+1}}}. \tag{3.16}$$

This formula expresses the distribution of $\{n_{ij}\}$ in terms of only n_{11}. Given the marginal totals, n_{11} determines the other three cell counts. The range of possible values for n_{11} is $m_- \leq n_{11} \leq m_+$, where $m_- = \max(0, n_{1+} + n_{+1} - n)$ and $m_+ = \min(n_{1+}, n_{+1})$.

For 2×2 tables, independence is equivalent to the odds ratio $\theta = 1$. To test H_0: $\theta = 1$, the P-value is the sum of certain hypergeometric probabilities. To illustrate, consider H_a: $\theta > 1$. For the given marginal totals, tables having larger n_{11} have larger sample odds ratios and hence stronger evidence in favor of H_a. Thus, the P-value equals $P(n_{11} \geq t_o)$, where t_o denotes the observed value of n_{11}. This test for 2×2 tables is called *Fisher's exact test* (Fisher 1934, 1935a,c; Irwin 1935; Yates 1934).

3.5.2 Fisher's Tea Drinker

R. A. Fisher (1935a) described the following experiment from his days at Rothamsted Experiment Station, an agriculture research lab north of London. Muriel Bristol, a colleague of Fisher's, claimed that when drinking tea she could distinguish whether milk or tea was added to the cup first (she preferred milk first). To test her claim, Fisher asked her to taste eight cups of tea, four of which had milk added first and four of which had tea added first. She knew there were four cups of each type and had to predict which four had the milk added first. The order of presenting the cups to her was randomized.

Table 3.8 shows a possible result. Distinguishing the order of pouring better than with pure guessing corresponds to $\theta > 1$, reflecting a positive association between order of pouring and the prediction. We conduct Fisher's exact test of H_0: $\theta = 1$ against H_a: $\theta > 1$.

The experimental design fixed both marginal distributions, since Dr. Bristol had to predict which four cups had milk added first. Thus, the hypergeometric applies naturally for the null distribution of n_{11}. The P-value for Fisher's exact test is the null probability of Table 3.8 and of tables having even more evidence in favor of her claim. The observed table, $t_o = 3$ correct choices of the cups having milk added first, has null probability

$$\frac{\binom{4}{3}\binom{4}{1}}{\binom{8}{4}} = 0.229.$$

The only table that is more extreme in the direction of H_a has $n_{11} = 4$ correct. It has a probability of 0.014. The P-value is $P(n_{11} \geq 3) = 0.243$. This result does not establish an association between the actual order of pouring and her predictions. It is difficult to do so with such a small sample. According to Fisher's daughter (Box 1978, p. 134), in reality Bristol did convince Fisher of her ability.

TABLE 3.8 Fisher's Tea Tasting Experiment

	Guess Poured First		
Poured First	Milk	Tea	Total
Milk	3	1	4
Tea	1	3	4
Total	4	4	

Source: Based on experiment described by Fisher (1935a).

3.5.3 Two-Sided *P*-Values for Fisher's Exact Test

For the one-sided alternative, the same *P*-value results using tables ordered according to larger n_{11}, larger odds ratio, or larger difference of proportions (Davis 1986a). For the two-sided alternative, different criteria can have different *P*-values.

For a two-sided *P*-value, a popular approach sums $P(n_{11} = t)$ in (3.16) for counts t such that $p(t) \leq p(t_o)$; that is, the *P*-value is $P = P[p(n_{11}) \leq p(t_0)]$ for the observed value t_o. Another possibility sums $p(t)$ for tables that are farther from H_0; that is,

$$P = P[|n_{11} - E(n_{11})| \geq |t_0 - E(n_{11})|],$$

where the hypergeometric $E(n_{11}) = n_{1+}n_{+1}/n$. This is identical to $P(X^2 \geq X_o^2)$ for observed Pearson statistic X_o^2. A third approach takes $P = 2\min[P(n_{11} \geq t_o), P(n_{11} \leq t_o)]$, but this can exceed 1. A fourth approach takes $P = \min[P(n_{11} \geq t_o), P(n_{11} \leq t_o)]$ plus an attainable probability in the other tail that is as close as possible to, but not greater than, that one-tailed probability.

Each approach has advantages and disadvantages (Blaker 2000; Davis 1986a; Dupont 1986; Lloyd 1988b; Mantel 1987b; Yates and discussants 1984). They can provide different results because of the discreteness and potential skewness. The approach of ordering tables by a distance measure from H_0, such as X^2, extends naturally to $I \times J$ tables.

In practice, two-sided tests are much more common than one-sided. Partly this is so that researchers can avoid charges of bias in giving evidence that supports their predicted direction for an effect. To conduct a test of size 0.05 when one truly believes that the effect has a particular direction, it is safest to conduct the one-sided test at the 0.025 level to guard against criticism. For instance, in the 1998 document *Biostatistical Principles for Clinical Trials*, the International Conference on Harmonization (ICH E9) stated: "The approach of setting type I errors for one-sided tests at half the conventional type I error used in two-sided tests is preferable in regulatory settings. This promotes consistency with two-sided confidence intervals that are generally appropriate for estimating the possible size of the difference between two treatments."

3.5.4 Discreteness and Conservatism Issues

The hypergeometric distribution (3.16) is highly discrete for small samples, as n_{11} and hence the *P*-value can assume relatively few values. It is usually not possible to achieve a fixed significance level (size) such as 0.05.

In the tea-tasting experiment, for instance, n_{11} can equal only 4, 3, 2, 1, 0. The one-sided *P*-values are restricted to 0.014, 0.243, 0.757, 0.986, and 1.0. If one rejects H_0 when the P-value does not exceed 0.05, then 0.05 is not the probability of type I error. Only the P-value of 0.014 does not exceed 0.05; thus, when H_0 is true, the probability of falsely rejecting it is 0.014, not 0.05. In this sense, the traditional approach to hypothesis testing is conservative: The true probability of type I error is less than the nominal level.

It *is* possible to achieve any fixed significance level by data-unrelated randomization on the boundary of the critical region, in deciding whether to reject H_0. For the tea-tasting experiment, suppose that we reject H_0 when $n_{11} = 4$, we reject H_0 with probability 0.157 when $n_{11} = 3$, and we do not reject H_0 otherwise; that is, when $n_{11} = 3$, we generate a uniform random variable U over $[0, 1]$ and reject H_0 if $U < 0.157$. For expectation taken with respect to the null hypergeometric distribution of n_{11}, the significance level equals

$$P(\text{reject } H_0) = E\big[P(\text{reject } H_0 | n_{11})\big]$$
$$= 1.0(0.014) + 0.157(0.229) + 0.0 \times P(n_{11} \leq 2) = 0.05.$$

With the randomization extension, Tocher (1950) showed that Fisher's test is uniformly most powerful unbiased (UMPU).

In practice, randomization having nothing to do with the data is unacceptable. We recommend simply reporting the P-value. To reduce conservativeness, report the mid-P-value (Section 1.4.5). The test is no longer guaranteed to have true P(type I error) no greater than the nominal value, but in practice it is rarely much greater. For the one-sided test with the tea-tasting data,

$$\text{mid-}P\text{-value} = (1/2)P(n_{11} = 3) + P(n_{11} > 3) = 0.129.$$

3.5.5 Small-Sample Unconditional Test of Independence*

A common sampling assumption for analyses comparing two groups on a binary response is that the rows are independent binomial samples. Then, only $\{n_{i+}\}$ are naturally fixed. For Poisson and multinomial sampling schemes, neither marginal distribution is fixed. For such cases it may seem artificial to condition on *both* sets of marginal counts. An alternative small-sample test, designed for independent binomial samples, conditions on only the row totals.

Under binomial sampling with parameter π_i in row i, consider testing H_0: $\pi_1 = \pi_2$ using some test statistic T, such as the Pearson X^2. For fixed $\{n_{i+}\}$, T can take a discrete set of values, one of which is the observed value t_o. Given $\pi_1 = \pi_2 = \pi$, the P-value is $P_\pi(T \geq t_o)$, calculated using the product of the two binomial probability mass functions. This is the sum of the product binomial probabilities for those pairs of binomial samples that have $T \geq t_o$. Since π is unknown, the actual P-value is defined as

$$P = \sup_{0 \leq \pi \leq 1} P_\pi(T \geq t_o).$$

This is an *unconditional* small-sample test of independence. Like Fisher's exact test, the true size is no greater than the nominal value (e.g., if we reject when $P \leq 0.05$, the actual P(type I error) is no greater than 0.05).

We illustrate using test statistic X^2 for the 2×2 table having entries (3, 0/0, 3), by row, with fixed row totals (3, 3) as binomial sample sizes. The sample $X^2 = 6.0$. This X^2 value for the observed table and for table (0, 3/3, 0) is the maximum possible. For a given value π for $\pi_1 = \pi_2$, the probability of the first table is $[\pi^3(1 - \pi)^0][\pi^0(1 - \pi)^3] = \pi^3(1 - \pi)^3$ (3 successes and 0 failures in the first row and 0 successes and 3 failures in the second), the product of two binomial probabilities. Similarly, the probability of the second table is $(1 - \pi)^3\pi^3$. Thus, the P-value is $P_\pi(X^2 \geq 6) = 2\pi^3(1 - \pi)^3$, the sum of the product binomial probabilities for those two tables. The supremum of this over $0 \leq \pi \leq 1$ occurs at $\pi = \frac{1}{2}$, giving overall P-value equal to $2(0.5)^3(0.5)^3 = 0.031$. By contrast, the two-sided Fisher's exact test has P-value equal to $2\binom{3}{0}\binom{3}{3}/\binom{6}{3} = 0.100$.

Barnard (1945, 1947) first proposed an unconditional test comparing binomial parameters, although he later (1949) refuted it in favor of Fisher's exact test. Several authors have since proposed related tests (e.g., Haber 1986; Suissa and Shuster 1985).

3.5.6 Conditional versus Unconditional Tests*

Since Barnard introduced the unconditional test, statisticians have debated the proper way to conduct small-sample analyses of 2×2 tables. Fisher criticized the unconditional approach, arguing that possible samples with quite different numbers of successes than observed were not relevant. In Fisher's (1945) view, "...the existence of these less informative possibilities should not affect our judgment of significance based on the series actually observed.... The fact that such an unhelpful outcome as these might occur...is surely no reason for enhancing our judgment of significance in cases where it has not occurred; ...it is only the sampling distribution of samples of the same type that can supply a rational test of significance." Sprott (2000, Sec. 6.4.4) recently provided a similar argument.

An adaptation of the unconditional approach by Berger and Boos (1994) addresses this criticism somewhat. They took the supremum for the P-value over a confidence interval of values for the nuisance parameter π rather than over all possible values. Their unconditional P-value is

$$P = \sup_{\pi \in C_\gamma} P_\pi(T \geq t_o) + \gamma,$$

where C_γ is a $100(1 - \gamma)\%$ confidence interval for π. Here, γ is taken to be very small (e.g., 0.001), and the test maintains the guaranteed upper bound on size.

Other arguments in favor of conditioning on both sets of marginal totals are that the conditional approach provides a simple way to eliminate nuisance parameters in a variety of problems (e.g., generalizing to other contingency table problems), and the margins contain little information about the association (Haber 1989; Yates 1984). Zhu and Reid (1994) noted that some information loss occurs in conditioning on the margins except when $\theta = 1$. Arguments against conditioning partly concern the increased discreteness that occurs. The few possible values for n_{11} make it difficult to obtain a small P-value. In repeated use with a nominal significance level, the actual type I error probability may be much smaller than the nominal value and the power may suffer. Finally, for inference about nonnull values (e.g, confidence intervals), we will see that the conditional approach applies only with the odds ratio and not other measures.

The conservatism problem is partly unavoidable. Statistics having discrete distributions are necessarily conservative in terms of achieving nominal significance levels. Because an unconditional test fixes only one margin, however, it has many more tables in the reference set for its sampling distribution. That distribution is less discrete, and a richer array of possible P-values occurs than with Fisher's exact test. An unconditional test tends to be less conservative and more powerful than Fisher's exact test. A disadvantage is that computations are very intensive for more complex problems, such as larger tables.

If a table truly has two independent binomial samples, the unconditional approach seems sensible. See Kempthorne (1979) for a cogent argument. The conditional approach is useful for other cases. In a randomized clinical trial a convenience sample of n subjects is randomly allocated to two treatments. The samples are not binomials, as they are not random samples from two populations of interest. One could focus on the sample alone and consider the probability of a result at least as extreme as observed if there truly is no treatment effect. For instance, out of all possible ways of choosing n_{1+} of the n subjects for treatment 1, for what proportion would n_{11} be at least as large as observed? Under the null hypothesis of no treatment effect, the same overall response distribution (n_{+1}, n_{+2}) of successes and failures occurs regardless of the allocation of subjects to treatments. Thus, the column margin is also naturally fixed. This argument leads to hypergeometric null probabilities and Fisher's exact test (Greenland 1981). This argument does not extend, however, to nonnull effect values and hence to confidence intervals.

When both sets of marginal totals are naturally fixed, such as in Table 3.8, the high degree of discreteness is unavoidable and Fisher's exact test is the best procedure. Regardless of which margins are naturally fixed, using the mid-P-value helps reduce conservative effects of discreteness.

3.5.7 Derivation of Exact Conditional Distribution*

We now show how the conditional test for independence yields the hypergeometric distribution. We do this for $I \times J$ tables, since we next discuss extensions of Fisher's exact test for them. We assume independent multinomial sampling within rows, as often applies in comparing I treatment groups. Then row totals $\{n_{i+}\}$ are fixed, and we estimate the I conditional distributions $\{\pi_{j|i}, \ j = 1, \ldots, J\}$. Under H_0: independence, $\pi_{j|1} = \pi_{j|2} = \cdots = \pi_{j|I} = \pi_{+j}$, for $j = 1, \ldots, J$. The product of the I multinomial probability functions then simplifies to

$$\prod_i \left(\frac{n_{i+}!}{\prod_j n_{ij}!} \prod_j \pi_{j|i}^{n_{ij}} \right) = \frac{\left(\prod_i n_{i+}! \right) \left(\prod_j \pi_{+j}^{n_{+j}} \right)}{\prod_i \prod_j n_{ij}!}. \tag{3.17}$$

This distribution for $\{n_{ij}\}$ depends on $\{\pi_{+j}\}$. These are nuisance parameters, since they do not describe the association. Fisher introduced the standard way of eliminating nuisance parameters, by conditioning on their sufficient statistics. From the definition of sufficiency, the resulting conditional distribution does not depend on those parameters.

The contribution of $\{\pi_{+j}\}$ to the product multinomial distribution (3.17) depends on the data only through $\{n_{+j}\}$, which are their sufficient statistics. The $\{n_{+j}\}$ have the multinomial $(n, \{\pi_{+j}\})$ distribution, namely

$$\frac{n!}{\prod_j n_{+j}!} \prod_j \pi_{+j}^{n_{+j}}. \tag{3.18}$$

The joint probability function of $\{n_{ij}\}$ and $\{n_{+j}\}$ is identical to the probability function of $\{n_{ij}\}$, since $\{n_{ij}\}$ determines $\{n_{+j}\}$. Thus, the probability function of $\{n_{ij}\}$, conditional on $\{n_{+j}\}$, equals the probability function (3.17) of $\{n_{ij}\}$ divided by the probability function (3.18) evaluated at $\{n_{+j}\}$, or

$$\frac{\left(\prod_i n_{i+}! \right) \left(\prod_j n_{+j}! \right)}{n! \prod_i \prod_j n_{ij}!}. \tag{3.19}$$

This is the *multiple hypergeometric* distribution. It applies to the set of $\{n_{ij}\}$ having the same $\{n_{i+}\}$ and $\{n_{+j}\}$ as the observed table. For 2×2 tables, it is the hypergeometric distribution (3.16).

When a table has a single multinomial sample, the unknown parameters are $\{\pi_{ij}\}$. For testing independence ($\pi_{ij} = \pi_{i+} \pi_{+j}$ all i and j), distribution (3.19) results from conditioning on the row and column totals. These are sufficient statistics for $\{\pi_{i+}\}$ and $\{\pi_{+j}\}$, which determine the null distribution. For either sampling model, both sets of margins are fixed after the conditioning. The end result (3.19) does not depend on unknown parameters and thus permits exact probability calculations.

3.5.8 Exact Tests of Independence for $I \times J$ Tables*

Exact tests for $I \times J$ tables utilize the multiple hypergeometric distribution. Freeman and Halton (1951) defined the P-value as the probability of the set

TABLE 3.9   Example for Exact Conditional Test

	Smoking Level (cigarettes/day)		
	0	1–24	> 25
Control	25	25	12
Myocardial infarction	0	1	3

Source: Reprinted with permission, based on Table 5 in S. Shapiro et al., *Lancet* 743–746 (1979).

of tables with the given margins that are no more likely to occur than the table observed. Other exact tests order the tables using a statistic describing distance from H_0. Yates (1934) used X^2. The *P*-value is then the null value of $P(X^2 \geq X_o^2)$ for observed value X_o^2. When classifications have ordered categories, an ordinal statistic is more relevant. For the alternative hypothesis of a positive association, we could use $P(T \geq t_o)$, where T is the correlation or gamma and where t_o denotes its observed value.

We illustrate an exact test for ordered categories with Table 3.9, which cross-classifies level of smoking and myocardial infarction for a sample of young women in a case–control study. The second row contains small counts, and large-sample tests may be inappropriate. Given the marginal counts, the only table having greater evidence of positive association between smoking and myocardial infarction has counts (25,26,11) for row 1 and (0,0,4) in row 2. Conditional on both sets of margins, the null probability of the observed table and this more extreme table [based on formula (3.19)] equals 0.018. Although the sample contains only four myocardial infarction patients, evidence exists of a positive association. The evidence is stronger than using X^2, which ignores the ordering of categories. The exact $P(X^2 \geq X_o^2) = P(X^2 \geq 6.96) = 0.052$.

Special algorithms and software for computing exact tests for $I \times J$ tables are widely available (e.g., Mehta and Patel 1983; see also Appendix A). We recommend these tests when asymptotic approximations may be invalid. Computing time increases exponentially as n, I, or J increase. However, one can use Monte Carlo to sample randomly from the set of tables with the given margins. The estimated *P*-value is then the sample proportion of tables having test statistic value at least as large as the value observed.

As I and/or J increase, the number of possible values for any test statistic T tends to increase. Thus, the conservativeness issue for conditional tests becomes less problematic.

3.6   SMALL-SAMPLE CONFIDENCE INTERVALS FOR 2×2 TABLES*

Small-sample methods also apply to estimation. Exact distributions depending only on the parameter of interest result from the same arguments. These distributions are the basis of confidence intervals for measures such as the odds ratio.

3.6.1 Small-Sample Inference for the Odds Ratio

For multinomial sampling, the distribution of $\{n_{ij}\}$ depends on n and cell probabilities $\{\pi_{ij}\}$. For 2×2 tables, the odds ratio is

$$\theta = \frac{\pi_{11}\pi_{22}}{\pi_{12}\pi_{21}} = \frac{\pi_{11}(1 - \pi_{1+} - \pi_{+1} + \pi_{11})}{(\pi_{1+} - \pi_{11})(\pi_{+1} - \pi_{11})}.$$

Hence, π_{11} is a function of θ and $\{\pi_{1+}, \pi_{+1}\}$. The same argument applies to any π_{ij}, so the multinomial distribution of $\{n_{ij}\}$ can use parameters $\{\theta, \pi_{1+}, \pi_{+1}\}$. Conditional on $\{n_{1+}, n_{+1}\}$, the distribution of $\{n_{ij}\}$ depends only on θ. Since n_{11} determines all other cell counts, given the marginal totals, the conditional distribution of $\{n_{ij}\}$ is specified by some function $P(n_{11} = t) = f(t; n_{1+}, n_{+1}, n, \theta)$. This distribution (Fisher 1935c) is the *noncentral hypergeometric*,

$$f(t; n_{1+}, n_{+1}, n, \theta) = \frac{\binom{n_{1+}}{t}\binom{n - n_{1+}}{n_{+1} - t}\theta^t}{\sum\limits_{u=m_-}^{m_+} \binom{n_{1+}}{u}\binom{n - n_{1+}}{n_{+1} - u}\theta^u} \qquad (3.20)$$

for $m_- \le t \le m_+$.

A confidence interval for θ results from inverting the test of $H_0: \theta = \theta_0$, having observed $n_{11} = t_o$. For $H_a: \theta > \theta_0$, the P-value is

$$P = \sum_{t \ge t_o} f(t; n_{1+}, n_{+1}, n, \theta_0).$$

For testing against $H_0: \theta < \theta_0$,

$$P = \sum_{t \le t_o} f(t; n_{1+}, n_{+1}, n, \theta_0).$$

When $\theta_0 = 1$, these are one-sided Fisher's exact tests. Cornfield (1956) constructed a confidence interval using the *tail method*. The lower endpoint is θ_0 for which $P = \alpha/2$ in testing against $H_a: \theta > \theta_0$. The upper endpoint is θ_0 for which $P = \alpha/2$ for $H_a: \theta < \theta_0$. The interval is the set of θ_0 for which both one-sided P-values $\ge \alpha/2$.

As in Fisher's exact test, the conditional approach to interval estimation is necessarily conservative because of discreteness. The actual confidence coefficient, defined as the infimum of the coverage probabilities for all possible θ, has the nominal confidence level as a lower bound. Less conservative behavior and shorter intervals result from inverting a single two-sided test rather than inverting two one-sided tests (Agresti and Min 2001; Baptista and Pike 1977). An alternative approach with independent binomial samples inverts nonnull unconditional small-sample tests. Because of the reduced discreteness, such intervals are also usually shorter.

The *conditional ML estimate* of θ is the value of θ that maximizes probability (3.20). Differentiating the log likelihood with respect to θ shows that this estimate satisfies the equation $n_{11} = E(n_{11})$ in θ, where the expectation refers to distribution (3.20). This equation has a unique solution $\hat{\theta}$ and is solved using iterative methods (Cornfield 1956). This estimator differs from the *unconditional ML estimator* $\hat{\theta} = n_{11}n_{22}/n_{12}n_{21}$, which uses the ML estimates of $\{\pi_{ij}\}$ for the multinomial distribution of $\{n_{ij}\}$. Using statistical software, we can calculate conditional ML estimates and small-sample confidence intervals for odds ratios (e.g., for SAS, see Table A.2).

3.6.2 Tea Tasting Example

We illustrate with Table 3.8 from Fisher's tea-tasting experiment. The conditional ML estimate of θ is 6.4. Software provides the Cornfield tail-method interval (0.2, 626.2) with confidence coefficient guaranteed ≥ 0.95. Not surprisingly, it is very wide because of the small sample. Inverting a family of two-sided "exact" conditional score tests gives a more precise interval, (0.3, 306.2). The unconditional approach is not appropriate here because of the sampling design. [If the table were two binomial samples, that approach gives interval (0.4, 234.4) by inverting "exact" unconditional score tests.]

3.6.3 Impact of Discreteness on Exact Confidence Intervals

Small-sample inference is "exact" in the sense that the conditional distribution is free of nuisance parameters. Confidence intervals and tests use exact probability calculations rather than approximate ones. However, their operating characteristics are conservative because of discreteness.

Large-sample methods do not have the guarantee of bounds on error probabilities. They can be conservative or liberal, and thus their results can appear quite different from exact methods. For example, for the tea-tasting data (Table 3.8), the P-value for the Pearson chi-squared test equals 0.157, compared to 0.486 for the two-sided exact test. The 95% large-sample confidence interval (3.2) for the odds ratio is (0.4, 220.9), compared to Cornfield's exact interval of (0.2, 626.2). Normally, one would prefer an exact method over an approximate one. When the conditional distribution is highly discrete, however, the choice is not so obvious. Exact methods then can be quite conservative, especially with small samples.

For highly discrete data, it seems sensible to use adjustments of exact methods based on the mid-P-value. Confidence intervals with the conditional approach then invert hypergeometric tests of $\theta = \theta_0$ using the mid-P-value. Although not guaranteed to have error probabilities no greater than the nominal level, this method usually comes closer than the exact method to the desired level. Compared to large-sample methods, it has the advantage of working well as the degree of discreteness diminishes, since it then is essentially the same as the corresponding exact method using an ordinary P-value.

Inference based on the mid-P-value compromises between the conservativeness of exact methods and the uncertain adequacy of large-sample methods. For interval estimation of the odds ratio, this method tends to be a bit conservative, but for small samples can yield much shorter intervals than the Cornfield exact interval. For the tea-tasting data, for instance, the 95% confidence interval based on inverting two one-sided hypergeometric tests using the mid-P-value is (0.31, 309), compared to the Cornfield interval of (0.21, 626).

3.6.4 Small-Sample Inference for Difference of Proportions

The conditional approach to eliminating nuisance parameters works when those parameters have sufficient statistics. However, we'll see (Section 6.7.9) that reduced sufficient statistics occur only for certain models. For binary data, such models must have odds ratios as parameters. For 2×2 tables, the conditional approach cannot yield confidence intervals for differences or ratios of proportions. The unconditional approach is more complex but does not require sufficient statistics. We used it in Section 3.5.5 for testing $\pi_1 - \pi_2 = 0$ with independent binomial samples.

A small-sample confidence interval inverts the corresponding unconditional test of H_0: $\pi_1 - \pi_2 = \delta_0$, for any fixed $-1 < \delta_0 < 1$. The probability function for the table is the product of $\text{bin}(n_1, \pi_1)$ and $\text{bin}(n_2, \pi_2)$ mass functions. One can express this in terms of $\delta = \pi_1 - \pi_2$ and a nuisance parameter λ. For instance, if $\lambda = \pi_1 + \pi_2$, one substitutes $\pi_1 = (\lambda + \delta)/2$ and $\pi_2 = (\lambda - \delta)/2$. For $\delta = \delta_0$ and a fixed value of λ, one then uses this binomial product to calculate the probability that the test statistic is at least as large as observed. The P-value is the supremum of such probabilities calculated over all possible values for λ. This provides a family of tests for the various values of δ_0. The confidence interval for $\pi_1 - \pi_2$ is the set of δ_0 for which this P-value exceeds α.

This approach can be quite conservative. For details regarding various test statistics, see Agresti and Min (2001), Coe and Tamhane (1993), Santner and Snell (1980), and Santner and Yamagami (1993). It is better to invert a single two-sided test, as in Coe and Tamhane (1993), than to invert two separate one-sided tests.

3.7 EXTENSIONS FOR MULTIWAY TABLES AND NONTABULATED RESPONSES

The methods of this chapter extend to multiway contingency tables. For instance, tests of independence for two-way tables extend to tests of conditional independence in three-way tables. In future chapters we present such methods with models that provide a basis for defining relevant parameters and their statistical inferences. The methods then apply in a greater variety of situations, such as when some explanatory variables are continuous rather than categorical.

3.7.1 Categorical Data Need Not Be Contingency Tables

Examples so far have presented categorical data in the format of contingency tables. However, this book has broader focus than contingency table analysis. Models for categorical response variables can have continuous as well as categorical explanatory variables. Even when all or most variables are categorical, source data files are not usually contingency tables but have the form of a line of data for each subject. The first three lines in a data file containing responses of a survey of subjects measuring gender, race, education (1 = less than high school, 2 = high school or some college, 3 = college graduate), and opinion about homosexuality (1 = tolerant, 2 = homophobic) might be:

subject	gender	race	education	opinion
1	f	w	2	1
2	m	b	3	1
3	m	w	1	2

Software can read data files of this type and then conduct analyses that may involve forming contingency tables.

In the next chapter we introduce the modeling framework used in the rest of the book. All the methods that we've studied in this chapter result from inferences for parameters in simple versions of these models.

NOTES

Section 3.1: Confidence Intervals for Association Parameters

3.1. Adaptations of Woolf's interval (3.2) for $\log \theta$ to handle zero cell counts include Agresti (1999) and Gart (1966, 1971). Goodman (1964a) presented simultaneous confidence intervals for all odds ratios in an $I \times J$ table. Brown and Benedetti (1977) and Goodman and Kruskal (1963, 1972) provided standard errors for many association measures. Goodman and Kruskal (1963, 1972) extended (3.9) for independent multinomial sampling.

3.2. Agresti and Caffo (2000) showed that as in the single-sample case (Problem 1.24), the Wald interval (3.4) for $\pi_1 - \pi_2$ behaves much better after adding two pseudo-observations of each type (one of each type in each sample).

Section 3.2: Testing Independence in Two-Way Contingency Tables

3.3. For hypergeometric sampling, $\{\hat{\mu}_{ij}\}$ in tests of independence are *exact* (rather than *estimated*) expected values. Specifically,

$$E(n_{11}) = \frac{n_{1+}n_{+1}}{n} \quad \text{and} \quad \text{var}(n_{11}) = \frac{n_{1+}n_{+1}n_{2+}n_{+2}}{n^2(n-1)}.$$

Haldane (1940) derived $E(X^2) = (I-1)(J-1)n/(n-1)$ and a complex formula for $\text{var}(X^2)$; Dawson (1954) provided a simplified expression. Lewis et al. (1984) derived the third central moment. Watson (1959) showed that the conditional distribution of X^2 also has the limiting chi-squared distribution.

3.4. Diaconis and Efron (1985) presented inference based on a uniform distribution over all possible tables of the same I, J, and n; their *volume test* considers the proportion of such tables having $X^2 \leq X_o^2$.

3.5. Specialized methods are necessary for complex sampling designs. Sequential methods are useful in biomedical applications (Jennison and Turnbull 2000, Chap. 12). Social science applications often incorporate clustering and/or stratification. LaVange et al. (2001) and Rao and Thomas (1988) surveyed analyses of categorical data for complex sampling methods. Gleser and Moore (1985) showed that positive dependence causes null distributions of Pearson statistics to stochastically increase. See also Bedrick (1983), Clogg and Eliason (1987), Fay (1985), Holt et al. (1980), Koehler and Wilson (1986), Rao and Scott (1987), Scott and Wild (2001), Shuster and Downing (1976), Tavaré and Altham (1983), and methods of Chapter 12.

Other modifications are necessary when some data are missing. Watson (1956) was perhaps the first to study this. Lipsitz and Fitzmaurice (1996) derived score tests of independence and conditional independence for contingency tables, assuming ignorable nonresponse, and showed that the test statistics have the usual asymptotic chi-squared null distributions. See Schafer (1997, Chap. 7) for a survey of methods.

Section 3.4: Two-Way Tables with Ordered Classifications

3.6. Bhapkar (1968) and Yates (1948) proposed statistics similar to M^2 and also proposed statistics for singly-ordered tables. Graubard and Korn (1987) listed 14 tests for $2 \times J$ tables that utilize a correlation-type statistic. See also Nair (1987) and Williams (1952). Cohen and Sackrowitz (1991, 1992) evaluated decision-theoretic aspects, such as admissibility, of tests based on gamma and local log odds ratios. Rayner and Best (2001) considered nonparametrics methods in a contingency table format.

Section 3.5: Small-Sample Tests of Independence

3.7. Yates (1934) mentioned that Fisher suggested the hypergeometric to him for an exact test. He proposed a continuity-corrected version of X^2,

$$X_c^2 = \sum\sum \frac{\left(|n_{ij} - \hat{\mu}_{ij}| - 0.5\right)^2}{\hat{\mu}_{ij}},$$

to approximate the exact test. Haber (1980, 1982), Plackett (1964), and Yates (1984) discussed its appropriateness. Since software now makes Fisher's exact test feasible even with large samples, this correction is no longer needed.

3.8. The UMPU property of Fisher's exact test follows from conditioning on a sufficient statistic that is complete and has distribution in the exponential family (Lehmann 1986, Secs. 4.5–4.7). Fleiss (1981), Gail and Gart (1973), and Suissa and Shuster (1985) studied sample size for obtaining fixed power in Fisher's test. The controversy over conditioning includes Barnard (1945, 1947, 1949, 1979), Berkson (1978), Fisher (1956), Howard (1998), Kempthorne (1979), Lloyd (1988a), Pearson (1947), Rice (1988), Routledge (1992), Suissa and Shuster (1984, 1985), and Yates (1984). Yates and discussants also addressed the choice of two-sided P-value. Discussion of unconditional methods includes Chan (1998), Martín Andrés and Silva Mato (1994), and Røhmel and Mansmann (1999). Altham (1969) and Howard (1998) discussed Bayesian analyses for 2×2 tables (see Section 15.2.3). Agresti (1992, 2001) surveyed small-sample methods.

3.9. For discussion of inference using the mid-P-value, see Berry and Armitage (1995), Hirji (1991), Hwang and Wells (2002), Hwang and Yang (2001), Mehta and Walsh (1992), and Routledge (1994). Similar benefits can accrue from alternative proposed P-values. One approach, useful when several tables have the same value for a test statistic, uses the table probability to create a more finely partitioned sample space; for tables having the observed test statistic value, only those contribute to the P-value that are no more likely than the observed table (Cohen and Sackrowitz 1992; Kim and Agresti 1995). This depends on more than the sufficient statistic, and in some cases a Rao–Blackwellized version is the mid-P-value (Hwang and Wells 2002). Ordinary P-values obtained with higher-order asymptotic methods without continuity corrections for discreteness yield performance similar to that of the mid-P-value (Pierce and Peters 1999; Strawderman and Wells 1998).

3.10. For exact treatment of $I \times J$ tables, see Mehta and Patel (1983). For ordered categories, see also Agresti et al. (1990). For Monte Carlo estimation of exact P-values, see Agresti et al. (1979), Booth and Butler (1999), Diaconis and Sturmfels (1998), Forster et al. (1996), Mehta et al. (1988), and Patefield (1982). Gail and Mantel (1977) and Good (1976) gave approximate formulas for the number of tables having certain fixed margins. Freidlin and Gastwirth (1999) extended the unconditional approach to a test for trend in $I \times 2$ tables and a test of conditional independence with several 2×2 tables.

Section 3.6: Small-Sample Confidence Intervals for 2 × 2 Tables

3.11. Suppose that (θ, λ) has minimal sufficient statistic (T, U), where λ is a nuisance parameter. Cox and Hinkley (1974, p. 35) defined U to be *ancillary* for θ if its distribution depends only on λ, and the distribution of T given U depends only on θ. For 2×2 tables with odds ratio θ and $\lambda = (\pi_{1+}, \pi_{+1})$, let $T = n_{11}$ and $U = (n_{1+}, n_{+1})$. Then U is not ancillary, because its distribution depends on θ as well as λ. Using a definition due to Godambe, Bhapkar (1989) referred to the marginals U as *partial ancillary* for θ. This means that the distribution of the data, given U, depends only on θ, and that for fixed θ, the family of distributions of U for various λ is complete. Liang (1984) gave an alternative definition referring to conditional and unconditional inference being equally efficient.

PROBLEMS

Applications

3.1 Refer to Table 2.9. Construct and interpret a 95% confidence interval for the population **(a)** odds ratio, **(b)** difference of proportions, and **(c)** relative risk between seat-belt use and type of injury.

3.2 Refer to Table 2.5 on lung cancer and smoking. Construct a confidence interval for a relevant measure of association. Interpret.

3.3 In professional basketball games during 1980–1982, when Larry Bird of the Boston Celtics shot a pair of free throws, 5 times he missed both, 251 times he made both, 34 times he made only the first, and 48 times he made only the second (Wardrop 1995). Is it plausible that the successive free throws are independent?

3.4 Refer to Table 3.10.
  a. Using X^2 and G^2, test the hypothesis of independence between party identification and race. Report the P-values and interpret.
  b. Use residuals to describe the evidence of association.
  c. Partition chi-squared into components regarding the choice between Democrat and Independent and between these two combined and Republican. Interpret.
  d. Summarize association by constructing a 95% confidence interval for the odds ratio between race and whether a Democrat or Republican. Interpret.

TABLE 3.10   Data for Problem 3.4

Race	Party Identification		
	Democrat	Independent	Republican
Black	103	15	11
White	341	105	405

Source: 1991 General Social Survey, National Opinion Research Center.

3.5 Refer to Table 3.10. In the same survey, gender was cross-classified with party identification. Table 3.11 shows some results. Explain how to interpret all the results on this printout.

3.6 In a study of the relationship between stage of breast cancer at diagnosis (local or advanced) and a woman's living arrangement, of 144 women living alone, 41.0% had an advanced case; of 209 living with spouse, 52.2% were advanced; of 89 living with others, 59.6% were advanced. The authors reported the P-value for the relationship as 0.02 (D. J. Moritz and W. A. Satariano, *J. Clin. Epidemiol.* **46**: 443–454, 1993). Reconstruct the analysis performed to obtain this P-value.

TABLE 3.11   Results for Problem 3.5

Frequency Expected	dem	indep	repub
female	279	73	225
	261.42	70.653	244.93
male	165	47	191
	182.58	49.347	171.07

Statistic	DF	Value	Prob
Chi-Square	2	7.0095	0.0301
Likelihood Ratio Chi-Square	2	7.0026	0.0302

Observ	Resraw	Reschi	StReschi	Observ	Resraw	Reschi	StReschi
1	17.584	1.088	2.293	4	−17.584	−1.301	−2.293
2	2.347	0.279	0.465	5	−2.347	−0.334	−0.464
3	−19.931	−1.274	−2.618	6	19.931	1.524	2.618

3.7   Refer to Table 2.1. Partition G^2 for testing whether the incidence of heart attacks is independent of aspirin intake into two components. Interpret.

3.8   *Project Blue Book: Analysis of Reports of Unidentified Aerial Objects* was published by the U.S. Air Force (Air Technical Intelligence Center at Wright-Patterson Air Force Base) in May 1955 to analyze reports of unidentified flying objects (UFOs). In its Table II, the report classified 1765 sightings later regarded as known objects and 434 sightings later regarded as unknown, according to the object color (nine categories). The report states: "The chi-square test is applicable only to distributions which have the same number of elements," so the investigators multiplied all counts in the known category by (434/1765), so each row has 434 observations, before computing X^2. They reported $X^2 = 26.15$ with df = 8. Explain why this is incorrect. What should X^2 equal? (*Hint:* For their adjusted table, first show that the contribution to X^2 is the same for each cell in a column, and then show the effect on those contributions of multiplying each count in one row by a constant.)

3.9   Table 3.12 classifies a sample of psychiatric patients by their diagnosis and by whether their treatment prescribed drugs.

   a. Obtain standardized Pearson residuals for independence, and interpret.

   b. Partition chi-squared into three components to describe differences and similarities among the diagnoses, by comparing (**i**) the first two rows, (**ii**) the third and fourth rows, and (**iii**) the last row to the first and second rows combined and the third and fourth rows combined.

TABLE 3.12 Data for Problem 3.9

Diagnosis	Drugs	No Drugs
Schizophrenia	105	8
Affective disorder	12	2
Neurosis	18	19
Personality disorder	47	52
Special symptoms	0	13

Source: Reprinted with permission from E. Helmes and G. C. Fekken, *J. Clin. Psychol.* **42:** 569–576 (1986).

3.10 Refer to Table 7.8. For the combined data for the two genders, yielding a single 4 × 4 table, $X^2 = 11.5$ ($P = 0.24$), whereas using row scores (3, 10, 20, 35) and column scores (1, 3, 4, 5), $M^2 = 7.04$ ($P = 0.008$). Explain why the results are so different.

3.11 A study on educational aspirations of high school students (S. Crysdale, *Internat. J. Compar. Sociol.* **16:** 19–36, 1975) measured aspirations with the scale (some high school, high school graduate, some college, college graduate). The student counts in these categories were (11, 52, 23, 22) when family income was low, (9, 44, 13, 10) when family income was middle, and (9, 41, 12, 27) when family income was high.

   a. Test independence of educational aspirations and family income using X^2 or G^2. Explain the deficiency of this test for these data.

   b. Find the standardized Pearson residuals. Do they suggest any association pattern?

   c. Conduct an alternative test that may be more powerful. Interpret.

3.12 Refer to Table 8.15. Obtain a 95% confidence interval for gamma. Interpret the association between schooling and attitude toward abortion.

3.13 Table 3.13 shows the results of a retrospective study comparing radiation therapy with surgery in treating cancer of the larynx. The response

TABLE 3.13 Data for Problem 3.13

	Cancer Controlled	Cancer Not Controlled
Surgery	21	2
Radiation therapy	15	3

Source: Reprinted with permission from W. M. Mendenhall, R. R. Million, D. E. Sharkey, and N. J. Cassisi, *Internat. J. Radiat. Oncol. Biol. Phys.* **10:** 357–363 (1984), Pergamon Press plc.

TABLE 3.14   SAS Output for Problem 3.13

```
              Fisher's Exact Test
- - - - - - - - - - - - - - - - - - - - - - - - - - -
Cell (1,1) Frequency (F)              21
Left-sided Pr <= F                0.8947
Right-sided Pr >= F               0.3808
Table Probability (P)            0.2755
Two-sided Pr<= P                  0.6384
- - - - - - - - - - - - - - - - - - - - - - - - - - -

Odds Ratio                        2.1000

Asymptotic Conf Limits:     95% Lower Conf Limit   0.3116
                            95% Upper Conf Limit  14.1523
Exact Conf Limits:          95% Lower Conf Limit   0.2089
                            95% Upper Conf Limit  27.5522
```

indicates whether the cancer was controlled for at least two years following treatment. Table 3.14 shows SAS output.

  a. Report and interpret the P-value for Fisher's exact test with **(i)** H_a: $\theta > 1$, and **(ii)** H_a: $\theta \neq 1$. Explain how the P-values are calculated.

  b. Interpret the confidence intervals for θ. Explain the difference between them and how they were calculated.

  c. Find and interpret the one-sided mid-P-value. Give advantages and disadvantages of this type of P-value.

3.14 A study considered the effect of prednisolone on severe hypercalcaemia in women with metastatic breast cancer (B. Kristensen et al., *J. Intern. Med.* **232**: 237–245, 1992). Of 30 patients, 15 were randomly selected to receive prednisolone. The other 15 formed a control group. Normalization in their level of serum-ionized calcium was achieved by 7 of the treated patients and none of the control group. Analyze whether results were significantly better for treatment than for control. Interpret.

3.15 For Problem 3.14, obtain a 95% confidence interval for the odds ratio using **(a)** the Woolf (i.e., Wald) interval, **(b)** Cornfield's "exact" approach, **(c)** the profile likelihood. In each case, note the effect of the zero cell count. Summarize advantages and disadvantages of each approach.

3.16 Refer to the tea-tasting data (Table 3.8). Construct the null distributions of the ordinary P-value and the mid-P-value for Fisher's exact test with H_a: $\theta > 1$. Find and compare their expected values.

3.17 Consider a 3×3 table having entries, by row, of (4, 2, 0 / 2, 2, 2 / 0, 2, 4). Conduct an exact test of independence, using X^2. Assuming ordered rows and columns and using equally spaced scores, conduct an ordinal exact test. Explain why results differ so much.

3.18 An advertisement by Schering Corp. in 1999 for the allergy drug Claritin mentioned that in a pediatric randomized clinical trial, symptoms of nervousness were shown by 4 of 188 patients on loratadine (Claritin), 2 of 262 patients taking placebo, and 2 of 170 patients on choropheniramine. In each part below, explain which method you used, and why.
  a. Is there inferential evidence that nervousness depends on drug?
  b. For the Claritin and placebo groups, construct and interpret a 95% confidence interval for the (**i**) odds ratio and (**ii**) difference of proportions suffering nervousness.

3.19 Refer to Problem 2.19 on sexual fun. Analyze these data. Present a short report summarizing results and interpretations.

Theory and Methods

3.20 Is $\hat{\theta}$ the midpoint of large- and small-sample confidence intervals for θ? Why or why not?

3.21 For comparing two binomial samples, show that the standard error (3.1) of a log odds ratio increases as the absolute difference of proportions of successes and failures for a given sample increases.

3.22 Using the delta method, show that the Wald confidence interval for the logit of a binomial parameter π is

$$\log[\hat{\pi}/(1 - \hat{\pi})] \pm z_{\alpha/2}/\sqrt{n\hat{\pi}(1 - \hat{\pi})} .$$

Explain how to use this interval to obtain one for π itself. [Newcombe (2001) noted that the sample logit is also the midpoint of the score interval for π, on the logit scale. He showed that this logit interval contains the score interval.]

3.23 For two parameters, a confidence interval for $\theta_1 - \theta_2$ based on single-sample estimate $\hat{\theta}_i$ and interval (ℓ_i, u_i) for θ_i, $i = 1, 2$, is

$$\left(\hat{\theta}_1 - \hat{\theta}_2 - \sqrt{\left(\hat{\theta}_1 - \ell_1\right)^2 + \left(u_2 - \hat{\theta}_2\right)^2}, \ \hat{\theta}_1 - \hat{\theta}_2 + \sqrt{\left(u_1 - \hat{\theta}_1\right)^2 + \left(\hat{\theta}_2 - \ell_2\right)^2} \right).$$

Newcombe (1998b) proposed an interval for $\pi_1 - \pi_2$ using the score interval (ℓ_i, u_i) for π_i that performs much better than the Wald interval (3.4). It is $(\hat{\pi}_1 - \hat{\pi}_2 - z_{\alpha/2}s_L, \hat{\pi}_1 - \hat{\pi}_2 + z_{\alpha/2}s_U)$, with

$$s_L = \sqrt{\frac{\ell_1(1 - \ell_1)}{n_1} + \frac{u_2(1 - u_2)}{n_2}}, \qquad s_U = \sqrt{\frac{u_1(1 - u_1)}{n_1} + \frac{\ell_2(1 - \ell_2)}{n_2}}.$$

Show that it has the general form above of an interval for $\theta_1 - \theta_2$.

3.24 For multinomial sampling, use the asymptotic variance of $\log \hat{\theta}$ to show that for Yule's Q (Problem 3.26) the asymptotic variance of $\sqrt{n}(\hat{Q} - Q)$ is $\sigma^2 = (\Sigma_i\Sigma_j\pi_{ij}^{-1})(1 - Q^2)^2/4$ (Yule 1900, 1912).

3.25 Refer to Problem 2.23. For multinomial sampling, show how to obtain a confidence interval for AR by first finding one for $\log(1 - AR)$ (Fleiss 1981, p. 76).

3.26 For multinomial probabilities $\boldsymbol{\pi} = (\pi_1, \pi_2, \ldots)$ with a contingency table of arbitrary dimensions, suppose that a measure $g(\boldsymbol{\pi}) = \nu/\delta$. Show that the asymptotic variance of $\sqrt{n}[g(\hat{\boldsymbol{\pi}}) - g(\boldsymbol{\pi})]$ is $\sigma^2 = [\Sigma_i\pi_i\eta_i^2 - (\Sigma_i\pi_i\eta_i)^2]/\delta^4$, where $\eta_i = \delta(\partial\nu/\partial\pi_i) - \nu(\partial\delta/\partial\pi_i)$ (Goodman and Kruskal, 1972).

3.27 For ordinal variables, consider gamma (2.14). Let

$$\pi_{ij}^{(c)} = \sum_{a<i}\sum_{b<j}\pi_{ab} + \sum_{a>i}\sum_{b>j}\pi_{ab}, \qquad \pi_{ij}^{(d)} = \sum_{a<i}\sum_{b>j}\pi_{ab} + \sum_{a>i}\sum_{b<j}\pi_{ab},$$

where i and j are fixed in the summations. Show that $\Pi_c = \Sigma_i\Sigma_j\pi_{ij}\pi_{ij}^{(c)}$ and $\Pi_d = \Sigma_i\Sigma_j\pi_{ij}\pi_{ij}^{(d)}$. Use the delta method to show that the large-sample normality (3.9) applies for $\hat{\gamma}$, with (Goodman and Kruskal 1963)

$$\phi_{ij} = 4[\Pi_d\pi_{ij}^{(c)} - \Pi_c\pi_{ij}^{(d)}]/(\Pi_c + \Pi_d)^2, \qquad \sum_i\sum_j\pi_{ij}\phi_{ij} = 0,$$

$$\sigma^2 = \frac{16}{(\Pi_c + \Pi_d)^4}\sum_i\sum_j\pi_{ij}[\Pi_d\pi_{ij}^{(c)} - \Pi_c\pi_{ij}^{(d)}]^2.$$

3.28 An $I \times J$ table has ordered columns and unordered rows. *Ridits* (Bross 1958) are data-based column scores. The jth sample ridit is the average cumulative proportion within category j,

$$\hat{r}_j = \sum_{k=1}^{j-1} p_{+k} + \left(\frac{1}{2}\right) p_{+j}.$$

The sample mean ridit in row i is $\hat{R}_i = \sum_j \hat{r}_j p_{j|i}$. Show that $\sum_j p_{+j} \hat{r}_j = 0.50$ and $\sum_i p_{i+} \hat{R}_i = 0.50$. [For ridit analyses, see Agresti (1984, Secs. 9.3 and 10.2), Bross (1958), Fleiss (1981, Sec. 9.4), and Landis et al. (1978).]

3.29 Show that $X^2 = n \sum\sum (p_{ij} - p_{i+} p_{+j})^2 / p_{i+} p_{+j}$. Thus, X^2 can be large when n is large, regardless of whether the association is practically important. Explain why this test, like other tests, simply indicates the degree of evidence against H_0 and does not describe strength of association. ("Like fire, the chi-square test is an excellent servant and a bad master," Sir Austin Bradford Hill, *Proc. Roy. Soc. Med.* **58:** 295–300, 1965.)

3.30 For testing $H_0: \pi_1 = \pi_2$ using independent binomial variates y_1 and y_2 with n_1 and n_2 trials, the score statistic is

$$z = \frac{\hat{\pi}_1 - \hat{\pi}_2}{\sqrt{\hat{\pi}(1 - \hat{\pi})(1/n_1 + 1/n_2)}},$$

where $\hat{\pi} = (y_1 + y_2)/(n_1 + n_2)$ is the *pooled estimate* of $\pi_1 = \pi_2$ under H_0. Show that $z^2 = X^2$.

3.31 For a 2×2 table, consider $H_0: \pi_{11} = \theta^2, \pi_{12} = \pi_{21} = \theta(1 - \theta), \pi_{22} = (1 - \theta)^2$.

   a. Show that the marginal distributions are identical and that independence holds.

   b. For a multinomial sample, under H_0 show that $\hat{\theta} = (p_{1+} + p_{+1})/2$.

   c. Explain how to test H_0. Show that df = 2 for the test statistic.

   d. Refer to Problem 3.3. Are Larry Bird's pairs of free throws plausibly independent *and* identically distributed?

3.32 For a 2×2 table, show that:

   a. The four Pearson residuals may take different values.

b. All four standardized Pearson residuals have the same absolute value. (This is sensible, since df = 1.)

c. The square of each standardized Pearson residual equals X^2. [*Note:* $X^2 = n(n_{11}n_{22} - n_{12}n_{21})^2/(n_{1+}n_{2+}n_{+1}n_{+2})$ for 2×2 tables. See Mirkin (2001) for alternative X^2 formulas for $I \times J$ tables.]

3.33 For testing independence, show that $X^2 \leq n \min(I - 1, J - 1)$. Hence $V^2 = X^2/[n\min(I - 1, J - 1)]$ falls between 0 and 1 (Cramér 1946). For 2×2 tables, X^2/n is often called *phi-squared*; it equals Goodman and Kruskal's tau (Problem 2.38). Other measures based on X^2 include the *contingency coefficient* $[X^2/(X^2 + n)]^{1/2}$ (Pearson 1904).

3.34 For counts $\{n_i\}$, the *power divergence statistic* for testing goodness of fit (Cressie and Read 1984; Read and Cressie 1988) is

$$\frac{2}{\lambda(\lambda + 1)} \sum n_i \left[(n_i/\hat{\mu}_i)^\lambda - 1 \right] \qquad \text{for } -\infty < \lambda < \infty.$$

a. For $\lambda = 1$, show that this equals X^2.

b. As $\lambda \to 0$, show that it converges to G^2. [*Hint:* $\log t = \lim_{h \to 0} (t^h - 1)/h$.]

c. As $\lambda \to -1$, show that it converges to $2\sum \hat{\mu}_i \log(\hat{\mu}_i/n_i)$, the *minimum discrimination information* statistic (Gokhale and Kullback 1978).

d. For $\lambda = -2$, show that it equals $\sum(n_i - \hat{\mu}_i)^2/n_i$, the *Neyman modified chi-squared* statistic (Neyman 1949).

e. For $\lambda = -\frac{1}{2}$, show that it equals $4\sum(\sqrt{n_i} - \sqrt{\hat{\mu}_i})^2$, the *Freeman–Tukey* statistic (Freeman and Tukey 1950).

[Under regularity conditions, their asymptotic distributions are identical (see Drost et al. 1989). The chi-squared null approximation works best for λ near $\frac{2}{3}$.]

3.35 Use a partitioning argument to explain why G^2 for testing independence cannot increase after combining two rows (or two columns) of a contingency table. (*Hint:* Argue that G^2 for full table = G^2 for collapsed table + G^2 for table of the two rows that are combined in the collapsed table.)

3.36 Motivate partitioning (3.14) by showing that the multiple hypergeometric distribution (3.19) for $\{n_{ij}\}$ factors as the product of hypergeometric distributions for the separate component tables (Lancaster, 1949).

3.37 Explain why $\{n_{+j}\}$ are sufficient for $\{\pi_{+j}\}$ in (3.17).

3.38 Assume independence, and let $p_{ij} = n_{ij}/n$ and $\hat{\pi}_{ij} = p_{i+}p_{+j}$.
  a. Show that p_{ij} and $\hat{\pi}_{ij}$ are unbiased for $\pi_{ij} = \pi_{i+}\pi_{+j}$.
  b. Show that $\text{var}(p_{ij}) = \pi_{i+}\pi_{+j}(1 - \pi_{i+}\pi_{+j})/n$.
  c. Using $E(p_{i+}p_{+j})^2 = E(p_{i+}^2)E(p_{+j}^2)$ and $E(p_{i+}^2) = \text{var}(p_{i+}) + [E(p_{i+})]^2$, show that

$$\text{var}(\hat{\pi}_{ij}) = \left\{ \pi_{i+}\pi_{+j}\left[\pi_{i+}(1 - \pi_{+j}) + \pi_{+j}(1 - \pi_{i+})\right]\right\}/n$$

$$+ \pi_{i+}(1 - \pi_{i+})\pi_{+j}(1 - \pi_{+j})/n^2.$$

  d. As $n \to \infty$, show that $\lim \text{var}(\sqrt{n}\,\hat{\pi}_{ij}) \leq \lim \text{var}(\sqrt{n}\,p_{ij})$, with equality only if $\pi_{ij} = 1$ or 0. Hence, if the model holds or if it nearly holds, the model estimator is better than the sample proportion.

3.39 Show that the sample value of the uncertainty coefficient (2.13) satisfies $\hat{U} = -G^2/2n(\Sigma p_{+j} \log p_{+j})$. [Haberman (1982) gave its standard error.]

3.40 When a test statistic has a continuous distribution, the P-value has a null uniform distribution, $P(P\text{-value} \leq \alpha) = \alpha$ for $0 < \alpha < 1$. For Fisher's exact test, explain why under the null, $P(P\text{-value} \leq \alpha) \leq \alpha$ for $0 < \alpha < 1$. (*Hint:* $P(P\text{-value} \leq \alpha) = E[P(P\text{-value} \leq \alpha | n_{1+}, n_{+1}, n)]$.)

3.41 Refer to Note 3.3 about moments of the hypergeometric distribution (3.16). Letting $\rho = n_{+1}/n$, show that n_{11} has the same mean as a binomial random variable for n_{1+} trials with success probability ρ, and that it has its variance multiplied by a finite population correction factor $(n - n_{1+})/(n - 1)$. (The hypergeometric is similar to the binomial when n_{1+} is small compared to n.)

3.42 A contingency table for two independent binomial variables has counts $(3, 0 \,/\, 0, 3)$ by row. For $H_0: \pi_1 = \pi_2$ and $H_a: \pi_1 > \pi_2$, show that the P-value equals $\frac{1}{64}$ for the exact unconditional test and $\frac{1}{20}$ for Fisher's exact test. [For discussion of this example, see Little (1989), G. Barnard's remarks at the end of Yates (1984), and Sprott (2000, Sec. 6.4.4).]

3.43 Refer to Problem 3.42 and exact tests using X^2 with H_a: $\pi_1 \neq \pi_2$. Explain why the unconditional P-value, evaluated at $\pi = 0.5$, is related to Fisher conditional P-values for various tables by

$$P(X^2 \geq 6) = \sum_{k=0}^{6} P(X^2 \geq 6|n_{+1} = k)P(n_{+1} = k) .$$

Thus, the unconditional P-value of $\frac{1}{32}$ is a weighted average of the Fisher P-value for the observed column margins and P-values of 0 corresponding to the impossibility of getting results as extreme as observed if other margins had occurred (i.e., $\frac{1}{32} = 0.10\left[\binom{6}{3}(1/2)^6\right]$). The Fisher quote in Section 3.5.6 gave his view about this.

3.44 Consider exact tests of independence, given the marginals, for the $I \times I$ table having $n_{ii} = 1$ for $i = 1,\ldots, I$, and $n_{ij} = 0$ otherwise. Show that (a) tests that order tables by their probabilities, X^2, or G^2 have P-value = 1.0, and (b) the one-sided test that orders tables by an ordinal statistic such as r or $C - D$ has P-value = $(1/I!)$.

3.45 A Monte Carlo scheme randomly samples M separate $I \times J$ tables having the observed margins to approximate $P_o = P(X^2 \geq X_o^2)$ for an exact test. Let $\hat{P}$ be the sample proportion of the M tables with $X^2 \geq X_o^2$. Show that $P(|\hat{P} - P_o| \leq B) = 1 - \alpha$ requires that $M \approx z_{\alpha/2}^2 P_o(1 - P_o)/B^2$.

3.46 Show that the conditional ML estimate of θ satisfies $n_{11} = E(n_{11})$ for distribution (3.18).

CHAPTER 4

Introduction to Generalized
Linear Models

In Chapters 2 and 3 we focused on methods for two-way contingency tables. Most studies, however, have several explanatory variables, and they may be continuous as well as categorical. The goal is usually to describe their effects on response variables. *Modeling* the effects helps us do this efficiently. A good-fitting model evaluates effects, includes relevant interactions, and provides smoothed estimates of response probabilities.

The rest of the book focuses on model building for categorical response variables. In this chapter we introduce a family of *generalized linear models* that contains the most important models for categorical responses as well as standard models for continuous responses. Section 4.1 covers three components common to all generalized linear models. Section 4.2 illustrates with models for binary responses. The most important case is *logistic regression*, a linear model for the *logit* transformation of a binomial parameter. In Chapters 5 through 7 we study these models in detail.

In Section 4.3 we present generalized linear models for counts. A *Poisson regression model* called a *loglinear model* is a linear model for the log of a Poisson mean. In Chapters 8 and 9 we study them for modeling counts in contingency tables.

Sections 4.4 through 4.8 are more technical. Readers wanting mainly an overview of methods can skip them or read them lightly. For generalized linear models, Section 4.4 covers likelihood equations and the asymptotic covariance matrix of ML model parameter estimates, and Section 4.5 summarizes inferential methods. Methods of solving the likelihood equations are presented in Section 4.6. In the final two sections we introduce generalizations, *quasi-likelihood* and *generalized additive models*, that further extend the scope of models.

4.1 GENERALIZED LINEAR MODEL

Generalized linear models (GLMs) extend ordinary regression models to encompass nonnormal response distributions and modeling functions of the mean. Three components specify a generalized linear model: A *random component* identifies the response variable Y and its probability distribution; a *systematic component* specifies explanatory variables used in a linear predictor function; and a *link function* specifies the function of $E(Y)$ that the model equates to the systematic component. Nelder and Wedderburn (1972) introduced the class of GLMs, although many models in the class were well established by then.

4.1.1 Components of Generalized Linear Models

The *random component* of a GLM consists of a response variable Y with independent observations $(y_1, \ldots, y_N)$ from a distribution in the natural exponential family. This family has probability density function or mass function of form

$$f(y_i; \theta_i) = a(\theta_i)b(y_i) \exp[y_i Q(\theta_i)]. \tag{4.1}$$

Several important distributions are special cases, including the Poisson and binomial. The value of the parameter θ_i may vary for $i = 1, \ldots, N$, depending on values of explanatory variables. The term $Q(\theta)$ is called the *natural parameter*. In Section 4.4 we present a more general formula that also has a dispersion parameter, but (4.1) is sufficient for basic discrete data models.

The *systematic component* of a GLM relates a vector $(\eta_1, \ldots, \eta_N)$ to the explanatory variables through a linear model. Let x_{ij} denote the value of predictor j $(j = 1, 2, \ldots, p)$ for subject i. Then

$$\eta_i = \sum_j \beta_j x_{ij}, \qquad i = 1, \ldots, N.$$

This linear combination of explanatory variables is called the *linear predictor*. Usually, one $x_{ij} = 1$ for all i, for the coefficient of an intercept (often denoted by α) in the model.

The third component of a GLM is a *link function* that connects the random and systematic components. Let $\mu_i = E(Y_i)$, $i = 1, \ldots, N$. The model links μ_i to η_i by $\eta_i = g(\mu_i)$, where the link function g is a monotonic, differentiable function. Thus, g links $E(Y_i)$ to explanatory variables through the formula

$$g(\mu_i) = \sum_j \beta_j x_{ij}, \qquad i = 1, \ldots, N. \tag{4.2}$$

The link function $g(\mu) = \mu$, called the *identity link*, has $\eta_i = \mu_i$. It specifies a linear model for the mean itself. This is the link function for ordinary regression with normally distributed Y. The link function that transforms the mean to the natural parameter is called the *canonical link*. For it, $g(\mu_i) = Q(\theta_i)$, and $Q(\theta_i) = \sum_j \beta_j x_{ij}$. The following subsections show examples.

In summary, a GLM is a linear model for a transformed mean of a response variable that has distribution in the natural exponential family. We now illustrate the three components by introducing the key GLMs for discrete response variables.

4.1.2 Binomial Logit Models for Binary Data

Many response variables are binary. Represent the success and failure outcomes by 1 and 0. The *Bernoulli distribution* for this *Bernoulli trial* specifies probabilities $P(Y = 1) = \pi$ and $P(Y = 0) = 1 - \pi$, for which $E(Y) = \pi$. This is the special case of the binomial (1.1) with $n = 1$. The probability mass function is

$$f(y; \pi) = \pi^y (1 - \pi)^{1-y} = (1 - \pi)[\pi/(1 - \pi)]^y$$

$$= (1 - \pi) \exp\left(y \log \frac{\pi}{1 - \pi} \right) \tag{4.3}$$

for $y = 0$ and 1. This is in the natural exponential family (4.1), identifying θ with π, $a(\pi) = 1 - \pi$, $b(y) = 1$, and $Q(\pi) = \log[\pi/(1 - \pi)]$. The natural parameter $\log[\pi/(1 - \pi)]$ is the log odds of response 1, the *logit* of π. This is the canonical link. GLMs using the logit link are often called *logit models*.

4.1.3 Poisson Loglinear Models for Count Data

Some response variables have counts as their possible outcomes. For a sample of silicon wafers used in manufacturing computer chips, each observation might be the number of imperfections on a wafer. Counts also occur as entries in contingency tables.

The simplest distribution for count data is the Poisson. Like counts, Poisson variates can take any nonnegative integer value. Let Y denote a count and let $\mu = E(Y)$. The Poisson probability mass function (1.4) for Y is

$$f(y; \mu) = \frac{e^{-\mu} \mu^y}{y!} = \exp(-\mu)\left(\frac{1}{y!} \right) \exp(y \log \mu), \qquad y = 0, 1, 2, \dots .$$

This has natural exponential form (4.1) with $\theta = \mu$, $a(\mu) = \exp(-\mu)$, $b(y) = 1/y!$, and $Q(\mu) = \log \mu$. The natural parameter is $\log \mu$, so the canonical

TABLE 4.1　Types of Generalized Linear Models for Statistical Analysis

Random Component	Link	Systematic Component	Model	Chapters
Normal	Identity	Continuous	Regression	
Normal	Identity	Categorical	Analysis of variance	
Normal	Identity	Mixed	Analysis of covariance	
Binomial	Logit	Mixed	Logistic regression	5 and 6
Poisson	Log	Mixed	Loglinear	8 and 9
Multinomial	Generalized logit	Mixed	Multinomial response	7

link function is the log link, $\eta = \log \mu$. The model using this link is

$$\log \mu_i = \sum_j \beta_j x_{ij}, \qquad i = 1, \ldots, N. \qquad (4.4)$$

This model is called a *Poisson loglinear model.*

4.1.4　Generalized Linear Models for Continuous Responses

The class of GLMs also includes models for continuous responses. The normal distribution is in a natural exponential family that includes dispersion parameters. Its natural parameter is the mean. Therefore, an ordinary regression model for $E(Y)$ is a GLM using the identity link. Table 4.1 lists this and other standard models for a normal random component. The table also lists GLMs for discrete responses that are presented in the next six chapters.

A traditional way to analyze data transforms Y so that it has approximately a normal distribution with constant variance; then, ordinary least-squares regression is applicable. With GLMs, by contrast, the choice of link function is separate from the choice of random component. If a link is useful in the sense that a linear model for the predictors is plausible for that link, it is not necessary that it also stabilizes variance or produces normality. This is because the fitting process maximizes the likelihood for the choice of distribution for Y, and that choice is not restricted to normality.

4.1.5　Deviance

For a particular GLM for observations $\mathbf{y} = (y_1, \ldots, y_N)$, let $L(\boldsymbol{\mu}; \mathbf{y})$ denote the log-likelihood function expressed in terms of the means $\boldsymbol{\mu} = (\mu_1, \ldots, \mu_N)$. Let $L(\hat{\boldsymbol{\mu}}; \mathbf{y})$ denote the maximum of the log likelihood for the model. Considered for all possible models, the maximum achievable log likelihood is

$L(\mathbf{y}; \mathbf{y})$. This occurs for the most general model, having a separate parameter for each observation and the perfect fit $\hat{\boldsymbol{\mu}} = \mathbf{y}$. Such a model is called the *saturated model*. This model is not useful, since it does not provide data reduction. However, it serves as a baseline for comparison with other model fits.

The *deviance* of a Poisson or binomial GLM is defined to be

$$-2[L(\hat{\boldsymbol{\mu}}; \mathbf{y}) - L(\mathbf{y}; \mathbf{y})] .$$

This is the likelihood-ratio statistic for testing the null hypothesis that the model holds against the general alternative (i.e., the saturated model). For some Poisson and binomial GLMs, the number of observations N stays fixed as the individual counts increase in size. Then the deviance has a chi-squared asymptotic null distribution. The df $= N - p$, where p is the number of model parameters; that is, df equals the difference between the numbers of parameters in the saturated and unsaturated models. The deviance then provides a test of model fit.

An example is binomial counts at N fixed settings of predictors when the number of trials at each setting increases. Let Y_i be $\text{bin}(n_i, \pi_i)$, $i = 1, \ldots,$ N. Consider the simple model of homogeneity, $\pi_i = \alpha$ all i. It has $p = 1$ parameter. The saturated model makes no assumption about $\{\pi_i\}$, letting them be any N values between 0 and 1.0. It has N parameters. The deviance for the homogeneity model has df $= N - 1$. In fact, it equals the G^2 likelihood-ratio statistic (3.11) for testing independence in the $N \times 2$ table that these samples form. Under independence, it has approximately a chi-squared distribution as the $\{n_i\}$ increase, for fixed N.

We use the deviance throughout the book for model checking and for inferential comparisons of models. Components of the deviance are residual measures of lack of fit. Methods for analyzing the deviance generalize analysis of variance methods for normal linear models.

4.1.6 Advantages of the GLM Formulation

GLMs provide a unified theory of modeling that encompasses the most important models for continuous and discrete variables. Models studied in this text are GLMs with binomial or Poisson random component, or multivariate extensions of GLMs. The ML parameter estimates are computed with an algorithm, presented in Section 4.6, that iteratively uses a weighted version of least squares. The reason for restricting GLMs to the exponential family of distributions for Y is that the same algorithm applies to this entire family, for any choice of link function.

Most statistical software has the facility to fit GLMs. Appendix A gives details.

4.2   GENERALIZED LINEAR MODELS FOR BINARY DATA

Let Y denote a binary response variable. For instance, Y might indicate vote in a British election (Labour, Conservative), choice of automobile (domestic, import), or diagnosis of breast cancer (present, absent). Each observation has one of two outcomes, denoted by 0 and 1, binomial for a single trial. The mean $E(Y) = P(Y = 1)$. We denote $P(Y = 1)$ by $\pi(\mathbf{x})$, reflecting its dependence on values $\mathbf{x} = (x_1, \ldots, x_p)$ of predictors. The variance of Y is

$$\text{var}(Y) = \pi(\mathbf{x})[1 - \pi(\mathbf{x})],$$

the binomial variance for one trial. In introducing GLMs for binary data, for simplicity we use a single explanatory variable.

4.2.1   Linear Probability Model

For a binary response, the regression model

$$\pi(x) = \alpha + \beta x \tag{4.5}$$

is called a *linear probability model*. With independent observations it is a GLM with binomial random component and identity link function.

The linear probability model has a major structural defect. Probabilities fall between 0 and 1, but linear functions take values over the entire real line. Model (4.5) has $\pi(x) < 0$ and $\pi(x) > 1$ for sufficiently large or small x values. For its extension with multiple predictors, difficulties often occur fitting this model because during the fitting process, $\hat{\pi}(\mathbf{x})$ falls outside the $[0, 1]$ range for some subjects' $\mathbf{x}$ values. The model can be valid over a restricted range of x values. When it is plausible, an advantage is its simple interpretation: β is the change in $\pi(x)$ for a one-unit increase in x.

We defer to Section 4.6 the technical details of fitting this and other GLMs. One should assume a binomial distribution for Y and use maximum likelihood (ML) rather than ordinary least squares. Least squares is ML for a normal distribution with constant variance. For binary responses, the constant variance condition that makes least squares estimators optimal (i.e., minimum variance in the class of linear unbiased estimators) is not satisfied. Since $\text{var}(Y) = \pi(x)[1 - \pi(x)]$, the variance depends on x through its influence on $\pi(x)$. As $\pi(x)$ moves toward 0 or 1, the distribution of Y is more nearly concentrated at a single point, and the variance moves toward 0. Because of the nonconstant variance, the binomial ML estimator is more efficient than least squares. Also Y, being binary, is very far from normally distributed. Thus, the usual sampling distributions for the least squares estimators do not apply. The estimates and standard errors for ML and least squares are usually similar, however, when $\hat{\pi}(x)$ for the sample x values falls in the range within which the variance is relatively stable (about 0.3 to 0.7).

TABLE 4.2   Relationship between Snoring and Heart Disease

Snoring	Heart Disease Yes	Heart Disease No	Proportion Yes	Linear Fit[a]	Logit Fit[a]
Never	24	1355	0.017	0.017	0.021
Occasionally	35	603	0.055	0.057	0.044
Nearly every night	21	192	0.099	0.096	0.093
Every night	30	224	0.118	0.116	0.132

[a]Model fits refer to proportion of yes responses.

Source: P. G. Norton and E. V. Dunn, *British Med. J.* **291**: 630–632 (1985), BMJ Publishing Group.

4.2.2   Snoring and Heart Disease Example

We illustrate the linear probability model with Table 4.2, from an epidemiological survey of 2484 subjects to investigate snoring as a risk factor for heart disease. Those surveyed were classified according to their spouses' report of how much they snored. The model states that the probability of heart disease is linearly related to the level of snoring x. We treat the rows of the table as independent binomial samples. No obvious choice of scores exists for categories of x. We used (0, 2, 4, 5), treating the last two levels as closer than the other adjacent pairs (Problem 4.4 uses equally spaced scores). ML estimates and standard errors are the same if we use a data file of 2484 binary observations or if we enter the four binomial totals of yes and no responses listed in Table 4.2.

Software (see, e.g., Table A.3 for SAS) reports the ML fit, $\hat{\pi}(x) = 0.0172 + 0.0198x$, with a standard error SE $= 0.0028$ for $\hat{\beta} = 0.0198$. For nonsnorers ($x = 0$), the estimated proportion of subjects having heart disease is 0.0172. We refer to the estimated values of $E(Y)$ for a GLM as *fitted values*. Table 4.2 shows the sample proportions and the fitted values for this model. Figure 4.1 graphs the sample and fitted values. The table and graph suggest that the model fits well. (In Section 5.2.3 we discuss formal goodness-of-fit analyses for binary-response GLMs.) The model interpretation is simple. The estimated probability of heart disease is about 0.02 for nonsnorers; it increases 2(0.0198) = 0.04 for occasional snorers, another 0.04 for those who snore nearly every night, and another 0.02 for those who always snore.

4.2.3   Logistic Regression Model

Usually, binary data result from a *nonlinear* relationship between $\pi(x)$ and x. A fixed change in x often has less impact when $\pi(x)$ is near 0 or 1 than when $\pi(x)$ is near 0.5. In the purchase of an automobile, consider the choice between buying new or used. Let $\pi(x)$ denote the probability of selecting new when annual family income $= x$. An increase of $50,000 in annual

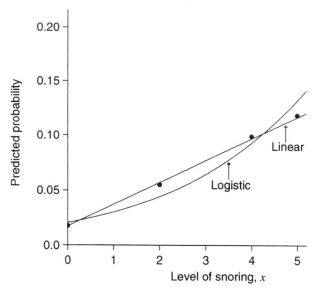

FIGURE 4.1 Predicted probabilities for linear probability and logistic regression models.

income would have less effect when $x = \$1,000,000$ [for which $\pi(x)$ is near 1] than when $x = \$50,000$.

In practice, nonlinear relationships between $\pi(x)$ and x are often monotonic, with $\pi(x)$ increasing continuously or $\pi(x)$ decreasing continuously as x increases. The S-shaped curves in Figure 4.2 are typical. The most important curve with this shape has the model formula

$$\pi(x) = \frac{\exp(\alpha + \beta x)}{1 + \exp(\alpha + \beta x)}. \tag{4.6}$$

This is the *logistic regression* model. As $x \to \infty$, $\pi(x)\downarrow 0$ when $\beta < 0$ and $\pi(x)\uparrow 1$ when $\beta > 0$.

Let's find the link function for which logistic regression is a GLM. For (4.6) the odds are

$$\frac{\pi(x)}{1 - \pi(x)} = \exp(\alpha + \beta x).$$

The log odds has the linear relationship

$$\log\frac{\pi(x)}{1 - \pi(x)} = \alpha + \beta x. \tag{4.7}$$

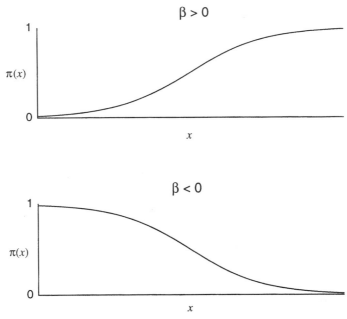

FIGURE 4.2   Logistic regression functions.

Thus, the appropriate link is the log odds transformation, the *logit*. Logistic regression models are GLMs with binomial random component and logit link function. Logistic regression models are also called *logit models*.

The logit is the natural parameter of the binomial distribution, so the logit link is its canonical link. Whereas $\pi(x)$ must fall in the $(0, 1)$ range, the logit can be any real number. The real numbers are also the range for linear predictors (such as $\alpha + \beta x$) that form the systematic component of a GLM. So this model does not have the structural problem that is true of the linear probability model.

For the snoring data in Table 4.2, software reports the logistic regression ML fit

$$\text{logit}[\hat{\pi}(x)] = -3.87 + 0.40x.$$

The positive $\hat{\beta} = 0.40$ reflects the increased incidence of heart disease at higher snoring levels. In Chapters 5 and 6 we study logistic regression in detail and interpret such equations. Estimated probabilities result from substituting x values into the estimate of probability formula (4.6). Table 4.2 also reports these fitted values. Figure 4.1 displays the fit. The fit is close to linear over this narrow range of estimated probabilities, and results are similar to those for the linear probability model.

4.2.4 Binomial GLM for 2 × 2 Contingency Tables

Among the simplest GLMs for a binary response is the one having a single explanatory variable X that is also binary. Label its values by 0 and 1. For a given link function, the GLM

$$\text{link}[\pi(x)] = \alpha + \beta x$$

has the effect of X described by

$$\beta = \text{link}[\pi(1)] - \text{link}[\pi(0)] .$$

For the identity link, $\beta = \pi(1) - \pi(0)$ is the difference between proportions. For the log link, $\beta = \log[\pi(1)] - \log[\pi(0)] = \log[\pi(1)/\pi(0)]$ is the log relative risk. For the logit link,

$$\beta = \text{logit}[\pi(1)] - \text{logit}[\pi(0)] = \log \frac{\pi(1)}{1 - \pi(1)} - \log \frac{\pi(0)}{1 - \pi(0)}$$

$$= \log \frac{\pi(1)/(1 - \pi(1))}{\pi(0)/(1 - \pi(0))}$$

is the log odds ratio. Measures of association for 2 × 2 tables are effect parameters in GLMs for binary data.

4.2.5 Probit and Inverse CDF Link Functions*

A monotone regression curve such as the first one in Figure 4.2 has the shape of a cumulative distribution function (cdf) for a continuous random variable. This suggests a model for a binary response having form $\pi(x) = F(x)$ for some cdf F.

Using an entire class of location-scale cdf's, such as normal cdf's with their variety of means and variances, permits the curve $\pi(x) = F(x)$ to have flexibility in the rate of increase and in the location where most of that increase occurs. Let $\Phi(\cdot)$ denote the standard cdf of the class, such as the $N(0, 1)$ cdf. Using Φ but writing the model as

$$\pi(x) = \Phi(\alpha + \beta x) \tag{4.8}$$

provides the same flexibility. Shapes of different cdf's in the class occur as α and β vary. Replacing x by βx permits the curve to increase at a different rate than the standard cdf (or even to decrease if $\beta < 0$); varying α moves the curve to the left or right.

When Φ is strictly increasing over the entire real line, its inverse function $\Phi^{-1}(\cdot)$ exists and (4.8) is, equivalently,

$$\Phi^{-1}[\pi(x)] = \alpha + \beta x . \tag{4.9}$$

For this class of cdf shapes, the link function for the GLM is Φ^{-1}. The link function maps the $(0, 1)$ range of probabilities onto $(-\infty, \infty)$, the range of linear predictors. The curve has the shape of a normal cdf when Φ is the standard normal cdf. Model (4.9) is then called the *probit* model. This curve has similar appearance to the logistic regression curve. Probit models are discussed in Section 6.6.

When $\beta > 0$, the logistic regression curve (4.6) is a cdf for the *logistic distribution*. When $\beta < 0$, the curve for $1 - \pi(x)$, the probability $Y = 0$, has that appearance. The cdf of the logistic distribution with mean μ and dispersion parameter $\tau > 0$ is

$$F(x) = \frac{\exp[(x - \mu)/\tau]}{1 + \exp[(x - \mu)/\tau]}, \qquad -\infty < x < \infty.$$

The corresponding probability density function is symmetric and bell-shaped, with standard deviation $\tau\pi/\sqrt{3}$ (here, π is the mathematical constant $3.14\ldots$). It looks much like the normal density with the same mean and standard deviation but with slightly thicker tails. (Its kurtosis equals that of a t distribution with df $= 9$.)

The standardized form of the logistic cdf has $\mu = 0$ and $\tau = 1$, so $\Phi(x) = e^x/(1 + e^x)$. For that function, the logistic regression curve (4.6) has form $\pi(x) = \Phi(\alpha + \beta x)$. By (4.9) the logit transformation is simply the inverse function for the standard logistic cdf; that is, when $\Phi(x) = \pi(x) = e^x/(1 + e^x)$, then $x = \Phi^{-1}[\pi(x)] = \log[\pi(x)/(1 - \pi(x))]$.

4.3 GENERALIZED LINEAR MODELS FOR COUNTS

The best known GLMs for count data assume a Poisson distribution for Y. We introduced this distribution in Section 1.2.3. In Chapters 8 and 9 we present Poisson GLMs for counts in contingency tables with categorical response variables. In this section we introduce Poisson GLMs using an alternative application: modeling count or rate data for a single discrete response variable.

4.3.1 Poisson Loglinear Models

The Poisson distribution has a positive mean μ. Although a GLM can model a positive mean using the identity link, it is more common to model the log of the mean. Like the linear predictor $\alpha + \beta x$, the log mean can take any real value. The log mean is the natural parameter for the Poisson distribution, and the log link is the canonical link for a Poisson GLM. A Poisson loglinear GLM assumes a Poisson distribution for Y and uses the log link.

The Poisson loglinear model with explanatory variable X is

$$\log \mu = \alpha + \beta x . \tag{4.10}$$

For this model, the mean satisfies the exponential relationship

$$\mu = \exp(\alpha + \beta x) = e^{\alpha}(e^{\beta})^{x}. \qquad (4.11)$$

A 1-unit increase in x has a multiplicative impact of e^{β} on μ: The mean at $x + 1$ equals the mean at x multiplied by e^{β}.

4.3.2 Horseshoe Crab Mating Example

We illustrate Poisson GLMs for Table 4.3 from a study of nesting horseshoe crabs. Each female horseshoe crab had a male crab resident in her nest. The study investigated factors affecting whether the female crab had any other males, called *satellites*, residing nearby. Explanatory variables are the female crab's color, spine condition, weight, and carapace width. The response outcome for each female crab is her number of satellites. For now, we use width alone as a predictor. Table 4.3 lists width in centimeters. The sample mean width equals 26.3 and the standard deviation equals 2.1.

Figure 4.3 plots the response counts of satellites against width, with numbered symbols indicating the number of observations at each point. The substantial variability makes it difficult to discern a clear trend. To get a clearer picture, we grouped the female crabs into width categories (≤ 23.25, 23.25–24.25, 24.25–25.25, 25.25–26.25, 26.25–27.25, 27.25–28.25, 28.25–29.25, > 29.25) and calculated the sample mean number of satellites for female crabs in each category. Figure 4.4 plots these sample means against the sample mean width for crabs in each category.

More sophisticated ways of portraying the trend smooth the data without grouping the width values or assuming a particular functional relationship. Figure 4.4 also shows a smoothed curve based on an extension of the GLM introduced in Section 4.8. The sample means and the smoothed curve both show a strong increasing trend. (The means tend to fall above the curve, since the response counts in a category tend to be skewed to the right; the smoothed curve is less susceptible to outlying observations.) The trend seems approximately linear, and we discuss next models for the ungrouped data for which the mean or the log of the mean is linear in width.

For a female crab, let μ be the expected number of satellites and $x = $ width. From GLM software (e.g., for SAS, see Table A.4), the ML fit of the Poisson loglinear model (4.10) is

$$\log \hat{\mu} = \hat{\alpha} + \hat{\beta} x = -3.305 + 0.164x.$$

The effect $\hat{\beta} = 0.164$ of width is positive, with SE $= 0.020$. The model fitted value at any width level is an estimated mean number of satellites $\hat{\mu}$. For instance, the fitted value at the mean width of $x = 26.3$ is

$$\hat{\mu} = \exp(\hat{\alpha} + \hat{\beta} x) = \exp[-3.305 + 0.164(26.3)] = 2.74.$$

TABLE 4.3   Number of Crab Satellites by Female's Characteristics[a]

C	S	W	Wt	Sa	C	S	W	Wt	Sa	C	S	W	Wt	Sa	C	S	W	Wt	Sa
2	3	28.3	3.05	8	3	3	22.5	1.55	0	1	1	26.0	2.30	9	3	3	24.8	2.10	0
3	3	26.0	2.60	4	2	3	23.8	2.10	0	3	2	24.7	1.90	0	2	1	23.7	1.95	0
3	3	25.6	2.15	0	3	3	24.3	2.15	0	2	3	25.8	2.65	0	2	3	28.2	3.05	11
4	2	21.0	1.85	0	2	1	26.0	2.30	14	1	1	27.1	2.95	8	2	3	25.2	2.00	1
2	3	29.0	3.00	1	4	3	24.7	2.20	0	2	3	27.4	2.70	5	2	3	23.2	1.95	4
1	2	25.0	2.30	3	2	1	22.5	1.60	1	3	3	26.7	2.60	2	4	3	25.8	2.00	3
4	3	26.2	1.30	0	2	3	28.7	3.15	3	2	1	26.8	2.70	5	4	3	27.5	2.60	0
2	3	24.9	2.10	0	1	1	29.3	3.20	4	1	3	25.8	2.60	0	2	2	25.7	2.00	0
2	1	25.7	2.00	8	2	1	26.7	2.70	5	4	3	23.7	1.85	0	2	3	26.8	2.65	0
2	3	27.5	3.15	6	4	3	23.4	1.90	0	2	3	27.9	2.80	6	3	3	27.5	3.10	3
1	1	26.1	2.80	5	1	1	27.7	2.50	6	2	1	30.0	3.30	5	3	1	28.5	3.25	9
3	3	28.9	2.80	4	2	3	28.2	2.60	6	2	3	25.0	2.10	4	2	3	28.5	3.00	3
2	1	30.3	3.60	3	4	3	24.7	2.10	5	2	3	27.7	2.90	5	1	1	27.4	2.70	6
2	3	22.9	1.60	4	2	1	25.7	2.00	5	2	3	28.3	3.00	15	2	3	27.2	2.70	3
3	3	26.2	2.30	3	2	1	27.8	2.75	0	4	3	25.5	2.25	0	3	3	27.1	2.55	0
3	3	24.5	2.05	5	3	1	27.0	2.45	3	2	3	26.0	2.15	5	2	3	28.0	2.80	1
2	3	30.0	3.05	8	2	3	29.0	3.20	10	2	3	26.2	2.40	0	2	1	26.5	1.30	0
2	3	26.2	2.40	3	3	3	25.6	2.80	7	3	3	23.0	1.65	1	3	3	23.0	1.80	0
2	3	25.4	2.25	6	3	3	24.2	1.90	0	2	2	22.9	1.60	0	3	2	26.0	2.20	3
2	3	25.4	2.25	4	3	3	25.7	1.20	0	2	3	25.1	2.10	5	3	2	24.5	2.25	0
4	3	27.5	2.90	0	3	3	23.1	1.65	0	3	1	25.9	2.55	4	2	3	25.8	2.30	0
4	3	27.0	2.25	3	2	3	28.5	3.05	0	4	1	25.5	2.75	0	4	3	23.5	1.90	0
2	2	24.0	1.70	0	2	1	29.7	3.85	5	2	1	26.8	2.55	0	4	3	26.7	2.45	0
2	1	28.7	3.20	0	3	3	23.1	1.55	0	2	1	29.0	2.80	1	3	3	25.5	2.25	0
3	3	26.5	1.97	1	3	3	24.5	2.20	1	3	3	28.5	3.00	1	2	3	28.2	2.87	1
2	3	24.5	1.60	1	2	3	27.5	2.55	1	2	2	24.7	2.55	4	2	1	25.2	2.00	1
3	3	27.3	2.90	1	2	3	26.3	2.40	1	2	3	29.0	3.10	1	2	3	25.3	1.90	2
2	3	26.5	2.30	4	2	3	27.8	3.25	3	2	3	27.0	2.50	6	3	3	25.7	2.10	0
2	3	25.0	2.10	2	2	3	31.9	3.33	2	4	3	23.7	1.80	0	4	3	29.3	3.23	12
3	3	22.0	1.40	0	2	3	25.0	2.40	5	3	3	27.0	2.50	6	3	3	23.8	1.80	6
1	1	30.2	3.28	2	3	3	26.2	2.22	0	2	3	24.2	1.65	2	2	3	27.4	2.90	3
2	2	25.4	2.30	0	3	3	28.4	3.20	3	4	3	22.5	1.47	4	2	3	26.2	2.02	2
2	1	24.9	2.30	6	1	2	24.5	1.95	6	2	3	25.1	1.80	0	2	1	28.0	2.90	4
4	3	25.8	2.25	10	2	3	27.9	3.05	7	2	3	24.9	2.20	0	2	1	28.4	3.10	5
3	3	27.2	2.40	5	2	2	25.0	2.25	6	2	3	27.5	2.63	6	2	1	33.5	5.20	7
2	3	30.5	3.32	3	3	3	29.0	2.92	3	2	1	24.3	2.00	0	2	3	25.8	2.40	0
4	3	25.0	2.10	8	2	1	31.7	3.73	4	2	3	29.5	3.02	4	3	3	24.0	1.90	10
2	3	30.0	3.00	9	2	3	27.6	2.85	4	2	3	26.2	2.30	0	2	1	23.1	2.00	0
2	1	22.9	1.60	0	4	3	24.5	1.90	0	2	3	24.7	1.95	4	2	3	28.3	3.20	0
2	3	23.9	1.85	2	3	3	23.8	1.80	0	3	2	29.8	3.50	4	2	3	26.5	2.35	4
2	3	26.0	2.28	3	2	3	28.2	3.05	8	4	3	25.7	2.15	0	2	3	26.5	2.75	7
2	3	25.8	2.20	0	3	3	24.1	1.80	0	3	3	26.2	2.17	2	3	3	26.1	2.75	3
3	3	29.0	3.28	4	1	1	28.0	2.62	0	4	3	27.0	2.63	0	2	2	24.5	2.00	0
1	1	26.5	2.35	0															

[a]C, color (1, light medium; 2, medium; 3, dark medium; 4, dark); S, spine condition (1, both good; 2, one worn or broken; 3, both worn or broken); W, carapace width (cm); Wt, weight (kg); Sa, number of satellites.

Source: Data courtesy of Jane Brockmann, Zoology Department, University of Florida; study described in *Ethology* **102**:1–21 (1996).

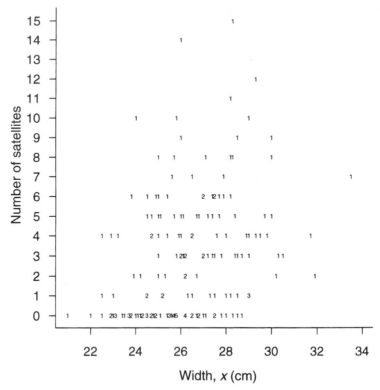

FIGURE 4.3   Number of satellites by width of female crab.

For this model, $\exp(\hat{\beta}) = \exp(0.164) = 1.18$ is the multiplicative effect on $\hat{\mu}$ for a 1-cm increase in x. For instance, the fitted value at $x = 27.3 = 26.3 + 1$ is $\exp[-3.305 + 0.164(27.3)] = 3.23$, which equals 1.18×2.74. A 1-cm increase in width yields an 18% increase in the estimated mean.

Figure 4.4 shows that $E(Y)$ may grow approximately linearly with width. This suggests the Poisson GLM with identity link. It has ML fit

$$\hat{\mu} = \hat{\alpha} + \hat{\beta}x = -11.53 + 0.55x .$$

This model has an additive rather than a multiplicative effect of X on μ. A 1-cm increase in x has an estimated increase of $\hat{\beta} = 0.55$ in $\hat{\mu}$. The fitted values are positive at all sampled x, and the model describes simply the effect: On the average, about a 2-cm increase in width is associated with an extra satellite.

Figure 4.5 plots $\hat{\mu}$ against width for the models with log link and identity link. Although they diverge somewhat for relatively small and large widths, they provide similar predictions over the width range in which most observations occur. We now study whether either model fits adequately.

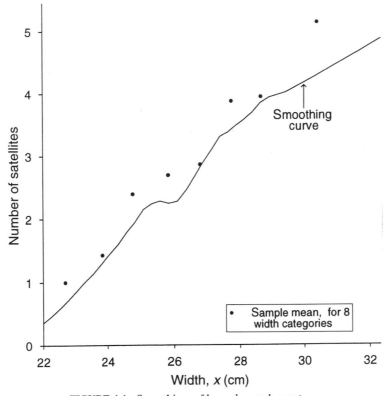

FIGURE 4.4 Smoothings of horseshoe crab counts.

TABLE 4.4 Sample Mean and Variance of Number of Satellites

Width (cm)	Number of Cases	Number of Satellites	Sample Mean	Sample Variance
< 23.25	14	14	1.00	2.77
23.25–24.25	14	20	1.43	8.88
24.25–25.25	28	67	2.39	6.54
25.25–26.25	39	105	2.69	11.38
26.25–27.25	22	63	2.86	6.88
27.25–28.25	24	93	3.87	8.81
28.25–29.25	18	71	3.94	16.88
> 29.25	14	72	5.14	8.29

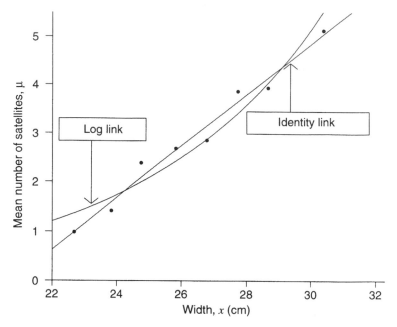

FIGURE 4.5   Estimated mean number of satellites for log and identity links.

4.3.3   Overdispersion for Poisson GLMs

In Section 1.2.4 we noted that count data often show greater variability than
the Poisson allows. For the grouped horseshoe crab data, Table 4.4 shows
the sample mean and variance for the counts of number of satellites for the
female crabs in each width category. The variances are much larger than the
means, whereas Poisson distributions have identical mean and variance.
The greater variability than predicted by the GLM random component
reflects *overdispersion.*

A common cause of overdispersion is subject heterogeneity. For instance,
suppose that width, weight, color, and spine condition are the four predictors
that affect a female crab's number of satellites. Suppose that Y has a Poisson
distribution at each fixed combination of those predictors. Our model uses
width alone as a predictor. Crabs having a certain width are then a mixture of
crabs of various weights, colors, and spine conditions. Thus, the population of
crabs having that width is a mixture of several Poisson populations, each
having its own mean for the response. This heterogeneity results in an overall
response distribution at that width having greater variation than the Poisson
predicts. If the variance equals the mean when all relevant variables are
controlled, it exceeds the mean when only one is controlled.

Overdispersion is not an issue in ordinary regression with normally distributed Y, because that distribution has a separate parameter (the variance)

to describe variability. For binomial and Poisson distributions, however, the variance is a function of the mean. Overdispersion is common in the modeling of counts. When the model for the mean is correct but the true distribution is not Poisson, the ML estimates of model parameters are still consistent but standard errors are incorrect. We next introduce an extension of the Poisson GLM that has an extra parameter and accounts better for overdispersion. In Section 4.7 we present another approach for this, quasi-likelihood inference.

4.3.4 Negative Binomial GLMs

The *negative binomial distribution* has probability mass function

$$f(y; k, \mu) = \frac{\Gamma(y + k)}{\Gamma(k)\Gamma(y + 1)} \left(\frac{k}{\mu + k} \right)^k \left(1 - \frac{k}{\mu + k} \right)^y, \qquad y = 0, 1, 2, \ldots,$$

$$(4.12)$$

where k and μ are parameters. This distribution has

$$E(Y) = \mu, \qquad \text{var}(Y) = \mu + \mu^2/k .$$

The index k^{-1} is called a *dispersion parameter*. As $k^{-1} \to 0$, $\text{var}(Y) \to \mu$ and the negative binomial distribution converges to the Poisson (Cameron and Trivedi 1998, p. 75). Usually, k^{-1} is unknown. Estimating it helps summarize the extent of overdispersion.

For k fixed, one can express (4.12) in natural exponential family form (4.1). Then, a model with negative binomial random component is a GLM. For simplicity, such models let k be the same constant for all observations but treat it as unknown. As in GLMs for binary data, a variety of link functions are possible. Most common is the log link, as in Poisson loglinear models, but sometimes the identity link is adequate.

In Section 13.4 we discuss negative binomial GLMs. We illustrate it here for the crab data analyzed above with Poisson GLMs. With the identity link and width as predictor, the Poisson GLM has $\hat{\mu} = -11.53 + 0.55x$ (SE = 0.06 for $\hat{\beta}$). For the negative binomial GLM, $\hat{\mu} = -11.15 + 0.53x$ (SE = 0.11). Moreover, $\hat{k}^{-1} = 0.98$, so at a predicted $\hat{\mu}$, the estimated variance is roughly $\hat{\mu} + \hat{\mu}^2$, compared to $\hat{\mu}$ for the Poisson GLM. Although fitted values are similar, the greater SE for $\hat{\beta}$ and the greater estimated variance in the negative binomial model reflect the overdispersion uncaptured with the Poisson GLM.

4.3.5 Poisson Regression for Rates

When events of a certain type occur over time, space, or some other index of size, it is usually more relevant to model the *rate* at which they occur than the number of them. For instance, a study of homicides in a given year for a sample of cities might model the homicide rate, defined for a city as its number of homicides that year divided by its population size. The model might describe how the rate depends on the city's unemployment rate, its residents' median income, and the percentage of residents having completed high school. In Section 9.7 we discuss Poisson regression for modeling rates.

4.3.6 Poisson GLM of Independence in $I \times J$ Contingency Tables

One use of Poisson loglinear models is in modeling counts in contingency tables. We illustrate for two-way tables with independent counts $\{Y_{ij}\}$ having Poisson distributions with means $\{\mu_{ij}\}$. Suppose that $\{\mu_{ij}\}$ satisfy

$$\mu_{ij} = \mu \alpha_i \beta_j,$$

where $\{\alpha_i\}$ and $\{\beta_j\}$ are positive constants satisfying $\Sigma_i \alpha_i = \Sigma_j \beta_j = 1$. This is a multiplicative model, but a linear predictor for a GLM results using the log link,

$$\log \mu_{ij} = \lambda + \alpha_i^* + \beta_j^*, \tag{4.13}$$

where $\lambda = \log \mu$, $\alpha_i^* = \log \alpha_i$, $\beta_j^* = \log \beta_j$. This Poisson loglinear model has additive main effects of the two classifications but no interaction.

Since the $\{Y_{ij}\}$ are independent, the total sample size $\Sigma_i \Sigma_j Y_{ij}$ has a Poisson distribution with mean $\Sigma_i \Sigma_j \mu_{ij} = \mu$. Conditional on $\Sigma_i \Sigma_j Y_{ij} = n$, the cell counts have a multinomial distribution with probabilities $\{\pi_{ij} = \mu_{ij}/\mu = \alpha_i \beta_j\}$. Similarly, you can check that conditional on n, the row totals $\{Y_{i+}\}$ have a multinomial distribution with probabilities $\{\pi_{i+} = \alpha_i\}$ and the column totals $\{Y_{+j}\}$ have a multinomial distribution with probabilities $\{\pi_{+j} = \beta_j\}$.

Conditional on n, the model is a multinomial one that satisfies $\pi_{ij} = \alpha_i \beta_j = \pi_{i+} \pi_{+j}$. This is independence of the two classifications. In fact, in Poisson form independence is the loglinear model (4.13). The inferences conducted in Chapter 3 about independence in two-way contingency tables relate to GLMs, either Poisson loglinear models or corresponding multinomial models that fix n or the row or column totals. In Chapters 8 and 9 we present more complex loglinear models for contingency tables.

4.4 MOMENTS AND LIKELIHOOD FOR GENERALIZED LINEAR MODELS*

Having introduced GLMs for binary and count data, we now turn our attention to details such as likelihood equations and methods for fitting them. The remainder of this chapter is somewhat technical, providing general results applying to most modeling methods presented in subsequent chapters. See McCullagh and Nelder (1989) for further details.

It is helpful to extend the notation for a GLM so that it can handle many distributions that have a second parameter. The random component of the GLM specifies that the N observations $(y_1, \ldots, y_N)$ on Y are independent, with probability mass or density function for y_i of form

$$f(y_i; \theta_i, \phi) = \exp\{[y_i\theta_i - b(\theta_i)]/a(\phi) + c(y_i, \phi)\}. \qquad (4.14)$$

This is called the *exponential dispersion family* and ϕ is called the *dispersion parameter* (Jørgensen 1987). The parameter θ_i is the *natural parameter*.

When ϕ is known, (4.14) simplifies to the form (4.1) for the natural exponential family, which is

$$f(y_i; \theta_i) = a(\theta_i)b(y_i)\exp[y_iQ(\theta_i)].$$

We identify $Q(\theta)$ here with $\theta/a(\phi)$ in (4.14), $a(\theta)$ with $\exp[-b(\theta)/a(\phi)]$ in (4.14), and $b(y)$ with $\exp[c(y, \phi)]$ in (4.14). The more general formula (4.14) is not needed for one-parameter families such as the binomial and Poisson. Usually, $a(\phi)$ has form $a(\phi) = \phi/\omega_i$ for a known weight ω_i. For instance, when y_i is a mean of n_i independent readings, such as a sample proportion for n_i Bernoulli trials, $\omega_i = n_i$ (Section 4.4.2).

4.4.1 Mean and Variance Functions for the Random Component

General expressions for $E(Y_i)$ and $\text{var}(Y_i)$ use terms in (4.14). Let $L_i = \log f(y_i; \theta_i, \phi)$ denote the contribution of y_i to the log likelihood; that is, the log-likelihood function is $L = \sum_i L_i$. Then, from (4.14),

$$L_i = [y_i\theta_i - b(\theta_i)]/a(\phi) + c(y_i, \phi). \qquad (4.15)$$

Therefore,

$$\partial L_i/\partial\theta_i = [y_i - b'(\theta_i)]/a(\phi), \qquad \partial^2 L_i/\partial\theta_i^2 = -b''(\theta_i)/a(\phi),$$

where $b'(\theta_i)$ and $b''(\theta_i)$ denote the first two derivatives of $b(\cdot)$ evaluated at θ_i. We now apply the general likelihood results

$$E\left(\frac{\partial L}{\partial\theta}\right) = 0 \quad \text{and} \quad -E\left(\frac{\partial^2 L}{\partial\theta^2}\right) = E\left(\frac{\partial L}{\partial\theta}\right)^2,$$

which hold under regularity conditions satisfied by the exponential family (Cox and Hinkley 1974, Sec. 4.8). From the first formula applied with a single observation, $E[Y_i - b'(\theta_i)]/a(\phi) = 0$, or

$$\mu_i = E(Y_i) = b'(\theta_i). \qquad (4.16)$$

From the second formula,

$$b''(\theta_i)/a(\phi) = E[(Y_i - b'(\theta_i))/a(\phi)]^2 = \operatorname{var}(Y_i)/[a(\phi)]^2,$$

so that

$$\operatorname{var}(Y_i) = b''(\theta_i)a(\phi). \tag{4.17}$$

In summary, the function $b(\cdot)$ in (4.14) determines moments of Y_i.

4.4.2 Mean and Variance Functions for Poisson and Binomial

We illustrate the mean and variance expressions for Poisson and binomial distributions. When Y_i is Poisson,

$$f(y_i; \mu_i) = \frac{e^{-\mu_i}\mu_i^{y_i}}{y_i!} = \exp(y_i \log \mu_i - \mu_i - \log y_i!)$$

$$= \exp[y_i\theta_i - \exp(\theta_i) - \log y_i!],$$

where $\theta_i = \log \mu_i$. This has exponential dispersion form (4.14) with $b(\theta_i) = \exp(\theta_i)$, $a(\phi) = 1$, and $c(y_i, \phi) = -\log y_i!$. The natural parameter is $\theta_i = \log \mu_i$. From (4.16) and (4.17),

$$E(Y_i) = b'(\theta_i) = \exp(\theta_i) = \mu_i,$$

$$\operatorname{var}(Y_i) = b''(\theta_i) = \exp(\theta_i) = \mu_i.$$

Next, suppose that n_iY_i has a $\operatorname{bin}(n_i, \pi_i)$ distribution; that is, here y_i is the sample *proportion* (rather than *number*) of successes, so $E(Y_i)$ is independent of n_i. Let $\theta_i = \log[\pi_i/(1 - \pi_i)]$. Then, $\pi_i = \exp(\theta_i)/[1 + \exp(\theta_i)]$ and $\log(1 - \pi_i) = -\log[1 + \exp(\theta_i)]$. Extending (4.3), one can show that

$$f(y_i; \pi_i, n_i) = \binom{n_i}{n_iy_i}\pi_i^{n_iy_i}(1 - \pi_i)^{n_i - n_iy_i}$$

$$= \exp\left[\frac{y_i\theta_i - \log[1 + \exp(\theta_i)]}{1/n_i} + \log\binom{n_i}{n_iy_i}\right]. \tag{4.18}$$

This has exponential dispersion form (4.14) with $b(\theta_i) = \log[1 + \exp(\theta_i)]$, $a(\phi) = 1/n_i$, and $c(y_i, \phi) = \log\binom{n_i}{n_iy_i}$. The natural parameter is the logit, $\theta_i = \log[\pi_i/(1 - \pi_i)]$. From (4.16) and (4.17),

$$E(Y_i) = b'(\theta_i) = \exp(\theta_i)/[1 + \exp(\theta_i)] = \pi_i,$$

$$\operatorname{var}(Y_i) = b''(\theta_i)a(\phi) = \exp(\theta_i)/\left\{[1 + \exp(\theta_i)]^2 n_i\right\} = \pi_i(1 - \pi_i)/n_i.$$

4.4.3 Systematic Component and Link Function

Let $(x_{i1}, \ldots, x_{ip})$ denote values of explanatory variables for observation i. The systematic component of a GLM relates parameters $\{\eta_i\}$ to these variables using a linear predictor

$$\eta_i = \sum_j \beta_j x_{ij}, \qquad i = 1, \ldots, N.$$

In matrix form,

$$\boldsymbol{\eta} = \mathbf{X}\boldsymbol{\beta},$$

where $\boldsymbol{\eta} = (\eta_1, \ldots, \eta_N)'$, $\boldsymbol{\beta} = (\beta_1, \ldots, \beta_p)'$ are column vectors of model parameters, and $\mathbf{X}$ is the $N \times p$ matrix of values of the explanatory variables for the N subjects. In ordinary linear models, $\mathbf{X}$ is called the *design matrix*. It need not refer to an experimental design, however, and the GLM literature calls it the *model matrix*.

The GLM links η_i to $\mu_i = E(Y_i)$ by a link function $g(\cdot)$. Thus, μ_i relates to the explanatory variables by

$$\eta_i = g(\mu_i) = \sum_j \beta_j x_{ij}, \qquad i = 1, \ldots, N.$$

The link function g for which $g(\mu_i) = \theta_i$ in (4.14) is the *canonical link*. For it, the direct relationship

$$\theta_i = \sum_j \beta_j x_{ij}$$

occurs between the natural parameter and the linear predictor.

Since $\mu_i = b'(\theta_i)$, the natural parameter is the function of the mean, $\theta_i = (b')^{-1}(\mu_i)$, where $(b')^{-1}(\cdot)$ denotes the inverse function to b'. Thus, the canonical link is the inverse of b'. In the Poisson case, for instance, $b(\theta_i) = \exp(\theta_i)$, so $b'(\theta_i) = \exp(\theta_i) = \mu_i$. Thus, $(b')^{-1}(\cdot)$ is the inverse of the exponential function, which is the log function (i.e., $\theta_i = \log \mu_i$). The canonical link is the log link.

4.4.4 Likelihood Equations for a GLM

For N independent observations, from (4.15) the log likelihood is

$$L(\boldsymbol{\beta}) = \sum_i L_i = \sum_i \log f(y_i; \theta_i, \phi) = \sum_i \frac{y_i \theta_i - b(\theta_i)}{a(\phi)} + \sum_i c(y_i, \phi).$$

$$(4.19)$$

The notation $L(\boldsymbol{\beta})$ reflects the dependence of $\boldsymbol{\theta}$ on the model parameters $\boldsymbol{\beta}$.

The likelihood equations are

$$\partial L(\boldsymbol{\beta})/\partial \beta_j = \sum_i \partial L_i/\partial \beta_j = 0$$

for all j. To differentiate the log likelihood (4.19), we use the chain rule,

$$\frac{\partial L_i}{\partial \beta_j} = \frac{\partial L_i}{\partial \theta_i} \frac{\partial \theta_i}{\partial \mu_i} \frac{\partial \mu_i}{\partial \eta_i} \frac{\partial \eta_i}{\partial \beta_j}. \qquad (4.20)$$

Since $\partial L_i/\partial \theta_i = [y_i - b'(\theta_i)]/a(\phi)$, and since $\mu_i = b'(\theta_i)$ and $\text{var}(Y_i) = b''(\theta_i)a(\phi)$ from (4.16) and (4.17),

$$\partial L_i/\partial \theta_i = (y_i - \mu_i)/a(\phi), \qquad \partial \mu_i/\partial \theta_i = b''(\theta_i) = \text{var}(Y_i)/a(\phi).$$

Also, since $\eta_i = \sum_j \beta_j x_{ij}$,

$$\partial \eta_i/\partial \beta_j = x_{ij}.$$

Finally, since $\eta_i = g(\mu_i)$, $\partial \mu_i/\partial \eta_i$ depends on the link function for the model. In summary, substituting into (4.20) gives us

$$\frac{\partial L_i}{\partial \beta_j} = \frac{y_i - \mu_i}{a(\phi)} \frac{a(\phi)}{\text{var}(Y_i)} \frac{\partial \mu_i}{\partial \eta_i} x_{ij} = \frac{(y_i - \mu_i)x_{ij}}{\text{var}(Y_i)} \frac{\partial \mu_i}{\partial \eta_i}. \qquad (4.21)$$

The likelihood equations are

$$\sum_{i=1}^{N} \frac{(y_i - \mu_i)x_{ij}}{\text{var}(Y_i)} \frac{\partial \mu_i}{\partial \eta_i} = 0, \qquad j = 1, \ldots, p. \qquad (4.22)$$

Although $\boldsymbol{\beta}$ does not appear in these equations, it is there implicitly through μ_i, since $\mu_i = g^{-1}(\sum_j \beta_j x_{ij})$. Different link functions yield different sets of equations.

Interestingly, the likelihood equations (4.22) depend on the distribution of Y_i only through μ_i and $\text{var}(Y_i)$. The variance itself depends on the mean through a particular functional form

$$\text{var}(Y_i) = v(\mu_i)$$

for some function v, such as $v(\mu_i) = \mu_i$ for the Poisson, $v(\mu_i) = \mu_i(1 - \mu_i)$ for the Bernoulli, and $v(\mu_i) = \sigma^2$ (i.e., constant) for the normal. When Y_i has distribution in the natural exponential family, the relationship between the mean and the variance characterizes the distribution (Jørgensen 1987). For instance, if Y_i has distribution in the natural exponential family and if $v(\mu_i) = \mu_i$, then necessarily Y_i has the Poisson distribution.

4.4.5 Likelihood Equations for Binomial GLMs

Using notation from Section 4.4.2, suppose that $n_i Y_i$ has a $\text{bin}(n_i, \pi_i)$ distribution. Then y_i is a sample proportion of successes for n_i trials. The binomial GLM (4.8) for a single predictor extends with several predictors to

$$\pi_i = \Phi\left(\sum_j \beta_j x_{ij}\right), \tag{4.23}$$

where Φ is the standard cdf of some class of continuous distributions. Since $\pi_i = \mu_i = \Phi(\eta_i)$ with $\eta_i = \sum_j \beta_j x_{ij}$,

$$\partial\mu_i / \partial\eta_i = \phi(\eta_i) = \phi\left(\sum_j \beta_j x_{ij}\right),$$

where $\phi(u) = \partial\Phi(u)/\partial u$ (i.e., the probability density function corresponding to the cdf Φ). Since $\text{var}(Y_i) = \pi_i(1 - \pi_i)/n_i$, the likelihood equations (4.22) simplify to

$$\sum_i \frac{n_i(y_i - \pi_i)x_{ij}}{\pi_i(1 - \pi_i)}\,\phi\left(\sum_j \beta_j x_{ij}\right) = 0, \tag{4.24}$$

where $\pi_i = \Phi(\sum_j \beta_j x_{ij})$. These depend on the link function Φ^{-1} through the derivative of its inverse.

For the logit link, $\eta_i = \log[\pi_i/(1 - \pi_i)]$, so $\partial\eta_i / \partial\pi_i = 1/[\pi_i(1 - \pi_i)]$ and $\partial\mu_i / \partial\eta_i = \partial\pi_i / \partial\eta_i = \pi_i(1 - \pi_i)$. Then the likelihood equations (4.22) and (4.24) simplify to

$$\sum_i n_i(y_i - \pi_i)x_{ij} = 0, \tag{4.25}$$

where π_i satisfies (4.23) with Φ the standard logistic cdf.

4.4.6 Asymptotic Covariance Matrix of Model Parameter Estimators

The likelihood function for the GLM also determines the asymptotic covariance matrix of the ML estimator $\hat{\boldsymbol{\beta}}$. This matrix is the inverse of the information matrix $\mathcal{I}$, which has elements $E[-\partial^2 L(\boldsymbol{\beta})/\partial\beta_h \partial\beta_j]$. To find this, for the contribution L_i to the log likelihood we use the helpful result

$$E\left(\frac{\partial^2 L_i}{\partial\beta_h \partial\beta_j}\right) = -E\left(\frac{\partial L_i}{\partial\beta_h}\right)\left(\frac{\partial L_i}{\partial\beta_j}\right),$$

which holds for exponential families (Cox and Hinkley 1974, Sec. 4.8). Thus,

$$E\left(\frac{\partial^2 L_i}{\partial \beta_h \partial \beta_j}\right) = -E\left[\frac{(Y_i - \mu_i)x_{ih}}{\text{var}(Y_i)}\frac{\partial \mu_i}{\partial \eta_i}\frac{(Y_i - \mu_i)x_{ij}}{\text{var}(Y_i)}\frac{\partial \mu_i}{\partial \eta_i}\right] \qquad \text{from (4.21)}$$

$$= \frac{-x_{ih}x_{ij}}{\text{var}(Y_i)}\left(\frac{\partial \mu_i}{\partial \eta_i}\right)^2.$$

Since $L(\boldsymbol{\beta}) = \Sigma_i L_i$,

$$E\left(-\frac{\partial^2 L(\boldsymbol{\beta})}{\partial \beta_h \partial \beta_j}\right) = \sum_{i=1}^{N}\frac{x_{ih}x_{ij}}{\text{var}(Y_i)}\left(\frac{\partial \mu_i}{\partial \eta_i}\right)^2.$$

Generalizing from this typical element to the entire matrix, the information matrix has the form

$$\mathcal{J} = \mathbf{X'WX}, \qquad (4.26)$$

where $\mathbf{W}$ is the diagonal matrix with main-diagonal elements

$$w_i = (\partial \mu_i / \partial \eta_i)^2 / \text{var}(Y_i). \qquad (4.27)$$

The asymptotic covariance matrix of $\hat{\boldsymbol{\beta}}$ is estimated by

$$\widehat{\text{cov}}(\hat{\boldsymbol{\beta}}) = \hat{\mathcal{J}}^{-1} = (\mathbf{X'\hat{W}X})^{-1}, \qquad (4.28)$$

where $\hat{\mathbf{W}}$ is $\mathbf{W}$ evaluated at $\hat{\boldsymbol{\beta}}$. From (4.27), the form of $\mathbf{W}$ also depends on the link function. We'll see an example for Poisson GLMs next and for binomial GLMs in Section 5.5.

4.4.7 Likelihood Equations and Covariance Matrix for Poisson Loglinear Model

The general Poisson loglinear model (4.4) has the matrix form

$$\log \boldsymbol{\mu} = \mathbf{X}\boldsymbol{\beta}.$$

For the log link, $\eta_i = \log \mu_i$, so $\mu_i = \exp(\eta_i)$ and $\partial \mu_i / \partial \eta_i = \exp(\eta_i) = \mu_i$. Since $\text{var}(Y_i) = \mu_i$, the likelihood equations (4.22) simplify to

$$\sum_i (y_i - \mu_i)x_{ij} = 0. \qquad (4.29)$$

These equate the sufficient statistics $\Sigma_i y_i x_{ij}$ for $\boldsymbol{\beta}$ to their expected values.

Also, since

$$w_i = (\partial \mu_i / \partial \eta_i)^2 / \text{var}(Y_i) = \mu_i$$

the estimated covariance matrix (4.28) of $\hat{\boldsymbol{\beta}}$ is $(\mathbf{X}'\hat{\mathbf{W}}\mathbf{X})^{-1}$, where $\hat{\mathbf{W}}$ is the diagonal matrix with elements of $\hat{\boldsymbol{\mu}}$ on the main diagonal.

4.5 INFERENCE FOR GENERALIZED LINEAR MODELS

For most GLMs the likelihood equations (4.22) are nonlinear functions of $\boldsymbol{\beta}$. For now, we put off details about solving them for the ML estimator $\hat{\boldsymbol{\beta}}$ and focus instead on using the fit for statistical inference.

The Wald, score, and likelihood-ratio methods introduced in Section 1.3.3 for significance testing and interval estimation apply to any GLM. In this section we concentrate on likelihood-ratio inference, through the *deviance* of the GLM.

4.5.1 Deviance and Goodness of Fit

From Section 4.1.5, the *saturated* GLM has a separate parameter for each observation. It gives a perfect fit. This sounds good, but it is not a helpful model. It does not smooth the data or have the advantages that a simpler model has, such as parsimony. Nonetheless, it serves as a baseline for other models, such as for checking model fit.

A saturated model explains all variation by the systematic component of the model. Let $\tilde{\theta}$ denote the estimate of θ for the saturated model, corresponding to estimated means $\tilde{\mu}_i = y_i$ for all i. For a particular unsaturated model, denote the corresponding ML estimates by $\hat{\theta}$ and $\hat{\mu}_i$. For maximized log likelihood $L(\hat{\boldsymbol{\mu}}; \mathbf{y})$ for that model and maximized log likelihood $L(\mathbf{y}; \mathbf{y})$ in the saturated case,

$$-2 \log \frac{\text{maximum likelihood for model}}{\text{maximum likelihood for saturated model}} = -2[L(\hat{\boldsymbol{\mu}}; \mathbf{y}) - L(\mathbf{y}; \mathbf{y})]$$

describes lack of fit. It is the likelihood-ratio statistic for testing the null hypothesis that the model holds against the alternative that a more general model holds. From (4.19),

$$-2[L(\hat{\boldsymbol{\mu}}; \mathbf{y}) - L(\mathbf{y}; \mathbf{y})]$$
$$= 2 \sum_i \left[y_i \tilde{\theta}_i - b(\tilde{\theta}_i) \right] / a(\phi) - 2 \sum_i \left[y_i \hat{\theta}_i - b(\hat{\theta}_i) \right] / a(\phi).$$

Usually, $a(\phi)$ in (4.14) has the form $a(\phi) = \phi/\omega_i$, and this statistic equals

$$2\sum_i \omega_i \left[y_i\left(\tilde{\theta}_i - \hat{\theta}_i\right) - b\left(\tilde{\theta}_i\right) + b\left(\hat{\theta}_i\right)\right]/\phi = D(\mathbf{y}; \hat{\boldsymbol{\mu}})/\phi. \qquad (4.30)$$

This is called the *scaled deviance* and $D(\mathbf{y}; \hat{\boldsymbol{\mu}})$ is called the *deviance*. The greater the scaled deviance, the poorer the fit. For some GLMs the scaled deviance has an approximate chi-squared distribution.

4.5.2 Deviance for Poisson Models

For Poisson GLMs, by Section 4.4.2, $\hat{\theta}_i = \log \hat{\mu}_i$ and $b(\hat{\theta}_i) = \exp(\hat{\theta}_i) = \hat{\mu}_i$. Similarly, $\tilde{\theta}_i = \log y_i$ and $b(\tilde{\theta}_i) = y_i$ for the saturated model. Also $a(\phi) = 1$, so the deviance and scaled deviance (4.30) equal

$$D(\mathbf{y}; \hat{\boldsymbol{\mu}}) = 2\sum_i \left[y_i\log(y_i/\hat{\mu}_i) - y_i + \hat{\mu}_i\right]. \qquad (4.31)$$

When a model with log link contains an intercept term, the likelihood equation (4.29) implied by that parameter is $\sum y_i = \sum \hat{\mu}_i$. Then the deviance simplifies to

$$D(\mathbf{y}; \hat{\boldsymbol{\mu}}) = 2\sum_i y_i \log(y_i/\hat{\mu}_i). \qquad (4.32)$$

For two-way contingency tables, this reduces to the G^2 statistic (3.11) in Section 3.2.1, substituting cell count n_{ij} for y_i and the independence fitted value $\hat{\mu}_{ij}$ for $\hat{\mu}_i$. For a Poisson or multinomial model applied to a contingency table with a fixed number of cells N, we will see in Section 14.3 that the deviance has an approximate chi-squared distribution for large $\{\mu_i\}$.

4.5.3 Deviance for Binomial Models: Grouped and Ungrouped Data

Now consider binomial GLMs with sample proportions $\{y_i\}$ based on $\{n_i\}$ trials. By Section 4.4.2, $\hat{\theta}_i = \log[\hat{\pi}_i/(1 - \hat{\pi}_i)]$ and $b(\hat{\theta}_i) = \log[1 + \exp(\hat{\theta}_i)] = -\log(1 - \hat{\pi}_i)$. Similarly, $\tilde{\theta}_i = \log[y_i/(1 - y_i)]$ and $b(\tilde{\theta}_i) = -\log(1 - y_i)$ for the saturated model. Also, $a(\phi) = 1/n_i$, so $\phi = 1$ and $\omega_i = n_i$. The deviance (4.30) equals

$$2\sum_i n_i \left\{ y_i\left(\log\frac{y_i}{1 - y_i} - \log\frac{\hat{\pi}_i}{1 - \hat{\pi}_i}\right) + \log(1 - y_i) - \log(1 - \hat{\pi}_i)\right\}$$

$$= 2\sum_i n_i y_i \log\frac{n_i y_i}{n_i - n_i y_i} - 2\sum_i n_i y_i \log\frac{n_i \hat{\pi}_i}{n_i - n_i \hat{\pi}_i} + 2\sum_i n_i \log\frac{1 - y_i}{1 - \hat{\pi}_i}$$

$$= 2\sum_i n_i y_i \log\frac{n_i y_i}{n_i \hat{\pi}_i} + 2\sum_i (n_i - n_i y_i) \log\frac{n_i - n_i y_i}{n_i - n_i \hat{\pi}_i}.$$

At setting i, $n_i y_i$ is the number of successes and $(n_i - n_i y_i)$ is the number of failures, $i = 1, \ldots, N$. Thus, the deviance is a sum over the $2N$ cells of successes and failures and has the same form,

$$D(y; \hat{\mu}) = 2 \sum \text{observed} \times \log(\text{observed}/\text{fitted}), \qquad (4.33)$$

as the deviance (4.32) for Poisson loglinear models with intercept term.

With binomial responses, it is possible to construct the data file as expressed here with the counts of successes and failures at each setting for the predictors, or with the individual Bernoulli 0–1 observations at the subject level. The deviance differs in the two cases. In the first case the saturated model has a parameter at each setting for the predictors, whereas in the second case it has a parameter for each subject. We refer to these as *grouped data* and *ungrouped data* cases. The approximate chi-squared distribution for the deviance occurs for grouped data but not for ungrouped data (see Problems 4.22 and 5.37). With grouped data, the sample size increases for a fixed number of settings of the predictors and hence a fixed number of parameters for the saturated model.

4.5.4 Likelihood-Ratio Model Comparison Using the Deviance

For a Poisson or binomial model M, $\phi = 1$, so the deviance (4.30) equals

$$D(y; \hat{\mu}) = -2[L(\hat{\mu}; y) - L(y; y)]. \qquad (4.34)$$

Consider two models, M_0 with fitted values $\hat{\mu}_0$ and M_1 with fitted values $\hat{\mu}_1$, with M_0 a special case of M_1. Model M_0 is said to be *nested* within M_1.

Since M_0 is simpler than M_1, a smaller set of parameter values satisfies M_0 than satisfies M_1. Maximizing the log likelihood over a smaller space cannot yield a larger maximum. Thus, $L(\hat{\mu}_0; y) \le L(\hat{\mu}_1; y)$, and it follows from (4.34) with the same $L(y; y)$ for each model that

$$D(y; \hat{\mu}_1) \le D(y; \hat{\mu}_0).$$

Simpler models have larger deviances. Assuming that model M_1 holds, the likelihood-ratio test of the hypothesis that M_0 holds uses the test statistic

$$-2[L(\hat{\mu}_0; y) - L(\hat{\mu}_1; y)]$$
$$= -2[L(\hat{\mu}_0; y) - L(y; y)] - \{-2[L(\hat{\mu}_1; y) - L(y; y)]\}$$
$$= D(y; \hat{\mu}_0) - D(y; \hat{\mu}_1).$$

The likelihood-ratio statistic comparing the two models is simply the difference between the deviances. This statistic is large when M_0 fits poorly compared to M_1.

In fact, since the part in (4.30) involving the saturated model cancels, the difference between deviances,

$$D(\mathbf{y}; \hat{\boldsymbol{\mu}}_0) - D(\mathbf{y}; \hat{\boldsymbol{\mu}}_1) = 2\sum \omega_i \left[y_i(\hat{\theta}_{1i} - \hat{\theta}_{0i}) - b(\hat{\theta}_{1i}) + b(\hat{\theta}_{0i}) \right],$$

also has the form of the deviance. Under regularity conditions, this difference has approximately a chi-squared null distribution with df equal to the difference between the numbers of parameters in the two models.

For binomial GLMs and Poisson loglinear GLMs with intercept, from expression (4.33) for the deviance, the difference in deviances uses the observed counts and the two sets of fitted values in the form

$$D(\mathbf{y}; \hat{\boldsymbol{\mu}}_0) - D(\mathbf{y}; \hat{\boldsymbol{\mu}}_1) = 2\sum \text{observed} \times \log(\text{fitted}_1/\text{fitted}_0).$$

With binomial responses, the test comparing models does not depend on whether the data file has grouped or ungrouped form. The saturated model differs in the two cases, but its log likelihood cancels when one forms the difference between the deviances.

4.5.5 Residuals for GLMs

When a GLM fits poorly according to an overall goodness-of-fit test, examination of residuals highlights where the fit is poor. One type of residual uses components of the deviance. In (4.30) let $D(\mathbf{y}; \hat{\boldsymbol{\mu}}) = \sum d_i$, where

$$d_i = 2\omega_i \left[y_i(\tilde{\theta}_i - \hat{\theta}_i) - b(\tilde{\theta}_i) + b(\hat{\theta}_i) \right].$$

The *deviance residual* for observation i is

$$\sqrt{d_i} \times \text{sign}(y_i - \hat{\mu}_i), \qquad (4.35)$$

An alternative is the *Pearson residual*,

$$e_i = \frac{y_i - \hat{\mu}_i}{\left[\widehat{\text{var}(Y_i)}\right]^{1/2}}. \qquad (4.36)$$

For instance, for a Poisson GLM, $\text{var}(Y_i) = \mu_i$ and the Pearson residual is

$$e_i = (y_i - \hat{\mu}_i)/\sqrt{\hat{\mu}_i}.$$

For two-way contingency tables identifying y_i with cell count n_{ij} and $\hat{\mu}_i$ with the independence fitted value $\hat{\mu}_{ij}$, this has the form (3.12); then $\sum e_{ij}^2 = X^2$, the Pearson X^2 statistic. Similarly, the sum of squared deviance residuals $\sum d_{ij} = G^2$, the likelihood-ratio statistic for testing independence.

When the model holds, Pearson and deviance residuals are less variable than standard normal because they compare y_i to the fitted means rather than the true mean (e.g., the denominator of (4.36) estimates $[\text{var}(Y_i)]^{1/2} = [\text{var}(Y_i - \mu_i)]^{1/2}$ rather than $[\text{var}(Y_i - \hat{\mu}_i)]^{1/2}$). Standardized residuals divide the ordinary residuals by their asymptotic standard errors. For GLMs the asymptotic covariance matrix of the vector of the raw residuals $\{y_i - \hat{\mu}_i\}$ is

$$\text{cov}(\mathbf{Y} - \hat{\boldsymbol{\mu}}) = \text{cov}(\mathbf{Y})[\mathbf{I} - \mathbf{Hat}].$$

Here, $\mathbf{I}$ is the identity matrix and $\mathbf{Hat}$ is the *hat matrix*,

$$\mathbf{Hat} = \mathbf{W}^{1/2}\mathbf{X}(\mathbf{X}'\mathbf{WX})^{-1}\mathbf{X}'\mathbf{W}^{1/2}, \tag{4.37}$$

where $\mathbf{W}$ is the diagonal matrix with elements (4.27) (Pregibon 1981). Let $\hat{h}_i$ denote the estimated diagonal element of $\mathbf{Hat}$ for observation i, called its *leverage*. Then, standardizing by dividing $y_i - \hat{\mu}_i$ by its estimated SE yields the standardized Pearson residual

$$r_i = \frac{y_i - \hat{\mu}_i}{\left\{[\text{var}(Y_i)](1 - \hat{h}_i)\right\}^{1/2}} = \frac{e_i}{\sqrt{1 - \hat{h}_i}}. \tag{4.38}$$

For Poisson GLMs, for instance, $r_i = (y_i - \hat{\mu}_i)/\sqrt{\hat{\mu}_i(1 - \hat{h}_i)}$. Pierce and Schafer (1986) presented standardized deviance residuals.

In linear models the hat matrix is so-named because $\mathbf{Hat} \times \mathbf{y}$ projects the data to the fitted values, $\hat{\boldsymbol{\mu}} = $ "mu-hat." For GLMs, applying the estimated hat matrix to a linearized approximation for $g(\mathbf{y})$ yields $\hat{\boldsymbol{\eta}} = g(\hat{\boldsymbol{\mu}})$, the model's estimated linear predictor values. The greater an observation's leverage, the greater its potential influence on the fit. As in ordinary regression, the leverages fall between 0 and 1 and sum to the number of model parameters. Unlike ordinary regression, the hat values depend on the fit as well as the model matrix, and points that have extreme predictor values need not have high leverage.

4.6 FITTING GENERALIZED LINEAR MODELS

Finally, we study how to find the ML estimators $\hat{\boldsymbol{\beta}}$ of GLM parameters. The likelihood equations (4.22) are usually nonlinear in $\hat{\boldsymbol{\beta}}$. We describe a general-purpose iterative method for solving nonlinear equations and apply it two ways to determine the maximum of a likelihood function.

4.6.1 Newton–Raphson Method

The *Newton–Raphson method* is an iterative method for solving nonlinear equations, such as equations whose solution determines the point at which a function takes its maximum. It begins with an initial guess for the solution. It obtains a second guess by approximating the function to be maximized in a neighborhood of the initial guess by a second-degree polynomial and then finding the location of that polynomial's maximum value. It then approximates the function in a neighborhood of the second guess by another second-degree polynomial, and the third guess is the location of its maximum. In this manner, the method generates a sequence of guesses. These converge to the location of the maximum when the function is suitable and/or the initial guess is good.

In more detail, here's how Newton–Raphson determines the value $\hat{\boldsymbol{\beta}}$ at which a function $L(\boldsymbol{\beta})$ is maximized. Let $\mathbf{u}' = (\partial L(\boldsymbol{\beta})/\partial \beta_1, \partial L(\boldsymbol{\beta})/\partial \beta_2, \ldots)$. Let $\mathbf{H}$ denote the matrix having entries $h_{ab} = \partial^2 L(\boldsymbol{\beta})/\partial \beta_a \partial \beta_b$, called the *Hessian matrix*. Let $\mathbf{u}^{(t)}$ and $\mathbf{H}^{(t)}$ be $\mathbf{u}$ and $\mathbf{H}$ evaluated at $\boldsymbol{\beta}^{(t)}$, the guess t for $\hat{\boldsymbol{\beta}}$. Step t in the iterative process $(t = 0, 1, 2, \ldots)$ approximates $L(\boldsymbol{\beta})$ near $\boldsymbol{\beta}^{(t)}$ by the terms up to second order in its Taylor series expansion,

$$L(\boldsymbol{\beta}) \approx L(\boldsymbol{\beta}^{(t)}) + \mathbf{u}^{(t)'} (\boldsymbol{\beta} - \boldsymbol{\beta}^{(t)}) + (\tfrac{1}{2})(\boldsymbol{\beta} - \boldsymbol{\beta}^{(t)})' \mathbf{H}^{(t)}(\boldsymbol{\beta} - \boldsymbol{\beta}^{(t)}).$$

Solving $\partial L(\boldsymbol{\beta})/\partial \boldsymbol{\beta} \approx \mathbf{u}^{(t)} + \mathbf{H}^{(t)}(\boldsymbol{\beta} - \boldsymbol{\beta}^{(t)}) = 0$ for $\boldsymbol{\beta}$ yields the next guess. That guess can be expressed as

$$\boldsymbol{\beta}^{(t+1)} = \boldsymbol{\beta}^{(t)} - (\mathbf{H}^{(t)})^{-1}\mathbf{u}^{(t)}, \tag{4.39}$$

assuming that $\mathbf{H}^{(t)}$ is nonsingular. (However, computing routines use standard methods for solving the linear equations rather than explicitly calculating the inverse.)

Iterations proceed until changes in $L(\boldsymbol{\beta}^{(t)})$ between successive cycles are sufficiently small. The ML estimator is the limit of $\boldsymbol{\beta}^{(t)}$ as $t \to \infty$; however, this need not happen if $L(\boldsymbol{\beta})$ has other local maxima at which the derivative of $L(\boldsymbol{\beta})$ equals 0. In that case, a good initial estimate is crucial. To help understand the Newton–Raphson process, work through these steps when β has a single element (Problem 4.34). Then, Figure 4.6 illustrates a cycle of the method, showing the parabolic (second-order) approximation at a given step.

In the next chapter we use Newton–Raphson for logistic regression models. For now, we illustrate it with a simpler problem for which we know the answer, maximizing the log likelihood based on an observation y from a $\mathrm{bin}(n, \pi)$ distribution. From Section 1.3.2, the first two derivatives of $L(\pi)$ $= y \log \pi + (n - y)\log(1 - \pi)$ are

$$u = (y - n\pi)/\pi(1 - \pi), \qquad H = -\left[y/\pi^2 + (n - y)/(1 - \pi)^2\right].$$

Each Newton–Raphson step has the form

$$\pi^{(t+1)} = \pi^{(t)} + \left[\frac{y}{(\pi^{(t)})^2} + \frac{n - y}{(1 - \pi^{(t)})^2}\right]^{-1} \frac{y - n\pi^{(t)}}{\pi^{(t)}(1 - \pi^{(t)})}.$$

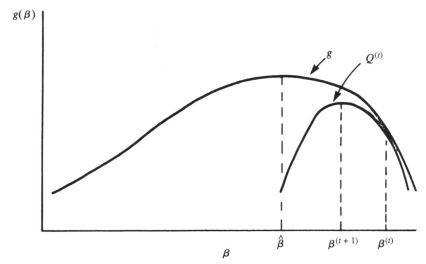

FIGURE 4.6 Cycle of Newton–Raphson method.

This adjusts $\pi^{(t)}$ up if $y/n > \pi^{(t)}$ and down if $y/n < \pi^{(t)}$. For instance, with $\pi^{(0)} = \frac{1}{2}$, you can check that $\pi^{(1)} = y/n$. When $\pi^{(t)} = y/n$, no adjustment occurs and $\pi^{(t+1)} = y/n$, which is the correct answer for $\hat{\pi}$. For starting values other than $\frac{1}{2}$, adequate convergence usually takes four or five iterations.

The convergence of $\boldsymbol{\beta}^{(t)}$ to $\hat{\boldsymbol{\beta}}$ for the Newton–Raphson method is usually fast. For large t, the convergence satisfies, for each j,

$$\left| \beta_j^{(t+1)} - \hat{\beta}_j \right| \leq c \left| \beta_j^{(t)} - \hat{\beta}_j \right|^2 \qquad \text{for some } c > 0$$

and is referred to as *second-order*. This implies that the number of correct decimals in the approximation roughly doubles after sufficiently many iterations. In practice, it often takes relatively few iterations for satisfactory convergence.

4.6.2 Fisher Scoring Method

Fisher scoring is an alternative iterative method for solving likelihood equations. It resembles the Newton–Raphson method, the distinction being with the Hessian matrix. Fisher scoring uses the *expected value* of this matrix, called the *expected information*, whereas Newton–Raphson uses the matrix itself, called the *observed information*.

Let $\mathcal{J}^{(t)}$ denote the approximation t for the ML estimate of the expected information matrix; that is, $\mathcal{J}^{(t)}$ has elements $-E\left(\partial^2 L(\boldsymbol{\beta})/\partial\beta_a\partial\beta_b \right)$, evaluated at $\beta^{(t)}$. The formula for Fisher scoring is

$$\beta^{(t+1)} = \beta^{(t)} + \left(\mathcal{J}^{(t)}\right)^{-1}\mathbf{u}^{(t)}$$

or

$$\mathcal{J}^{(t)}\beta^{(t+1)} = \mathcal{J}^{(t)}\beta^{(t)} + \mathbf{u}^{(t)}. \tag{4.40}$$

For estimating a binomial parameter, from Section 1.3.2 the information is $n/[\pi(1 - \pi)]$. A step of Fisher scoring gives

$$\pi^{(t+1)} = \pi^{(t)} + \left[\frac{n}{\pi^{(t)}(1 - \pi^{(t)})}\right]^{-1}\frac{y - n\pi^{(t)}}{\pi^{(t)}(1 - \pi^{(t)})}$$

$$= \pi^{(t)} + \frac{y - n\pi^{(t)}}{n} = \frac{y}{n}.$$

This gives the answer for $\hat{\pi}$ after a single iteration and stays at that value for successive iterations.

Formula (4.26) showed that $\mathcal{J} = \mathbf{X'WX}$. Similarly, $\mathcal{J}^{(t)} = \mathbf{X'W}^{(t)}\mathbf{X}$, where $\mathbf{W}^{(t)}$ is $\mathbf{W}$ [see (4.27)] evaluated at $\beta^{(t)}$. The estimated asymptotic covariance matrix $\mathcal{J}^{-1}$ of $\hat{\beta}$ [see (4.28)] occurs as a by-product of this algorithm as $\left(\mathcal{J}^{(t)}\right)^{-1}$ for t at which convergence is adequate. From (4.22), for both Fisher scoring and Newton–Raphson, $\mathbf{u}$ has elements

$$u_j = \frac{\partial L(\beta)}{\partial \beta_j} = \sum_{i=1}^{N}\frac{(y_i - \mu_i)x_{ij}}{\text{var}(Y_i)}\frac{\partial \mu_i}{\partial \eta_i}. \tag{4.41}$$

For GLMs with a canonical link, we'll see (Section 4.6.4) that the observed and expected information are the same. For noncanonical link models, Fisher scoring has the advantages that it produces the asymptotic covariance matrix as a by-product, the expected information is necessarily nonnegative definite, and as seen next, it is closely related to weighted least squares methods for ordinary linear models. However, it need not have second-order convergence, and for complex models the observed information is often easier to calculate. Efron and Hinkley (1978), developing arguments of R. A. Fisher, gave reasons for preferring observed information. They argued that its variance estimates better approximate a relevant conditional variance (conditional on statistics not relevant to the parameter being estimated), it is "closer to the data," and it tends to agree more closely with Bayesian analyses.

4.6.3 ML as Iterative Reweighted Least Squares*

A relation exists between *weighted least squares estimation* and using Fisher scoring to find ML estimates. We refer here to the general linear model of form

$$z = X\beta + \epsilon.$$

When the covariance matrix of ϵ is V, the weighted least squares (WLS) estimator of β is

$$(X'V^{-1}X)^{-1}X'V^{-1}z.$$

From $\mathcal{J} = X'WX$, expression (4.41) for elements of u, and since diagonal elements of W are $w_i = (\partial\mu_i/\partial\eta_i)^2/\mathrm{var}(Y_i)$, it follows that in (4.40),

$$\mathcal{J}^{(t)}\beta^{(t)} + u^{(t)} = X'W^{(t)}z^{(t)},$$

where $z^{(t)}$ has elements

$$z_i^{(t)} = \sum_j x_{ij}\beta_j^{(t)} + \left(y_i - \mu_i^{(t)}\right)\frac{\partial\eta_i^{(t)}}{\partial\mu_i^{(t)}}$$

$$= \eta_i^{(t)} + \left(y_i - \mu_i^{(t)}\right)\frac{\partial\eta_i^{(t)}}{\partial\mu_i^{(t)}}.$$

Equations (4.40) for Fisher scoring then have the form

$$(X'W^{(t)}X)\beta^{(t+1)} = X'W^{(t)}z^{(t)}.$$

These are the normal equations for using weighted least squares to fit a linear model for a response variable $z^{(t)}$, when the model matrix is X and the inverse of the covariance matrix is $W^{(t)}$. The equations have solution

$$\beta^{(t+1)} = (X'W^{(t)}X)^{-1}X'W^{(t)}z^{(t)}.$$

The vector z in this formulation is a linearized form of the link function g, evaluated at y,

$$g(y_i) \approx g(\mu_i) + (y_i - \mu_i)g'(\mu_i) = \eta_i + (y_i - \mu_i)(\partial\eta_i/\partial\mu_i) = z_i. \quad (4.42)$$

This *adjusted* (or "working") *response variable* z has element i approximated by $z_i^{(t)}$ for cycle t of the iterative scheme. That cycle regresses $z^{(t)}$ on X with weight (i.e., inverse covariance) $W^{(t)}$ to obtain a new estimate $\beta^{(t+1)}$. This estimate yields a new linear predictor value $\eta^{(t+1)} = X\beta^{(t+1)}$ and a new adjusted response value $z^{(t+1)}$ for the next cycle. The ML estimator results from iterative use of weighted least squares, in which the weight matrix changes at each cycle. The process is called *iterative reweighted least squares*.

A simple way to begin the iterative process uses the data y as the initial estimate of μ. This determines the first estimate of the weight matrix W and hence the initial estimate of $\boldsymbol{\beta}$. It may be necessary to alter some observations slightly for this first cycle only so that $g(\mathbf{y})$, the initial value of $\mathbf{z}$, is finite. For instance, when g is the log link applied to counts, a count of $y_i = 0$ is problematic, so one could set $y_i = \frac{1}{2}$. This is not a problem with the model itself, since the log applies to the mean, and fitted means are usually strictly positive in successive iterations.

4.6.4 Simplifications for Canonical Links*

Certain simplifications result with GLMs using the canonical link. For that link,

$$\eta_i = \theta_i = \sum_j \beta_j x_{ij}.$$

Often, $a(\phi)$ in the density or mass function (4.14) is identical for all observations, such as for Poisson GLMs $[a(\phi) = 1]$ and binomial GLMs with each $n_i = 1$ [for which $a(\phi) = 1/n_i = 1$]. Then the part of the log likelihood (4.19) involving both parameters and data is $\sum y_i \theta_i$, which simplifies to

$$\sum_i y_i \left(\sum_j \beta_j x_{ij} \right) = \sum_j \beta_j \left(\sum_i y_i x_{ij} \right).$$

Sufficient statistics for estimating $\boldsymbol{\beta}$ in the GLM are then

$$\sum_i y_i x_{ij}, \qquad j = 1, \ldots, p.$$

For the canonical link,

$$\partial \mu_i / \partial \eta_i = \partial \mu_i / \partial \theta_i = \partial b'(\theta_i) / \partial \theta_i = b''(\theta_i).$$

Thus, the contribution (4.21) to the likelihood equation for β_j simplifies to

$$\frac{\partial L_i}{\partial \beta_j} = \frac{y_i - \mu_i}{\text{var}(Y_i)} b''(\theta_i) x_{ij} = \frac{(y_i - \mu_i) x_{ij}}{a(\phi)}. \tag{4.43}$$

When $a(\phi)$ is identical for all observations, the likelihood equations are

$$\sum_i x_{ij} y_i = \sum_i x_{ij} \mu_i, \qquad j = 1, \ldots, p. \tag{4.44}$$

These equations equate the sufficient statistics for the model parameters to their expected values (Nelder and Wedderburn 1972). For a normal distribution with identity link, these are the *normal equations*. We obtained these for Poisson loglinear models in (4.29) and for binomial logistic regression models (when each $n_i = 1$) in (4.25).

From expression (4.43) for $\partial L_i / \partial \beta_j$, with the canonical link the second derivatives of the log likelihood have components

$$\frac{\partial^2 L_i}{\partial \beta_j \partial \beta_h} = - \frac{x_{ij}}{a(\phi)} \left(\frac{\partial \mu_i}{\partial \beta_h} \right).$$

This does not depend on the observation y_i, so

$$\partial^2 L(\boldsymbol{\beta}) / \partial \beta_h \partial \beta_j = E\left[\partial^2 L(\boldsymbol{\beta}) / \partial \beta_h \partial \beta_j \right].$$

That is, $\mathbf{H} = -\mathcal{J}$, and the Newton–Raphson and Fisher scoring algorithms are identical for canonical link models (Nelder and Wedderburn 1972).

4.7   QUASI-LIKELIHOOD AND GENERALIZED LINEAR MODELS*

A GLM $g(\mu_i) = \sum_j \beta_j x_{ij}$ specifies μ_i using a link function g and linear predictor. From (4.22) and (4.41), the ML estimates $\hat{\boldsymbol{\beta}}$ are the solutions of the likelihood equations

$$u_j(\boldsymbol{\beta}) = \sum_{i=1}^{N} \frac{(y_i - \mu_i) x_{ij}}{v(\mu_i)} \left(\frac{\partial \mu_i}{\partial \eta_i} \right) = 0, \qquad j = 1, \ldots, p, \qquad (4.45)$$

where $\mu_i = g^{-1}(\sum_j \beta_j x_{ij})$ and $v(\mu_i) = \mathrm{var}(Y_i)$. These equations set the *score functions* $\{u_j(\boldsymbol{\beta})\}$, which are derivatives of the log likelihood with respect to $\{\beta_j\}$, equal to 0. As we noted in Section 4.4.4, the likelihood equations depend on the assumed distribution for Y_i only through μ_i and $v(\mu_i)$. The choice of distribution determines the mean–variance relationship $v(\mu_i)$.

4.7.1   Mean–Variance Relationship Determines Quasi-likelihood Estimates

Wedderburn (1974) proposed an alternative approach, *quasi-likelihood estimation*, which assumes only a mean–variance relationship rather than a specific distribution for Y_i. It has a link function and linear predictor of the usual GLM form, but instead of assuming a distributional type for Y_i it assumes only

$$\mathrm{var}(Y_i) = v(\mu_i)$$

for some chosen variance function v. The equations that determine quasi-likelihood estimates are the same as the likelihood equations (4.45) for GLMs. They are not likelihood equations, however, without the additional assumption that $\{Y_i\}$ has distribution in the natural exponential family.

To illustrate, suppose we assume that the $\{Y_i\}$ are independent with

$$v(\mu_i) = \mu_i.$$

The quasi-likelihood (QL) estimates are the solution of (4.45) with $v(\mu_i)$ replaced by μ_i. Under the additional assumption that $\{Y_i\}$ have distribution in the exponential dispersion family (4.14), these estimates are also ML estimates. That case is simply the Poisson distribution. Thus, for $v(\mu) = \mu$, quasi-likelihood estimates are also ML estimates when the random component has a Poisson distribution.

Wedderburn suggested using the estimating equations (4.45) for *any* variance function, even if it does not occur for a member of the natural exponential family. In fact, the purpose of the quasi-likelihood method was to encompass a greater variety of cases, such as discussed in Section 4.7.2. The QL estimates have asymptotic covariance matrix of the same form (4.28) as in GLMs, namely $(\mathbf{X}'\hat{\mathbf{W}}\mathbf{X})^{-1}$ with $w_i = (\partial\mu_i/\partial\eta_i)^2/\mathrm{var}(Y_i)$.

4.7.2 Overdispersion for Poisson GLMs and Quasi-likelihood

For count data, we've seen (Section 4.3.3) that the Poisson assumption is often unrealistic because of overdispersion—the variance exceeds the mean. One cause for this is heterogeneity among subjects. This suggests an alternative to a Poisson GLM in which the mean–variance relationship has the form

$$v(\mu_i) = \phi\mu_i$$

for some constant ϕ. The case $\phi > 1$ represents overdispersion for the Poisson model.

In the estimating equations (4.45) with $v(\mu_i) = \phi\mu_i$, ϕ drops out. Thus, the equations are identical to likelihood equations for Poisson models, and model parameter estimates are also identical. Also,

$$w_i = (\partial\mu_i/\partial\eta_i)^2\mathrm{var}(Y_i) = (\partial\mu_i/\partial\eta_i)^2/\phi\mu_i,$$

so the estimated $\mathrm{cov}(\hat{\boldsymbol{\beta}}) = (\mathbf{X}'\hat{\mathbf{W}}\mathbf{X})^{-1}$ is ϕ times that for the Poisson model.

When a variance function has the form $v(\mu_i) = \phi v^*(\mu_i)$, usually ϕ is also unknown. However, ϕ is not in the estimating equations. Let $X^2 = \Sigma(y_i - \hat{\mu}_i)^2/v^*(\hat{\mu}_i)$, a Pearson-type statistic for the simpler model with $\phi = 1$. Then X^2/ϕ is a sum of squares of N standardized terms. When X^2/ϕ is approximately chi-squared or when μ_i is approximately linear in $\boldsymbol{\beta}$ with $v^*(\hat{\mu}_i)$ close to $v^*(\mu_i)$, then $E(X^2/\phi) \approx N - p$, the number of observations minus the number of model parameters p. Hence, $E[X^2/(N-p)] \approx \phi$. Using the motivation of moment estimation, Wedderburn (1974) suggested taking $\hat{\phi} = X^2/(N-p)$ as the estimated multiple of the covariance matrix.

In summary, this quasi-likelihood approach for count data is simple: Fit the ordinary Poisson model and use its p parameter estimates. Multiply the ordinary standard error estimates by $\sqrt{X^2/(N-p)}$.

We illustrate for the horseshoe crab data analyzed with Poisson GLMs in Section 4.3.2. With the log link, the fit using width to predict number of satellites was $\log \hat{\mu} = -3.305 + 0.164x$, with SE $= 0.020$ for $\hat{\beta} = 0.164$. To improve the adequacy of using a chi-squared statistic to summarize fit, we use the satellite totals and fit for all female crabs at a given width, to increase the counts and fitted values relative to those for individual female crabs. The $N = 66$ distinct width levels each have a total count y_i for the number of satellites and a fitted total $\hat{\mu}_i$. The Pearson statistic comparing these is $X^2 = 174.3$. The quasi-likelihood adjustment for standard errors equals $\sqrt{174.3/(66 - 2)} = 1.65$. Thus, SE $= 1.65(0.020) = 0.033$ is a more plausible standard error for $\hat{\beta} = 0.164$ in this prediction equation.

Alternative ways of handling overdispersion include mixture models that allow heterogeneity in the mean at fixed settings of predictors. For count data these include Poisson GLMs having random effects (Section 13.5) and negative binomial GLMs that result when a Poisson parameter itself has a gamma distribution (Section 4.3.4 and 13.4).

4.7.3 Overdispersion for Binomial GLMs and Quasi-likelihood

The quasi-likelihood approach can also handle overdispersion for counts based on binary data. When y_i is the sample mean of n_i independent binary observations with parameter π_i, $i = 1, \ldots, N$, then binomial sampling has $E(Y_i) = \pi_i$ and $\text{var}(Y_i) = \pi_i(1 - \pi_i)/n_i$. A simple quasi-likelihood approach uses the alternative variance function

$$v(\pi_i) = \phi\pi_i(1 - \pi_i)/n_i. \tag{4.46}$$

Overdispersion occurs when $\phi > 1$. The quasi-likelihood estimates are the same as ML estimates for the binomial model, since ϕ drops out of the estimating equations (4.45). As in the overdispersed Poisson case, ϕ enters the denominator of w_i. Thus, the asymptotic covariance matrix multiplies by ϕ, and standard errors multiply by $\sqrt{\phi}$. An estimate of ϕ using the X^2 fit statistic for the ordinary binomial model is $X^2/(N - p)$ (Finney 1947).

Methods like these that use estimates from ordinary models but inflate their standard errors are appropriate only if the model chosen describes well the structural relationship between the mean of Y and the predictors. If a large goodness-of-fit statistic is due to some other type of lack of fit, such as failing to include a relevant interaction term, making an adjustment for overdispersion will not address the inadequacy.

For counts with binary data, alternative mechanisms for handling overdispersion include mixture models such as binomial GLMs with random effects (Section 12.3) and models for which a binomial parameter itself has a beta distribution (Section 13.3).

4.7.4 Teratology Overdispersion Example

Table 4.5 shows results of a teratology experiment in which female rats on iron-deficient diets were assigned to four groups. Rats in group 1 were given placebo injections, and rats in other groups were given injections of an iron

TABLE 4.5   Response Counts of (Litter Size, Number Dead) for 58 Litters of Rats in Low-Iron Teratology Study

Group 1: Untreated (low iron)
   (10, 1) (11, 4) (12, 9) (4, 4) (10, 10) (11, 9) (9, 9) (11, 11) (10, 10) (10, 7) (12, 12)
   (10, 9) (8, 8) (11, 9) (6, 4) (9, 7) (14, 14) (12, 7) (11, 9) (13, 8) (14, 5) (10, 10)
   (12, 10) (13, 8) (10, 10) (14, 3) (13, 13) (4, 3) (8, 8) (13, 5) (12, 12)

Group 2: Injections days 7 and 10
   (10, 1) (3, 1) (13, 1) (12, 0) (14, 4) (9, 2) (13, 2) (16, 1) (11, 0) (4, 0) (1, 0)(12, 0)

Group 3: Injections days 0 and 7
   (8, 0) (11, 1) (14, 0) (14, 1) (11, 0)

Group 4: Injections weekly
   (3, 0) (13, 0) (9, 2) (17, 2) (15, 0) (2, 0) (14, 1) (8, 0) (6, 0) (17, 0)

Source: Moore and Tsiatis (1991).

supplement; this was done weekly in group 4, only on days 7 and 10 in group 2, and only on days 0 and 7 in group 3. The 58 rats were made pregnant, sacrificed after three weeks, and then the total number of dead fetuses was counted in each litter. In teratology experiments, due to unmeasured covariates and genetic variability the probability of death may vary from litter to litter within a particular treatment group.

Let $y_{i(g)}$ denote the proportion of dead fetuses out of the $n_{i(g)}$ in litter i in treatment group g. Let $\pi_{i(g)}$ denote the probability of death for a fetus in that litter. Consider the model with $n_{i(g)}y_{i(g)}$ a $bin(n_{i(g)}, \pi_{i(g)})$ variate, where

$$\pi_{i(g)} = \pi_g, \qquad g = 1, 2, 3, 4.$$

That is, the model treats all litters in a particular group g as having the same probability of death π_g. The ML fit has estimate $\hat{\pi}_g$ equal to the sample proportion of deaths for all fetuses from litters in that group. These equal $\hat{\pi}_1 = 0.758$ (SE = 0.024), $\hat{\pi}_2 = 0.102$ (SE = 0.028), $\hat{\pi}_3 = 0.034$ (SE = 0.024), and $\hat{\pi}_4 = 0.048$ (SE = 0.021), where for group g, SE = $\sqrt{\hat{\pi}_g(1 - \hat{\pi}_g)/(\sum_i n_{i(g)})}$. The estimated probability of death is considerably higher for the placebo group.

For litter i in group g, $n_{i(g)}\hat{\pi}_g$ is a fitted number of deaths and $n_{i(g)}(1 - \hat{\pi}_g)$ is a fitted number of nondeaths. Comparing these fitted values to the observed counts of deaths and nondeaths in the $N = 58$ litters using the Pearson statistic gives $X^2 = 154.7$ with df = $58 - 4 = 54$. There is considerable evidence of overdispersion. With the quasi-likelihood approach, $\{\hat{\pi}_g\}$ are the same as the binomial ML estimates; however, $\hat{\phi} = X^2/(N - p)$ = $154.7/(58 - 4) = 2.86$, so standard errors multiply by $\hat{\phi}^{1/2} = 1.69$.

Even with this adjustment for overdispersion, strong evidence remains that the probability of death is substantially higher for the placebo group. For instance, a 95% confidence interval for $\pi_1 - \pi_2$ is

$$(0.758 - 0.102) \pm 1.96\left[(1.69 \times 0.024)^2 + (1.69 \times 0.028)^2\right]^{1/2}$$

$$\text{or} \quad (0.54, 0.78).$$

This is wider, however, than the Wald interval of (0.59, 0.73) for comparing independent proportions, which ignores the overdispersion.

4.8 GENERALIZED ADDITIVE MODELS*

The GLM generalizes the ordinary linear model to permit nonnormal distributions and modeling functions of the mean. Quasi-likelihood provides a further generalization, specifying how the variance depends on the mean without assuming a given distribution. Another generalization replaces the linear predictor by smooth functions of the predictors.

4.8.1 Smoothing Data

The GLM structure $g(\mu_i) = \Sigma_j \beta_j x_{ij}$ generalizes to

$$g(\mu_i) = \sum_j s_j(x_{ij}),$$

where $s_j(\cdot)$ is an unspecified smooth function of predictor j. A useful smooth function is the *cubic spline*. It has separate cubic polynomials over sets of disjoint intervals, joined together smoothly at boundaries of those intervals.

Like GLMs, this model specifies a distribution for the random component and a link function g. The resulting model is called a *generalized additive model*, symbolized by GAM (Hastie and Tibshirani 1990). The GLM is the special case in which each s_j is a linear function. Also possible is taking some s_j as smooth functions and others as linear functions or as dummy variables for qualitative predictors.

The details for fitting GAMs are beyond our scope. The fitting algorithm employs a generalization of the Newton–Raphson method that utilizes local smoothing. This corresponds to subtracting from the log-likelihood function a penalty function that increases as the smooth function gets more wiggly. The model fit assigns a deviance and an approximate df value to each s_j in the additive predictor, enabling inference about those terms. For instance, a smooth function having df = 5 is similar in overall complexity to a fourth-degree polynomial, which has five parameters. One's choice of a df value (or smoothing parameter) determines how smooth the resulting GAM fit looks.

It is usually worth trying a variety of degrees of smoothing to find one that smooths the data sufficiently so that the trend is not too irregular but does not smooth so much that it suppresses interesting patterns. This approach may suggest that a linear model is adequate with a particular link or suggest ways to improve on linearity. Some software packages that do not have GAMs can smooth the data by employing a type of regression that gives greater weight to nearby observations in predicting the value at a given point; such *locally weighted least squares regression* is often referred to as *lowess*. We prefer GAMs because they recognize explicitly the form of the response. For instance, with a binary response, lowess can give predicted values below 0 or above 1, which cannot happen with a GAM.

Even when one plans to use GLMs, a GAM can be helpful for exploratory analysis. For instance, for continuous X with continuous responses, scatter diagrams provide visual information about the dependence of Y on X. For binary responses, the following example shows that such diagrams are not very informative. Plotting the fitted smooth function for a predictor may reveal a general trend without assuming a particular functional relationship.

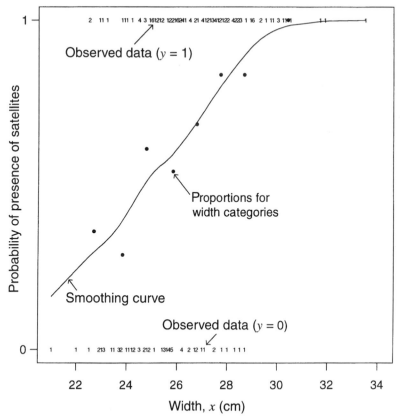

FIGURE 4.7 Whether satellites are present (1, yes; 0, no), by width of female crab, with smoothing fit of generalized additive model.

4.8.2 GAMs for Horseshoe Crab Example

In Section 4.3.2, Figure 4.4 showed the trend relating number of satellites for horseshoe crabs to their width. This smooth curve is the fit of a generalized additive model, assuming a Poisson distribution and using the log link.

In the next chapter we'll use logistic regression to model the probability that a crab has at least one satellite. For crab i, let $y_i = 1$ if she has at least one satellite and $y_i = 0$ otherwise. Figure 4.7 plots these data against $x =$ crab width. It consists of a set of points with $y_i = 1$ and a second set of points with $y_i = 0$. The numbered symbols indicate the number of observations at each point. It appears that $y_i = 1$ tends to occur relatively more often at higher x values. Figure 4.7 also shows a curve based on smoothing the data using a GAM, assuming a binomial response and logit link. This curve shows a roughly increasing trend and is more informative than viewing the binary data alone. It suggests that an S-shaped regression function may describe this relationship relatively well.

NOTES

Section 4.1: Generalized Linear Model

4.2. Distribution (4.1) is called a *natural* (or *linear*) exponential family to distinguish it from a more general exponential family that replaces y by $r(y)$ in the exponential term. For other generalizations, see Jørgensen (1987). Books on GLMs and related models, in approximate order of technical level from highest to lowest, are McCullagh and Nelder (1989), Fahrmeir and Tutz (2001), Aitkin et al. (1989), Dobson (2002), and Gill (2000). See also Firth (1991).

Section 4.3: Generalized Linear Models for Counts

4.2. For further discussion of Poisson regression and related models for count data, see Breslow (1984), Cameron and Trivedi (1998), Frome (1983), Hinde (1982), Lawless (1987), and Seeber (1998) and references therein.

Section 4.4: Moments and Likelihood for Generalized Linear Models

4.3. The function $b(\cdot)$ in (4.14) is called the *cumulant function*, since when $a(\phi) = 1$ its derivatives yield the cumulants of the distribution (Jørgensen 1987).

For many GLMs, including Poisson models with log link and binary models with logit link, with full-rank model matrix the Hessian is negative definite and the log likelihood is a strictly concave function. Then ML estimates of model parameters exist and are unique under quite general conditions (Wedderburn 1976).

Section 4.5: Inference for Generalized Linear Models

4.4. The matrix $\mathbf{W}$ used in cov($\hat{\boldsymbol{\beta}}$) [see (4.28)], in the hat matrix for standardized Pearson residuals [see (4.38)], and in Fisher scoring [see (4.40)] is the inverse of the covariance matrix of the linearized form of $g(\mathbf{y})$ (see Section 4.6.3).

McCullagh and Nelder (1989, Chap. 12) discussed model checking for GLMs. For discussions about residuals, see also Green (1984), Pierce and Schafer (1986), Pregibon (1980, 1981), and Williams (1987). Pregibon (1982) showed that the squared standardized Pearson residual is the score statistic for testing whether the observation is an outlier. Davison and Hinkley (1997, Sec. 7.2) discussed bootstrapping in GLMs.

Section 4.6: Fitting Generalized Linear Models

4.5. Fisher (1935b) introduced the Fisher scoring method to calculate ML estimates for probit models. For further discussion of GLM model fitting and the relationship between iterative reweighted least squares and ML estimation, see Green (1984), Jørgensen (1983), McCullagh and Nelder (1989), and Nelder and Wedderburn (1972). Green (1984), Jørgensen (1983), and Palmgren and Ekholm (1987) also discussed this relation for exponential family nonlinear models.

Section 4.7: Quasi-likelihood and Generalized Linear Models

4.6. For more on quasi-likelihood, see Sections 11.4, 12.6.4, and 13.3, Breslow (1984), Cox (1983), Firth (1987), Hinde and Demétrio (1998), McCullagh (1983), McCullagh and Nelder (1989), Nelder and Pregibon (1987), and Wedderburn (1974, 1976). See Heyde (1997) for a theoretical perspective.

Section 4.8: Generalized Additive Models

4.7. Besides GAMs, other nonparametric smoothing methods can describe the dependence of a binary response on a predictor. For instance, see Copas (1983), Lloyd (1999, Chap. 5), and Section 15.3.3 for kernel smoothing and Kauermann and Tutz (2001) for models with random effects.

PROBLEMS

Applications

4.1 In the 2000 U.S. presidential election, Palm Beach County in Florida was the focus of unusual voting patterns (including a large number of illegal double votes) apparently caused by a confusing "butterfly ballot." Many voters claimed that they voted mistakenly for the Reform Party candidate, Pat Buchanan, when they intended to vote for Al Gore. Figure 4.8 shows the total number of votes for Buchanan plotted against the number of votes for the Reform Party candidate in 1996 (Ross Perot), by county in Florida. (For details, see A. Agresti and B. Presnell, *J. Law Public Policy*, Volume 13, Fall 2001, 117–134.)

a. In county i, let π_i denote the proportion of the vote for Buchanan and let x_i denote the proportion of the vote for Perot in 1996. For the linear probability model fitted to all counties except Palm Beach County, $\hat{\pi}_i = -0.0003 + 0.0304x_i$. Give the value of P in the

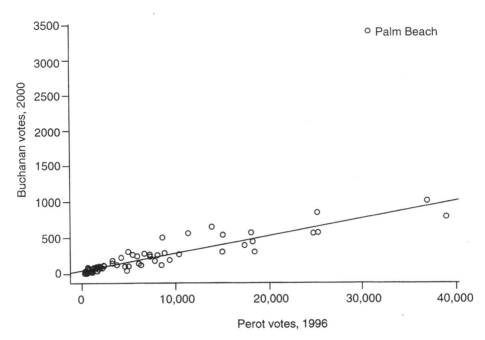

FIGURE 4.8 Total vote, by county in Florida, for Reform Party candidates Buchanan in 2000 and Perot in 1996.

interpretation: The estimated proportion vote for Buchanan in 2000 was roughly $P\%$ of that for Perot in 1996.

b. For Palm Beach County, $\pi_i = 0.0079$ and $x_i = 0.0774$. Does this result appear to be an outlier? Explain.

c. For logistic regression, $\log[\hat{\pi}_i/(1 - \hat{\pi}_i)] = -7.164 + 12.219x_i$. Find $\hat{\pi}_i$ in Palm Beach County. Is that county an outlier for this model?

4.2 For games in baseball's National League during nine decades, Table 4.6 shows the percentage of times that the starting pitcher pitched a complete game.

TABLE 4.6   Data for Problem 4.2

Decade	Percent Complete	Decade	Percent Complete	Decade	Percent Complete
1900–1909	72.7	1930–1939	44.3	1960–1969	27.2
1910–1919	63.4	1940–1949	41.6	1970–1979	22.5
1920–1929	50.0	1950–1959	32.8	1980–1989	13.3

Source: Data from George Will, *Newsweek*, Apr. 10, 1989.

a. Treating the number of games as the same in each decade, the ML fit of the linear probability model is $\hat{\pi} = 0.7578 - 0.0694x$, where x = decade $(x = 1, 2, \ldots, 9)$. Interpret 0.7578 and -0.0694.

b. Substituting $x = 10, 11, 12$, predict the percentages of complete games for the next three decades. Are these predictions plausible? Why?

c. The ML fit with logistic regression is $\hat{\pi} = \exp(1.148 - 0.315x)/[1 + \exp(1.148 - 0.315x)]$. Obtain $\hat{\pi}_i$ for $x = 10, 11, 12$. Are these more plausible?

4.3 For Table 3.7 with scores (0, 0.5, 1.5, 4.0, 7.0) for alcohol consumption, ML fitting of the linear probability model for malformation has output.

Parameter	Estimate	Std Error	Wald 95% Conf Limits	
Intercept	0.0025	0.0003	0.0019	0.0032
Alcohol	0.0011	0.0007	−0.0003	0.0025

Interpret the model fit. Use it to estimate the relative risk of malformation for alcohol consumption levels 0 and 7.0.

4.4 For Table 4.2, refit the linear probability model or the logistic regression model using the scores (**a**) (0, 2, 4, 6), (**b**) (0, 1, 2, 3), and (**c**) (1, 2, 3, 4). Compare $\hat{\beta}$ for the three choices. Compare fitted values. Summarize the effect of linear transformations of scores, which preserve relative sizes of spacings between scores.

4.5 For Table 4.3, let $Y = 1$ if a crab has at least one satellite, and $Y = 0$ otherwise. Using x = weight, fit the linear probability model.

a. Use ordinary least squares. Interpret the parameter estimates. Find the estimated probability at the highest observed weight (5.20 kg). Comment.

b. Try to fit the model using ML, treating Y as binomial. [The failure is due to a fitted probability falling outside the (0, 1) range. The fit in part (a) is ML for a normal random component, for which fitted values outside this range are permissible.]

c. Fit the logistic regression model. Show that the fitted probability at a weight of 5.20 kg equals 0.9968.

d. Fit the probit model. Find the fitted probability at 5.20 kg.

4.6 An experiment analyzes imperfection rates for two processes used to fabricate silicon wafers for computer chips. For treatment A applied to 10 wafers, the numbers of imperfections are 8, 7, 6, 6, 3, 4, 7, 2, 3, 4. Treatment B applied to 10 other wafers has 9, 9, 8, 14, 8, 13, 11, 5, 7, 6 imperfections. Treat the counts as independent Poisson variates having means μ_A and μ_B.

a. Fit the model $\log \mu = \alpha + \beta x$, where $x = 1$ for treatment B and $x = 0$ for treatment A. Show that $\exp(\beta) = \mu_B/\mu_A$, and interpret its estimate.

b. Test $H_0: \mu_A = \mu_B$ with the Wald or likelihood ratio test of $H_0: \beta = 0$. Interpret.

c. Construct a 95% confidence interval for μ_B/μ_A. (*Hint:* First construct one for β.)

d. Test $H_0: \mu_A = \mu_B$ based on this result: If Y_1 and Y_2 are independent Poisson with means μ_1 and μ_2, then $(Y_1|Y_1 + Y_2)$ is binomial with $n = Y_1 + Y_2$ and $\pi = \mu_1/(\mu_1 + \mu_2)$.

4.7 For Table 4.3, Table 4.7 shows SAS output for a Poisson loglinear model fit using X = weight and Y = number of satellites.

a. Estimate $E(Y)$ for female crabs of average weight, 2.44 kg.

b. Use $\hat{\beta}$ to describe the weight effect. Show how to construct the reported confidence interval.

c. Construct a Wald test that Y is independent of X. Interpret.

d. Can you conduct a likelihood-ratio test of this hypothesis? If not, what else do you need?

e. Is there evidence of overdispersion? If necessary, adjust standard errors and interpret.

TABLE 4.7   SAS Output for Problem 4.7

Criterion	DF	Value
Deviance	171	560.8664
Pearson Chi-Square	171	535.8957
Log Likelihood		71.9524

Parameter	Estimate	Std Error	Wald 95% Conf Limits		Chi-Sq	Pr > ChiSq
Intercept	−0.4284	0.1789	−0.7791	−0.0777	5.73	0.0167
weight	0.5893	0.0650	0.4619	0.7167	82.15	<.0001

4.8 Refer to Problem 4.7. Using the identity link with x = weight, $\hat{\mu} = -2.60 + 2.264x$, where $\hat{\beta} = 2.264$ has SE = 0.228. Repeat parts (a) through (c).

4.9 Refer to Table 4.3.

a. Fit a Poisson loglinear model using both W = weight and C = color to predict Y = number of satellites. Assigning dummy variables, treat C as a nominal factor. Interpret parameter estimates.

    b. Estimate $E(Y)$ for female crabs of average weight (2.44 kg) that are (i) medium light, and (ii) dark.

    c. Test whether color is needed in the model. (*Hint:* From Section 4.5.4, the likelihood-ratio statistic comparing models is the difference in deviances.)

    d. The estimated color effects are monotone across the four categories. Fit a simpler model that treats C as quantitative and assumes a linear effect. Interpret its color effect and repeat the analyses of parts (b) and (c). Compare the fit to the model in part (a). Interpret.

    e. Add width to the model. What effect does the strong positive correlation between width and weight have? Are both needed in the model?

4.10 In Section 4.3.2, refer to the Poisson model with identity link. The fit using least squares is $\hat{\mu} = -10.42 + 0.51x$ (SE = 0.11). Explain why the parameter estimates differ and why the SE values are so different.

4.11 For the negative binomial model fitted to the crab satellite counts with log link and width predictor, $\hat{\alpha} = -4.05$, $\hat{\beta} = 0.192$ (SE = 0.048), $\hat{k}^{-1} = 1.106$ (SE = 0.197). Interpret. Why is SE for $\hat{\beta}$ so different from SE = 0.020 for the corresponding Poisson GLM in Sec 4.3.2? Which is more appropriate? Why?

4.12 Refer to Problem 4.6. The sample mean and variance are 5.0 and 4.2 for treatment A and 9.0 and 8.4 for treatment B.

    a. Is there evidence of overdispersion for the Poisson model having a dummy variable for treatment? Explain.

    b. Fit the negative binomial loglinear model. Note that the estimated dispersion parameter is 0 and that estimates of treatment means and standard errors are the same as with the Poisson loglinear GLM.

    c. For the overall sample of 20 observations, the sample mean and variance are 7.0 and 10.2. Fit the loglinear model having only an intercept term under Poisson and negative binomial assumptions. Compare results, and compare confidence intervals for the overall mean response. Why do they differ? (*Note:* This shows how the Poisson model can deteriorate when an important covariate is unmeasured.)

4.13 Table 4.8 shows the free-throw shooting, by game, of Shaq O'Neal of the Los Angeles Lakers during the 2000 NBA (basketball) playoffs. Commentators remarked that his shooting varied dramatically from game to game. In game i, suppose that Y_i = number of free throws

TABLE 4.8 Data for Problem 4.13

Game	Number Made	Number of Attempts	Game	Number Made	Number of Attempts	Game	Number Made	Number of Attempts
1	4	5	9	4	12	17	8	12
2	5	11	10	1	4	18	1	6
3	5	14	11	13	27	19	18	39
4	5	12	12	5	17	20	3	13
5	2	7	13	6	12	21	10	17
6	7	10	14	9	9	22	1	6
7	6	14	15	7	12	23	3	12
8	9	15	16	3	10			

Source: www.nba.com.

made out of n_i attempts is a $\text{bin}(n_i, \pi_i)$ variate and the $\{Y_i\}$ are independent.

a. Fit the model, $\pi_i = \alpha$, and find and interpret $\hat{\alpha}$ and its standard error. Does the model appear to fit adequately? (*Note:* You could check this with a small-sample test of independence of the 23×2 table of game and the binary outcome.)

b. Adjust the standard error for overdispersion. Using the original SE and its correction, find and compare 95% confidence intervals for α. Interpret.

4.14 Refer to Table 13.6. Fit a loglinear model with a dummy variable for race, (**a**) assuming a Poisson distribution, and (**b**) allowing overdispersion with a quasi-likelihood approach. Compare results.

4.15 Refer to Problem 4.6. The wafers are also classified by thickness of silicon coating ($z = 0$, low; $z = 1$, high). The first five imperfection counts reported for each treatment refer to $z = 0$ and the last five refer to $z = 1$. Analyze these data.

14.6 Refer to Table 13.9 on frequency of sexual intercourse. Analyze these data.

Theory and Methods

4.17 Describe the purpose of the link function of a GLM. What is the identity link? Explain why it is not often used with binomial or Poisson responses.

4.18 For known k, show that the negative binomial distribution (4.12) has exponential family form (4.1) with natural parameter $\log[\mu/(\mu + k)]$.

4.19 For binary data, define a GLM using the log link. Show that effects refer to the relative risk. Why do you think this link is not often used? (*Hint:* What happens if the linear predictor takes a positive value?)

4.20 For the logistic regression model (4.6) with $\beta > 0$, show that (**a**) as $x \to \infty$, $\pi(x)$ is monotone increasing, and (**b**) the curve for $\pi(x)$ is the cdf of a logistic distribution having mean $-\alpha/\beta$ and standard deviation $\pi/(|\beta|\sqrt{3})$.

4.21 Show representation (4.18) for the binomial distribution.

4.22 Let Y_i be a bin(n_i, π_i) variate for group i, $i = 1, \ldots, N$, with $\{Y_i\}$ independent. Consider the model that $\pi_1 = \cdots = \pi_N$. Denote that common value by π. For observations $\{y_i\}$, show that $\hat{\pi} = (\Sigma y_i)/(\Sigma n_i)$. When all $n_i = 1$, for testing this model's fit in the $N \times 2$ table, show that $X^2 = n$. Thus, goodness-of-fit statistics can be completely uninformative for ungrouped data. (See also Problem 5.37.)

4.23 Suppose that Y_i is Poisson with $g(\mu_i) = \alpha + \beta x_i$, where $x_i = 1$ for $i = 1, \ldots, n_A$ from group A and $x_i = 0$ for $i = n_A + 1, \ldots, n_A + n_B$ from group B. Show that for any link function g, the likelihood equations (4.22) imply that fitted means $\hat{\mu}_A$ and $\hat{\mu}_B$ equal the sample means.

4.24 For binary data with sample proportion y_i based on n_i trials, we use quasi-likelihood to fit a model using variance function (4.46). Show that parameter estimates are the same as for the binomial GLM but that the covariance matrix multiplies by ϕ.

4.25 A binomial GLM $\pi_i = \Phi(\Sigma_j \beta_j x_{ij})$ with arbitrary inverse link function Φ assumes that $n_i Y_i$ has a bin(n_i, π_i) distribution. Find w_i in (4.27) and hence $\widehat{\text{cov}}(\hat{\beta})$. For logistic regression, show that $w_i = n_i \pi_i(1 - \pi_i)$.

4.26 A GLM has parameter β with sufficient statistic S. A goodness-of-fit test statistic T has observed value t_o. If β were known, a P-value is $P = P(T \geq t_o; \beta)$. Explain why $P(T \geq t_o | S)$ is the uniform minimum variance unbiased estimator of P.

4.27 Let y_{ij} be observation j of a count variable for group $i, i = 1, \ldots, I$, $j = 1, \ldots, n_i$. Suppose that $\{Y_{ij}\}$ are independent Poisson with $E(Y_{ij}) = \mu_i$.
  a. Show that the ML estimate of μ_i is $\hat{\mu}_i = \bar{y}_i = \Sigma_j y_{ij}/n_i$.
  b. Simplify the expression for the deviance for this model. [For testing this model, it follows from Fisher (1970, p. 58, originally published

in 1925) that the deviance and the Pearson statistic $\Sigma_i \Sigma_j (y_{ij} - \bar{y}_i)^2 / \bar{y}_i$ have approximate chi-squared distributions with df $= \Sigma_i(n_i - 1)$. For a single group, Cochran (1954) referred to $\Sigma_j(y_{1j} - \bar{y}_1)^2 / \bar{y}_1$ as the *variance test* for the fit of a Poisson distribution, since it compares the sample variance to the estimated Poisson variance $\bar{y}_1$.]

4.28 Conditional on λ, Y has a Poisson distribution with mean λ. Values of λ vary according to gamma density (13.12), which has $E(\lambda) = \mu$, $\text{var}(\lambda) = \mu^2/k$. Show that marginally Y has the negative binomial distribution (4.12). Explain why the negative binomial model is a way to handle overdispersion for the Poisson.

4.29 Consider the class of binary models (4.8) and (4.9). Suppose that the standard cdf Φ corresponds to a probability density function ϕ that is symmetric around 0.
   a. Show that x at which $\pi(x) = 0.5$ is $x = -\alpha/\beta$.
   b. Show that the rate of change in $\pi(x)$ when $\pi(x) = 0.5$ is $\beta\phi(0)$. Show this is 0.25β for the logit link and $\beta/\sqrt{2\pi}$ (where $\pi = 3.14\ldots$) for the probit link.
   c. Show that the probit regression curve has the shape of a normal cdf with mean $-\alpha/\beta$ and standard deviation $1/|\beta|$.

4.30 Show the normal distribution $N(\mu, \sigma^2)$ with fixed σ satisfies family (4.1), and identify the components. Formulate the ordinary regression model as a GLM.

4.31 In Problem 4.30, when σ is also a parameter, show that it satisfies the exponential dispersion family (4.14).

4.32 For binary observations, consider the model $\pi(x) = \frac{1}{2} + (1/\pi)\tan^{-1}(\alpha + \beta x)$. Which distribution has cdf of this form? Explain when a GLM using this curve might be more appropriate than logistic regression.

4.33 Find the form of the deviance residual (4.35) for an observation in a **(a)** binomial GLM, and **(b)** Poisson GLM. Illustrate part (b) for a cell count in a two-way contingency table for the model of independence.

4.34 Consider the value $\hat{\beta}$ that maximizes a function $L(\beta)$. Let $\beta^{(0)}$ denote an initial guess.
   a. Using $L'(\hat{\beta}) = L'(\beta^{(0)}) + (\hat{\beta} - \beta^{(0)})L''(\beta^{(0)}) + \cdots$, argue that for $\beta^{(0)}$ close to $\hat{\beta}$, approximately $0 = L'(\beta^{(0)}) + (\hat{\beta} - \beta^{(0)})L''(\beta^{(0)})$. Solve this equation to obtain an approximation $\beta^{(1)}$ for $\hat{\beta}$.

b. Let $\beta^{(t)}$ denote approximation t for $\hat{\beta}$, $t = 0, 1, 2, \ldots$. Justify that the next approximation is

$$\beta^{(t+1)} = \beta^{(t)} - L'(\beta^{(t)})/L''(\beta^{(t)}).$$

4.35 For n independent observations from a Poisson distribution, show that Fisher scoring gives $\mu^{(t+1)} = \bar{y}$ for all $t > 0$. By contrast, what happens with Newton–Raphson?

4.36 Write a computer program using the Newton–Raphson algorithm to maximize the likelihood for a binomial sample. For $\hat{\pi} = 0.3$ based on $n = 10$, print out results of the first six iterations when the starting value $\pi^{(0)}$ is **(a)** 0.1, **(b)** 0.2, $\ldots$, **(i)** 0.9. Summarize the effects of the starting value on speed of convergence. What happens if it is 0 or 1?

4.37 In a GLM, suppose that var$(Y) = v(\mu)$ for $\mu = E(Y)$. Show that the link g satisfying $g'(\mu) = [v(\mu)]^{-1/2}$ has the same weight matrix $\mathbf{W}^{(t)}$ at each cycle. Show this link for a Poisson random component is $g(\mu) = 2\sqrt{\mu}$.

4.38 For noncanonical links in a GLM, show that the observed information matrix may depend on the data and hence differs from the expected information. Illustrate using the probit model.

CHAPTER 5

Logistic Regression

In introducing generalized linear models for binary data in Chapter 4 we highlighted logistic regression. This is the most important model for categorical response data. It is used increasingly in a wide variety of applications. Early uses were in biomedical studies but the past 20 years have also seen much use in social science research and marketing.

Recently, logistic regression has become a popular tool in business applications. Some *credit-scoring* applications use logistic regression to model the probability that a subject is credit worthy. For instance, the probability that a subject pays a bill on time may use predictors such as the size of the bill, annual income, occupation, mortgage and debt obligations, percentage of bills paid on time in the past, and other aspects of an applicant's credit history. A company that relies on catalog sales may determine whether to send a catalog to a potential customer by modeling the probability of a sale as a function of indices of past buying behavior.

Another area of increasing application is genetics. For instance, one recent article (J. M. Henshall and M. E. Goddard, *Genetics* 151:885–894, 1999) used logistic regression to estimate quantitative trait loci effects, modeling the probability that an offspring inherits an allele of one type instead of another type as a function of phenotypic values on various traits for that offspring. Another recent article (D. F. Levinson et al., *Amer. J. Hum. Genet.*, 67:652–663, 2000) used logistic regression for analysis of the genotype data of affected sibling pairs (ASPs) and their parents from several research centers. The model studied the probability that ASPs have identity-by-descent allele sharing and tested its heterogeneity among the centers.

In this chapter we study logistic regression more closely. Section 5.1 covers parameter interpretation. In Section 5.2 we present inferential methods for those parameters. Sections 5.3 and 5.4 generalize to multiple predictors, some of which may be qualitative. Finally, in Section 5.5 we apply GLM model-fitting methods to determine and solve likelihood equations for logistic regression.

165

5.1 INTERPRETING PARAMETERS IN LOGISTIC REGRESSION

For a binary response variable Y and an explanatory variable X, let $\pi(x) = P(Y = 1 \mid X = x) = 1 - P(Y = 0 \mid X = x)$. The logistic regression model is

$$\pi(x) = \frac{\exp(\alpha + \beta x)}{1 + \exp(\alpha + \beta x)}. \tag{5.1}$$

Equivalently, the log odds, called the *logit*, has the linear relationship

$$\text{logit}[\pi(x)] = \log \frac{\pi(x)}{1 - \pi(x)} = \alpha + \beta x. \tag{5.2}$$

This equates the logit link function to the linear predictor.

5.1.1 Interpreting β: Odds, Probabilities, and Linear Approximations

How can we interpret β in (5.2)? Its sign determines whether $\pi(x)$ is increasing or decreasing as x increases. The rate of climb or descent increases as $|\beta|$ increases; as $\beta \to 0$ the curve flattens to a horizontal straight line. When $\beta = 0$, Y is independent of X. For quantitative x with $\beta > 0$, the curve for $\pi(x)$ has the shape of the cdf of the logistic distribution (recall Section 4.2.5). Since the logistic density is symmetric, $\pi(x)$ approaches 1 at the same rate that it approaches 0.

Exponentiating both sides of (5.2) shows that the odds are an exponential function of x. This provides a basic interpretation for the magnitude of β: The odds increase multiplicatively by e^β for every 1-unit increase in x. In other words, e^β is an odds ratio, the odds at $X = x + 1$ divided by the odds at $X = x$.

Most scientists are not familiar with odds or logits, so the interpretation of a multiplicative effect of e^β on the odds scale or an additive effect of β on the logit scale is not helpful to them. A simpler, although approximate slope interpretation uses a linearization argument (Berkson 1951). Since it has a curved rather than a linear appearance, the logistic regression function (5.1) implies that the rate of change in $\pi(x)$ per unit change in x varies. A straight line drawn tangent to the curve at a particular x value, shown in Figure 5.1, describes the rate of change at that point. Calculating $\partial\pi(x)/\partial x$ using (5.1) yields a fairly complex function of the parameters and x, but it simplifies to the form $\beta\pi(x)[1 - \pi(x)]$.

For instance, the line tangent to the curve at x for which $\pi(x) = \frac{1}{2}$ has slope $\beta(\frac{1}{2})(\frac{1}{2}) = \beta/4$; when $\pi(x) = 0.9$ or 0.1, it has slope 0.09β. The slope approaches 0 as $\pi(x)$ approaches 1.0 or 0. The steepest slope occurs at x for which $\pi(x) = \frac{1}{2}$; that x value is $x = -\alpha/\beta$. [To check that $\pi(x) = \frac{1}{2}$ at this

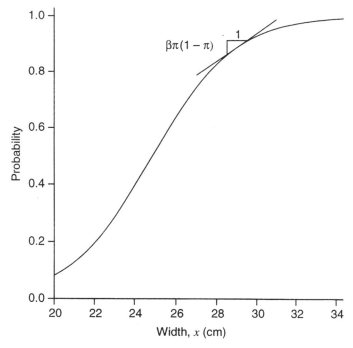

FIGURE 5.1   Linear approximation to logistic regression curve.

point, substitute $-\alpha/\beta$ for x in (5.1), or substitute $\pi(x) = \frac{1}{2}$ in (5.2) and solve for x.] This x value is sometimes called the *median effective level* and denoted EL_{50}. In toxicology studies it is called LD_{50} (LD = lethal dose), the dose with a 50% chance of a lethal result.

From this linear approximation, near x where $\pi(x) = \frac{1}{2}$, a change in x of $1/\beta$ corresponds to a change in $\pi(x)$ of roughly $(1/\beta)(\beta/4) = \frac{1}{4}$; that is, $1/\beta$ approximates the distance between x values where $\pi(x) = 0.25$ or 0.75 (in reality, 0.27 and 0.73) and where $\pi(x) = 0.50$. The linear approximation works better for smaller changes in x, however.

An alternative way to interpret the effect reports the values of $\pi(x)$ at certain x values, such as their quartiles. This entails substituting those quartiles for x into formula (5.1) for $\pi(x)$. The change in $\pi(x)$ over the middle half of x values, from the lower quartile to the upper quartile of x, then describes the effect. It can be compared to the corresponding change over the middle half of values of other predictors.

The intercept parameter α is not usually of particular interest. However, by centering the predictor about 0 [i.e., replacing x by $(x - \bar{x})$], α becomes the logit at that mean, and thus $e^{\alpha}/(1 + e^{\alpha}) = \pi(\bar{x})$. (As in ordinary regression, centering is also helpful in complex models containing quadratic or interaction terms to reduce correlations among model parameter estimates.)

5.1.2 Looking at the Data

In practice, these interpretations use formula (5.1) with ML estimates substituted for parameters. Before fitting the model and making such interpretations, look at the data to check that the logistic regression model is appropriate. Since Y takes only values 0 and 1, it is difficult to check this by plotting Y against x.

It can be helpful to plot sample proportions or logits against x. Let n_i denote the number of observations at setting i of x. Of them, let y_i denote the number of "1" outcomes, with $p_i = y_i/n_i$. Sample logit i is $\log[p_i/(1 - p_i)] = \log[y_i/(n_i - y_i)]$. This is not finite when $y_i = 0$ or n_i. An ad hoc adjustment adds a positive constant to the number of outcomes of the two types. The adjustment

$$\log \frac{y_i + \frac{1}{2}}{n_i - y_i + \frac{1}{2}}$$

is the least-biased estimator of this form of the true logit (Note 5.2). The plot of sample logits should be roughly linear.

When X is continuous and all $n_i = 1$, or when it is essentially continuous and all n_i are small, this is unsatisfactory. One could group the data with nearby x values into categories before calculating sample proportions and sample logits. A better approach that does not require choosing arbitrary categories uses a smoothing mechanism to reveal trends. One such smoothing approach fits a generalized additive model (Section 4.8), which replaces the linear predictor of a GLM by a smooth function. Inspect a plot of the fit to see if severe discrepancies occur from the S-shaped trend predicted by logistic regression.

5.1.3 Horseshoe Crabs Revisited

To illustrate logistic regression, we reanalyze the horseshoe crab data introduced in Section 4.3.2. The binary response is whether a female crab has any male crabs residing nearby (satellites): $Y = 1$ if she has at least one satellite, and $Y = 0$ if she has none. We first use as a predictor the female crab's width.

Figure 4.7 plotted the data and showed the smoothed prediction of the mean provided by a generalized additive model (GAM), assuming a binomial response and logit link. The logistic regression model appears to be adequate. This is also suggested by the grouping of the data used to investigate the adequacy of Poisson regression models in Section 4.3.2 (Table 4.4). In each of the eight width categories, we computed the sample proportion of crabs having satellites and the mean width for the crabs in that category. Figure 5.2 shows eight dots representing the sample proportions of female crabs having satellites plotted against the mean widths for the eight cate-

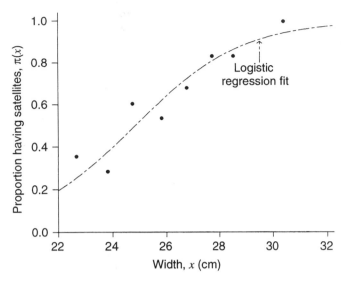

FIGURE 5.2 Observed and fitted proportions of satellites by width of female crab.

gories. The eight plotted sample proportions and the GAM smoothing curve both show a roughly increasing trend, so we proceed with fitting the logistic regression model with linear width predictor.

We defer to Section 5.5 details about ML fitting. Software (e.g., for SAS see Table A.8) reports output such as Table 5.1 exhibits. For the ungrouped data from Table 4.3, let $\pi(x)$ denote the probability that a female horseshoe crab of width x has a satellite. The ML fit is

$$\hat{\pi}(x) = \frac{\exp(-12.351 + 0.497x)}{1 + \exp(-12.351 + 0.497x)}.$$

TABLE 5.1 Computer Output for Logistic Regression Model with Horseshoe Crab Data

Criteria For Assessing Goodness Of Fit		
Criterion	DF	Value
Deviance	171	194.4527
Pearson Chi-Square	171	165.1434
Log Likelihood		-97.2263

Parameter	Estimate	Std Error	Likelihood-Ratio 95% Conf Limits		Wald Chi-Sq	P>ChiSq
Intercept	-12.3508	2.6287	-17.8097	-7.4573	22.07	<.0001
width	0.4972	0.1017	0.3084	0.7090	23.89	<.0001

Substituting $x = 26.3$ cm, the mean width level in this sample, $\hat{\pi}(x) = 0.674$. The estimated probability equals $\frac{1}{2}$ when $x = -\hat{\alpha}/\hat{\beta} = 12.351/0.497 = 24.8$. Figure 5.2 plots $\hat{\pi}(x)$ against width.

The estimated odds of a satellite multiply by $\exp(\hat{\beta}) = \exp(0.497) = 1.64$ for each 1-cm increase in width; that is, there is a 64% increase. To convey the effect less technically, we could report the incremental rate of change in the probability of a satellite. At the mean width, $\hat{\pi}(x) = 0.674$, and $\hat{\pi}(x)$ increases by about $\hat{\beta}[\hat{\pi}(x)(1 - \hat{\pi}(x))] = 0.497(0.674)(0.326) = 0.11$ for a 1-cm increase in width. Or, we could report $\hat{\pi}(x)$ at the quartiles of x. The lower quartile, median, and upper quartile for width are 24.9, 26.1, and 27.7; $\hat{\pi}(x)$ at those values equals 0.51, 0.65, and 0.81, increasing by 0.30 over the x values for the middle half of the sample.

The latter summary is useful for comparing the effects of predictors having different units. For instance, with crab weight as the predictor, $\text{logit}[\hat{\pi}(x)] = -3.695 + 1.815x$. A 1-kg increase in weight is not comparable to a 1-cm increase in width, so $\hat{\beta} = 0.497$ for $x = $ width is not comparable to $\hat{\beta} = 1.815$ for $x = $ weight. The quartiles for weight are 2.00, 2.35, and 2.85; $\hat{\pi}(x)$ at those values are 0.48, 0.64, and 0.81, increasing by 0.33 over the middle half of the sampled weights. The effect is similar to that of width.

5.1.4 Logistic Regression with Retrospective Studies

Another property of logistic regression relates to situations in which the explanatory variable X rather than the response variable Y is random. This occurs with retrospective sampling designs, such as case–control biomedical studies (Section 2.1.6). For samples of subjects having $Y = 1$ (cases) and having $Y = 0$ (controls), the value of X is observed. Evidence exists of an association if the distribution of X values differs between cases and controls. In retrospective studies, one can estimate odds ratios (Section 2.2.4). Effects in the logistic regression model refer to odds ratios. Thus, one can fit such models and estimate effects in case–control studies.

Here is a justification for this. Let Z indicate whether a subject is sampled $(1 = \text{yes}, 0 = \text{no})$. Let $\rho_1 = P(Z = 1 | y = 1)$ denote the probability of sampling a case, and let $\rho_0 = P(Z = 1 | y = 0)$ denote the probability of sampling a control. Even though the conditional distribution of Y given $X = x$ is not sampled, we need a model for $P(Y = 1 | z = 1, x)$, assuming that $P(Y = 1 | x)$ follows the logistic model. By Bayes' theorem,

$$P(Y = 1 | z = 1, x) = \frac{P(Z = 1 | y = 1, x) P(Y = 1 | x)}{\sum_{j=0}^{1} P(Z = 1 | y = j, x) P(Y = j | x)}. \quad (5.3)$$

Now, suppose that $P(Z = 1 | y, x) = P(Z = 1 | y)$ for $y = 0$ and 1; that is, for each y, the sampling probabilities do not depend on x. For instance, often x

refers to exposure of some type, such as whether someone has been a smoker. Then, for cases and for controls, the probability of being sampled is the same for smokers and nonsmokers. Under this assumption, substituting ρ_1 and ρ_0 in (5.3) and dividing numerator and denominator by $P(Y = 0|x)$, (5.3) simplifies to

$$P(Y = 1|z = 1, x) = \frac{\rho_1 \exp(\alpha + \beta x)}{\rho_0 + \rho_1 \exp(\alpha + \beta x)}.$$

Then, dividing numerator and denominator by ρ_0 and using $\rho_1/\rho_0 = \exp[\log(\rho_1/\rho_0)]$ yields

$$\text{logit}[P(Y = 1|z = 1, x)] = \alpha^* + \beta x$$

with $\alpha^* = \alpha + \log(\rho_1/\rho_0)$.

Thus, the logistic regression model holds with the same effect parameter β as in the model for $P(Y = 1|x)$. If the sampling rate for cases is 10 times that for controls, the intercept estimated is $\log(10) = 2.3$ larger than the one estimated with a prospective study. For related comments, see Anderson (1972), Breslow and Day (1980, p. 203), Breslow and Powers (1978), Carroll et al. (1995), Farewell (1979), Mantel (1973), Prentice (1976a), and Prentice and Pyke (1979).

With case–control studies, one cannot estimate β in other binary-response models. Unlike the odds ratio, the effect for the conditional distribution of X given Y does not then equal that for Y given X. This is an important advantage of the logit link and is a major reason why logit models have surpassed other models in popularity in biomedical studies.

Many case–control studies employ matching. Each case is matched with one or more control subjects. The controls are like the case on key character-istics such as age. The model and subsequent analysis should take the matching into account. In Section 10.2.5 we discuss logistic regression for matched case–control studies.

Regardless of the sampling mechanism, logistic regression may or may not describe a relationship well. In one special case, it necessarily holds. Given that $Y = i$, suppose that X has $N(\mu_i, \sigma^2)$ distribution, $i = 0, 1$. Then, by Bayes' theorem, $P(Y = 1|X = x)$ equals (5.1) with $\beta = (\mu_1 - \mu_0)/\sigma^2$ (Cornfield 1962). When a population is a mixture of two types of subjects, one type with $Y = 1$ that is approximately normally distributed on X and the other type with $Y = 0$ that is approximately normal on X with similar variance, the logistic regression function (5.1) approximates well the curve for $\pi(x)$. If the distributions are normal but with different variances, the model applies also having a quadratic term (Anderson 1975). In that case, the relationship is nonmonotone, with $\pi(x)$ increasing and then decreasing, or the reverse (Problem 5.33).

5.2   INFERENCE FOR LOGISTIC REGRESSION

By Wald's (1943) asymptotic results for ML estimators, parameter estimators in logistic regression models have large-sample normal distributions. Thus, inference can use the (Wald, likelihood-ratio, score) triad of methods (Section 1.3.3).

5.2.1   Types of Inference

For the model with a single predictor,

$$\text{logit}[\pi(x)] = \alpha + \beta x,$$

significance tests focus on H_0: $\beta = 0$, the hypothesis of independence. The Wald test uses the log likelihood at $\hat{\beta}$, with test statistic $z = \hat{\beta}/\text{SE}$ or its square; under H_0, z^2 is asymptotically χ_1^2. The likelihood-ratio test uses twice the difference between the maximized log likelihood at $\hat{\beta}$ and at $\beta = 0$ and also has an asymptotic χ_1^2 null distribution. The score test uses the log likelihood at $\beta = 0$ through the derivative of the log likelihood (i.e., the score function) at that point. The test statistic compares the sufficient statistic for β to its null expected value, suitably standardized [$N(0,1)$ or χ_1^2]. In Section 5.3.5 present this test of H_0: $\beta = 0$.

   For large samples, the three tests usually give similar results. The likelihood-ratio test is preferred over the Wald. It uses more information, since it incorporates the log likelihood at H_0 as well as at $\hat{\beta}$. When $|\beta|$ is relatively large, the Wald test is not as powerful as the likelihood-ratio test and can even show aberrant behavior [see Hauck and Donner (1977) and Problem 5.38].

   Confidence intervals are more informative than tests. An interval for β results from inverting a test of H_0: $\beta = \beta_0$. The interval is the set of β_0 for which the chi-squared test statistic is no greater than $\chi_1^2(\alpha) = z_{\alpha/2}^2$. For the Wald approach, this means $[(\hat{\beta} - \beta_0)/\text{SE}]^2 \leq z_{\alpha/2}^2$; the interval is $\hat{\beta} \pm z_{\alpha/2}(\text{SE})$.

   For summarizing the relationship, other characteristics may have greater importance than β, such as $\pi(x)$ at various x values. For fixed $x = x_0$, $\text{logit}[\hat{\pi}(x_0)] = \hat{\alpha} + \hat{\beta}x_0$ has a large-sample SE given by the estimated square root of

$$\text{var}(\hat{\alpha} + \hat{\beta}x_0) = \text{var}(\hat{\alpha}) + x_0^2\,\text{var}(\hat{\beta}) + 2x_0\,\text{cov}(\hat{\alpha}, \hat{\beta}).$$

A 95% confidence interval for $\text{logit}[\pi(x_0)]$ is $(\hat{\alpha} + \hat{\beta}x_0) \pm 1.96\,\text{SE}$. Substituting each endpoint into the inverse transformation $\pi(x_0) = \exp(\text{logit})/[1 + \exp(\text{logit})]$ gives a corresponding interval for $\pi(x_0)$.

   Each method of inference can also produce small-sample confidence intervals and tests. We defer discussion of this until Section 6.7.

5.2.2 Inference for Horseshoe Crab Data

We illustrate logistic regression inferences with the model for the probability a horseshoe crab has a satellite, with width as the predictor. Table 5.1 showed the fit and standard errors. The statistic $z = \hat{\beta}/\text{SE} = 0.497/0.102 = 4.9$ provides strong evidence of a positive width effect ($P < 0.0001$). The equivalent Wald chi-squared statistic, $z^2 = 23.9$, has df = 1. The maximized log likelihoods equal -112.88 under H_0: $\beta = 0$ and -97.23 for the full model. The likelihood-ratio statistic equals $-2(-112.88 - 97.23) = 31.3$, with df = 1. This provides even stronger evidence than the Wald test.

The Wald 95% confidence interval for β is $0.497 \pm 1.96(0.102)$, or $(0.298, 0.697)$. Table 5.1 reports a likelihood-ratio confidence interval of $(0.308, 0.709)$, based on the profile likelihood function. The confidence interval for the effect on the odds per 1-cm increase in width equals $(e^{0.308}, e^{0.709}) = (1.36, 2.03)$. We infer that a 1-cm increase in width has at least a 36% increase and at most a doubling in the odds of a satellite.

Most software for logistic regression also reports estimates and confidence intervals for $\pi(x)$ (e.g., PROC GENMOD in SAS with the OBSTATS option). Consider this for crabs of width $x = 26.5$, near the mean width. The estimated logit is $-12.351 + 0.497(26.5) = 0.825$, and $\hat{\pi}(x) = 0.695$. Software reports

$$\widehat{\text{var}}(\hat{\alpha}) = 6.910, \qquad \widehat{\text{var}}(\hat{\beta}) = 0.01035, \qquad \widehat{\text{cov}}(\hat{\alpha}, \hat{\beta}) = -0.2668,$$

from which

$$\widehat{\text{var}}\{\text{logit}[\hat{\pi}(x)]\} = 6.910 + x^2(0.01035) + 2x(-0.2668).$$

At $x = 26.5$ this is 0.038, so the 95% confidence interval for $\text{logit}[\pi(26.5)]$ equals $0.825 \pm (1.96)\sqrt{0.038}$, or $(0.44, 1.21)$. This translates to the interval $(0.61, 0.77)$ for the probability of satellites (e.g., $\exp(0.44)/[1 + \exp(0.44)] = 0.61$). (Alternatively, for the model fit using predictor $x^* = x - 26.5$, $\hat{\alpha}$ and its SE are the estimated logit and its SE.) Figure 5.3 plots the confidence bands around the prediction equation for $\pi(x)$ as a function of x. Hauck (1983) gave alternative bands for which the confidence coefficient applies simultaneously to all possible predictor values.

One could ignore the model fit and simply use sample proportions (i.e., the saturated model) to estimate such probabilities. Six female crabs in the sample had $x = 26.5$, and four of them had satellites. The sample proportion estimate at $x = 26.5$ is $\hat{\pi} = 4/6 = 0.67$, similar to the model-based estimate. The 95% score confidence interval (Section 1.4.2) based on these six observations alone equals $(0.30, 0.90)$.

When the logistic regression model truly holds, the model-based estimator of a probability is considerably better than the sample proportion. The model has only two parameters to estimate, whereas the saturated model has a

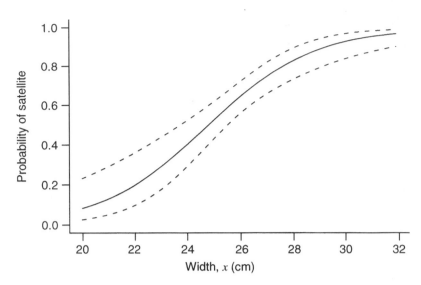

FIGURE 5.3 Prediction equation and 95% confidence bands for probability of satellite as a function of width.

separate parameter for every distinct value of x. For instance, at $x = 26.5$, software reports SE = 0.04 for the model-based estimate 0.695, whereas the SE is $\sqrt{\hat{\pi}(1 - \hat{\pi})/n} = \sqrt{(0.67)(0.33)/6} = 0.19$ for the sample proportion of 0.67 with only 6 observations. The 95% confidence intervals are (0.61, 0.77) using the model versus (0.30, 0.90) using the sample proportion. Instead of using only 6 observations, the model uses the information that all 173 observations provide in estimating the two model parameters. The result is a much more precise estimate.

Reality is a bit more complicated. In practice, the model is not *exactly* the true relationship between $\pi(x)$ and x. However, if it approximates the true probabilities decently, its estimator still tends to be closer than the sample proportion to the true value. The model smooths the sample data, somewhat dampening the observed variability. The resulting estimators tend to be better unless each sample proportion is based on an extremely large sample. Section 6.4.5 discusses this advantage of using models.

5.2.3 Checking Goodness of Fit: Ungrouped and Grouped Data

In practice, there is no guarantee that a certain logistic regression model fits the data well. For any type of binary data, one way to detect lack of fit uses a likelihood-ratio test to compare the model to more complex ones. A more complex model might contain a nonlinear effect, such as a quadratic term. Models with multiple predictors would consider interaction. If more complex models do not fit better, this provides some assurance that the model chosen is reasonable.

Other approaches to detecting lack of fit search for *any* way that the model fails. This is simplest when the explanatory variables are solely categorical, as we'll illustrate in Section 5.4.3. At each setting of x, one can multiply the estimated probabilities of the two outcomes by the number of subjects at that setting to obtain estimated expected frequencies for $y = 0$ and $y = 1$. These are *fitted values*. The test of the model compares the observed counts and fitted values using a Pearson X^2 or likelihood-ratio G^2 statistic. For a fixed number of settings, as the fitted counts increase, X^2 and G^2 have limiting chi-squared null distributions. The degrees of freedom, called the *residual* df for the model, subtract the number of parameters in the model from the number of parameters in the saturated model (i.e., the number of settings of x).

The reason for the restriction to categorical predictors for a global test of fit relates to the distinction in Section 4.5.3 that we mentioned between grouped and ungrouped data for binomial models. The saturated model differs in the two cases. An asymptotic chi-squared distribution for the deviance results as $n \to \infty$ with a fixed number of parameters in that model and hence a fixed number of settings of predictor values.

5.2.4 Goodness of Fit of Model for Horseshoe Crabs

We illustrate with a goodness-of-fit analysis for the model using $x = $ width to predict the probability that a female crab has a satellite. One way to check it compares it to a more complex model, such as the model containing a quadratic term. With width centered at 0 by subtracting its mean of 26.3, that model has fit

$$\text{logit}[\hat{\pi}(x)] = 0.618 + 0.533x + 0.040x^2.$$

The quadratic estimate has SE $= 0.046$. There is not much evidence to support adding that term. The likelihood-ratio statistic for testing that the true coefficient of x^2 is 0 equals 0.83 (df $= 1$).

We next consider overall goodness of fit. Width takes 66 distinct values for the 173 crabs, with few observations at most widths. One can view the data as a 66×2 contingency table. The two cells in each row count the number of crabs with satellites and the number of crabs without satellites, at that width. The chi-squared theory for X^2 and G^2 applies when the number of levels of x is fixed, and the number of observations at each level grows. Although we grouped the data using the distinct width values rather than using 173 separate binary responses, this theory is violated here in two ways. First, most fitted counts are very small. Second, when more data are collected, additional width values would occur, so the contingency table would contain more cells rather than a fixed number. Because of this, X^2 and G^2 for logistic regression models with continuous or nearly continuous predictors do not have approximate chi-squared distributions. (Normal approximations can be

TABLE 5.2   Grouping of Observed and Fitted Values for Fit of Logistic
Regression Model to Horseshoe Crab Data

Width (cm)	Number Yes	Number No	Fitted Yes	Fitted No
< 23.25	5	9	3.64	10.36
23.25–24.25	4	10	5.31	8.69
24.25–25.25	17	11	13.78	14.22
25.25–26.25	21	18	24.23	14.77
26.25–27.25	15	7	15.94	6.06
27.25–28.25	20	4	19.38	4.62
28.25–29.25	15	3	15.65	2.35
> 29.25	14	0	13.08	0.92

more appropriate, but no single method has received much attention; see
Section 9.8.6 for references.)

One could use X^2 and G^2 to compare the observed and fitted values in
grouped form. Table 5.2 uses the groupings of Table 4.4, giving an 8×2
table. In each width category, the fitted value for a yes response is the sum of
the estimated probabilities $\hat{\pi}(x)$ for all crabs having width in that category;
the fitted value for a no response is the sum of $1 - \hat{\pi}(x)$ for those crabs. The
fitted values are then much larger. Then, X^2 and G^2 have better validity,
although the chi-squared theory still is not perfect since $\pi(x)$ is not constant
in each category. Their values are $X^2 = 5.3$ and $G^2 = 6.2$. Table 5.2 has
eight binomial samples, one for each width setting; the model has two
parameters, so df = $8 - 2 = 6$. Neither X^2 nor G^2 shows evidence of lack of
fit ($P > 0.4$). Thus, we can feel more comfortable about using the model for
the original ungrouped data.

5.2.5   Checking Goodness of Fit with Ungrouped Data by Grouping

As just noted, with ungrouped data or with continuous or nearly continuous
predictors, X^2 and G^2 do not have limiting chi-squared distributions. They
are still useful for comparing models, as done above for checking a quadratic
term and as we will discuss in Sections 5.4.3 and 9.8.5. Also, as just noted,
one can apply them in an approximate manner to grouped observed and
fitted values for a partition of the space of x values. As the number of
explanatory variables increases, however, simultaneous grouping of values for
each variable can produce a contingency table with a large number of cells,
most of which have small counts.

Regardless of the number of predictors, one can partition observed and
fitted values according to the estimated probabilities of success using the
original ungrouped data. One common approach forms the groups in the
partition so they have approximately equal size. With 10 groups, the first pair

of observed counts and corresponding fitted counts refers to the $n/10$ observations having the highest estimated probabilities, the next pair refers to the $n/10$ observations having the second decile of estimated probabilities, and so on. Each group has an observed count of subjects with each outcome and a fitted value for each outcome. The fitted value for an outcome is the sum of the estimated probabilities for that outcome for all observations in that group.

This construction is the basis of a test due to Hosmer and Lemeshow (1980). They proposed a Pearson statistic comparing the observed and fitted counts for this partition. Let y_{ij} denote the binary outcome for observation j in group i of the partition, $i = 1, \ldots, g$, $j = 1, \ldots, n_i$. Let $\hat{\pi}_{ij}$ denote the corresponding fitted probability for the model fitted to the ungrouped data. Their statistic equals

$$\sum_{i=1}^{g} \frac{\left(\Sigma_j y_{ij} - \Sigma_j \hat{\pi}_{ij}\right)^2}{\left(\Sigma_j \hat{\pi}_{ij}\right)\left[1 - \left(\Sigma_j \hat{\pi}_{ij}\right)/n_i\right]}.$$

When many observations have the same estimated probability, there is some arbitrariness in forming the groups, and different software may report somewhat different values. This statistic does not have a limiting chi-squared distribution, because the observations in a group are not identical trials, since they do not share a common success probability. However, Hosmer and Lemeshow noted that when the number of distinct patterns of covariate values equals the sample size, the null distribution is approximated by chi-squared with df $= g - 2$.

For the logistic regression fit to the horseshoe crab data with continuous width predictor, the Hosmer–Lemeshow statistic with $g = 10$ groups equals 3.5, with df $= 8$. It also indicates a decent fit.

Unfortunately, like other proposed global fit statistics, the Hosmer–Lemeshow statistic does not have good power for detecting particular types of lack of fit (Hosmer et al. 1997). In any case, a large value of a global fit statistic merely indicates *some* lack of fit but provides no insight about its nature. The approach of comparing the working model to a more complex one is more useful from a scientific perspective, since it searches for lack of fit of a particular type. For either approach, when the fit is poor, diagnostic measures describe the influence of individual observations on the model fit and highlight reasons for the inadequacy. We discuss these in Section 6.2.1.

5.3 LOGIT MODELS WITH CATEGORICAL PREDICTORS

Like ordinary regression, logistic regression extends to include qualitative explanatory variables, often called *factors*. In this section we use dummy variables to do this.

5.3.1   ANOVA-Type Representation of Factors

For simplicity, we first consider a single factor X, with I categories. In row i of the $I \times 2$ table, y_i is the number of outcomes in the first column (successes) out of n_i trials. We treat y_i as binomial with parameter π_i.

The logit model with a factor is

$$\log \frac{\pi_i}{1 - \pi_i} = \alpha + \beta_i. \tag{5.4}$$

The higher β_i is, the higher the value of π_i. The right-hand side of (5.4) resembles the model formula for cell means in one-way ANOVA. As in ANOVA, the factor has as many parameters $\{\beta_i\}$ as categories, but one is redundant. With I categories, X has $I - 1$ nonredundant parameters. One parameter can be set to 0, say $\beta_I = 0$. If the values do not satisfy this, we can recode so that it is true. For instance, set $\tilde{\beta}_i = \beta_i - \beta_I$ and $\tilde{\alpha} = \alpha + \beta_I$, which satisfy $\tilde{\beta}_I = 0$. Then

$$\text{logit}(\pi_i) = \alpha + \beta_i = (\tilde{\alpha} - \beta_I) + (\tilde{\beta}_i + \beta_I) = \tilde{\alpha} + \tilde{\beta}_i,$$

where the newly defined parameters satisfy the constraint. When $\beta_I = 0$, α equals the logit in row I, and β_i is the difference between the logits in rows i and I. Thus, β_i equals the log odds ratio for that pair of rows.

For *any* $\{\pi_i > 0\}$, $\{\beta_i\}$ exist such that model (5.4) holds. The model has as many parameters (I) as binomial observations and is *saturated*. When a factor has *no* effect, $\beta_1 = \beta_2 = \cdots = \beta_I$. Since this is equivalent to $\pi_1 = \cdots = \pi_I$, this model with only an intercept term specifies statistical independence of X and Y.

5.3.2   Dummy Variables in Logit Models

An equivalent expression of model (5.4) uses *dummy variables*. Let $x_i = 1$ for observations in row i and $x_i = 0$ otherwise, $i = 1, \ldots, I - 1$. The model is

$$\text{logit}(\pi_i) = \alpha + \beta_1 x_1 + \beta_2 x_2 + \cdots + \beta_{I-1} x_{I-1}.$$

This accounts for parameter redundancy by not forming a dummy variable for category I. The constraint $\beta_I = 0$ in (5.4) corresponds to this form of dummy variable. The choice of category to exclude for the dummy variable is arbitrary. Some software sets $\beta_1 = 0$; this corresponds to a model with dummy variables for categories 2 through I, but not category 1.

Another way to impose constraints sets $\sum_i \beta_i = 0$. Suppose that X has $I = 2$ categories, so $\beta_1 = -\beta_2$. This results from *effect coding* for a dummy variable, $x = 1$ in category 1 and $x = -1$ in category 2.

The same substantive results occur for any coding scheme. For model (5.4), regardless of the constraint for $\{\beta_i\}$, $\{\hat{\alpha} + \hat{\beta}_i\}$ and hence $\{\hat{\pi}_i\}$ are the same. The differences $\hat{\beta}_a - \hat{\beta}_b$ for pairs (a, b) of categories of X are identical and represent estimated log odds ratios. Thus, $\exp(\hat{\beta}_a - \hat{\beta}_b)$ is the estimated odds of success in category a of X divided by the estimated odds of success in category b of X. Reparameterizing a model may change parameter estimates but does not change the model fit or the effects of interest.

The value β_i or $\hat{\beta}_i$ for a single category is irrelevant. Different constraint systems result in different values. For a binary predictor, for instance, using dummy variables with reference value $\beta_2 = 0$, the log odds ratio equals $\beta_1 - \beta_2 = \beta_1$; by contrast, for effect coding with ± 1 dummy variable and hence $\beta_1 + \beta_2 = 0$, the log odds ratio equals $\beta_1 - \beta_2 = \beta_1 - (-\beta_1) = 2\beta_1$. A parameter or its estimate makes sense only by comparison with one for another category.

5.3.3   Alcohol and Infant Malformation Example Revisited

We return now to Table 3.7 from the study of maternal alcohol consumption and child's congenital malformations, shown again in Table 5.3. For model (5.4), we treat malformations as the response and alcohol consumption as an explanatory factor. Regardless of the constraint for $\{\beta_i\}$, $\{\hat{\alpha} + \hat{\beta}_i\}$ are the sample logits, reported in Table 5.3. For instance,

$$\text{logit}(\hat{\pi}_1) = \hat{\alpha} + \hat{\beta}_1 = \log(48/17{,}066) = -5.87.$$

For the coding that constrains $\beta_5 = 0$, $\hat{\alpha} = -3.61$ and $\hat{\beta}_1 = -2.26$. For the coding $\beta_1 = 0$, $\hat{\alpha} = -5.87$. Table 5.3 shows that except for the slight reversal between the first and second categories of alcohol consumption, the logits and hence the sample proportions of malformation cases increase as alcohol consumption increases.

The simpler model with all $\beta_i = 0$ specifies independence. For it, $\hat{\alpha}$ equals the logit for the overall sample proportion of malformations, or $\log(93/32481) = -5.86$. To test H_0: independence (df = 4), the Pearson

TABLE 5.3   Logits and Proportion of Malformation for Table 3.7

Alcohol Consumption	Present	Absent	Logit	Proportion Malformed Observed	Proportion Malformed Fitted
0	48	17,066	−5.87	0.0028	0.0026
< 1	38	14,464	−5.94	0.0026	0.0030
1–2	5	788	−5.06	0.0063	0.0041
3–5	1	126	−4.84	0.0079	0.0091
≥ 6	1	37	−3.61	0.0263	0.0231

statistic (3.10) is $X^2 = 12.1$ ($P = 0.02$), and the likelihood-ratio statistic (3.11) is $G^2 = 6.2$ ($P = 0.19$). These provide mixed signals. Table 5.3 has a mixture of very small, moderate, and extremely large counts. Even though $n = 32,574$, the null sampling distributions of X^2 or G^2 may not be close to chi-squared. The P-values using the exact conditional distributions of X^2 and G^2 are 0.03 and 0.13. These are closer, but still give differing evidence. In any case, these statistics ignore the ordinality of alcohol consumption. The sample suggests that malformations may tend to be more likely with higher alcohol consumption. The first two percentages are similar and the next two are also similar, however, and any of the last three percentages changes substantially with the addition or deletion of one malformation case.

5.3.4   Linear Logit Model for $I \times 2$ Tables

Model (5.4) treats the explanatory factor as nominal, since it is invariant to the ordering of categories. For ordered factor categories, other models are more parsimonious than this, yet more complex than the independence model. For instance, let scores $\{x_1, x_2, \ldots, x_I\}$ describe distances between categories of X. When one expects a monotone effect of X on Y, it is natural to fit the *linear logit model*

$$\text{logit}(\pi_i) = \alpha + \beta x_i. \tag{5.5}$$

The independence model is the special case $\beta = 0$.

The near-monotone increase in sample logits in Table 5.3 indicates that the linear logit model (5.5) may fit better than the independence model. As measured, alcohol consumption groups a naturally continuous variable. With scores $\{x_1 = 0, x_2 = 0.5, x_3 = 1.5, x_4 = 4.0, x_5 = 7.0\}$, the last score being somewhat arbitrary, Table 5.4 shows results. The estimated multiplicative

TABLE 5.4   Computer Output for Logistic Regression Model with Infant Malformation Data

Criteria For Assessing Goodness Of Fit		
Criterion	DF	Value
Deviance	3	1.9487
Pearson Chi-Square	3	2.0523
Log Likelihood		-635.5968

Parameter	Estimate	Std Error	Likelihood-Ratio 95% Conf Limits		Wald Chi-Sq	Pr>ChiSq
Intercept	-5.9605	0.1154	-6.1930	-5.7397	2666.41	<.0001
alcohol	0.3166	0.1254	0.0187	0.5236	6.37	0.0116

effect of a unit increase in daily alcohol consumption on the odds of malformation is $\exp(0.317) = 1.37$. Table 5.3 shows the observed and fitted proportions of malformation. The model seems to fit well, as statistics comparing observed and fitted counts are $G^2 = 1.95$ and $X^2 = 2.05$, with df $= 3$.

5.3.5 Cochran–Armitage Trend Test

Armitage (1955) and Cochran (1954) were among the first to emphasize the importance of utilizing ordered categories in a contingency table. For $I \times 2$ tables with ordered rows and I independent $\text{bin}(n_i, \pi_i)$ variates $\{y_i\}$, they proposed a trend statistic for testing independence by partitioning the Pearson statistic for that hypothesis. They used a linear probability model,

$$\pi_i = \alpha + \beta x_i, \tag{5.6}$$

fitted by ordinary least squares. For this model, the null hypothesis of independence is $H_0: \beta = 0$. Let $\bar{x} = \Sigma_i n_i x_i / n$. Let $p_i = y_i / n_i$, and let $p = (\Sigma_i y_i)/n$ denote the overall proportion of successes. The prediction equation is

$$\hat{\pi}_i = p + b(x_i - \bar{x}),$$

where

$$b = \frac{\Sigma_i n_i (p_i - p)(x_i - \bar{x})}{\Sigma_i n_i (x_i - \bar{x})^2}.$$

Denote the Pearson statistic for testing independence by $X^2(I)$. For $I \times 2$ tables with ordered rows, it satisfies

$$X^2(I) = \frac{1}{p(1-p)} \sum_i n_i (p_i - p)^2 = z^2 + X^2(L),$$

where

$$X^2(L) = \frac{1}{p(1-p)} \sum_i n_i (p_i - \hat{\pi}_i)^2$$

$$z^2 = \frac{b^2}{p(1-p)} \sum_i n_i (x_i - \bar{x})^2 = \left[\frac{\Sigma_i (x_i - \bar{x}) y_i}{\sqrt{p(1-p)\Sigma_i n_i (x_i - \bar{x})^2}} \right]^2. \tag{5.7}$$

When the linear probability model holds, $X^2(L)$ is asymptotically chi-squared with df $= I - 2$. It tests the fit of the model. The statistic z^2, with df $= 1$, tests $H_0: \beta = 0$ for the linear trend in the proportions (5.6). The test of independence using this statistic is called the *Cochran–Armitage trend test*.

This analysis seems unrelated to the linear logit model. However, the Cochran–Armitage statistic is equivalent to the score statistic for testing $H_0: \beta = 0$ in that model. Moreover, this statistic relates to the statistic M^2 in (3.15) used to test for a linear trend in an $I \times J$ table; namely, it equals M^2 applied when $J = 2$, except with $(n - 1)$ replaced by n. When $I = 2$, $X^2(L) = 0$ and $z^2 = X^2(I)$.

For Table 5.3 on alcohol consumption and malformation, $X^2(I) = 12.1$. Using the same scores as in the linear logit model, the Cochran–Armitage trend test has $z^2 = 6.6$ (*P*-value $= 0.010$). The test suggests strong evidence of a positive slope. In addition,

$$X^2(I) = 12.1 = 6.6 + 5.5,$$

where $X^2(L) = 5.5$ (df $= 3$) shows only slight evidence of departure of the proportions from linearity. The trend test agrees with M^2 for the sample correlation of $r = 0.014$ for $n = 32{,}573$ (Section 3.4.5). For the chosen scores, the correlation seems weak. However, r has limited use as a descriptive measure for tables that are highly discrete and unbalanced.

The Cochran–Armitage trend test (i.e., the score test) usually gives results similar to the Wald or likelihood-ratio test of $H_0: \beta = 0$ in the linear logit model. The asymptotics work well even for quite small n when $\{n_i\}$ are equal and $\{x_i\}$ are equally spaced. With Table 5.3, the Wald statistic equals $(\hat{\beta}/\text{SE})^2 = (0.317/0.125)^2 = 6.4$ ($P = 0.012$) and the likelihood-ratio statistic equals 4.25 ($P = 0.039$). The highly unbalanced counts suggest that it is safest to use the likelihood function through the likelihood-ratio approach. This is also true for estimation. The profile likelihood 95% confidence interval of $(0.02, 0.52)$ for β reported in Table 5.4 is preferable to the Wald interval of $0.317 \pm 1.96(0.125) = (0.07, 0.56)$. Even though n is very large, exact inference based on small-sample methods presented in Section 6.7.4 is relevant here.

5.4 MULTIPLE LOGISTIC REGRESSION

Like ordinary regression, logistic regression extends to models with multiple explanatory variables. For instance, the model for $\pi(\mathbf{x}) = P(Y = 1)$ at values $\mathbf{x} = (x_1, \ldots, x_p)$ of p predictors is

$$\text{logit}[\pi(\mathbf{x})] = \alpha + \beta_1 x_1 + \beta_2 x_2 + \cdots + \beta_p x_p. \tag{5.8}$$

The alternative formula, directly specifying $\pi(\mathbf{x})$, is

$$\pi(\mathbf{x}) = \frac{\exp(\alpha + \beta_1 x_1 + \beta_2 x_2 + \cdots + \beta_p x_p)}{1 + \exp(\alpha + \beta_1 x_1 + \beta_2 x_2 + \cdots + \beta_p x_p)}. \tag{5.9}$$

The parameter β_i refers to the effect of x_i on the log odds that $Y = 1$, controlling the other x_j. For instance, $\exp(\beta_i)$ is the multiplicative effect on the odds of a 1-unit increase in x_i, at fixed levels of other x_j. An explanatory variable can be qualitative, using dummy variables for categories.

5.4.1 Logit Models for Multiway Contingency Tables

When all variables are categorical, a multiway contingency table displays the data. We illustrate ideas with binary predictors X and Z. We treat the sample size at given combinations (i, k) of X and Z as fixed and regard the two counts on Y at each setting as binomial, with different binomials treated as independent. Denote the two categories for each variable by $(0, 1)$, and let dummy variables for X and Z have $x_1 = z_1 = 1$ and $x_2 = z_2 = 0$. The model

$$\text{logit}[P(Y = 1)] = \alpha + \beta_1 x_i + \beta_2 z_k \tag{5.10}$$

has main effects for X and Z but assumes an absence of interaction. The effect of one factor is the same at each level of the other.

At a fixed level z_k of Z, the effect on the logit of changing categories of X is

$$[\alpha + \beta_1(1) + \beta_2 z_k] - [\alpha + \beta_1(0) + \beta_2 z_k] = \beta_1. \tag{5.11}$$

This logit difference equals the difference of log odds, which is the log odds ratio between X and Y, fixing Z. Thus, $\exp(\beta_1)$ is the conditional odds ratio between X and Y. Controlling for Z, the odds of success when $X = 1$ equal $\exp(\beta_1)$ times the odds when $X = 0$. This conditional odds ratio is the same at each level of Z; that is, there is *homogeneous XY association* (Section 2.3.5). The lack of an interaction term in (5.10) implies a common odds ratio for the partial tables. When $\beta_1 = 0$, that common odds ratio equals 1. Then X and Y are independent in each partial table, or *conditionally independent*, *given Z* (Section 2.3.4).

Additivity on the logit scale is the generally accepted definition of no interaction for categorical variables. However, one could, instead, define it as additivity on some other scale, such as with probit or identity link. Significant interaction can occur on one scale when there is none on another scale. In some applications, a particular definition may be natural. For instance, theory might assume an underlying normal distribution and predict that the probit is an additive function of predictor effects.

A factor with I categories needs $I - 1$ dummy variables, as we showed in Section 5.3.2. An alternative representation of such factors resembles the way that ANOVA models often express them. The model formula

$$\text{logit}[P(Y = 1)] = \alpha + \beta_i^X + \beta_k^Z \tag{5.11}$$

represents effects of X with parameters $\{\beta_i^X\}$ and effects of Z with parameters $\{\beta_k^Z\}$. (The X and Z superscripts are merely labels and do not represent powers.) Model form (5.11) applies for any number of categories for X and Z. The parameter β_i^X denotes the effect on the logit of classification in category i of X. Conditional independence between X and Y, given Z, corresponds to $\beta_1^X = \beta_2^X = \cdots = \beta_I^X$, whereby $P(Y = 1)$ does not change as i changes.

For each factor, one parameter in (5.11) is redundant. Fixing one at 0, such as $\beta_I^X = \beta_K^Z = 0$, represents the category not having its own dummy variable. When X and Z have two categories, the parameterization in model (5.11) then corresponds to that in model (5.10) with $\beta_1^X = \beta_1$ and $\beta_2^X = 0$, and with $\beta_1^Z = \beta_2$ and $\beta_2^Z = 0$.

5.4.2 AIDS and AZT Example

Table 5.5 is from a study on the effects of AZT in slowing the development of AIDS symptoms. In the study, 338 veterans whose immune systems were beginning to falter after infection with the AIDS virus were randomly assigned either to receive AZT immediately or to wait until their T cells showed severe immune weakness. Table 5.5 cross-classifies the veterans' race, whether they received AZT immediately, and whether they developed AIDS symptoms during the 3-year study.

In model (5.10), we identify X with AZT treatment ($x_1 = 1$ for immediate AZT use, $x_2 = 0$ otherwise) and Z with race ($z_1 = 1$ for whites, $z_2 = 0$ for blacks), for predicting the probability that AIDS symptoms developed. Thus, α is the log odds of developing AIDS symptoms for black subjects without immediate AZT use, β_1 is the increment to the log odds for those with immediate AZT use, and β_2 is the increment to the log odds for white

TABLE 5.5 Development of AIDS Symptoms by AZT Use and Race

| Race | AZT Use | Symptoms | |
		Yes	No
White	Yes	14	93
	No	32	81
Black	Yes	11	52
	No	12	43

Source: New York Times, Feb. 15, 1991.

TABLE 5.6 Computer Output for Logit Model with AIDS Symptoms Data

```
                        Goodness-of-Fit Statistics
                Criterion      DF      Value     Pr > ChiSq
                Deviance        1     1.3835       0.2395
                Pearson         1     1.3910       0.2382
```

```
                Analysis of Maximum Likelihood Estimates
   Parameter   Estimate   Std Error   Wald Chi-Square   Pr > ChiSq
   Intercept    -1.0736     0.2629        16.6705         < .0001
   azt          -0.7195     0.2790         6.6507          0.0099
   race          0.0555     0.2886         0.0370          0.8476
```

```
                        Odds Ratio Estimates
           Effect    Estimate    95% Wald Confidence Limits
           azt        0.487         0.282         0.841
           race       1.057         0.600         1.861
```

```
        Profile Likelihood Confidence Interval for Odds Ratios
           Effect    Estimate      95% Confidence Limits
           azt        0.487          0.279         0.835
           race       1.057          0.605         1.884
```

```
Obs   race   azt    y     n     pi_hat     lower      upper
 1     1      1     14    107   0.14962    0.09897    0.21987
 2     1      0     32    113   0.26540    0.19668    0.34774
 3     0      1     11     63   0.14270    0.08704    0.22519
 4     0      0     12     55   0.25472    0.16953    0.36396
```

subjects. Table 5.6 shows output. The estimated odds ratio between immediate AZT use and development of AIDS symptoms equals $\exp(-0.7195) = 0.487$. For each race, the estimated odds of symptoms are half as high for those who took AZT immediately. The Wald confidence interval for this effect is $\exp[-0.720 \pm 1.96(0.279)] = (0.28, 0.84)$. Similar results occur for the likelihood-based interval.

The hypothesis of conditional independence of AZT treatment and development of AIDS symptoms, controlling for race, is $H_0: \beta_1 = 0$ in (5.10). The likelihood-ratio statistic comparing model (5.10) with the simpler model having $\beta_1 = 0$ equals 6.9 (df = 1), showing evidence of association ($P = 0.01$). The Wald statistic ($\hat{\beta}_1/\mathrm{SE})^2 = (-0.720/0.279)^2 = 6.65$ provides similar results.

Table 5.7 shows parameter estimates for three ways of defining factor parameters in (5.11): (1) setting the last parameter equal to 0, (2) setting the first parameter equal to 0, and (3) having parameters sum to zero. For each coding scheme, at a given combination of AZT use and race, the estimated probability of developing AIDS symptoms is the same. For instance, the intercept estimate plus the estimate for immediate AZT use plus the estimate for being white is -1.738 for each scheme, so the estimated probability

TABLE 5.7   Parameter Estimates for Logit Model Fitted to Table 5.5

	Definition of Parameters		
Parameter	Last = Zero	First = Zero	Sum = Zero
Intercept	−1.074	−1.738	−1.406
AZT Yes	−0.720	0.000	−0.360
No	0.000	0.720	0.360
Race White	0.055	0.000	0.028
Black	0.000	−0.055	−0.028

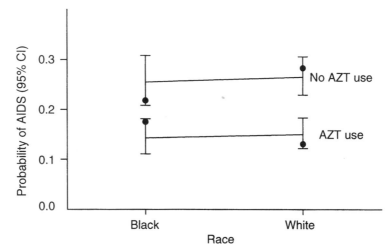

FIGURE 5.4 Estimated effects of AZT use and race on probability of developing AIDS symptoms (dots are sample proportions).

that white veterans with immediate AZT use develop AIDS symptoms equals $\exp(-1.738)/[1 + \exp(-1.738)] = 0.15$. The bottom of Table 5.6 shows point and interval estimates of the probabilities. Figure 5.4 shows a graphical representation of the sample proportions (the four dots) and the point estimates enclosed in 95% confidence intervals.

Similarly, for each coding scheme, $\beta_1^X - \beta_2^X$ is identical and represents the conditional log odds ratio of X with the response, given Z. Here, $\exp(\hat{\beta}_1^X - \hat{\beta}_2^X) = \exp(-0.720) = 0.49$ estimates the common odds ratio between immediate AZT use and AIDS symptoms, for each race.

5.4.3   Goodness of Fit as a Likelihood-Ratio Test

The likelihood-ratio statistic $-2(L_0 - L_1)$ tests whether certain model parameters are zero by comparing the log likelihood L_1 for the fitted model M_1 with L_0 for a simpler model M_0. Denote this statistic for testing M_0, given that M_1 holds, by $G^2(M_0|M_1)$. The goodness-of-fit statistic $G^2(M)$ is a special case in which $M_0 = M$ and M_1 is the saturated model. In testing whether M fits, we test whether *all* parameters in the saturated model but not in M equal zero. The asymptotic df is the difference in the number of parameters in the two models, which is the number of binomials modeled minus the number of parameters in M.

We illustrate by checking the fit of model (5.10) for the AIDS data. For its fit, white veterans with immediate AZT use had estimated probability 0.150 of developing AIDS symptoms during the study. Since 107 white veterans took AZT, the fitted value is $107(0.150) = 16.0$ for developing symptoms and $107(0.850) = 91.0$ for not developing them. Similarly, one can obtain fitted values for all eight cells in Table 5.5. The goodness-of-fit statistics comparing these with the cell counts are $G^2 = 1.38$ and $X^2 = 1.39$. The model has four binomials, one at each combination of AZT use and race. Since it has three parameters, residual df $= 4 - 3 = 1$. The small G^2 and X^2 values suggest that the model fits decently ($P > 0.2$).

For model (5.10), the odds ratio between X and Y is the same at each level of Z. The goodness-of-fit test checks this structure. That is, the test also provides a test of homogeneous odds ratios. For Table 5.5, homogeneity is plausible. Since residual df $= 1$, the more complex model that adds an interaction term and permits the two odds ratios to differ is saturated.

Let L_S denote the maximized log likelihood for the saturated model. As discussed in Section 4.5.4, the likelihood-ratio statistic for comparing models M_1 and M_0 is

$$G^2(M_0|M_1) = -2(L_0 - L_1)$$
$$= -2(L_0 - L_S) - [-2(L_1 - L_S)]$$
$$= G^2(M_0) - G^2(M_1).$$

The test statistic comparing two models is identical to the difference in G^2 goodness-of-fit statistics (deviances) for the two models. To illustrate, consider H_0: $\beta_2 = 0$ for the race effect with the AIDS data. The likelihood-ratio statistic equals 0.04, suggesting that the simpler model is adequate. But this equals $G^2(M_0) - G^2(M_1) = 1.42 - 1.38$, where M_0 is the simpler model with $\beta_2 = 0$.

The model comparison statistic often has an approximate chi-squared null distribution even when separate $G^2(M_i)$ do not. For instance, when a predictor is continuous or a contingency table has very small fitted values, the sampling distribution of $G^2(M_i)$ may be far from chi-squared. Nonetheless, if df for the comparison statistic is modest (as in comparing two models that differ by a few parameters), the null distribution of $G^2(M_0|M_1)$ is approximately chi-squared.

5.4.4 Horseshoe Crab Example Revisited

Like ordinary regression, logistic regression can have a mixture of quantitative and qualitative predictors. We illustrate with the horseshoe crab data (Section 5.1.3), using the female crab's width and color as predictors. Color has five categories: light, medium light, medium, medium dark, dark. It is a surrogate for age, older crabs tending to be darker. The sample contained no light crabs, so our models use only the other four categories.

We first treat color as qualitative. The four categories use three dummy variables. The model is

$$\text{logit}(\pi) = \alpha + \beta_1 c_1 + \beta_2 c_2 + \beta_3 c_3 + \beta_4 x, \qquad (5.12)$$

where $\pi = P(Y = 1)$, $x =$ width in centimeters, and

$$c_1 = 1 \text{ for medium-light color, and } 0 \text{ otherwise,}$$

$$c_2 = 1 \text{ for medium color, and } 0 \text{ otherwise,}$$

$$c_3 = 1 \text{ for medium-dark color, and } 0 \text{ otherwise.}$$

The crab color is dark (category 4) when $c_1 = c_2 = c_3 = 0$. Table 5.8 shows the ML parameter estimates. For instance, for dark crabs, $\text{logit}(\hat{\pi}) = -12.715 + 0.468x$; by contrast, for medium-light crabs, $c_1 = 1$, and $\text{logit}(\hat{\pi}) = (-12.715 + 1.330) + 0.468x = -11.385 + 0.468x$. At the average width of 26.3 cm, $\hat{\pi} = 0.399$ for dark crabs and 0.715 for medium-light crabs.

The model assumes a lack of interaction between color and width in their effects. Width has the same coefficient (0.468) for all colors, so the shapes of the curves relating width to π are identical. For each color, a 1-cm increase in width has a multiplicative effect of $\exp(0.468) = 1.60$ on the odds that $Y = 1$. Figure 5.5 displays the fitted model. Any one curve equals any other

TABLE 5.8 Computer Output for Model with Width and Color Predictors

		Criteria For Assessing Goodness Of Fit				
		Criterion		DF	Value	
		Deviance		168	187.4570	
		Pearson Chi-Square		168	168.6590	
		Log Likelihood			−93.7285	

Parameter	Estimate	Standard Error	Likelihood-Ratio 95% Confidence Limits		Chi-Square	Pr>ChiSq
intercept	−12.7151	2.7618	−18.4564	−7.5788	21.20	<.0001
c1	1.3299	0.8525	−0.2738	3.1354	2.43	0.1188
c2	1.4023	0.5484	0.3527	2.5260	6.54	0.0106
c3	1.1061	0.5921	−0.0279	2.3138	3.49	0.0617
width	0.4680	0.1055	0.2713	0.6870	19.66	<.0001

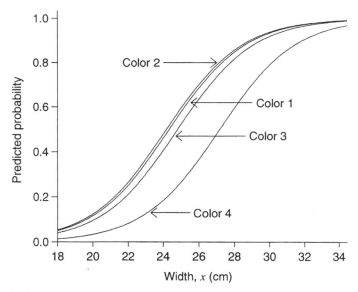

FIGURE 5.5 Logistic regression model using width and color predictors of satellite presence for horseshoe crabs.

curve shifted to the right or left. The parallelism of curves in the horizontal dimension implies that any two curves never cross. At all width values, color 4 (dark) has a lower estimated probability of a satellite than the other colors. There is a noticeable positive effect of width.

The exponentiated difference between two color parameter estimates is an odds ratio comparing those colors. For instance, the difference for medium-light crabs and dark crabs equals 1.330. At any given width, the estimated odds that a medium-light crab has a satellite are $\exp(1.330) = 3.8$ times the estimated odds for a dark crab. At width $x = 26.3$, the odds equal $0.715/0.285 = 2.51$ for a medium-light crab and $0.399/0.601 = 0.66$ for a dark crab, for which $2.51/0.66 = 3.8$.

5.4.5 Model Comparison

To test whether color contributes significantly to model (5.12), we test $H_0: \beta_1 = \beta_2 = \beta_3 = 0$. This states that controlling for width, the probability of a satellite is independent of color. We compare the maximized log-likelihood L_1 for the full model (5.12) to L_0 for the simpler model. The test statistic $-2(L_0 - L_1) = 7.0$ has df $= 3$, the difference between the numbers of parameters in the two models. The chi-squared P-value of 0.07 provides slight evidence of a color effect.

The more complex model allowing color $\times$ width interaction has three additional terms, the cross-products of width with the color dummy variables.

Fitting this model is equivalent to fitting logistic regression with width predictor separately for crabs of each color. Each color then has a different-shaped curve relating width to $P(Y = 1)$, so a comparison of two colors varies according to the width value. The likelihood-ratio statistic comparing the models with and without the interaction terms equals 4.4, with df = 3. The evidence of interaction is weak ($P = 0.22$).

5.4.6 Quantitative Treatment of Ordinal Predictor

Color has ordered categories, from lightest to darkest. A simpler model yet treats this predictor as quantitative. Color may have a linear effect, for a set of monotone scores. To illustrate, for scores $c = \{1, 2, 3, 4\}$ for the color categories, the model

$$\text{logit}(\pi) = \alpha + \beta_1 c + \beta_2 x \tag{5.13}$$

has $\hat{\beta}_1 = -0.509$ (SE = 0.224) and $\hat{\beta}_2 = 0.458$ (SE = 0.104). This shows strong evidence of an effect for each. At a given width, for every one-category increase in color darkness, the estimated odds of a satellite multiply by $\exp(-0.509) = 0.60$.

The likelihood-ratio statistic comparing this fit to the more complex model (5.12) having a separate parameter for each color equals 1.7 (df = 2). This statistic tests that the simpler model (5.13) is adequate, given that model (5.12) holds. It tests that when plotted against the color scores, the color parameters in (5.12) follow a linear trend. The simplification seems permissible ($P = 0.44$).

The color parameter estimates in the qualitative-color model (5.12) are $(1.33, 1.40, 1.11, 0)$, the 0 value for the dark category reflecting its lack of a dummy variable. Although these values do not depart significantly from a linear trend, the first three are quite similar compared to the last one. Thus, another potential color scoring for model (5.13) is $\{1, 1, 1, 0\}$; that is, score = 0 for dark-colored crabs, and score = 1 otherwise. The likelihood-ratio statistic comparing model (5.13) with these binary scores to model (5.12) equals 0.5 (df = 2), showing that this simpler model is also adequate. Its fit is

$$\text{logit}(\hat{\pi}) = -12.980 + 1.300c + 0.478x, \tag{5.14}$$

with standard errors 0.526 and 0.104. At a given width, the estimated odds that a lighter-colored crab has a satellite are $\exp(1.300) = 3.7$ times the estimated odds for a dark crab.

In summary, the qualitative-color model, the quantitative-color model with scores $\{1, 2, 3, 4\}$, and the model with binary color scores $\{1, 1, 1, 0\}$ all suggest that dark crabs are least likely to have satellites. A much larger sample is needed to determine which color scoring is most appropriate. It is advantageous to treat ordinal predictors in a quantitative manner when such models fit well. The model is simpler and easier to interpret, and tests of the predictor effect are more powerful when it has a single parameter rather than several parameters. In Section 6.4 we discuss this issue further.

5.4.7 Standardized and Probability-Based Interpretations

To compare effects of quantitative predictors having different units, it can be helpful to report standardized coefficients. One approach fits the model to standardized predictors, replacing each x_j by $(x_j - \bar{x}_j)/s_{x_j}$. Then, each regression coefficient represents the effect of a standard deviation change in a predictor, controlling for the other variables. Equivalently, for each j one can multiply unstandardized estimate $\hat{\beta}_j$ by s_{x_j} (see also Note 5.9).

Regardless of the units, many find it difficult to understand odds or odds ratio effects. The simpler interpretation of the approximate change in the probability based on a linearization of the model (Section 5.1.1) applies also to multiple predictors. Consider a setting of predictors at which $\hat{P}(Y = 1) = \hat{\pi}$. Then, controlling for the other predictors, a 1-unit increase in x_j corresponds approximately to a $\hat{\beta}_j \hat{\pi}(1 - \hat{\pi})$ change in $\hat{\pi}$. For instance, at predictor settings at which $\hat{\pi} = 0.5$ for fit (5.14), the approximate effect of a 1-cm increase in width is $(0.478)(0.5)(0.5) = 0.12$. This is considerable, since a 1-cm change in width is less than half a standard deviation.

This linear approximation deteriorates as the change in the predictor increases. More precise interpretations use the probability formula directly. To describe the effect of x_j, one could set the other predictors at their sample means and compute the estimated probabilities at the smallest and largest x_j values. These are sensitive to outliers, however. It is often more sensible to use the quartiles.

For fit (5.14), the sample means are 26.3 for x and 0.873 for c. The lower and upper quartiles of x are 24.9 and 27.7. At $x = 24.9$ and $c = \bar{c}$, $\hat{\pi} = 0.51$. At $x = 27.7$ and $c = \bar{c}$, $\hat{\pi} = 0.80$. The change in $\hat{\pi}$ from 0.51 to 0.80 over the middle 50% of the range of width values reflects a strong width effect. Since c takes only values 0 and 1, one could instead report this effect separately for each. Also, when an explanatory variable is a dummy, it makes sense to report the estimated probabilities at its two values rather than at quartiles, which could be identical. At $\bar{x} = 26.3$, $\hat{\pi} = 0.40$ when $c = 0$ and $\hat{\pi} = 0.71$ when $c = 1$. This color effect, differentiating dark crabs from others, is also substantial.

Table 5.9 shows a way to present effects that can be understandable to those not familiar with odds ratios. It also shows results of the extension of model (5.14), permitting interaction. The estimated width effect is then greater for the lighter-colored crabs. However, the interaction is not significant.

TABLE 5.9 Summary of Effects in Model (5.14) with Crab Width and Color as Predictors of Presence of Satellites

Variable	Estimate	SE	Comparison	Change in Probability
No interaction model				
Intercept	− 12.980	2.727		
Color (0 = dark,				
1 = other)	1.300	0.526	$(1, 0)$ at $\bar{x}$	$0.31 = 0.71 - 0.40$
Width, x (cm)	0.478	0.104	(UQ, LQ) at $\bar{c}$	$0.29 = 0.80 - 0.51$
Interaction model				
Intercept	− 5.854	6.694		
Color (0 = dark,				
1 = other)	− 6.958	7.318		
Width, x (cm)	0.200	0.262	(UQ, LQ) at $c = 0$	$0.13 = 0.43 - 0.30$
Width × color	0.322	0.286	(UQ, LQ) at $c = 1$	$0.29 = 0.84 - 0.55$

5.5 FITTING LOGISTIC REGRESSION MODELS

The mechanics of ML estimation and model fitting for logistic regression are special cases of the GLM fitting results of Section 4.6. With n subjects, we treat the n binary responses as independent. Let $\mathbf{x}_i = (x_{i1}, \ldots, x_{ip})$ denote setting i of values of p explanatory variables, $i = 1, \ldots, N$. When explanatory variables are continuous, a different setting may occur for each subject, in which case $N = n$. The logistic regression model (5.8), regarding α as a regression parameter with unit coefficient, is

$$\pi(\mathbf{x}_i) = \frac{\exp\left(\sum_{j=1}^{p} \beta_j x_{ij}\right)}{1 + \exp\left(\sum_{j=1}^{p} \beta_j x_{ij}\right)}. \tag{5.15}$$

5.5.1 Likelihood Equations

When more than one observation occurs at a fixed x_i value, it is sufficient to record the number of observations n_i and the number of successes. We then let y_i refer to this success count rather than to an individual binary response. Then $\{Y_1, \ldots, Y_N\}$ are independent binomials with $E(Y_i) = n_i \pi(\mathbf{x}_i)$, where $n_1 + \cdots + n_N = n$. Their joint probability mass function is proportional to the product of N binomial functions,

$$\prod_{i=1}^{N} \pi(\mathbf{x}_i)^{y_i} [1 - \pi(\mathbf{x}_i)]^{n_i - y_i}$$

$$= \left\{ \prod_{i=1}^{N} \exp\left[\log\left(\frac{\pi(\mathbf{x}_i)}{1 - \pi(\mathbf{x}_i)} \right)^{y_i} \right] \right\} \left\{ \prod_{i=1}^{N} [1 - \pi(\mathbf{x}_i)]^{n_i} \right\}$$

$$= \left\{ \exp\left[\sum_i y_i \log\frac{\pi(\mathbf{x}_i)}{1 - \pi(\mathbf{x}_i)} \right] \right\} \left\{ \prod_{i=1}^{N} [1 - \pi(\mathbf{x}_i)]^{n_i} \right\}.$$

For model (5.15), the ith logit is $\sum_j \beta_j x_{ij}$, so the exponential term in the last expression equals $\exp[\sum_i y_i(\sum_j \beta_j x_{ij})] = \exp[\sum_j(\sum_i y_i x_{ij})\beta_j]$. Also, since $[1 - \pi(\mathbf{x}_i)] = [1 + \exp(\sum_j \beta_j x_{ij})]^{-1}$, the log likelihood equals

$$L(\boldsymbol{\beta}) = \sum_j \left(\sum_i y_i x_{ij}\right)\beta_j - \sum_i n_i \log\left[1 + \exp\left(\sum_j \beta_j x_{ij}\right)\right]. \quad (5.16)$$

This depends on the binomial counts only through the sufficient statistics $\{\sum_i y_i x_{ij}, \ j = 1, \ldots, p\}$.

The likelihood equations result from setting $\partial L(\boldsymbol{\beta})/\partial \boldsymbol{\beta} = 0$. Since

$$\frac{\partial L(\boldsymbol{\beta})}{\partial \beta_j} = \sum_i y_i x_{ij} - \sum_i n_i x_{ij} \frac{\exp(\sum_k \beta_k x_{ik})}{1 + \exp(\sum_k \beta_k x_{ik})},$$

the likelihood equations are

$$\sum_i y_i x_{ij} - \sum_i n_i \hat{\pi}_i x_{ij} = 0, \qquad j = 1, \ldots, p, \quad (5.17)$$

where $\hat{\pi}_i = \exp(\sum_k \hat{\beta}_k x_{ik})/[1 + \exp(\sum_k \hat{\beta}_k x_{ik})]$ is the ML estimate of $\pi(\mathbf{x}_i)$. We observed these equations as a special case of those for binomial GLMs in (4.25) (but there y_i is the *proportion* of successes). The equations are nonlinear and require iterative solution.

Let $\mathbf{X}$ denote the $N \times p$ matrix of values of $\{x_{ij}\}$. The likelihood equations (5.17) have form

$$\mathbf{X}'\mathbf{y} = \mathbf{X}'\hat{\boldsymbol{\mu}}, \quad (5.18)$$

where $\hat{\mu}_i = n_i \hat{\pi}_i$. This equation illustrates a fundamental result: For GLMs with canonical link, the likelihood equations equate the sufficient statistics to the estimates of their expected values. Equation (4.44) showed this result in the GLM context, and (5.18) are the normal equations in ordinary regression.

5.5.2 Asymptotic Covariance Matrix of Parameter Estimators

The ML estimators $\hat{\boldsymbol{\beta}}$ have a large-sample normal distribution with covariance matrix equal to the inverse of the information matrix. The observed information matrix has elements

$$-\frac{\partial^2 L(\boldsymbol{\beta})}{\partial \beta_a \partial \beta_b} = \sum_i \frac{x_{ia} x_{ib} n_i \exp(\sum_j \beta_j x_{ij})}{\left[1 + \exp(\sum_j \beta_j x_{ij})\right]^2} = \sum_i x_{ia} x_{ib} n_i \pi_i (1 - \pi_i). \quad (5.19)$$

This is not a function of $\{y_i\}$, so the observed and expected information are identical. This happens for all GLMs that use canonical links (Section 4.6.4).

The estimated covariance matrix is the inverse of the matrix having elements (5.19), substituting $\hat{\boldsymbol{\beta}}$. This has form

$$\widehat{\text{cov}}(\hat{\boldsymbol{\beta}}) = \left\{\mathbf{X}' \, \mathbf{diag}\big[n_i\hat{\pi}_i(1 - \hat{\pi}_i)\big]\mathbf{X}\right\}^{-1}, \tag{5.20}$$

where $\mathbf{diag}[n_i\hat{\pi}_i(1 - \hat{\pi}_i)]$ denotes the $N \times N$ diagonal matrix having $\{n_i\hat{\pi}_i(1 - \hat{\pi}_i)\}$ on the main diagonal. This is the special case of the GLM covariance matrix (4.28) with estimated diagonal weight matrix $\hat{\mathbf{W}}$ having elements $\hat{w}_i = n_i\hat{\pi}_i(1 - \hat{\pi}_i)$. The square roots of the main diagonal elements of (5.20) are estimated standard errors of $\hat{\boldsymbol{\beta}}$.

5.5.3 Distribution of Probability Estimators

Using $\widehat{\text{cov}}(\hat{\boldsymbol{\beta}})$, one can conduct inference about $\boldsymbol{\beta}$ and related effects such as odds ratios. One can also construct confidence intervals for response probabilities $\pi(\mathbf{x})$ at particular settings $\mathbf{x}$.

The estimated variance of $\text{logit}[\hat{\pi}(\mathbf{x})] = \mathbf{x}\hat{\boldsymbol{\beta}}$ is $\mathbf{x}\,\widehat{\text{cov}}(\hat{\boldsymbol{\beta}})\mathbf{x}'$. For large samples, $\text{logit}[\hat{\pi}(\mathbf{x})] \pm z_{\alpha/2}\sqrt{\mathbf{x}\,\widehat{\text{cov}}(\hat{\boldsymbol{\beta}})\mathbf{x}'}$ is a confidence interval for the true logit. The endpoints invert to a corresponding interval for $\pi(\mathbf{x})$ using the transform $\pi = \exp(\text{logit})/[1 + \exp(\text{logit})]$.

5.5.4 Newton–Raphson Method Applied to Logistic Regression

We refer back to Section 4.6.1 for the Newton–Raphson iterative method. Let

$$u_j^{(t)} = \frac{\partial L(\boldsymbol{\beta})}{\partial \beta_j}\bigg|_{\boldsymbol{\beta}^{(t)}} = \sum_i \big(y_i - n_i\pi_i^{(t)}\big)x_{ij}$$

$$h_{ab}^{(t)} = \frac{\partial^2 L(\boldsymbol{\beta})}{\partial \beta_a \, \partial \beta_b}\bigg|_{\boldsymbol{\beta}^{(t)}} = -\sum_i x_{ia}x_{ib}n_i\pi_i^{(t)}\big(1 - \pi_i^{(t)}\big).$$

Here, $\boldsymbol{\pi}^{(t)}$, approximation t for $\hat{\boldsymbol{\pi}}$, is obtained from $\boldsymbol{\beta}^{(t)}$ through

$$\pi_i^{(t)} = \frac{\exp\big(\sum_{j=1}^p \beta_j^{(t)}x_{ij}\big)}{1 + \exp\big(\sum_{j=1}^p \beta_j^{(t)}x_{ij}\big)}. \tag{5.21}$$

We use $\mathbf{u}^{(t)}$ and $\mathbf{H}^{(t)}$ with formula (4.39) to obtain the next value $\boldsymbol{\beta}^{(t+1)}$, which in this context is

$$\boldsymbol{\beta}^{(t+1)} = \boldsymbol{\beta}^{(t)} + \left\{\mathbf{X}'\mathbf{diag}\big[n_i\pi_i^{(t)}(1 - \pi_i^{(t)})\big]\mathbf{X}\right\}^{-1}\mathbf{X}'(\mathbf{y} - \boldsymbol{\mu}^{(t)}), \tag{5.22}$$

where $\mu_i^{(t)} = n_i\pi_i^{(t)}$. This is used to obtain $\boldsymbol{\pi}^{(t+1)}$, and so forth.

With an initial guess $\beta^{(0)}$, (5.21) yields $\pi^{(0)}$, and for $t > 0$ the iterations proceed as just described using (5.22) and (5.21). In the limit, $\pi^{(t)}$ and $\beta^{(t)}$ converge to the ML estimates $\hat{\pi}$ and $\hat{\beta}$ (Walker and Duncan 1967). The $\mathbf{H}^{(t)}$ matrices converge to $\hat{\mathbf{H}} = -\mathbf{X}'\mathbf{diag}[n_i\hat{\pi}_i(1 - \hat{\pi}_i)]\mathbf{X}$. By (5.20) the estimated asymptotic covariance matrix of $\hat{\beta}$ is a by-product of the Newton–Raphson method, namely $-\hat{\mathbf{H}}^{-1}$.

From the argument in Section 4.6.3, $\beta^{(t+1)}$ has the iterative reweighted least squares form $(\mathbf{X}'\mathbf{V}_t^{-1}\mathbf{X})^{-1}\mathbf{X}'\mathbf{V}_t^{-1}\mathbf{z}^{(t)}$, where $\mathbf{z}^{(t)}$ has elements

$$z_i^{(t)} = \log\frac{\pi_i^{(t)}}{1 - \pi_i^{(t)}} + \frac{y_i - n_i\pi_i^{(t)}}{n_i\pi_i^{(t)}\left(1 - \pi_i^{(t)}\right)}, \tag{5.23}$$

and where $\mathbf{V}_t$ is a diagonal matrix with elements $\{1/n_i\pi_i^{(t)}(1 - \pi_i^{(t)})\}$. In this expression, $\mathbf{z}^{(t)}$ is the linearized form of the logit link function for the sample data, evaluated at $\pi^{(t)}$ [see (4.42)]. From Section 3.1.6 the elements of $\mathbf{V}_t$ are estimated asymptotic variances of the sample logits. The ML estimate is the limit of a sequence of weighted least squares estimates, where the weight matrix changes at each cycle.

5.5.5 Convergence and Existence of Finite Estimates

The log-likelihood function for logistic regression models is strictly concave. ML estimates exist and are unique except in certain boundary cases (Haberman 1974a; Wedderburn 1976; Albert and Anderson 1984). Estimates do not exist or may be infinite when there is no overlap in the sets of explanatory variable values having $y = 0$ and having $y = 1$; that is, when a hyperplane can pass through the space of predictor values such that on one side of that hyperplane $y = 0$ for all observations, whereas on the other side, $y = 1$ always. There is then *perfect discrimination*, as one can predict the sample outcomes perfectly by knowing the predictor values (except possibly at a boundary point). When there is overlap, ML estimates exist and are unique. Similar results occur for the probit and some other links (Silvapulle 1981).

Figure 5.6 illustrates for a single explanatory variable. Here, $y = 0$ at $x = 10, 20, 30, 40$, and $y = 1$ at $x = 60, 70, 80, 90$. An ideal fit has $\hat{\pi} = 0$ for $x \leq 40$ and $\hat{\pi} = 1$ for $x \geq 60$. By letting $\hat{\beta} \to \infty$ and, for fixed $\hat{\beta}$, letting $\hat{\alpha} = -\hat{\beta}(50)$ so that $\hat{\pi} = 0.5$ at $x = 50$, one generates a sequence with ever-increasing value of the likelihood that comes successively closer to a perfect fit.

In practice, most software fails to recognize that $\hat{\beta} = \infty$. After a few cycles of iterative fitting, the log likelihood looks flat at the working estimate, and convergence criteria are satisfied. Because the log likelihood is so flat and because variances come from the inverse of the matrix of negative second derivatives, software typically reports huge standard errors. For these data, for instance, PROC GENMOD in SAS reports $\mathrm{logit}(\hat{\pi}) = -192.2 + 3.8x$ with standard errors of 8.0×10^8 and 1.5×10^7.

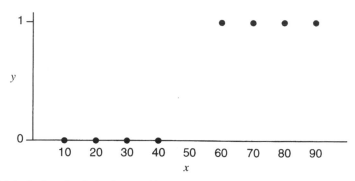

FIGURE 5.6　Perfect discrimination resulting in infinite logistic regression parameter estimate.

NOTES

Section 5.1: Interpreting Parameters in Logistic Regression

5.1. Books focusing on applied logistic regression include Collett (1991) and Hosmer and Lemeshow (2000). Books having major components on logistic regression include Christensen (1997), Cox and Snell (1989), and Morgan (1992). Prentice (1976b) and Stukel (1988) extended the scope by introducing shape parameters that modify the behavior of the curve in extreme probability regions and allow for asymmetric treatment of the two tails.

5.2. Haldane (1956) recommended adding $\frac{1}{2}$ to the numerator and denominator of the sample logit. With this modification, the bias is on the order of only $1/n_i^2$, for large n_i (see Firth 1993a and Problem 14.4).

5.3. The Cornfield (1962) result about normal distributions for $(X \mid Y = i)$ implying the logistic curve for $P(Y = 1 \mid x)$ suggests that logistic regression is useful in *discrimination* and *classification* problems. These use a subject's x value to predict to which of two populations they belong. Anderson (1975), Bull and Donner (1987), Efron (1975), and Press and Wilson (1978) compared logistic regression favorably to discriminant analysis, which assumes that explanatory variables have a normal distribution at each level of Y.

5.4. Rosenbaum and Rubin (1983) used logistic regression to adjust for bias in comparing two groups in observational studies. They defined the *propensity* as the probability of being in one group, for a given setting of the explanatory variables $\mathbf{x}$, and they used logistic regression to estimate how propensity depends on $\mathbf{x}$. In comparing the groups on the response variable, they showed that one can control for differing distributions of the groups on $\mathbf{x}$ by adjusting for the estimated propensity. This is done by using the propensity to match samples from the groups or to subclassify subjects into several strata consisting of intervals of propensity scores or to adjust directly by entering the propensity in the model. See D'Agostino (1998) for a tutorial.

5.5. Adelbasit and Plackett (1983), Chaloner and Larntz (1988), Minkin (1987), and Wu (1985) discussed design problems for binary response experiments, such as choosing settings for a predictor to optimize a criterion for estimating parameter values or estimating the setting at which the response probability equals some fixed value. The nonconstant variance makes this challenging.

Section 5.2: Inference for Logistic Regression

5.6. Albert and Anderson (1984), Berkson (1951, 1953, 1955), Cox (1958a), Hodges (1958), and Walker and Duncan (1967) discussed ML estimation for logistic regression. For adjustments with complex sample surveys, see Hosmer and Lemeshow (2000, Sec. 6.4) and LaVange et al. (2001). Scott and Wild (2001) discussed the analyses of case–control studies with complex sampling designs.

5.7. Tsiatis (1980) suggested an alternative goodness-of-fit test that partitions values for the explanatory variables into a set of regions and adds a dummy variable to the model for each region. The test statistic compares the fit of this model to the simpler one, testing that the extra parameters are not needed. The idea of grouping values to check model fit by comparing observed and fitted counts extends to any GLM (Pregibon 1982). Hosmer et al. (1997) compared various ways of doing this.

Section 5.3: Logit Models with Categorical Predictors

5.8. The Cochran–Armitage trend test is locally asymptotically efficient for both linear and logistic alternatives for $P(Y = 1)$. Its efficiency against linear alternatives follows from the approximate normality of the sample proportions, with constant Bernoulli variance when $\beta = 0$. For the linear logit model (5.5), its efficiency follows from its equivalence with the score test. See Problem 9.35 and Cox (1958a) for related remarks. Tarone and Gart (1980) showed that the score test for a binary linear trend model does not depend on the link function. Gross (1981) noted that for the linear logit model, the local asymptotic relative efficiency for testing independence using the statistic with an incorrect set of scores equals the square of the Pearson correlation between the true and incorrect scores. Simon (1978) gave related asymptotic results. Corcoran et al. (2001), Mantel (1963), and Podgor et al. (1996) extended the trend test.

Section 5.4: Multiple Logistic Regression

5.9. Since the standardized logistic cdf has standard deviation $\pi/\sqrt{3}$, some software (e.g., PROC LOGISTIC in SAS) defines a standardized estimate by multiplying the unstandardized estimate by $s_{x_j}\sqrt{3}/\pi$.

PROBLEMS

Applications

5.1 For a study using logistic regression to determine characteristics associated with remission in cancer patients, Table 5.10 shows the most important explanatory variable, a labeling index (LI). This index measures proliferative activity of cells after a patient receives an injection of tritiated thymidine, representing the percentage of cells that are "labeled." The response Y measured whether the patient achieved remission (1 = yes). Software reports Table 5.11 for a logistic regression model using LI to predict the probability of remission.

TABLE 5.10 Data for Problem 5.1

LI	Number of Cases	Number of Remissions	LI	Number of Cases	Number of Remissions	LI	Number of Cases	Number of Remissions
8	2	0	18	1	1	28	1	1
10	2	0	20	3	2	32	1	0
12	3	0	22	2	1	34	1	1
14	3	0	24	1	0	38	3	2
16	3	0	26	1	1			

Source: Data reprinted with permission from E. T. Lee, *Comput. Prog. Biomed.* **4**: 80–92 (1974).

TABLE 5.11 Computer Output for Problem 5.1

Criterion	Intercept Only	Intercept and Covariates
$-2 \operatorname{Log} L$	34.372	26.073

Testing Global Null Hypothesis: BETA = 0

Test	Chi-Square	DF	Pr > ChiSq
Likelihood Ratio	8.2988	1	0.0040
Score	7.9311	1	0.0049
Wald	5.9594	1	0.0146

Parameter	Estimate	Standard Error	Chi-Square	Pr > ChiSq
Intercept	−3.7771	1.3786	7.5064	0.0061
li	0.1449	0.0593	5.9594	0.0146

Odds Ratio Estimates

Effect	Point Estimate	95% Wald Confidence Limits	
li	1.156	1.029	1.298

Estimated Covariance Matrix

Variable	Intercept	li
Intercept	1.900616	−0.07653
li	−0.07653	0.003521

Obs	li	remiss	n	pi_hat	lower	upper
1	8	0	2	0.06797	0.01121	0.31925
2	10	0	2	0.08879	0.01809	0.34010

a. Show how software obtained $\hat{\pi} = 0.068$ when LI = 8.

b. Show that $\hat{\pi} = 0.5$ when LI = 26.0.

c. Show that the rate of change in $\hat{\pi}$ is 0.009 when LI = 8 and 0.036 when LI = 26.

d. The lower quartile and upper quartile for LI are 14 and 28. Show that $\hat{\pi}$ increases by 0.42, from 0.15 to 0.57, between those values.

e. For a unit change in LI, show that the estimated odds of remission multiply by 1.16.

    f. Explain how to obtain the confidence interval reported for the odds ratio. Interpret.

    g. Construct a Wald test for the effect. Interpret.

    h. Conduct a likelihood-ratio test for the effect, showing how to construct the test statistic using the $-2\log L$ values reported.

    i. Show how software obtained the confidence interval for π reported at LI $= 8$. (*Hint:* Use the reported covariance matrix.)

TABLE 5.12 Data for Problem 5.2[a]

Ft	Temp	TD	Ft	Temp	TD	Ft	Temp	TD	Ft	Temp	TD	Ft	Temp	TD
1	66	0	2	70	1	3	69	0	4	68	0	5	67	0
6	72	0	7	73	0	8	70	0	9	57	1	10	63	1
11	70	1	12	78	0	13	67	0	14	53	1	15	67	0
16	75	0	17	70	0	18	81	0	19	76	0	20	79	0
21	75	1	22	76	0	23	58	1						

[a]Ft, flight number; Temp, temperature (°F); TD, thermal distress (1, yes; 0, no).
Source: Data based on Table 1 in *J. Amer. Statist. Assoc.*, **84**: 945–957, (1989), by S. R. Dalal, E. B. Fowlkes, and B. Hoadley. Reprinted with permission from the *Journal of the American Statistical Association*.

5.2 For the 23 space shuttle flights before the *Challenger* mission disaster in 1986, Table 5.12 shows the temperature at the time of the flight and whether at least one primary O-ring suffered thermal distress.

    a. Use logistic regression to model the effect of temperature on the probability of thermal distress. Plot a figure of the fitted model, and interpret.

    b. Estimate the probability of thermal distress at 31°F, the temperature at the place and time of the *Challenger* flight.

    c. Construct a confidence interval for the effect of temperature on the odds of thermal distress, and test the statistical significance of the effect.

    d. Check the model fit by comparing it to a more complex model.

5.3 Refer to Table 4.2. Using scores $\{0, 2, 4, 5\}$ for snoring, fit the logistic regression model. Interpret using fitted probabilities, linear approximations, and effects on the odds. Analyze the goodness of fit.

5.4 Hastie and Tibshirani (1990, p. 282) described a study to determine risk factors for kyphosis, severe forward flexion of the spine following corrective spinal surgery. The age in months at the time of the operation for the 18 subjects for whom kyphosis was present were 12, 15, 42, 52, 59, 73, 82, 91, 96, 105, 114, 120, 121, 128, 130, 139, 139, 157 and for 22 of the subjects for whom kyphosis was absent were 1, 1, 2, 8, 11, 18, 22, 31, 37, 61, 72, 81, 97, 112, 118, 127, 131, 140, 151, 159, 177, 206.

a. Fit a logistic regression model using age as a predictor of whether kyphosis is present. Test whether age has a significant effect.

b. Plot the data. Note the difference in dispersion on age at the two levels of kyphosis. Fit the model $logit[\pi(x)] = \alpha + \beta_1 x + \beta_2 x^2$. Test the significance of the squared age term, plot the fit, and interpret. (Note also Problem 5.33.)

5.5 Refer to Table 6.11. The Pearson test of independence has $X^2(I) = 6.88$ $(P = 0.14)$. For equally spaced scores, the Cochran–Armitage trend test has $z^2 = 6.67$ $(P = 0.01)$. Interpret, and explain why results differ so. Analyze the data using a linear logit model. Test independence using the Wald and likelihood-ratio tests, and compare results to the Cochran–Armitage test. Check the fit of the model, and interpret.

5.6 For Table 5.3, conduct the trend test using alcohol consumption scores (1, 2, 3, 4, 5) instead of (0.0, 0.5, 1.5, 4.0, 7.0). Compare results, noting the sensitivity to the choice of scores for highly unbalanced data.

5.7 Refer to Table 2.11. Using scores (0, 3, 9.5, 19.5, 37, 55) for cigarette smoking, analyze these data using a logit model. Is the intercept estimate meaningful? Explain.

5.8 A study used the 1998 Behavioral Risk Factors Social Survey to consider factors associated with women's use of oral contraceptives in the United States. Table 5.13 summarizes effects for a logistic regression model for the probability of using oral contraceptives. Each predictor uses a dummy variable, and the table lists the category having dummy outcome 1. Interpret effects. Construct and interpret a confidence interval for the conditional odds ratio between contraceptive use and education.

TABLE 5.13   Data for Problem 5.8

Variable	Coding = 1 if:	Estimate	SE
Age	35 or younger	−1.320	0.087
Race	White	0.622	0.098
Education	≥ 1 year college	0.501	0.077
Marital status	Married	−0.460	0.073

Source: Data courtesy of Debbie Wilson, College of Pharmacy, University of Florida.

TABLE 5.14   Computer Output for Problem 5.9

```
            Criteria For Assessing Goodness Of Fit
            Criterion                 DF          Value
            Deviance                   1          0.3798
            Pearson Chi-Square         1          0.1978
            Log Likelihood                     -209.4783

                         Standard    Likelihood Ratio      Chi-
Parameter    Estimate     Error       95% Conf Limits     Square
Intercept    -3.5961     0.5069     -4.7754    -2.7349     50.33
def          -0.8678     0.3671     -1.5633    -0.1140      5.59
vic           2.4044     0.6006      1.3068     3.7175     16.03

                         LR Statistics
        Source    DF    Chi-Square      Pr > ChiSq
        def        1        5.01          0.0251
        vic        1       20.35          <.0001
```

5.9 Refer to Table 2.6. Table 5.14 shows the results of fitting a logit model, treating death penalty as the response (1 = yes) and defendant's race (1 = white) and victims' race (1 = white) as dummy predictors.

  a. Interpret parameter estimates. Which group is most likely to have the yes response? Find the estimated probability in that case.

  b. Interpret 95% confidence intervals for conditional odds ratios.

  c. Test the effect of defendant's race, controlling for victims' race, using a **(i)** Wald test, and **(ii)** likelihood-ratio test. Interpret.

  d. Test the goodness of fit. Interpret.

5.10 Model the effects of victim's race and defendant's race for Table 2.13. Interpret.

5.11 Table 5.15 appeared in a national study of 15- and 16-year-old adolescent. The event of interest is ever having sexual intercourse. Analyze,

TABLE 5.15   Data for Problem 5.11

Race	Gender	Intercourse	
		Yes	No
White	Male	43	134
	Female	26	149
Black	Male	29	23
	Female	22	36

Source: S. P. Morgan and J. D. Teachman, *J. Marriage Fam.* **50**: 929–936 (1988).   Reprinted with permission from the National Council on Family Relations.

including description and inference about the effects of gender and race, goodness of fit, and summary interpretations.

5.12 According to the *Independent* newspaper (London, Mar. 8, 1994), the Metropolitan Police in London reported 30,475 people as missing in the year ending March 1993. For those of age 13 or less, 33 of 3271 missing males and 38 of 2486 missing females were still missing a year later. For ages 14 to 18, the values were 63 of 7256 males and 108 of 8877 females; for ages 19 and above, the values were 157 of 5065 males and 159 of 3520 females. Analyze and interpret. (Thanks to Pat Altham for showing me these data.)

5.13 The National Collegiate Athletic Association studied graduation rates for freshman student athletes during the 1984–1985 academic year. The (sample size, number graduated) totals were $(796, 498)$ for white females, $(1625, 878)$ for white males, $(143, 54)$ for black females, and $(60, 197)$ for black males (J. J. McArdle and F. Hamagami, *J. Amer. Statist. Assoc.* **89**: 1107–1123, 1994). Analyze and interpret.

5.14 In a study designed to evaluate whether an educational program makes sexually active adolescents more likely to obtain condoms, adolescents were randomly assigned to two experimental groups. The educational program, involving a lecture and videotape about transmission of the HIV virus, was provided to one group but not the other. Table 5.16 summarizes results of a logistic regression model for factors observed to influence teenagers to obtain condoms.

  a. Find the parameter estimates for the fitted model, using $(1, 0)$ dummy variables for the first three predictors. Based on the corresponding confidence interval for the log odds ratio, determine the standard error for the group effect.

  b. Explain why either the estimate of 1.38 for the odds ratio for gender or the corresponding confidence interval is incorrect. Show that if the reported interval is correct, 1.38 is actually the *log* odds ratio, and the estimated odds ratio equals 3.98.

TABLE 5.16   Data for Problem 5.14

Variable	Odds Ratio	95% Confidence Interval
Group (education vs. none)	4.04	(1.17, 13.9)
Gender (males vs. females)	1.38	(1.23. 12.88)
SES (high vs. low)	5.82	(1.87, 18.28)
Lifetime number of partners	3.22	(1.08, 11.31)

Source: V. I. Rickert et al., *Clin. Pediatr.* **31**: 205–210 (1992).

TABLE 5.17   Data for Problem 5.15

Variable	Effect	P-value
Intercept	−7.00	< 0.01
Alcohol use	0.10	0.03
Smoking	1.20	< 0.01
Race	0.30	0.02
Race × smoking	0.20	0.04

5.15   Table 5.17 shows estimated effects for a logistic regression model with squamous cell esophageal cancer ($Y = 1$, yes; $Y = 0$, no) as the response. Smoking status (S) equals 1 for at least one pack per day and 0 otherwise, alcohol consumption (A) equals the average number of alcoholic drinks consumed per day, and race (R) equals 1 for blacks and 0 for whites. To describe the race × smoking interaction, construct the prediction equation when $R = 1$ and again when $R = 0$. Find the fitted YS conditional odds ratio for each case. Similarly, construct the prediction equation when $S = 1$ and again when $S = 0$. Find the fitted YR conditional odds ratios. Note that for each association, the coefficient of the cross-product term is the difference between the log odds ratios at the two fixed levels for the other variable. Explain why the coefficient of S represents the log odds ratio between Y and S for whites. To what hypotheses do the P-values for R and S refer?

5.16   A survey of high school students on Y = whether the subject has driven a motor vehicle after consuming a substantial amount of alcohol (1 = yes), s = gender (1 = female), r = race (1 = black; 0 = white), and g = grade ($g_1 = 1$, grade 9; $g_2 = 1$, grade 10; $g_3 = 1$, grade 11; $g_1 = g_2 = g_3 = 0$, grade 12) has prediction equation

$$\text{logit}\left[\hat{P}(Y = 1)\right] = -0.88 - 0.40s - 0.72r - 2.22g_1 - 1.43g_2 - 0.58g_3$$

$$+ 0.74rg_1 + 0.38rg_2 + 0.01rg_3.$$

a.   Carefully interpret effects. Explain the interaction by describing the race effect at each grade and the grade effect for each race.

b.   Replace r above by r_1 (1 = black, 0 = other). The study also measured r_2 (1 = Hispanic, 0 = other), with $r_1 = r_2 = 0$ for white. Suppose that the prediction equation is as above but with additional terms $-0.29\, r_2 + 0.53\, r_2g_1 + 0.25\, r_2g_2 - 0.06\, r_2g_3$. Interpret the effects.

TABLE 5.18   Data for Problem 5.17

Patient	D	T	Y	Patient	D	T	Y	Patient	D	T	Y
1	45	0	0	13	50	1	0	25	20	1	0
2	15	0	0	14	75	1	1	26	45	0	1
3	40	0	1	15	30	0	0	27	15	1	0
4	83	1	1	16	25	0	1	28	25	0	1
5	90	1	1	17	20	1	0	29	15	1	0
6	25	1	1	18	60	1	1	30	30	0	1
7	35	0	1	19	70	1	1	31	40	0	1
8	65	0	1	20	30	0	1	32	15	1	0
9	95	0	1	21	60	0	1	33	135	1	1
10	35	0	1	22	61	0	0	34	20	1	0
11	75	0	1	23	65	0	1	35	40	1	0
12	45	1	1	24	15	1	0				

Source: Data from D. Collett, in *Encyclopedia of Biostatistics* (New York: Wiley: 1998), pp. 350–358.

5.17 Table 5.18 shows the results of a study about Y = whether a patient having surgery with general anesthesia experienced a sore throat on waking (0 = no, 1 = yes) as a function of the D = duration of the surgery (in minutes) and the T = type of device used to secure the airway (0 = laryngeal mask airway, 1 = tracheal tube). Fit a logit model using these predictors, interpret parameter estimates, and conduct inference about the effects.

5.18 Refer to model (5.2) for the horseshoe crabs using x = width.
   a. Show that (i) at the mean width (26.3), the estimated odds of a satellite equal 2.07; (ii) at x = 27.3, the estimated odds equal 3.40; and (iii) since $\exp(\hat{\beta})$ = 1.64, 3.40 = (1.64)2.07, and the odds increase by 64%.
   b. Based on the 95% confidence interval for β, show that for x near where $\pi = 0.5$, the rate of increase in the probability of a satellite per 1-cm increase in x falls between about 0.07 and 0.17.

5.19 For Table 4.3, fit a logistic regression model for the probability of a satellite, using color alone as the predictor.
   a. Treat color as nominal. Explain why this model is saturated. Express its parameter estimates in terms of the sample logits for each color.
   b. Conduct a likelihood-ratio test that color has no effect.
   c. Fit a model that treats color as quantitative. Interpret the fit, and test that color has no effect.
   d. Test the goodness of fit of the model in part (c). Interpret.

5.20 Refer to model (5.14). Describe the effect of width by finding the estimated probabilities of a satellite at its lower and upper quartiles, separately for $c = 1$ and $c = 0$.

5.21 Refer to the prediction equation $\text{logit}(\hat{\pi}) = -10.071 - 0.509c + 0.458x$ for model (5.13). The means and standard deviations are $\bar{c} = 2.44$ and $s = 0.80$ for color, and $\bar{x} = 26.30$ and $s = 2.11$ for width. For standardized predictors [e.g., $x = (\text{width} - 26.3)/2.11$], explain why the estimated coefficients of c and x equal -0.41 and 0.97. Interpret these by comparing the partial effects of a 1 standard deviation increase in each predictor on the odds. Describe the color effect by estimating the change in $\hat{\pi}$ between the first and last color categories at the mean score for width.

5.22 Refer to model (5.12).
  a. Fit the model using $x = $ weight. Interpret effects of weight and color.
  b. Does the model permitting interaction provide an improved fit? Interpret.
  c. For part (b), construct a confidence interval for a difference between the slope parameters for medium-light and dark crabs. Interpret.
  d. Using models that treat color as quantitative, repeat the analyses in parts (a) to (c).

5.23 Fowlkes et al. (1988) reported results of a survey of employees of a large national corporation to determine how satisfaction depends on race, gender, age, and regional location. The data are at the book's Web site (*www.stat.ufl.edu / ~ aa /cda /cda.html*). Fit a logit model to these data and carefully interpret the parameter estimates. Fowlkes et al. (1988) reported "The least-satisfied employees are less than 35 years of age, female, other (race), and work in the Northeast; .... . The most satisfied group is greater than 44 years of age, male, other, and working in the Pacific or Mid-Atlantic regions; the odds of such employees being satisfied are about 3.5 to 1." Show how these interpretations result from the fit of this model.

5.24 Let Y denote a subject's opinion about current laws legalizing abortion ($1 = $ support), for gender h ($h = 1$, female; $h = 2$, male), religious affiliation i ($i = 1$, Protestant; $i = 2$, Catholic; $i = 3$, Jewish), and political party affiliation j ($j = 1$, Democrat; $j = 2$, Republican; $j = 3$, Independent). For survey data, software for fitting the model

$$\text{logit}[P(Y = 1)] = \alpha + \beta_h^G + \beta_i^R + \beta_j^P$$

reports $\hat{\alpha} = 0.62$, $\hat{\beta}_1^G = 0.08$, $\hat{\beta}_2^G = -0.08$, $\hat{\beta}_1^R = -0.16$, $\hat{\beta}_2^R = -0.25$, $\hat{\beta}_3^R = 0.41$, $\hat{\beta}_1^P = 0.87$, $\hat{\beta}_2^P = -1.27$, $\hat{\beta}_3^P = 0.40$.

a. Interpret how the odds of support depends on religion.

b. Estimate the probability of support for the group most (least) likely to support current laws.

c. If, instead, parameters used constraints $\beta_1^G = \beta_1^R = \beta_1^P = 0$, report the estimates.

5.25 Table 5.19 refers to a sample of subjects randomly selected for an Italian study on the relation between income and whether one possesses a travel credit card. At each level of annual income in millions of lira, the table indicates the number of subjects sampled and the number possessing at least one travel credit card. Analyze these data.

TABLE 5.19 Data for Problem 5.25

Income (millions of lira)	Number of Cases	Credit Cards	Income (millions of lira)	Number of Cases	Credit Cards	Income (millions of lira)	Number of Cases	Credit Cards
24	1	0	39	2	0	65	6	6
27	1	0	40	5	0	68	3	3
28	5	2	41	2	0	70	5	3
29	3	0	42	2	0	79	1	0
30	9	1	45	1	1	80	1	0
31	5	1	48	1	0	84	1	0
32	8	0	49	1	0	94	1	0
33	1	0	50	10	2	120	6	6
34	7	1	52	1	0	130	1	1
35	1	1	59	1	0			
38	3	1	60	5	2			

Source: Categorical Data Analysis, Quaderni del Corso Estivo di Statistica e Calcolo delle Probabilità, n. 4., Istituto di Metodi Quantitativi, Università Luigi Bocconi, by R. Piccarreta.

5.26 Refer to Table 9.1, treating marijuana use as the response variable. Analyze these data.

5.27 The book's Web site (*www.stat.ufl.edu/~aa/cda/cda.html*) contains a five-way table relating occupational aspirations (high, low) to gender, residence, IQ, and socioeconomic status. Analyze these data.

Theory and Methods

5.28 For model (5.1), show that $\partial \pi(x)/\partial x = \beta \pi(x)[1 - \pi(x)]$.

5.29 For model (5.1), when $\pi(x)$ is small, explain why you can interpret $\exp(\beta)$ approximately as $\pi(x + 1)/\pi(x)$.

5.30 Prove that the logistic regression curve (5.1) has the steepest slope where $\pi(x) = \frac{1}{2}$. Generalize to model (5.8).

5.31 The calibration problem is that of estimating x at which $\pi(x) = \pi_0$. For the linear logit model, argue that a confidence interval is the set of x values for which

$$\left|\hat\alpha + \hat\beta x - \text{logit}(\pi_0)\right| / \left[\text{var}(\hat\alpha) + x^2 \,\text{var}(\hat\beta) + 2x \,\text{cov}(\hat\alpha,\hat\beta)\right]^{1/2} < z_{\alpha/2}.$$

[Morgan (1992, Sec. 2.7) surveyed other approaches.]

5.32 A study for several professional sports of the effect of a player's draft position d $(d = 1, 2, 3, \dots)$ of selection from the pool of potential players in a given year on the probability π of eventually being named an all star used the model $\text{logit}(\pi) = \alpha + \beta \log d$ (S. M. Berry, *Chance*, **14**:53–57, 2001).

    a. Show that $\pi/(1 - \pi) = e^\alpha d^\beta$. Show that $e^\alpha =$ odds for the first draft pick.

    b. In the United States, Berry reported $\hat\alpha = 2.3$ and $\hat\beta = -1.1$ for pro basketball and $\hat\alpha = 0.7$ and $\hat\beta = -0.6$ for pro baseball. This suggests that in basketball a first draft pick is more crucial and picks with high d are relatively less likely to be all-stars. Explain why.

5.33 For the population of subjects having $Y = j$, X has a $N(\mu_j, \sigma^2)$ distribution, $j = 0,1$.

    a. Using Bayes theorem, show that $P(Y = 1|x)$ satisfies the logistic regression model with $\beta = (\mu_1 - \mu_0)/\sigma^2$.

    b. Suppose that $(X|Y = j)$ is $N(\mu_j,\sigma_j^2)$ with $\sigma_0 \neq \sigma_1$. Show that the logistic model holds with a quadratic term (Anderson 1975). [Problem 5.4 showed that a quadratic term is helpful when x values have quite different dispersion at $y = 0$ and $y = 1$. This result also suggests that to test equality of means of normal distributions when the variances differ, one can fit a quadratic logistic regression with the two groups as the response and test the quadratic term; see O'Brien (1988).]

    c. Suppose that $(X|Y = j)$ has exponential dispersion family density $f(x; \theta_j) = \exp\{[x\theta_j - b(\theta_j)]/a(\phi) + c(x, \phi)\}$. Find the relevant logistic model.

d. For multiple predictors, suppose that $(\mathbf{X} \mid Y = j)$ has a multivariate $N(\boldsymbol{\mu}_j, \boldsymbol{\Sigma})$ distribution, $j = 0, 1$. Show that $P(Y = 1 \mid \mathbf{x})$ satisfies logistic regression with effect parameters $\boldsymbol{\Sigma}^{-1}(\boldsymbol{\mu}_1 - \boldsymbol{\mu}_0)$ (Cornfield 1962).

5.34 Suppose that $\pi(x) = F(x)$ for some strictly increasing cdf F. Explain why a monotone transformation of x exists such that the logistic regression model holds. Generalize to alternative link functions.

5.35 For an $I \times 2$ contingency table, consider logit model (5.4).
   a. Given $\{\pi_i > 0\}$, show how to find $\{\beta_i\}$ satisfying $\beta_I = 0$.
   b. Prove that $\beta_1 = \beta_2 = \cdots = \beta_I$ is the independence model. Find its likelihood equation, and show that $\hat{\alpha} = \text{logit}[(\sum_i y_i)/(\sum_i n_i)]$.

5.36 Construct the log-likelihood function for the model $\text{logit}[\pi(x)] = \alpha + \beta x$ with independent binomial outcomes of y_0 successes in n_0 trials at $x = 0$ and y_1 successes in n_1 trials at $x = 1$. Derive the likelihood equations, and show that $\hat{\beta}$ is the sample log odds ratio.

5.37 A study has n_i independent binary observations $\{y_{i1}, \ldots, y_{in_i}\}$ when $X = x_i$, $i = 1, \ldots, N$, with $n = \sum_i n_i$. Consider the model $\text{logit}(\pi_i) = \alpha + \beta x_i$, where $\pi_i = P(Y_{ij} = 1)$.
   a. Show that the kernel of the likelihood function is the same treating the data as n Bernoulli observations or N binomial observations.
   b. For the saturated model, explain why the likelihood function is different for these two data forms. (*Hint:* The number of parameters differs.) Hence, the deviance reported by software depends on the form of data entry.
   c. Explain why the difference between deviances for two unsaturated models does not depend on the form of data entry.
   d. Suppose that each $n_i = 1$. Show that the deviance depends on $\hat{\pi}_i$ but not y_i. Hence, it is not useful for checking model fit (see also Problem 4.22).

5.38 Suppose that Y has a $\text{bin}(n, \pi)$ distribution. For the model, $\text{logit}(\pi) = \alpha$, consider testing H_0: $\alpha = 0$ (i.e., $\pi = 0.5$). Let $\hat{\pi} = y/n$.
   a. From Section 3.1.6, the asymptotic variance of $\hat{\alpha} = \text{logit}(\hat{\pi})$ is $[n\pi(1 - \pi)]^{-1}$. Compare the estimated SE for the Wald test and the SE using the null value of π, using test statistic $[\text{logit}(\hat{\pi})/\text{SE}]^2$. Show that the ratio of the Wald statistic to the statistic with null SE equals $4\hat{\pi}(1 - \hat{\pi})$. What is the implication about performance of the Wald test if $|\alpha|$ is large and $\hat{\pi}$ tends to be near 0 or 1?

b. Wald inference depends on the parameterization. How does the comparison of tests change with the scale $[(\hat{\pi} - 0.5)/\text{SE}]^2$, where SE is now the estimated or null SE of $\hat{\pi}$?

c. Suppose that $y = 0$ or $y = n$. Show that the Wald test in part (a) cannot reject $H_0: \pi = \pi_0$ for any $0 < \pi_0 < 1$, whereas the Wald test in part (b) rejects every such π_0. [*Note:* Analogous results apply for inference about the Poisson mean versus the log mean; see Mantel (1987a).]

5.39 Find the likelihood equations for model (5.10). Show that they imply the fitted values and that the sample values are identical in the marginal two-way tables.

5.40 Consider the linear logit model (5.5) for an $I \times 2$ table, with y_i a $\text{bin}(n_i, \pi_i)$ variate.

a. Show that the log likelihood is

$$L(\boldsymbol{\beta}) = \sum_{i=1}^{I} y_i(\alpha + \beta x_i) - \sum_{i=1}^{I} n_i \log[1 + \exp(\alpha + \beta x_i)].$$

b. Show that the sufficient statistic for β is $\sum_i y_i x_i$, and explain why this is essentially the variable utilized in the Cochran–Armitage test. (Hence that test is a score test of $H_0: \beta = 0$.)

c. Letting $S = \sum_i y_i$, show that the likelihood equations are

$$S = \sum_i n_i \frac{\exp(\alpha + \beta x_i)}{1 + \exp(\alpha + \beta x_i)}$$

$$\sum_i y_i x_i = \sum_i n_i x_i \frac{\exp(\alpha + \beta x_i)}{1 + \exp(\alpha + \beta x_i)}.$$

d. Let $\{\hat{\mu}_i = n_i \hat{\pi}_i\}$. Explain why $\sum_i \hat{\mu}_i = \sum_i y_i$ and

$$\sum_i x_i \frac{y_i}{S} = \sum_i x_i \frac{\hat{\mu}_i}{\sum_a \hat{\mu}_a}.$$

Explain why this implies that the mean score on x across the rows in the first column is the same for the model fit as for the observed data. They are also identical for the second column.

5.41 Let Y_i be $\text{bin}(n_i, \pi_i)$ at x_i, and let $p_i = y_i/n_i$. For binomial GLMs with logit link:

a. For p_i near π_i, show that

$$\log \frac{p_i}{1 - p_i} \approx \log \frac{\pi_i}{1 - \pi_i} + \frac{p_i - \pi_i}{\pi_i(1 - \pi_i)}.$$

b. Show that $z_i^{(t)}$ in (5.23) is a linearized version of the ith sample logit, evaluated at approximation $\pi_i^{(t)}$ for $\hat{\pi}_i$.

c. Verify the formula (5.20) for $\widehat{\text{cov}(\hat{\boldsymbol{\beta}})}$.

5.42 Using graphs or tables, explain what is meant by *no interaction* in modeling response Y and explanatory X and Z when:

a. All variables are continuous (multiple regression).

b. Y and X are continuous, Z is categorical (analysis of covariance).

c. Y is continuous, X and Z are categorical (two-way ANOVA).

d. Y is binary, X and Z are categorical (logit model).

CHAPTER 6

Building and Applying Logistic Regression Models

Having studied the basics of fitting and interpreting logistic regression models, we now turn our attention to building and applying them. With several explanatory variables, there are many potential models. In Section 6.1 we discuss strategies for model selection. After choosing a preliminary model, model checking addresses whether systematic lack of fit exists. Section 6.2 covers diagnostics, such as residuals, for model checking.

In practice, a common application compares two groups on a binary response, with data stratified by control variables. In Section 6.3 we present logit-related analyses of such data. In Section 6.4 we show the advantages of a well-chosen model in enhancing inferential power for detecting and estimating associations. Section 6.5 covers power and sample size determination for logistic regression. Although the logit is the most popular link function for probabilities, other links are sometimes more appropriate. In Section 6.6 we present models using the probit link and links making a double log transform.

For small samples or models with many parameters, ordinary large-sample ML inference may perform poorly. In Section 6.7 we discuss *conditional logistic regression*. Like small-sample methods for 2×2 tables, this uses conditioning arguments to eliminate nuisance parameters.

6.1 STRATEGIES IN MODEL SELECTION

Model selection for logistic regression faces the same issues as for ordinary regression. The selection process becomes harder as the number of explanatory variables increases, because of the rapid increase in possible effects and interactions. There are two competing goals: The model should be complex enough to fit the data well. On the other hand, it should be simple to interpret, smoothing rather than overfitting the data.

Most studies are designed to answer certain questions. Those questions guide the choice of model terms. Confirmatory analyses then use a restricted set of models. For instance, a study hypothesis about an effect may be tested by comparing models with and without that effect. For studies that are exploratory rather than confirmatory, a search among possible models may provide clues about the dependence structure and raise questions for future research.

In either case, it is helpful first to study the effect on Y of each predictor by itself using graphics (incorporating smoothing) for a continuous predictor or a contingency table for a discrete predictor. This gives a "feel" for the marginal effects. Unbalanced data, with relatively few responses of one type, limit the number of predictors for the model. One guideline suggests at least 10 outcomes of each type should occur for every predictor (Peduzzi et al. 1996). If $y = 1$ only 30 times out of $n = 1000$, for instance, the model should contain no more than about three x terms. Such guidelines are approximate, and this does *not* mean that if you have 500 outcomes of each type you are well served by a model with 50 predictors.

Many model selection procedures exist, no one of which is always best. Cautions that apply to ordinary regression hold for any generalized linear model. For instance, a model with several predictors may suffer from *multicollinearity*—correlations among predictors making it seem that no one variable is important when all the others are in the model. A variable may seem to have little effect because it overlaps considerably with other predictors in the model, itself being predicted well by the other predictors. Deleting such a redundant predictor can be helpful, for instance to reduce standard errors of other estimated effects.

6.1.1 Horseshoe Crab Example Revisited

The horseshoe crab data set in Table 4.3 has four predictors: color (four categories), spine condition (three categories), weight, and width of the carapace shell. We now fit a logistic regression model using all these to predict whether the female crab has satellites ($y = 1$).

We start by fitting a model containing main effects,

$$\text{logit}[P(Y = 1)] = \alpha + \beta_1 \text{weight} + \beta_2 \text{width} + \beta_3 c_1$$
$$+ \beta_4 c_2 + \beta_5 c_3 + \beta_6 s_1 + \beta_7 s_2,$$

treating color (c_i) and spine condition (s_j) as qualitative (factors), with dummy variables for the first three colors and the first two spine conditions. Table 6.1 shows results. A likelihood-ratio test that Y is jointly independent of these predictors simultaneously tests H_0: $\beta_1 = \cdots = \beta_7 = 0$. The test statistic equals 40.6 with df $= 7$ ($P < 0.0001$). This shows extremely strong evidence that at least one predictor has an effect.

**TABLE 6.1   Computer Output from Fitting Model with All Main
Effects to Horseshoe Crab Data**

```
                 Testing Global Null Hypothesis: BETA = 0
        Test                    Chi-Square     DF      Pr > ChiSq
        Likelihood Ratio          40.5565       7        <.0001

                 Analysis of Maximum Likelihood Estimates
Parameter      Estimate     Std Error     Chi-Square     Pr > ChiSq
Intercept      −9.2734       3.8378         5.8386         0.0157
weight          0.8258       0.7038         1.3765         0.2407
width           0.2631       0.1953         1.8152         0.1779
color  1        1.6087       0.9355         2.9567         0.0855
color  2        1.5058       0.5667         7.0607         0.0079
color  3        1.1198       0.5933         3.5624         0.0591
spine  1       −0.4003       0.5027         0.6340         0.4259
spine  2       −0.4963       0.6292         0.6222         0.4302
```

Although the overall test is highly significant, the Table 6.1 results are discouraging. The estimates for weight and width are only slightly larger than their SE values. The estimates for the factors compare each category to the final one as a baseline. For color, the largest difference is less than two standard errors; for spine condition, the largest difference is less than a standard error.

The small P-value for the overall test, yet the lack of significance for individual effects, is a warning sign of multicollinearity. In Section 5.2.2 we showed strong evidence of a width effect. Controlling for weight, color, and spine condition, little evidence remains of a partial width effect. However, weight and width have a strong correlation (0.887). For practical purposes they are equally good predictors, but it is nearly redundant to use them both. Our further analysis uses width (W) with color (C) and spine condition (S) as predictors. For simplicity, we symbolize models by their highest-order terms, regarding C and S as factors. For instance, $(C + S + W)$ denotes a model with main effects, whereas $(C + S*W)$ denotes a model that has those main effects plus an $S \times W$ interaction. It is not usually sensible to consider a model with interaction but not the main effects that make up that interaction.

6.1.2   Stepwise Procedures

In exploratory studies, an algorithmic method for searching among models can be informative if we use results cautiously. Goodman (1971a) proposed methods analogous to forward selection and backward elimination in ordinary regression.

Forward selection adds terms sequentially until further additions do not improve the fit. At each stage it selects the term giving the greatest improvement in fit. The minimum *P*-value for testing the term in the model is a sensible criterion, since reductions in deviance for different terms may have different df values. A stepwise variation of this procedure retests, at each stage, terms added at previous stages to see if they are still significant.

Backward elimination begins with a complex model and sequentially removes terms. At each stage, it selects the term for which its removal has the least damaging effect on the model (e.g., largest *P*-value). The process stops when any further deletion leads to a significantly poorer fit. With either approach, for qualitative predictors with more than two categories, the process should consider the entire variable at any stage rather than just individual dummy variables. Add or drop the entire variable rather than just one of its dummies. Otherwise, the result depends on the coding. The same remark applies to interactions containing that variable.

Many statisticians prefer backward elimination over forward selection, feeling it safer to delete terms from an overly complex model than to add terms to an overly simple one. Forward selection can stop prematurely because a particular test in the sequence has low power. Neither strategy necessarily yields a meaningful model. Use variable selection procedures with caution! When you evaluate many terms, one or two that are not important may look impressive simply due to chance. For instance, when all the true effects are weak, the largest sample effect may substantially overestimate its true effect. See Westfall and Wolfinger (1997) and Westfall and Young (1993) for ways to adjust *P*-values to take multiple tests into account.

Some software has additional options for selecting a model. One approach attempts to determine the best model with some fixed number of terms, according to some criterion. If such a method and backward and forward selection procedures yield quite different models, this is an indication that such results are of dubious use. Another such indication would be when a quite different model results from applying a given procedure to a bootstrap sample of the same size from the sample distribution.

Finally, statistical significance should not be the sole criterion for inclusion of a term in a model. It is sensible to include a variable that is central to the purposes of the study and report its estimated effect even if it is not statistically significant. Keeping it in the model may help reduce bias in estimated effects of other predictors and may make it possible to compare results with other studies where the effect is significant (perhaps because of a larger sample size.) Algorithmic selection procedures are no substitute for careful thought in guiding the formulation of models.

6.1.3   Backward Elimination for Horseshoe Crab Example

Table 6.2 summarizes results of fitting and comparing several logit models to the horseshoe crab data with predictors width, color, and spine condition. The deviance (G^2) test of fit compares the model to the saturated model. As noted in Sections 5.2.4 and 5.2.5, this is not approximately chi-squared when a predictor is continuous, as width is. However, the difference of deviances

TABLE 6.2 Results of Fitting Several Logistic Regression Models to Horseshoe Crab Data

Model	Predictors[a]	Deviance G^2	df	AIC	Models Compared	Deviance Difference	Corr. $r(y, \hat{\mu})$
1	$(C*S*W)$	170.44	152	212.4	—	—	
2	$(C*S + C*W + S*W)$	173.68	155	209.7	(2)–(1)	3.2 (df = 3)	
3a	$(C*S + S*W)$	177.34	158	207.3	(3a)–(2)	3.7 (df = 3)	
3b	$(C*W + S*W)$	181.56	161	205.6	(3b)–(2)	7.9 (df = 6)	
3c	$(C*S + C*W)$	173.69	157	205.7	(3c)–(2)	0.0 (df = 2)	
4a	$(S + C*W)$	181.64	163	201.6	(4a)–(3c)	8.0 (df = 6)	
4b	$(W + C*S)$	177.61	160	203.6	(4b)–(3c)	3.9 (df = 3)	
5	$(C + S + W)$	186.61	166	200.6	(5)–(4b)	9.0 (df = 6)	
6a	$(C + S)$	208.83	167	220.8	(6a)–(5)	22.2 (df = 1)	
6b	$(S + W)$	194.42	169	202.4	(6b)–(5)	7.8 (df = 3)	
6c	$(C + W)$	187.46	168	197.5	(6c)–(5)	0.8 (df = 2)	0.452
7a	(C)	212.06	169	220.1	(7a)–(6c)	24.5 (df = 1)	0.285
7b	(W)	194.45	171	198.5	(7b)–(6c)	7.0 (df = 3)	0.402
8	$(C = \text{dark} + W)$	187.96	170	194.0	(8)–(6c)	0.5 (df = 2)	0.447
9	None	225.76	172	227.8	(9)–(8)	37.8 (df = 2)	0.000

[a]C, color; S, spine condition; W, width.

between two models that differ by a modest number of parameters is relevant. That difference is the likelihood-ratio statistic $-2(L_0 - L_1)$ comparing the models, and it has an approximate null chi-squared distribution..

To select a model, we use backward elimination. We test only the highest-order terms for each variable. It is inappropriate, for instance, to remove a main effect term if the model has interactions involving that term.

We begin with the most complex model, symbolized by $(C*S*W)$, model 1 in Table 6.2. This model uses main effects for each term as well as the three two-factor interactions and the three-factor interaction. It allows a separate width effect at each CS combination. (In fact, at some of those combinations y outcomes of only one type occur, so effects are not estimable.) The likelihood-ratio statistic comparing this model to the simpler model $(C*S + C*W + S*W)$ removing the three-factor interaction term equals 3.2 (df = 3). This suggests that the three-factor term is not needed $(P = 0.36)$, thank goodness, so we continue the simplification process.

In the next stage we consider the three models that remove a two-factor interaction. Of these, $(C*S + C*W)$ gives essentially the same fit as the more complex model, so we drop the $S \times W$ interaction. Next, we consider dropping one of the other two-factor interactions. The model $(S + C*W)$, dropping the $C \times S$ interaction, has an increased deviance of 8.0 on df = 6 $(P = 0.24)$; the model $(W + C*S)$, dropping the $C \times W$ interaction, has an increased deviance of 3.9 on df = 3 $(P = 0.27)$. Neither increase is important, suggesting that we can drop either and proceed. In either case, dropping next the remaining interaction also seems permissible. For instance, dropping the $C \times S$ interaction from model $(W + C^*S)$, leaving model $(C + S + W)$, increases the deviance by 9.0 on df = 6 ($P = 0.17$).

The working model now has the main effects alone. In the next stage we consider dropping one of them. Table 6.2 shows little consequence of removing S. Both remaining variables (C and W) then have nonnegligible effects. For instance, removing C increases the deviance (comparing models 7b and 6c) by 7.0 on df = 3 ($P = 0.07$). The analysis in Section 5.4.6 revealed a noticeable difference between dark crabs (category 4) and the others. The simpler model that has a single dummy variable for color, equaling 0 for dark crabs and 1 otherwise, fits essentially as well. (The deviance difference between models 8 and 6c equals 0.5, with df = 2.) Further simplification results in large increases in deviance and is unjustified.

6.1.4 AIC, Model Selection, and the Correct Model

In selecting a model, we are mistaken if we think that we have found the true one. Any model is a simplification of reality. For instance, width does not exactly have a linear effect on the probability of satellites, whether we use the logit link or the identity link.

What is the logic of testing the fit of a model when we know that it does not truly hold? A simple model that fits adequately has the advantages of model parsimony. If a model has relatively little bias, describing reality well, it tends to provide more accurate estimates of the quantities of interest. This was discussed in Sections 3.3.7 and 5.2.2 and is examined further in Section 6.4.5.

Other criteria besides significance tests can help select a good model in terms of estimating quantities of interest. The best known is the *Akaike information criterion* (AIC). It judges a model by how close its fitted values tend to be to the true values, in terms of a certain expected value. Even though a simple model is farther from the true model than is a more complex model, it may be preferred because it tends to provide better estimates of certain characteristics of the true model, such as cell probabilities. Thus, the optimal model is the one that tends to have fit closest to reality. Given a sample, Akaike showed that this criterion selects the model that minimizes

$$\text{AIC} = -2(\text{maximized log likelihood} - \text{number of parameters in model}).$$

This penalizes a model for having many parameters. With models for categorical Y, this ordering is equivalent to one based on an adjustment of the deviance, $[G^2 - 2(\text{df})]$, by twice its residual df. For cogent arguments supporting this criterion, see Burnham and Anderson (1998).

We illustrate AIC for model selection using the models Table 6.2 lists. That table also shows the AIC values. Of models using the three basic variables, AIC is smallest (AIC = 197.5) for $C + W$, having main effects of color and width. The simpler model having a dummy variable for whether a crab is dark fares better yet (AIC = 194.0). Either model seems reasonable.

We should balance the lower AIC for the simpler model against its having been suggested by the fit of $C + W$.

6.1.5 Using Causal Hypotheses to Guide Model Building

Although selection procedures are helpful exploratory tools, the model-building process should utilize theory and common sense. Often, a time ordering among the variables suggests possible causal relationships. Analyzing a certain sequence of models helps to investigate those relationships (Goodman 1973).

We illustrate with Table 6.3, from a British study. A sample of men and women who had petitioned for divorce and a similar number of married people were asked: (a) "Before you married your (former) husband/wife, had you ever made love with anyone else?"; (b) "During your (former) marriage, (did you have) have you had any affairs or brief sexual encounters with another man/woman?" The $2 \times 2 \times 2 \times 2$ table has variables $G =$ gender, $E =$ whether reported extramarital sex, $P =$ whether reported premarital sex, and $M =$ marital status.

The time points at which responses on the four variables occur suggests the following ordering of the variables:

$$
\begin{array}{ccccccc}
G & \rightarrow & P & \rightarrow & E & \rightarrow & M \\
\text{gender} & & \text{premarital} & & \text{extramarital} & & \text{marital} \\
& & \text{sex} & & \text{sex} & & \text{status}
\end{array}
$$

Any of these is an explanatory variable when a variable listed to its right is the response. Figure 6.1 shows one possible causal structure. In this figure, a variable at the tip of an arrow is a response for a model at some stage. The explanatory variables have arrows pointing to the response, directly or indirectly.

We first treat P as a response. Figure 6.1 predicts that G has a direct effect on P, so the model of independence of these variables is inadequate.

TABLE 6.3 Marital Status by Report of Pre- and Extramarital Sex (PMS and EMS)

		Gender							
		Women				Men			
	PMS:	Yes		No		Yes		No	
Marital Status	EMS:	Yes	No	Yes	No	Yes	No	Yes	No
Divorced		17	54	36	214	28	60	17	68
Still married		4	25	4	322	11	42	4	130

Source: G. N. Gilbert, *Modelling Society* (London: George Allen & Unwin, 1981). Reprinted with permission from Unwin Hyman Ltd.

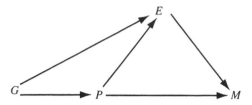

FIGURE 6.1   Causal diagram for Table 6.3.

At the second stage, E is the response. Figure 6.1 predicts that P and G have direct effects on E. It also suggests that G has an indirect effect on E, through its effect on P. These effects on E can be analyzed using the logit model for E with additive G and P effects. If G has only an indirect effect on E, the model with P alone as a predictor is adequate; that is, controlling for P, E and G are conditionally independent. At the third stage, M is the response. Figure 6.1 predicts that E has a direct effect on M, P has direct effects and indirect effects through its effects on E, and G has indirect effects through its effects on P and E. This suggests the logit model for M having additive E and P effects. For this model, G and M are independent, given P and E.

Table 6.4 shows results. The first stage, having P as the response, shows strong evidence of a GP association. The sample odds ratio for their marginal table is 0.27; the estimated odds of premarital sex for females are 0.27 times that for males. The second stage has E as the response. Only weak evidence occurs that G had a direct as well as an indirect effect on E, as G^2 drops by 2.9 (df = 1) after adding G to a model already containing P as a predictor. For this model, the estimated EP conditional odds ratio is 4.0.

The third stage has M as the response. Figure 6.1 specifies the logit model with main effects of E and P, but it fits poorly. The model that allows an

TABLE 6.4   Goodness of Fit of Various Models for Table 6.3[a]

Stage	Response Variable	Potential Explanatory	Actual Explanatory	G^2	df
1	P	G	None	75.3	1
			(G)	0.0	0
2	E	G, P	None	48.9	3
			(P)	2.9	2
			$(G + P)$	0.0	1
3	M	G, P, E	$(E + P)$	18.2	5
			$(E*P)$	5.2	4
			$(E*P + G)$	0.7	3

[a]P, premarital sex; E, extramarital sex; M, marital status; G, gender.

$E \times P$ interaction in their effects on M but assumes conditional independence of G and M fits much better (G^2 decrease of 13.0, df = 1). The model that also has a main effect for G fits slightly better yet. Either model is more complicated than Figure 6.1 predicted, since the effects of E on M vary according to the level of P. However, some preliminary thought about causal relationships suggested a model similar to one giving a good fit. We leave it to the reader to estimate and interpret effects for the third stage.

6.1.6   New Model-Building Strategies for Data Mining

As computing power continues to explode, enormous data sets are more common. A financial institution that markets credit cards may have observations for millions of subjects to whom they sent advertising, on whether they applied for a card. For their customers, they have monthly data on whether they paid their bill on time plus information on many variables measured on the credit card application. The analysis of huge data sets is called *data mining*.

Model building for huge data sets is challenging. There is currently considerable study of alternatives to traditional statistical methods, including automated algorithms that ignore concepts such as sampling error or modeling. Significance tests are usually irrelevant, as nearly any variable has a significant effect if n is sufficiently large. Model-building strategies view some models as useful for prediction even if they have complex structure. Nonetheless, a point of diminishing returns still occurs in adding predictors to models. After a point, new predictors tend to be so correlated with a linear combination of ones already in the model that they do not improve predictive power. For large n, inference is less relevant than summary measures of predictive power. This is a topic of the next section.

6.2   LOGISTIC REGRESSION DIAGNOSTICS

In Section 5.2.3 we introduced statistics for checking model fit in a global sense. After selecting a preliminary model, we obtain further insight by switching to a microscopic mode of analysis. In contingency tables, for instance, the pattern of lack of fit revealed in cell-by-cell comparisons of observed and fitted counts may suggest a better model. For continuous predictors, graphical displays are also helpful. Such diagnostic analyses may suggest a reason for the lack of fit, such as nonlinearity in the effect of an explanatory variable.

6.2.1   Pearson, Deviance, and Standardized Residuals

With categorical predictors, it is useful to form residuals to compare observed and fitted counts. Let y_i denote the binomial variate for n_i trials at setting i of the explanatory variables, $i = 1, \ldots, N$. Let $\hat{\pi}_i$ denote the model estimate of $P(Y = 1)$. Then $n_i \hat{\pi}_i$ is the fitted number of successes. For a GLM with binomial random component, the Pearson residual (4.36) for this fit is

$$e_i = \frac{y_i - n_i \hat{\pi}_i}{\left[\widehat{\mathrm{var}(Y_i)}\right]^{1/2}} = \frac{y_i - n_i \hat{\pi}_i}{\sqrt{\left[n_i \hat{\pi}_i (1 - \hat{\pi}_i)\right]}}. \tag{6.1}$$

This divides the raw residual $(y_i - \hat{\mu}_i)$ by the estimated binomial standard deviation of y_i. The Pearson statistic for testing the model fit satisfies

$$X^2 = \sum_{i=1}^{N} e_i^2.$$

Each squared Pearson residual is a component of X^2.

With $\hat{\pi}_i$ replaced by π_i in the numerator of (6.1), e_i is the difference between a binomial random variable and its expectation, divided by its estimated standard deviation. For large n_i, e_i then has an approximate $N(0, 1)$ distribution, when the model holds. Since π_i is estimated by $\hat{\pi}_i$ and the $\{\hat{\pi}_i\}$ depend on $\{y_i\}$, however, $\{y_i - n_i \hat{\pi}_i\}$ tend to be smaller than $\{y_i - n_i \pi_i\}$ and the $\{e_i\}$ are less variable than $N(0, 1)$. If X^2 has df $= \nu$, $X^2 = \Sigma_i e_i^2$ is asymptotically comparable to the sum of squares of ν (rather than N) independent standard normal random variables. Thus, when the model holds, $E(\Sigma_i e_i^2)/N \approx \nu/N < 1$.

The standardized Pearson residual is slightly larger in absolute value and is approximately $N(0, 1)$ when the model holds. In Section 4.5.5 we showed the adjustment uses the leverage from an estimated hat matrix. For observation i with leverage $\hat{h}_i$, the standardized residual is

$$r_i = \frac{e_i}{\sqrt{1 - \hat{h}_i}} = \frac{y_i - n_i \hat{\pi}_i}{\sqrt{\left[n_i \hat{\pi}_i (1 - \hat{\pi}_i)(1 - \hat{h}_i)\right]}}.$$

Absolute values larger than roughly 2 or 3 provide evidence of lack of fit.

An alternative residual uses components of the G^2 fit statistic. These are the *deviance residuals*, introduced for GLMs in (4.35). The deviance residual for observation i is

$$\sqrt{d_i} \times \mathrm{sign}(y_i - n_i \hat{\pi}_i), \tag{6.2}$$

where

$$d_i = 2\left(y_i \log \frac{y_i}{n_i \hat{\pi}_i} + (n_i - y_i) \log \frac{n_i - y_i}{n_i - n_i \hat{\pi}_i}\right).$$

This also tends to be less variable then $N(0, 1)$ and can be standardized.

Plots of residuals against explanatory variables or linear predictor values may detect a type of lack of fit. When fitted values are very small, however, just as X^2 and G^2 lose relevance, so do residuals. When explanatory variables are continuous, often $n_i = 1$ at each setting. Then y_i can equal only 0 or 1, and e_i can assume only two values. One must then be cautious about regarding either outcome as extreme, and a single residual is usually uninformative. Plots of residuals also then have limited use, consisting simply of two parallel lines of dots. The deviance itself is then completely uninformative (Problem 5.37). When data can be grouped into sets of observations having common predictor values, it is better to compute residuals for the grouped data than for individual subjects.

6.2.2 Heart Disease Example

A sample of male residents of Framingham, Massachusetts, aged 40 through 59, were classified on several factors, including blood pressure (Table 6.5). The response variable is whether they developed coronary heart disease during a six-year follow-up period.

Let π_i be the probability of heart disease for blood pressure category i. The table shows the fit and the standardized Pearson residuals for two logistic regression models. The first model,

$$\text{logit}(\pi_i) = \alpha,$$

treats the response as independent of blood pressure. Some residuals for that model are large. This is not surprising, since the model fits poorly ($G^2 = 30.0$, $X^2 = 33.4$, df $= 7$).

TABLE 6.5 Standardized Pearson Residuals for Logit Models Fitted to Data on Blood Pressure and Heart Disease

Blood Pressure	Sample Size	Observed Heart Disease	Fitted Indep. Model	Fitted Linear Logit	Residual Indep. Model	Residual Linear Logit
< 117	156	3	10.8	5.2	−2.62	−1.11
117–126	252	17	17.4	10.6	−0.12	2.37
127–136	284	12	19.7	15.1	−2.02	−0.95
137–146	271	16	18.8	18.1	−0.74	−0.57
147–156	139	12	9.6	11.6	0.84	0.13
157–166	85	8	5.9	8.9	0.93	−0.33
167–186	99	16	6.9	14.2	3.76	0.65
> 186	43	8	3.0	8.4	3.07	−0.18

Source: Data from Cornfield (1962).

TABLE 6.6   Residuals Reported in SAS for Heart Disease Data of Table 6.5[a]

			Observation Statistics			
Observ	disease	n	blood	Reschi	Resdev	StReschi
1	3	156	111.5	−0.9794	−1.0617	−1.1058
2	17	252	121.5	2.0057	1.8501	2.3746
3	12	284	131.5	−0.8133	−0.8420	−0.9453
4	16	271	141.5	−0.5067	−0.5162	−0.5727
5	12	139	151.5	0.1176	0.1170	0.1261
6	8	85	161.5	−0.3042	−0.3088	−0.3261
7	16	99	176.5	0.5135	0.5050	0.6520
8	8	43	191.5	−0.1395	−0.1402	−0.1773

[a]Reschi, Pearson residual; StReschi, adjusted residual.

A plot of the residuals show an increasing trend. This suggests the linear logit model,

$$\text{logit}(\pi_i) = \alpha + \beta x_i,$$

with scores $\{x_i\}$ for blood pressure level. We used scores (111.5, 121.5, 131.5, 141.5, 151.5, 161.5, 176.5, 191.5). The nonextreme scores are midpoints for the intervals of blood pressure. The trend in residuals disappears for this model, and only the second category shows some evidence of lack of fit.

Table 6.6 reports residuals for the linear logit model, as reported by SAS. The Pearson residuals (Reschi), deviance residuals (Resdev), and standardized Pearson residuals (StReschi) show similar results. Each is somewhat large in the second category. One relatively large residual is not surprising, however. With many residuals, some may be large purely by chance. Here the

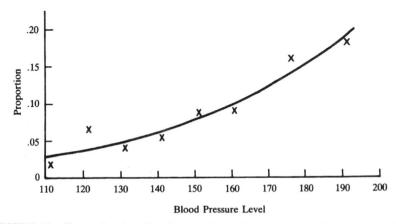

FIGURE 6.2   Observed and predicted proportions of heart disease for linear logit model.

overall fit statistics ($G^2 = 5.9$, $X^2 = 6.3$ with df = 6) do not indicate problems. In analyzing residual patterns, we should be cautious about attributing patterns to what might be chance variation from a model.

Another useful graphical display for showing lack of fit compares observed and fitted proportions by plotting them against each other or by plotting both of them against explanatory variables. For the linear logit model, Figure 6.2 plots both the observed proportions and the estimated probabilities of heart disease against blood pressure. The fit seems decent.

Studying residuals helps us understand either why a model fits poorly or where there is lack of fit in a generally good-fitting model. The next example illustrates the second case.

6.2.3 Graduate Admissions Example

Table 6.7 refers to graduate school applications to the 23 departments in the College of Liberal Arts and Sciences at the University of Florida during the 1997–1998 academic year. It cross-classifies applicant's gender (G), whether admitted (A), and department (D) to which the prospective students applied. We consider logit models with A as the response variable. Let y_{ik} denote the number admitted and let π_{ik} denote the probability of admission for gender i in department k. We treat $\{Y_{ik}\}$ as independent bin(n_{ik}, π_{ik}). Other things being equal, one would hope the admissions decision is independent of gender. However, the model with no gender effect, given the department,

$$\text{logit}(\pi_{ik}) = \alpha + \beta_k^D,$$

fits rather poorly ($G^2 = 44.7$, $X^2 = 40.9$, df = 23).

TABLE 6.7 Data Relating Admission to Gender and Department for Model with No Gender Effect

	Females		Males		Std. Res		Females		Males		Std. Res
Dept	Yes	No	Yes	No	(Fem,Yes)	Dept	Yes	No	Yes	No	(Fem,Yes)
anth	32	81	21	41	−0.76	ling	21	10	7	8	1.37
astr	6	0	3	8	2.87	math	25	18	31	37	1.29
chem	12	43	34	110	−0.27	phil	3	0	9	6	1.34
clas	3	1	4	0	−1.07	phys	10	11	25	53	1.32
comm	52	149	5	10	−0.63	poli	25	34	39	49	−0.23
comp	8	7	6	12	1.16	psyc	2	123	4	41	−2.27
engl	35	100	30	112	0.94	reli	3	3	0	2	1.26
geog	9	1	11	11	2.17	roma	29	13	6	3	0.14
geol	6	3	15	6	−0.26	soci	16	33	7	17	0.30
germ	17	0	4	1	1.89	stat	23	9	36	14	−0.01
hist	9	9	21	19	−0.18	zool	4	62	10	54	−1.76
lati	26	7	25	16	1.65						

Source: Data courtesy of James Booth.

Table 6.7 also reports standardized Pearson residuals for the number of females who were admitted for this model. For instance, the astronomy department admitted 6 females, which was 2.87 standard deviations higher than the model predicted. Each department has only a single nonredundant standardized residual, because of marginal constraints for the model. The model has fit $\hat{\pi}_{ik} = (y_{1k} + y_{2k})/n_{+k}$, corresponding to an independence fit ($\hat{\pi}_{1k} = \hat{\pi}_{2k}$) in each partial table. Now, $y_{1k} - n_{1k}\hat{\pi}_{1k} = y_{1k} - n_{1k}(y_{1k} + y_{2k})/n_{+k} = (n_{2k}/n_{+k})y_{1k} - (n_{1k}/n_{+k})y_{2k} = -(y_{2k} - n_{2k}\hat{\pi}_{2k})$. Thus, standard errors of $(y_{1k} - n_{1k}\hat{\pi}_{1k})$ and $(y_{2k} - n_{2k}\hat{\pi}_{2k})$ are identical. The standardized residuals are identical in absolute value for males and females but of different sign. Astronomy admitted 3 males, and their standardized residual was -2.87; the number admitted was 2.87 standard deviations fewer than predicted. This is another advantage of standardized over ordinary Pearson residuals. The model of independence in a partial table has df = 1. Only one bit of information exists about how the data depart from independence, yet the ordinary Pearson residual for males need not equal the ordinary Pearson residual for females.

Departments with large standardized Pearson residuals reveal the reason for the lack of fit. Significantly more females were admitted than the model predicts in the astronomy and geography departments, and fewer in the psychology department. Without these three departments, the model fits reasonably well ($G^2 = 24.4$, $X^2 = 22.8$, df = 20).

For the complete data, adding a gender effect to the model does not provide an improved fit ($G^2 = 42.4$, $X^2 = 39.0$, df = 22), because the departments just described have associations in different directions and of greater magnitude than other departments. This model has an ML estimate of 1.19 for the GA conditional odds ratio, the odds of admission being 19% higher for females than males, given department. By contrast, the marginal table collapsed over department has a GA sample odds ratio of 0.94, the overall odds of admission being 6% lower for females. This illustrates Simpson's paradox (Section 2.3.2), the conditional association having different direction than the marginal association.

6.2.4 Influence Diagnostics for Logistic Regression

Other regression diagnostic tools are also helpful in assessing fit. These include plots of ordered residuals against normal percentiles (Haberman 1973a) and analyses that describe an observation's influence on parameter estimates and fit statistics. Whenever a residual indicates that a model fits an observation poorly, it can be informative to delete the observation and refit the model to remaining ones. This is equivalent to adding a parameter to the model for that observation, forcing a perfect fit for it.

As in ordinary regression, an observation may be relatively influential in determining parameter estimates. The greater an observation's leverage, the greater its potential influence. The fit could be quite different if an observation that appears to be an outlier on y and has large leverage is deleted. However, a single observation can have a more exorbitant influence in ordinary regression than a single binary observation in logistic regression, since there is no bound on the distance of y_i from its expected value. Also, in Section 4.5.5 we observed that the GLM estimated hat matrix

$$\widehat{\text{Hat}} = \hat{\mathbf{W}}^{1/2} \mathbf{X} (\mathbf{X}'\hat{\mathbf{W}}\mathbf{X})^{-1} \mathbf{X}'\hat{\mathbf{W}}^{1/2}$$

depends on the fit as well as the model matrix $\mathbf{X}$. For logistic regression, in Section 5.5.2 we showed that the weight matrix $\hat{\mathbf{W}}$ is diagonal with element $\hat{w}_i = n_i \hat{\pi}_i (1 - \hat{\pi}_i)$ for the n_i observations at setting i of predictors. Points that have extreme predictor values need not have high leverage. In fact, the leverage can be small if $\hat{\pi}_i$ is close to 0 or 1.

Several measures that describe the effect on parameter estimates and fit statistics of removing an observation from the data set are related algebraically to the observation's leverage (Pregibon 1981; Williams 1987). In logistic regression, the observation could be a single binary response or a binomial response for a set of subjects all having the same predictor values. Influence measures for each observation include:

1. For each model parameter, the change in the parameter estimate when the observation is deleted. This change, divided by its standard error, is called *Dfbeta*.

2. A measure of the change in a joint confidence interval for the parameters produced by deleting the observation. This confidence interval displacement diagnostic is denoted by c.

3. The change in X^2 or G^2 goodness-of-fit statistics when the observation is deleted.

For each measure, the larger the value, the greater the influence. We illustrate them using the linear logit model with blood pressure as a predictor for heart disease in Table 6.5. Table 6.8 contains simple approximations (due to Pregibon 1981) for the *Dfbeta* measure for the coefficient of blood pressure, the confidence interval diagnostic c, the change in G^2, and the change in X^2. (This is the square of the standardized Pearson residual, r_i^2.) All their values show that deleting the second observation has the greatest effect. This is not surprising, as that observation has the only relatively large residual. By contrast, Table 6.8 also contains the changes in X^2 and G^2 for deleting observations in fitting the independence model. At the low and high ends of the blood pressure values, several changes are very large. However, these all relate to removing an entire binomial sample at a blood pressure level instead of removing a single subject's binary observation. Such subject-level deletions have little effect even for this model.

With continuous or multiple predictors, it can be informative to plot these diagnostics, for instance against the estimated probabilities. See Cook and

TABLE 6.8 Diagnostic Measures for Logistic Regression Models Fitted to Heart Disease Data

Blood Pressure	Dfbeta	c	Pearson X^2 Diff.	Likelihood-Ratio G^2 Diff.	Pearson X^2 Diff.[a]	Likelihood-Ratio G^2 Diff.[a]
111.5	0.49	0.34	1.22	1.39	6.86	9.13
121.5	−1.14	2.26	5.64	5.04	0.02	0.02
131.5	0.33	0.31	0.89	0.94	4.08	4.56
141.5	0.08	0.09	0.33	0.34	0.55	0.57
151.5	0.01	0.00	0.02	0.02	0.70	0.66
161.5	−0.07	0.02	0.11	0.11	0.87	0.80
176.5	0.40	0.26	0.42	0.42	14.17	10.83
191.5	−0.12	0.02	0.03	0.03	9.41	6.73

[a]Independence model; other values refer to model with blood pressure predictor.
Source: Data from Cornfield (1962).

Weisberg (1999, Chap. 22), Fowlkes (1987), and Landwehr et al. (1984) for examples of useful diagnostic plots.

6.2.5 Summarizing Predictive Power: R and R-Squared Measures

In ordinary regression, R^2 describes the proportional reduction in variation in comparing the conditional variation of the response to the marginal variation. It and the multiple correlation R describe the power of the explanatory variables to predict the response, with $R = 1$ for perfect prediction. Despite various attempts to define analogs for categorical response models, no proposed measure is as widely useful as R and R^2. We present a few proposed measures in this section.

For any GLM, the correlation $r(y, \hat{\mu})$ between the observed responses $\{y_i\}$ and the model's fitted values $\{\hat{\mu}_i\}$ measures predictive power. For least squares regression, this is the multiple correlation between Y and the predictors. An advantage of the correlation relative to its square is the appeal of working on the original scale and its approximate proportionality to effect size: For a small effect with a single predictor, doubling the slope corresponds roughly to doubling the correlation. This measure can be useful for comparing fits of different models to the same data set.

In logistic regression, $\hat{\mu}_i$ for a particular model is the estimated probability $\hat{\pi}_i$ for binary observation i. Table 6.2 shows $r(y, \hat{\mu})$ for a few models fitted to the horseshoe crab data. Width alone has $r = 0.402$, and adding color to the model increases r to 0.452. The simpler model that uses color merely to indicate whether a crab is dark does essentially as well, with $r = 0.447$. The complex model containing color, spine condition, width, and all their two- and three-way interactions has $r = 0.526$. This seems considerably higher, but with multiple predictors the r estimates become more highly biased in estimating the true correlation. It can be misleading to compare r values for models with greatly different df values. After a jackknife adjustment designed to reduce bias, there is little difference between r for this overly complex model and the simpler model (Zheng and Agresti 2000). Little is lost and much is gained by using the simpler model.

Another way to measure the association between the binary responses $\{y_i\}$ and their fitted values $\{\hat{\pi}_i\}$ uses the proportional reduction in squared error

$$1 - \frac{\Sigma_i(y_i - \hat{\pi}_i)^2}{\Sigma_i(y_i - \bar{y})^2},$$

obtained by using $\hat{\pi}_i$ instead of $\bar{y} = \Sigma y_i/n$ as a predictor of y_i (Efron 1978). Amemiya (1981) suggested a related measure that weights squared deviations by inverse predicted variances. For logistic regression, unlike normal GLMs, these and $r(y, \hat{\mu})$ need not be nondecreasing as the model gets more complex. Like any correlation-type measure, they can depend strongly on the range of observed values of explanatory variables.

Other measures directly use the likelihood function. Denote the maximized log likelihood by L_M for a given model, L_S for the saturated model, and L_0 for the null model containing only an intercept term. Probabilities are no greater than 1.0, so log likelihoods are nonpositive. As the model complexity increases, the parameter space expands, so the maximized log likelihood increases. Thus, $L_0 \le L_M \le L_S \le 0$. The measure

$$\frac{L_M - L_0}{L_S - L_0} \tag{6.3}$$

falls between 0 and 1. It equals 0 when the model provides no improvement in fit over the null model, and it equals 1 when the model fits as well as the saturated model. A weakness is the log likelihood is not an easily interpretable scale. Interpreting the numerical value is difficult, other than in a comparative sense for different models.

For n independent Bernoulli observations, the maximized log likelihood is

$$\log \prod_{i=1}^{n} \left[\hat{\pi}_i^{y_i}(1 - \hat{\pi}_i)^{1-y_i} \right] = \sum_{i=1}^{n} \left[y_i \log \hat{\pi}_i + (1 - y_i) \log (1 - \hat{\pi}_i) \right].$$

The null model gives $\hat{\pi}_i = (\Sigma y_i)/n = \bar{y}$, so that

$$L_0 = n[\bar{y}(\log \bar{y}) + (1 - \bar{y}) \log (1 - \bar{y})].$$

The saturated model has a parameter for each subject and implies that

$\hat{\pi}_i = y_i$ for all i. Thus, $L_S = 0$ and (6.3) simplifies to

$$D = \frac{L_0 - L_M}{L_0}.$$

McFadden (1974) proposed this measure.

With multiple observations at each setting of explanatory variables, the data file can take the grouped-data form of N binomial counts rather than n Bernoulli indicators. The saturated model then has a parameter for each count. It gives N fitted proportions equal to the N sample proportions of success. Then L_S is nonzero and (6.3) takes a different value than when calculated using individual subjects. For N binomial counts, the maximized likelihoods are related to the G^2 goodness-of-fit statistic by $G^2(M) = -2(L_M - L_S)$, so (6.3) becomes

$$D^* = \frac{G^2(0) - G^2(M)}{G^2(0)}.$$

Goodman (1971a) and Theil (1970) discussed this and related partial associa-tion measures.

With grouped data D^* can be large even when predictive power is weak at the subject level. For instance, a model can fit much better than the null model even though fitted probabilities are close to 0.5 for the entire sample. In particular, $D^* = 1$ when it fits perfectly, regardless of how well one can predict individual subject's responses on Y with that model. Also, suppose that the population satisfies the given model, but not the null model. As the sample size n increases with number of settings N fixed, $G^2(M)$ behaves like a chi-squared random variable but $G^2(0)$ grows unboundedly. Thus, $D^* \to 1$ as $n \to \infty$, and its magnitude tends to depend on n. This measure confounds model goodness of fit with predictive power. Similar behavior occurs for R^2 in regression analyses when calculated using *means* of Y values (rather than individual subjects) at N different x settings. It is more sensible to use D for binary, ungrouped data.

6.2.6 Summarizing Predictive Power: Classification Tables and ROC Curves

A *classification table* cross-classifies the binary response with a prediction of whether $y = 0$ or 1. The prediction is $\hat{y} = 1$ when $\hat{\pi}_i > \pi_0$ and $\hat{y} = 0$ when $\hat{\pi}_i \le \pi_0$, for some cutoff π_0. Most classification tables use $\pi_0 = 0.5$ and summarize predictive power by

$$\text{sensitivity} = P(\hat{y} = 1 | y = 1) \quad \text{and} \quad \text{specificity} = P(\hat{y} = 0 | y = 0)$$

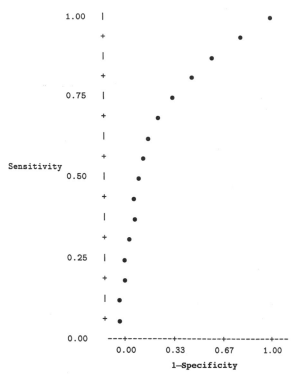

FIGURE 6.3   ROC curve for logistic regression model with horseshoe crab data.

(Recall Sections 2.1.2.) Limitations of this table are that it collapses continuous predictive values $\hat{\pi}$ into binary ones, the choice of π_0 is arbitrary, and it is highly sensitive to the relative numbers of times $y = 1$ and $y = 0$.

A *receiver operating characteristic* (ROC) curve is a plot of sensitivity as a function of $(1 - $ specificity) for the possible cutoffs π_0. This curve usually has a concave shape connecting the points $(0, 0)$ and $(1, 1)$. The higher the area under the curve, the better the predictions. The ROC curve is more informative than the classification table, since it summarizes predictive power for all possible π_0. Figure 6.3 shows how PROC LOGISTIC in SAS reports the ROC curve for the model for the horseshoe crabs using width and color as predictors.

The area under a ROC curve is identical to the value of another measure of predictive power, the *concordance index*. Consider all pairs of observations (i, j) such that $y_i = 1$ and $y_j = 0$. The concordance index c estimates the probability that the predictions and the outcomes are concordant, the observation with the larger y also having the larger $\hat{\pi}$ (Harrell et al. 1982). A value $c = 0.5$ means predictions were no better than random guessing. This corresponds to a model having only an intercept term and an ROC curve that is a straight line connecting points $(0, 0)$ and $(1, 1)$. For the horseshoe crab data, $c = 0.639$ with color alone as a predictor, 0.742 with width alone, 0.771 with width and color, and 0.772 with width and a dummy for whether a crab has dark color.

ROC curves are a popular way of evaluating diagnostic tests. Sometimes such tests have $J > 2$ ordered response categories rather than (positive, negative). The ROC curve then refers to the various possible cutoffs for defining a result to be positive. It plots sensitivity against $1 -$ specificity for the possible collapsings of the J categories to a (positive, negative) scale [see Toledano and Gatsonis (1996)].

6.3 INFERENCE ABOUT CONDITIONAL ASSOCIATIONS IN $2 \times 2 \times K$ TABLES

The analysis of the graduate admissions data in Sections 6.2.3 used the model of conditional independence. This model is an important one in biomedical studies that investigate whether an association exists between a treatment variable and a disease outcome after controlling for a possibly confounding variable that might influence that association. In this section we review the test of conditional independence as a logit model analysis for a $2 \times 2 \times K$ contingency table. We also present a test (Mantel and Haenszel 1959) that seems non-model-based but relates to the logit model.

We illustrate using Table 6.9, showing results of a clinical trial with eight centers. The study compared two cream preparations, an active drug and a

TABLE 6.9   Clinical Trial Relating Treatment to Response for Eight Centers

| Center | Treatment | Response | | Odds Ratio | μ_{11k} | var(n_{11k}) |
		Success	Failure			
1	Drug	11	25	1.19	10.36	3.79
	Control	10	27			
2	Drug	16	4	1.82	14.62	2.47
	Control	22	10			
3	Drug	14	5	4.80	10.50	2.41
	Control	7	12			
4	Drug	2	14	2.29	1.45	0.70
	Control	1	16			
5	Drug	6	11	∞	3.52	1.20
	Control	0	12			
6	Drug	1	10	∞	0.52	0.25
	Control	0	10			
7	Drug	1	4	2.0	0.71	0.42
	Control	1	8			
8	Drug	4	2	0.33	4.62	0.62
	Control	6	1			

Source: Beitler and Landis (1985).

control, on their success in curing an infection. This table illustrates a common pharmaceutical application, comparing two treatments on a binary response with observations from several strata. The strata are often medical centers or clinics; or they may be levels of age or severity of the condition being treated or combinations of levels of several control variables; or they may be different studies of the same sort evaluated in a meta analysis.

6.3.1 Using Logit Models to Test Conditional Independence

For a binary response Y, we study the effect of a binary predictor X, controlling for a qualitative covariate Z. Let $\pi_{ik} = P(Y = 1 \,|\, X = i, Z = k)$. Consider the model

$$\text{logit}(\pi_{ik}) = \alpha + \beta x_i + \beta_k^Z, \qquad i = 1, 2, \quad k = 1, \ldots, K, \qquad (6.4)$$

where $x_1 = 1$ and $x_2 = 0$. This model assumes that the XY conditional odds ratio is the same at each category of Z, namely $\exp(\beta)$. The null hypothesis of XY conditional independence is H_0: $\beta = 0$. The Wald statistic is $(\hat{\beta}/\text{SE})^2$. The likelihood-ratio statistic is the difference between G^2 statistics for the reduced model

$$\text{logit}(\pi_{ik}) = \alpha + \beta_k^Z \qquad (6.5)$$

and the full model. These tests are sensible when X has a similar effect at each category of Z. They have df $= 1$.

Alternatively, since the reduced model (6.5) is equivalent to conditional independence of X and Y, one could test conditional independence using a goodness-of-fit test of that model. That test has df $= K$ when X is binary. This corresponds to comparing model (6.5) and the saturated model, which permits $\beta \neq 0$ and contains XZ interaction parameters. When no interaction exists or when interaction exists but it has minor substantive importance, it follows from results to be presented in Section 6.4.2 that this approach is less powerful, especially when K is large. However, when the direction of the XY association varies among categories of Z, it can be more powerful.

6.3.2 Cochran–Mantel–Haenszel Test of Conditional Independence

Mantel and Haenszel (1959) proposed a non-model-based test of H_0: conditional independence in $2 \times 2 \times K$ tables. Focusing on retrospective studies of disease, they treated response (column) marginal totals as fixed. Thus, in each partial table k of cell counts $\{n_{ijk}\}$, their analysis conditions on both the predictor totals (n_{1+k}, n_{2+k}) and the response outcome totals (n_{+1k}, n_{+2k}). The usual sampling schemes then yield a hypergeometric distribution (3.16) for the first cell count n_{11k} in each partial table. That count determines $\{n_{12k}, n_{21k}, n_{22k}\}$, given the marginal totals.

Under H_0, the hypergeometric mean and variance of n_{11k} are

$$\mu_{11k} = E(n_{11k}) = n_{1+k} n_{+1k}/n_{++k}$$

$$\text{var}(n_{11k}) = n_{1+k} n_{2+k} n_{+1k} n_{+2k}/n_{++k}^2(n_{++k} - 1).$$

Cell counts from different partial tables are independent. The test statistic combines information from the K tables by comparing $\Sigma_k n_{11k}$ to its null expected value. It equals

$$\text{CMH} = \frac{[\Sigma_k(n_{11k} - \mu_{11k})]^2}{\Sigma_k \text{var}(n_{11k})}. \tag{6.6}$$

This statistic has a large-sample chi-squared null distribution with df = 1.

When the odds ratio $\theta_{XY(k)} > 1$ in partial table k, we expect that $(n_{11k} - \mu_{11k}) > 0$. When $\theta_{XY(k)} > 1$ in every partial table or $\theta_{XY(k)} < 1$ in each table, $\Sigma_k(n_{11k} - \mu_{11k})$ tends to be relatively large in absolute value. This test works best when the XY association is similar in each partial table. In this sense it is similar to the tests of H_0: $\beta = 0$ in logit model (6.4). When the sample sizes in the strata are moderately large, this test usually gives similar results. In fact, it is a score test (Section 1.3.3) of H_0: $\beta = 0$ in that model (Day and Byar 1979).

Cochran (1954) proposed a similar statistic. He treated the rows in each 2×2 table as two independent binomials rather than a hypergeometric. Cochran's statistic is (6.6) with $\text{var}(n_{11k})$ replaced by

$$\text{var}(n_{11k}) = n_{1+k} n_{2+k} n_{+1k} n_{+2k}/n_{++k}^3.$$

Because of the similarity in their approaches, we call (6.6) the *Cochran–Mantel–Haenszel* (CMH) *statistic*. The Mantel and Haenszel approach using the hypergeometric is more general in that it also applies to some cases in which the rows are not independent binomial samples from two populations. Examples are retrospective studies and randomized clinical trials with the available subjects randomly allocated to two treatments. In the first case the column totals are naturally fixed. In the second, under the null hypothesis the column margins are the same regardless of how subjects were assigned to treatments, and randomization arguments lead to the hypergeometric in each 2×2 table.

Mantel and Haenszel (1959) proposed (6.6) with a continuity correction. The P-value from the test then better approximates an exact conditional test (Section 6.7.5) but it tends to be conservative. The CMH statistic generalizes for $I \times J \times K$ tables (Section 7.5.3).

6.3.3   Multicenter Clinical Trial Example

For the multicenter clinical trial, Table 6.9 reports the sample odds ratio for each table and the expected value and variance of the number of successes for the drug treatment (n_{11k}) under H_0: conditional independence. In each table except the last, the sample odds ratio shows a positive association. Thus, it makes sense to combine results with CMH = 6.38, with df = 1. There is considerable evidence against H_0 ($P = 0.012$).

Similar results occur in testing H_0: $\beta = 0$ in logit model (6.4). The model fit has $\hat{\beta} = 0.777$ with SE = 0.307. The Wald statistic is $(0.777/0.307)^2 = 6.42$ ($P = 0.011$). The likelihood-ratio statistic equals 6.67 ($P = 0.010$).

6.3.4 CMH Test and Sparse Data*

In summary, for logit model (6.4), CMH is the score statistic alternative to the likelihood-ratio or Wald test of H_0: $\beta = 0$. As $n \to \infty$ with fixed K, the tests have the same asymptotic chi-squared behavior under H_0. An advantage of CMH is that its chi-squared limit also applies with an alternative asymptotic scheme in which $K \to \infty$ as $n \to \infty$. The asymptotic theory for likelihood-ratio and Wald tests requires the number of parameters (and hence K) to be fixed, so it does not apply to this scheme. An application of this type is when each stratum has a single matched pair of subjects, one in each group.

With strata of matched pairs, $n_{1+k} = n_{2+k} = 1$ for each k. Then $n = 2K$, so $K \to \infty$ as $n \to \infty$. Table 6.10 shows the data layout for this situation. When both subjects in stratum k make the same response (as in the first case in Table 6.10), $n_{+1k} = 0$ or $n_{+2k} = 0$. Given the marginal counts, the internal counts are then completely determined, and $\mu_{11k} = n_{11k}$ and var(n_{11k}) = 0. When the subjects make differing responses (as in the second case), $n_{+1k} = n_{+2k} = 1$, so that $\mu_{11k} = 0.5$ and var(n_{11k}) = 0.25. Thus, a matched pair contributes to the CMH statistic only when the two subjects' responses differ. Let K^* denote the number of the K tables that satisfy this. Although each n_{11k} can take only two values, the central limit theorem implies that $\sum_k n_{11k}$ is approximately normal for large K^*. Thus, the distribution of CMH is approximately chi-squared.

Usually, when K grows with n, each stratum has few observations. There may be more than two observations, such as case–control studies that match several controls with each case. Contingency tables with relatively few observations are referred to as *sparse*. The nonstandard setting in which $K \to \infty$ as $n \to \infty$ is called *sparse-data asymptotics*. Ordinary ML estimation then breaks down because the number of parameters is not fixed, instead having the same order as the sample size. In particular, an approximate chi-squared distribution holds for the likelihood-ratio and Wald statistics for testing conditional

TABLE 6.10   Stratum Containing a Matched Pair

Element of Pair	Response		Response	
	Success	Failure	Success	Failure
First	1	0	1	0
Second	1	0	0	1

independence only when the strata marginal totals generally exceed about 5 to 10 and K is fixed and small relative to n.

6.3.5 Estimation of Common Odds Ratio

It is more informative to estimate the strength of association than to test hypotheses about it. When the association seems stable among partial tables, it is helpful to combine the K sample odds ratios into a summary measure of conditional association. The logit model (6.4) implies homogeneous association, $\theta_{XY(1)} = \cdots = \theta_{XY(K)} = \exp(\beta)$. The ML estimate of the common odds ratio is $\exp(\hat{\beta})$.

Other estimators of a common odds ratio are not model-based. Woolf (1955) proposed an exponentiated weighted average of the K sample log odds ratios. Mantel and Haenszel (1959) proposed that

$$\hat{\theta}_{MH} = \frac{\sum_k (n_{11k}n_{22k}/n_{++k})}{\sum_k (n_{12k}n_{21k}/n_{++k})} = \frac{\sum_k p_{11|k}p_{22|k}n_{++k}}{\sum_k p_{12|k}p_{21|k}n_{++k}}, \tag{6.7}$$

where $p_{ij|k} = n_{ijk}/n_{++k}$. This gives more weight to strata with larger sample sizes. It is preferred over the ML estimator when K is large and the data are sparse. The ML estimator $\hat{\beta}$ of the log odds ratio then tends to be too large in absolute value. For sparse-data asymptotics with only a single matched pair in each stratum, for instance, $\hat{\beta} \xrightarrow{p} 2\beta$. [This *convergence in probability* means that for any $\epsilon > 0$, $P(|\hat{\beta} - 2\beta| < \epsilon) \to 1$ as $n \to \infty$; see Problem 10.24.]

Hauck (1979) gave an asymptotic variance for $\log(\hat{\theta}_{MH})$ that applies for a fixed number of strata. In that case $\log(\hat{\theta}_{MH})$ is slightly less efficient than the ML estimator $\hat{\beta}$ unless $\beta = 0$ (Tarone et al. 1983). Robins et al. (1986) derived an estimated variance that applies both for these standard asymptotics with large n and fixed K and for sparse asymptotics in which K is also large. Expressing $\hat{\theta}_{MH} = R/S = (\sum_k R_k)/(\sum_k S_k)$ with $R_k = n_{11k}n_{22k}/n_{++k}$, their derivation showed that $(\log \hat{\theta}_{MH} - \log \theta)$ is approximately proportional to $(R - \theta S)$. They also showed that $E(R - \theta S) = 0$ and derived the variance of $(R - \theta S)$. Their result is

$$\hat{\sigma}^2 \left[\log \hat{\theta}_{MH} \right] = \frac{1}{2R^2} \sum_k n_{++k}^{-1}(n_{11k} + n_{22k}) R_k$$

$$+ \frac{1}{2S^2} \sum_k n_{++k}^{-1}(n_{12k} + n_{21k}) S_k$$

$$+ \frac{1}{2RS} \sum_k n_{++k}^{-1} [(n_{11k} + n_{22k}) S_k + (n_{12k} + n_{21k}) R_k].$$

For the eight-center clinical trial summarized by Table 6.9,

$$\hat{\theta}_{MH} = \frac{(11 \times 27)/73 + \cdots + (4 \times 1)/13}{(25 \times 10)/73 + \cdots + (2 \times 6)/13} = 2.13.$$

For $\log \hat{\theta}_{MH} = 0.758$, $\hat{\sigma}[\log \hat{\theta}_{MH}] = 0.303$. A 95% confidence interval for the common odds ratio is $\exp(0.758 \pm 1.96 \times 0.303)$ or $(1.18, 3.87)$. Similar results occur using model (6.4). The 95% confidence interval for $\exp(\beta)$ is $\exp(0.777 \pm 1.96 \times 0.307)$, or $(1.19, 3.97)$, using the Wald interval, and $(1.20, 4.02)$ using the likelihood-ratio interval. Although the evidence of an effect is considerable, inference about its size is rather imprecise. The odds of success may be as little as 20% higher with the drug, or they may be as much as four times as high.

If the true odds ratios are not identical but do not vary drastically, $\hat{\theta}_{MH}$ still is a useful summary of the conditional associations. Similarly, the CMH test is a powerful summary of evidence against H_0: conditional independence, as long as the sample associations fall primarily in a single direction. It is not necessary to assume equality of odds ratios to use the CMH test.

6.3.6 Testing Homogeneity of Odds Ratios

The homogeneous association condition $\theta_{XY(1)} = \cdots = \theta_{XY(K)}$ for $2 \times 2 \times K$ tables is equivalent to logit model (6.4). A test of homogeneous association is implicitly a goodness-of-fit test of this model. The usual G^2 and X^2 test statistics provide this, with df $= K - 1$. They test that the $K - 1$ parameters in the saturated model that are the coefficients of interaction terms [cross products of the dummy variable for x with $(K - 1)$ dummy variables for categories of Z] all equal 0. Breslow and Day (1980, p. 142) proposed an alternative large-sample test (Note 6.5).

For the eight-center clinical trial data in Table 6.9, $G^2 = 9.7$ and $X^2 = 8.0$ (df $= 7$) do not contradict the hypothesis of equal odds ratios. It is reasonable to summarize the conditional association by a single odds ratio (e.g., $\hat{\theta}_{MH} = 2.1$) for all eight partial tables. In fact, even with a small P-value in a test of homogeneous association, if the variability in the sample odds ratios is not substantial, a summary measure such as $\hat{\theta}_{MH}$ is useful. A test of homogeneity is not a prerequisite for this measure or for testing conditional independence.

6.3.7 Summarizing Heterogeneity in Odds Ratios

In practice, a predictor effect is often similar from stratum to stratum. In multicenter clinical trials comparing a new drug to a standard, for example, if the new drug is truly more beneficial, the true effect is usually positive in each stratum.

In strict terms, however, a model with homogeneous effects is unrealistic. First, we rarely expect the true odds ratio to be *exactly* the same in each stratum, because of unmeasured covariates that affect it. Breslow (1976) discussed modeling of the log odds ratio using a set of explanatory variables. Second, the model regards the strata effects $\{\beta_k^Z\}$ as fixed effects, treating them as the only strata of interest. Often the strata are merely a sampling of the possible ones. Multicenter clinical trials have data for certain centers but many other centers could have been used. Scientists would like their conclusions to apply to all such centers, not only those in the study.

A somewhat different logit model treats the true log odds ratios in partial tables as a random sample from a $N(\mu, \sigma^2)$ distribution. Fitting the model yields an estimated mean log odds ratio and an estimated variability about that mean. The inference applies to the population of strata rather than only those sampled. This type of model uses *random effects* in the linear predictor to induce this extra type of variability. In Chapter 12 we discuss GLMs with random effects, and in Section 12.3.4 we fit such a model to Table 6.9.

6.4 USING MODELS TO IMPROVE INFERENTIAL POWER

When contingency tables have ordered categories, in Section 3.4 we showed that tests that utilize the ordering can have improved power. Testing independence against a linear trend alternative in a linear logit model (Sections 5.3.4, and 5.4.6) is a way to do this. In this section we present the reason for these power improvements.

6.4.1 Directed Alternatives

Consider an $I \times 2$ contingency table for I binomial variates with parameters $\{\pi_i\}$. H_0: independence states

$$\text{logit}(\pi_i) = \alpha.$$

The ordinary X^2 and G^2 statistics of Section 3.2.1 refer to the general alternative,

$$\text{logit}(\pi_i) = \alpha + \beta_i,$$

which is saturated. They test H_0: $\beta_1 = \beta_2 = \cdots = \beta_I = 0$ in that model, with df $= (I - 1)$. Their general alternative treats both classifications as nominal. Denote these test statistics as $G^2(I)$ and $X^2(I)$. Recall that $G^2(I)$ is the likelihood-ratio statistic $G^2(M_0|M_1) = -2(L_0 - L_1)$ for comparing the saturated model M_1 with the independence (I) model M_0.

Ordinal test statistics refer to narrower, usually more relevant, alternatives. With ordered rows, an example is a test of H_0: $\beta = 0$ in the linear logit model, $\text{logit}(\pi_i) = \alpha + \beta x_i$. The likelihood-ratio statistic $G^2(I|L) = G^2(I) - G^2(L)$ compares the linear logit model and the independence model. When a test statistic focuses on a single parameter, such as β in that model, it has df = 1. Now, df equals the mean of the chi-squared distribution. A large test statistic with df = 1 falls farther out in its right-hand tail than a comparable value of $X^2(I)$ or $G^2(I)$ with df = $(I - 1)$. Thus, it has a smaller P-value.

6.4.2 Noncentral Chi-Squared Distribution

To compare power of $G^2(I|L)$ and $G^2(I)$, it is necessary to compare their nonnull sampling distributions. When H_0 is false, their distributions are approximately *noncentral chi-squared*. This distribution, introduced by R. A. Fisher in 1928, arises from the following construction: If $Z_i \sim N(\mu_i, 1)$, $i = 1, \ldots, \nu$, and if $Z_1, \ldots, Z_\nu$ are independent, ΣZ_i^2 has the noncentral chi-squared distribution with df = ν and *noncentrality parameter* $\lambda = \Sigma \mu_i^2$. Its mean is $\nu + \lambda$ and its variance is $2(\nu + 2\lambda)$. The ordinary (central) chi-squared distribution, which occurs when H_0 is true, has $\lambda = 0$.

Let $X_{\nu,\lambda}^2$ denote a noncentral chi-squared random variable with df = ν and noncentrality λ. A fundamental result for chi-squared analyses is that, for fixed λ,

$$P[X_{\nu,\lambda}^2 > \chi_\nu^2(\alpha)] \text{ increases as } \nu \text{ decreases}.$$

That is, the power for rejecting H_0 at a fixed α-level increases as the df of the test decreases (e.g., Das Gupta and Perlman 1974). For fixed ν, the power equals α when $\lambda = 0$, and it increases as λ increases. The inverse relation between power and df suggests that focusing the noncentrality on a statistic having a small df value can improve power.

6.4.3 Increased Power for Narrower Alternatives

Suppose that X has, at least approximately, a linear effect on $\text{logit}[P(Y = 1)]$. To test independence, it is then sensible to use a statistic having strong power for that effect. This is the purpose of the tests based on the linear logit model, using the likelihood-ratio statistic $G^2(I|L)$, the Wald statistic $z = \hat{\beta}/SE$, and the Cochran–Armitage (score) statistic.

When is $G^2(I|L)$ more powerful than $G^2(I)$? The statistics satisfy

$$G^2(I) = G^2(I|L) + G^2(L),$$

where $G^2(L)$ tests goodness of fit of the linear logit model. When the linear logit model holds, $G^2(L)$ has an asymptotic chi-squared distribution with df $= I - 2$; then if $\beta \neq 0$, $G^2(I)$ and $G^2(I|L)$ both have approximate noncentral chi-squared distributions with the same noncentrality. Whereas df $= I - 1$ for $G^2(I)$, df $= 1$ for $G^2(I|L)$. Thus, $G^2(I|L)$ is more powerful, since it uses fewer degrees of freedom.

When the linear logit model does not hold, $G^2(I)$ has greater noncentrality than $G^2(I|L)$, the discrepancy increasing as the model fits more poorly. However, when the model approximates reality fairly well, usually $G^2(I|L)$ is still more powerful. That test's df value of 1 more than compensates for its loss in noncentrality. The closer the true relationship is to the linear logit, the more nearly $G^2(I|L)$ captures the same noncentrality as $G^2(I)$, and the more powerful it is compared to $G^2(I)$. To illustrate, Figure 6.4 plots power as a function of noncentrality when df $= 1$ and 7. When the noncentrality of a test having df $= 1$ is at least about half that of a test having df $= 7$, the test with df $= 1$ is more powerful. The linear logit model then helps detect a key component of an association. As Mantel (1963) argued in a similar context, "that a linear regression is being tested does not mean that an assumption of linearity is being made. Rather it is that test of a linear component of regression provides power for detecting any progressive association which may exist."

The improved power results from sacrificing power in other cases. The $G^2(I)$ test can have greater power than $G^2(I|L)$ when the linear logit model describes reality very poorly.

The remark about the desirability of focusing noncentrality holds for nominal variables also. For instance, consider testing conditional independence in $2 \times 2 \times K$ tables. One approach tests $\beta = 0$ in model (6.4), using df $= 1$. Another approach tests goodness of fit of model (6.5), using df $= K$

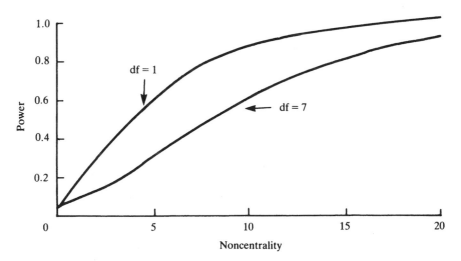

FIGURE 6.4   Power and noncentrality, for df $= 1$ and df $= 7$, when $\alpha = 0.05$.

TABLE 6.11   Change in Clinical Condition by Degree of Infiltration

Clinical Change	Degree of Infiltration		Proportion High
	High	Low	
Worse	1	11	0.08
Stationary	13	53	0.20
Slight improvement	16	42	0.28
Moderate improvement	15	27	0.36
Marked improvement	7	11	0.39

Source: Reprinted with permission from the Biometric Society (Cochran 1954).

(Section 6.3.1). When model (6.4) holds, both tests have the same noncentrality. Thus, the test of $\beta = 0$ is more powerful, since is has fewer degrees of freedom.

6.4.4   Treatment of Leprosy Example

Table 6.11 refers to an experiment on the use of sulfones and streptomycin drugs in the treatment of leprosy. The degree of infiltration at the start of the experiment measures a type of skin damage. The response is the change in the overall clinical condition of the patient after 48 weeks of treatment. We use response scores $\{-1, 0, 1, 2, 3\}$. The question of interest is whether subjects with high infiltration changed differently from those with low infiltration.

Here, the clinical change response variable is ordinal. It seems natural to compare the mean change for the two infiltration levels. Cochran (1954) and Yates (1948) noted that this analysis is identical to a trend test treating the binary variable as the response. That test is sensitive to linearity between clinical change and the proportion of cases with high infiltration.

The test $G^2(I) = 7.28$ (df = 4) does not show much evidence of association ($P = 0.12$), but it ignores the row ordering. The sample proportion of high infiltration increases monotonically as the clinical change improves. The test of H_0: $\beta = 0$ in the linear logit model has $G^2(I|L) = 6.65$, with df = 1 ($P = 0.01$). It gives strong evidence of more positive clinical change at the higher level of infiltration. Using the ordering by decreasing df from 4 to 1 pays a strong dividend. In addition, $G^2(L) = 0.63$ with df = 3 suggests that the linear trend model fits well.

6.4.5   Model Smoothing Improves Precision of Estimation

Using directed alternatives can improve not only *test power*, but also *estimation* of cell probabilities and summary measures. In generic form, let π be true cell probabilities in a contingency table, let **p** denote sample proportions, and let $\hat{\pi}$ denote model-based ML estimates of π.

When π satisfy a certain model, both $\hat{\pi}$ for that model and $\mathbf{p}$ are consistent estimators of π. The model-based estimator $\hat{\pi}$ is better, as its true asymptotic standard error cannot exceed that of $\mathbf{p}$. This happens because of model parsimony: The unsaturated model, on which $\hat{\pi}$ is based, has fewer parameters than the saturated model, on which $\mathbf{p}$ is based. In fact, model-based estimators are also more efficient in estimating functions $g(\pi)$ of cell probabilities. For any differentiable function g,

$$\text{asymp. var}\left[\sqrt{n}\, g(\hat{\pi})\right] \leq \text{asymp. var}\left[\sqrt{n}\, g(\mathbf{p})\right].$$

In Section 14.2.2 we prove this result. It holds more generally than for categorical data models (Altham 1984). This is one reason that statisticians prefer parsimonious models.

In reality, of course, a chosen model is unlikely to hold exactly. However, when the model approximates π well, unless n is extremely large, $\hat{\pi}$ is still better than $\mathbf{p}$. Although $\hat{\pi}_i$ is biased, it has smaller variance than p_i, and $\text{MSE}(\hat{\pi}_i) < \text{MSE}(p_i)$ when its variance plus squared bias is smaller than $\text{var}(p_i)$. In Section 3.3.7 we showed that in two-way tables, independence-model estimates of cell probabilities can be better than sample proportions even when that model does not hold.

6.5 SAMPLE SIZE AND POWER CONSIDERATIONS*

In any statistical procedure, the sample size n influences the results. Strong effects are likely to be detected even when n is small. By contrast, detection of weak effects requires large n. A study design should reflect the sample size needed to provide good power for detecting the effect.

6.5.1 Sample Size and Power for Comparing Two Proportions

For test statistics having large-sample normal distributions, power calculations can use ordinary methods. To illustrate, consider a test comparing binomial parameters π_1 and π_2 for two medical treatments. An experiment plans independent samples of size $n_i = n/2$ receiving each treatment. The researchers expect $\pi_i \approx 0.6$ for each, and a difference of at least 0.10 is important. In testing $H_0: \pi_1 = \pi_2$, the variance of the difference $\hat{\pi}_1 - \hat{\pi}_2$ in sample proportions is $\pi_1(1 - \pi_1)/(n/2) + \pi_2(1 - \pi_2)/(n/2) \approx 0.6 \times 0.4 \times (4/n) = 0.96/n$. In particular,

$$z = \frac{(\hat{\pi}_1 - \hat{\pi}_2) - (\pi_1 - \pi_2)}{(0.96/n)^{1/2}}$$

has approximately a standard normal distribution for π_1 and π_2 near 0.6.

The power of an α-level test of H_0 is approximately

$$P\left[\frac{|\hat{\pi}_1 - \hat{\pi}_2|}{(0.96/n)^{1/2}} \geq z_{\alpha/2}\right].$$

When $\pi_1 - \pi_2 = 0.10$, for $\alpha = 0.05$, this equals

$$P\left[\frac{(\hat{\pi}_1 - \hat{\pi}_2) - 0.10}{(0.96/n)^{1/2}} > 1.96 - 0.10(n/0.96)^{1/2}\right]$$

$$+P\left[\frac{(\hat{\pi}_1 - \hat{\pi}_2) - 0.10}{(0.96/n)^{1/2}} < -1.96 - 0.10(n/0.96)^{1/2}\right]$$

$$= P\left[z > 1.96 - 0.10(n/0.96)^{1/2}\right] + P\left[z < -1.96 - 0.10(n/0.96)^{1/2}\right]$$

$$= 1 - \Phi\left[1.96 - 0.10(n/0.96)^{1/2}\right] + \Phi\left[-1.96 - 0.10(n/0.96)^{1/2}\right],$$

where Φ is the standard normal cdf. The power is approximately 0.11 when $n = 50$ and 0.30 when $n = 200$. It is not easy to attain significance when effects are small and the sample is not very large. Figure 6.5 shows how the power increases in n when $\pi_1 - \pi_2 = 0.1$. By contrast, it shows how the power improves when $\pi_1 - \pi_2 = 0.2$.

For a given $P(\text{type I error}) = \alpha$ and $P(\text{type II error}) = \beta$ (and hence power $= 1 - \beta$), one can determine the sample size needed to attain those

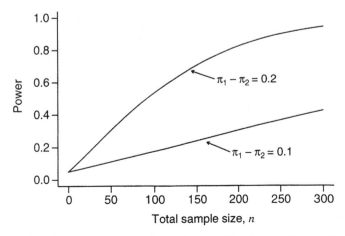

FIGURE 6.5 Approximate power for testing equality of proportions, with true values near middle of range and $\alpha = 0.05$.

values. A study using $n_1 = n_2$ requires approximately

$$n_1 = n_2 = (z_{\alpha/2} + z_\beta)^2 [\pi_1(1 - \pi_1) + \pi_2(1 - \pi_2)]/(\pi_1 - \pi_2)^2.$$

For a test with $\alpha = 0.05$ and $\beta = 0.10$ when π_1 and π_2 are truly about 0.60 and 0.70, $n_1 = n_2 = 473$. This formula also provides the sample sizes needed for a comparable confidence interval for $\pi_1 - \pi_2$. With about 473 subjects in each group, a 95% confidence interval has only a 0.10 chance of containing 0 when actually, $\pi_1 = 0.60$ and $\pi_2 = 0.70$.

This sample-size formula is approximate and may underestimate slightly the actual values required. It is adequate for most practical work, though, in which only rough conjectures are available for π_1 and π_2. Fleiss (1981) showed more precise formulas.

6.5.2 Sample Size Determination in Logistic Regression

Consider now the model $\text{logit}[\pi(x_i)] = \alpha + \gamma x_i$, $i = 1, \ldots, n$, in which x is quantitative. [We use γ so as not to confuse with $\beta = P(\text{type II error})$.] The sample size needed to achieve a certain power for testing $H_0: \gamma = 0$ depends on the variance of $\hat{\gamma}$. This depends on $\{\pi(x_i)\}$, and formulas for n use a guess for $\hat{\pi} = \pi(\bar{x})$ and the distribution of X. The effect size is the log odds ratio τ comparing $\pi(\bar{x})$ to $\pi(\bar{x} + s_x)$, the probability for a standard deviation above the mean of x. For a one-sided test when X is approximately normal, Hsieh (1989) derived

$$n = \left[z_\alpha + z_\beta \exp(-\tau^2/4) \right]^2 (1 + 2\hat{\pi}\delta)/(\hat{\pi}\tau^2),$$

where

$$\delta = \left[1 + (1 + \tau^2)\exp(5\tau^2/4) \right] / \left[1 + \exp(-\tau^2/4) \right].$$

The value n decreases as $\hat{\pi} \to 0.5$ and as $|\tau|$ increases.

We illustrate for modeling the effect of $x =$ cholesterol level on the probability of severe heart disease for a population for which that probability at an average level of cholesterol is about 0.08. Researchers want the test to be sensitive to a 50% increase in this probability, for a standard deviation increase in cholesterol. The odds of severe heart disease at the mean cholesterol level equal $0.08/0.92 = 0.087$, and the odds one standard deviation above the mean equal $0.12/0.88 = 0.136$. The odds ratio equals $0.136/0.087 = 1.57$, and $\tau = \log(1.57) = 0.450$. For $\alpha = 0.05$ and $\beta = 0.10$, $\delta = 1.306$ and $n = 612$.

6.5.3 Sample Size in Multiple Logistic Regression

A multiple logistic regression model requires larger n to detect effects. Let R denote the multiple correlation between the predictor X of interest and the others in the model. The formula for n above divides by $(1 - R^2)$. In that formula, $\hat{\pi}$ is evaluated at the mean of all the explanatory variables, and the odds ratio refers to the effect of X at the mean level of the other predictors.

Consider the example in Section 6.5.2 when blood pressure is also a predictor. If the correlation between cholesterol and blood pressure is 0.40, we need $n \approx 612/[1 - (0.40)^2] = 729$.

These formulas provide, at best, rough indications of sample size. Most applications have only a crude guess for $\hat{\pi}$ and R, and X may be far from normally distributed. For other work on this problem, see Hsieh et al. (1998) and Whittemore (1981).

6.5.4 Power for Chi-Squared Tests in Contingency Tables

When hypotheses are false, squared normal and X^2 and G^2 statistics have large-sample noncentral chi-squared distributions (Section 6.4.2). Suppose that H_0 is equivalent to model M for a contingency table. Let π_i denote the true probability in cell i, and let $\pi_i(M)$ denote the value to which the ML estimate $\hat{\pi}_i$ for model M converges, where $\Sigma\pi_i = \Sigma\pi_i(M) = 1$. For a multinomial sample of size n, the noncentrality parameter for X^2 equals

$$\lambda = n \sum_i \frac{[\pi_i - \pi_i(M)]^2}{\pi_i(M)}. \qquad (6.8)$$

This has the same form as X^2, with π_i in place of the sample proportion p_i and $\pi_i(M)$ in place of $\hat{\pi}_i$. The noncentrality parameter for G^2 equals

$$\lambda = 2n \sum_i \pi_i \log \frac{\pi_i}{\pi_i(M)}. \qquad (6.9)$$

TABLE 6.12   Power of Chi-Squared Test for $\alpha = 0.05$

						Noncentrality							
df	0.0	0.2	0.4	0.6	0.8	1.0	2.0	3.0	4.0	5.0	7.0	10.0	15.0	25.0
1	.050	.073	.097	.121	.146	.170	.293	.410	.516	.609	.754	.885	.972	.998
2	.050	.065	.081	.098	.115	.133	.226	.322	.415	.504	.655	.815	.944	.996
3	.050	.062	.075	.088	.102	.116	.192	.275	.358	.440	.590	.761	.917	.993
4	.050	.060	.071	.082	.093	.106	.172	.244	.320	.396	.540	.716	.891	.989
6	.050	.058	.066	.075	.084	.094	.146	.206	.270	.336	.468	.644	.843	.980
8	.050	.057	.064	.071	.079	.087	.131	.182	.238	.296	.417	.588	.799	.968
10	.050	.056	.062	.068	.075	.082	.121	.166	.215	.268	.379	.542	.760	.956
20	.050	.053	.056	.060	.063	.066	.096	.125	.158	.193	.273	.402	.611	.883
50	.050	.052	.054	.056	.059	.061	.076	.092	.110	.129	.173	.250	.398	.687

Source: Reprinted with permission from G. E. Haynam, Z. Govindarajulu, and F. C. Leone, in *Selected Tables in Mathematical Statistics*, eds. H. L. Harter and D. B. Owen (Chicago: Markham, 1970).

When H_0 is true, all $\pi_i = \pi_i(M)$. Then, for either statistic, $\lambda = 0$ and the central chi-squared distribution applies.

To determine the approximate power for a chi-squared test with df = ν, (1) choose a hypothetical set of true values $\{\pi_i\}$, (2) calculate $\{\pi_i(M)\}$ by fitting to $\{\pi_i\}$ the model M for H_0, (3) calculate the noncentrality parameter λ, and (4) calculate $P[X_{\nu,\lambda}^2 > \chi_\nu^2(\alpha)]$. Table 6.12 shows an excerpt from a table of noncentral chi-squared probabilities for step 4 with $\alpha = 0.05$.

6.5.5 Power for Testing Conditional Independence

We use an example based on one in O'Brien (1986). A standard fetal heart rate monitoring test predicts whether a fetus will require nonroutine care following delivery. The standard test has categories (worrisome, reassuring). The response Y is whether the newborn required some nonroutine medical care during the first week after birth (1 = yes, 0 = no). A new fetal heart rate monitoring test is developed, having categories (very worrisome, somewhat worrisome, reassuring). A physician plans to study whether this new test can help make predictions about the outcome; that is, given the result of the standard test, is there an association between the response and the result of the new test? A relevant statistic tests the effect of the new monitoring test in the logit model having the new test (N) and standard test (S) as qualitative predictors.

To help select n, a statistician asks the physician to conjecture about the joint distribution of the explanatory variables, with questions such as "What proportion of the cases do you think will be scored 'reassuring' by both tests?" For each NS combination, the physician also guessed $P(Y = 1)$. Table 6.13 shows one scenario for marginal and conditional probabilities. These yield a joint distribution $\{\pi_{ijk}\}$ from their product, such as $0.04 \times 0.40 = 0.016$ for the proportion of cases judged worrisome by the standard test and very worrisome by the new test and requiring nonroutine medical care. These joint probabilities yield fitted probabilities $\pi(M_0)$ and $\pi(M_1)$ for the null and alternative logit models. (One can get these by entering $\{\pi_{ijk}\}$ in

TABLE 6.13 Scenario for Power Computation

Standard	New	Joint Probability	P (nonroutine care)
Worrisome	Very worrisome	0.04	0.40
	Somewhat worrisome	0.08	0.32
	Reassuring	0.04	0.27
Reassuring	Very worrisome	0.02	0.30
	Somewhat worrisome	0.18	0.22
	Reassuring	0.64	0.15

Source: Reprinted with permission from O'Brien (1986).

percentage form as counts in software for logistic regression, fit the relevant model, and divide the fitted counts by 100 to get the fitted joint probabilities.) The likelihood-ratio test comparing these models has noncentrality (6.9) with $\pi(M_1)$ playing the role of π and $\pi(M_0)$ playing the role of $\pi(M)$.

For the scenario in Table 6.13, the noncentrality equals $0.00816n$, with df = 2. For n = 400, 600, and 1000, the approximate powers when α = 0.05 are 0.35, 0.49, and 0.73. This scenario predicts 64% of the observations to occur at only one combination of the factors. The lack of dispersion for the factors weakens the power.

6.5.6   Effects of Sample Size on Model Selection and Inference

The effects of sample size suggest some cautions for model selection. For small n, the most parsimonious model accepted in a goodness-of-fit test may be quite simple. By contrast, larger samples usually require more complex models to pass goodness-of-fit tests. Then, some effects that are statistically significant may be weak and substantively unimportant. With large n it may be adequate to use a model that is simpler than models that pass goodness-of-fit tests. An analysis that focuses solely on goodness-of-fit tests is incomplete. It is also necessary to estimate model parameters and describe strengths of effects.

These remarks merely reflect limitations of significance testing. Null hypotheses are rarely true. With large enough n, they will be rejected. A more relevant concern is whether the difference between true parameter values and null hypothesis values is sufficient to be important. Many methodologists overemphasize testing and underutilize estimation methods such as confidence intervals. When the P-value is small, a confidence interval specifies the extent to which H_0 may be false, thus helping us determine whether rejecting it has practical importance. When the P-value is not small, the confidence interval indicates whether some plausible parameter values are far from H_0. A wide confidence interval containing the H_0 value indicates that the test had weak power at important alternatives.

6.6   PROBIT AND COMPLEMENTARY LOG-LOG MODELS*

For binary responses, in this section we discuss two alternatives to logit models. Like the logit model, these models have form (4.8),

$$\pi(x) = \Phi(\alpha + \beta x) \qquad (6.10)$$

for a continuous cdf Φ. The following argument motivates this class.

6.6.1   Tolerance Motivation for Binary Response Models

In toxicology, binary response models describe the effect of dosage of a toxin on whether a subject dies. The *tolerance distribution* provides justification for model (6.10). Let x denote the dosage level. For a randomly selected subject, let $Y = 1$ if the subject dies. Suppose that the subject has tolerance T for the dosage, with $(Y = 1)$ equivalent to $(T \le x)$. For instance, an insect survives if the dosage x is less than T and dies if the dosage is at least T. Tolerances vary among subjects, and let $F(t) = P(T \le t)$. For fixed dosage x, the probability a randomly selected subject dies is

$$\pi(x) = P(Y = 1 \mid X = x) = P(T \le x) = F(x).$$

That is, the appropriate binary model is the one having the shape of the cdf F of the tolerance distribution. Let Φ denote the standard cdf for the family to which F belongs. A common standardization uses the mean and standard deviation of T, so that

$$\pi(x) = F(x) = \Phi[(x - \mu)/\sigma].$$

Then, the model has form $\pi(x) = \Phi(\alpha + \beta x)$.

6.6.2 Probit Models

Toxicological experiments often measure dosage as the log concentration (Bliss 1935). Often, the tolerance distribution for the dosage is approximately $N(\mu, \sigma^2)$ for unknown μ and σ. If F is the $N(\mu, \sigma^2)$ cdf, then $\pi(x)$ has the form $\pi(x) = \Phi(\alpha + \beta x)$, where Φ is the standard normal cdf, $\alpha = -\mu/\sigma$ and $\beta = 1/\sigma$. In GLM form,

$$\Phi^{-1}[\pi(x)] = \alpha + \beta x \qquad (6.11)$$

is the *probit model*. The probit link function is $\Phi^{-1}(\cdot)$. Whereas the cdf maps the real line onto the $(0, 1)$ probability scale, the inverse cdf maps the $(0, 1)$ scale for $\pi(x)$ onto the real line values for linear predictors in binary response models.

The response curve for $\pi(x)$ [or for $1 - \pi(x)$, when $\beta < 0$] has the appearance of the normal cdf with mean $\mu = -\alpha/\beta$ and standard deviation $\sigma = 1/|\beta|$. Since 68% of the normal density falls within a standard deviation of the mean, $1/|\beta|$ is the distance between x values where $\pi(x) = 0.16$ or 0.84 and where $\pi(x) = 0.50$. The rate of change in $\pi(x)$ is $\partial\pi(x)/\partial x = \beta\phi(\alpha + \beta x)$, where $\phi(\cdot)$ is the standard normal density function. The rate is highest when $\alpha + \beta x = 0$ (i.e., at $x = -\alpha/\beta$), where it equals $\beta/(2\pi)^{1/2} = 0.40\beta$ (for $\pi = 3.14\ldots$). At that point, $\pi(x) = \frac{1}{2}$.

By comparison, in logistic regression with parameter β, the curve for $\pi(x)$ is a logistic cdf with standard deviation $\pi/|\beta|\sqrt{3}$. Its rate of change in $\pi(x)$ at $x = -\alpha/\beta$ is 0.25β. The rates of change where $\pi(x) = \frac{1}{2}$ are the same for the cdf's corresponding to the probit and logistic curves when the logistic β is $0.40/0.25 = 1.6$ times the probit β. The standard deviations are the same when the logistic β is $\pi/\sqrt{3} = 1.8$ times the probit β. When both models fit well, parameter estimates in logistic regression are about 1.6 to 1.8 times those in probit models.

The likelihood equations that (4.24) showed for binomial regression models apply to probit models (see also Problem 6.32). One can solve them using the Fisher scoring algorithm for GLMs (Bliss 1935, Fisher 1935b). Newton–Raphson yields the same ML estimates but slightly different standard errors. For the information matrix inverted to obtain the asymptotic covariance matrix, Newon–Raphson uses observed information, whereas Fisher scoring uses expected information. These differ for binary links other than the logit.

6.6.3   Beetle Mortality Example

Table 6.14 reports the number of beetles killed after 5 hours of exposure to gaseous carbon disulfide at various concentrations. Figure 6.6 plots (as dots) the proportion killed against the log concentration. The proportion jumps up at about $x = 1.8$, and it is close to 1 above there.

The ML fit of the probit model is

$$\Phi^{-1}[\hat{\pi}(x)] = -34.96 + 19.74x.$$

For this fit, $\hat{\pi}(x) = 0.5$ at $x = 34.96/19.74 = 1.77$. The fit corresponds to a normal tolerance distribution with $\mu = 1.77$ and $\sigma = 1/19.74 = 0.05$. The curve for $\hat{\pi}(x)$ is that of a $N(1.77, 0.05^2)$ cdf.

At dosage x_i with n_i beetles, $n_i\hat{\pi}(x_i)$ is the fitted count for death, $i = 1, \ldots, 8$. Table 6.14 reports the fitted values and Figure 6.6 shows the fit. The table also shows fitted values for the linear logit model. These models fit similarly and rather poorly. The G^2 goodness-of-fit statistic equals 11.1 for the logit model and 10.0 for the probit model, with df = 6.

TABLE 6.14   Beetles Killed after Exposure to Carbon Disulfide

Log Dose	Number of Beetles	Number Killed	Fitted Values		
			Comp. Log-Log	Probit	Logit
1.691	59	6	5.7	3.4	3.5
1.724	60	13	11.3	10.7	9.8
1.755	62	18	20.9	23.4	22.4
1.784	56	28	30.3	33.8	33.9
1.811	63	52	47.7	49.6	50.0
1.837	59	53	54.2	53.4	53.3
1.861	62	61	61.1	59.7	59.2
1.884	60	60	59.9	59.2	58.8

Source: Data reprinted with permission from Bliss (1935).

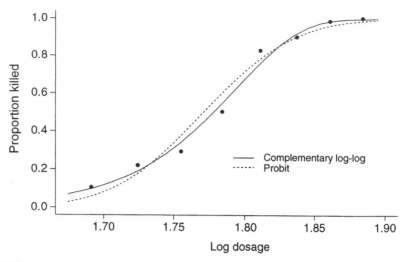

FIGURE 6.6 Proportion of beetles killed versus log dosage, with fits of probit and complementary log-log models.

6.6.4   Complementary Log-Log Link Models

The logit and probit links are symmetric about 0.5, in the sense that

$$\text{link}[\pi(x)] = -\text{link}[1 - \pi(x)].$$

To illustrate,

$$\begin{aligned} \text{logit}[\pi(x)] &= \log[\pi(x)/(1 - \pi(x))] \\ &= -\log[(1 - \pi(x))/\pi(x)] = -\text{logit}[1 - \pi(x)]. \end{aligned}$$

This means that the response curve for $\pi(x)$ has a symmetric appearance about the point where $\pi(x) = 0.5$, so $\pi(x)$ approaches 0 at the same rate it approaches 1. Logit and probit models are inappropriate when this is badly violated.

The response curve

$$\pi(x) = 1 - \exp[-\exp(\alpha + \beta x)] \tag{6.12}$$

has the shape shown in Figure 6.7. It is asymmetric, $\pi(x)$ approaching 0 fairly slowly but approaching 1 quite sharply. For this model,

$$\log[-\log(1 - \pi(x))] = \alpha + \beta x.$$

The link for this GLM is called the *complementary log-log* link, since the log-log link applies to the complement of $\pi(x)$.

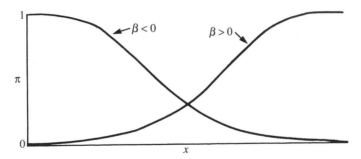

FIGURE 6.7 Model with complementary log–log link.

To interpret model (6.12), we note that at x_1 and x_2,

$$\log\left[-\log(1 - \pi(x_2))\right] - \log\left[-\log(1 - \pi(x_1))\right] = \beta(x_2 - x_1),$$

so that

$$\frac{\log[1 - \pi(x_2)]}{\log[1 - \pi(x_1)]} = \exp\left[\beta(x_2 - x_1)\right]$$

and

$$1 - \pi(x_2) = \left[1 - \pi(x_1)\right]^{\exp[\beta(x_2 - x_1)]}.$$

For $x_2 - x_1 = 1$, the complement probability at x_2 equals the complement probability at x_1 raised to the power $\exp(\beta)$.

A related model to (6.12) is

$$\pi(x) = \exp\left[-\exp(\alpha + \beta x)\right]. \tag{6.13}$$

For it, $\pi(x)$ approaches 0 sharply but approaches 1 slowly. As x increases, the curve is monotone decreasing when $\beta > 0$, and monotone increasing when $\beta < 0$. In GLM form it uses the *log-log* link

$$\log\left[-\log(\pi(x))\right] = \alpha + \beta x.$$

When the complementary log-log model holds for the probability of a success, the log-log model holds for the probability of a failure.

Model (6.13) with log-log link is the special case of (6.10) with cdf of the *extreme value* (or *Gumbel*) distribution. The cdf equals

$$F(x) = \exp\left\{-\exp\left[-(x - a)/b\right]\right\}$$

for parameters $b > 0$ and $-\infty < a < \infty$. It has mean $a + 0.577b$ and standard deviation $\pi b / \sqrt{6}$. Models with log-log links can be fitted using the Fisher scoring algorithm for GLMs.

6.6.5 Beetle Mortality Example Revisited

For the beetle mortality data (Table 6.14), the complementary log-log model has ML estimates $\hat{\alpha} = -39.52$ and $\hat{\beta} = 22.01$. At dosage $x = 1.7$, the fitted probability of survival is $1 - \hat{\pi}(x) = \exp\{-\exp[-39.52 + 22.01(1.7)]\} = 0.885$, whereas at $x = 1.8$ it is 0.332 and at $x = 1.9$ it is 5×10^{-5}. The probability of survival at dosage $x + 0.1$ equals the probability at dosage x raised to the power $\exp(22.01 \times 0.1) = 9.03$. For instance, $0.332 = (0.885)^{9.03}$.

Table 6.14 shows the fitted values and Figure 6.6 shows the fit. They are close to the observed death counts ($G^2 = 3.5$, df $= 6$). The fit seems adequate. Aranda-Ordaz (1981) and Stukel (1988) discussed these data further.

6.7 CONDITIONAL LOGISTIC REGRESSION AND EXACT DISTRIBUTIONS*

ML estimators of logistic model parameters work best when the sample size n is large compared to the number of parameters. When n is small or when the number of parameters grows as n does, improved inference results using *conditional maximum likelihood*. In this section we present this approach and in Section 10.2 apply it with matched case–control studies.

6.7.1 Conditional Likelihood

This conditional likelihood approach eliminates nuisance parameters by conditioning on their sufficient statistics. This generalizes Fisher's method for 2×2 tables (Section 3.5). The conditional likelihood refers to a conditional distribution defined for potential samples that provide the same information about the nuisance parameters that occurs in the observed sample.

We begin with a general exposition and then discuss special cases. Let y_i denote the binary response for subject i, $i = 1, \ldots, N$. (For now, each y_i refers to a single trial, so $n_i = 1$.) Let x_{ij} be the value of predictor j for that subject, $j = 1, \ldots, p$. The model is

$$P(Y_i = y_i) = \frac{\exp\left[y_i\left(\alpha + \sum_{j=1}^{p} \beta_j x_{ij}\right)\right]}{1 + \exp\left(\alpha + \sum_{j=1}^{p} \beta_j x_{ij}\right)}, \tag{6.14}$$

where substituting $y_i = 1$ gives the usual expression, such as (5.15). Here, we explicitly separate the intercept from the coefficients of the p predictors. For N independent observations,

$$P(Y_1 = y_1, \ldots, Y_N = y_N) = \frac{\exp\left[\left(\sum_i y_i\right)\alpha + \sum_{j=1}^{p}\left(\sum_i y_i x_{ij}\right)\beta_j\right]}{\prod_i\left[1 + \exp\left(\alpha + \sum_{j=1}^{p} \beta_j x_{ij}\right)\right]}. \tag{6.15}$$

From this likelihood function, the sufficient statistic for β_j is $\sum_i y_i x_{ij}$, $j = 1, \ldots, p$. The sufficient statistic for α is $\sum_i y_i$, the total number of successes.

Usually, some parameters refer to effects of primary interest. Others may be there to adjust for relevant effects, but their values are not of special interest. We can eliminate the latter parameters from the likelihood by conditioning on their sufficient statistics. We illustrate by eliminating α. (In Section 10.2.5 we show that for models for matched case–control studies, intercept terms cause difficulties with inference about the primary parameters, so it can be helpful to eliminate them.) Since the sufficient statistic for α is $\sum_i y_i$, we condition on $\sum_i y_i$. Suppose that $\sum_i y_i = t$. Denote the conditional reference set of samples having the same value of $\sum_i y_i$ as observed by

$$ S(t) = \left\{ (y_1^*, \ldots, y_N^*) : \sum_i y_i^* = t \right\}. $$

With $\{y_i\}$ such that $\sum_i y_i = t$, the conditional likelihood function equals

$$ P\left(Y_1 = y_1, \ldots, Y_N = y_N \Big| \sum_i y_i = t\right) = \frac{P(Y_1 = y_1, \ldots, Y_N = y_N)}{\sum_{S(t)} P(Y_1 = y_1^*, \ldots, Y_N = y_N^*)} $$

$$ = \frac{\exp\left[t\alpha + \sum_{j=1}^p (\sum_i y_i x_{ij})\beta_j\right] \prod_i \left[1 + \exp(\alpha + \sum_{j=1}^p \beta_j x_{ij})\right]}{\sum_{S(t)} \exp\left[t\alpha + \sum_{j=1}^p (\sum_i y_i^* x_{ij})\beta_j\right] / \prod_i \left[1 + \exp(\alpha + \sum_{j=1}^p \beta_j x_{ij})\right]} $$

$$ = \frac{\exp\left[\sum_{j=1}^p (\sum_i y_i x_{ij})\beta_j\right]}{\sum_{S(t)} \exp\left[\sum_{j=1}^p (\sum_i y_i^* x_{ij})\beta_j\right]}. $$

This does not depend on α.

A conditional likelihood is used just like an ordinary likelihood. For the parameters in it, their conditional ML estimates are the values maximizing it. Calculated using iterative methods, the estimators are asymptotically normal with covariance matrix equal to the negative inverse of the matrix of second partial derivatives of the conditional log likelihood.

6.7.2 Small-Sample Conditional Inference for Logistic Regression

For small samples, inference for a parameter uses the conditional distribution after eliminating all other parameters. With it, one can calculate probabilities such as P-values exactly rather than with crude approximations (Cox 1970).

For instance, suppose that inference focuses on β_p in model (6.14). To eliminate other parameters, we condition on their sufficient statistics $T_j = \sum_i y_i x_{ij}$, $j = 0, \ldots, p - 1$ (where $x_{i0} = 1$). With an argument like that just shown, one obtains the conditional distribution

$$P\left(Y_1 = y_1, \ldots, Y_N = y_N \mid T_j = t_j, j = 0, \ldots, p - 1\right)$$

$$= \frac{\exp\left[\left(\Sigma_i y_i x_{ip}\right)\beta_p\right]}{\Sigma_{S(t_0, \ldots, t_{p-1})} \exp\left[\left(\Sigma_i y_i^* x_{ip}\right)\beta_p\right]} = \frac{\exp\left(t_p \beta_p\right)}{\Sigma_{S(t_0, \ldots, t_{p-1})} \exp\left(t_p^* \beta_p\right)},$$

where

$$S\left(t_0, \ldots, t_{p-1}\right) = \left\{\left(y_1^*, \ldots, y_N^*\right) : \sum_i y_i^* x_{ij} = t_j, j = 0, \ldots, p - 1\right\}.$$

This depends only on β_p. Inference for β_p uses the conditional distribution of its sufficient statistic, $T_p = \Sigma_i y_i x_{ip}$, given the others. Let $c(t_0, \ldots, t_{p-1}, t)$ denote the number of data vectors in $S(t_0, \ldots, t_{p-1})$ for which $T_p = t$. The conditional distribution of T_p is

$$P\left(T_p = t \mid T_j = t_j, j = 0, \ldots, p - 1\right) = \frac{c\left(t_0, \ldots, t_{p-1}, t\right)\exp\left(t\beta_p\right)}{\Sigma_u c\left(t_0, \ldots, t_{p-1}, u\right)\exp\left(u\beta_p\right)},$$

$$(6.16)$$

where the denominator summation refers to the possible values u of T_p.

For testing H_0: $\beta_p = 0$, the conditional distribution simplifies. For H_a: $\beta_p > 0$ and observed $T_p = t_{\text{obs}}$, the exact conditional P-value is

$$\sum_{t \geq t_{\text{obs}}} P\left(T_p = t \mid T_j = t_j, j = 0, \ldots, p - 1\right) = \frac{\Sigma_{t \geq t_{\text{obs}}} c\left(t_0, \ldots, t_{p-1}, t\right)}{\Sigma_u c\left(t_0, \ldots, t_{p-1}, u\right)},$$

the proportion of data configurations in the conditional set that have the sufficient statistic for β_p at least as large as observed. Implementing this inference requires calculating $\{c(t_0, \ldots, t_{p-1}, u)\}$. For all but the simplest problems, computations are intensive and require specialized software (e.g., LogXact of Cytel Software or PROC LOGISTIC in SAS). In the remainder of this section we consider special cases for small-sample inference.

6.7.3 Small-Sample Conditional Inference for 2 × 2 Contingency Tables

First, consider logistic regression with a single predictor x,

$$\text{logit}\left[P(Y_i = 1)\right] = \alpha + \beta x_i, \quad i = 1, \ldots, N, \quad (6.17)$$

when x_i takes only two values. The model applies to 2 × 2 tables, where $x_i = 1$ denotes row 1 and $x_i = 0$ denotes row 2. The sufficient statistic for α

is $\Sigma_i y_i$, which is the first column total. The sufficient statistic for β is $T = \Sigma_i y_i x_i$, which simplifies to the number of successes in the first row. Equivalently, the sufficient statistics for the model are the numbers of successes in the two rows. Let s_1 and s_2 denote these binomial variates. The row totals n_1 and n_2 are their indices.

To eliminate α, we condition on $s = s_1 + s_2$, the first column total. Since $N = n_1 + n_2$ is fixed, so then is the other column marginal total. Fixing both sets of marginal totals yields hypergeometric probabilities for s_1 that depend only on β [see (3.20), identifying $\theta = \exp(\beta)$]. In that case the conditional distribution satisfies (6.16) with $c(t_0, t) = \binom{n_1}{t}\binom{N - n_1}{t_0 - t}$ and with $t_0 = s$ and $t = s_1$. The resulting exact conditional test that $\beta = 0$ is Fisher's exact test for 2×2 tables (Section 3.5.1).

6.7.4 Small-Sample Conditional Inference for Linear Logit Model

The linear logit model, $\text{logit}(\pi_i) = \alpha + \beta x_i$, applies to $I \times 2$ tables with ordered rows. We discussed this model in Section 5.3.4. For it, the data $\{y_i\}$ are I independent $\{\text{bin}(n_i, \pi_i)\}$ counts, with fixed row totals $\{n_i\}$. Conditioning on $s = \Sigma y_i$ and hence the column totals yields a conditional likelihood free of α. Exact inference about β uses its sufficient statistic, $T = \Sigma x_i y_i$. From (6.16) its distribution has the form

$$P\left(T = t \,\middle|\, \sum_i y_i = s; \beta\right) = \frac{c(s, t)e^{\beta t}}{\Sigma_u c(s, u)e^{\beta u}}. \tag{6.18}$$

Here, $c(s, u)$ equals the sum of $\left[\Pi_i \binom{n_i}{y_i}\right]$ for all tables with the given marginal totals that have $T = u$.

When $\beta = 0$, the cell counts have the multiple hypergeometric distribution (3.19). To test this, ordering the tables with the given margins by T is equivalent to ordering them by the Cochran–Armitage statistic (Section 5.3.5). Thus, this test for the linear logit model is an exact trend test.

In Section 5.3.5 we applied the Cochran–Armitage test to Table 5.3 on maternal alcohol consumption and infant malformation. Even though $n = 32,573$, the table is highly unbalanced, with both very small and very large counts. It is safer to use small-sample methods. For the exact conditional trend test with the same scores, the one-sided P-value for H_a: $\beta > 0$ is 0.0168. The two-sided P-value is 0.0172, reflecting asymmetry of the conditional distribution, given the marginal counts. This is not much different from the two-sided P-value of 0.010 obtained with the large-sample Cochran–Armitage test.

6.7.5 Small-Sample Tests of Conditional Independence in $2 \times 2 \times K$ Tables

For $2 \times 2 \times K$ tables $\{n_{ijk}\}$, the Cochran–Mantel–Haenszel test uses $\Sigma_k n_{11k}$. For logit model (6.4), this is the sufficient statistic for β, the effect of X. To conduct a small-sample test of $\beta = 0$, one needs to eliminate the other model parameters. Constructing the likelihood reveals that the sufficient statistics for $\{\beta_k^Z\}$ are the column marginal totals $\{n_{+jk}\}$ in each partial table. When X and Z are predictors, it is natural to treat the numbers of trials $\{n_{i+k}\}$ at each combination of XZ values as fixed. Thus, exact inference about β conditions on the row and column totals in each stratum.

Conditional on the strata margins, an exact test uses $\Sigma_k n_{11k}$. Hypergeometric probabilities occur in each partial table for the independent null distributions of $\{n_{11k}, k = 1, \ldots, K\}$. The product of the K mass functions gives the null joint distribution of $\{n_{11k}, k = 1, \ldots, K\}$. [This is (6.19) below, setting $\theta = 1$.] This determines the null distribution of $\Sigma_k n_{11k}$. For H_a: $\beta > 0$, the P-value is the null probability that $\Sigma_k n_{11k}$ is at least as large as observed, for the fixed strata marginal totals. Mehta et al. (1985) presented a fast algorithm. The test simplifies to Fisher's exact test when $K = 1$.

6.7.6 Promotion Discrimination Example

Table 6.15 refers to U.S. government computer specialists of similar seniority considered for promotion. The table cross-classifies promotion decision by employee's race, considered for three separate months. We test conditional independence of promotion decision and race, or H_0: $\beta = 0$, in model (6.4). The table contains several small counts. The overall sample size is not small ($n = 74$), but one marginal count (collapsing over month of decision) equals zero, so we might be wary of using the CMH test.

For H_a: $\beta < 0$ (i.e., odds ratio < 1), the probability of promotion was lower for black employees than for white employees. For the margins of the partial tables in Table 6.15, n_{111} can range between 0 and 4, n_{112} can range between 0 and 4, and n_{113} can range between 0 and 2. The total $\Sigma_k n_{11k}$ can range between 0 and 10. The sample data are the most extreme possible

TABLE 6.15   Promotion Decisions by Race and by Month

	July Promotions		August Promotions		September Promotions	
Race	Yes	No	Yes	No	Yes	No
Black	0	7	0	7	0	8
White	4	16	4	13	2	13

Source: J. Gastwirth, *Statistical Reasoning in Law and Public Policy* (San Diego, CA: Academic Press, 1988), p. 266.

result in each case. The observed $\Sigma_k n_{11k} = 0$, and the P-value is the null probability of this outcome. Software provides $P = 0.026$. A two-sided P-value, based on summing the probabilities of all tables no more likely than the observed table, equals 0.056.

6.7.7 Exact Conditional Estimation and Comparison of Odds Ratios

For model (6.4) of homogeneous association in $2 \times 2 \times K$ tables, the ordinary ML estimator of the odds ratio $\theta = \exp(\beta)$ behaves poorly for sparse-data asymptotics. The conditional ML estimator maximizes the conditional likelihood function after reducing the parameter space by conditioning on sufficient statistics for the other parameters (Andersen 1970; Birch 1964b).

For cell counts $\{n_{ijk}\}$, given $\{n_{i+k}, n_{+jk}\}$ for all k, the conditional probability mass function that $(n_{111} = t_1, \ldots, n_{11K} = t_K)$ is the product of the functions (3.20) from the separate strata, or

$$\prod_k P(n_{11k} = t_k \mid n_{1+k}, n_{+1k}, n_{++k}; \theta) = \prod_k \frac{\dbinom{n_{1+k}}{t_k}\dbinom{n_{++k} - n_{1+k}}{n_{+1k} - t_k}\theta^{t_k}}{\Sigma_u \dbinom{n_{1+k}}{u}\dbinom{n_{++k} - n_{1+k}}{n_{+1k} - u}\theta^u}.$$

$$(6.19)$$

The conditional ML estimator $\hat{\theta}$ maximizes (6.19). Like the Mantel–Haenszel estimator $\hat{\theta}_{\text{MH}}$, it has good properties for both standard and sparse-data asymptotic cases (Andersen 1970; Breslow 1981), since the number of parameters does not change as K does. It can be slightly more efficient than $\hat{\theta}_{\text{MH}}$, except when $\theta = 1.0$, where they are equally efficient, or for matched pairs, where they are identical (Breslow 1981).

The conditional distribution (6.19) propagates one for $\Sigma_k n_{11k}$, which is used to test $H_0: \theta = \theta_0$ for an arbitrary value. Then, a 95% confidence interval for θ consists of all θ_0 for which the P-value exceeds 0.05. Such an interval is guaranteed to have at least the nominal coverage probability (Gart 1970; Kim and Agresti 1995; Mehta et al. 1985). This extends the interval for a single 2×2 table (Section 3.6.1). For the promotion discrimination case (Table 6.15), $\Sigma_k n_{11k} = 0$, so the lower bound of any confidence interval for θ should be 0. For the generalization to several strata of Cornfield's tail-method interval, StatXact reports a 95% confidence interval of $(0, 1.01)$.

Zelen (1971) presented a small-sample test of homogeneity of the odds ratios. See Agresti (1992) for discussion of this and other small-sample methods for contingency tables.

TABLE 6.16   Example for Exact Conditional Logistic Regression

Cephalexin[a]	Age[a]	Length of Stay[a]	Cases of Diarrhea	Sample Size
0	0	0	0	385
0	0	1	5	233
0	1	0	3	789
0	1	1	47	1081
1	1	1	5	5

[a]See the text for an explanation of 0 and 1.

Source: Based on study by E. Jaffe and V. Chang, Cornell Medical Center, reported in the Manual for *LogXact* (Cambridge, MA: CYTEL Software, 1999), p. 259.

6.7.8   Diarrhea Example

The final example deals with a larger number of variables. Table 6.16 refers to 2493 patients having stays in a hospital. The response is whether they suffered an acute form of diarrhea during their stay. The three predictors are age (1 for over 50 years old, 0 for under 50), length of stay in hospital (1 for more than 1 week, 0 for less than 1 week), and exposure to an antibiotic called Cephalexin (1 for yes, 0 for no). We discuss estimation of the effect of Cephalexin, controlling for age and length of stay, using a model containing only main-effect terms.

The sample size is large, yet relatively few cases of acute diarrhea occurred. Moreover, all subjects having exposure to Cephalexin were also diarrhea cases. Such boundary outcomes in which none or all responses fall in one category cause infinite ML estimates of some model parameters. An ML estimate of ∞ for the Cephalexin effect means that the likelihood function increases continually as the parameter estimate for Cephalexin increases indefinitely.

To study the Cephalexin effect, we use an exact distribution, conditioning on sufficient statistics for the other predictors. Although the estimate of the log-odds-ratio parameter for the effect of Cephalexin is infinite, it is possible to construct a confidence interval by inverting the family of tests for the parameter, using the conditional distribution. Doing this, a 95% confidence interval is $(19, \infty)$ for the odds ratio. Assuming that the main-effects model is valid, Cephalexin appears to have a strong effect. Similarly, $P < 0.0001$ for testing that the log odds ratio equals zero.

Results must be qualified somewhat because no Cephalexin cases occurred at the first three combinations of levels of age and length of stay. In fact, the first three rows of Table 6.16 make no contribution to the analysis (Problem 6.18). The data actually provide evidence about the effect of Cephalexin only for older subjects having a long stay.

6.7.9 Complications from Discreteness

Like Fisher's exact test, exact conditional inference for contingency tables is conservative because of discreteness. This is especially true when n is small or the data are unbalanced, with most observations falling in a single column or row. Using mid-P-values or P-values based on a finer partitioning of the sample space (Note 3.9) in tests and related confidence intervals reduces conservativeness. For the promotion discrimination data (Table 6.15), we reported a 95% confidence interval for the common odds ratio of (0, 1.01). Inverting exact tests of H_0: $\theta = \theta_0$ with the mid-P-value yields the interval (0, 0.78). However, this approach cannot guarantee that the actual coverage probability is bounded below by 0.95.

A particular problem occurs when no other set of $\{y_i^*\}$ values has the same value of a given sufficient statistic $\sum_i y_i x_{ij}$ as the observed data. In that case the conditional distribution of the sufficient statistic for the parameter of interest is degenerate. The P-value for the exact test then equals 1.0. This commonly happens when at least one explanatory variable x_j whose effect is conditioned out for the inference is continuous, with unequally spaced observed values.

Finally, a limitation of the conditional approach is requiring sufficient statistics for the nuisance parameters. This happens only with GLMs that use the canonical link. Thus, for instance, the conditional approach works for logit models but not probit models.

NOTES

Section 6.1: Strategies in Model Selection

    6.1. A Bayesian argument motivates the Bayesian information criterion BIC = $[G^2 - (\log n)(\text{df})]$, an alternative to AIC. It takes sample size into account. Compared to AIC, BIC gravitates less quickly toward more complex models as n increases. For details and critiques, see Raftery (1986) and the February 1999 issue of *Sociological Methods and Research*.

    6.2. Tree-structured methods such as CART are alternatives to logistic regression that formalize a decision process using a sequential set of questions that branch in different directions depending on a subject's responses. An example is deciding whether a subject with chest pains may be suffering a heart attack. Zhang et al. (1998) surveyed such methods.

Section 6.2: Logistic Regression Diagnostics

    6.3. For logistic regression diagnostics, see Copas (1988), Fowlkes (1987), Hosmer and Lemeshow (2000, Chap. 5), Johnson (1985), Landwehr et al. (1984), and Pregibon (1981). Separate diagnostics are useful for checking the adequacy of each component of a GLM (McCullagh and Nelder 1989, Chap. 12). For a family $g(\mu; \gamma)$ of link functions indexed by parameter γ, Pregibon (1980) showed how to estimate γ giving the link with best fit and how to check the adequacy of a given link $g(\mu; \gamma_0)$.

6.4. Amemiya (1981), Efron (1978), Maddala (1983), and Zheng and Agresti (2000) and references therein reviewed R^2 measures for binary regression. Hosmer and Lemeshow (2000, Sec. 5.2.3) discussed classification tables and their limitations. Pepe (2000) and references therein surveyed ROC methodology.

Section 6.3: Inference about Conditional Associations in $2 \times 2 \times K$ Tables

6.5. Analogs of $\hat{\theta}_{MH}$ summarize differences of proportions or relative risks from several strata (Greenland and Robins 1985). Breslow and Day (1980, p. 142) proposed an alternative large-sample test of homogeneity of odds ratios. In each partial table let $\{\hat{\mu}_{ijk}\}$ have the same marginals as the data observed, yet have odds ratio equal to $\hat{\theta}_{MH}$. Their test statistic has the Pearson form comparing $\{n_{ijk}\}$ to $\{\hat{\mu}_{ijk}\}$. Tarone (1985) showed that because of the inefficiency of $\hat{\theta}_{MH}$ one must adjust the Breslow–Day statistic for it to have a limiting chi-squared null distribution with df $= K - 1$. This adjustment is usually minor. Jones et al. (1989) reviewed and compared several tests of homogeneity in sparse and nonsparse settings. Other work on comparing odds ratios and estimating a common value include Breslow and Day (1980, Sec. 4.4), Donner and Hauck (1986), Gart (1970), and Liang and Self (1985). For modeling the odds ratio, see Breslow (1976), Breslow and Day (1980, Sec. 7.5), and Prentice (1976a). Breslow emphasized retrospective studies, in which the conditional approach is natural since the outcome totals are fixed.

Section 6.5: Sample Size and Power Considerations

6.6. For sample-size determination for comparing proportions, Fleiss (1981, Sec. 3.2) provided tables. See Lachin (1977) for the $I \times J$ case. Chapman and Meng (1966), Drost et al. (1989), Haberman (1974a, pp. 109–112), Harkness and Katz (1964), Mitra (1958), and Patnaik (1949) derived theory for asymptotic nonnull behavior of chi-squared statistics; see also Section 14.3.5. O'Brien's (1986) simulation results suggested that the noncentral chi-squared approximation for G^2 holds well for a wide range of powers. Read and Cressie (1988, pp. 147–148) listed other articles that studied the nonnull behavior of X^2 and G^2.

Section 6.6: Probit and Complementary Log-Log Models

6.7. Finney (1971) is the standard reference on probit modeling. Chambers and Cox (1967) showed that it is difficult to distinguish between probit and logit models unless n is extremely large. Ashford and Sowden (1970) generalized the probit model for multivariate binary responses; see also Lesaffre and Molenberghs (1991) and Ochi and Prentice (1984). Wedderburn (1976) showed that the log likelihood is concave for probit and complementary log-log links.

Section 6.7: Conditional Logistic Regression

6.8. For details about conditional logistic regression, see Section 10.2, Breslow and Day (1980, Chap. 7), Cox (1970), and Hosmer and Lemeshow (2000, Chap. 5). Liang (1984) showed that conditional ML estimators and conditional score tests are asymptotically equivalent to their unconditional counterparts under sampling from exponential families. For exact inference using the conditional likelihood, see Hirji et al. (1987), Mehta and Patel (1995), and the LogXact manual (Cytel Software). Mehta et al. (2000) discussed Monte Carlo approximations.

PROBLEMS

Applications

6.1 For the horseshoe crab data, fit a model using weight and width as predictors. Conduct (**a**) a likelihood-ratio test of H_0: $\beta_1 = \beta_2 = 0$, and (**b**) separate tests for the partial effects. Why does neither test in part (b) show evidence of an effect when the test in part (a) shows strong evidence?

6.2 Refer to the data for Problem 8.13. Treating opinion about premarital sex as the response variable, use backward elimination to select a model. Interpret.

6.3 Refer to Table 6.4. Fit the stage 3 model denoted there by $(E*P + G)$. Use parameter estimates to interpret the G effect and the dependence of the E effect on P.

6.4 Discern the reasons that Simpson's paradox occurs for Table 6.7.

6.5 Refer to Problem 2.12.
   a. Fit the model with G and D main effects. Using it, estimate the AG conditional odds ratio. Compare to the marginal odds ratio, and explain why they are so different. Test its goodness of fit.
   b. Fit the model of no G effect, given the department. Use X^2 to test fit. Obtain residuals, and interpret the lack of fit. (Each department has a single nonredundant standardized Pearson residual. They satisfy $\sum_{i=1}^{6} r_i^2 = X^2$, their squares giving six df = 1 components.)
   c. Fit the two models excluding department A. Again consider lack of fit, and interpret.

6.6 Conduct a residual analysis for the independence model with Table 6.11. What type of lack of fit is indicated?

6.7 Table 6.17, refers to the effectiveness of immediately injected or $1\frac{1}{2}$-hour-delayed penicillin in protecting rabbits against lethal injection with β-hemolytic streptococci.
   a. Let X = delay, Y = whether cured, and Z = penicillin level. Fit the logit model (6.4). Argue that the pattern of 0 cell counts suggests that (with no intercept) $\hat{\beta}_1^Z = -\infty$ and $\hat{\beta}_5^Z = \infty$. What does your software report?
   b. Using the logit model, conduct the likelihood-ratio test of XY conditional independence. Interpret.

TABLE 6.17   Data for Problem 6.7

Penicillin Level	Delay	Response Cured	Response Died
$\frac{1}{8}$	None	0	6
	$1\frac{1}{2}$h	0	5
$\frac{1}{4}$	None	3	3
	$1\frac{1}{2}$h	0	6
$\frac{1}{2}$	None	6	0
	$1\frac{1}{2}$h	2	4
1	None	5	1
	$1\frac{1}{2}$h	6	0
4	None	2	0
	$1\frac{1}{2}$h	5	0

Source: Reprinted with permission from Mantel (1963).

c. Test *XY* conditional independence using the Cochran–Mantel–Haenszel test. Interpret.

d. Estimate the *XY* conditional odds ratio using (**i**) ML with the logit model, and (**ii**) the Mantel–Haenszel estimate. Interpret.

e. The small cell counts make large-sample analyses questionnable. Conduct small-sample inference, and interpret.

6.8 Refer to Table 2.6. Use the CMH statistic to test independence of death penalty verdict and victim's race, controlling for defendant's race. Show another test of this hypothesis, and compare results.

6.9 Treatments A and B were compared on a binary response for 40 pairs of subjects matched on relevant covariates. For each pair, treatments were assigned to the subjects randomly. Twenty pairs of subjects made the same response for each treatment. Six pairs had a success for the subject receiving A and a failure for the subject receiving B, whereas the other 14 pairs had a success for B and a failure for A. Use the Cochran–Mantel–Haenszel procedure to test independence of response and treatment. (In Section 10.1 we present an equivalent test, McNemar's test.)

6.10 Refer to Section 6.5.1. Suppose that $\pi_1 = 0.7$ and $\pi_2 = 0.6$. What sample size is needed for the test to have approximate power 0.80, when $\alpha = 0.05$, for (**a**) H_a: $\pi_1 \neq \pi_2$, and (**b**) H_a: $\pi_1 > \pi_2$?

6.11 Refer to Section 6.5.1. Suppose that $\pi_1 = 0.63$ and $\pi_2 = 0.57$. When treatment sample sizes are equal, explain why the joint probabilities in the 2×2 table are 0.315 and 0.185 in the row for treatment A and 0.285 and 0.215 in the row for treatment B. For the model of independence, explain why the fitted joint probabilities are 0.30 for success and 0.20 for failure, in each row. Show that X^2 has noncentrality parameter $0.00375n$ and df $= 1$. For $n = 200$ and $\alpha = 0.05$, find the power.

6.12 In an experiment designed to compare two treatments on a three-category response, a researcher expects the conditional distributions to be approximately (0.2, 0.2, 0.6) and (0.3, 0.3, 0.4).

    a. With $\alpha = 0.05$, find the approximate power using **(i)** X^2, and **(ii)** G^2 to compare the distributions with 100 observations for each treatment. Compare results.

    b. What sample size is needed for each treatment for the tests in part (a) to have approximate power 0.90?

6.13 The horseshoe crab width values in Table 4.3 have $\bar{x} = 26.3$ and $s_x = 2.1$. If the true relationship were similar to the fitted equation in Section 5.1.3, about how large a sample yields $P(\text{type II error}) = 0.10$, with $\alpha = 0.05$, for testing $H_0: \beta = 0$ against $H_a: \beta > 0$?

6.14 Refer to Problem 5.1. Table 6.18 shows output for fitting a probit model. Interpret the parameter estimates **(a)** using characteristics of the normal cdf response curve, **(b)** finding the estimated rate of change in the probability of remission where it equals 0.5, and **(c)** finding the difference between the estimated probabilities of remission at the upper and lower quartiles of the labeling index, 14 and 28.

TABLE 6.18   Data for Problem 6.14

Parameter	Estimate	Standard Error	Likelihood Ratio 95% Confidence Limits		Chi-Square	Pr > ChiSq
Intercept	-2.3178	0.7795	-4.0114	-0.9084	8.84	0.0029
LI	0.0878	0.0328	0.0275	0.1575	7.19	0.0073

6.15 Use probit models to describe the effects of width and color on the probability of a satellite for Table 4.3. Interpret.

6.16 Refer to Table 6.14. Fit the model having log-log link rather than complementary log-log. Test the fit. Why does it fit so poorly?

6.17 For the linear logit model with Table 3.9 and scores (0, 15, 30), conduct the exact test of H_0: $\beta = 0$ and find a point and interval estimate of β using the conditional likelihood. Interpret.

6.18 Refer to Table 6.16. Apply conditional logistic regression to the model discussed in Section 6.7.8.

     a. Obtain an exact P-value for testing no C effect against the alternative of a positive effect. Construct a 95% confidence interval for the conditional CD odds ratio.

     b. Construct the partial tables relating C to D for the combinations of levels of (A, L). Note that three tables have no data when $C = 1$. For the sole partial table having data at both C levels, find a 95% exact confidence interval for the odds ratio and find an exact one-sided P-value. Compare to results using the entire data set. Comment about the contribution to inference of tables having only a single positive row total or a single positive column total.

     c. Obtain the ordinary ML fit of the logistic regression model. To investigate the sensitivity of the estimated C effect, find the change in the estimate and SE after adding one observation to the data set, a case with no diarrhea when $(C, A, L) = (1, 1, 1)$.

6.19 Consider Table 6.19, from a study of nonmetastatic osteosarcoma (A. M. Goorin, *J. Clin Oncol.* **5**: 1178–1184, 1987, and the manual for *LogXact*). The response is whether the subject achieved a three-year disease-free interval.

     a. Show that each predictor has a significant effect when used individually without the others.

     b. Try to fit a main-effects logistic regression model containing all three predictors. Explain why the ML estimate for the effect of lymphocytic infiltration is infinite.

TABLE 6.19   Data for Problem 6.19

Lymphocytic Infiltration	Gender	Osteoblastic Pathology	Disease-Free Yes	Disease-Free No
High	Female	No	3	0
		Yes	2	0
	Male	No	4	0
		Yes	1	0
Low	Female	No	5	0
		Yes	3	2
	Male	No	5	4
		Yes	6	11

Source: LogXact 4 for Windows (Cambridge, MA: CYTEL Software, 1999).

    c. Using conditional logistic regression, **(i)** conduct an exact test for the effect of lymphocytic infiltration, controlling for the other variables; and **(ii)** find a 95% confidence interval for the effect. Interpret results.

6.20 Use the methods discussed in this chapter to select a model for Table 5.5.

6.21 Logistic regression is applied increasingly to large financial databases, such as for credit scoring to model the influence of predictors on whether a consumer is creditworthy. The data archive found under the index at *www.stat.uni-muenchen.de* contains such a data set that includes 20 covariates for 1000 observations. Build a model for creditworthiness using the predictors running account, duration of credit, payment of previous credits, intended use, gender, and marital status.

Theory and Methods

6.22 For a sequence of s nested models $M_1, \ldots, M_s$, model M_s is the most complex. Let ν denote the difference in residual df between M_1 and M_s.

    a. Explain why for $j < k$, $G^2(M_j | M_k) \leq G^2(M_j | M_s)$.

    b. Assume model M_j, so that M_k also holds when $k > j$. For all $k > j$, as $n \to \infty$, $P[G^2(M_j | M_k) > \chi_\nu^2(\alpha)] \leq \alpha$. Explain why.

    c. Gabriel (1966) suggested a simultaneous testing procedure in which, for each pair of models, the critical value for differences between G^2 values is $\chi_\nu^2(\alpha)$. The final model accepted must be more complex than any model rejected in a pairwise comparison. Since part (b) is true for all $j < k$, argue that Gabriel's procedure has type I error probability no greater than α.

6.23 Prove that the Pearson residuals for the linear logit model applied to a $I \times 2$ contingency table satisfy $X^2 = \sum_{i=1}^{I} e_i^2$. Note that this holds for a binomial GLM with any link.

6.24 Refer to logit model (6.4) for a $2 \times 2 \times K$ contingency table $\{n_{ijk}\}$.

    a. Using dummy variables, write the log-likelihood function. Identify the sufficient statistics for the various parameters. Explain how to conduct exact conditional inference about the effect of X, controlling for Z.

    b. Using a basic result for testing in exponential families, explain why uniformly most powerful unbiased tests of conditional XY independence are based on $\sum_k n_{11k}$ (Birch 1964b; Lehmann 1986, Sec. 4.8).

6.25 Suppose that $\{\pi_{ijk}\}$ in a $2 \times 2 \times 2$ table are, by row, $(0.15, 0.10 \,/\, 0.10, 0.15)$ when $Z = 1$ and $(0.10, 0.15 \,/\, 0.15, 0.10)$ when $Z = 2$. For testing conditional XY independence with logit models having Y as a response, explain why the likelihood-ratio test comparing models $X + Z$ and Z is not consistent but the likelihood-ratio test of fit of the XY conditional independence model is.

6.26 Refer to Section 6.4.1. When Y is $N(\mu_i, \sigma^2)$, consider the comparison of $(\mu_1, \ldots, \mu_I)$ based on independent samples at the I categories of X. When approximately $\mu_i = \alpha + \beta x_i$, explain why the t or F test of H_0: $\beta = 0$ is more powerful than the one-way ANOVA F test. Describe a pattern for $\{\mu_i\}$ for which the ANOVA test would be more powerful.

6.27 For a multinomial distribution, let $\gamma = \Sigma_i b_i \pi_i$, and suppose that $\pi_i = f_i(\theta) > 0$, $i = 1, \ldots, I$. For sample proportions $\{p_i\}$, let $S = \Sigma_i b_i p_i$. Let $T = \Sigma_i b_i \hat{\pi}_i$, where $\hat{\pi}_i = f_i(\hat{\theta})$, for the ML estimator $\hat{\theta}$ of θ.
  a. Show that $\mathrm{var}(S) = [\Sigma_i b_i^2 \pi_i - (\Sigma_i b_i \pi_i)^2]/n$.
  b. Using the delta method, show $\mathrm{var}(T) \approx [\mathrm{var}(\hat{\theta})][\Sigma_i b_i f_i'(\theta)]^2$.
  c. By computing the information for $L(\theta) = \Sigma_i n_i \log[f_i(\theta)]$, show that $\mathrm{var}(\hat{\theta})$ is approximately $[n\Sigma_i(f_i'(\theta))^2/f_i(\theta)]^{-1}$.
  d. Asymptotically, show that $\mathrm{var}[\sqrt{n}\,(T - \gamma)] \leq \mathrm{var}[\sqrt{n}\,(S - \gamma)]$. [*Hint*: Show that $\mathrm{var}(T)/\mathrm{var}(S)$ is a squared correlation between two random variables, where with probability π_i the first equals b_i and the second equals $f_i'(\theta)/f_i(\theta)$.]

6.28 A *threshold model* can also motivate the probit model. For it, there is an unobserved continuous response Y^* such that the observed $y_i = 0$ if $y_i^* \leq \tau$ and $y_i = 1$ if $y_i^* > \tau$. Suppose that $y_i^* = \mu_i + \epsilon_i$, where $\mu_i = \alpha + \beta x_i$ and where $\{\epsilon_i\}$ are independent from a $N(0, \sigma^2)$ distribution. For identifiability one can set $\sigma = 1$ and the threshold $\tau = 0$. Show that the probit model holds and explain why β represents the expected number of standard deviation change in Y^* for a 1-unit increase in x.

6.29 Consider the choice between two options, such as two product brands. Let U_0 denote the *utility* of outcome $y = 0$ and U_1 the utility of $y = 1$. For $y = 0$ and 1, suppose that $U_y = \alpha_y + \beta_y x + \epsilon_y$, using a scale such that ϵ_y has some standardized distribution. A subject selects $y = 1$ if $U_1 > U_0$ for that subject.
  a. If ϵ_0 and ϵ_1 are independent $N(0, 1)$ random variables, show that $P(Y = 1)$ satisfies the probit model.
  b. If ϵ_y are independent extreme-value random variables, with cdf $F(\epsilon) = \exp[-\exp(-\epsilon)]$, show that $P(Y = 1)$ satisfies the logistic regression model (Maddala 1983, p. 60; McFadden 1974).

6.30 Consider model (6.12) with complementary log-log link.

    a. Find x at which $\pi(x) = \frac{1}{2}$.

    b. Show the greatest rate of change of $\pi(x)$ occurs at $x = -\alpha/\beta$. What does $\pi(x)$ equal at that point? Give the corresponding result for the model with log-log link, and compare to the logit and probit models.

6.31 Suppose that log-log model (6.13) holds. Explain how to interpret β.

6.32 Let y_i, $i = 1, \ldots, n$, denote n independent binary random variables.

    a. Derive the log likelihood for the probit model $\Phi^{-1}[\pi(\mathbf{x}_i)] = \sum_j \beta_j x_{ij}$.

    b. Show that the likelihood equations for the logistic and probit regression models are

$$\sum_i (y_i - \hat{\pi}_i) z_i x_{ij} = 0, \qquad j = 0, \ldots, p,$$

    where $z_i = 1$ for the logistic case and $z_i = \phi(\sum_j \hat{\beta}_j x_{ij})/\hat{\pi}_i(1 - \hat{\pi}_i)$ for the probit case. (When the link is not canonical, there is no reduction of the data in sufficient statistics.)

6.33 Sometimes, sample proportions are continuous rather than of the binomial form (number of successes)/(number of trials). Each observation is any real number between 0 and 1, such as the proportion of a tooth surface that is covered with plaque. For independent responses $\{y_i\}$, Aitchison and Shen (1980) and Bartlett (1937) modeled logit$(Y_i) \sim N(\beta_i, \sigma^2)$. Then Y_i itself is said to have a *logistic-normal distribution*.

    a. Expressing a $N(\beta, \sigma^2)$ variate as $\beta + \sigma Z$, where Z is standard normal, show that $Y_i = \exp(\beta_i + \sigma Z)/[1 + \exp(\beta_i + \sigma Z)]$.

    b. Show that for small σ,

$$Y_i = \frac{e^{\beta_i}}{1 + e^{\beta_i}} + \frac{e^{\beta_i}}{1 + e^{\beta_i}} \frac{1}{1 + e^{\beta_i}} \sigma Z + \frac{e^{\beta_i}(1 - e^{\beta_i})}{2(1 + e^{\beta_i})^3} \sigma^2 Z^2 + \cdots .$$

    c. Letting $\mu_i = e^{\beta_i}/(1 + e^{\beta_i})$, when σ is close to 0 show that

$$E(Y_i) \approx \mu_i, \qquad \mathrm{var}(Y_i) \approx [\mu_i(1 - \mu_i)]^2 \sigma^2.$$

    d. For independent continuous proportions $\{y_i\}$, let $\mu_i = E(Y_i)$. For a GLM, it is sensible to use an inverse cdf link for μ_i, but it is unclear how to choose a distribution for Y_i. The approximate moments for the logistic-normal motivate a quasi-likelihood approach (Wedderburn 1974) with variance function $v(\mu_i) = \phi[\mu_i(1 - \mu_i)]^2$ for unknown ϕ. Explain why this provides similar results as fitting a normal regression model to the sample logits assuming constant variance. (The QL approach has the advantage of not requiring adjustment of 0 or 1 observations, for which sample logits don't exist.)

e. Wedderburn (1974) gave an example with response the proportion of a leaf showing a type of blotch. Envision an approximation of binomial form based on cutting each leaf into a large number of small regions of the same size and observing for each region whether it is mostly covered with blotch. Explain why this suggests that $v(\mu_i) = \phi\mu_i(1 - \mu_i)$. What violation of the binomial assumptions might make this questionnable? [The parametric family of beta distributions has variance function of this form (see Section 13.3.1). Barndorff-Nielsen and Jørgensen (1991) proposed a distribution having $v(\mu_i) = \phi[\mu_i(1 - \mu_i)]^3$; see also Cox (1996).]

6.34 For independent binomial sampling, construct the log likelihood and identify the sufficient statistics to be conditioned out to perform exact inference about β in model (6.4).

6.35 Let $\hat{\boldsymbol{\pi}}^{(-)} = (\hat{\pi}^{(-1)}, \ldots, \hat{\pi}^{(-n)})$, where $\hat{\pi}^{(-i)}$ denotes the estimate of $E(Y_i)$ for binary observation i after fitting the model without that observation. Cross-validation declares a model to have good predictive power if corr$(\hat{\boldsymbol{\pi}}^{(-)}, \mathbf{y})$ is high. Consider the model logit$(\pi_i) = \alpha$ for all i. Show that $\hat{\pi}_i = \bar{y}$ and hence $\hat{\pi}^{(-i)} = [n/(n - 1)][\bar{y} - (1/n)y_i]$, and hence corr$(\hat{\boldsymbol{\pi}}^{(-)}, \mathbf{y}) = -1$ regardless of how well the model fits. Thus, cross-validation can be misleading with binary data (Zheng and Agresti 2000).

CHAPTER 7

Logit Models for Multinomial Responses

In Chapters 5 and 6 we discussed modeling binary response variables with binomial GLMs. Multicategory responses use multinomial GLMs. In this chapter we generalize logistic regression for multinomial (nominal and ordinal) response variables.

In Section 7.1 we present a model for nominal responses that uses a separate binary logit model for each pair of response categories. In Section 7.2 we present a model for ordinal responses that uses logits of cumulative response probabilities. In Section 7.3 we use other link functions for those cumulative probabilities. Section 7.4 covers alternative ordinal-response models.

In Section 7.5 we discuss tests of conditional independence with multinomial responses using models and using generalizations of the Cochran–Mantel–Haenszel statistic. In the final section we introduce a multinomial logit model for *discrete-choice modeling* of a subject's choice from one of several options when values of predictors may depend on the option.

7.1 NOMINAL RESPONSES: BASELINE-CATEGORY LOGIT MODELS

Let Y be a categorical response with J categories. Multicategory (also called *polytomous*) logit models for nominal response variables simultaneously describe log odds for all $\binom{J}{2}$ pairs of categories. Given a certain choice of $J - 1$ of these, the rest are redundant.

7.1.1 Baseline-Category Logits

Let $\pi_j(\mathbf{x}) = P(Y = j \mid \mathbf{x})$ at a fixed setting $\mathbf{x}$ for explanatory variables, with $\sum_j \pi_j(\mathbf{x}) = 1$. For observations at that setting, we treat the counts at the J categories of Y as multinomial with probabilities $\{\pi_1(\mathbf{x}), \ldots, \pi_J(\mathbf{x})\}$.

Logit models pair each response category with a baseline category, often the last one or the most common one. The model

$$\log \frac{\pi_j(\mathbf{x})}{\pi_J(\mathbf{x})} = \alpha_j + \boldsymbol{\beta}_j' \mathbf{x}, \qquad j = 1, \ldots, J - 1, \qquad (7.1)$$

simultaneously describes the effects of $\mathbf{x}$ on these $J - 1$ logits. The effects vary according to the response paired with the baseline. These $J - 1$ equations determine parameters for logits with other pairs of response categories, since

$$\log \frac{\pi_a(\mathbf{x})}{\pi_b(\mathbf{x})} = \log \frac{\pi_a(\mathbf{x})}{\pi_J(\mathbf{x})} - \log \frac{\pi_b(\mathbf{x})}{\pi_J(\mathbf{x})}.$$

With categorical predictors, X^2 and G^2 goodness-of-fit statistics provide a model check when data are not sparse. When an explanatory variable is continuous or the data are sparse, such statistics are still valid for comparing nested models differing by relatively few terms (Haberman 1974a, pp. 372–373; 1977a).

7.1.2 Alligator Food Choice Example

Table 7.1 is from a study of factors influencing the primary food choice of alligators. It used 219 alligators captured in four Florida lakes. The nominal response variable is the primary food type, in volume, found in an alligator's stomach. This had five categories: fish, invertebrate, reptile, bird, other. The invertebrates included apple snails, aquatic insects, and crayfish. The reptiles were primarily turtles, although one stomach contained the tags of 23 baby alligators released in the lake the previous year! The "other" category consisted of amphibian, mammal, plant material, stones or other debris, or no food or dominant type. Table 7.1 also classifies the alligators according to L = lake of capture (Hancock, Oklawaha, Trafford, George), G = gender (male, female), and S = size (≤ 2.3 meters long, > 2.3 meters long).

Baseline-category logit models can investigate the effects of L, G, and S on primary food type. Table 7.2 contains fit statistics for several models. We denote a model by its predictors: for instance, $(L + S)$ having additive lake and size effects and $()$ having no predictors. The data are sparse, 219 observations scattered among 80 cells. Thus, G^2 is more reliable for comparing models than for testing fit. The statistics $G^2[()|(G)] = 2.1$ and $G^2 = [(L + S)|(G + L + S)] = 2.2$, each based on df = 4, suggest simplifying by collapsing the table over gender. (Other analyses, not presented here, show that adding interaction terms including G do not improve the fit significantly.) The G^2 and X^2 values for the collapsed table indicate that both L and S have effects. Table 7.3 exhibits fitted values for model $(L + S)$ for the

TABLE 7.1   Primary Food Choice of Alligators

Lake	Gender	Size (m)	Fish	Invertebrate	Reptile	Bird	Other
Hancock	Male	≤ 2.3	7	1	0	0	5
		> 2.3	4	0	0	1	2
	Female	≤ 2.3	16	3	2	2	3
		> 2.3	3	0	1	2	3
Oklawaha	Male	≤ 2.3	2	2	0	0	1
		> 2.3	13	7	6	0	0
	Female	≤ 2.3	3	9	1	0	2
		> 2.3	0	1	0	1	0
Trafford	Male	≤ 2.3	3	7	1	0	1
		> 2.3	8	6	6	3	5
	Female	≤ 2.3	2	4	1	1	4
		> 2.3	0	1	0	0	0
George	Male	≤ 2.3	13	10	0	2	2
		> 2.3	9	0	0	1	2
	Female	≤ 2.3	3	9	1	0	1
		> 2.3	8	1	0	0	1

Source: Data courtesy of Clint Moore, from an unpublished manuscript by M. F. Delaney and C. T. Moore.

TABLE 7.2   Goodness of Fit of Baseline-Category Logit Models for Table 7.1

Model[a]	G^2	X^2	df
( )	116.8	106.5	60
(G)	114.7	101.2	56
(S)	101.6	86.9	56
(L)	73.6	79.6	48
$(L + S)$	52.5	58.0	44
$(G + L + S)$	50.3	52.6	40
Collapsed over G			
( )	81.4	73.1	28
(S)	66.2	54.3	24
(L)	38.2	32.7	16
$(L + S)$	17.1	15.0	12

[a] G, gender; S, size; L, lake of capture. See the text for details.

TABLE 7.3 Observed and Fitted Values for Study of Alligator's Primary Food Choice

Lake	Size of alligator (meters)	Fish	Invertebrate	Reptile	Bird	Other
			Primary Food Choice			
Hancock	≤ 2.3	23	4	2	2	8
		(20.9)	(3.6)	(1.9)	(2.7)	(9.9)
	> 2.3	7	0	1	3	5
		(9.1)	(0.4)	(1.1)	(2.3)	(3.1)
Oklawaha	≤ 2.3	5	11	1	0	3
		(5.2)	(12.0)	(1.5)	(0.2)	(1.1)
	> 2.3	13	8	6	1	0
		(12.8)	(7.0)	(5.5)	(0.8)	(1.9)
Trafford	≤ 2.3	5	11	2	1	5
		(4.4)	(12.4)	(2.1)	(0.9)	(4.2)
	> 2.3	89	7	6	3	5
		(8.6)	(5.6)	(5.9)	(3.1)	(5.8)
George	≤ 2.3	16	19	1	2	3
		(18.5)	(16.9)	(0.5)	(1.2)	(3.8)
	> 2.3	17	1	0	1	3
		(14.5)	(3.1)	(0.5)	(1.8)	(2.2)

collapsed table. Absolute values of standardized Pearson residuals comparing observed and fitted values exceed 2 in only two of the 40 cells and exceed 3 in none of the cells. The fit seems adequate.

Fish was the most common food choice. We now estimate the effects of lake and size on the odds that alligators select other primary food types instead of fish. With fish as the baseline category, Table 7.4 contains ML estimates of effect parameters. These result from models using dummy variables for the first three lakes and for size. The table uses letter subscripts to denote the food choice categories. For example, the prediction equation for the log odds of selecting invertebrates instead of fish is

$$\log(\hat{\pi}_I/\hat{\pi}_F) = -1.55 + 1.46s - 1.66z_H + 0.94z_O + 1.12z_T,$$

TABLE 7.4 Estimated Parameters in Logit Model for Alligator Food Choice, Based on Dummy Variable for First Size Category and Each Lake Except Lake George[a]

Logit[b]	Intercept	Size ≤ 2.3	Hancock	Oklawaha	Trafford
				Lake	
$\log(\pi_I/\pi_F)$	−1.55	1.46 (0.40)	−1.66 (0.61)	0.94 (0.47)	1.12 (0.49)
$\log(\pi_R/\pi_F)$	−3.31	−0.35 (0.58)	1.24 (1.19)	2.46 (1.12)	2.94 (1.12)
$\log(\pi_B/\pi_F)$	−2.09	−0.63 (0.64)	0.70 (0.78)	−0.65 (1.20)	1.09 (0.84)
$\log(\pi_O/\pi_F)$	−1.90	0.33 (0.45)	0.83 (0.56)	0.01 (0.78)	1.52 (0.62)

[a]SE values in parentheses.
I, invertebrate; *R*, reptile; *B*, bird; *O*, other; *F*, fish.

where $s = 1$ for size ≤ 2.3 meters and 0 otherwise, z_H is a dummy variable for Lake Hancock ($z_H = 1$ for alligators in that lake and 0 otherwise), and z_O and z_T are dummy variables for lakes Oklawaha and Trafford. Size of alligator has a noticeable effect. For a given lake, for small alligators the estimated odds that primary food choice was invertebrates instead of fish are $\exp(1.46) = 4.3$ times the estimated odds for large alligators; the Wald 95% confidence interval is $\exp[1.46 \pm 1.96(0.396)] = (2.0, 9.3)$. The lake effects indicate that the estimated odds that the primary food choice was invertebrates instead of fish are relatively higher at Lakes Trafford and Oklawaha and relatively lower at Lake Hancock than they are at Lake George.

The equations in Table 7.4 determine those for other food-choice pairs. For instance, for (invertebrate, other),

$$\log(\hat{\pi}_I/\hat{\pi}_O) = \log(\hat{\pi}_I/\hat{\pi}_F) - \log(\hat{\pi}_O/\hat{\pi}_F)$$

$$= (-1.55 + 1.46s - 1.66z_H + 0.94z_O + 1.12z_T)$$

$$- (-1.90 + 0.33s + 0.83z_H + 0.01z_O + 1.52z_T)$$

$$= 0.35 + 1.13s - 2.48z_H + 0.93z_O - 0.39z_T.$$

7.1.3 Estimating Response Probabilities

The equation that expresses multinomial logit models directly in terms of response probabilities $\{\pi_j(\mathbf{x})\}$ is

$$\pi_j(\mathbf{x}) = \frac{\exp(\alpha_j + \boldsymbol{\beta}_j'\mathbf{x})}{1 + \sum_{h=1}^{J-1}\exp(\alpha_h + \boldsymbol{\beta}_h'\mathbf{x})} \tag{7.2}$$

with $\alpha_J = 0$ and $\boldsymbol{\beta}_J = \mathbf{0}$. This follows from (7.1), using the fact that (7.1) also holds with $j = J$ by setting $\alpha_J = 0$ and $\boldsymbol{\beta}_J = \mathbf{0}$. (Also, the parameters equal zero for a baseline category for identifiability reasons; see Problem 7.26.) The denominator of (7.2) is the same for each j. The numerators for various j sum to the denominator, so $\sum_j \pi_j(\mathbf{x}) = 1$. For $J = 2$, (7.2) simplifies to the formula of type (5.1) used for binary logistic regression.

From Table 7.4 the estimated probability that a large alligator in Lake Hancock has invertebrates as the primary food choice is

$$\hat{\pi}_I = \frac{e^{-1.55-1.66}}{1 + e^{-1.55-1.66} + e^{-3.31+1.24} + e^{-2.09+0.70} + e^{-1.90+0.83}} = 0.023.$$

The estimated probabilities for reptile, bird, other, and fish are 0.072, 0.141, 0.194, and 0.570.

This example used qualitative predictors. Multinomial logit models can also contain quantitative predictors. In this study, the biologists used the size dummy variable to distinguish between adult and subadult alligators. However, the alligators' actual length was measured and is quantitative. With quantitative predictors, it is informative to plot the estimated probabilities.

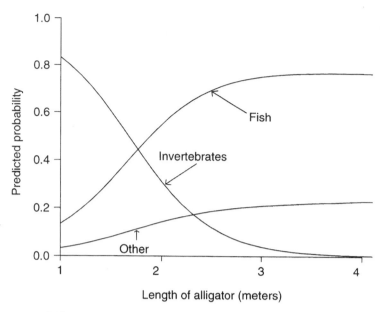

FIGURE 7.1   Estimated probabilities for primary food choice.

To illustrate, for alligators at one lake, Figure 7.1 plots the estimated probabilities that primary food choice is fish, invertebrate, or other (which combines the other, bird, and reptile categories) as a function of length. With more than two response categories, the probability for a given category need not continuously increase or decrease (Problem 7.27).

7.1.4   Fitting of Baseline-Category Logit Models*

ML fitting of multinomial logit models maximizes the likelihood subject to $\{\pi_j(\mathbf{x})\}$ simultaneously satisfying the $J - 1$ equations that specify the model. For $i = 1, \ldots, n$, let $\mathbf{y}_i = (y_{i1}, \ldots, y_{iJ})$ represent the multinomial trial for subject i, where $y_{ij} = 1$ when the response is in category j and $y_{ij} = 0$ otherwise. Thus, $\sum_j y_{ij} = 1$. Let $\mathbf{x}_i = (x_{i1}, \ldots, x_{ip})'$ denote explanatory variable values for subject i. Let $\boldsymbol{\beta}_j = (\beta_{j1}, \ldots, \beta_{jp})'$ denote parameters for the jth logit.

Since $\pi_J = 1 - (\pi_1 + \cdots + \pi_{J-1})$ and $y_{iJ} = 1 - (y_{i1} + \cdots + y_{i, J-1})$, the contribution to the log likelihood by subject i is

$$\log\left[\prod_{j=1}^{J} \pi_j(\mathbf{x}_i)^{y_{ij}}\right] = \sum_{j=1}^{J-1} y_{ij}\log\pi_j(\mathbf{x}_i) + \left(1 - \sum_{j=1}^{J-1} y_{ij}\right)\log\left[1 - \sum_{j=1}^{J-1} \pi_j(\mathbf{x}_i)\right]$$

$$= \sum_{j=1}^{J-1} y_{ij}\log\frac{\pi_j(\mathbf{x}_i)}{1 - \sum_{j=1}^{J-1}\pi_j(\mathbf{x}_i)} + \log\left[1 - \sum_{j=1}^{J-1} \pi_j(\mathbf{x}_i)\right].$$

Thus, the baseline-category logits are the natural parameters for the multinomial distribution.

Now assume n independent observations. In the last expression above, substituting $\alpha_j + \boldsymbol{\beta}'_j \mathbf{x}_i$ for the logit in the first term and $\pi_J(\mathbf{x}_i) = 1/[1 + \sum_{j=1}^{J-1} \exp(\alpha_j + \boldsymbol{\beta}'_j \mathbf{x}_i)]$ in the second term, the log likelihood is

$$
\log \prod_{i=1}^{n} \left[\prod_{j=1}^{J} \pi_j(\mathbf{x}_i)^{y_{ij}} \right]
$$

$$
= \sum_{i=1}^{n} \left\{ \sum_{j=1}^{J-1} y_{ij}(\alpha_j + \boldsymbol{\beta}'_j \mathbf{x}_i) - \log\left[1 + \sum_{j=1}^{J-1} \exp(\alpha_j + \boldsymbol{\beta}'_j \mathbf{x}_i)\right] \right\}
$$

$$
= \sum_{j=1}^{J-1} \left[\alpha_j \left(\sum_{i=1}^{n} y_{ij} \right) + \sum_{k=1}^{p} \beta_{jk} \left(\sum_{i=1}^{n} x_{ik} y_{ij} \right) \right]
$$

$$
- \sum_{i=1}^{n} \log\left[1 + \sum_{j=1}^{J-1} \exp(\alpha_j + \boldsymbol{\beta}'_j \mathbf{x}_i) \right].
$$

The sufficient statistic for β_{jk} is $\sum_i x_{ik} y_{ij}$, $j = 1, \ldots, J-1$, $k = 1, \ldots, p$. The sufficient statistic for α_j is $\sum_i y_{ij} = \sum_i x_{i0} y_{ij}$ for $x_{i0} = 1$; this is the total number of outcomes in category j.

The likelihood equations equate the sufficient statistics to their expected values. The log likelihood is concave, and the Newton–Raphson method yields the ML parameter estimates. The estimators have large-sample normal distributions. Their asymptotic standard errors are square roots of diagonal elements of the inverse information matrix.

Most statistical software can fit multinomial logit models, but some can fit only binary logistic regression models. An alternative fitting approach fits binary logit models separately for the $J - 1$ pairings of responses: model (7.1) for $j = 1$ alone, using only observations in category 1 or J of the response variable to obtain estimates of α_1 and $\boldsymbol{\beta}_1$; model (7.1) using only categories 2 and J to obtain estimates of α_2 and $\boldsymbol{\beta}_2$; in this manner, obtaining $J - 1$ separate fits of logit models. A logit model fitted using data from only two response categories is the same as a regular logit model fitted *conditional* on classification into one of those categories. For instance, the jth baseline-category logit is a logit of conditional probabilities

$$
\log \frac{\pi_j(\mathbf{x})/\big(\pi_j(\mathbf{x}) + \pi_J(\mathbf{x})\big)}{\pi_J(\mathbf{x})/\big(\pi_j(\mathbf{x}) + \pi_J(\mathbf{x})\big)} = \log \frac{\pi_j(\mathbf{x})}{\pi_J(\mathbf{x})}.
$$

The separate-fitting estimates differ from the ML estimates for simultaneous fitting of the $J - 1$ logits. They are less efficient, tending to have larger standard errors. However, Begg and Gray (1984) showed that the efficiency loss is minor when the response category having highest prevalence is the baseline. To illustrate this approach, we used the data for the categories invertebrate and fish alone. The fit is $\log(\hat{\pi}_I/\hat{\pi}_F) = -1.69 + 1.66s - 1.78z_H + 1.05z_O + 1.22z_T$, with standard errors (0.43, 0.62, 0.49, 0.52) for the effects. The effects are similar to those from simultaneous fitting with all five response categories—see the first row of Table 7.4. The estimated standard errors are only slightly larger, since 155 of the 219 observations were in the fish or invertebrate categories of food type.

7.1.5 Multicategory Logit Model as Multivariate GLM*

For a univariate response variable in the natural exponential family, a GLM has form $g(\mu_i) = \mathbf{x}'_i\boldsymbol{\beta}$ for a link function g, expected response $\mu_i = E(Y_i)$, vector of values $\mathbf{x}_i$ of p explanatory variables for observation i, and parameter vector $\boldsymbol{\beta} = (\beta_1, \ldots, \beta_p)'$. This extends to a multivariate GLM for distributions in the multivariate exponential family (Problem 7.24), such as the multinomial.

Let $\mathbf{y}_i = (y_{i1}, y_{i2}, \ldots)'$ be a vector response for subject i, with $\boldsymbol{\mu}_i = E(\mathbf{Y}_i)$. Let $\mathbf{g}$ be a vector of link functions. The multivariate GLM has the form

$$\mathbf{g}(\boldsymbol{\mu}_i) = \mathbf{X}_i\boldsymbol{\beta}, \tag{7.3}$$

where row h of the model matrix $\mathbf{X}_i$ for observation i contains values of explanatory variables for y_{ih}. For details, see Fahrmeir and Tutz (2001, Chap. 3).

The baseline-category logit model is a multivariate GLM. Here $\mathbf{y}_i = (y_{i1}, \ldots, y_{i,J-1})'$, since y_{iJ} is redundant. Then, $\boldsymbol{\mu}_i = (\pi_1(\mathbf{x}_i), \ldots, \pi_{J-1}(\mathbf{x}_i))'$ and

$$g_j(\boldsymbol{\mu}_i) = \log\{\mu_{ij}/[1 - (\mu_{i1} + \cdots + \mu_{i,J-1})]\}.$$

The model matrix for observation i is

$$\mathbf{X}_i = \begin{pmatrix} 1 & \mathbf{x}'_i & & & \\ & & 1 & \mathbf{x}'_i & \\ & & & \cdots & \\ & & & & 1 & \mathbf{x}'_i \end{pmatrix}.$$

with 0 entries in other locations, and $\boldsymbol{\beta}' = (\alpha_1, \boldsymbol{\beta}'_1, \ldots, \alpha_{J-1}, \boldsymbol{\beta}'_{J-1})$. One can also formulate it for grouped data using sample proportions in the categories.

7.2 ORDINAL RESPONSES: CUMULATIVE LOGIT MODELS

In Section 6.4.1 we showed the benefits of utilizing the ordinality of a variable by focusing inferences on a single parameter. These benefits extend to models for ordinal responses. Models with terms that reflect ordinal characteristics such as monotone trend have improved model parsimony and power. In this section we introduce the most popular logit model for ordinal responses.

7.2.1 Cumulative Logits

One way to use category ordering forms logits of cumulative probabilities,

$$P(Y \le j \mid \mathbf{x}) = \pi_1(\mathbf{x}) + \cdots + \pi_j(\mathbf{x}), \quad j = 1, \ldots, J.$$

The *cumulative logits* are defined as

$$\text{logit}[P(Y \le j \mid \mathbf{x})] = \log \frac{P(Y \le j \mid \mathbf{x})}{1 - P(Y \le j \mid \mathbf{x})}$$

$$= \log \frac{\pi_1(\mathbf{x}) + \cdots + \pi_j(\mathbf{x})}{\pi_{j+1}(\mathbf{x}) + \cdots + \pi_J(\mathbf{x})}, \quad j = 1, \ldots, J - 1. \quad (7.4)$$

Each cumulative logit uses all J response categories.

A model for logit$[P(Y \le j)]$ alone is an ordinary logit model for a binary response in which categories 1 to j form one outcome and categories $j + 1$ to J form the second. Better, models can use all $J - 1$ cumulative logits in a single parsimonious model.

7.2.2 Proportional Odds Model

A model that simultaneously uses all cumulative logits is

$$\text{logit}[P(Y \le j \mid \mathbf{x})] = \alpha_j + \boldsymbol{\beta}' \mathbf{x}, \quad j = 1, \ldots, J - 1. \quad (7.5)$$

Each cumulative logit has its own intercept. The $\{\alpha_j\}$ are increasing in j, since $P(Y \le j \mid \mathbf{x})$ increases in j for fixed $\mathbf{x}$, and the logit is an increasing function of this probability.

This model has the same effects $\boldsymbol{\beta}$ for each logit. For a continuous predictor x, Figure 7.2 depicts the model when $J = 4$. For fixed j, the

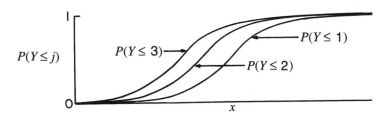

FIGURE 7.2 Cumulative logit model with effect independent of cutpoint.

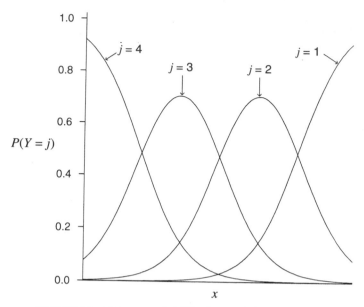

FIGURE 7.3   Category probabilities in cumulative logit model.

response curve is a logistic regression curve for a binary response with outcomes $Y \leq j$ and $Y > j$. The response curves for $j = 1$, 2, and 3 have the same shape. They share exactly the same rate of increase or decrease but are horizontally displaced from each other. For $j < k$, the curve for $P(Y \leq k)$ is the curve for $P(Y \leq j)$ translated by $(\alpha_k - \alpha_j)/\beta$ units in the x direction; that is,

$$P(Y \leq k \mid X = x) = P\big(Y \leq j \mid X = x + (\alpha_k - \alpha_j)/\beta\big) .$$

Figure 7.3 portrays the curves for the category probabilities.

The cumulative logit model (7.5) satisfies

$$\text{logit}\big[P(Y \leq j \mid \mathbf{x}_1)\big] - \text{logit}\big[P(Y \leq j \mid \mathbf{x}_2)\big]$$

$$= \log \frac{P(Y \leq j \mid \mathbf{x}_1)/P(Y > j \mid \mathbf{x}_1)}{P(Y \leq j \mid \mathbf{x}_2)/P(Y > j \mid \mathbf{x}_2)} = \boldsymbol{\beta}'(\mathbf{x}_1 - \mathbf{x}_2).$$

An odds ratio of cumulative probabilities is called a *cumulative odds ratio*. The odds of making response $\leq j$ at $\mathbf{x} = \mathbf{x}_1$ are $\exp[\boldsymbol{\beta}'(\mathbf{x}_1 - \mathbf{x}_2)]$ times the odds at $\mathbf{x} = \mathbf{x}_2$. The log cumulative odds ratio is proportional to the distance between $\mathbf{x}_1$ and $\mathbf{x}_2$. The same proportionality constant applies to each logit. Because of this property, McCullagh (1980) called (7.5) a *proportional odds model*.

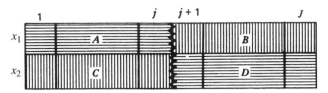

FIGURE 7.4 Uniform odds ratios AD/BC whenever $x_1 - x_2 = 1$, for all response cutpoints with proportional odds model.

   With a single predictor, the cumulative odds ratio equals e^{β} whenever $x_1 - x_2 = 1$. Figure 7.4 illustrates the constant cumulative odds ratio this model then implies for all j. It shows the J-category response collapsed into the binary outcome $(\leq j, > j)$ and shows the sets of cells that determine the cumulative odds ratio AD/BC that takes the same value e^{β} for each such collapsing.

   Model (7.5) constrains the $J - 1$ response curves to have the same shape. Thus, its fit is not the same as fitting separate logit models for each j. Again let $(y_{i1}, \ldots, y_{iJ})$ be binary indicators of the response for subject i. The likelihood function is

$$\prod_{i=1}^{n}\left[\prod_{j=1}^{J}\pi_j(\mathbf{x}_i)^{y_{ij}}\right] = \prod_{i=1}^{n}\left[\prod_{j=1}^{J}(P(Y \leq j \mid \mathbf{x}_i) - P(Y \leq j - 1 \mid \mathbf{x}_i))^{y_{ij}}\right]$$

$$= \prod_{i=1}^{n}\left[\prod_{j=1}^{J}\left(\frac{\exp(\alpha_j + \boldsymbol{\beta}'\mathbf{x}_i)}{1 + \exp(\alpha_j + \boldsymbol{\beta}'\mathbf{x}_i)} - \frac{\exp(\alpha_{j-1} + \boldsymbol{\beta}'\mathbf{x}_i)}{1 + \exp(\alpha_{j-1} + \boldsymbol{\beta}'\mathbf{x}_i)}\right)^{y_{ij}}\right],$$

$$(7.6)$$

viewed as a function of $(\{\alpha_j\}, \boldsymbol{\beta})$. McCullagh (1980) and Walker and Duncan (1967) used Fisher scoring algorithms to obtain ML estimates.

7.2.3 Latent Variable Motivation*

A regression model for a continuous variable assumed to underlie Y motivates the common effect $\boldsymbol{\beta}$ for different j in the proportional odds model (Anderson and Philips 1981). Let Y^* denote this underlying variable. In statistics, such an unobserved variable is called a *latent variable*. Suppose that it has cdf $G(y^* - \eta)$, where values of y^* vary around a location parameter η (such as a mean) that depends on $\mathbf{x}$ through $\eta(\mathbf{x}) = \boldsymbol{\beta}'\mathbf{x}$. Suppose that $-\infty = \alpha_0 < \alpha_1 < \cdots < \alpha_J = \infty$ are *cutpoints* of the continuous scale such

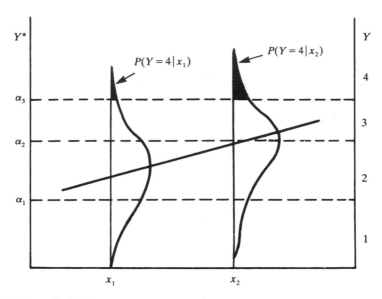

FIGURE 7.5   Ordinal measurement and underlying regression model for a latent variable.

that the observed response Y satisfies

$$Y = j \quad \text{if } \alpha_{j-1} < Y^* \leq \alpha_j.$$

That is, Y falls in category j when the latent variable falls in the jth interval of values (Figure 7.5). Then

$$P(Y \leq j \mid \mathbf{x}) = P(Y^* \leq \alpha_j \mid \mathbf{x}) = G(\alpha_j - \boldsymbol{\beta}'\mathbf{x}).$$

The appropriate model for Y implies that the link G^{-1}, the inverse of the cdf for Y^*, applies to $P(Y \leq j \mid \mathbf{x})$. If $Y^* = \boldsymbol{\beta}'\mathbf{x} + \epsilon$, where the cdf G of ϵ is the logistic (Section 4.2.5), then G^{-1} is the logit link and a proportional odds model results. Normality for ϵ implies a probit link for cumulative probabilities (Section 7.3.1).

In this derivation, the same parameters $\boldsymbol{\beta}$ occur for the effects on Y regardless of how the cutpoints $\{\alpha_j\}$ chop up the scale for the latent variable. The effect parameters are invariant to the choice of categories for Y. If a continuous variable measuring political philosophy has a linear regression with some predictor variables, then the same effect parameters apply to a discrete version of political philosophy with the categories (liberal, moderate, conservative) or (very liberal, slightly liberal, moderate, slightly conservative, very conservative). This feature makes it possible to compare estimates from studies using different response scales.

Note that the use of a cdf of form $G(y^* - \eta)$ for the latent variable results in linear predictor $\alpha_j - \beta'\mathbf{x}$ rather than $\alpha_j + \beta'\mathbf{x}$. When $\beta > 0$, as x increases each cumulative logit then decreases, so each cumulative probability decreases and relatively less probability mass falls at the low end of the Y scale. Thus, Y tends to be larger at higher values of x. With this parameterization the sign of β has the usual meaning. However, most software (e.g., SAS) uses form (7.5).

7.2.4 Mental Impairment Example

Table 7.5 comes from a study of mental health for a random sample of adult residents of Alachua County, Florida. It relates mental impairment to two explanatory variables. Mental impairment is an ordinal response, with categories (well, mild symptom formation, moderate symptom formation, impaired). The life events index x_1 is a composite measure of the number and severity of important life events such as birth of child, new job, divorce, or death in family that occurred to the subject within the past 3 years. Socioeconomic status (x_2 = SES) is measured here as binary (1 = high, 0 = low).

TABLE 7.5   Mental Impairment by SES and Life Events

Subject	Mental Impairment	SES[a] x_2	Life Events x_1	Subject	Mental Impairment	SES[a] x_2	Life Events x_1
1	Well	1	1	21	Mild	1	9
2	Well	1	9	22	Mild	0	3
3	Well	1	4	23	Mild	1	3
4	Well	1	3	24	Mild	1	1
5	Well	0	2	25	Moderate	0	0
6	Well	1	0	26	Moderate	1	4
7	Well	0	1	27	Moderate	0	3
8	Well	1	3	28	Moderate	0	9
9	Well	1	3	29	Moderate	1	6
10	Well	1	7	30	Moderate	0	4
11	Well	0	1	31	Moderate	0	3
12	Well	0	2	32	Impaired	1	8
13	Mild	1	5	33	Impaired	1	2
14	Mild	0	6	34	Impaired	1	7
15	Mild	1	3	35	Impaired	0	5
16	Mild	0	1	36	Impaired	0	4
17	Mild	1	8	37	Impaired	0	4
18	Mild	1	2	38	Impaired	1	8
19	Mild	0	5	39	Impaired	0	8
20	Mild	1	5	40	Impaired	0	9

[a]0, low; 1, high.

TABLE 7.6   Output for Fitting Cumulative Logit Model to Table 7.5

Score Test for the Proportional Odds Assumption

Chi-Square	DF	Pr > ChiSq
2.3255	4	0.6761

Parameter	Estimate	Std Error	Like. Ratio 95% Conf Limits		Chi-Square	Pr > Chi Sq
Intercept1	−0.2819	0.6423	−1.5615	0.9839	0.19	0.6607
Intercept2	1.2128	0.6607	−0.0507	2.5656	3.37	0.0664
Intercept3	2.2094	0.7210	0.8590	3.7123	9.39	0.0022
life	−0.3189	0.1210	−0.5718	−0.0920	6.95	0.0084
ses	1.1112	0.6109	−0.0641	2.3471	3.31	0.0689

The main-effects model of form (7.5) is

$$\text{logit}[P(Y \le j \mid \mathbf{x})] = \alpha_j + \beta_1 x_1 + \beta_2 x_2.$$

Table 7.6 shows output. With $J = 4$ response categories, the model has three $\{\alpha_j\}$ intercepts. Usually, these are not of interest except for computing response probabilities. The parameter estimates yield estimated logits and hence estimates of $P(Y \le j)$, $P(Y > j)$, or $P(Y = j)$. We illustrate for subjects at the mean life events score of $x_1 = 4.275$ with low SES ($x_2 = 0$). Since $\hat{\alpha}_1 = -0.282$, the estimated probability of response *well* is

$$\hat{P}(Y = 1) = \hat{P}(Y \le 1) = \frac{\exp[-0.282 - 0.319(4.275)]}{1 + \exp[-0.282 - 0.319(4.275)]} = 0.16.$$

Figure 7.6 plots $\hat{P}(Y > 2)$ as a function of the life events index, at the two levels of SES.

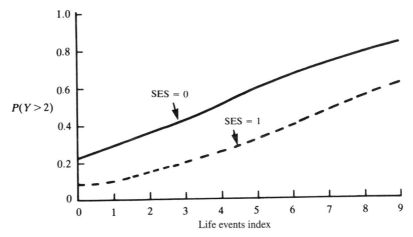

FIGURE 7.6   Estimated values of $P(Y > 2)$ for Table 7.5.

The effect estimates $\hat{\beta}_1 = -0.319$ and $\hat{\beta}_2 = 1.111$ suggest that the cumulative probability starting at the well end of the scale decreases as the life events score increases and increases at the higher level of SES. Given the life events score, at the high SES level the estimated odds of mental impairment below any fixed level are $e^{1.111} = 3.0$ times the estimated odds at the low SES level.

Descriptions of effects can compare cumulative probabilities rather than use odds ratios. These can be easier to understand. We describe effects of quantitative variables by comparing probabilities at their quartiles. We describe effects of qualitative variables by comparing probabilities for different categories. We control for quantitative variables by setting them at their mean. We control for qualitative variables by fixing the category, unless there are several, in which case we can set each at their dummy means. We illustrate again with $P(Y = 1)$, the *well* outcome. First, we describe the SES effect. At the mean life events of 4.275, $\hat{P}(Y = 1) = 0.37$ at high SES (i.e., $x_2 = 1$) and 0.16 at low SES ($x_2 = 0$). Next, we describe the life events effect. The lower and upper quartiles of the life events score are 2.0 and 6.5. For high SES, $\hat{P}(Y = 1)$ changes from 0.55 to 0.22 between these quartiles; for low SES, it changes from 0.28 to 0.09. (Note that comparing 0.55 to 0.28 at the lower quartile and 0.22 to 0.09 at the upper quartile provides further information about the SES effect.) The sample effect is substantial for both predictors.

The output in Table 7.6, taken from SAS, also presents a score test of the proportional odds property. This tests whether the effects are the same for each cumulative logit against the alternative of separate effects. It compares the model with one parameter for x_1 and one for x_2 to a more complex model with three parameters for each, allowing different effects for logit$[P(Y \leq 1)$, logit$[P(Y \leq 2)]$, and logit$[P(Y \leq 3)]$. Here, the score statistic equals 2.33. It has df = 4, since the more complex model has four additional parameters. The more complex model does not fit significantly better ($P = 0.68$).

7.2.5 More Complex Models

More complex cumulative logit models are formulated as in ordinary logistic regression. They simply require a set of intercept parameters rather than a single one. In the previous example, for instance, permitting interaction yields a model with ML fit

$$\text{logit}\left[\hat{P}(Y \leq j \mid \mathbf{x})\right] = \hat{\alpha}_j - 0.420x_1 + 0.371x_2 + 0.181x_1x_2,$$

where the coefficient of x_1x_2 has SE = 0.238. The estimated effect of life events on the cumulative logit is -0.420 for the low SES group and $(-0.420 + 0.181) = -0.239$ for the high SES group. The impact of life events seems more severe for the low SES group, but the difference in effects is not significant.

Models in this section used the proportional odds assumption of the same effects for different cumulative logits. An advantage is that effects are simple to summarize and interpret, requiring only a single parameter for each predictor. The models generalize to include separate effects, replacing $\boldsymbol{\beta}$ in (7.5) by $\boldsymbol{\beta}_j$. This implies nonparallelism of curves for different logits. However, curves for different cumulative probabilities then cross for some $\mathbf{x}$ values. Such models violate the proper order among the cumulative probabilities.

Even if such a model fits better over the observed range of $\mathbf{x}$, for reasons of parsimony the simple model might be preferable. One case is when effects $\{\hat{\boldsymbol{\beta}}_j\}$ with different logits are not substantially different in practical terms. Then the significance in a test of proportional odds may reflect primarily a large value of n. Even with smaller n, although effect estimators using the simple model are biased, they may have smaller MSE than estimators from a more complex model having many more parameters. So even if a test of proportional odds has a small P-value, don't discard this model automatically.

If a proportional odds model fits poorly in terms of practical as well as statistical significance, alternative strategies exist. These include (1) trying a link function for which the response curve is nonsymmetric (e.g., complementary log-log); (2) adding additional terms, such as interactions, to the linear predictor; (3) adding dispersion parameters; (4) permitting separate effects for each logit for some but not all predictors (i.e., *partial proportional odds*; and (5) fitting baseline-category logit models and using the ordinality in an informal way in interpreting the associations. For approach (4), see Peterson and Harrell (1990), Stokes et al. (2000, Sec. 15.13), and criticism by Cox (1995). In the next section we generalize the cumulative logit model to permit extensions (1) and (3).

7.3 ORDINAL RESPONSES: CUMULATIVE LINK MODELS

Cumulative logit models use the logit link. As in univariate GLMs, other link functions are possible. Let G^{-1} denote a link function that is the inverse of the continuous cdf G (recall Section 4.2.5). The *cumulative link* model

$$G^{-1}[P(Y \le j \mid \mathbf{x})] = \alpha_j + \boldsymbol{\beta}'\mathbf{x} \qquad (7.7)$$

links the cumulative probabilities to the linear predictor. The logit link function $G^{-1}(u) = \log[u/(1-u)]$ is the inverse of the standard logistic cdf.

As in the proportional odds model (7.5), effects of $\mathbf{x}$ in (7.7) are assumed the same for each cutpoint, $j = 1, \ldots, J - 1$. In Section 7.2.3 we showed that this assumption holds when a linear regression for a latent variable Y^* has standardized cdf G. Model (7.7) results from discrete measurement of Y^* from a location-parameter family having cdf $G(y^* - \beta'x)$. The parameters $\{\alpha_j\}$ are category cutpoints on a standardized version of the latent scale. In this sense, cumulative link models are regression models, using a linear predictor $\beta'x$ to describe effects of explanatory variables on crude ordinal measurement of Y^*. Using $-\beta$ rather than $+\beta$ in the linear predictor merely results in change of sign of $\hat{\beta}$. Most software (e.g., GENMOD and LOGISTIC in SAS) fits it in $+\beta$ form.

7.3.1 Types of Cumulative Links

Use of the standard normal cdf Φ for G gives the *cumulative probit model*. This generalizes the binary probit model (Section 6.6) to ordinal responses. It is appropriate when the distribution for Y^* is normal. Parameters in probit models can be interpreted in terms of the latent variable Y^*. For instance, consider the model $\Phi^{-1}[P(Y \leq j)] = \alpha_j - \beta x$. From Section 7.2.3, since $Y^* = \beta x + \epsilon$ where $\epsilon \sim N(0, 1)$ has cdf Φ, β has the interpretation that a 1-unit increase in x corresponds to a β increase in $E(Y^*)$. When ϵ need not be in standardized form with $\sigma = 1$, a 1-unit increase in x corresponds to a β standard deviation increase in $E(Y^*)$. Cumulative logit models provide fits similar to those for cumulative probit models, and their parameter interpretation is simpler.

An underlying extreme value distribution for Y^* implies a model of the form

$$\log\{-\log[1 - P(Y \leq j \mid x)]\} = \alpha_j + \beta'x .$$

In section 6.6 we introduced this *complementary log-log link* for binary data. The ordinal model using this link is sometimes called a *proportional hazards* model since it results from a generalization of the proportional hazards model for survival data to handle grouped survival times (Prentice and Gloeckler 1978). It has the property

$$P(Y > j \mid x_1) = [P(Y > j \mid x_2)]^{\exp[\beta'(x_1 - x_2)]} .$$

With this link, $P(Y \leq j)$ approaches 1.0 at a faster rate than it approaches 0.0. The related *log-log link* $\log\{-\log[P(Y \leq j)]\}$ is appropriate when the complementary log-log link holds for the categories listed in reverse order.

7.3.2 Estimation for Cumulative Link Models

McCullagh (1980) and Thompson and Baker (1981) treated cumulative link models as multivariate GLMs. McCullagh presented a Fisher scoring algorithm for ML estimation, expressing the likelihood in the form (7.6) using cumulative probabilities. McCullagh showed that sufficiently large n guarantees a unique maximum of the likelihood. Burridge (1981) and Pratt (1981) showed that the log likelihood is concave for many cumulative link models, including the logit, probit, and complementary log-log. Iterative algorithms usually converge rapidly to the ML estimates.

7.3.3 Life Table Example

Table 7.7 shows the life-length distribution for U.S. residents in 1981, by race and gender. Life length uses five ordered categories. The underlying continuous cdf of life length increases slowly at small to moderate ages but increases sharply at older ages. This suggests the complementary log-log link. This link also results from assuming that the hazard rate increases exponentially with age, which happens for an extreme value distribution (the Gompertz).

For gender G (1 = female; 0 = male), race R (1 = black; 0 = white), and life length Y, Table 7.7 contains fitted distributions for the model

$$\log\{-\log[1 - P(Y \leq j \mid G = g, R = r)]\} = \alpha_j + \beta_1 g + \beta_2 r.$$

Goodness-of-fit statistics are irrelevant, since the table contains population distributions. The model describes well the four distributions. Its parameter values are $\beta_1 = -0.658$ and $\beta_2 = 0.626$. The fitted cdf's satisfy

$$P(Y > j \mid G = 0, R = r) = [P(Y > j \mid G = 1, R = r)]^{\exp(0.658)}.$$

Given race, the proportion of men living longer than a fixed time equaled the proportion for women raised to the $\exp(0.658) = 1.93$ power. Given gender, the proportion of blacks living longer than a fixed time equaled the proportion for whites to the $\exp(0.626) = 1.87$ power. The β_1 and β_2 values indicate that white men and black women had similar distributions, that white women tended to have longest lives and black men tended to have shortest lives. If the probability of living longer than some fixed time equaled π for white women, that probability was about π^2 for white men and black women and π^4 for black men.

TABLE 7.7 Life-Length Distribution of U.S. Residents (Percent),[a] 1981

| Life Length | Males | | Females | |
	White	Black	White	Black
0–20	2.4 (2.4)	3.6 (4.4)	1.6 (1.2)	2.7 (2.3)
20–40	3.4 (3.5)	7.5 (6.4)	1.4 (1.9)	2.9 (3.4)
40–50	3.8 (4.4)	8.3 (7.7)	2.2 (2.4)	4.4 (4.3)
50–60	17.5 (16.7)	25.0 (26.1)	9.9 (9.6)	16.3 (16.3)
Over 65	72.9 (73.0)	55.6 (55.4)	84.9 (84.9)	73.7 (73.7)

[a] Values in parentheses are fit of proportional hazards (i.e., complementary log-log link) model.

Source: Data from *Statistical Abstract of the United States* (Washington, DC: U.S. Bureau of the Census, 1984), p. 69.

7.3.4 Incorporating Dispersion Effects*

For cumulative link models, settings of the explanatory variables are *stochastically ordered* on the response: For any pair x_1 and x_2, either $P(Y \leq j|x_1) \leq P(Y \leq j|x_2)$ for all j or $P(Y \leq j|x_1) \geq P(Y \leq j|x_2)$ for all j. Figure 7.7a illustrates for underlying continuous density functions and cdf's at two settings of x. When this is violated and such models fit poorly, often it is because the dispersion also varies with x. For instance, perhaps responses tend to concentrate around the same location but more dispersion occurs at x_1 than at x_2. Then perhaps $P(Y \leq j|x_1) > P(Y \leq j|x_2)$ for small j but $P(Y \leq j|x_1) < P(Y \leq j|x_2)$ for large j. In other words, at x_1 the responses concentrate more at the extreme categories than at x_2. Figure 7.7b illustrates for underlying continuous distributions.

A cumulative link model that incorporates dispersion effects is

$$G^{-1}[P(Y \leq j|x)] = \frac{\alpha_j + \beta'x}{\exp(\gamma'x)}. \qquad (7.8)$$

(Again, one can replace + by − to more closely mimic a location–scale family for an underlying continuous variable.) The denominator contains

When X and Y are independent,

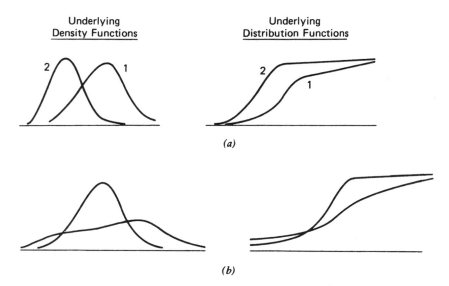

FIGURE 7.7 (*a*) Distribution 1 stochastically higher than distribution 2; (*b*) distributions not stochastically ordered.

scale parameters $\boldsymbol{\gamma}$ that describe the dispersion's dependence on $\mathbf{x}$. The ordinary model (7.7) is the special case $\boldsymbol{\gamma} = 0$. Otherwise, the cumulative probabilities tend to shrink toward each other when $\boldsymbol{\gamma}'\mathbf{x} > 0$. This creates higher probabilities in the end categories and overall greater dispersion. The cumulative probabilities tend to move apart (creating less dispersion) when $\boldsymbol{\gamma}'\mathbf{x} < 0$.

To illustrate, we use this model to compare two groups on an ordinal scale. Suppose that x is a dummy variable with $x = 1$ for the first group. With cumulative logits, model (7.8) is

$$\text{logit}[P(Y \le j)] = \alpha_j, \qquad x = 0,$$

$$\text{logit}[P(Y \le j)] = (\alpha_j + \beta)/\exp(\gamma), \qquad x = 1.$$

The case $\gamma = 0$ is the usual model, in which β is a location shift that determines a common cumulative log odds ratio for all 2×2 collapsings of the $2 \times J$ table. When $\gamma \ne 0$ the difference between the logits for the two groups, and hence the cumulative odds ratio, varies as j does. When $\gamma > 0$, responses at $x = 1$ tend to be more disperse than at $x = 0$. See Cox (1995) and McCullagh (1980) for model fitting and examples.

7.4   ALTERNATIVE MODELS FOR ORDINAL RESPONSES*

Models for ordinal responses need not use cumulative probabilities. In this section we discuss alternative logit models and a simpler model that resembles ordinary regression.

7.4.1   Adjacent-Categories Logits

The adjacent-categories logits are

$$\text{logit}[P(Y = j | Y = j \text{ or } j + 1)] = \log\frac{\pi_j}{\pi_{j+1}}, \quad j = 1, \dots, J - 1. \quad (7.9)$$

These logits are a basic set equivalent to the baseline-category logits. The connections are

$$\log\frac{\pi_j}{\pi_J} = \log\frac{\pi_j}{\pi_{j+1}} + \log\frac{\pi_{j+1}}{\pi_{j+2}} + \cdots + \log\frac{\pi_{J-1}}{\pi_J}, \qquad (7.10)$$

and

$$\log\frac{\pi_j}{\pi_{j+1}} = \log\frac{\pi_j}{\pi_J} - \log\frac{\pi_{j+1}}{\pi_J}, \quad j = 1, \dots, J - 1.$$

Either set determines logits for all $\binom{J}{2}$ pairs of response categories.

Models using adjacent-categories logits can be expressed as baseline-category logit models. For instance, consider the adjacent-categories logit model

$$\log \frac{\pi_j(\mathbf{x})}{\pi_{j+1}(\mathbf{x})} = \alpha_j + \boldsymbol{\beta}'\mathbf{x}, \quad j = 1, \ldots, J - 1, \tag{7.11}$$

with common effect $\boldsymbol{\beta}$. From adding $(J - j)$ terms as in (7.10), the equivalent baseline-category logit model is

$$\log \frac{\pi_j(\mathbf{x})}{\pi_J(\mathbf{x})} = \sum_{k=j}^{J-1} \alpha_k + \boldsymbol{\beta}'(J - j)\mathbf{x}, \quad j = 1, \ldots, J - 1$$

$$= \alpha_j^* + \boldsymbol{\beta}'\mathbf{u}_j, \quad j = 1, \ldots, J - 1$$

with $\mathbf{u}_j = (J - j)\mathbf{x}$. The adjacent-categories logit model corresponds to a baseline-category logit model with adjusted model matrix but also a single parameter for each predictor. With some software one can fit model (7.11) by fitting the equivalent baseline-category logit model.

The construction of the adjacent-categories logits recognizes the ordering of Y categories. To benefit from this in model parsimony requires appropriate specification of the linear predictor. For instance, if an explanatory variable has similar effect for each logit, advantages accrue from having a single parameter instead of $(J - 1)$ parameters describing that effect. When used with this proportional odds form, model (7.11) with adjacent-categories logits fit well in similar situations as model (7.5) with cumulative logits. They both imply stochastically ordered distributions for Y at different predictor values.

The choice of model should depend less on goodness of fit than on whether one prefers effects to refer to individual response categories, as the adjacent-categories logits provide, or instead to groupings of categories using the entire scale or an underlying latent variable, which cumulative logits provide. Since effects in cumulative logit models refer to the entire scale, they are usually larger. The ratio of estimate to standard error, however, is usually similar for the two model types. An advantage of the cumulative logit model is the approximate invariance of effect estimates to the choice and number of response categories. This does not happen with the adjacent-categories logits.

7.4.2 Job Satisfaction Example

Table 7.8 refers to the relationship between job satisfaction (Y) and income, stratified by gender, for black Americans. For simplicity, we use income scores $(1, 2, 3, 4)$. For income x and gender g ($1 =$ females, $0 =$ males), consider the model

$$\log(\pi_j / \pi_{j+1}) = \alpha_j + \beta_1 x + \beta_2 g, \quad j = 1, 2, 3.$$

TABLE 7.8   Job Satisfaction and Income, Controlling for Gender

Gender	Income (dollars)	Very Dissatisfied	A Little Satisfied	Moderately Satisfied	Very Satisfied
Female	< 5000	1	3	11	2
	5000–15,000	2	3	17	3
	15,000–25,000	0	1	8	5
	> 25,000	0	2	4	2
Male	< 5000	1	1	2	1
	5000–15,000	0	3	5	1
	15,000–25,000	0	0	7	3
	> 25,000	0	1	9	6

Source: 1991, General Social Survey, National Opinion Research Center.

It describes the odds of being very dissatisfied instead of a little satisfied, a little instead of moderately satisfied, and moderately instead of very satisfied. This model is equivalent to the baseline-category logit model

$$\log(\pi_j/\pi_4) = \alpha_j^* + \beta_1(4-j)x + \beta_2(4-j)g, \quad j = 1, 2, 3.$$

The value of the first predictor in this model is set equal to $3x$ in the equation for $\log(\pi_1/\pi_4)$, $2x$ in the equation for $\log(\pi_2/\pi_4)$, and x in the equation for $\log(\pi_3/\pi_4)$. Some software (e.g., PROC CATMOD in SAS; see Table A.12) allows one to enter a row of a model matrix for each baseline-category logit at a given setting of predictors. Then, after fitting the baseline-category logit model that constrains the effects to be the same for each logit, the estimated regression parameters are the ML estimates of parameters for the adjacent-categories logit model. The ML fit gives $\hat{\beta}_1 = -0.389$ (SE = 0.155) and $\hat{\beta}_2 = 0.045$ (SE = 0.314). For this parameterization, $\hat{\beta}_1 < 0$ means the odds of lower job satisfaction decrease as income increases. Given gender, the estimated odds of response in the lower of two adjacent categories multiplies by $\exp(-0.389) = 0.68$ for each category increase in income. The model describes 24 logits (three for each income $\times$ gender combination) with five parameters. Its deviance $G^2 = 12.6$ with $df = 19$. This model with a linear trend for the income effect and a lack of interaction between income and gender seems adequate.

Similar substantive results occur with a cumulative logit model. Its deviance $G^2 = 13.3$ with $df = 19$. The income effect is larger ($\hat{\beta}_1 = -0.51$, SE = 0.20), since it refers to the entire response scale rather than adjacent categories. However, significance is similar, with $\hat{\beta}_1/\text{SE} \approx -2.5$ for each model.

7.4.3 Continuation-Ratio Logits

Continuation-ratio logits are defined as

$$\log\frac{\pi_j}{\pi_{j+1} + \cdots + \pi_J}, \quad j = 1, \ldots, J - 1 \tag{7.12}$$

or as

$$\log\frac{\pi_{j+1}}{\pi_1 + \cdots + \pi_j}, \quad j = 1, \ldots, J - 1. \tag{7.13}$$

The continuation-ratio logit model form is useful when a sequential mechanism, such as survival through various age periods, determines the response outcome (e.g., Tutz 1991). Let $\omega_j = P(Y = j \mid Y \geq j)$. With explanatory variables,

$$\omega_j(\mathbf{x}) = \frac{\pi_j(\mathbf{x})}{\pi_j(\mathbf{x}) + \cdots + \pi_J(\mathbf{x})}, \quad j = 1, \ldots, J - 1. \tag{7.14}$$

The continuation-ratio logits (7.12) are ordinary logits of these conditional probabilities: namely, $\log[\omega_j(\mathbf{x})/(1 - \omega_j(\mathbf{x}))]$.

At the ith setting $\mathbf{x}_i$ of $\mathbf{x}$, let $\{y_{ij}, j = 1, \ldots, J\}$ denote the response counts, with $n_i = \sum_j y_{ij}$. When $n_i = 1$, y_{ij} indicates whether the response is in category j, as in Section 7.1.4. Let $b(n, y; \omega)$ denote the binomial probability of y successes in n trials with parameter ω for each trial. By expressing the multinomial probability of $(y_{i1}, \ldots, y_{iJ})$ in the form $p(y_{i1})p(y_{i2}|y_{i1}) \cdots p(y_{iJ}|y_{i1}, \ldots, y_{i,J-1})$, one can show that the multinomial mass function has factorization

$$b[n_i, y_{i1}; \omega_1(\mathbf{x}_i)]b[n_i - y_{i1}, y_{i2}; \omega_2(\mathbf{x}_i)] \cdots$$
$$b[n_i - y_{i1} - \cdots - y_{i,J-2}, y_{i,J-1}; \omega_{J-1}(\mathbf{x}_i)]. \tag{7.15}$$

The full likelihood is the product of multinomial mass functions from the different $\mathbf{x}_i$ values. Thus, the log likelihood is a sum of terms such that different ω_j enter into different terms. When parameters in the model specification for logit(ω_j) are distinct from those for logit(ω_k) whenever $j \neq k$, maximizing each term separately maximizes the full log likelihood. Thus, separate fitting of models for different continuation-ratio logits gives the same results as simultaneous fitting. The sum of the $J - 1$ separate G^2 statistics provides an overall goodness-of-fit statistic pertaining to the simultaneous fitting of $J - 1$ models.

Because these logits refer to a binary response in which one category combines levels of the original scale, separate fitting can use methods for binary logit models. Similar remarks apply to continuation-ratio logits (7.13), although those logits and the subsequent analysis do not give equivalent results. Sometimes, simpler models with the same effects for each logit are plausible (McCullagh and Nelder 1989, p. 164; Tutz 1991).

7.4.4 Developmental Toxicity Study with Pregnant Mice

We illustrate continuation-ratio logits using Table 7.9 from a developmental toxicity study. Such experiments with rodents test substances posing potential danger to developing fetuses. Diethylene glycol dimethyl ether (diEGdiME), one such substance, is an industrial solvent used in the manufacture of protective coatings such as lacquer and metal coatings.

This study administered diEGdiME in distilled water to pregnant mice. Each mouse was exposed to one of five concentration levels for 10 days early in the pregnancy. The mice exposed to level 0 formed a control group. Two days later, the uterine contents of the pregnant mice were examined for defects. Each fetus has three possible outcomes (nonlive, malformation, normal). The outcomes are ordered, with nonlive the least desirable result. We use continuation-ratio logits to model (1) the probability π_1 of a nonlive fetus, and (2) the conditional probability $\pi_2/(\pi_2 + \pi_3)$ of a malformed fetus, given that the fetus was live.

We fitted the continuation-ratio logit models

$$
\log\frac{\pi_1(x_i)}{\pi_2(x_i) + \pi_3(x_i)} = \alpha_1 + \beta_1 x_i, \quad \log\frac{\pi_2(x_i)}{\pi_3(x_i)} = \alpha_2 + \beta_2 x_i,
$$

using x_i scores $\{0, 62.5, 125, 250, 500\}$ for concentration level. The ML estimates are $\hat{\beta}_1 = 0.0064$ (SE $= 0.0004$) and $\hat{\beta}_2 = 0.0174$ (SE $= 0.0012$). In each case, the less desirable outcome is more likely as the concentration increases. For instance, given that a fetus was live, the estimated odds that it was malformed rather than normal multiplies by $\exp(1.74) = 5.7$ for every 100-unit increase in the concentration of diEGdiME. The likelihood-ratio fit

TABLE 7.9 Outcomes for Pregnant Mice in Developmental Toxicity Study

Concentration (mg/kg per day)	Response		
	Nonlive	Malformation	Normal
0 (controls)	15	1	281
62.5	17	0	225
125	22	7	283
250	38	59	202
500	144	132	9

[a]Based on results in C. J. Price et al., *Fund. Appl. Toxicol.* **8**:115–126 (1987). I thank Louise Ryan for showing me these data.

statistics are $G^2 = 5.78$ for $j = 1$ and $G^2 = 6.06$ for $j = 2$, each based on $df = 3$. Their sum, $G^2 = 11.84$ (or similarly $X^2 = 9.76$), with df $= 6$, summarizes the fit.

This analysis treats pregnancy outcomes for different fetuses as independent, identical observations. In fact, each pregnant mouse had a litter of fetuses, and statistical dependence may exist among different fetuses in the same litter. Different litters at a given concentration level may also have different response probabilities. Heterogeneity of various sorts among the litters (e.g., due to varying physical characteristics among different pregnant mice) would cause these probabilities to vary somewhat. Either statistical dependence or heterogeneous probabilities violates the binomial assumption and causes overdispersion. At a fixed concentration level, the number of fetuses in a litter that die may vary among pregnant mice more than if the counts were independent and identical binomial variates. The total G^2 shows some evidence of lack of fit ($P = 0.07$) but may reflect overdispersion caused by these factors rather than an inappropriate choice of response curve.

To account for overdispersion, we could adjust standard errors using the quasi-likelihood approach (Section 4.7). This multiplies standard errors by $\sqrt{X^2/df} = \sqrt{9.76/6} = 1.28$. For each logit, strong evidence remains that $\beta_j > 0$. In Chapters 12 and 13 we present other methods that account for the clustering of fetuses in litters.

7.4.5 Mean Response Models for Ordered Response

We now present a model that resembles ordinary regression for a continuous response variable. For scores $v_1 \leq v_2 \leq \cdots \leq v_J$, let

$$M(\mathbf{x}) = \sum_j v_j \pi_j(\mathbf{x})$$

denote the mean response. The model

$$M(\mathbf{x}) = \alpha + \boldsymbol{\beta}'\mathbf{x} \qquad (7.16)$$

assumes a linear relationship between the mean and the explanatory variables. With $J = 2$, it is the linear probability model (Section 4.2.1). With $J > 2$, it does not structurally specify the response probabilities but merely describes the dependence of the mean on $\mathbf{x}$.

Assuming independent multinomial sampling at different $\mathbf{x}_i$, Bhapkar (1968), Grizzle et al. (1969), and Williams and Grizzle (1972) presented weighted least squares (WLS) fits for mean response models. The WLS approach, described in Section 15.1, applies when all explanatory variables are categorical. The ML approach for maximizing the product multinomial likelihood applies for categorical or continuous explanatory variables. Haber (1985) and Lipsitz (1992) presented algorithms for ML fitting of a family, including mean response models. This is somewhat complex, since the probabilities in the multinomial likelihood are not direct functions of the parameters in (7.16). Specialized software is available (see Appendix A).

7.4.6 Job Satisfaction Example Revisited

We illustrate for Table 7.8, modeling the mean of Y = job satisfaction using income x and gender g (1 = females, 0 = males). For simplicity, we use job satisfaction scores and income scores (1, 2, 3, 4). The model has ML fit,

$$\hat{M} = 2.59 + 0.181x - 0.030g,$$

with SE = 0.069 for income and 0.145 for gender. Given gender, the estimated increase in mean job satisfaction is about 0.2 response category for each category increase of income. Although the evidence is strong of a positive effect [e.g., Wald statistic $(0.181/0.069)^2 = 6.8$, df = 1, $P = 0.009$], the strength of the effect is weak. Job satisfaction at the highest income level is estimated to average about half a category higher than at the lowest income level, since $3(0.181) = 0.54$. Similar results occur with the WLS solution, for which the estimated income effect of 0.182 has SE = 0.068 (Table A.12 shows the use of CATMOD in SAS).

The deviance for testing the model fit equals 5.1. Since means occur at eight income $\times$ gender settings and the model has three parameters, residual df = 5. The fit seems adequate.

7.4.7 Advantages and Disadvantages of Mean Response Models

Treating ordinal variables in a quantitative manner is sensible if their categorical nature reflects crude measurement of an inherently continuous variable. Mean response models have the advantage of closely resembling ordinary regression.

With $J = 2$, in Section 4.2.1 we noted that linear probability models have a structural difficulty because of the restriction of probabilities to $(0, 1)$. A similar difficulty occurs here, since a linear model can have predicted means outside the range of assigned scores. This happens less frequently when J is large and reasonable dispersion of responses occurs throughout the domain of interest for the explanatory variables. The notion of an underlying latent variable makes more sense for an ordinal variable than for a strictly binary response, so this difficulty has less relevance here.

Unlike logit models, mean response models do not uniquely determine cell probabilities. Thus, mean response models do not specify structural aspects such as stochastic orderings. These models do not represent the categorical response structure as fully as do models for probabilities, and conditions such as independence do not occur as special cases. However, they provide simpler descriptions than odds ratios or summaries from cumulative link models. As J increases, they also interface with ordinary regression models. For large J, they are a simple mechanism for approximating results for a regression model we would use if we could measure Y continuously.

7.5 TESTING CONDITIONAL INDEPENDENCE IN $I \times J \times K$ TABLES*

In Section 6.3.2 we introduced the Cochran–Mantel–Haenszel (CMH) test of conditional independence for $2 \times 2 \times K$ tables. This section presents related tests with multicategory responses for $I \times J \times K$ tables. Likelihood-ratio tests compare the fit of a model specifying XY conditional independence with a model having dependence. Alternatively, generalizations of the CMH statistic are score statistics for certain models.

7.5.1 Using Multinomial Models to Test Conditional Independence

Treating Z as a nominal control factor, we discuss four cases with (Y, X) as (ordinal, ordinal), (ordinal, nominal), (nominal, ordinal), (nominal, nominal). For ordinal Y we use cumulative logit models, but other ordinal links yield analogous tests. As we noted in Section 6.3.2 when the XY association is similar in the partial tables, the power benefits from basing a test statistic on a model of homogeneous association.

1. *Y ordinal, X ordinal.* Let $\{x_i\}$ be ordered scores. The model

$$\text{logit}[P(Y \leq j | X = i, Z = k)] = \alpha_j + \beta x_i + \beta_k^Z \quad (7.17)$$

   has the same linear trend for the X effect in each partial table. For it, XY conditional independence is H_0: $\beta = 0$. Likelihood-ratio, score, or Wald statistics for H_0 provide large-sample chi-squared tests with df = 1 that are sensitive to the trend alternative.

2. *Y ordinal, X nominal.* An alternative to conditional independence that treats X as a factor is

$$\text{logit}[P(Y \leq j | X = i, Z = k)] = \alpha_j + \beta_i + \beta_k^Z,$$

   with constraint such as $\beta_I = 0$. For this model, XY conditional independence is H_0: $\beta_1 = \cdots = \beta_I$. Large-sample chi-squared tests have df = $I - 1$.

3. *Y nominal, X ordinal.* When Y is nominal, analogous tests use baseline-category logit models. The model of XY conditional independence is

$$\log \frac{P(Y = j | X = i, Z = k)}{P(Y = J | X = i, Z = k)} = \alpha_{jk}. \quad (7.18)$$

For ordered scores $\{x_i\}$, a test that is sensitive to the same linear trend alternatives in each partial table compares this model to

$$\log \frac{P(Y = j \mid X = i, Z = k)}{P(Y = J \mid X = i, Z = k)} = \alpha_{jk} + \beta_j x_i.$$

Conditional independence is H_0: $\beta_1 = \cdots = \beta_{J-1} = 0$. Large-sample chi-squared tests have df $= J - 1$.

4. *Y nominal, X nominal.* An alternative to *XY* conditional independence that treats *X* as a factor is

$$\log \frac{P(Y = j \mid X = i, Z = k)}{P(Y = J \mid X = i, Z = k)} = \alpha_{jk} + \beta_{ij} \qquad (7.19)$$

with constraint such as $\beta_{Ij} = 0$ for each j. For each j, X and Z have additive effects of form $\alpha_k + \beta_i$. Conditional independence is H_0: $\beta_{1j} = \cdots = \beta_{Ij}$ for $j = 1, \ldots, J - 1$. Large-sample chi-squared tests have df $= (I - 1)(J - 1)$.

Table 7.10 summarizes the four tests. They work well when the model describes at least a major component of the departure from conditional independence. This does not mean that one must test the fit of the model to use the test (see the remarks at the end of Section 6.3.2).

Occasionally, the association may change dramatically across the K partial tables. When Z is ordinal, an alternative by which a log odds ratio changes linearly across levels of Z is sometimes of use. For instance, when $Z =$ age of subject, the association between a risk factor X (e.g., level of smoking) and a response Y (e.g., severity of heart disease) may tend to increase with Z. When Z is nominal, one can test the conditional independence models

TABLE 7.10 Summary of Models for Testing Conditional Independence

Y-X	Model	Conditional Independence	df
Ord-Ord	logit$[P(Y \leq j)] = \alpha_j + \beta x_i + \beta_k^Z$	$\beta = 0$	1
-Nom	logit$[P(Y \leq j)] = \alpha_j + \beta_i + \beta_k^Z$	$\beta_1 = \cdots = \beta_I$	$I - 1$
Nom-Ord	$\log\left[\dfrac{P(Y = j)}{P(Y = J)}\right] = \alpha_{jk} + \beta_j x_i$	$\beta_1 = \cdots = \beta_{J-1} = 0$	$J - 1$
-Nom	$\log\left[\dfrac{P(Y = j)}{P(Y = J)}\right] = \alpha_{jk} + \beta_{ij}$	all $\beta_{ij} = 0$	$(I - 1)(J - 1)$

against a more general alternative with separate effect parameters at each level of Z. Allowing effects to vary across levels of Z, however, results in the test df being multiplied by K, which handicaps power.

7.5.2 Job Satisfaction Example Revisited

We now revisit the job satisfaction data (Table 7.8). Table 7.11 summarizes the fit of several models. The model treating income as an ordinal predictor uses scores $\{3, 10, 20, 35\}$, approximate midpoints of categories in thousands of dollars. Each likelihood-ratio test compares a given model to the model deleting the income effect, controlling for gender.

Testing conditional independence with the cumulative logit model (7.17) yields likelihood-ratio statistic $19.62 - 13.95 = 5.7$ with df $= 20 - 19 = 1$, strong evidence of an effect. Models that treat either or both variables as nominal do not provide such strong evidence. Focusing the test on a linear trend alternative yields a smaller P-value. However, we learn more from estimating parameters than from significance tests, as in Sections 7.4.2 and 7.4.6.

7.5.3 Generalized Cochran–Mantel–Haenszel Tests for $I \times J \times K$ Tables

Birch (1965), Landis et al. (1978), and Mantel and Byar (1978) generalized the CMH statistic (Section 6.3.2). The tests treat X and Y symmetrically, so the three cases correspond to treating both as nominal, both as ordinal, or one of each. Conditional on row and column totals, each stratum has $(I - 1)(J - 1)$ nonredundant cell counts. Let

$$\mathbf{n}_k = \left(n_{11k}, n_{12k}, \ldots, n_{1,J-1,k}, \ldots, n_{I-1,J-1,k} \right)'.$$

TABLE 7.11 Summary of Model-Based Likelihood-Ratio Tests of Conditional Independence for Table 7.8

Satisfaction	Income	G^2 Fit	df	Test Statistic	df	P-value
Ordinal	Ordinal	13.95	19	5.7	1	0.017
	Nominal	10.51	17	9.1	3	0.028
	Not in model	19.62	20	—	—	—
Nominal	Ordinal	11.74	15	7.6	3	0.054
	Nominal	7.09	9	12.3	9	0.198
	Not in model	19.37	18	—	—	—

Let $\boldsymbol{\mu}_k = E(\mathbf{n}_k)$ under H_0: conditional independence, namely

$$\boldsymbol{\mu}_k = \left(n_{1+k}n_{+1k}, n_{1+k}n_{+2k}, \ldots, n_{I-1,+,k}n_{+,J-1,k}\right)'/n_{++k}.$$

Let $\mathbf{V}_k$ denote the null covariance matrix of $\mathbf{n}_k$, where

$$\text{cov}(n_{ijk}, n_{i'j'k}) = \frac{n_{i+k}(\delta_{ii'}n_{++k} - n_{i'+k})n_{+jk}(\delta_{jj'}n_{++k} - n_{+j'k})}{n_{++k}^2(n_{++k} - 1)}$$

with $\delta_{ab} = 1$ when $a = b$ and $\delta_{ab} = 0$ otherwise.

The most general statistic treats rows and columns as unordered. Summing over the K strata, let

$$\mathbf{n} = \sum \mathbf{n}_k, \qquad \boldsymbol{\mu} = \sum \boldsymbol{\mu}_k, \qquad \mathbf{V} = \sum \mathbf{V}_k.$$

The generalized CMH statistic for nominal X and Y is

$$\text{CMH} = (\mathbf{n} - \boldsymbol{\mu})'\mathbf{V}^{-1}(\mathbf{n} - \boldsymbol{\mu}). \qquad (7.20)$$

Its large-sample chi-squared distribution has df $= (I - 1)(J - 1)$. The df value equals that for the statistics comparing logit models (7.18) and (7.19). Both statistics are sensitive to detecting a conditional association that is similar in each stratum. For $K = 1$ stratum with n observations, CMH $= [(n - 1)/n]X^2$, where X^2 is the Pearson statistic (3.10).

Mantel (1963) introduced a generalized statistic for ordinal X and Y. Using ordered scores $\{u_i\}$ and $\{v_j\}$, it is sensitive to a correlation of common sign in each stratum. Evidence of a positive trend occurs if in each stratum $T_k = \sum_i \sum_j u_i v_j n_{ijk}$ exceeds its null expectation. Given the marginal totals in each stratum, under conditional independence

$$E(T_k) = \left[\sum_i u_i n_{i+k}\right]\left[\sum_j v_j n_{+jk}\right]\bigg/n_{++k},$$

$$\text{var}(T_k) = \frac{1}{n_{++k} - 1}\left[\sum_i u_i^2 n_{i+k} - \frac{(\sum_i u_i n_{i+k})^2}{n_{++k}}\right]$$

$$\times \left[\sum_j v_j^2 n_{+jk} - \frac{(\sum_j v_j n_{+jk})^2}{n_{++k}}\right].$$

The statistic $[T_k - E(T_k)]/[\text{var}(T_k)]^{1/2}$ equals the correlation between X and Y in stratum k multiplied by $\sqrt{n_{++k} - 1}$. To summarize across the K strata,

Mantel (1963) proposed

$$M^2 = \frac{\left\{ \sum_k \left[\sum_i \sum_j u_i v_j n_{ijk} - E\left(\sum_i \sum_j u_i v_j n_{ijk} \right) \right] \right\}^2}{\sum_k \text{var}\left(\sum_i \sum_j u_i v_j n_{ijk} \right)}. \qquad (7.21)$$

This has an approximate χ_1^2 null distribution, the same as for testing H_0: $\beta = 0$ in ordinal model (7.17). For $K = 1$, this is the M^2 statistic (3.15).

Landis et al. (1978) presented a statistic that has (7.20) and (7.21) as special cases. His statistic also can treat X as nominal and Y as ordinal, summarizing information about how I row means compare to their null expected values, with df $= I - 1$ (see Note 7.7).

7.5.4 Job Satisfaction Example Revisited

Table 7.12 shows output from conducting generalized CMH tests for Table 7.8. Statistics treating a variable as ordinal used scores $\{3, 10, 20, 35\}$ for income and scores $\{1, 3, 4, 5\}$ for job satisfaction. (Table A.12 shows the use of PROC FREQ in SAS, but with different scores.

The *general association* alternative treats X and Y as nominal and uses (7.20). It is sensitive to any association that is similar in each level of Z. The *row mean scores differ* alternative treats rows as nominal and columns as ordinal. It is sensitive to variation among the I row mean scores on Y, when that variation is similar in each level of Z. Finally, the *nonzero correlation* alternative treats X and Y as ordinal and uses (7.21). It is sensitive to a similar linear trend in each level of Z. As in the model-based analyses that Table 7.11 summarized, the evidence is stronger using the df $= 1$ ordinal test.

7.5.5 Related Score Tests for Multinomial Logit Models

The generalized CMH tests seem to be non-model-based alternatives to those of Section 7.5.1 using multinomial logit models. However, a close connection exists between them. For various multinomial logit models, the generalized CMH tests are score tests.

TABLE 7.12 Output for Generalized Cochran–Mantel–Haenszel Tests with Job Satisfaction and Income Data

	Summary Statistics for income by satisf Controlling for gender			
Cochran-Mantel-Haenszel Statistics (Based on Table Scores)				
Statistic	Alternative Hypothesis	DF	Value	Prob
1	Nonzero Correlation	1	6.1563	0.0131
2	Row Mean Scores Differ	3	9.0342	0.0288
3	General Association	9	10.2001	0.3345

The generalized CMH test (7.20) that treats X and Y as nominal is the score test that the $(I - 1)(J - 1)\{\beta_{ij}\}$ parameters in logit model (7.19) equal 0. The generalized CMH test using M^2 that treats X and Y as ordinal is the score test of $\beta = 0$ in model (7.17). For the cumulative logit model, the equivalence has the same $\{x_i\}$ scores in the model as in M^2, and the $\{v_j\}$ scores in M^2 are average rank scores. For the adjacent-categories logit model analog of (7.17), the $\{v_j\}$ scores in M^2 are any equally spaced scores.

With large samples in each stratum, the generalized CMH tests give similar results as likelihood-ratio tests comparing the relevant models. An advantage of the model-based approach is providing estimates of effects. An advantage of the generalized CMH tests is maintaining good performance under sparse asymptotics whereby K grows as n does. Remarks in Section 6.3.4 apply here also.

7.5.6 Exact Tests of Conditional Independence

In principle, exact tests of conditional independence can use the generalized CMH statistics, generalizing Section 6.7.5 for $2 \times 2 \times K$ tables. To eliminate nuisance parameters, one conditions on row and column totals in each stratum. The distribution of counts in each stratum is the multiple hypergeometric (Section 3.5.7), and this propagates an exact conditional distribution for the statistic of interest. The P-value is the probability of those tables having the same strata margins as observed but test statistic at least as large as observed (see Birch 1965; Kim and Agresti 1997; Mehta et al. 1988).

7.6 DISCRETE-CHOICE MULTINOMIAL LOGIT MODELS*

An important application of multinomial logit models is determining effects of explanatory variables on a subject's choice from a discrete set of options—for instance, the choice of transportation system to take to work (drive, bus, subway, walk, bicycle), housing (buy house, buy condominium, rent), primary shopping location (downtown, mall, catalogs, Internet), or product brand. Models for response variables consisting of a discrete set of choices are called *discrete-choice models*.

7.6.1 Discrete-Choice Modeling

In many discrete-choice applications, an explanatory variable takes different values for different response choices. As predictors of choice of transportation system, cost and time to reach destination take different values for each option. As a predictor of choice of product brand, price varies according to the option. Explanatory variables of this type are *characteristics of the choices*. They differ from the usual ones, for which values remain constant across the choice set. Such variables, *characteristics of the chooser*, include income, education, and other demographic characteristics.

McFadden (1974) proposed a discrete-choice model for explanatory variables that are characteristics of the choices. His model also permits the choice set to vary among subjects. For instance, some subjects may not have the subway as an option for travel to work. For subject i and response choice j, let $\mathbf{x}_{ij} = (x_{ij1}, \ldots, x_{ijp})'$ denote the values of the p explanatory variables, and let $\mathbf{x}_i = (\mathbf{x}_{i1}, \ldots, \mathbf{x}_{ip})$. Conditional on the choice set C_i for subject i, the model for the probability of selecting option j is

$$\pi_j(\mathbf{x}_i) = \frac{\exp(\boldsymbol{\beta}'\mathbf{x}_{ij})}{\sum_{h \in C_i} \exp(\boldsymbol{\beta}'\mathbf{x}_{ih})}. \tag{7.22}$$

For each pair of choices a and b, this model has the logit form

$$\log[\pi_a(\mathbf{x}_i)/\pi_b(\mathbf{x}_i)] = \boldsymbol{\beta}'(\mathbf{x}_{ia} - \mathbf{x}_{ib}). \tag{7.23}$$

Conditional on the choice being a or b, a variable's influence depends on the distance between the subject's values of that variable for those choices. If the values are the same, the model asserts that the variable has no influence on the choice between a and b. Reflecting this property, McFadden originally referred to model (7.22) as a *conditional logit* model.

From (7.23), the odds of choosing a over b do not depend on the other alternatives in the choice set or on their values of the explanatory variables. Luce (1959) called this property *independence from irrelevant alternatives*. It is unrealistic in some applications. For instance, for travel options auto and red bus, suppose that 80% choose auto, an odds of 4.0. Now suppose that the options are auto, red bus, and blue bus. According to (7.23), the odds are still 4.0 of choosing auto instead of red bus, but intuitively, we expect them to be about 8.0 (10% choosing each bus option), McFadden (1974) stated: "Application of the model should be limited to situations where the alternatives can plausibly be assumed to be distinct and weighed independently in the eyes of each decision-maker."

7.6.2 Discrete-Choice and Multinomial Logit Models

Model (7.22) can also incorporate explanatory variables that are characteristics of the chooser. This may seem surprising, since (7.22) has a single parameter for each explanatory variable; that is, the parameter vector is the same for each pair of choices. However, multinomial logit model (7.2) has discrete-choice form (7.22) after replacing such an explanatory variable by J artificial variables; the jth is the product of the explanatory variable with a dummy variable that equals 1 when the response choice is j. For instance, for a single explanatory variable, let x_i denote its value for subject i. For $j = 1, \ldots, J$, let δ_{jk} equal 1 when $k = j$ and 0 otherwise, and let

$$\mathbf{z}_{ij} = \left(\delta_{j1}, \ldots, \delta_{jJ}, \delta_{j1}x_i, \ldots, \delta_{jJ}x_i\right)'.$$

Let $\boldsymbol{\beta} = (\alpha_1, \ldots, \alpha_J, \beta_1, \ldots, \beta_J)'$. Then $\boldsymbol{\beta}'\mathbf{z}_{ij} = \alpha_j + \beta_j x_i$, and (7.2) is (with $\alpha_J = \beta_J = 0$ for identifiability)

$$
\pi_j(x_i) = \frac{\exp(\alpha_j + \beta_j x_i)}{\exp(\alpha_1 + \beta_1 x_i) + \cdots + \exp(\alpha_J + \beta_J x_i)}
$$

$$
= \frac{\exp(\boldsymbol{\beta}'\mathbf{z}_{ij})}{\exp(\boldsymbol{\beta}'\mathbf{z}_{i1}) + \cdots + \exp(\boldsymbol{\beta}'\mathbf{z}_{iJ})}.
$$

This has form (7.22).

With this approach, discrete-choice models can contain characteristics of the chooser and the choices. Thus, model (7.22) is very general. The ordinary multinomial logit model (7.2) using baseline-category logits is a special case.

7.6.3 Shopping Choice Example

McFadden (1974) used multinomial logit models to describe how residents of Pittsburgh, Pennsylvania chose a shopping destination. The five possible destinations were different city zones. One explanatory variable measured shopping opportunities, defined to be the retail employment in the zone as a percentage of total retail employment in the region. The other explanatory variable was price of the trip, defined from a separate analysis using auto in-vehicle time and auto operating cost.

The ML estimates of model parameters were -1.06 (SE $= 0.28$) for price of trip and 0.84 (SE $= 0.23$) for shopping opportunity. From (7.23),

$$
\log(\hat{\pi}_a / \hat{\pi}_b) = -1.06(P_a - P_b) + 0.84(S_a - S_b),
$$

where $P =$ price and $S =$ shopping opportunity. Not surprisingly, a destination is relatively more attractive as the trip price decreases and as the shopping opportunity increases. Given values of P and S for each destination, the sample analog of (7.22) provides estimated probabilities of choosing each destination.

NOTES

Section 7.1: Nominal Responses: Baseline-Category Logit Models

7.1. Multicategory models derive from latent variable constructions that generalize those for binary responses. One approach uses the principle of selecting the category having maximum utility (Problem 6.29). Fahrmeir and Tutz (2001, Chap. 3) gave discussion and references. Baseline-category logit models were developed in Bock (1970), Haberman (1974a, pp. 352–373), Mantel (1966), Nerlove and Press (1973), and Theil (1969, 1970). Lesaffre and Albert (1989) presented regression diagnostics. Amemiya (1981), Haberman (1982), and Theil (1970) presented R-squared measures.

Section 7.2: Ordinal Responses: Cumulative Logit Models

7.2. Early uses of cumulative logit models include Bock and Jones (1968), Simon (1974), Snell (1964), Walker and Duncan (1967), and Williams and Grizzle (1972). McCullagh (1980) popularized the proportional odds case. Later articles include Agresti and Lang (1993a), Hastie and Tibshirani (1987), Peterson and Harrell (1990), and Tutz (1989). See also Section 11.3.3, Note 11.3, and Section 12.4.1. McCullagh and Nelder (1989, Sec. 5.6) suggested using cumulative totals in forming residuals.

7.3. McCullagh (1980) noted that score tests for model (7.5) are equivalent to nonparametric tests using average ranks. For instance, for $2 \times J$ tables assume that $\text{logit}[P(Y \leq j)] = \alpha_j + \beta x$, with x an indicator. The score test of H_0: $\beta = 0$ is equivalent to a discrete version of the Wilcoxon–Mann–Whitney test. Whitehead (1993) gave sample size formulas for this case. The sample size n_J needed for a certain power decreases as J increases: When response categories have equal probabilities, $n_J \approx 0.75 n_2/(1 - 1/J^2)$. Thus, for large J, $n_J \approx 0.75 n_2$, and $1 - 1/J^2$ is a type of efficiency measure of using J categories instead of a continuous response. The efficiency loss is minor with $J \approx 5$, but major in collapsing to $J = 2$. Edwardes (1997) innovatively adapted the test by treating the cutpoints as random. This relates to random effects models of Section 12.4.1.

Section 7.3: Ordinal Responses: Cumulative Link Models

7.4. Aitchison and Silvey (1957) and Bock and Jones (1968, Chap. 8) studied cumulative probit models. Farewell (1982) generalized the complementary log-log model to allow variation among the sample in the category boundaries for the underlying scale; this relates to random effects models (Section 12.4). Genter and Farewell (1985) introduced a generalized link function that permits comparison of fits provided by probit, complementary log-log, and other links. Yee and Wild (1996) defined generalized additive models for nominal and ordinal responses. Hamada and Wu (1990) and Nair (1987) presented alternatives to model (7.8) for detecting dispersion effects.

7.5. Some authors have considered inference relating generally to stochastic ordering; see, for instance, Dardanoni and Forcina (1998) and survey articles in a 2002 issue of *J. Statist. Plann. Inference* (Vol. 107, Nos. 1–2).

Section 7.4: Alternative Models for Ordinal Responses

7.6. The ratio of a pdf to the complement of the cdf is the *hazard function* (Section 9.7.3). For discrete variables, this is the ratio found in continuation-ratio logits. Hence, continuation-ratio logits are sometimes interpreted as log hazards. Thompson (1977) used them in modeling discrete survival-time data. When lengths of time intervals approach 0, his model converges to the Cox proportional hazards model. Other applications of continuation-ratio logits include Läärä and Matthews (1985) and Tutz (1991).

Section 7.5: Testing Conditional Independence in $I \times J \times K$ Tables

7.7. Let $\mathbf{B}_k = \mathbf{u}_k \otimes \mathbf{v}_k$ denote a matrix of constants based on row scores $\mathbf{u}_k$ and column scores $\mathbf{v}_k$ for stratum k, where $\otimes$ denotes the Kronecker product. The Landis et al.

(1978) generalized statistic is

$$L^2 = \left[\sum_k \mathbf{B}_k(\mathbf{n}_k - \boldsymbol{\mu}_k) \right]' \left[\sum_k \mathbf{B}_k \mathbf{V}_k \mathbf{B}'_k \right]^{-1} \left[\sum_k \mathbf{B}_k(\mathbf{n}_k - \boldsymbol{\mu}_k) \right].$$

When $\mathbf{u}_k = (u_1, \ldots, u_I)$ and $\mathbf{v}_k = (v_1, \ldots, v_J)$ for all strata, $L^2 = M^2$. When $\mathbf{u}_k$ is an $(I - 1) \times I$ matrix $(\mathbf{I}, -\mathbf{1})$, where $\mathbf{I}$ is an identity matrix of size $(I - 1)$ and $\mathbf{1}$ denotes a column vector of $I - 1$ ones, and $\mathbf{v}_k$ is the analogous matrix of size $(J - 1) \times J$, L^2 simplifies to (7.20) with df $= (I - 1)(J - 1)$. With this $\mathbf{u}_k$ and $\mathbf{v}_k = (v_1, \ldots, v_J)$, L^2 sums over the strata information about how I row means compare to their null expected values, and it has df $= I - 1$. Rank score versions are analogs for ordered categorical responses of strata-adjusted Spearman correlation and Kruskal–Wallis tests. Landis et al. (1998) and Stokes et al. (2000) reviewed CMH methods. Koch et al. (1982) reviewed related methods.

Section 7.6: Discrete-Choice Multinomial Logit Models

7.8. McFadden's model relates to models proposed by Bradley and Terry (1952) (see Section 10.6) and Luce (1959). See Train (1986) for a text treatment. McFadden (1982) discussed hierarchical models having a nesting of choices in a tree-like structure. For other discussion, see Maddala (1983) and Small (1987). Models that do not assume independence from irrelevant alternatives result with probit link (Amemiya 1981) or with the logit link but including random effects (Brownstone and Train 1999). Methods in Section 12.6 for random effects models are useful for fitting such models. These include Monte Carlo methods for approximating integrals that determine the likelihood function. See Stern (1997) for a review.

PROBLEMS

Applications

7.1 For Table 7.13, let $Y =$ belief in life after death, $x_1 =$ gender (1 = females, 0 = males), and $x_2 =$ race (1 = whites, 0 = blacks). Table 7.14 shows the fit of the model

$$\log(\pi_j / \pi_3) = \alpha_j + \beta_j^G x_1 + \beta_j^R x_2, \qquad j = 1, 2,$$

with SE values in parentheses.

TABLE 7.13   Data for Problem 7.1

		\multicolumn{3}{c}{Belief in Afterlife}		
Race	Gender	Yes	Undecided	No
White	Female	371	49	74
	Male	250	45	71
Black	Female	64	9	15
	Male	25	5	13

Source: 1991 General Social Survey, National Opinion Research Center.

TABLE 7.14   Fit of Model for Problem 7.1

	Belief Categories for Logit	
Parameter	Yes/No	Undecided/No
Intercept	0.883 (0.243)	−0.758 (0.361)
Gender	0.419 (0.171)	0.105 (0.246)
Race	0.342 (0.237)	0.271 (0.354)

a. Find the prediction equation for $\log(\pi_1/\pi_2)$.

b. Using the yes and no response categories, interpret the conditional gender effect using a 95% confidence interval for an odds ratio.

c. Show that for white females, $\hat{\pi}_1 = \hat{P}(Y = \text{yes}) = 0.76$.

d. Without calculating estimated probabilities, explain why the intercept estimates indicate that for black males $\hat{\pi}_1 > \hat{\pi}_3 > \hat{\pi}_2$. Use the intercept and gender estimates to show that the same ordering applies for black females.

e. Without calculating estimated probabilities, explain why the estimates in the gender and race rows indicate that $\hat{\pi}_3$ is highest for black males.

f. For this fit, $G^2 = 0.9$. Explain why residual df = 2. Deleting the gender effect, $G^2 = 8.0$. Test whether opinion is independent of gender, given race. Interpret.

7.2 A model fit predicting preference for U.S. President (Democrat, Republican, Independent) using $x =$ annual income (in \$10,000) is $\log(\hat{\pi}_D/\hat{\pi}_I) = 3.3 - 0.2x$ and $\log(\hat{\pi}_R/\hat{\pi}_I) = 1.0 + 0.3x$.

a. Find the prediction equation for $\log(\hat{\pi}_R/\hat{\pi}_D)$ and interpret the slope. For what range of x is $\hat{\pi}_R > \hat{\pi}_D$?

b. Find the prediction equation for $\hat{\pi}_I$.

c. Plot $\hat{\pi}_D$, $\hat{\pi}_I$, and $\hat{\pi}_R$ for x between 0 and 10, and interpret.

7.3 Table 7.15 refers to the effect on political party identification of gender and race. Find a baseline-category logit model that fits well.

TABLE 7.15   Data for Problem 7.3

		Party Identification		
Gender	Race	Democrat	Republican	Independent
Male	White	132	176	127
	Black	42	6	12
Female	White	172	129	130
	Black	56	4	15

Interpret estimated effects on the odds that party identification is Democrat instead of Republican.

TABLE 7.16 Data for Problem 7.4[a]

Males				Females			
Length (m)	Choice	Length (m)	Choice	Length (m)	Choice	Length (m)	Choice
1.30	I	1.80	F	1.24	I	2.56	O
1.32	F	1.85	F	1.30	I	2.67	F
1.32	F	1.93	I	1.45	I	2.72	I
1.40	F	1.93	F	1.45	O	2.79	F
1.42	I	1.98	I	1.55	I	2.84	F
1.42	F	2.03	F	1.60	I		
1.47	I	2.03	F	1.60	I		
1.47	F	2.31	F	1.65	F		
1.50	I	2.36	F	1.78	I		
1.52	I	2.46	F	1.78	O		
1.63	I	3.25	O	1.80	I		
1.65	O	3.28	O	1.88	I		
1.65	O	3.33	F	2.16	F		
1.65	I	3.56	F	2.26	F		
1.65	F	3.58	F	2.31	F		
1.68	F	3.66	F	2.36	F		
1.70	I	3.68	O	2.39	F		
1.73	O	3.71	F	2.41	F		
1.78	F	3.89	F	2.44	F		
1.78	O						

[a]*I*, invertebrates; *F*, fish; *O*, other.

7.4 For 63 alligators caught in Lake George, Florida, Table 7.16 classifies primary food choice as (fish, invertebrate, other) and shows length in meters. Alligators are called subadults if length < 1.83 meters (6 feet) and adults if length > 1.83 meters.

a. Measuring length as (adult, subadult), find a model that adequately describes effects of gender and length on food choice. Interpret the effects. For adult females, find the estimated probabilities of the food-choice categories.

b. Using only observations for which primary food choice was fish or invertebrate, find a model that adequately describes effects of gender and binary length. Compare parameter estimates and standard errors for this separate-fitting approach to those obtained with simultaneous fitting, including the other category.

c. Treating length as binary loses information. Adapt the model in part (a) to use the continuous measurements. Interpret, explaining how the estimated outcome probabilities vary with length. Find the estimated length at which the invertebrate and other categories are equally likely.

7.5 For recent data from a General Social Survey, the cumulative logit model (7.5) with Y = political ideology (very liberal, slightly liberal, moderate, slightly conservative, very conservative) and $x = 1$ for the 428 Democrats and $x = 0$ for the 407 Republicans has $\hat{\beta} = 0.975$ (SE = 0.129) and $\hat{\alpha}_1 = -2.469$. Interpret $\hat{\beta}$. Find the estimated probability of a very liberal response for each group.

7.6 Refer to Problem 7.5. With adjacent-categories logits, $\hat{\beta} = 0.435$. Interpret using odds ratios for adjacent categories and for the (very liberal, very conservative) pair of categories.

7.7 Table 7.17 is an expanded version of a data set analyzed in Section 8.4.2. The response categories are (1) not injured, (2) injured but not transported by emergency medical services, (3) injured and transported by emergency medical services but not hospitalized, (4) injured and hospitalized but did not die, and (5) injured and died. Table 7.18 shows output for a model of form (7.5), using dummy variables for predictors.

  a. Why are there four intercepts? Explain how they determine the estimated response distribution for males in urban areas wearing seat belts.
  b. Construct a confidence interval for the effect of gender, given seat-belt use and location. Interpret.
  c. Find the estimated cumulative odds ratio between the response and seat-belt use for those in rural locations and for those in urban locations, given gender. Based on this, explain how the effect of seat-belt use varies by region, and explain how to interpret the interaction estimate, -0.1244.

TABLE 7.17   Data for Problem 7.7

Gender	Location	Seat Belt	Response 1	2	3	4	5
Female	Urban	No	7,287	175	720	91	10
		Yes	11,587	126	577	48	8
	Rural	No	3,246	73	710	159	31
		Yes	6,134	94	564	82	17
Male	Urban	No	10,381	136	566	96	14
		Yes	10,969	83	259	37	1
	Rural	No	6,123	141	710	188	45
		Yes	6,693	74	353	74	12

Source: Data courtesy of Cristanna Cook, Medical Care Development, Augusta, Maine.

TABLE 7.18 Output for Problem 7.7

Parameter			DF	Estimate	Std Error
Intercept1			1	3.3074	0.0351
Intercept2			1	3.4818	0.0355
Intercept3			1	5.3494	0.0470
Intercept4			1	7.2563	0.0914
gender	female		1	-0.5463	0.0272
gender	male		0	0.0000	0.0000
location	rural		1	-0.6988	0.0424
location	urban		0	0.0000	0.0000
seatbelt	no		1	-0.7602	0.0393
seatbelt	yes		0	0.0000	0.0000
location*seatbelt	rural	no	1	-0.1244	0.0548
location*seatbelt	rural	yes	0	0.0000	0.0000
location*seatbelt	urban	no	0	0.0000	0.0000
location*seatbelt	urban	yes	0	0.0000	0.0000

7.8 Refer to the cumulative logit model for Table 7.8.
  a. Compare the estimated income effect $\hat{\beta}_1 = -0.510$ to the estimate after collapsing the response to three categories by combining categories **(i)** very satisfied and moderately satisfied, and **(ii)** very dissatisfied and a little satisfied. What property of the model does this reflect?
  b. Consider $\hat{\beta}_1/SE$ using the full scale to $\hat{\beta}_1/SE$ for the collapsing in part (a(i)). Usually, a disadvantage of collapsing multinomial responses is that the significance of effects diminishes.
  c. Check whether an improved model results from permitting interaction between income and gender. Interpret.

7.9 Table 7.19 refers to a clinical trial for the treatment of small-cell lung cancer. Patients were randomly assigned to two treatment groups. The sequential therapy administered the same combination of chemotherapeutic agents in each treatment cycle; the alternating therapy had three different combinations, alternating from cycle to cycle.

TABLE 7.19 Data for Problem 7.9

		Response to Chemotherapy			
Therapy	Gender	Progressive Disease	No Change	Partial Remission	Complete Remission
Sequential	Male	28	45	29	26
	Female	4	12	5	2
Alternating	Male	41	44	20	20
	Female	12	7	3	1

Source: W. Holtbrugge and M. Schumacher, *Appl. Statist.* **40**: 249–259 (1991).

a. Fit a cumulative logit model with main effects for treatment and gender. Interpret.

b. Fit the model that also contains an interaction term. Interpret. Does it fit better? Explain why it is equivalent to using the four gender–treatment combinations as levels of a single factor.

7.10 Refer to Table 7.13. Treating belief in an afterlife as ordinal, fit and interpret an ordinal model.

7.11 Table 9.7 displays associations among smoking status (S), breathing test results (B), and age (A) for workers in certain industrial plants. Treat B as a response.

a. Specify a baseline-category logit model with additive factor effects of S and A. This model has deviance $G^2 = 25.9$. Show that df = 4, and explain why this model treats all variables as nominal.

b. Treat B as ordinal and S as ordinal in terms of how recently one was a smoker, with scores $\{s_i\}$. Consider the model

$$\log\frac{P(B = k + 1 \mid S = i, A = j)}{P(B = k \mid S = i, A = j)} = \alpha_k + \beta_1 s_i + \beta_2 a_j + \beta_3 s_i a_j$$

with $a_1 = 0$ and $a_2 = 1$. Show that this assumes a linear effect of S with slope β_1 for age < 40 and $\beta_1 + \beta_3$ for age 40–59. Using $\{s_i = i\}$, $\hat{\beta}_1 = 0.115$, $\hat{\beta}_2 = 0.311$, and $\hat{\beta}_3 = 0.663$ (SE = 0.164). Interpret the interaction.

c. From part (b), for age 40–59 show that the estimated odds of abnormal rather than borderline breathing for current smokers are 2.18 times those for former smokers and $\exp(2 \times 0.778) = 4.74$ times those for never smokers. Explain why the squares of these values are estimated odds of abnormal rather than normal breathing.

7.12 The book's Web site (*www.stat.ufl.edu/~aa/cda/cda.html*) has a 7×2 table that refers to subjects who graduated from high school in 1965. They were classified as protestors if they took part in at least one demonstration, protest march, or sit-in, and classified according to their party identification in 1982. Analyze the data, using response **(a)** party identification, **(b)** whether a protestor. Compare interpretations.

7.13 For Table 7.5, the cumulative probit model has fit $\Phi^{-1}[\hat{P}(Y \leq j)] = \hat{\alpha}_j - 0.195x_1 + 0.683x_2$, with $\hat{\alpha}_1 = -0.161$, $\hat{\alpha}_2 = 0.746$, and $\hat{\alpha}_3 = 1.339$. Find the means and standard deviation for the two normal cdf's that provide the curves for $\hat{P}(Y > 2)$ as a function of $x_1 = $ life events index, at the two levels of $x_2 = $ SES. Interpret effects.

7.14 Analyze Table 7.8 with a cumulative probit model. Compare interpretations to those in the text with other ordinal models.

7.15 Fit a model with complementary log-log link to Table 7.20, which shows family income distributions by percent for families in the northeast U.S. Interpret the difference between the income distributions.

TABLE 7.20 Data for Problem 7.15

Year	\multicolumn{7}{c}{Income ($1000)}
	0–3	3–5	5–7	7–10	10–12	12–15	15 +
1960	6.5	8.2	11.3	23.5	15.6	12.7	22.2
1970	4.3	6.0	7.7	13.2	10.5	16.3	42.1

Source: Reproduced with permission from the Royal Statistical Society, London (McCullagh 1980).

7.16 Table 7.21 shows results of fitting the mean response model to Table 7.8 using scores $\{3, 10, 20, 35\}$ for income and $\{1, 3, 4, 5\}$ for job satisfaction. Interpret the income effect, provide a confidence interval for the difference in mean satisfaction at income levels 35 and 3, controlling for gender, and check the model fit.

TABLE 7.21 Results for Problem 7.16

Source	DF	Chi-Square	Pr > ChiSq
Residual	5	6.99	0.2211

Analysis of Weighted Least Squares Estimates

Effect	Parameter	Estimate	Std Error	Chi-Square	Pr > ChiSq
Intercept	1	3.8076	0.1796	449.47	<.0001
gender	2	−0.0687	0.1419	0.23	0.6283
income	3	0.0160	0.0066	5.97	0.0146

7.17 The book's Web site (*www.stat.ufl.edu/ ~aa/cda/cda.html*) has a 3 × 4 × 4 table that cross-classifies dumping severity (Y) and operation (X) for four hospitals (H). The four operations refer to treatments for duodenal ulcer patients and have a natural ordering. Dumping severity describes a possible undesirable side effect of the operation. Its three categories are also ordered.

    a. Table 7.22 shows results of generalized CMH tests. Interpret, explaining how one test can be much more significant than the others.

TABLE 7.22   Results for Problem 7.17

	Summary Statistics for dumping by operate Controlling for hospital			
Statistic	Alternative Hypothesis	DF	Value	Prob
1	Nonzero Correlation	1	6.3404	0.0118
2	Row Mean Scores Differ	3	6.5901	0.0862
3	General Association	6	10.5983	0.1016

b. Let $\{x_i = i\}$. Fit the model

$$\text{logit}\left[P(Y \le j \mid H = h, X = i) \right] = \alpha_j + \mu_h + \beta x_i.$$

Test conditional independence of X and Y using it, and interpret $\hat{\beta}$. Which generalized CMH test has the same spirit as this?

c. Does an improved fit result from allowing the operation effect to vary by hospital? Interpret.

d. Find a mean response model that fits well. Interpret.

7.18 Table 7.23 refers to a study that randomly assigned subjects to a control or treatment group. Daily during the study, treatment subjects ate cereal containing psyllium. The study analyzed the effect on LDL cholesterol.

a. Model the ending cholesterol level as a function of treatment, using the beginning level as a covariate. Interpret the treatment effect.

b. Repeat part (a), now treating the beginning level as qualitative. Compare results.

c. An alternative to part (b) uses a generalized CMH test relating treatment to the ending response for partial tables defined by beginning cholesterol level. Apply such a test, taking into account the response ordering, to compare treatments. Interpret, and compare to part (b).

TABLE 7.23   Data for Problem 7.18

	Ending LDL Cholesterol Level							
	Control				Treatment			
Beginning	≤ 3.4	3.4–4.1	4.1–4.9	> 4.9	3.4	3.4–4.1	4.1–4.9	> 4.9
≤ 3.4	18	8	0	0	21	4	2	0
3.4–4.1	16	30	13	2	17	25	6	0
4.1–4.9	0	14	28	7	11	35	36	6
> 4.9	0	2	15	22	1	5	14	12

Source: Data courtesy of Sallee Anderson, Kellogg Co.

7.19   Analyze Table 7.5 with each type of model studied in this chapter. Write a report summarizing results and advantages and disadvantages of each modeling strategy.

7.20   The book's Web site (*www.stat.ufl.edu/~aa/cda/cda.html*) has a $4 \times 4 \times 5$ table that cross-classifies assessment of cognitive impairment, Alzheimer's disease, and age. Analyze these data, treating **(a)** Alzheimer's disease, and **(b)** cognitive impairment, as the response variable.

7.21   Analyze Table 9.5 using logit models that treat **(a)** party affiliation, and **(b)** ideology, as the response variable.

7.22   The book's Web site (*www.stat.ufl.edu/~aa/cda/cda.html*) has a $4 \times 2 \times 3 \times 3$ table that refers to a sample of residents of Copenhagen. The variables are type of housing (H), degree of contact with other residents (C), feeling of influence on apartment management (I), and satisfaction with housing conditions (S). Treating S as the response variable, analyze these data.

7.23   Refer to Table 7.17. Analyze these data.

Theory and Methods

7.24   A multivariate generalization of the exponential dispersion family (4.14) is

$$f(\mathbf{y}_i; \boldsymbol{\theta}_i, \phi) = \exp\{[\mathbf{y}_i'\boldsymbol{\theta}_i - b(\boldsymbol{\theta}_i)]/a(\phi) + c(\mathbf{y}_i, \phi)\},$$

where $\boldsymbol{\theta}_i$ is the natural parameter. Show that the multinomial variate $\mathbf{y}_i$ defined in Section 7.1.5 for a single trial with parameters $\{\pi_j, j = 1, \ldots, J - 1\}$ is in the $(J - 1)$-parameter exponential family, with baseline-category logits as natural parameters.

7.25   Cell counts $\{y_{ij}\}$ in an $I \times J$ contingency table have a multinomial $(n; \{\pi_{ij}\})$ distribution. Show that $\{P(Y_{ij} = n_{ij}), i = 1, \ldots, I, j = 1, \ldots, J\}$ can be expressed as

$$d^n \, n! \prod_i \prod_j (n_{ij}!)^{-1} \exp\left[\sum_{i=1}^{I-1} \sum_{j=1}^{J-1} n_{ij}\log(\alpha_{ij}) \right.$$

$$\left. + \sum_{i=1}^{I-1} n_{i+}\log(\pi_{iJ}/\pi_{IJ}) + \sum_{j=1}^{J-1} n_{+j}\log(\pi_{Ij}/\pi_{IJ}) \right]$$

where $\alpha_{ij} = \pi_{ij}\pi_{IJ}/\pi_{iJ}\pi_{Ij}$ and d is a constant independent of the data. Find an alternative expression using local odds ratios $\{\theta_{ij}\}$, by showing that

$$\sum_i \sum_j n_{ij}\log\alpha_{ij} = \sum_i \sum_j s_{ij}\log\theta_{ij}, \quad \text{where} \quad s_{ij} = \sum_{a\le i}\sum_{b\le j} n_{ab}.$$

7.26 Suppose that we express (7.2) as

$$\pi_j(\mathbf{x}) = \frac{\exp(\alpha_j + \boldsymbol{\beta}'_j\mathbf{x})}{\sum_{h=1}^J \exp(\alpha_h + \boldsymbol{\beta}'_h\mathbf{x})}.$$

Show that dividing numerator and denominator by $\exp(\alpha_J + \boldsymbol{\beta}'_J\mathbf{x})$ yields new parameters $\alpha_j^* = \alpha_j - \alpha_J$ and $\beta_j^* = \beta_j - \beta_J$ that satisfy $\alpha_J = 0$ and $\boldsymbol{\beta}_J = \mathbf{0}$. Thus, without loss of generality, $\alpha_J = 0$ and $\boldsymbol{\beta}_J = \mathbf{0}$.

7.27 When $J = 3$, suppose that

$$\pi_j(x) = \exp(\alpha_j + \beta_j x)/[1 + \exp(\alpha_1 + \beta_1 x) + \exp(\alpha_2 + \beta_2 x)],$$

$j = 1, 2$. Show that $\pi_3(x)$ is (**a**) decreasing in x if $\beta_1 > 0$ and $\beta_2 > 0$, (**b**) increasing in x if $\beta_1 < 0$ and $\beta_2 < 0$, and (**c**) nonmonotone when β_1 and β_2 have different signs.

7.28 Refer to the log-likelihood function for the baseline-category logit model (Section 7.1.4). Denote the sufficient statistics by $np_j = \sum_i y_{ij}$ and $S_{jk} = \sum_i x_{ik}y_{ij}$, $j = 1, \ldots J - 1$, $k = 1, \ldots, p$. Let $\mathbf{S} = (S_{11}, \ldots, S_{1t}, \ldots S_{J1}, \ldots, S_{Jt})'$. Condition on $\sum_i y_{ij}$, $j = 1, \ldots, J$. Under the null hypothesis that explanatory variables have no effect, show that

$$E(\mathbf{S}) = n(\mathbf{p} \otimes \mathbf{m}), \qquad \text{var}(\mathbf{S}) = n(\mathbf{V} \otimes \boldsymbol{\Sigma}),$$

where $\mathbf{p} = (p_1, \ldots, p_J)'$; $\mathbf{m} = (\bar{x}_1, \ldots, \bar{x}_t)'$, where $\bar{x}_k = (\sum_i x_{ik})/n$; $\boldsymbol{\Sigma}$ has elements (s_{kv}^2), where $s_{kv}^2 = [\sum_i(x_{ik} - \bar{x}_k)(x_{iv} - \bar{x}_v)]/(n - 1)$; $\mathbf{V}$ has elements $v_{ii} = p_i(1 - p_i)$ and $v_{ij} = -p_ip_j$, and $\otimes$ denotes the Kronecker product (Zelen 1991).

7.29 Is the proportional odds model a special case of a baseline-category logit model? Explain why or why not.

7.30 Prove factorization (7.15) for the multinomial distribution.

7.31   Show that for the model, logit$[P(Y \leq j)] = \alpha_j + \beta_j x$, cumulative probabilities may be misordered for some x values.

7.32   For an $I \times J$ contingency table with ordinal Y and scores $\{x_i = i\}$ for x, consider the model

$$\text{logit}\big[P(Y \leq j \mid X = x_i)\big] = \alpha_j + \beta x_i. \qquad (7.24)$$

a. Show that logit$[P(Y \leq j \mid X = x_{i+1})] - \text{logit}[P(Y \leq j \mid X = x_i)] = \beta$. Show that this difference in logits is a log cumulative odds ratio for the 2×2 table consisting of rows i and $i + 1$ and the binary response having cutpoint following category j. Thus, (7.24) is a *uniform association model* in cumulative odds ratios.

b. Show that residual df $= IJ - I - J$.

c. Show that independence of X and Y is the special case $\beta = 0$.

d. Using the same linear predictor but with adjacent-categories logits, show that uniform association applies to the local odds ratios (2.10).

e. A generalization of (7.24) replaces $\{\beta x_i\}$ by unordered parameters $\{\mu_i\}$, hence treating X as nominal. For rows a and b, show that the log cumulative odds ratio equals $\mu_a - \mu_b$ for all $J - 1$ cutpoints.

7.33   Suppose that model (7.24) holds for a $2 \times J$ table with $J > 2$, and let $x_2 - x_1 = 1$. Explain why local log odds ratios are typically smaller in absolute value than the cumulative log odds ratio β. [In fact, on p. 122 of their first edition, McCullagh and Nelder (1989) noted that local odds ratios $\{\theta_{1j}\}$ relate to β by

$$\log \theta_{1j} = \beta\big[P(Y \leq j + 1) - P(Y \leq j - 1)\big] + o(\beta), \quad j = 1, \ldots, J - 1,$$

where $o(\beta)/\beta \rightarrow 0$ as $\beta \rightarrow 0$.]

7.34   A response scale has the categories (strongly agree, mildly agree, mildly disagree, strongly disagree, don't know). One way to model such a scale uses a logit model for the probability of a don't know response and uses a separate ordinal model for the ordered categories conditional on response in one of those categories. Explain how to construct a likelihood to do this simultaneously.

7.35   For the cumulative probit model $\Phi^{-1}[P(Y \leq j)] = \alpha_j - \boldsymbol{\beta}' \mathbf{x}$, explain why a 1-unit increase in x_i corresponds to a β_i standard deviation increase in the expected underlying latent response, controlling for other predictors.

7.36 For cumulative link model (7.7), show that for $1 \leq j < k \leq J - 1$, $P(Y \leq k \mid \mathbf{x}) = P(Y \leq j \mid \mathbf{x}^*)$, where $\mathbf{x}^*$ is obtained by increasing the ith component of $\mathbf{x}$ by $(\alpha_k - \alpha_j)/\beta_i$. Interpret.

7.37 A cumulative link model for an $I \times J$ contingency table with a qualitative predictor is

$$G^{-1}[P(Y \leq j)] = \alpha_j + \mu_i, \quad i = 1, \ldots, I, \, j = 1, \ldots, J - 1.$$

  a. Show that the residual df $= (I - 1)(J - 2)$.
  b. When this model holds, show that independence corresponds to $\mu_1 = \cdots = \mu_I$ and the test of independence has df $= I - 1$.
  c. When this model holds, show that the rows are stochastically ordered on Y.

7.38 $F_1(y) = 1 - \exp(-\lambda y)$ for $y > 0$ is a negative exponential cdf with parameter λ, and $F_2(y) = 1 - \exp(-\mu y)$ for $y > 0$. Show that the difference between the cdf's on a complementary log-log scale is identical for all y. Give implications for categorical data analysis.

7.39 Consider the model Link$[\omega_j(\mathbf{x})] = \alpha_j + \boldsymbol{\beta}_j'\mathbf{x}$, where $\omega_j(\mathbf{x})$ is (7.14).
  a. Explain why this model can be fitted separately for $j = 1, \ldots, J - 1$.
  b. For the complementary log-log link, show that this model is equivalent to one using the same link for cumulative probabilities (Läärä and Matthews 1985).

7.40 Why is it not optimal to fit mean response models for ordinal responses using ordinary least squares as is done for normal regression?

7.41 When X and Y are ordinal, explain how to test conditional independence by allowing a different trend in each partial table. [*Hint*: Generalize model (7.17) by replacing β by β_k.]

7.42 A cafe has four entrées: chicken, beef, fish, vegetarian. Specify a model of form (7.22) for the selection of an entrée using $x =$ gender ($1 =$ female, $0 =$ male) and $u =$ cost of entrée, which is a characteristic of the choices. Interpret the model parameters.

CHAPTER 8

Loglinear Models for Contingency Tables

In Section 4.3 we introduced loglinear models as generalized linear models (GLMs) using the log link function with a Poisson response. A common use is modeling cell counts in contingency tables. The models specify how the expected count depends on levels of the categorical variables for that cell as well as associations and interactions among those variables. The purpose of loglinear modeling is the analysis of association and interaction patterns.

In Section 8.1 we introduce loglinear models for two-way contingency tables. In Sections 8.2 and 8.3 we extend them to three-way tables, and in Section 8.4 discuss models for multiway tables. Loglinear models are of use primarily when at least two variables are response variables. With a single categorical response, it is simpler and more natural to use logit models. When one variable is treated as a response and the others as explanatory variables, logit models for that response variable are equivalent to certain loglinear models. Section 8.5 covers this connection. In Sections 8.6 and 8.7 we discuss ML loglinear model fitting.

8.1 LOGLINEAR MODELS FOR TWO-WAY TABLES

Consider an $I \times J$ contingency table that cross-classifies a multinomial sample of n subjects on two categorical responses. The cell probabilities are $\{\pi_{ij}\}$ and the expected frequencies are $\{\mu_{ij} = n\pi_{ij}\}$. Loglinear model formulas use $\{\mu_{ij}\}$ rather than $\{\pi_{ij}\}$, so they also apply with Poisson sampling for $N = IJ$ independent cell counts $\{Y_{ij}\}$ having $\{\mu_{ij} = E(Y_{ij})\}$. In either case we denote the observed cell counts by $\{n_{ij}\}$.

8.1.1 Independence Model

Under statistical independence, in Section 4.3.6 we noted that the $\{\mu_{ij}\}$ have the structure

$$\mu_{ij} = \mu \alpha_i \beta_j.$$

For multinomial sampling, for instance, $\mu_{ij} = n\pi_{i+}\pi_{+j}$. Denote the row variable by X and the column variable by Y. The formula expressing independence is multiplicative. Thus, $\log \mu_{ij}$ has additive form

$$\log \mu_{ij} = \lambda + \lambda_i^X + \lambda_j^Y \tag{8.1}$$

for a row effect λ_i^X and a column effect λ_j^Y. This is the *loglinear model of independence*. As usual, identifiability requires constraints such as $\lambda_I^X = \lambda_J^Y = 0$.

The ML fitted values are $\{\hat{\mu}_{ij} = n_{i+}n_{+j}/n\}$, the estimated expected frequencies for chi-squared tests of independence. The tests using X^2 and G^2 (Section 3.2.1) are also goodness-of-fit tests of this loglinear model.

8.1.2 Interpretation of Parameters

Loglinear models for contingency tables are GLMs that treat the N cell counts as independent observations of a Poisson random component. Loglinear GLMs identify the data as the N cell counts rather than the individual classifications of the n subjects. The expected cell counts link to the explanatory terms using the log link. As (8.1) illustrates, of the cross-classified variables, the model does not distinguish between response and explanatory variables. It treats both jointly as responses, modeling $\{\mu_{ij}\}$ for combinations of their levels. To interpret parameters, however, it is helpful to treat the variables asymmetrically.

We illustrate with the independence model for $I \times 2$ tables. In row i, the logit equals

$$\text{logit}[P(Y = 1 | X = i)] = \log \frac{P(Y = 1 | X = i)}{P(Y = 2 | X = i)}$$

$$= \log \frac{\mu_{i1}}{\mu_{i2}} = \log \mu_{i1} - \log \mu_{i2}$$

$$= \left(\lambda + \lambda_i^X + \lambda_1^Y\right) - \left(\lambda + \lambda_i^X + \lambda_2^Y\right) = \lambda_1^Y - \lambda_2^Y.$$

The final term does not depend on i; that is, $\text{logit}[P(Y = 1 | X = i)]$ is identical at each level of X. Thus, independence implies a model of form, $\text{logit}[P(Y = 1 | X = i)] = \alpha$. In each row, the odds of response in column 1 equal $\exp(\alpha) = \exp(\lambda_1^Y - \lambda_2^Y)$.

An analogous property holds when $J > 2$. Differences between two parameters for a given variable relate to the log odds of making one response, relative to the other, on that variable. Of course, with a single response variable, logit models apply directly and loglinear models are unneeded.

8.1.3 Saturated Model

Statistically dependent variables satisfy a more complex loglinear model,

$$\log \mu_{ij} = \lambda + \lambda_i^X + \lambda_j^Y + \lambda_{ij}^{XY}. \tag{8.2}$$

The $\{\lambda_{ij}^{XY}\}$ are association terms that reflect deviations from independence. The right-hand side of (8.2) resembles the formula for cell means in two-way ANOVA, allowing interaction. The $\{\lambda_{ij}^{XY}\}$ represent interactions between X and Y, whereby the effect of one variable on μ_{ij} depends on the level of the other. The independence model (8.1) results when all $\lambda_{ij}^{XY} = 0$.

With constraints $\lambda_I^X = \lambda_J^Y = 0$ in (8.1) and (8.2), $\{\lambda_i^X\}$ and $\{\lambda_j^Y\}$ are, equivalently, coefficients of dummy variables for the first $(I-1)$ categories of X and the first $(J-1)$ categories of Y. Thus, λ_{ij}^{XY} is the coefficient of the product of dummy variables for λ_i^X and λ_j^Y. Since there are $(I-1)(J-1)$ such cross products, $\lambda_{Ij}^{XY} = \lambda_{iJ}^{XY} = 0$, and only $(I-1)(J-1)$ of these parameters are nonredundant. Tests of independence analyze whether these $(I-1)(J-1)$ parameters equal zero, so they have residual df $= (I-1)(J-1)$.

The number of parameters in model (8.2) equals $1 + (I-1) + (J-1) + (I-1)(J-1) = IJ$, the number of cells. Hence, this model describes perfectly any $\{\mu_{ij} > 0\}$ (see Problem 8.16). It is the most general model for two-way contingency tables, the *saturated model*. For it, direct relationships exist between log odds ratios and $\{\lambda_{ij}^{XY}\}$. For instance, for 2×2 tables,

$$\log \theta = \log \frac{\mu_{11} \mu_{22}}{\mu_{12} \mu_{21}} = \log \mu_{11} + \log \mu_{22} - \log \mu_{12} - \log \mu_{21}$$

$$= \left(\lambda + \lambda_1^X + \lambda_1^Y + \lambda_{11}^{XY} \right) + \left(\lambda + \lambda_2^X + \lambda_2^Y + \lambda_{22}^{XY} \right)$$

$$- \left(\lambda + \lambda_1^X + \lambda_2^Y + \lambda_{12}^{XY} \right) - \left(\lambda + \lambda_2^X + \lambda_1^Y + \lambda_{21}^{XY} \right)$$

$$= \lambda_{11}^{XY} + \lambda_{22}^{XY} - \lambda_{12}^{XY} - \lambda_{21}^{XY}. \tag{8.3}$$

Thus, $\{\lambda_{ij}^{XY}\}$ determine the association.

In practice, unsaturated models are preferable, since their fit smooths the sample data and has simpler interpretations. For tables with at least three variables, unsaturated models can include association terms. Then, loglinear models are more commonly used to describe associations (through two-factor terms) than to describe odds (through single-factor terms).

Like others in this book, model (8.2) is *hierarchical*. This means that the model includes all lower-order terms composed from variables contained in a higher-order model term. When the model contains λ_{ij}^{XY}, it also contains λ_i^X and λ_j^Y. A reason for including lower-order terms is that, otherwise, the statistical significance and the interpretation of a higher-order term depends on how variables are coded. This is undesirable, and with hierarchical models the same results occur no matter how variables are coded.

An example of a nonhierarchical model is

$$\log \mu_{ij} = \lambda + \lambda_i^X + \lambda_{ij}^{XY}.$$

This model permits association but forces unnatural behavior of expected frequencies, with the pattern depending on constraints used for parameters. For instance, with constraints whereby parameters are zero at the last level, $\log \mu_{Ij} = \lambda$ in every column. Nonhierarchical models are rarely sensible in practice. Using them is analogous to using ANOVA or regression models with interaction terms but without the corresponding main effects.

When a model has two-factor terms, interpretations focus on them rather than on the single-factor terms. By analogy with two-way ANOVA with two-factor interaction, it can be misleading to report main effects. The estimates of the main-effect terms depend on the coding scheme used for the higher-order effects, and the interpretation also depends on that scheme (see Problem 8.16). Normally, we restrict our attention to the highest-order terms for a variable, as we illustrate in Section 8.2.

8.1.4 Alternative Parameter Constraints

As with the independence model, the parameter constraints for the saturated model are arbitrary. Instead of setting all $\lambda_{Ij}^{XY} = \lambda_{iJ}^{XY} = 0$, one could set $\sum_i \lambda_{ij}^{XY} = \sum_j \lambda_{ij}^{XY} = 0$ for all i and j. Different software uses different constraints. What *is* unique are contrasts such as $\lambda_{11}^{XY} + \lambda_{22}^{XY} - \lambda_{12}^{XY} - \lambda_{21}^{XY}$ in (8.3) that determine odds ratios.

For instance, suppose that a log odds ratio equals 2.0 in a 2×2 table. With the first set of constraints, 2.0 is the coefficient of a product of a dummy variable indicating the first category of X and a dummy variable indicating the first category of Y. With it, $\lambda_{11}^{XY} = 2.0$ and $\lambda_{12}^{XY} = \lambda_{21}^{XY} = \lambda_{22}^{XY} = 0$. For sum-to-zero constraints, $\lambda_{11}^{XY} = \lambda_{22}^{XY} = 0.5$, $\lambda_{12}^{XY} = \lambda_{21}^{XY} = -0.5$. For either set, the log odds ratio (8.3) equals 2.0. For a set of parameters, an advantage of setting a baseline parameter equal to 0 instead of the sum equal to 0 is that some parameters in a set can have infinite estimates.

8.1.5 Multinomial Models for Cell Probabilities

Conditional on the sum n of the cell counts, Poisson loglinear models for $\{\mu_{ij}\}$ become multinomial models for cell probabilities $\{\pi_{ij} = \mu_{ij}/(\sum\sum \mu_{ab})\}$. To illustrate, for the saturated model,

$$\pi_{ij} = \frac{\exp\left(\lambda + \lambda_i^X + \lambda_j^Y + \lambda_{ij}^{XY}\right)}{\sum_a \sum_b \exp\left(\lambda + \lambda_a^X + \lambda_b^Y + \lambda_{ab}^{XY}\right)}. \tag{8.4}$$

This representation implies the usual constraints for probabilities, $\{\pi_{ij} \geq 0\}$ and $\Sigma_i \Sigma_j \pi_{ij} = 1$. The λ intercept parameter cancels in the multinomial model (8.4). This parameter relates purely to the total sample size, which is random in the Poisson model but not in the multinomial model.

8.2 LOGLINEAR MODELS FOR INDEPENDENCE AND INTERACTION IN THREE-WAY TABLES

In Section 2.3 we introduced three-way contingency tables and related structure such as conditional independence and homogeneous association. Loglinear models for three-way tables describe their independence and association patterns.

8.2.1 Types of Independence

A three-way $I \times J \times K$ cross-classification of response variables X, Y, and Z has several potential types of independence. We assume a multinomial distribution with cell probabilities $\{\pi_{ijk}\}$, and $\Sigma_i \Sigma_j \Sigma_k \pi_{ijk} = 1.0$. The models also apply to Poisson sampling with means $\{\mu_{ijk}\}$.

The three variables are *mutually independent* when

$$\pi_{ijk} = \pi_{i++} \pi_{+j+} \pi_{++k} \qquad \text{for all } i, j, \text{ and } k. \tag{8.5}$$

For expected frequencies $\{\mu_{ijk}\}$, mutual independence has loglinear form

$$\log \mu_{ijk} = \lambda + \lambda_i^X + \lambda_j^Y + \lambda_k^Z. \tag{8.6}$$

Variable Y is *jointly independent* of X and Z when

$$\pi_{ijk} = \pi_{i+k} \pi_{+j+} \qquad \text{for all } i, j, \text{ and } k. \tag{8.7}$$

This is ordinary two-way independence between Y and a variable composed of the IK combinations of levels of X and Z. The loglinear model is

$$\log \mu_{ijk} = \lambda + \lambda_i^X + \lambda_j^Y + \lambda_k^Z + \lambda_{ik}^{XZ}. \tag{8.8}$$

Similarly, X could be jointly independent of Y and Z, or Z could be jointly independent of X and Y. Mutual independence (8.5) implies joint independence of any one variable from the others.

From Section 2.3, X and Y are *conditionally independent, given* Z when independence holds for each partial table within which Z is fixed. That is, if $\pi_{ij|k} = P(X = i, Y = j | Z = k)$, then

$$\pi_{ij|k} = \pi_{i+|k} \pi_{+j|k} \qquad \text{for all } i, j, \text{ and } k.$$

For joint probabilities over the entire table, equivalently

$$\pi_{ijk} = \pi_{i+k}\pi_{+jk}/\pi_{++k} \qquad \text{for all } i, j, \text{ and } k. \tag{8.9}$$

Conditional independence of X and Y, given Z, is the loglinear model

$$\log \mu_{ijk} = \lambda + \lambda_i^X + \lambda_j^Y + \lambda_k^Z + \lambda_{ik}^{XZ} + \lambda_{jk}^{YZ}. \tag{8.10}$$

This is a weaker condition than mutual or joint independence. Mutual independence implies that Y is jointly independent of X and Z, which itself implies that X and Y are conditionally independent. Table 8.1 summarizes these three types of independence.

In Section 2.3.2 we showed that partial associations can be quite different from marginal associations. For instance, conditional independence does not imply marginal independence. Conditional independence and marginal independence both hold when one of the stronger types of independence studied above applies. Figure 8.1 summarizes relationships among the four types of independence.

8.2.2 Homogeneous Association and Three-Factor Interaction

Loglinear models (8.6), (8.8), and (8.10) have three, two, and one pair of conditionally independent variables, respectively. In the latter two models,

TABLE 8.1   Summary of Loglinear Independence Models

Model	Probabilistic Form for π_{ijk}	Association Terms in Loglinear Model	Interpretation
(8.6)	$\pi_{i++}\pi_{+j+}\pi_{++k}$	None	Variables mutually independent
(8.8)	$\pi_{i+k}\pi_{+j+}$	λ_{ik}^{XZ}	Y independent of X and Z
(8.10)	$\pi_{i+k}\pi_{+jk}/\pi_{++k}$	$\lambda_{ik}^{XZ} + \lambda_{jk}^{YZ}$	X and Y independent, given Z

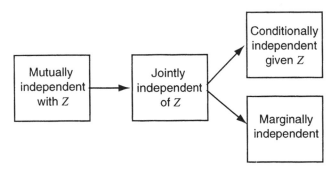

FIGURE 8.1 Relationships among types of XY independence.

the doubly subscripted terms (such as λ_{ij}^{XY}) pertain to conditionally dependent variables. A model that permits all three pairs to be conditionally dependent is

$$\log \mu_{ijk} = \lambda + \lambda_i^X + \lambda_j^Y + \lambda_k^Z + \lambda_{ij}^{XY} + \lambda_{ik}^{XZ} + \lambda_{jk}^{YZ}. \qquad (8.11)$$

From exponentiating both sides, the cell probabilities have form

$$\pi_{ijk} = \psi_{ij} \phi_{jk} \omega_{ik}.$$

No closed-form expression exists for the three components in terms of margins of $\{\pi_{ijk}\}$ except in certain special cases (see Note 9.2).

For this model, in the next section we show that conditional odds ratios between any two variables are identical at each category of the third variable. That is, each pair has *homogeneous association* (Section 2.3.5). Model (8.11) is called the loglinear model of *homogeneous association* or of *no three-factor interaction*.

The general loglinear model for a three-way table is

$$\log \mu_{ijk} = \lambda + \lambda_i^X + \lambda_j^Y + \lambda_k^Z + \lambda_{ij}^{XY} + \lambda_{ik}^{XZ} + \lambda_{jk}^{YZ} + \lambda_{ijk}^{XYZ}. \qquad (8.12)$$

With dummy variables, λ_{ijk}^{XYZ} is the coefficient of the product of the ith dummy variable for X, jth dummy variable for Y, and kth dummy variable for Z. The total number of nonredundant parameters is

$$1 + (I - 1) + (J - 1) + (K - 1) + (I - 1)(J - 1) + (I - 1)(K - 1)$$
$$+ (J - 1)(K - 1) + (I - 1)(J - 1)(K - 1) = IJK,$$

the total number of cell counts. This model has as many parameters as observations and is saturated. It describes all possible positive $\{\mu_{ijk}\}$. Each pair of variables may be conditionally dependent, and an odds ratio for any pair may vary across categories of the third variable.

Setting certain parameters equal to zero in (8.12) yields the models introduced previously. Table 8.2 lists some of these models. To ease referring to models, Table 8.2 assigns to each model a symbol that lists the highest-order

TABLE 8.2 Loglinear Models for Three-Dimensional Tables

Loglinear Model	Symbol
$\log \mu_{ijk} = \lambda + \lambda_i^X + \lambda_j^Y + \lambda_k^Z$	(X, Y, Z)
$\log \mu_{ijk} = \lambda + \lambda_i^X + \lambda_j^Y + \lambda_k^Z + \lambda_{ij}^{XY}$	(XY, Z)
$\log \mu_{ijk} = \lambda + \lambda_i^X + \lambda_j^Y + \lambda_k^Z + \lambda_{ij}^{XY} + \lambda_{jk}^{YZ}$	(XY, YZ)
$\log \mu_{ijk} = \lambda + \lambda_i^X + \lambda_j^Y + \lambda_k^Z + \lambda_{ij}^{XY} + \lambda_{jk}^{YZ} + \lambda_{ik}^{XZ}$	(XY, YZ, XZ)
$\log \mu_{ijk} = \lambda + \lambda_i^X + \lambda_j^Y + \lambda_k^Z + \lambda_{ij}^{XY} + \lambda_{jk}^{YZ} + \lambda_{ik}^{XZ} + \lambda_{ijk}^{XYZ}$	(XYZ)

term(s) for each variable. For instance, the model (8.10) of conditional independence between X and Y has symbol (XZ, YZ), since its highest-order terms are λ_{ik}^{XZ} and λ_{jk}^{YZ}. In the notation we used for logit models in Sections 6.1 and 7.1.2 this stands for $(X*Z + Y*Z)$, which is itself shorthand for notation $(X + Y + Z + X \times Z + Y \times Z)$ that has the main effects as well as interactions.

8.2.3 Interpreting Model Parameters

Interpretations of loglinear model parameters use their highest-order terms. For instance, interpretations for model (8.11) use the two-factor terms to describe conditional odds ratios. At a fixed level k of Z, the *conditional association* between X and Y uses $(I - 1)(J - 1)$ odds ratios, such as the local odds ratios

$$\theta_{ij(k)} = \frac{\pi_{ijk}\pi_{i+1,j+1,k}}{\pi_{i,j+1,k}\pi_{i+1,j,k}}, \qquad 1 \le i \le I - 1, \quad 1 \le j \le J - 1. \quad (8.13)$$

Similarly, $(I - 1)(K - 1)$ odds ratios $\{\theta_{i(j)k}\}$ describe XZ conditional association, and $(J - 1)(K - 1)$ odds ratios $\{\theta_{(i)jk}\}$ describe YZ conditional association. Loglinear models have characterizations using constraints on conditional odds ratios. For instance, conditional independence of X and Y is equivalent to $\{\theta_{ij(k)} = 1, i = 1, \ldots, I - 1, j = 1, \ldots, J - 1, k = 1, \ldots, K\}$.

The two-factor parameters relate directly to the conditional odds ratios. To illustrate, substituting (8.11) for model (XY, XZ, YZ) into $\log \theta_{ij(k)}$ yields

$$\log \theta_{ij(k)} = \log \frac{\mu_{ijk}\mu_{i+1,j+1,k}}{\mu_{i+1,jk}\mu_{1,j+1,k}} = \lambda_{ij}^{XY} + \lambda_{i+1,j+1}^{XY} - \lambda_{i,j+1}^{XY} - \lambda_{i+1,j}^{XY}. \quad (8.14)$$

Since the right-hand side is the same for all k, an absence of three-factor interaction is equivalent to

$$\theta_{ij(1)} = \theta_{ij(2)} = \cdots = \theta_{ij(K)} \quad \text{for all } i \text{ and } j.$$

The same argument for the other conditional odds ratios shows that model (XY, XZ, YZ) is also equivalent to

$$\theta_{i(1)k} = \theta_{i(2)k} = \cdots = \theta_{i(J)k} \quad \text{for all } i \text{ and } k,$$

and to

$$\theta_{(1)jk} = \theta_{(2)jk} = \cdots = \theta_{(I)jk} \quad \text{for all } j \text{ and } k.$$

Any model not having the three-factor interaction term has a homogeneous association for each pair of variables.

When X and Y have two categories, only one nonredundant λ_{ij}^{XY} parameter occurs. Thus, expression (8.14) is simplified depending on the constraints. By the same argument as in Section 8.1.3 for 2×2 tables, the conditional log odds ratio simplifies to λ_{11}^{XY} with dummy-variable constraints setting parameters at the second level of X or Y equal to 0.

The λ_{ijk}^{XYZ} term in the general model (8.12) refers to three-factor interaction. It describes how the odds ratio between two variables changes across categories of the third. We illustrate for $2 \times 2 \times 2$ tables. By direct substitution of the general model formula,

$$\log\frac{\theta_{11(1)}}{\theta_{11(2)}} = \log\frac{(\mu_{111}\,\mu_{221})/(\mu_{121}\,\mu_{211})}{(\mu_{112}\,\mu_{222})/(\mu_{122}\,\mu_{212})}$$

$$= \left(\lambda_{111}^{XYZ} + \lambda_{221}^{XYZ} - \lambda_{121}^{XYZ} - \lambda_{211}^{XYZ}\right)$$

$$- \left(\lambda_{112}^{XYZ} + \lambda_{222}^{XYZ} - \lambda_{122}^{XYZ} - \lambda_{212}^{XYZ}\right).$$

Only one parameter is nonredundant. For constraints setting the second-category parameters equal to 0, this log ratio of odds ratios equals λ_{111}^{XYZ}. When $\lambda_{111}^{XYZ} = 0$, $\theta_{11(1)} = \theta_{11(2)}$, giving homogeneous XY association.

8.2.4 Alcohol, Cigarette, and Marijuana Use Example

Table 8.3 refers to a 1992 survey by the Wright State University School of Medicine and the United Health Services in Dayton, Ohio. The survey asked 2276 students in their final year of high school in a nonurban area near Dayton, Ohio whether they had ever used alcohol, cigarettes, or marijuana. Denote the variables in this $2 \times 2 \times 2$ table by A for alcohol use, C for cigarette use, and M for marijuana use.

Section 8.7 covers the fitting of loglinear models. For now, we emphasize interpretation. Table 8.4 shows fitted values for several loglinear models. The

TABLE 8.3 Alcohol, Cigarette, and Marijuana Use for High School Seniors

Alcohol Use	Cigarette Use	Marijuana Use	
		Yes	No
Yes	Yes	911	538
	No	44	456
No	Yes	3	43
	No	2	279

Source: Data courtesy of Harry Khamis, Wright State University.

TABLE 8.4   Fitted Values for Loglinear Models Applied to Table 8.3

Alcohol Use	Cigarette Use	Marijuana Use	Loglinear Model[a]				
			(A, C, M)	(AC, M)	(AM, CM)	(AC, AM, CM)	(ACM)
Yes	Yes	Yes	540.0	611.2	909.24	910.4	911
		No	740.2	837.8	438.84	538.6	538
	No	Yes	282.1	210.9	45.76	44.6	44
		No	386.7	289.1	555.16	455.4	456
No	Yes	Yes	90.6	19.4	4.76	3.6	3
		No	124.2	26.6	142.16	42.4	43
	No	Yes	47.3	118.5	0.24	1.4	2
		No	64.9	162.5	179.84	279.6	279

[a]A, alcohol use; C, cigarette use; M, marijuana use.

fit for model (AC, AM, CM) is close to the observed data, which are the fitted values for the saturated model (ACM). The other models fit poorly.

Table 8.5 illustrates model association patterns by presenting estimated conditional and marginal odds ratios. For example, the entry 1.0 for the AC conditional association for the model (AM, CM) of AC conditional independence is the common value of the AC fitted odds ratios at the two levels of M,

$$1.0 = \frac{909.24 \times 0.24}{45.76 \times 4.76} = \frac{438.84 \times 179.84}{555.16 \times 142.16}.$$

The entry 2.7 for the AC marginal association for this model is the odds ratio for the marginal AC fitted table. The odds ratios for the observed data are those reported for the saturated model (ACM).

Table 8.5 shows that estimated conditional odds ratios equal 1.0 for each pairwise term not appearing in a model, such as the AC association in model (AM, CM). For that model, the estimated marginal AC odds ratio differs from 1.0, since conditional independence does not imply marginal independence. Some models have conditional associations that are necessarily the

TABLE 8.5   Estimated Odds Ratios for Loglinear Models in Table 8.5

Model	Conditional Association			Marginal Association		
	AC	AM	CM	AC	AM	CM
(A, C, M)	1.0	1.0	1.0	1.0	1.0	1.0
(AC, M)	17.7	1.0	1.0	17.7	1.0	1.0
(AM, CM)	1.0	61.9	25.1	2.7	61.9	25.1
(AC, AM, CM)	7.8	19.8	17.3	17.7	61.9	25.1
(ACM) level 1	13.8	24.3	17.5	17.7	61.9	25.1
(ACM) level 2	7.7	13.5	9.7			

same as the corresponding marginal associations. In Section 9.1.2 we present a condition guaranteeing this.

Model (AC, AM, CM) permits all pairwise associations but maintains homogeneous odds ratios between two variables at each level of the third. The AC fitted conditional odds ratios for this model equal 7.8. One can calculate this odds ratio using the model's fitted values at either level of M, or [from (8.14)] using $\exp(\hat{\lambda}_{11}^{AC} + \hat{\lambda}_{22}^{AC} - \hat{\lambda}_{12}^{AC} - \hat{\lambda}_{21}^{AC})$.

Table 8.5 shows that estimated odds ratios are very dependent on the model. This highlights the importance of good model selection. An estimate from this table is informative only to the extent that its model fits well. In the next section we discuss goodness of fit.

8.3 INFERENCE FOR LOGLINEAR MODELS

A good-fitting loglinear model provides a basis for describing and making inferences about associations among categorical responses. Standard methods apply for checking fit and making inference about model parameters.

8.3.1 Chi-Squared Goodness-of-Fit Tests

As usual, X^2 and G^2 test whether a model holds by comparing cell fitted values to observed counts. Here df equals the number of cell counts minus the number of model parameters.

For the student survey (Table 8.3), Table 8.6 shows results of testing fit for several loglinear models. Models that lack any association term fit poorly. The model (AC, AM, CM) that has all pairwise associations fits well ($P = 0.54$). It is suggested by other criteria also, such as minimizing

$$\text{AIC} = -2(\text{maximized log likelihood} - \text{number of parameters in model})$$

or equivalently, minimizing $[G^2 - 2(\text{df})]$.

TABLE 8.6 Goodness-of-Fit Tests for Loglinear Models in Table 8.4

Model	G^2	X^2	df	P-value[a]
(A, C, M)	1286.0	1411.4	4	< 0.001
(A, CM)	534.2	505.6	3	< 0.001
(C, AM)	939.6	824.2	3	< 0.001
(M, AC)	843.8	704.9	3	< 0.001
(AC, AM)	497.4	443.8	2	< 0.001
(AC, CM)	92.0	80.8	2	< 0.001
(AM, CM)	187.8	177.6	2	< 0.001
(AC, AM, CM)	0.4	0.4	1	0.54
(ACM)	0.0	0.0	0	—

[a]P-value for G^2 statistic.

8.3.2   Inference about Conditional Associations

Tests about conditional associations compare loglinear models. The likelihood-ratio statistic $-2(L_0 - L_1)$ is identical to the difference $G^2(M_0 | M_1) = G^2(M_0) - G^2(M_1)$ between deviances for models without that term and with it. For model (XY, XZ, YZ), consider the hypothesis of XY conditional independence. This is H_0: $\lambda_{ij}^{XY} = 0$ for the $(I - 1)(J - 1)$ XY association parameters. The test statistic is $G^2(XZ, YZ) - G^2(XY, XZ, YZ)$, with df = $(I - 1)(J - 1)$. This has the same purpose as the generalized CMH and model-based tests for nominal variables presented in Section 7.5.

For instance, the test of conditional independence between alcohol use and cigarette smoking compares model (AM, CM) with the alternative (AC, AM, CM). The test statistic is

$$G^2[(AM, CM) | (AC, AM, CM)] = 187.8 - 0.4 = 187.4,$$

with df = $2 - 1 = 1$ ($P < 0.001$). The statistics comparing (AC, CM) and (AC, AM) with (AC, AM, CM) also provide strong evidence of AM and CM conditional associations. Further analyses of Table 8.3 use model (AC, AM, CM).

With large sample sizes, statistically significant effects can be weak and unimportant. A more relevant concern is whether the associations are strong enough to be important. Confidence intervals are more useful than tests for assessing this. Table 8.7 shows output from fitting model (AC, AM, CM) with

TABLE 8.7   Output for Fitting Loglinear Model to Table 8.3

```
                  Criteria For Assessing Goodness Of Fit
         Criterion                DF       Value       Value / DF
         Deviance                  1      0.3740         0.3740
         Pearson Chi-Square        1      0.4011         0.4011

                                       Standard      Wald
Parameter               Estimate        Error     Chi-Square    Pr>ChiSq
Intercept                 5.6334       0.0597      8903.96       <.0001
a          1              0.4877       0.0758        41.44       <.0001
c          1             -1.8867       0.1627       134.47       <.0001
m          1             -5.3090       0.4752       124.82       <.0001
a*m        1    1         2.9860       0.4647        41.29       <.0001
a*c        1    1         2.0545       0.1741       139.32       <.0001
c*m        1    1         2.8479       0.1638       302.14       <.0001

         LR Statistics
              Source       DF     Chi-Square     Pr>ChiSq
               a*m          1        91.64         <.0001
               a*c          1       187.38         <.0001
               c*m          1       497.00         <.0001
```

parameters in the last row and in the last column equal to zero, such as by using $(1, 0)$ dummy variables for each classification. Consider the conditional AC odds ratio, assuming model (AC, AM, CM). Table 8.7 reports $\hat{\lambda}_{11}^{AC} = 2.054$, with SE $= 0.174$. For these constraints, this is the estimated conditional log odds ratio. A 95% Wald confidence interval for the true conditional AC odds ratio is $\exp[2.054 \pm 1.96(0.174)]$, or $(5.5, 11.0)$. Strong positive association exists between cigarette use and alcohol use, both for users and nonusers of marijuana.

For model (AC, AM, CM), the 95% Wald confidence intervals are $(8.0, 49.2)$ for the AM conditional odds ratio and $(12.5, 23.8)$ for the CM conditional odds ratio. The intervals are wide, but these associations also are strong. Table 8.5 shows that estimated marginal associations are even stronger. Controlling for outcome on one response moderates the association somewhat between the other two.

The analyses in this section pertain to associations. A different analysis pertains to comparing single-variable marginal distributions, for instance to determine if students used cigarettes more than alcohol or marijuana. That type of analysis is presented in Section 10.1.

8.4 LOGLINEAR MODELS FOR HIGHER DIMENSIONS

Loglinear models for three-way tables are more complex than for two-way tables, because of the variety of potential association terms. Loglinear models for three-way tables extend readily, however, to multiway tables. As the number of dimensions increases, some complications arise. One is the increase in the number of possible association and interaction terms, making model selection more difficult. Another is the increase in number of cells. In Section 9.8 we show that this can cause difficulties with existence of estimates and appropriateness of asymptotic theory.

8.4.1 Four-Way Contingency Tables

We illustrate models for higher dimensions using a four-way table with variables W, X, Y, and Z. Interpretations are simplest when the model has no three-factor interaction terms. Such models are special cases of

$$\log \mu_{hijk} = \lambda + \lambda_h^W + \lambda_i^X + \lambda_j^Y + \lambda_k^Z$$
$$+ \lambda_{hi}^{WX} + \lambda_{hj}^{WY} + \lambda_{hk}^{WZ} + \lambda_{ij}^{XY} + \lambda_{ik}^{XZ} + \lambda_{jk}^{YZ},$$

denoted by (WX, WY, WZ, XY, XZ, YZ). Each pair of variables is conditionally dependent, with the same odds ratios at each combination of categories of the other two variables. An absence of a two-factor term implies conditional independence, given the other two variables.

A variety of models exhibit three-factor interaction. A model could contain any of WXY, WXZ, WYZ, or XYZ terms. For model (WXY, WZ, XZ, YZ), each pair of variables is conditionally dependent, but at each level of Z the WX association, the WY association, and the XY association may vary across categories of the remaining variable. The conditional association between Z and another variable is homogeneous. The saturated model contains all the three-factor terms plus a four-factor interaction term.

8.4.2 Automobile Accident Example

Table 8.8 summarizes observations of 68,694 passengers in autos and light trucks involved in accidents in the state of Maine in 1991. The table classifies passengers by gender (G), location of accident (L), seat-belt use (S), and injury (I). Table 8.8 reports the sample proportion of passengers who were injured. For each GL combination, the proportion of injuries was about halved for passengers wearing seat belts.

Table 8.9 displays tests of fit for several loglinear models. To investigate the complexity of model needed, we consider models (G, I, L, S),

TABLE 8.8   Loglinear Models for Injury, Seat-Belt Use, Gender, and Location[a]

| | | Seat | Injury | | (GI, GL, GS, IL, IS, LS) | | (GLS, GI, IL, IS) | | Sample Proportion |
Gender	Location	Belt	No	Yes	No	Yes	No	Yes	Yes
Female	Urban	No	7,287	996	7,166.4	993.0	7,273.2	1,009.8	0.12
		Yes	11,587	759	11,748.3	721.3	11,632.6	713.4	0.06
	Rural	No	3,246	973	3,353.8	988.8	3,254.7	964.3	0.23
		Yes	6,134	757	5,985.5	781.9	6,093.5	797.5	0.11
Male	Urban	No	10,381	812	10,471.5	845.1	10,358.9	834.1	0.07
		Yes	10,969	380	10,837.8	387.6	10,959.2	389.8	0.03
	Rural	No	6,123	1,084	6,045.3	1,038.1	6,150.2	1,056.8	0.15
		Yes	6,693	513	6,811.4	518.2	6,697.6	508.4	0.07

[a]G, gender; I, injury; L, location; S, seat-belt use.

Source: Data courtesy of Cristanna Cook, Medical Care Development, Augusta, Maine.

TABLE 8.9   Goodness-of-Fit Tests for Loglinear Models in Table 8.8

Model	G^2	df	P-Value
(G, I, L, S)	2792.8	11	< 0.0001
(GI, GL, GS, IL, IS, LS)	23.4	5	< 0.001
(GIL, GIS, GLS, ILS)	1.3	1	0.25
(GIL, GS, IS, LS)	18.6	4	0.001
(GIS, GL, IL, LS)	22.8	4	< 0.001
(GLS, GI, IL, IS)	7.5	4	0.11
(ILS, GI, GL, GS)	20.6	4	< 0.001

TABLE 8.10   Estimated Conditional Odds Ratios for Models of Table 8.8

| Odds Ratio | Loglinear Model | |
	(GI, GL, GS, IL, IS, LS)	(GLS, GI, IL, IS)
GI	0.58	0.58
IL	2.13	2.13
IS	0.44	0.44
GL S = no	1.23	1.33
S = yes	1.23	1.17
GS L = urban	0.63	0.66
L = rural	0.63	0.58
LS G = female	1.09	1.17
G = male	1.09	1.03

(GI, GL, GS, IL, IS, LS), and (GIL, GIS, GLS, ILS) having all terms of varying complexity. Model (G, I, L, S) of mutual independence fits very poorly. Model (GI, GL, GS, IL, IS, LS) fits much better but still has a lack of fit $(P < 0.001)$. Model (GIL, GIS, GLS, ILS) fits well $(G^2 = 1.3, \text{df} = 1)$ but is complex and difficult to interpret. This suggests studying models more complex than (GI, GL, GS, IL, IS, LS) but simpler than (GIL, GIS, GLS, ILS).

First, however, we analyze model (GI, GL, GS, IL, IS, LS), which focuses on pairwise associations. Table 8.8 displays its fitted values. Table 8.10 reports the model-based estimated conditional odds ratios. One can obtain them directly using the fitted values for partial tables relating two variables at any combination of levels of the other two. They also follow directly from parameter estimates; for instance, $0.44 = \exp(\hat{\lambda}_{11}^{IS} + \hat{\lambda}_{22}^{IS} - \hat{\lambda}_{12}^{IS} - \hat{\lambda}_{21}^{IS})$.

Since the sample size is large, the estimates of odds ratios are quite precise. For instance, the standard error of the estimated IS conditional log odds ratio of -0.814 is 0.028. A 95% Wald confidence interval for the true odds ratio is $\exp[-0.814 \pm 1.96(0.028)]$ or $(0.42, 0.47)$. This model estimates that the odds of injury for passengers wearing seat belts were less than half the odds for passengers not wearing them, at each gender–location combination. The fitted odds ratios in Table 8.10 also suggest that other factors being fixed, injury was more likely in rural than urban accidents and more likely for females than for males. The estimated odds that males used seat belts were only 0.63 times the estimated odds for females.

Interpretations are more complex for models containing three-factor interaction terms. Table 8.9 shows results of adding a single three-factor term to model (GI, GL, GS, IL, IS, LS). Of the four possible models, (GLS, GI, IL, IS) appears to fit best. Table 8.8 also displays its fit. Given the large sample size, its G^2 value suggests that it fits quite well.

For model (GLS, GI, IL, IS), each pair of variables is conditionally dependent, and at each category of I the association between any two of the others varies across categories of the remaining variable. For this model, it is inappropriate to interpret the GL, GS, and LS two-factor terms on their own. Since I does not occur in a three-factor interaction, the conditional odds ratio between I and each variable (see the top portion of Table 8.10) is the same at each combination of categories of the other two variables.

When a model has a three-factor interaction term but no term of higher order than that, one can study the interaction by calculating fitted odds ratios between two variables at each level of the third. One can do this at any levels of remaining variables not involved in the interaction. The bottom portion of Table 8.10 illustrates this for model (GLS, GI, IL, IS). For instance, the fitted GS odds ratio of 0.66 for $(L = \text{urban})$ refers to four fitted values for urban accidents, both the four with (injury = no) and the four with (injury = yes); for example, $0.66 = (7273.2 \times 10{,}959.2)/(11{,}632.6 \times 10{,}358.9)$.

8.4.3 Large Samples and Statistical versus Practical Significance

Model (GLS, GI, IL, IS) seems to fit much better than (GI, GL, GS, IL, IS, LS). The difference in G^2 values of $23.4 - 7.5 = 15.9$ has df $= 5 - 4 = 1$ $(P = 0.0001)$. Table 8.10 indicates, however, that the degree of three-factor interaction is weak. The fitted odds ratio between any two of G, L, and S is similar at both levels of the third variable. The significantly better fit of model (GLS, GI, IL, IS) reflects mainly the enormous sample size.

As in any test, a statistically significant effect need not be practically important. With huge samples, it is crucial to focus on estimation rather than hypothesis testing. For instance, a comparison of fitted odds ratios for the two models in Table 8.10 suggests that the simpler model (GI, GL, GS, IL, IS, LS) is adequate for most purposes.

8.4.4 Dissimilarity Index

For a table of arbitrary dimension with cell counts $\{n_i = np_i\}$ and fitted values $\{\hat{\mu}_i = n\hat{\pi}_i\}$, one can summarize the closeness of a model fit to the data by the *dissimilarity index* (Gini 1914),

$$\hat{\Delta} = \sum_i |n_i - \hat{\mu}_i|/2n = \sum_i |p_i - \hat{\pi}_i|/2 \,.$$

This index falls between 0 and 1, with smaller values representing a better fit. It represents the proportion of sample cases that must move to different cells for the model to fit perfectly.

The dissimilarity index $\hat{\Delta}$ estimates a corresponding population index Δ describing model lack of fit. The value $\Delta = 0$ occurs when the model holds perfectly. In practice, this is unrealistic for unsaturated models, and $\Delta > 0$. The estimator $\hat{\Delta}$ helps study whether the lack of fit is important in a practical sense. When $\hat{\Delta} < 0.02$ or 0.03, the sample data follow the model pattern quite closely, even though the model is not perfect. When Δ is near 0, $\hat{\Delta}$ tends to overestimate Δ, substantially so for small n. Firth and Kuha (2000) provided an approximate variance for $\hat{\Delta}$ and studied ways to reduce its estimation bias.

For Table 8.8, model (GI, GL, GS, IL, IS, LS) has $\hat{\Delta} = 0.008$, and model (GLS, GI, IL, IS) has $\hat{\Delta} = 0.003$. For either model, moving less than 1% of the data yields a perfect fit. The relatively large G^2 value for (GI, GL, GS, IL, IS, LS) indicated that it does not truly hold. Nevertheless, the small $\hat{\Delta}$ value suggests that, in practical terms, it fits decently.

8.5 LOGLINEAR–LOGIT MODEL CONNECTION

Loglinear models treat categorical response variables symmetrically, focusing on associations and interactions in their joint distribution. Logit models, by contrast, describe how a single categorical response depends on explanatory variables. The model types seem distinct, but connections exist between them. For a loglinear model, forming logits on one response helps to interpret the model. Moreover, logit models with categorical explanatory variables have equivalent loglinear models.

8.5.1 Using Logit Models to Interpret Loglinear Models

To understand implications of a loglinear model formula, it can help to form a logit on one variable. We illustrate with the loglinear model (XY, XZ, YZ). When Y is binary, its logit is

$$\log \frac{P(Y = 1 \mid X = i, Z = k)}{P(Y = 2 \mid X = i, Z = k)} = \log \frac{\mu_{i1k}}{\mu_{i2k}} = \log \mu_{i1k} - \log \mu_{i2k}$$

$$= \left(\lambda + \lambda_i^X + \lambda_1^Y + \lambda_k^Z + \lambda_{i1}^{XY} + \lambda_{ik}^{XZ} + \lambda_{1k}^{YZ} \right)$$

$$- \left(\lambda + \lambda_i^X + \lambda_2^Y + \lambda_k^Z + \lambda_{i2}^{XY} + \lambda_{ik}^{XZ} + \lambda_{2k}^{YZ} \right)$$

$$= \left(\lambda_1^Y - \lambda_2^Y \right) + \left(\lambda_{i1}^{XY} - \lambda_{i2}^{XY} \right) + \left(\lambda_{1k}^{YZ} - \lambda_{2k}^{YZ} \right).$$

The first parenthetical term is a constant, not depending on i or k. The second parenthetical term depends on the category i of X. The third parenthetical term depends on the category k of Z. This logit has the additive form

$$\text{logit}[P(Y = 1 \mid X = i, Z = k)] = \alpha + \beta_i^X + \beta_k^Z. \tag{8.15}$$

Using the notation summarizing logit models by their predictors, we denote it by $(X + Z)$.

In Section 5.4.1 we discussed this logit model. When Y is binary, the loglinear model (XY, XZ, YZ) is equivalent to it. The λ_{ik}^{XZ} terms for association among explanatory variables cancel in the difference in logarithms the logit defines. The logit model does not study this association.

8.5.2 Auto Accident Example Revisited

For the Maine auto accidents (Table 8.8), in Section 8.4.2 we showed that the loglinear model (GLS, GI, LI, IS),

$$\log \mu_{gi\,\ell s} = \lambda + \lambda_g^G + \lambda_i^I + \lambda_\ell^L + \lambda_s^S + \lambda_{gi}^{GI} + \lambda_{g\ell}^{GL} + \lambda_{gs}^{GS}$$
$$+ \lambda_{i\ell}^{IL} + \lambda_{is}^{IS} + \lambda_{\ell s}^{LS} + \lambda_{g\,\ell s}^{GLS},$$

fits well. It is natural to treat injury (I) as a response variable and gender (G), location (L), and seat-belt use (S) as explanatory variables, or perhaps S as a response with G and L as explanatory. One can show that this loglinear model is equivalent to logit model $(G + L + S)$,

$$\text{logit}[P(I = 1 \mid G = g, L = \ell, S = s)] = \alpha + \beta_g^G + \beta_\ell^L + \beta_s^S. \quad (8.16)$$

For instance, the seat-belt effects in the two models satisfy $\beta_s^S = \lambda_{1s}^{IS} - \lambda_{2s}^{IS}$. In the logit calculation, all terms in the loglinear model not having the injury index i cancel. Fitted values, goodness-of-fit statistics, residual df, and standardized Pearson residuals for the logit model are identical to those for the loglinear model.

Odds ratios describing effects on I relate to two-factor loglinear parameters and main-effect logit parameters. In the logit model, the log odds ratio for the effect of S on I equals $\beta_1^S - \beta_2^S$. This equals $\lambda_{11}^{IS} + \lambda_{22}^{IS} - \lambda_{12}^{IS} - \lambda_{21}^{IS}$ in the loglinear model. Their estimates are the same no matter how software sets up constraints. For Table 8.8, $\hat{\beta}_1^S - \hat{\beta}_2^S = -0.817$ for the logit model, and $\hat{\lambda}_{11}^{IS} + \hat{\lambda}_{22}^{IS} - \hat{\lambda}_{12}^{IS} - \hat{\lambda}_{21}^{IS} = -0.817$ for the loglinear model.

Loglinear models are GLMs that treat the 16 cell counts in Table 8.8 as 16 independent Poisson variates. Logit models are GLMs that treat the table as binomial counts. Logit models with I as the response treat the marginal GLS table $\{n_{g+\ell s}\}$ as fixed and regard $\{n_{g1\,\ell s}\}$ as eight independent binomial variates on that response. Although the sampling models differ, the results from fits of corresponding models are identical.

8.5.3 Correspondence between Loglinear and Logit Models

In the derivation of the logit model $(X + Z)$ [see (8.15)] from loglinear model (XY, XZ, YZ), the λ_{ik}^{XZ} term cancels. It might seem as if the model (XY, YZ) omitting this term is also equivalent to that logit model. Indeed, forming the logit on Y for (XY, YZ) results in the same logit formula. The loglinear

TABLE 8.11 Equivalent Loglinear and Logit Models for a Three-Way Table with Binary Response Variable Y

Loglinear Symbol	Logit Model	Logit Symbol
(Y, XZ)	α	$(-)$
(XY, XZ)	$\alpha + \beta_i^X$	(X)
(YZ, XZ)	$\alpha + \beta_k^Z$	(Z)
(XY, YZ, XZ)	$\alpha + \beta_i^X + \beta_k^Z$	$(X + Z)$
(XYZ)	$\alpha + \beta_i^X + \beta_k^Z + \beta_{ik}^{XZ}$	$(X*Z)$

model that has the same fit as the logit model, however, contains a general interaction term for relationships among the explanatory variables. The logit model does not assume anything about relationships among explanatory variables, so it allows an arbitrary interaction pattern for them.

Table 8.11 summarizes equivalent logit and loglinear models for three-way tables when Y is a binary response. Each loglinear model contains the XZ association term relating the explanatory variables in the logit models. The simple loglinear model (Y, XZ) states that Y is jointly independent of both X and Z, and is equivalent to the logit model having only an intercept. The saturated loglinear model (XYZ) contains the three-factor interaction term. When Y is a binary response, this model is equivalent to a logit model with an interaction between the predictors X and Z. For instance, the effect of X on Y depends on Z, meaning that the XY odds ratio varies across its categories. That logit model is also saturated.

Analogous correspondences hold when Y has several categories, using baseline-category logit models. An advantage of the loglinear approach is its generality. It applies when more than one response variable exists. The alcohol–cigarette–marijuana example in Section 8.2.4, for instance, used loglinear models to study association patterns among three response variables. Loglinear models are most natural when at least two variables are response variables. When only one is a response, it is more sensible to use logit models directly.

8.5.4 Generalized Loglinear Model*

Let $\mathbf{n} = (n_1, \ldots, n_N)'$ and $\boldsymbol{\mu} = (\mu_1, \ldots, \mu_N)'$ denote column vectors of observed and expected counts for the N cells of a contingency table, with $n = \Sigma_i n_i$. For simplicity we use a single index, but the table may be multidimensional. Loglinear models for positive Poisson means have the form

$$\log \boldsymbol{\mu} = \mathbf{X}\boldsymbol{\beta} \qquad (8.17)$$

for model matrix $\mathbf{X}$ and column vector $\boldsymbol{\beta}$ of model parameters.

We illustrate with the independence model, $\log \mu_{ij} = \lambda + \lambda_i^X + \lambda_j^Y$, for a 2×2 table. With constraints $\lambda_2^X = \lambda_2^Y = 0$, it is

$$
\begin{bmatrix} \log \mu_{11} \\ \log \mu_{12} \\ \log \mu_{21} \\ \log \mu_{22} \end{bmatrix} = \begin{bmatrix} 1 & 1 & 1 \\ 1 & 1 & 0 \\ 1 & 0 & 1 \\ 1 & 0 & 0 \end{bmatrix} \begin{bmatrix} \lambda \\ \lambda_1^X \\ \lambda_1^Y \end{bmatrix}.
$$

A generalization of (8.17) allows many additional models. This *generalized loglinear model* is

$$
\mathbf{C} \log(\mathbf{A}\boldsymbol{\mu}) = \mathbf{X}\boldsymbol{\beta} \tag{8.18}
$$

for matrices $\mathbf{C}$ and $\mathbf{A}$. The ordinary loglinear model (8.17) results when $\mathbf{C}$ and $\mathbf{A}$ are identity matrices. Other special cases include logit models for binary or multicategory responses.

For instance, the loglinear model of independence for a 2×2 table is equivalent to a model by which the logit for Y is the same in each row of X (see Section 8.1.2). That logit model has form (8.18): $\mathbf{A}$ is a 4×4 identity matrix, so $\mathbf{A}\boldsymbol{\mu}$ is the 4×1 vector $\boldsymbol{\mu} = (\mu_{11}, \mu_{12}, \mu_{21}, \mu_{22})'$; the product $\mathbf{C} \log(\mathbf{A}\boldsymbol{\mu})$ forms the logit in row 1 and the logit in row 2 using

$$
\mathbf{C} = \begin{bmatrix} 1 & -1 & 0 & 0 \\ 0 & 0 & 1 & -1 \end{bmatrix};
$$

then $\mathbf{X} = (1, 1)'$ is a 2×1 matrix, and $\boldsymbol{\beta}$ is a single constant α, so $\mathbf{X}\boldsymbol{\beta}$ forms a common value for those two logits.

In Chapters 10 and 11 we use the generalized loglinear model for models outside the classes of GLMs studied thus far. An example is modeling marginal distributions of multivariate responses.

8.6 LOGLINEAR MODEL FITTING: LIKELIHOOD EQUATIONS AND ASYMPTOTIC DISTRIBUTIONS*

In discussing the fitting of loglinear models, we first derive sufficient statistics and likelihood equations. We then present large-sample normal distributions for ML estimators of model parameters and cell probabilities. We illustrate results with models for three-way tables. For simplicity, derivations use the Poisson sampling model, which does not require a constraint on parameters such as the multinomial does.

8.6.1 Minimal Sufficient Statistics

For three-way tables, the joint Poisson probability that cell counts $\{Y_{ijk} = n_{ijk}\}$ is

$$\prod_i \prod_j \prod_k \frac{e^{-\mu_{ijk}} \mu_{ijk}^{n_{ijk}}}{n_{ijk}!},$$

where the product refers to all cells of the table. The kernel of the log likelihood is

$$L(\boldsymbol{\mu}) = \sum_i \sum_j \sum_k n_{ijk} \log \mu_{ijk} - \sum_i \sum_j \sum_k \mu_{ijk}. \tag{8.19}$$

For the general loglinear model (8.12), this simplifies to

$$L(\boldsymbol{\mu}) = n\lambda + \sum_i n_{i++} \lambda_i^X + \sum_j n_{+j+} \lambda_j^Y + \sum_k n_{++k} \lambda_k^Z$$

$$+ \sum_i \sum_j n_{ij+} \lambda_{ij}^{XY} + \sum_i \sum_k n_{i+k} \lambda_{ik}^{XZ} + \sum_j \sum_k n_{+jk} \lambda_{jk}^{YZ}$$

$$+ \sum_i \sum_j \sum_k n_{ijk} \lambda_{ijk}^{XYZ} - \sum_i \sum_j \sum_k \exp\left(\lambda + \cdots + \lambda_{ijk}^{XYZ}\right). \tag{8.20}$$

Since the Poisson distribution is in the exponential family, coefficients of the parameters are sufficient statistics. For this saturated model, $\{n_{ijk}\}$ are coefficients of $\{\lambda_{ijk}^{XYZ}\}$, so there is no reduction of the data. For simpler models, certain parameters are zero and (8.20) simplifies. For instance, for the model (X, Y, Z) of mutual independence, sufficient statistics are the coefficients in (8.20) of $\{\lambda_i^X\}$, $\{\lambda_j^Y\}$, and $\{\lambda_k^Z\}$. These are $\{n_{i++}\}$, $\{n_{+j+}\}$, and $\{n_{++k}\}$.

Table 8.12 lists minimal sufficient statistics for several loglinear models. Each one is the coefficient of the highest-order term(s) in which a variable appears. In fact, they are the marginal distributions for terms in the model symbol. Simpler models use more condensed sample information. For instance, whereas (X, Y, Z) uses only the single-factor marginal distributions, (XY, XZ, YZ) uses the two-way marginal tables.

TABLE 8.12 Minimal Sufficient Statistics for Fitting Loglinear Models

Model	Minimal Sufficient Statistics
(X, Y, Z)	$\{n_{i++}\}, \{n_{+j+}\}, \{n_{++k}\}$
(XY, Z)	$\{n_{ij+}\}, \{n_{++k}\}$
(XY, YZ)	$\{n_{ij+}\}, \{n_{+jk}\}$
(XY, XZ, YZ)	$\{n_{ij+}\}, \{n_{i+k}\}, \{n_{+jk}\}$

8.6.2 Likelihood Equations for Loglinear Models

The fitted values for a model are solutions to the likelihood equations. We derive likelihood equations using general representation (8.17) for a loglinear model. For a vector of counts $\mathbf{n}$ with $\boldsymbol{\mu} = E(\mathbf{n})$, the model is $\log \boldsymbol{\mu} = \mathbf{X}\boldsymbol{\beta}$, for which $\log(\mu_i) = \sum_j x_{ij} \beta_j$ for all i.

Extending (8.19), for Poisson sampling the log likelihood is

$$L(\boldsymbol{\mu}) = \sum_i n_i \log \mu_i - \sum_i \mu_i$$

$$= \sum_i n_i \left(\sum_j x_{ij} \beta_j \right) - \sum_i \exp \left(\sum_j x_{ij} \beta_j \right). \tag{8.21}$$

The sufficient statistic for β_j is its coefficient, $\sum_i n_i x_{ij}$. Since

$$\frac{\partial}{\partial \beta_j} \left[\exp \left(\sum_j x_{ij} \beta_j \right) \right] = x_{ij} \exp \left(\sum_j x_{ij} \beta_j \right) = x_{ij} \mu_i,$$

$$\frac{\partial L(\boldsymbol{\mu})}{\partial \beta_j} = \sum_i n_i x_{ij} - \sum_i \mu_i x_{ij}, \quad j = 1, 2, \ldots, p.$$

The likelihood equations equate these derivatives to zero. They have the form

$$\mathbf{X}'\mathbf{n} = \mathbf{X}'\hat{\boldsymbol{\mu}}. \tag{8.22}$$

These equations equate the sufficient statistics to their expected values, a result obtained with GLM theory in (4.29). For models considered so far, these sufficient statistics are the marginal tables in the model symbol.

To illustrate, consider model (XZ, YZ). Its log likelihood is (8.20) with $\lambda^{XY} = \lambda^{XYZ} = 0$. The log-likelihood derivatives

$$\frac{\partial L}{\partial \lambda_{ik}^{XZ}} = n_{i+k} - \mu_{i+k} \quad \text{and} \quad \frac{\partial L}{\partial \lambda_{jk}^{YZ}} = n_{+jk} - \mu_{+jk}$$

yield the likelihood equations

$$\hat{\mu}_{i+k} = n_{i+k} \quad \text{for all } i \text{ and } k, \tag{8.23}$$

$$\hat{\mu}_{+jk} = n_{+jk} \quad \text{for all } j \text{ and } k. \tag{8.24}$$

Derivatives with respect to lower-order terms yield equations implied by these (Problem 8.30). For model (XZ, YZ), the fitted values have the same XZ and YZ marginal totals as the observed data.

8.6.3 Birch's Results for Loglinear Models

For model (XZ, YZ), from (8.23), (8.24), and Table 8.12, the minimal sufficient statistics are the ML estimates of the corresponding marginal distributions of expected frequencies. Equation (8.22) gives the corresponding result for any loglinear model. Birch (1963) showed that likelihood equations for loglinear models match minimal sufficient statistics to their expected values. Poisson GLM theory implied this result in (4.29) and (4.44). Thus, fitted values for loglinear models are smoothed versions of the cell counts that match them in certain marginal distributions but have associations and interactions satisfying the model-implied patterns.

Birch showed that a unique set of fitted values both satisfy the model and match the data in the minimal sufficient statistics. Hence, if we find such a solution, it must be the ML solution. To illustrate, the independence model for a two-way table

$$\log \mu_{ij} = \lambda + \lambda_i^X + \lambda_j^Y$$

has minimal sufficient statistics $\{n_{i+}\}$ and $\{n_{+j}\}$. The likelihood equations are

$$\hat{\mu}_{i+} = n_{i+}, \quad \hat{\mu}_{+j} = n_{+j}, \quad \text{for all } i \text{ and } j.$$

The fitted values $\{\hat{\mu}_{ij} = n_{i+} n_{+j}/n\}$ satisfy these equations and also satisfy the model. Birch's result implies that they are the ML estimates.

8.6.4 Direct versus Iterative Calculation of Fitted Values

To illustrate how to solve likelihood equations, we continue the analysis of model (XZ, YZ). From (8.9), the model satisfies

$$\pi_{ijk} = \frac{\pi_{i+k} \pi_{+jk}}{\pi_{++k}} \quad \text{for all } i, j, \text{ and } k.$$

For Poisson sampling, the related formula uses expected frequencies. Setting $\pi_{ijk} = \mu_{ijk}/n$, this is $\{\mu_{ijk} = \mu_{i+k} \mu_{+jk}/\mu_{++k}\}$. The likelihood equations (8.23) and (8.24) specify that ML estimates satisfy $\hat{\mu}_{i+k} = n_{i+k}$ and $\hat{\mu}_{+jk} = n_{+jk}$ and thus also $\hat{\mu}_{++k} = n_{++k}$. Since ML estimates of functions of parameters are the same functions of the ML estimates of those parameters,

$$\hat{\mu}_{ijk} = \frac{\hat{\mu}_{i+k} \hat{\mu}_{+jk}}{\hat{\mu}_{++k}} = \frac{n_{i+k} n_{+jk}}{n_{++k}}.$$

This solution satisfies the model and matches the data in the sufficient statistics. Thus, it is the unique ML solution.

TABLE 8.13 Fitted Values for Loglinear Models in Three-Way Tables

Model[a]	Probabilistic Form	Fitted Value
(X, Y, Z)	$\pi_{ijk} = \pi_{i++}\,\pi_{+j+}\,\pi_{++k}$	$\hat{\mu}_{ijk} = \dfrac{n_{i++}\,n_{+j+}\,n_{++k}}{n^2}$
(XY, Z)	$\pi_{ijk} = \pi_{ij+}\,\pi_{++k}$	$\hat{\mu}_{ijk} = \dfrac{n_{ij+}\,n_{++k}}{n}$
(XY, XZ)	$\pi_{ijk} = \dfrac{\pi_{ij+}\,\pi_{i+k}}{\pi_{i++}}$	$\hat{\mu}_{ijk} = \dfrac{n_{ij+}\,n_{i+k}}{n_{i++}}$
(XY, XZ, YZ)	$\pi_{ijk} = \psi_{ij}\,\phi_{jk}\,\omega_{ik}$	Iterative methods (Section 8.7)
(XYZ)	No restriction	$\hat{\mu}_{ijk} = n_{ijk}$

[a] Formulas for models not listed are obtained by symmetry; for example, for (XZ, Y), $\hat{\mu}_{ijk} = n_{i+k}\,n_{+j+}/n$.

Similar reasoning produces $\{\hat{\mu}_{ijk}\}$ for all except one model in Table 8.12. Table 8.13 shows formulas. That table also expresses $\{\pi_{ijk}\}$ in terms of marginal probabilities. These expressions and the likelihood equations determine the ML formulas, using the approach just described.

For models having explicit formulas for $\hat{\mu}_{ijk}$, the estimates are said to be *direct*. Many loglinear models do not have direct estimates. ML estimation then requires iterative methods. Of models in Tables 8.12 and 8.13, the only one not having direct estimates is (XY, XZ, YZ). Although the two-way marginal tables are its minimal sufficient statistics, it is not possible to express $\{\pi_{ijk}\}$ directly in terms of $\{\pi_{ij+}\}$, $\{\pi_{i+k}\}$, and $\{\pi_{+jk}\}$. Direct estimates do not exist for unsaturated models containing all two-factor associations. In practice, it is not essential to know which models have direct estimates. Iterative methods for models not having direct estimates also apply with models that have direct estimates. Statistical software for loglinear models uses such iterative methods for *all* cases.

8.6.5 Chi-Squared Goodness-of-Fit Tests

Model goodness-of-fit statistics compare fitted cell counts to sample counts. For Poisson GLMs, in Section 4.5.2 we showed that for models with an intercept term, the deviance equals the G^2 statistic. With a fixed number of cells, G^2 and X^2 have approximate chi-squared null distributions when expected frequencies are large. The df equal the difference in dimension between the alternative and null hypotheses. This equals the difference between the number of parameters in the general case and when the model holds.

We illustrate with model (X, Y, Z), for multinomial sampling with probabilities $\{\pi_{ijk}\}$. In the general case, the only constraint is $\sum_i\sum_j\sum_k \pi_{ijk} = 1$, so there are $IJK - 1$ parameters. For model (X, Y, Z), $\{\pi_{ijk} = \pi_{i++}\,\pi_{+j+}\,\pi_{++k}\}$ are determined by $I - 1$ of $\{\pi_{i++}\}$ (since $\sum_i \pi_{i++} = 1$), $J - 1$ of $\{\pi_{+j+}\}$, and $K - 1$ of $\{\pi_{++k}\}$. Thus,

$$\mathrm{df} = (IJK - 1) - [(I - 1) + (J - 1) + (K - 1)] = IJK - I - J - K + 2.$$

TABLE 8.14   **Residual Degrees of Freedom for Loglinear Models for Three-Way Tables**

Model	Degrees of Freedom
(X, Y, Z)	$IJK - I - J - K + 2$
(XY, Z)	$(K - 1)(IJ - 1)$
(XZ, Y)	$(J - 1)(IK - 1)$
(YZ, X)	$(I - 1)(JK - 1)$
(XY, YZ)	$J(I - 1)(K - 1)$
(XZ, YZ)	$K(I - 1)(J - 1)$
(XY, XZ)	$I(J - 1)(K - 1)$
(XY, XZ, YZ)	$(I - 1)(J - 1)(K - 1)$
(XYZ)	0

The same df formula applies for Poisson sampling. Then, the general case has IJK $\{\mu_{ijk}\}$ parameters. The residual df equal the number of cells in the table minus the number of parameters in the Poisson loglinear model for $\{\mu_{ijk}\}$. For instance, model (X, Y, Z) has residual df $= IJK - [1 + (I - 1) + (J - 1) + (K - 1)]$, reflecting the single intercept parameter λ and constraints such as $\lambda_I^X = \lambda_J^Y = \lambda_K^Z = 0$. This equals the number of linearly independent parameters equated to zero in the saturated model to obtain the given model. Table 8.14 shows df formulas for testing three-way loglinear models.

8.6.6   Covariance Matrix of ML Parameter Estimators

To present large-sample distributions of ML parameter estimators, we return to general expression $\log(\mu_i) = \sum_j x_{ij} \beta_j$, from which we obtained the log-likelihood derivatives

$$\frac{\partial L(\boldsymbol{\mu})}{\partial \beta_j} = \sum_i n_i x_{ij} - \sum_i \mu_i x_{ij}, \qquad j = 1, 2, \ldots, p.$$

The Hessian matrix of second partial derivatives has elements

$$\frac{\partial^2 L(\boldsymbol{\mu})}{\partial \beta_j \partial \beta_k} = -\sum_i x_{ij} \frac{\partial \mu_i}{\partial \beta_k}$$

$$= -\sum_i x_{ij} \left\{ \frac{\partial}{\partial \beta_k} \left[\exp\left(\sum_h x_{ih} \beta_h \right) \right] \right\} = -\sum_i x_{ij} x_{ik} \mu_i.$$

Like logistic regression models, loglinear models are GLMs using the canonical link; thus this matrix does not depend on the observed data. The information matrix, the negative of this matrix, is

$$\mathscr{I} = \mathbf{X}' \operatorname{diag}(\boldsymbol{\mu})\mathbf{X},$$

where $\operatorname{diag}(\boldsymbol{\mu})$ has the elements of $\boldsymbol{\mu}$ on the main diagonal.

For a fixed number of cells, as $n \to \infty$, the ML estimator $\hat{\boldsymbol{\beta}}$ is asymptotically normal with mean $\boldsymbol{\beta}$ and covariance matrix $\mathscr{I}^{-1}$. Thus, for Poisson sampling, the asymptotic covariance matrix

$$\operatorname{cov}(\hat{\boldsymbol{\beta}}) = [\mathbf{X}' \operatorname{diag}(\boldsymbol{\mu})\mathbf{X}]^{-1}. \tag{8.25}$$

Substituting ML fitted values and then taking square roots of diagonal elements yields standard errors for $\hat{\boldsymbol{\beta}}$. This also follows from the general expression (4.28) for GLMs, as noted in Section 4.4.7.

8.6.7   Connection between Multinomial and Poisson Loglinear Models

Similar asymptotic results hold with multinomial sampling. When $\{Y_i, i = 1, \ldots, N\}$ are independent Poisson random variables, the conditional distribution of $\{Y_i\}$ given $n = \Sigma_i Y_i$ is multinomial with parameters $\{\pi_i = \mu_i / (\Sigma_a \mu_a)\}$. Birch (1963) showed that ML estimates of loglinear model parameters are the same for multinomial sampling as for independent Poisson sampling. He showed that estimates are also the same for independent multinomial sampling, as long as the model contains a term for the marginal distribution fixed by the sampling design. To illustrate, suppose that at each combination of categories of X and Z, an independent multinomial sample occurs on Y. Then, $\{n_{i+k}\}$ are fixed. The model must contain λ_{ik}^{XZ}, so the fitted values satisfy $\{\hat{\mu}_{i+k} = n_{i+k}\}$.

That separate inferential theory is unnecessary for multinomial loglinear models follows from the following argument. Express the Poisson loglinear model for $\{\mu_i\}$ as

$$\log \mu_i = \lambda + \mathbf{x}_i\boldsymbol{\beta},$$

where $(1, \mathbf{x}_i)$ is row i of the model matrix $\mathbf{X}$ and $(\lambda, \boldsymbol{\beta}')'$ is the model parameter vector. The Poisson log likelihood is

$$L = L(\lambda, \boldsymbol{\beta}) = \sum_i n_i \log \mu_i - \sum_i \mu_i$$

$$= \sum_i n_i(\lambda + \mathbf{x}_i\boldsymbol{\beta}) - \sum_i \exp(\lambda + \mathbf{x}_i\boldsymbol{\beta}) = n\lambda + \sum_i n_i\mathbf{x}_i\boldsymbol{\beta} - \tau,$$

where $\tau = \Sigma_i \mu_i = \Sigma_i \exp(\lambda + \mathbf{x}_i\boldsymbol{\beta})$. Since $\log \tau = \lambda + \log[\Sigma_i \exp(\mathbf{x}_i\boldsymbol{\beta})]$, this log likelihood has the form

$$L = L(\tau, \boldsymbol{\beta}) = \left\{ \sum_i n_i\mathbf{x}_i\boldsymbol{\beta} - n\log\left[\sum_i \exp(\mathbf{x}_i\boldsymbol{\beta})\right] \right\} + (n\log \tau - \tau). \tag{8.26}$$

Now $\pi_i = \mu_i/(\Sigma_a \mu_a) = \exp(\lambda + \mathbf{x}_i\boldsymbol{\beta})/[\Sigma_a\exp(\lambda + \mathbf{x}_a\boldsymbol{\beta})]$, and $\exp(\lambda)$ cancels in the numerator and denominator. Thus, the first term (in braces) on the right-hand side in (8.26) is $\Sigma n_i \log \pi_i$, which is the multinomial log likelihood, conditional on the total cell count n. Unconditionally, $n = \Sigma_i n_i$ has a Poisson distribution with expectation $\Sigma_i \mu_i = \tau$, so the second term in (8.26) is the Poisson log likelihood for n. Since $\boldsymbol{\beta}$ enters only in the first term, the ML estimator $\hat{\boldsymbol{\beta}}$ and its covariance matrix for the Poisson log likelihood $L(\lambda, \boldsymbol{\beta})$ are identical to those for the multinomial log likelihood. The Poisson loglinear model has one more parameter (i.e., λ) than the multinomial loglinear model because of the random sample size. See Birch (1963), Lang (1996c), McCullagh and Nelder (1989, p. 211), and Palmgren (1981) for details.

For a multinomial sample, we show in Section 14.4.1 that the estimated covariance matrix of loglinear parameter estimators is

$$\widehat{\text{cov}}(\hat{\boldsymbol{\beta}}) = \{\mathbf{X}'[\text{diag}(\hat{\boldsymbol{\mu}}) - \hat{\boldsymbol{\mu}}\hat{\boldsymbol{\mu}}'/n]\mathbf{X}\}^{-1}. \tag{8.27}$$

The intercept λ from the Poisson model is not relevant, and $\mathbf{X}$ for the multinomial model deletes the column of $\mathbf{X}$ pertaining to it in the Poisson model.

A similar argument applies with several independent multinomial samples. Each log-likelihood term is a sum of components from different samples, but the Poisson log likelihood again decomposes into two parts. One part is a Poisson log likelihood for the independent sample sizes, and the other part is the sum of the independent multinomial log likelihoods. Palmgren (1981) showed that conditional on observed marginal totals for explanatory variables, the asymptotic covariances for estimators of parameters involving the response are the same as for Poisson sampling. For a single multinomial sample, Palmgren's result implies that (8.27) is identical to (8.25) with the row and column referring to λ deleted. Birch (1963) and Goodman (1970) gave related results. Lang (1996c) gave an elegant discussion of connections between multinomial and Poisson models. His results imply that the asymptotic variance of any linear contrast of estimated log means within a covariate level is identical for the two models.

8.6.8 Distribution of Probability Estimators

For multinomial sampling, the ML estimates of cell probabilities are $\hat{\boldsymbol{\pi}} = \hat{\boldsymbol{\mu}}/n$. We next give the asymptotic $\text{cov}(\hat{\boldsymbol{\pi}})$. Lang (1996c) showed the asymptotic covariance matrix for $\hat{\boldsymbol{\mu}}$ for Poisson sampling and its connection with $\text{cov}(\hat{\boldsymbol{\pi}})$.

The saturated model has $\hat{\boldsymbol{\pi}} = \mathbf{p}$, the sample proportions. Under multinomial sampling, from (3.7) and (3.8), their covariance matrix is

$$\text{cov}(\mathbf{p}) = [\text{diag}(\boldsymbol{\pi}) - \boldsymbol{\pi}\boldsymbol{\pi}']/n. \tag{8.28}$$

With I independent multinomial samples on a response variable with J categories, $\boldsymbol{\pi}$ and $\mathbf{p}$ consist of I sets of proportions, each having $J - 1$ nonredundant elements. Then, $\text{cov}(\mathbf{p})$ is a block diagonal matrix. Each of the independent samples has a $(J - 1) \times (J - 1)$ block of form (8.28), and the matrix contains zeros off the main diagonal of blocks.

Now assume an unsaturated model. Using the delta method we show in Sections 14.2.2 and 14.4.1 that $\hat{\boldsymbol{\pi}}$ has an asymptotic normal distribution about $\boldsymbol{\pi}$. The estimated covariance matrix equals

$$\widehat{\text{cov}}(\hat{\boldsymbol{\pi}}) = \left\{ \widehat{\text{cov}}(\mathbf{p})\mathbf{X}\left[\mathbf{X}'\widehat{\text{cov}}(\mathbf{p})\mathbf{X}\right]^{-1}\mathbf{X}'\widehat{\text{cov}}(\mathbf{p}) \right\}/n.$$

For a single multinomial sample, this expression equals

$$\widehat{\text{cov}}(\hat{\boldsymbol{\pi}}) = \left\{ \left[\text{diag}(\hat{\boldsymbol{\pi}}) - \hat{\boldsymbol{\pi}}\hat{\boldsymbol{\pi}}'\right]\mathbf{X}\left[\mathbf{X}'(\text{diag}(\hat{\boldsymbol{\pi}}) - \hat{\boldsymbol{\pi}}\hat{\boldsymbol{\pi}}')\mathbf{X}\right]^{-1} \right.$$
$$\left. \mathbf{X}'\left[\text{diag}(\hat{\boldsymbol{\pi}}) - \hat{\boldsymbol{\pi}}\hat{\boldsymbol{\pi}}'\right]\right\}/n.$$

For tables with many cells, it is not unusual to have a sample proportion of 0 in a cell. In this case the ordinary standard error is 0, which is unappealing. An advantage of fitting a model is that it typically has a positive fitted probability and standard error.

8.6.9 Uniqueness of ML Estimates

When all $\{n_i > 0\}$, the ML estimates exist and are unique. To show this, for simplicity we use Poisson sampling. Suppose that the model is parameterized so that $\mathbf{X}$ has full rank. Birch (1963) showed that the likelihood equations are soluble, by noting that the kernel of the Poisson log likelihood

$$L(\boldsymbol{\mu}) = \sum_i (n_i \log \mu_i - \mu_i)$$

has individual terms converging to $-\infty$ as $\log(\mu_i) \to \pm\infty$; thus, the log likelihood is bounded above and attains its maximum at finite values of the model parameters. It is stationary at this maximum, since it has continuous first partial derivatives.

Birch showed that the likelihood equations have a unique solution, and the likelihood is maximized at that point. He proved this by showing that the matrix of values $\{-\partial^2 L/\partial\beta_h\,\partial\beta_j\}$ [i.e., the information matrix $\mathbf{X}'\text{diag}(\boldsymbol{\mu})\mathbf{X}$] is nonsingular and nonnegative definite, and hence positive definite. Nonsingularity follows from $\mathbf{X}$ having full rank and the diagonal matrix having positive elements $\{\mu_i\}$. Any quadratic form $\mathbf{c}'\mathbf{X}'\text{diag}(\boldsymbol{\mu})\mathbf{X}\mathbf{c}$ equals $\sum_i [\sqrt{\mu_i}\,(\sum_j x_{ij}c_j)]^2$ ≥ 0, so the matrix is also nonnegative definite.

8.7 LOGLINEAR MODEL FITTING: ITERATIVE METHODS AND THEIR APPLICATION*

When a loglinear model does not have direct estimates, iterative algorithms such as Newton–Raphson can solve the likelihood equations. In this section we also present a simpler but more limited method, *iterative proportional fitting*.

8.7.1 Newton–Raphson Method

In Section 4.6.1 we introduced the Newton–Raphson method. Referring to notation there, we identify $L(\boldsymbol{\beta})$ as the log likelihood for Poisson loglinear models.

From (8.21), let

$$L(\boldsymbol{\beta}) = \sum_i n_i \left(\sum_h x_{ih} \beta_h \right) - \sum_i \exp\left(\sum_h x_{ih} \beta_h \right).$$

Then

$$u_j = \frac{\partial L(\boldsymbol{\beta})}{\partial \beta_j} = \sum_i n_i x_{ij} - \sum_i \mu_i x_{ij},$$

$$h_{jk} = \frac{\partial^2 L(\boldsymbol{\beta})}{\partial \beta_j \partial \beta_k} = -\sum_i \mu_i x_{ij} x_{ik},$$

so that

$$u_j^{(t)} = \sum_i \left(n_i - \mu_i^{(t)} \right) x_{ij} \quad \text{and} \quad h_{jk}^{(t)} = -\sum_i \mu_i^{(t)} x_{ij} x_{ik}.$$

The tth approximation $\boldsymbol{\mu}^{(t)}$ for $\hat{\boldsymbol{\mu}}$ derives from $\boldsymbol{\beta}^{(t)}$ through $\boldsymbol{\mu}^{(t)} = \exp(\mathbf{X}\boldsymbol{\beta}^{(t)})$. It generates the next value $\boldsymbol{\beta}^{(t+1)}$ using (4.39), which in this context is

$$\boldsymbol{\beta}^{(t+1)} = \boldsymbol{\beta}^{(t)} + \left[\mathbf{X}' \mathbf{diag}(\boldsymbol{\mu}^{(t)}) \mathbf{X} \right]^{-1} \mathbf{X}' \left(\mathbf{n} - \boldsymbol{\mu}^{(t)} \right).$$

This in turn produces $\boldsymbol{\mu}^{(t+1)}$, and so on.

Alternatively, $\boldsymbol{\beta}^{(t+1)}$ can be expressed as

$$\boldsymbol{\beta}^{(t+1)} = -\left(\mathbf{H}^{(t)} \right)^{-1} \mathbf{r}^{(t)}, \tag{8.29}$$

where $r_j^{(t)} = \sum \mu_i^{(t)} x_{ij} \left[\log \mu_i^{(t)} + \left(n_i - \mu_i^{(t)} \right) / \mu_i^{(t)} \right]$. The expression in brackets is the first term in the Taylor series expansion of $\log n_i$ at $\log \mu_i^{(t)}$.

The iterative process begins with all $\mu_i^{(0)} = n_i$, or with an adjustment such as $\mu_i^{(0)} = n_i + \frac{1}{2}$ if any $n_i = 0$. Then (8.29) produces $\boldsymbol{\beta}^{(1)}$, and for $t > 0$ the iterations proceed as just described with $\{n_i\}$. For loglinear models $L(\boldsymbol{\beta})$ is concave, and $\boldsymbol{\mu}^{(t)}$ and $\boldsymbol{\beta}^{(t)}$ usually converge rapidly to the ML estimates $\hat{\boldsymbol{\mu}}$ and $\hat{\boldsymbol{\beta}}$ as t increases. The $\mathbf{H}^{(t)}$ matrix converges to $\hat{\mathbf{H}} = -\mathbf{X}'\mathbf{diag}(\hat{\boldsymbol{\mu}})\mathbf{X}$. By (8.25), the estimated large-sample covariance matrix of $\hat{\boldsymbol{\beta}}$ is $-\hat{\mathbf{H}}^{-1}$, a by-product of the method.

As we discussed in Section 4.6.3 for GLMs, (8.29) has the iterative reweighted least squares form

$$\boldsymbol{\beta}^{(t+1)} = \left(\mathbf{X}'\hat{\mathbf{V}}_t^{-1}\mathbf{X}\right)^{-1}\mathbf{X}'\hat{\mathbf{V}}_t^{-1}\mathbf{z}^{(t)}.$$

Here, $\mathbf{z}^{(t)}$ has elements $n_i = \log \mu_i^{(t)} + (n_i - \mu_i^{(t)})/\mu_i^{(t)}$ and $\hat{\mathbf{V}}_t = [\mathbf{diag}(\boldsymbol{\mu}^{(t)})]^{-1}$. Thus, $\boldsymbol{\beta}^{(t+1)}$ is the weighted least squares solution for a model

$$\mathbf{z}^{(t)} = \mathbf{X}\boldsymbol{\beta} + \boldsymbol{\epsilon},$$

where $\{\epsilon_i\}$ are uncorrelated with variances $\{1/\mu_i^{(t)}\}$. With $\{\mu_i^{(0)} = n_i\}$, $\boldsymbol{\beta}^{(1)}$ is the weighted least squares estimate for model $\log(\mathbf{n}) = \mathbf{X}\boldsymbol{\beta} + \boldsymbol{\epsilon}$.

8.7.2 Iterative Proportional Fitting

The *iterative proportional fitting* (IPF) *algorithm* is a simple method for calculating $\{\hat{\mu}_i\}$ for loglinear models. Introduced by Deming and Stephan (1940), it has the following steps:

1. Start with $\{\mu_i^{(0)}\}$ satisfying a model no more complex than the one being fitted. For instance, $\{\mu_i^{(0)} \equiv 1.0\}$ are trivially adequate.
2. By multiplying by appropriate factors, adjust $\{\mu_i^{(0)}\}$ successively to match each marginal table in the set of minimal sufficient statistics.
3. Continue until the maximum difference between the sufficient statistics and their fitted values is sufficiently close to zero.

We illustrate using model (XY, XZ, YZ). Its minimal sufficient statistics are $\{n_{ij+}\}$, $\{n_{i+k}\}$, and $\{n_{+jk}\}$. Initial estimates must satisfy the model. The first cycle of the IPF algorithm has three steps:

$$\mu_{ijk}^{(1)} = \mu_{ijk}^{(0)}\frac{n_{ij+}}{\mu_{ij+}^{(0)}}, \quad \mu_{ijk}^{(2)} = \mu_{ijk}^{(1)}\frac{n_{i+k}}{\mu_{i+k}^{(1)}}, \quad \mu_{ijk}^{(3)} = \mu_{ijk}^{(2)}\frac{n_{+jk}}{\mu_{+jk}^{(2)}}.$$

Summing both sides of the first expression over k shows that $\mu_{ij+}^{(1)} = n_{ij+}$ for all i and j. After step 1, observed and fitted values match in the XY marginal table. After step 2, all $\mu_{i+k}^{(2)} = n_{i+k}$, but the XY marginal tables no longer match. After step 3, all $\mu_{+jk}^{(3)} = n_{+jk}$, but the XY and XZ marginal tables no longer match. A new cycle begins by again matching the XY marginal tables, using $\mu_{ijk}^{(4)} = \mu_{ijk}^{(3)}(n_{ij+}/\mu_{ij+}^{(3)})$, and so on.

At each step, the updated estimates continue to satisfy the model. For instance, step 1 uses the same adjustment factor $(n_{ij+}/\mu_{ij+}^{(0)})$ at different levels k of Z. Thus, XY odds ratios from different levels of Z have ratio equal to 1, and the homogeneous association pattern continues at each step.

As the cycles progress, the G^2 statistic comparing cell counts to the updated fit is monotone decreasing, and the process must converge (Fienberg 1970a; Haberman 1974a). The IPF algorithm produces ML estimates because it generates a sequence of fitted values converging to a solution that both satisfies the model and matches the sufficient statistics. By Birch's results (Section 8.6.3), only one such solution exists, and it is ML.

The IPF method works even for models having direct estimates. Then, IPF normally yields ML estimates within one cycle (Haberman 1974a, p. 197). We illustrate with the model of independence. The minimal sufficient statistics are $\{n_{i+}\}$ and $\{n_{+j}\}$. With $\{\mu_{ij}^{(0)} \equiv 1.0\}$, the first cycle gives

$$\mu_{ij}^{(1)} = \mu_{ij}^{(0)}\frac{n_{i+}}{\mu_{i+}^{(0)}} = \frac{n_{i+}}{J},$$

$$\mu_{ij}^{(2)} = \mu_{ij}^{(1)}\frac{n_{+j}}{\mu_{+j}^{(1)}} = \frac{n_{i+}n_{+j}}{n}.$$

The IPF algorithm then gives $\hat{\mu}_{ij}^{(t)} = n_{i+}n_{+j}/n$ for all $t > 2$.

8.7.3 Comparison of Iterative Methods

The IPF algorithm is simple and easy to implement. It converges to the ML fit even when the likelihood is poorly behaved, for instance with zero fitted counts and estimates on the boundary of the parameter space. The Newton–Raphson method is more complex, requiring solving a system of equations at each step. Newton–Raphson is sometimes not feasible when the model is of high dimensionality—for instance, when the contingency table and parameter vector are huge.

However, IPF has disadvantages. It is applicable primarily to models for which likelihood equations equate observed and fitted counts in marginal tables. By contrast, Newton–Raphson is a general-purpose method that can solve more complex likelihood equations. IPF sometimes converges slowly compared to Newton–Raphson. Unlike Newton–Raphson, IPF does not produce the model parameter estimates and their estimated covariance matrix as a by-product. Fitted values that IPF produces can generate this information. Model parameter estimates are contrasts of $\{\log \hat{\mu}_i\}$ (see Problems 8.16 and 8.17), and substituting fitted values into (8.25) yields $\text{cov}(\hat{\boldsymbol{\beta}})$.

Because Newton–Raphson applies to a wide variety of models and also yields standard errors, it is the fitting routine used by most software for loglinear models. IPF is increasingly viewed as primarily of historical interest. However, for some applications the analysis is more transparent using IPF, as the next example illustrates.

8.7.4 Contingency Table Standardization

Table 8.15 relates education and attitudes toward legalized abortion using a General Social Survey, conducted by the National Opinion Research Center. To make patterns of association clearer, Smith (1976) standardized the table so that all row and column marginal totals equal 100 while maintaining the sample odds ratio structure.

The IPF routine to standardize with margins of 100 is

$$\mu_{ij}^{(0)} = n_{ij}$$

and then for $t = 1, 3, 5, \ldots,$

$$\mu_{ij}^{(t)} = \mu_{ij}^{(t-1)} \frac{100}{\mu_{i+}^{(t-1)}}, \quad \mu_{ij}^{(t+1)} = \mu_{ij}^{(t)} \frac{100}{\mu_{+j}^{(t)}}.$$

At the end of each odd-numbered step, all row totals equal 100. At the end of each even-numbered step, all column totals equal 100. Odds ratios do not change at each odd (even) step, since all counts in a given row (column) multiply by the same constant.

The IPF algorithm converges to the entries in parentheses in Table 8.15. The association is clearer in this standardized table. A ridge appears down the main diagonal, with higher levels of education having more favorable attitudes about abortion. The other counts fall away smoothly on both sides.

Table standardization is useful for comparing tables having different marginal structures. Mosteller (1968) compared intergenerational occupa-

TABLE 8.15 Marginal Standardization of Attitudes toward Abortion by Years of Schooling

	Attitude toward Legalized Abortion			
Schooling	Generally Disapprove	Middle Position	Generally Approve	Total
Less than high school	209 (49.4)	101 (32.0)	237 (18.6)	(100)
High school	151 (32.8)	126 (36.6)	426 (30.6)	(100)
More than high school	16 (17.8)	21 (31.3)	138 (50.9)	(100)
Total	(100)	(100)	(100)	

Source: Smith (1976).

tional mobility tables from Britain and Denmark. Yule (1912) compared three hospitals on vaccination and recovery for smallpox patients. A modern application is adjusting sample data to match marginal distributions specified by census results.

The process of table standardization is called *raking* the table. Imrey et al. (1981) and Little and Wu (1991) derived the asymptotic covariance matrix for raked sample proportions. For sample counts $\{n_{ij}\}$ with $\{\mu_{ij} = E(n_{ij})\}$, let $\{E_{ij}\}$ denote expected frequencies for the standardized table and $\{\hat{E}_{ij}\}$ fitted values in the standardized table. The standardization process corresponds to fitting the model

$$\log(E_{ij}/\mu_{ij}) = \lambda + \lambda_i^E + \lambda_j^A.$$

That is, maintaining the odds ratios means that the two-way tables of $\{E_{ij}/\mu_{ij}\}$ and of $\{\hat{E}_{ij}/n_{ij}\}$ satisfy independence.

The fitted values $\{\hat{E}_{ij}\}$ in the standardized table satisfy

$$\log \hat{E}_{ij} - \log n_{ij} = \hat{\lambda} + \hat{\lambda}_i^E + \hat{\lambda}_j^A.$$

The adjustment term, $-\log n_{ij}$, to the log link of the fit is called an *offset*. The fit corresponds to using $\log n_{ij}$ as a predictor on the right-hand side and forcing its coefficient to equal 1.0. Standard GLM software can fit models having offsets. To rake a table, one enters as sample data pseudo-values that satisfy independence and have the desired margins, taking $\log n_{ij}$ as an offset. (For SAS, see Table A.14). In Section 9.7.1 we discuss further the use of model offsets.

NOTES

Section 8.2: Loglinear Models for Independence and Interaction in Three-Way Tables

8.1. Roy and Mitra (1956) discussed types of independence for three-way tables and their large-sample tests. Birch's (1963) article on ML estimation for loglinear models was part of substantial research on loglinear models in the 1960s, much due to L. A. Goodman (see Section 16.4). Haberman (1974a) presented an influential theoretical study of loglinear models.

Section 8.3: Inference for Loglinear Models

8.2. Goodman (1970, 1971b), Haberman (1974a, Chap. 5), Lauritzen (1996), Sundberg (1975), and Whittaker (1990, Sec. 12.4) discussed families of loglinear models that have direct ML estimates and interpretations in terms of independence, conditional independence, or equiprobability. Such models are called *decomposable*, since expected frequencies decompose into products and ratios of expected marginal sufficient statistics. Haberman proved conditions under which loglinear models have direct estimates. Baglivo et al. (1992), Forster et al. (1996), and Morgan and Blumenstein (1991) discussed exact inference.

8.3. For methods that allow for misclassification error, see Kuha and Skinner (1997) and Kuha et al. (1998) and references therein. For treatment of missing data, see Little (1998), Schafer (1997, Chap. 8), and their references.

Section 8.7: Loglinear Model Fitting: Iterative Methods and Their Application

8.4. Deming (1964, Chap. VII) described early work on IPF by Deming and Stephan. Darroch (1962) used IPF to obtain ML estimates in contingency tables. Bishop et al. (1975), Fienberg (1970a), and Speed (1998) presented other applications of IPF. Darroch and Ratcliff (1972) generalized IPF for models in which sufficient statistics are more complex than marginal tables.

8.5. For further discussion of table raking, see Bishop et al. (1975, pp. 76–102), Fleiss (1981, Chap. 14), Haberman (1979, Chap. 9), Hoem (1987), and Little and Wu (1991).

PROBLEMS

Applications

8.1 The 1988 General Social Survey compiled by the National Opinion Research Center asked: "Do you support or oppose the following measures to deal with AIDS? (1) Have the government pay all of the health care costs of AIDS patients; (2) Develop a government information program to promote safe sex practices, such as the use of condoms." Table 8.16 summarizes opinions about health care costs (H) and the information program (I), classified also by the respondent's gender (G).

  a. Fit loglinear models (GH, GI), (GH, HI), (GI, HI), and (GH, GI, HI). Show that models that lack the HI term fit poorly.

  b. For model (GH, GI, HI), show that 95% Wald confidence intervals equal $(0.55, 1.10)$ for the GH conditional odds ratio and $(0.99, 2.55)$ for the GI conditional odds ratio. Interpret. Is it plausible that gender has no effect on opinion for these issues?

TABLE 8.16   Data for Problem 8.1

Gender	Information Opinion	Health Opinion	
		Support	Oppose
Male	Support	76	160
	Oppose	6	25
Female	Support	114	181
	Oppose	11	48

Source: 1988 General Social Survey, National Opinion Research Center.

TABLE 8.17   Data for Problem 8.2[a]

President	Busing	Home 1	Home 2	Home 3
1	1	41	65	0
	2	71	157	1
	3	1	17	0
2	1	2	5	0
	2	3	44	0
	3	1	0	0
3	1	0	3	1
	2	0	10	0
	3	0	0	1

[a]1, Yes; 2, no; 3, don't know.

Source: 1991 General Social Survey, National Opinion Research Center.

8.2 Refer to Table 8.17 from the 1991 General Social Survey. White subjects were asked: (B) "Do you favor busing of (Negro/Black) and white school children from one school district to another?", (P) "If your party nominated a (Negro/Black) for President, would you vote for him if he were qualified for the job?", (D) "During the last few years, has anyone in your family brought a friend who was a (Negro/Black) home for dinner?" The response scale for each item was (yes, no, don't know). Fit model (BD, BP, DP).

 a. Using the yes and no categories, estimate the conditional odds ratio for each pair of variables. Interpret.

 b. Analyze the model's goodness of fit. Interpret.

 c. Conduct inference for the BP conditional association using a Wald or likelihood-ratio confidence interval and test. Interpret.

8.3 Refer to Section 8.3.2. Explain why software for which parameters sum to zero across levels of each index reports $\hat{\lambda}_{11}^{AC} = \hat{\lambda}_{22}^{AC} = 0.514$ and $\hat{\lambda}_{12}^{AC} = \hat{\lambda}_{21}^{AC} = -0.514$, with SE = 0.044 for each term.

8.4 Refer to Table 2.6. Let D = defendant's race, V = victims' race, and P = death penalty verdict. Fit the loglinear model (DV, DP, PV).

 a. Using the fitted values, estimate and interpret the odds ratio between D and P at each level of V. Note the common odds ratio property.

 b. Calculate the marginal odds ratio between D and P, **(i)** using the fitted values, and **(ii)** using the sample data. Why are they equal? Contrast the odds ratio with part (a). Explain why Simpson's paradox occurs.

c. Fit the corresponding logit model, treating P as the response. Show the correspondence between parameter estimates and fit statistics.

d. Is there a simpler model that fits well? Interpret, and show the logit–loglinear connection.

TABLE 8.18 Data for Problem 8.5

Safety Equipment in Use	Whether Ejected	Injury	
		Nonfatal	Fatal
Seat belt	Yes	1,105	14
	No	411,111	483
None	Yes	4,624	497
	No	157,342	1,008

Source: Florida Department of Highway Safety and Motor Vehicles.

8.5 Table 8.18 refers to automobile accident records in Florida in 1988.

a. Find a loglinear model that describes the data well. Interpret associations.

b. Treating whether killed as the response, fit an equivalent logit model. Interpret the effects.

c. Since n is large, goodness-of-fit statistics are large unless the model fits very well. Calculate the dissimilarity index for the model in part (a), and interpret.

8.6 Refer to Table 8.19. Subjects were asked their opinions about government spending on the environment (E), health (H), assistance to big cities (C), and law enforcement (L).

TABLE 8.19 Data for Problem 8.6[a]

			Cities								
			1			2			3		
		Law									
Environment	Health	Enforcement:	1	2	3	1	2	3	1	2	3
1	1		62	17	5	90	42	3	74	31	11
	2		11	7	0	22	18	1	19	14	3
	3		2	3	1	2	0	1	1	3	1
2	1		11	3	0	21	13	2	20	8	3
	2		1	4	0	6	9	0	6	5	2
	3		1	0	1	2	1	1	4	3	1
3	1		3	0	0	2	1	0	9	2	1
	2		1	0	0	2	1	0	4	2	0
	3		1	0	0	0	0	0	1	2	3

[a]1, Too little; 2, about right; 3, too much.
Source: 1989 General Social Survey, National Opinion Research Center.

TABLE 8.20   Output for Fitting Model to Table 8.19

Criteria For Assessing Goodness Of Fit

Criterion	DF	Value	Value / DF
Deviance	48	31.6695	0.6598
Pearson Chi-Square	48	26.5224	0.5526
Log Likelihood		1284.9404	

Parameter			DF	Estimate	Standard Error	Wald 95% Confidence	Limits	Chi-Square
e*h	1	1	1	2.1425	0.5566	1.0515	3.2335	14.81
e*h	1	2	1	1.4221	0.6034	0.2394	2.6049	5.55
e*h	2	1	1	0.7294	0.5667	-0.3813	1.8402	1.66
e*h	2	2	1	0.3183	0.6211	-0.8991	1.5356	0.26
e*l	1	1	1	-0.1328	0.6378	-1.3829	1.1172	0.04
e*l	1	2	1	0.3739	0.6975	-0.9931	1.7410	0.29
e*l	2	1	1	-0.2630	0.6796	-1.5949	1.0689	0.15
e*l	2	2	1	0.4250	0.7361	-1.0178	1.8678	0.33
e*c	1	1	1	1.2000	0.5177	0.1854	2.2147	5.37
e*c	1	2	1	1.3896	0.4774	0.4540	2.3253	8.47
e*c	2	1	1	0.6917	0.5605	-0.4068	1.7902	1.52
e*c	2	2	1	1.3767	0.5024	0.3921	2.3614	7.51
h*c	1	1	1	-0.1865	0.4547	-1.0777	0.7048	0.17
h*c	1	2	1	0.7464	0.4808	-0.1959	1.6886	2.41
h*c	2	1	1	-0.4675	0.4978	-1.4431	0.5081	0.88
h*c	2	2	1	0.7293	0.5023	-0.2553	1.7138	2.11
h*l	1	1	1	1.8741	0.5079	0.8786	2.8696	13.61
h*l	1	2	1	1.0366	0.5262	0.0052	2.0680	3.88
h*l	2	1	1	1.9371	0.6226	0.7168	3.1574	9.68
h*l	2	2	1	1.8230	0.6355	0.5775	3.0686	8.23
c*l	1	1	1	0.8735	0.4604	-0.0289	1.7760	3.60
c*l	1	2	1	0.5707	0.4863	-0.3824	1.5239	1.38
c*l	2	1	1	1.0793	0.4326	0.2314	1.9271	6.23
c*l	2	2	1	1.2058	0.4462	0.3312	2.0804	7.30

a. Table 8.20 shows some results, including the two-factor estimates, for the homogeneous association model. Check the fit, and interpret.

b. All estimates at category 3 of each variable equal 0. Report the estimated conditional odds ratios using the too much and too little categories for each pair of variables. Summarize the associations. Based on these results, which term(s) might you consider dropping from the model? Why?

c. Table 8.21 reports $\{\hat{\lambda}_{eh}^{EH}\}$ when parameters sum to zero within rows and within columns, and when parameters are zero in the first row and first column. Show how these yield the estimated EH conditional odds ratio for the too much and too little categories. Compare to part (b). Construct a confidence interval for that odds ratio. Interpret.

TABLE 8.21 Parameter Estimates for Problem 8.6

	Sum to Zero Constraints H			Zero for First Level H		
E	1	2	3	1	2	3
1	0.509	0.166	−0.676	0	0	0
2	−0.065	−0.099	0.163	0	0.309	1.413
3	−0.445	−0.068	0.513	0	0.720	2.142

8.7 Refer to the loglinear models for Table 8.8.
   a. Explain why the fitted odds ratios in Table 8.10 for model (GI, GL, GS, IL, IS, LS) suggest that the most likely accident case for injury is females not wearing seat belts in rural locations.
   b. Fit model (GLS, GI, IL, IS). Using model parameter estimates, show that the fitted IS conditional odds ratio equals 0.44. Show that for each injury level, the estimated conditional LS odds ratio is 1.17 for $(G = $ female) and 1.03 for $(G = $ male). How can you get these using the model parameter estimates?

8.8 Consider the following two-stage model for Table 8.8. The first stage is a logit model with S as the response for the three-way GLS table. The second stage is a logit model with these three variables as predictors for I in the four-way table. Explain why this composite model is sensible, fit the models, and interpret results.

8.9 Refer to the logit model in Problem 5.24. Let $A = $ opinion on abortion.
   a. Give the symbol for the loglinear model that is equivalent to this logit model.
   b. Which logit model corresponds to loglinear model (AR, AP, GRP)?
   c. State the equivalent loglinear and logit models for which (i) A is jointly independent of G, R, and P; (ii) there are main effects of R on A, but A is conditionally independent of G and P, given R; (iii) there is interaction between P and R in their effects on A, and G has main effects.

8.10 For a multiway contingency table, when is a logit model more appropriate than a loglinear model? When is a loglinear model more appropriate?

8.11 Using software, conduct the analyses described in this chapter for the student survey data (Table 8.3).

8.12 Standardize Table 10.6. Describe the migration patterns.

8.13 The book's Web site (*www.stat.ufl.edu/~aa/cda/cda.html*) has a $2 \times 3 \times 2 \times 2$ table relating responses on frequency of attending religious services, political views, opinion on making birth control available to teenagers, and opinion about a man and woman having sexual relations before marriage. Analyze these data using loglinear models.

Theory and Methods

8.14 Suppose that $\{\mu_{ij} = n\pi_{ij}\}$ satisfy the independence model (8.1).
    a. Show that $\lambda_a^Y - \lambda_b^Y = \log(\pi_{+a}/\pi_{+b})$.
    b. Show that {all $\lambda_j^Y = 0$} is equivalent to $\pi_{+j} = 1/J$ for all j.

8.15 Refer to the independence model, $\mu_{ij} = \mu\alpha_i\beta_j$. For the corresponding loglinear model (8.1):
    a. Show that one can constrain $\Sigma\lambda_i^X = \Sigma\lambda_j^Y = 0$ by setting

$$\lambda_i^X = \log \alpha_i - \left(\sum_h \log \alpha_h\right)\bigg/ I, \quad \lambda_j^Y = \log \beta_j - \left(\sum_h \log \beta_h\right)\bigg/ J,$$

$$\lambda = \log \mu + \left(\sum_h \log \alpha_h\right)\bigg/ I + \left(\sum_h \log \beta_h\right)\bigg/ J.$$

    b. Show that one can constrain $\lambda_1^X = \lambda_1^Y = 0$ by defining $\lambda_i^X = \log \alpha_i - \log \alpha_1$ and $\lambda_j^Y = \log \beta_j - \log \beta_1$. Then, what does λ equal?

8.16 For an $I \times J$ table, let $\eta_{ij} = \log \mu_{ij}$, and let a dot subscript denote the mean for that index (e.g., $\eta_{i.} = \Sigma_j\eta_{ij}/J$). Then, let $\lambda = \eta_{..}$, $\lambda_i^X = \eta_{i.} - \eta_{..}$, $\lambda_j^Y = \eta_{.j} - \eta_{..}$, and $\lambda_{ij}^{XY} = \eta_{ij} - \eta_{i.} - \eta_{.j} + \eta_{..}$.
    a. Show that $\log \mu_{ij} = \lambda + \lambda_i^X + \lambda_j^Y + \lambda_{ij}^{XY}$. Hence, any set of positive $\{\mu_{ij}\}$ satisfies the saturated model.
    b. Show that $\Sigma_i\lambda_i^X = \Sigma_j\lambda_j^Y = \Sigma_i\lambda_{ij}^{XY} = \Sigma_j\lambda_{ij}^{XY} = 0$.
    c. For 2×2 tables, show that $\log \theta = 4\lambda_{11}^{XY}$.
    d. For $2 \times J$ tables, show that $\lambda_{11}^{XY} = (\Sigma_j\log \alpha_j)/2J$, where $\alpha_j = \mu_{11}\mu_{2j}/\mu_{21}\mu_{1j}$, $j = 2,\ldots,J$.
    e. Alternative constraints have other odds ratio formulas. Let $\lambda = \eta_{11}$, $\lambda_i^X = \eta_{i1} - \eta_{11}$, $\lambda_j^Y = \eta_{1j} - \eta_{11}$, and $\lambda_{ij}^{XY} = \eta_{ij} - \eta_{i1} - \eta_{1j} + \eta_{11}$. Then, show that the saturated model holds with $\lambda_1^X = \lambda_1^Y = \lambda_{1j}^{XY} = \lambda_{i1}^{XY} = 0$ for all i and j, and $\lambda_{ij}^{XY} = \log(\mu_{11}\mu_{ij}/\mu_{1j}\mu_{i1})$.

8.17 Suppose that all $\mu_{ijk} > 0$. Let $\eta_{ijk} = \log \mu_{ijk}$, and consider model parameters with zero-sum constraints.

    a. For the general loglinear model (8.12), define parameters in the fashion of Problem 8.16 (e.g., $\lambda_{ij}^{XY} = \eta_{ij.} - \eta_{i..} - \eta_{.j.} + \eta_{...}$).

    b. For model (XY, XZ, YZ) with a $2 \times 2 \times 2$ table, show that $\lambda_{11}^{XY} = \frac{1}{4}\log \theta_{11(k)}$.

    c. For (XYZ) with a $2 \times 2 \times 2$ table, show that

$$\lambda_{111}^{XYZ} = \tfrac{1}{8}\log\left[\theta_{11(1)}/\theta_{11(2)}\right].$$

    Thus, $\lambda_{ijk}^{XYZ} = 0$ is equivalent to $\theta_{11(1)} = \theta_{11(2)}$.

8.18 Two balanced coins are flipped, independently. Let $X = $ whether the first flip resulted in a head (yes, no), $Y = $ whether the second flip resulted in a head, and $Z = $ whether both flips had the same result. Using this example, show that marginal independence for each pair of three variables does not imply that the variables are mutually independent.

8.19 For three categorical variables X, Y, and Z:

    a. When Y is jointly independent of X and Z, show that X and Y are conditionally independent, given Z.

    b. Prove that mutual independence of X, Y, and Z implies that X and Y are both marginally and conditionally independent.

    c. When X is independent of Y and Y is independent of Z, does it follow that X is independent of Z? Explain.

    d. When any pair of variables is conditionally independent, explain why there is no three-factor interaction.

8.20 Suppose that X and Y are conditionally independent, given Z, and X and Z are marginally independent.

    a. Show that X is jointly independent of Y and Z.

    b. Show X and Y are marginally independent.

    c. Show that if X and Z are conditionally (rather than marginally) independent, then X and Y are still marginally independent.

8.21 A $2 \times 2 \times 2$ table satisfies $\pi_{i++} = \pi_{+j+} = \pi_{++k} = \frac{1}{2}$, all i, j, k. Give an example of $\{\pi_{ijk}\}$ that satisfies model (**a**) (X, Y, Z), (**b**) (XY, Z), (**c**) (XY, YZ), (**d**) (XY, XZ, YZ), and (**e**) (XYZ), but in each case not a simpler model.

8.22 Suppose that model (XY, XZ, YZ) holds in a $2 \times 2 \times 2$ table, and the common XY conditional log odds ratio at the two levels of Z is positive. If the XZ and YZ conditional log odds ratios are both positive or both negative, show that the XY marginal odds ratio is larger than the XY conditional odds ratio. Hence, Simpson's paradox cannot occur for the XY association.

8.23 Show that the general loglinear model in T dimensions has 2^T terms. [*Hint:* It has an intercept, $\binom{T}{1}$ single-factor terms, $\binom{T}{2}$ two-factor terms, ...]

8.24 Each of T responses is binary. For dummy variables $\{z_1, \ldots, z_T\}$, the loglinear model of mutual independence has the form

$$\log \mu_{z_1, \ldots, z_T} = \lambda_1 z_1 + \cdots + \lambda_T z_T.$$

Show how to express the general loglinear model (Cox 1972).

8.25 Consider a cross-classification of W, X, Y, Z.
   a. Explain why (WXZ, WYZ) is the most general loglinear model for which X and Y are conditionally independent.
   b. State the model symbol for which X and Y are conditionally independent *and* there is no three-factor interaction.

8.26 For a four-way table with binary response Y, give the equivalent loglinear and logit models that have:
   a. Main effects of A, B, and C on Y.
   b. Interaction between A and B in their effects on Y, and C has main effects.
   c. Repeat part (a) for a nominal response Y with a baseline-category logit model.

8.27 For a 3×3 table with ordered rows having scores $\{x_i\}$, identify all terms in the generalized loglinear model (8.18) for models **(a)** $\text{logit}[P(Y \leq j)] = \alpha_j + \beta x_i$, and **(b)** $\log[P(Y = j)/P(Y = 3)] = \alpha_j + \beta_j x_i$.

8.28 For the independence model for a two-way table, derive minimal sufficient statistics, likelihood equations, fitted values, and residual df.

8.29 For the loglinear model for an $I \times J$ table, $\log \mu_{ij} = \lambda + \lambda_i^X$, show that $\hat{\mu}_{ij} = n_{i+}/J$ and residual df $= I(J - 1)$.

8.30 Write the log likelihood L for model (XZ, YZ). Calculate $\partial L/\partial \lambda$ and show that it implies $\hat{\mu}_{+++} = n$. Show that $\partial L/\partial \lambda_i^X = n_{i++} - \mu_{i++}$.

Similarly, differentiate with respect to each parameter to obtain likelihood equations. Show (8.23) and (8.24) imply the other equations, so those equations determine the ML estimates.

8.31 For model (XY, Z), derive (**a**) minimal sufficient statistics, (**b**) likelihood equations, (**c**) fitted values, and (**d**) residual df for tests of fit.

8.32 Consider the loglinear model with symbol (XZ, YZ).

    a. For fixed k, show that $\{\hat{\mu}_{ijk}\}$ equal the fitted values for testing independence between X and Y within level k of Z.

    b. Show that the Pearson and likelihood-ratio statistics for testing this model's fit have form $X^2 = \sum X_k^2$, where X_k^2 tests independence between X and Y at level k of Z.

8.33 Verify the df values shown in Table 8.14 for models (XY, Z), (XY, YZ), and (XY, XZ, YZ).

8.34 Verify that loglinear model (GLS, GI, LI, IS) implies logit model (8.16). Show that the conditional log odds ratio for the effect of S on I equals $\beta_1^S - \beta_2^S$ in the logit model and $\lambda_{11}^{IS} + \lambda_{22}^{IS} - \lambda_{12}^{IS} - \lambda_{21}^{IS}$ in the loglinear model.

8.35 Table 8.22 shows fitted values for models for four-way tables that have direct estimates.

    a. Use Birch's results to verify that the entry is correct for (W, X, Y, Z). Verify its residual df.

    b. Motivate the estimate and df formulas for (WX, YZ), (WXY, Z), (WXY, WZ), and (WXY, WXZ) using composite variables and the corresponding results for two-way tables [e.g., for (WXY, WZ), given W, Z is independent of the composite XY variable].

TABLE 8.22   Data for Problem 8.35[a]

Model	Expected Frequency Estimate	Residual DF
(W, X, Y, Z)	$n_{h+++}n_{+i++}n_{++j+}n_{+++k}/n^3$	$HIJK - H - I - J - K + 3$
(WX, Y, Z)	$n_{hi++}n_{++j+}n_{+++k}/n^2$	$HIJK - HI - J - K + 2$
(WX, WY, Z)	$n_{hi++}n_{h+j+}n_{+++k}/n_{h+++}n$	$HIJK - HI - HJ - K + H + 1$
(WX, YZ)	$n_{hi++}n_{++jk}/n$	$(HI - 1)(JK - 1)$
(WX, WY, XZ)	$n_{hi++}n_{h+j+}n_{+i+k}/n_{h+++}n_{+i++}$	$HIJK - HI - HJ - IK + H + I$
(WX, WY, WZ)	$n_{hi++}n_{h+j+}n_{h++k}/(n_{h+++})^2$	$HIJK - HI - HJ - HK + 2H$
(WXY, Z)	$n_{hij+}n_{+++k}/n$	$(HIJ - 1)(K - 1)$
(WXY, WZ)	$n_{hij+}n_{h++k}/n_{h+++}$	$H(IJ - 1)(K - 1)$
(WXY, WXZ)	$n_{hij+}n_{hi+k}/n_{hi++}$	$HI(J - 1)(K - 1)$

[a] Number of levels of W, X, Y, Z, denoted by H, I, J, K. Estimates for other models of each type are obtained by symmetry.

8.36 A T-dimensional table $\{n_{ab \ldots t}\}$ has I_i categories in dimension i.

    a. Find minimal sufficient statistics, ML estimates of cell probabilities, and residual df for the mutual independence model.

    b. Find the minimal sufficient statistics and residual df for the hierarchical model having all two-factor associations but no three-factor interactions.

8.37 Consider loglinear model (X, Y, Z) for a $2 \times 2 \times 2$ table.

    a. Express the model in the form $\log \boldsymbol{\mu} = \mathbf{X}\boldsymbol{\beta}$.

    b. Show that the likelihood equations $\mathbf{X}'\mathbf{n} = \mathbf{X}'\hat{\boldsymbol{\mu}}$ equate $\{n_{ijk}\}$ and $\{\hat{\mu}_{ijk}\}$ in the one-dimensional margins.

8.38 Apply IPF to model **(a)** (X, YZ), and **(b)** (XZ, YZ). Show that the ML estimates result within one cycle.

8.39 Given target row totals $\{r_i > 0\}$ and column totals $\{c_j > 0\}$:

    a. Explain how to use IPF to adjust sample proportions $\{p_{ij}\}$ to have these totals but maintain the sample odds ratios.

    b. Show how to find cell proportions that have these totals and for which all local odds ratios equal $\theta > 0$. (*Hint:* Take initial values of 1.0 in all cells in the first row and in the first column. This determines all other initial cell entries such that all local odds ratios equal θ.)

    c. Explain how cell proportions are determined by the marginal proportions and the local odds ratios.

8.40 Refer to Birch's results in Section 8.6.3. Show that L has individual terms converging to $-\infty$ as $\log \mu_i \to \pm\infty$. Explain why positive definiteness of the information matrix implies that the solution of the likelihood equations is unique, with likelihood maximized at that point.

CHAPTER 9

Building and Extending Loglinear/Logit Models

In Chapters 5 through 7 we presented logistic regression models, which use the logit link for binomial or multinomial responses. In Chapter 8 we presented loglinear models for contingency tables, which use the log link for Poisson cell counts. Equivalences between them were discussed in Section 8.5.3. In this chapter we discuss building and extending these models with contingency tables.

In Section 9.1 we present graphs that show a model's association and conditional independence patterns. In Section 9.2 we discuss selection and comparison of loglinear models. Diagnostics for checking models, such as residuals, are presented in Section 9.3.

The loglinear models of Chapter 8 treat all variables as nominal. In Section 9.4 we present loglinear models of association between ordinal variables. In Sections 9.5 and 9.6 we present generalizations that replace fixed scores by parameters. In the final section we discuss complications that occur with sparse contingency tables.

9.1 ASSOCIATION GRAPHS AND COLLAPSIBILITY

A graphical representation for associations in loglinear models indicates the pairs of conditionally independent variables. This representation helps reveal implications of models. Our presentation derives partly from Darroch et al. (1980), who used mathematical graph theory to represent certain loglinear models (called *graphical models*) having a conditional independence structure.

9.1.1 Association Graphs

An *association graph* has a set of vertices, each vertex representing a variable. An edge connecting two variables represents a conditional association be-

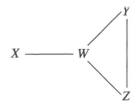

FIGURE 9.1   Association graph for model (WX, WY, WZ, YZ).

tween them. For instance, loglinear model (WX, WY, WZ, YZ) lacks XY and XZ terms. It assumes independence between X and Y and between X and Z, conditional on the remaining two variables. Figure 9.1 portrays this model's association graph. The four variables form the vertices. The four edges represent pairwise conditional associations. Edges do not connect X and Y or X and Z, the conditionally independent pairs.

Two loglinear models with the same pairwise associations have the same association graph. For instance, this association graph is also the one for model (WX, WYZ), which adds a three-factor WYZ interaction.

A *path* in an association graph is a sequence of edges leading from one variable to another. Two variables X and Y are said to be *separated* by a subset of variables if all paths connecting X and Y intersect that subset. For instance, in Figure 9.1, W separates X and Y, since any path connecting X and Y goes through W. The subset $\{W, Z\}$ also separates X and Y. A fundamental result states that two variables are conditionally independent given *any* subset of variables that separates them (Kreiner 1987; Whittaker 1990, p. 67). Thus, not only are X and Y conditionally independent given W and Z, but also given W alone. Similarly, X and Z are conditionally independent given W alone.

9.1.2   Collapsibility in Three-Way Contingency Tables

In Section 2.3.3 we showed that conditional associations in partial tables usually differ from marginal associations. Under certain *collapsibility conditions*, however, they are the same.

> For three-way tables, XY marginal and conditional odds ratios are identical if either Z and X are conditionally independent or if Z and Y are conditionally independent.

The conditions state that the variable treated as the control (Z) is conditionally independent of X or Y, or both. These conditions occur for loglinear models (XY, YZ) and (XY, XZ). Thus, the fitted XY odds ratio is identical in the partial tables and the marginal table for models with association graphs

$$X \text{——} Y \text{——} Z \quad \text{and} \quad Y \text{——} X \text{——} Z$$

or even simpler models, but not for the model with graph

$$X \text{——} Z \text{——} Y$$

in which an edge connects Z to both X and Y. The proof follows directly from the formulas for models (XY, YZ) and (XY, XZ) (Problem 9.26).

We illustrate for the student survey (Table 8.3) from Section 8.2.4, with A = alcohol use, C = cigarette use, and M = marijuana use. Model (AM, CM) specifies AC conditional independence, given M. It has association graph

$$A \text{——} M \text{——} C.$$

Consider the AM association. Since C is conditionally independent of A, the AM fitted conditional odds ratios are the same as the AM fitted marginal odds ratio collapsed over C. From Table 8.5, both equal 61.9. Similarly, the CM association is collapsible. The AC association is not, because M is conditionally dependent with both A and C in model (AM, CM). Thus, A and C may be marginally dependent, even though they are conditionally independent. In fact, from Table 8.5, the fitted AC marginal odds ratio for this model is 2.7.

For model (AC, AM, CM), no pair is conditionally independent. No collapsibility conditions are fulfilled. Table 8.5 showed that each pair has quite different fitted marginal and conditional associations for this model. When a model contains all two-factor effects, effects may change after collapsing over any variable.

9.1.3 Collapsibility and Logit Models

The collapsibility conditions apply also to logit models. For instance, suppose that a clinical trial studies the association between a binary treatment variable X ($x_1 = 1$, $x_2 = 0$) and a binary response Y, using data from K centers (Z). The logit model

$$\text{logit}[P(Y = 1 | X = i, Z = k)] = \alpha + \beta x_i + \beta_k^Z$$

has the same treatment effect β for each center. Since this model corresponds to loglinear model (XY, XZ, YZ), this effect may differ after collapsing the $2 \times 2 \times K$ table over centers. The estimated XY conditional odds ratio, $\exp(\hat{\beta})$, typically differs from the sample odds ratio in the marginal 2×2 table.

Next, consider the simpler model that lacks center effects,

$$\text{logit}[P(Y = 1 | X = i, Z = k)] = \alpha + \beta x_i.$$

For a given treatment, the success probability is identical for each center. The model satisfies a collapsibility condition, because it states that Z is conditionally independent of Y, given X. This logit model is equivalent to loglinear model (XY, XZ), for which the XY association is collapsible. So, when center effects are negligible and the simpler model fits nearly as well, the estimated treatment effect is approximately the marginal XY odds ratio.

9.1.4 Collapsibility and Association Graphs for Multiway Tables

Bishop et al. (1975, p. 47) provided a parametric collapsibility condition with multiway tables:

> Suppose that a model for a multiway table partitions variables into three mutually exclusive subsets, A, B, C, such that B separates A and C. After collapsing the table over the variables in C, parameters relating variables in A and parameters relating variables in A to variables in B are unchanged.

We illustrate using model (WX, WY, WZ, YZ) (Figure 9.1). Let $A = \{X\}$, $B = \{W\}$, and $C = \{Y, Z\}$. Since the XY and XZ terms do not appear, all parameters linking set A with set C equal zero, and B separates A and C. If we collapse over Y and Z, the WX association is unchanged. Next, identify $A = \{Y, Z\}$, $B = \{W\}$, $C = \{X\}$. Then, conditional associations among W, Y, and Z remain the same after collapsing over X.

This result also implies that when any variable is independent of all other variables, collapsing over it does not affect any other model terms. For instance, associations among W, X, and Y in model (WX, WY, XY, Z) are the same as in (WX, WY, XY).

When set B contains more than one variable, although parameter values are unchanged in collapsing over set C, the ML estimates of those parameters may differ slightly. A stronger collapsibility definition also requires that the estimates be identical. This condition of commutativity of fitting and collapsing holds if the model contains the highest-order term relating variables in B to each other. Asmussen and Edwards (1983) discussed this property, which relates to *decomposability* of tables (Note 8.2).

9.2   MODEL SELECTION AND COMPARISON

Strategies for selecting and comparing loglinear models are similar to those for logistic regression discussed in Section 6.1. A model should be complex enough to fit well but also relatively simple to interpret, smoothing rather than overfitting the data.

9.2.1   Considerations in Model Selection

The potentially useful models are usually a small subset of the possible models. A study designed to answer certain questions through confirmatory analyses may plan to compare models that differ only by the inclusion of certain terms. Also, models should recognize distinctions between response and explanatory variables. The modeling process should concentrate on terms linking responses and terms linking explanatory variables to responses. The model should contain the most general interaction term relating the explanatory variables. From the likelihood equations, this has the effect of equating the fitted totals to the sample totals at combinations of their levels. This is natural, since one normally treats such totals as fixed. Related to this, certain marginal totals are often fixed by the sampling design. Any potential model should include those totals as sufficient statistics, so likelihood equations equate them to the fitted totals.

Consider Table 8.8 with I = automobile injury and S = seat-belt use as responses and G = gender and L = location as explanatory variables. Then we treat $\{n_{g+\ell+}\}$ as fixed at each combination for G and L. For example, 20,629 women had accidents in urban locations, so the fitted counts should have 20,629 women in urban locations. To ensure this, a loglinear model should contain the GL term, which implies from its likelihood equations that $\{\hat{\mu}_{g+\ell+} = n_{g+\ell+}\}$. Thus, the model should be at least as complex as (GL, S, I) and focus on the effects of G and L on S and I as well as the SI association.

If S is also explanatory and only I is a response, $\{n_{g+\ell s}\}$ should be fixed. With a single categorical response, relevant loglinear models correspond to logit models for that response. One should then use logit rather than loglinear models, when the main focus is describing effects on that response.

For exploratory studies, a search among potential models may provide clues about associations and interactions. One approach first fits the model having single-factor terms, then the model having two-factor and single-factor terms, then the model having three-factor and lower terms, and so on. Fitting such models often reveals a restricted range of good-fitting models. In Section 8.4.2 we used this strategy with the automobile injury data set. Automatic search mechanisms among possible models, such as backward elimination, may also be useful but should be used with care and skepticism. Such a strategy need not yield a meaningful model.

9.2.2 Model Building for the Dayton Student Survey

In Sections 8.2.4 and 8.3.2 we analyzed the use of alcohol (A), cigarettes (C), and marijuana (M) by a sample of high school seniors. The study also classified students by gender (G) and race (R). Table 9.1 shows the five-dimensional contingency table. In selecting a model, we treat A, C, and M as responses and G and R as explanatory. Thus, a model should contain the GR term, which forces the GR fitted marginal totals to equal the sample marginal totals

Table 9.2 displays goodness-of-fit tests for several models. Because many cell counts are small, the chi-squared approximation for G^2 may be poor, but this index is useful for comparing models. The first model listed contains only the GR association and assumes conditional independence for the other nine pairs of associations. It fits horribly, which is no surprise. Model 2, with all two-factor terms, on the other hand, seems to fit well. Model 3, containing all

TABLE 9.1   Alcohol, Cigarette, and Marijuana Use for High School Seniors

Alcohol Use	Cigarette Use	Marijuana Use							
		Race = White				Race = Other			
		Female		Male		Female		Male	
		Yes	No	Yes	No	Yes	No	Yes	No
Yes	Yes	405	268	453	228	23	23	30	19
	No	13	218	28	201	2	19	1	18
No	Yes	1	17	1	17	0	1	1	8
	No	1	117	1	133	0	12	0	17

Source: Harry Khamis, Wright State University.

TABLE 9.2   Goodness-of-Fit Tests for Loglinear Models for Table 9.1

Model[a]	G^2	df
1. Mutual independence + GR	1325.1	25
2. Homogeneous association	15.3	16
3. All three-factor terms	5.3	6
4a. $(2)-AC$	201.2	17
4b. $(2)-AM$	107.0	17
4c. $(2)-CM$	513.5	17
4d. $(2)-AG$	18.7	17
4e. $(2)-AR$	20.3	17
4f. $(2)-CG$	16.3	17
4g. $(2)-CR$	15.8	17
4h. $(2)-GM$	25.2	17
4i. $(2)-MR$	18.9	17
5. $(AC, AM, CM, AG, AR, GM, GR, MR)$	16.7	18
6. $(AC, AM, CM, AG, AR, GM, GR)$	19.9	19
7. (AC, AM, CM, AG, AR, GR)	28.8	20

[a]G, gender; R, race; A, alcohol use; C, cigarette use; M, marijuana use.

the three-factor interaction terms, also fits well, but the improvement in fit is not great (difference in G^2 of $15.3 - 5.3 = 10.0$ based on df = $16 - 6 = 10$). Thus, we consider models without three-factor terms. Beginning with model 2, we eliminate two-factor terms. We use backward elimination, sequentially taking out terms for which the resulting increase in G^2 is smallest, when refitting the model.

Table 9.2 shows the start of this process. Nine pairwise associations are candidates for removal from model 2 (all except GR), shown in models 4a through 4i. The smallest increase in G^2, compared to model 2, occurs in removing the CR term (i.e., model 4g). The increase is $15.8 - 15.3 = 0.5$, with df = $17 - 16 = 1$, so this elimination seems sensible. After removing it, the smallest additional increase results from removing the CG term (model 5), resulting in $G^2 = 16.7$ with df = 18, and a change in G^2 of 0.9 based on df = 1. Removing next the MR term (model 6) yields $G^2 = 19.9$ with df = 19, a change in G^2 of 3.2 based on df = 1.

Further removals have a more severe effect. For instance, removing the AG term increases G^2 by 5.3, with df = 1, for a P-value of 0.02. One cannot take such P-values literally, since the data suggested these tests, but it seems safest not to drop additional terms. [See Westfall and Wolfinger (1997) and Westfall and Young (1993) for methods of adjusting P-values to account for multiple tests]. Model 6, denoted by $(AC, AM, CM, AG, AR, GM, GR)$, has association graph

$$M \frac{\quad\quad}{} G$$
$$C \frac{\quad\quad}{A} \frac{\quad\quad}{R}$$

Every path between C and $\{G, R\}$ involves a variable in $\{A, M\}$. Given the outcome on alcohol use and marijuana use, the model states that cigarette use is independent of both gender and race. Collapsing over the explanatory variables race and gender, the conditional associations between C and A and between C and M are the same as with the model (AC, AM, CM) fitted in Section 8.2.4.

Removing the GM term from this model yields model 7 in Table 9.2. Its association graph reveals that A separates $\{G, R\}$ from $\{C, M\}$. Thus, all pairwise conditional associations among A, C, and M in model 7 are identical to those in model (AC, AM, CM), collapsing over G and R. In fact, model 7 does not fit poorly ($G^2 = 28.8$ with df = 20) considering the large sample size. (Its sample dissimilarity index is $\hat{\Delta} = 0.036$.) Hence, one might collapse over gender and race in studying associations among the primary variables. An advantage of the full five-variable model is that it estimates effects of gender and race on these responses, in particular the effects of race and gender on alcohol use and the effect of gender on marijuana use.

9.2.3 Loglinear Model Comparison Statistics

Consider two loglinear models, M_1 and M_0, with M_0 a special case of M_1. By Sections 4.5.4 and 5.4.3, the likelihood-ratio statistic for testing M_0 against M_1 is $G^2(M_0 | M_1) = G^2(M_0) - G^2(M_1)$. We used this statistic above in comparing pairs of models.

Let **n** denote a column vector of the observed cell counts $\{n_i\}$. Let $\hat{\mu}_0$ and $\hat{\mu}_1$ denote vectors of the fitted values $\{\hat{\mu}_{0i}\}$ and $\{\hat{\mu}_{1i}\}$ for M_0 and M_1. The deviance $G^2(M_0)$ for the simpler model partitions into

$$G^2(M_0) = G^2(M_1) + G^2(M_0 | M_1). \tag{9.1}$$

Just as $G^2(M)$ measures the distance of fitted values for M from **n**, $G^2(M_0 | M_1)$ measures the distance of fit $\hat{\mu}_0$ from fit $\hat{\mu}_1$. In this sense, decomposition (9.1) expresses a certain orthogonality: The distance of $\mathbf{n}$ from $\hat{\boldsymbol{\mu}}_0$ equals the distance of $\mathbf{n}$ from $\hat{\boldsymbol{\mu}}_1$ plus the distance of $\hat{\boldsymbol{\mu}}_1$ from $\hat{\boldsymbol{\mu}}_0$.

The model comparison statistic equals

$$G^2(M_0 \mid M_1) = 2\sum_i n_i \log(n_i/\hat{\mu}_{0i}) - 2\sum_i n_i \log(n_i/\hat{\mu}_{1i})$$

$$= 2\sum_i n_i \log(\hat{\mu}_{1i}/\hat{\mu}_{0i}). \tag{9.2}$$

The two loglinear models have the matrix form (8.17), or

$$\log \boldsymbol{\mu}_0 = \mathbf{X}_0 \boldsymbol{\beta}_0 \quad \text{and} \quad \log \boldsymbol{\mu}_1 = \mathbf{X}_1 \boldsymbol{\beta}_1.$$

Since M_0 is simpler than M_1, one can express $\log \boldsymbol{\mu}_0 = \mathbf{X}_0 \boldsymbol{\beta}_0 = \mathbf{X}_1 \boldsymbol{\beta}_1^*$, where $\boldsymbol{\beta}_1^*$ equals $\boldsymbol{\beta}_0$ with 0 elements appended corresponding to the extra parameters in $\boldsymbol{\beta}_1$ but not in $\boldsymbol{\beta}_0$. Then, from (9.2),

$$G^2(M_0 \mid M_1) = 2\mathbf{n}'(\log \hat{\boldsymbol{\mu}}_1 - \log \hat{\boldsymbol{\mu}}_0) = 2\mathbf{n}'\left[\mathbf{X}_1 \hat{\boldsymbol{\beta}}_1 - \mathbf{X}_1 \hat{\boldsymbol{\beta}}_1^*\right]$$

$$= 2\hat{\boldsymbol{\mu}}_1'\left[\mathbf{X}_1 \hat{\boldsymbol{\beta}}_1 - \mathbf{X}_1 \hat{\boldsymbol{\beta}}_1^*\right] = 2\hat{\boldsymbol{\mu}}_1'(\log \hat{\boldsymbol{\mu}}_1 - \log \hat{\boldsymbol{\mu}}_0)$$

$$= 2\sum \hat{\mu}_{1i}\log(\hat{\mu}_{1i}/\hat{\mu}_{0i}), \tag{9.3}$$

where the replacement of $\mathbf{n}$ by $\hat{\boldsymbol{\mu}}_1$ follows from the likelihood equations $\mathbf{n}'\mathbf{X}_1 = \hat{\boldsymbol{\mu}}_1'\mathbf{X}_1$ for M_1 [Recall (8.22)]. Statistic (9.3) has the same form as $G^2(M_0)$, but with $\{\hat{\mu}_{1i}\}$ playing the role of the observed data. Note that $G^2(M_0)$ is the special case of $G^2(M_0 \mid M_1)$ with M_1 saturated.

The Pearson difference $X^2(M_0) - X^2(M_1)$ does not have Pearson form. It is not even necessarily nonnegative. A more appropriate Pearson statistic for comparing models is

$$X^2(M_0 \mid M_1) = \sum (\hat{\mu}_{1i} - \hat{\mu}_{0i})^2/\hat{\mu}_{0i}. \tag{9.4}$$

This has the usual form with $\{\hat{\mu}_{1i}\}$ in place of $\{n_i\}$. Statistics (9.3) and (9.4) depend on the data only through the fitted values and thus only through sufficient statistics for M_1.

When M_0 holds, $G^2(M_0)$ and $G^2(M_1)$ have asymptotic chi-squared distributions, and $G^2(M_0 \mid M_1)$ is asymptotically chi-squared with df equal to the difference between df for M_0 and M_1. Haberman (1977a) showed that $G^2(M_0 \mid M_1)$ and $X^2(M_0 \mid M_1)$ have the same null large-sample behavior, even for fairly sparse tables. (Under certain conditions, their difference converges in probability to 0 as n increases.) When M_1 holds but M_0 does not, $G^2(M_1)$ still has its asymptotic chi-squared distribution, but the other two statistics tend to grow unboundedly as n increases.

9.2.4 Partitioning Chi-Squared with Model Comparisons

Equation (9.1) utilizes the property by which a chi-squared statistic with df > 1 partitions into components. We used such partitionings in tests for trend with ordinal predictors in linear logit or linear probability models (Section 5.3.5) and with ordinal responses in cumulative logit models (Section 7.2). More generally, this property applies with a set of nested models to test a sequence of hypotheses. The separate tests for comparing pairs of models are asymptotically independent.

For example, a chi-squared decomposition with $J - 1$ models justifies the partitioning of G^2 stated in Section 3.3.3 for $2 \times J$ tables. For $j = 2, \ldots, J$, let M_j denote the model that satisfies

$$\theta_i = (\mu_{1i} \mu_{2, i+1})/(\mu_{1, i+1} \mu_{2i}) = 1, \quad i = 1, \ldots, j - 1.$$

For M_j, the $2 \times j$ table consisting of columns 1 through j satisfies independence. Model M_J is independence in the complete $2 \times J$ table. Model M_h is a special case of M_j whenever $h > j$. By (9.2),

$$\begin{aligned}
G^2(M_J) &= G^2(M_J | M_{J-1}) + G^2(M_{J-1}) \\
&= G^2(M_J | M_{J-1}) + G^2(M_{J-1} | M_{J-2}) + G^2(M_{J-2}) \\
&= \cdots = G^2(M_J | M_{J-1}) + \cdots + G^2(M_3 | M_2) + G^2(M_2).
\end{aligned}$$

From (9.3), $G^2(M_j | M_{j-1})$ has the G^2 form with the fitted values for model M_{j-1} playing the role of the observed data. Substitution of fitted values for the two models into (9.3) shows that $G^2(M_j | M_{j-1})$ is identical to G^2 for testing independence in a 2×2 table; the first column combines column 1 through $j - 1$ of the original table, and the second column is column j of the original table.

With several preplanned comparisons, simultaneous test procedures lessen the probability of attributing importance to sample effects that simply reflect chance variation. These procedures use adjusted significance levels. For a set of s tests for nested models, when each test has level $1 - (1 - \alpha)^{1/s}$, the overall asymptotic $P(\text{type I error}) \leq \alpha$ (Goodman 1969a). For instance, suppose that we test the fit of (WXZ, WY, XY, ZY), compare that model to (WX, WZ, XZ, WY, XY, ZY), and compare that model to (WX, WZ, XZ, WY, ZY). To ensure overall $\alpha = 0.05$ for the $s = 3$ tests, use level $1 - (0.95)^{1/3} = 0.017$ for each.

9.2.5 Identical Marginal and Conditional Tests of Independence

A test using $G^2(M_0 | M_1)$ simplifies dramatically when both models have direct estimates. In that case, the models have independence linkages necessary to ensure collapsibility. A test of conditional independence has the same result as the test of independence applied to the marginal table. Sundberg (1975) proved the following: When two direct models M_0 and M_1 are identical except for a pairwise association term, $G^2(M_0|M_1)$ is identical to G^2 for testing independence in the marginal table for that pair of variables. Bishop (1971) and Goodman (1970, 1971b) have related discussion.

For instance, $G^2[(X, Y, Z)|(XY, Z)]$ tests $\lambda^{XY} = 0$ in model (XY, Z). Thus, it tests XY conditional independence under the assumption that X and Y are jointly independent of Z. Using the two sets of fitted values, from (9.3), it equals

$$2 \sum_i \sum_j \sum_k \frac{n_{ij+}n_{++k}}{n} \log \frac{n_{ij+}n_{++k}/n}{n_{i++}n_{+j+}n_{++k}/n^2}$$

$$= 2 \sum_i \sum_j n_{ij+} \log \frac{n_{ij+}}{n_{i++}n_{+j+}/n},$$

which equals $G^2[(X, Y)]$ for testing independence in the marginal XY table. This is not surprising. The collapsibility conditions imply that for model (XY, Z), the marginal XY association is the same as the conditional XY association.

9.3 DIAGNOSTICS FOR CHECKING MODELS

The model comparison test using $G^2(M_0|M_1)$ is useful for detecting whether an extra term improves a model fit. Cell residuals provide a cell-specific indication of model lack of fit.

9.3.1 Residuals for Loglinear Models

In Section 4.5.5 we noted that residuals for the independence model (Section 3.3.1) extend to any Poisson GLM. For cell i in a contingency table with observed count n_i and fitted value $\hat{\mu}_i$, the *Pearson residual* is

$$e_i = \frac{n_i - \hat{\mu}_i}{\sqrt{\hat{\mu}_i}}. \tag{9.5}$$

These relate to the Pearson statistic by $\sum e_i^2 = X^2$.

Like the Pearson residual (6.1) for binomial models, the asymptotic variances of $\{e_i\}$ are less than 1.0. They average (residual df)/(number of cells). Haberman (1973a) defined the standardized Pearson residual,

$$r_i = e_i / \sqrt{1 - \hat{h}_i}\,,$$

where the leverage $\hat{h}_i$ is a diagonal element of the estimated hat matrix (Section 4.5.5). This has an asymptotic standard normal distribution and is preferable to the Pearson residual. A closed-form expression applies for loglinear models having direct estimates (Haberman 1978, p. 275). Alternative residuals use components of the deviance (Section 4.5.5).

9.3.2 Student Survey Example Revisited

For Table 9.1 cross-classifying alcohol, cigarette, and marijuana use by gender and race, we suggested in Section 9.2.2 that the model with all two-factor associations is plausible. For it, the only large standardized Pearson residual equals 3.2, resulting from a fitted value of 3.1 in the cell having a count of 8. Further comparisons suggested that the simpler model $(AC, AM, CM, AG, AR, GM, GR)$ is adequate. Its only large standardized residual equals 3.3, referring to a fitted value of 2.9 in that cell. The number of nonwhite males who did not use alcohol or marijuana but who smoked cigarettes is somewhat greater than either model predicts. The standardized Pearson residuals do not suggest problems with either model, considering the large sample size and many cells studied.

9.3.3 Correspondence between Loglinear and Logit Residuals

In Section 8.5 we showed that logit models in contingency tables are equivalent to certain loglinear models. However, a Pearson residual for a logit model differs from a Pearson residual for a loglinear model. The numerators comparing the ith observed and fitted binomial or Poisson count are the same, since the model fitted values are the same. However, the logit model uses a fitted binomial standard deviation in the denominator [see (6.1)], whereas the loglinear model uses a fitted Poisson standard deviation [see (9.5)]. Thus, the logit Pearson residual exceeds the loglinear Pearson residual (9.5).

Once standardized by dividing by estimated standard errors, the standardized Pearson residuals are identical for the two models. This is another reason for preferring standardized residuals over ordinary Pearson residuals.

9.4 MODELING ORDINAL ASSOCIATIONS

The loglinear models presented so far have a serious limitation—they treat all classifications as nominal. If the order of a variable's categories changes in

TABLE 9.3 Opinions about Premarital Sex and Availability of Teenage Birth Control

	Teenage Birth Control[a]			
Premarital Sex	Strongly Disagree	Disagree	Agree	Strongly Agree
Always wrong	81	68	60	38
	(42.4)[1]	(51.2)	(86.4)	(67.0)
	7.6[2]	3.1	−4.1	−4.8
	(80.9)[3]	(67.6)	(69.4)	(29.1)
Almost always wrong	24	26	29	14
	(16.0)	(19.3)	(32.5)	(25.2)
	2.3	1.8	−0.8	−2.8
	(20.8)	(23.1)	(31.5)	(17.6)
Wrong only sometimes	18	41	74	42
	(30.1)	(36.3)	(61.2)	(47.4)
	−2.7	1.0	2.2	−1.0
	(24.4)	(36.1)	(65.7)	(48.8)
Not wrong at all	36	57	161	157
	(70.6)	(85.2)	(143.8)	(111.4)
	−6.1	−4.6	2.4	6.8
	(33.0)	(65.1)	(157.4)	(155.5)

[a] [1] Independence model fit; [2] standardized Pearson residuals for the independence model fit; [3] linear-by-linear association model fit.

Source: 1991 General Social Survey, National Opinion Research Center.

any way, the fit is the same. For ordinal classifications, these models ignore important information.

Refer to Table 9.3. Subjects were asked their opinion about a man and woman having sexual relations before marriage (always wrong, almost always wrong, wrong only sometimes, not wrong at all). They were also asked whether methods of birth control should be available to teenagers between the ages of 14 and 16 (strongly disagree, disagree, agree, strongly agree). For the loglinear model of independence, denoted by I, $G^2(I) = 127.6$ with df = 9. The model fits poorly. Yet, adding the ordinary association term makes it saturated and unhelpful.

Table 9.3 also contains fitted values and standardized residuals for independence. The residuals in the corners stand out. Sample counts are much larger than independence predicts where both responses are the most negative possible or the most positive possible. By contrast, the counts are much smaller than fitted values where one response is the most positive and the other is the most negative. Cross-classifications of ordinal variables often exhibit their greatest deviations from independence in the corner cells. This pattern for Table 9.3 indicates lack of fit in the form of a positive trend.

Subjects who are more willing to make birth control available to teenagers also tend to feel more tolerant about premarital sex.

Models for ordinal variables use association terms that permit trends. The models are more complex than the independence model, yet unsaturated. Models with association and interaction terms exist in situations in which nominal models are saturated. Tests with ordinal models have improved power for detecting trends.

9.4.1 Linear-by-Linear Association in Two-Way Tables

For two-way tables, a simple model for two ordinal variables assigns ordered row scores $u_1 \leq u_2 \leq \cdots \leq u_I$ and column scores $v_1 \leq v_2 \leq \cdots \leq v_J$. The model is

$$\log \mu_{ij} = \lambda + \lambda_i^X + \lambda_j^Y + \beta u_i v_j, \qquad (9.6)$$

with constraints such as $\lambda_I^X = \lambda_J^Y = 0$. This is the special case of the saturated model (8.2) in which $\lambda_{ij}^{XY} = \beta u_i v_j$. It requires only one parameter to describe association, whereas the saturated model requires $(I - 1)(J - 1)$.

Independence occurs when $\beta = 0$. The term $\beta u_i v_j$ represents the deviation of $\log \mu_{ij}$ from independence. The deviation is linear in the Y scores at a fixed level of X and linear in the X scores at a fixed level of Y. In column j, for instance, the deviation is a linear function of X, having form (slope) $\times$ (score for X), with slope βv_j. Because of this property, (9.6) is called the *linear-by-linear association model* (abbreviated, $L \times L$). The model has its greatest departures from independence in the corners of the table. Birch (1965), Goodman (1979a), and Haberman (1974b) introduced special cases.

The direction and strength of the association depend on β. When $\beta > 0$, Y tends to increase as X increases. Expected frequencies are larger than expected (under independence) in cells where X and Y are both high or both low. When $\beta < 0$, Y tends to decrease as X increases. When the data display a positive or negative trend, the $L \times L$ model usually fits much better than the independence model.

For the 2×2 table using the cells intersecting rows a and c with columns b and d, direct substitution shows that the model has

$$\log \frac{\mu_{ab} \mu_{cd}}{\mu_{ad} \mu_{cb}} = \beta(u_c - u_a)(v_d - v_b). \qquad (9.7)$$

This log odds ratio is stronger as $|\beta|$ increases and for pairs of categories that are farther apart. Simple interpretations result when $u_2 - u_1 = \cdots = u_I - u_{I-1}$ and $v_2 - v_1 = \cdots = v_J - v_{J-1}$. When $\{u_i = i\}$ and $\{v_j = j\}$, for instance, the *local odds ratios* (2.10) for adjacent rows and adjacent columns have common value e^β. Goodman (1979a) called this case *uniform association*. Figure 9.2 portrays local odds ratios having uniform value.

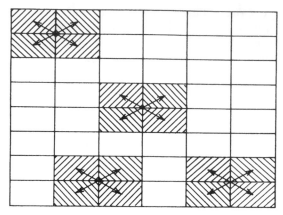

FIGURE 9.2 Constant odds ratio implied by uniform association model. (*Note:* β = the constant log odds ratio for adjacent rows and adjacent columns.)

The choice of scores affects the interpretation of β. Often, the response scale discretizes an inherently continuous scale. It is sensible to choose scores that approximate distances between midpoints of categories for the underlying scale, such as we did in measuring alcohol consumption for a linear logit model in Section 3.4.5. It is sometimes useful to standardize the scores, subtracting the mean and dividing by the standard deviation, so

$$\sum u_i \pi_{i+} = \sum v_j \pi_{+j} = 0$$
$$\sum u_i^2 \pi_{i+} = \sum v_j^2 \pi_{+j} = 1.$$

Then, β represents the log odds ratios for standard deviation distances in the X and Y directions. The $L \times L$ model tends to fit well when an underlying continuous distribution is approximately bivariate normal. For standardized scores, β is then comparable to $\rho/(1 - \rho^2)$, where ρ is the underlying correlation. For weak associations, $\beta \approx \rho$ (see Becker 1989b; Goodman 1981a, b, 1985).

9.4.2 Corresponding Logit Model for Adjacent Responses

A logit formulation of the $L \times L$ model treats Y as a response and X as explanatory. Let $\pi_{j|i} = P(Y = j | X = i)$. Using logits for adjacent response categories (Section 7.4.1),

$$\log \frac{\pi_{j+1|i}}{\pi_{j|i}} = \log \frac{\mu_{i,j+1}}{\mu_{ij}} = \left(\lambda_{j+1}^Y - \lambda_j^Y\right) + \beta(v_{j+1} - v_j)u_i.$$

For unit-spaced $\{v_j\}$, this simplifies to

$$\log \frac{\pi_{j+1|i}}{\pi_{j|i}} = \alpha_j + \beta u_i$$

where $\alpha_j = \lambda_{j+1}^Y - \lambda_j^Y$. The same linear logit effect β applies simultaneously for all $(J - 1)$ pairs of adjacent response categories: The odds $Y = j + 1$ instead of $Y = j$ multiply by e^β for each unit change in X. In using equal-interval response scores, we implicitly assume that the effect of X is the same on each of the $J - 1$ adjacent-categories logits for Y.

9.4.3 Likelihood Equations and Model Fitting

The Poisson log-likelihood $L(\mathbf{\mu}) = \Sigma_i \Sigma_j n_{ij} \log \mu_{ij} - \Sigma_i \Sigma_j \mu_{ij}$ simplifies for the $L \times L$ model (9.6) to

$$L(\mathbf{\mu}) = n\lambda + \sum_i n_{i+} \lambda_i^X + \sum_j n_{+j} \lambda_j^Y + \beta \sum_i \sum_j u_i v_j n_{ij}$$
$$- \sum_i \sum_j \exp\left(\lambda + \lambda_i^X + \lambda_j^Y + \beta u_i v_j\right).$$

Differentiating $L(\mathbf{\mu})$ with respect to $(\lambda_i^X, \lambda_j^Y, \beta)$ and setting the three partial derivatives equal to zero yields likelihood equations

$$\hat{\mu}_{i+} = n_{i+}, \, i = 1, \ldots, I, \qquad \hat{\mu}_{+j} = n_{+j}, \, j = 1, \ldots, J,$$
$$\sum_i \sum_j u_i v_j \hat{\mu}_{ij} = \sum_i \sum_j u_i v_j n_{ij}.$$

Iterative methods such as Newton–Raphson yield the ML fit.

Let $p_{ij} = n_{ij}/n$ and $\hat{\pi}_{ij} = \hat{\mu}_{ij}/n$. The third likelihood equation implies that

$$\sum_i \sum_j u_i v_j \hat{\pi}_{ij} = \sum_i \sum_j u_i v_j p_{ij}.$$

Since marginal distributions and hence marginal means and variances are identical for fitted and observed distributions, the third equation implies the correlation between the scores for X and Y is the same for both distributions. The fitted counts display the same positive or negative trend as the data.

Since $\{u_i\}$ and $\{v_j\}$ are fixed, the $L \times L$ model (9.6) has only one more parameter (β) than the independence model. Its residual

$$\text{df} = IJ - [1 + (I - 1) + (J - 1) + 1] = IJ - I - J,$$

unsaturated for all but 2×2 tables.

9.4.4 Sex Opinions Example

Table 9.3 also reports fitted values for the linear-by-linear association model applied to Table 9.3, using scores $\{1, 2, 3, 4\}$ for rows and columns. Table 9.4

TABLE 9.4　Output for Fitting Linear-by-Linear Association Model to Table 9.3

```
        Criteria For Assessing Goodness Of Fit
        Criterion                   DF      Value
        Deviance                     8     11.5337
        Pearson Chi-Square           8     11.5085
```

Parameter		Estimate	Standard Error	Wald 95% Conf. Limits		Chi-Square	Pr > ChiSq
Intercept		0.4735	0.4339	−0.3769	1.3239	1.19	0.2751
premar	1	1.7537	0.2343	1.2944	2.2129	56.01	<.0001
premar	2	0.1077	0.1988	−0.2820	0.4974	0.29	0.5880
premar	3	−0.0163	0.1264	−0.2641	0.2314	0.02	0.8972
premar	4	0.0000	0.0000	0.0000	0.0000	.	.
birth	1	1.8797	0.2491	1.3914	2.3679	56.94	<.0001
birth	2	1.4156	0.1996	1.0243	1.8068	50.29	<.0001
birth	3	1.1551	0.1291	0.9021	1.4082	80.07	<.0001
birth	4	0.0000	0.0000	0.0000	0.0000	.	.
linlin		0.2858	0.0282	0.2305	0.3412	102.46	<.0001

```
                       LR Statistics
          Source      DF      Chi-Square      Pr > ChiSq
          linlin       1        116.12          >.0001
```

shows software output. To get this, we added a variable (denoted "linlin") to the independence model having values equal to the product of row and column number. Compared to the independence model, for which $G^2(I) = 127.6$ with df = 9, the $L \times L$ model fits dramatically better $[G^2(L \times L) = 11.5$, df = 8]. This is especially noticeable in the corners, where it predicts the greatest departures from independence.

The ML estimate $\hat{\beta} = 0.286$ (SE = 0.028) indicates that subjects having more favorable attitudes about teen birth control also tend to have more tolerant attitudes about premarital sex. The estimated local odds ratio is $\exp(\hat{\beta}) = \exp(0.286) = 1.33$. A 95% Wald confidence interval is $\exp(0.286 \pm 1.96 \times 0.028)$, or $(1.26, 1.41)$. The strength of association seems weak. From (9.7), however, nonlocal odds ratios are stronger. The estimated odds ratio for the four corner cells equals

$$\exp\left[\hat{\beta}(u_4 - u_1)(v_4 - v_1)\right] = \exp[0.286(4 - 1)(4 - 1)] = 13.1.$$

This also results from the corner fitted values, $(80.9 \times 155.5)/(29.1 \times 33.0) = 13.1$.

Two sets of scores having the same spacings yield the same $\hat{\beta}$ and the same fit. Any other sets of equally spaced scores yield the same fit but an appropriately rescaled $\hat{\beta}$. For instance, using row scores $\{2, 4, 6, 8\}$ with $\{v_j = j\}$ also yields $G^2 = 11.5$, but $\hat{\beta} = 0.143$ with SE = 0.014 (both half as large). For Table 9.3, one might regard categories 2 and 3 as farther apart than categories 1 and 2, or categories 3 and 4. Scores such as $\{1, 2, 4, 5\}$ for rows and columns recognize this. The $L \times L$ model then has $G^2 = 8.8$ (df $= 8$) and $\hat{\beta} = 0.146$ (SE $= 0.014$).

One need not regard the scores as approximations for distances between categories or as reasonable scalings of ordinal variables in order for the models to be valid. They simply imply a certain pattern for the odds ratios. If the $L \times L$ model fits well with equally spaced row and column scores, the uniform local odds ratio describes the association regardless of whether the scores are sensible indexes of true distances between categories.

For scores $\{u_i = i\}$ with Table 9.3, the marginal mean and standard deviation for premarital sex are 2.81 and 1.26. The standardized scores are $\{(i - 2.81)/1.26\}$, or $(-1.44, -0.65, 0.15, 0.95)$. The standardized equal-interval scores for birth control are $(-1.65, -0.69, 0.27, 1.23)$. For these scores, $\hat{\beta} = 0.374$. By solving $\hat{\beta} = \hat{\rho}/(1 - \hat{\rho}^2)$ for $\hat{\rho}$, $\hat{\rho} = 0.333$. If there is an underlying bivariate normal distribution, we estimate the correlation to be 0.333.

9.4.5 Directed Ordinal Test of Independence

For the linear-by-linear association model, H_0: independence is H_0: $\beta = 0$. The likelihood-ratio test statistic equals

$$G^2(I | L \times L) = G^2(I) - G^2(L \times L).$$

Designed to detect positive or negative trends, it has df $= 1$. For Table 9.3, $G^2(I | L \times L) = 127.6 - 11.5 = 116.1$. This has $P < 0.0001$, extremely strong evidence of an association. The Wald statistic $z^2 = (\hat{\beta}/\text{SE})^2 = (0.286/0.0282)^2 = 102.5$ (df $= 1$) also shows strong evidence. The correlation statistic (3.15) presented in Section 3.4.1 for testing independence is the score statistic for H_0: $\beta = 0$ in this model. It equals 112.6 (df $= 1$).

When the $L \times L$ model holds, the ordinal test using $G^2(I | L \times L)$ is asymptotically more powerful than the test using $G^2(I)$. This is true for the same reason given in Section 6.4.2 for the linear logit model. The power of a chi-squared test increases when df decrease, for fixed noncentrality. When the $L \times L$ model holds, the noncentrality is the same for $G^2(I | L \times L)$ and $G^2(I)$; thus $G^2(I | L \times L)$ is more powerful, since its df $= 1$ compared to $(I - 1)(J - 1)$ for $G^2(I)$. The power advantage increases as I and J increase, since the noncentrality remains focused on df $= 1$ for $G^2(I | L \times L)$ but df also increases for $G^2(I)$.

9.5 ASSOCIATION MODELS*

Generalizations of the linear-by-linear association model apply to multiway tables or treat scores as parameters rather than fixed. The models are called *association models*, because they focus on the association structure.

9.5.1 Row and Column Effects Models

We first present a model that treats X as nominal and Y as ordinal. It is appropriate for two-way tables with ordered columns, using scores $v_1 \leq v_2 \leq \cdots \leq v_J$. Since the rows are unordered, they do not have scores. Replacing the ordered values $\{\beta u_i\}$ in the linear-by-linear term $\beta u_i v_j$ in model (9.6) by unordered parameters $\{\mu_i\}$ gives

$$\log \mu_{ij} = \lambda + \lambda_i^X + \lambda_j^Y + \mu_i v_j. \tag{9.8}$$

Constraints are needed such as $\lambda_I^X = \lambda_J^Y = \mu_I = 0$. The $\{\mu_i\}$ are called *row effects*. The model is called the *row effects model*.

Model (9.8) has $I - 1$ more parameters (the $\{\mu_i\}$) than the independence model. Independence is the special case $\mu_1 = \cdots = \mu_I$. A corresponding *column effects model* has association term $u_i v_j$. It treats X as ordinal with scores $\{u_i\}$ and Y as nominal with parameters $\{v_j\}$. The row effects and column effects models were developed by Goodman (1979a), Haberman (1974b), and Simon (1974).

9.5.2 Logit Model for Adjacent Responses

With $\{v_{j+1} - v_j = 1\}$, the row effects model has adjacent-categories logit form

$$\log \frac{P(Y = j + 1 \mid X = i)}{P(Y = j \mid X = i)} = \alpha_j + \mu_i. \tag{9.9}$$

The effect in row i is identical for each pair of adjacent responses. Plots of these logits against i $(i = 1, \ldots, I)$ for different j are parallel. Goodman (1983) referred to model (9.9) as the *parallel odds* model.

Differences among $\{\mu_i\}$ compare rows with respect to their conditional distributions on Y. When $\mu_i = \mu_h$, rows h and i have identical conditional distributions. If $\mu_i > \mu_h$, Y is stochastically higher in row i than row h.

The likelihood equations for the row effects model (9.8) are $\{\hat{\mu}_{i+} = n_{i+}\}$, $\{\hat{\mu}_{+j} = n_{+j}\}$, and

$$\sum_j v_j \hat{\mu}_{ij} = \sum v_j n_{ij}, \quad i = 1, \ldots, I.$$

Let $\hat{\pi}_{j|i} = \hat{\mu}_{ij}/\hat{\mu}_{i+}$ and $p_{j|i} = n_{ij}/n_{i+}$. Since $\hat{\mu}_{i+} = n_{i+}$, the third likelihood equation is $\sum_j v_j \hat{\pi}_{j|i} = \sum_j v_j p_{j|i}$. For the conditional distribution within each row, the mean column score is the same for the fitted and sample distributions. The likelihood equations are solved using iterative methods.

TABLE 9.5 Observed Frequencies and Fitted Values for Political Ideology Data

	Political Ideology[a]			
Party Affiliation	Liberal	Moderate	Conservative	Total
Democrat	143	156	100	399
	$(102.0)^1$	(161.4)	(135.6)	
	$(136.6)^2$	(168.7)	(93.6)	
Independent	119	210	141	470
	(120.2)	(190.1)	(159.7)	
	(123.8)	(200.4)	(145.8)	
Republican	15	72	127	214
	(54.7)	(86.6)	(72.7)	
	(16.6)	(68.9)	(128.6)	

[a][1] Independence model; [2] row effects model.

Source: Based on data in R. D. Hedlund, *Public Opinion Quart.* **41**: 498–514 (1978).

9.5.3 Political Ideology Example

Table 9.5 displays the relationship between political ideology and political party affiliation for a sample of voters in a presidential primary in Wisconsin. The table shows fitted values for the independence (I) model and the row effects (R) model with $\{v_j = j\}$.

Table 9.6 shows output. Goodness-of-fit tests show that independence is inadequate. Adding the row effects parameters much improves the fit $(G^2(I) = 105.7$, df = 4; $G^2(R) = 2.8$, df = 2). Also, testing $H_0: \mu_1 = \mu_2 = \mu_3$ using $G^2(I|R) = 102.9$ (df = 2) shows very strong evidence of an association. In Table 9.5, the improved fit is especially noticeable at the ends of the ordinal scale, where the model has greatest deviation from independence.

The output uses dummy variables for the first two categories of each classification. The interaction term equals the product of the score for ideology and a parameter for party. Thus, the row effect estimates satisfy $\hat{\mu}_3 = 0$, and the other two estimates contrast the first two parties with Republicans. The estimates are $\hat{\mu}_1 = -1.213$ and $\hat{\mu}_2 = -0.943$. The further $\hat{\mu}_i$ falls in the negative direction, the greater the tendency for the party i to locate at the liberal end of the ideology scale, relative to Republicans. In this sample the Republicans are much more conservative than the other two groups, and the Democrats (row 1) are the most liberal. From (9.9) the model predicts constant odds ratios for adjacent columns of political ideology. For instance, since $\hat{\mu}_3 - \hat{\mu}_1 = 1.213$, the estimated odds that Republicans were conservative instead of moderate, or moderate instead of liberal, were $\exp(1.213) = 3.36$ times the corresponding estimated odds for Democrats. Figure 9.3 shows the parallelism of the estimated logits for the row effects model.

The loglinear model does not distinguish between response and explanatory variables. Instead, one could use a cumulative logit model to describe

TABLE 9.6   Output for Fitting Row Effects Model to Table 9.5

```
           Criteria For Assessing Goodness Of Fit
           Criterion              DF        Value
           Deviance                2       2.8149
           Pearson Chi-Square      2       2.8039
```

Parameter		Estimate	Std Error	Wald 95% Conf. Limits		Chi-Square	Pr > ChiSq
Intercept		4.8565	0.0858	4.6883	5.0246	3204.02	<.0001
party	Democ	3.3230	0.3188	2.6981	3.9479	108.63	<.0001
party	Indep	2.9536	0.3149	2.3364	3.5707	87.98	<.0001
party	Repub	0.0000	0.0000	0.0000	0.0000	.	.
ideology	1	−2.0488	0.2216	−2.4831	−1.6145	85.50	<.0001
ideology	2	−0.6244	0.1139	−0.8476	−0.4013	30.08	<.0001
ideology	3	0.0000	0.0000	0.0000	0.0000	.	.
score*party	Democ	−1.2134	0.1304	−1.4690	−0.9577	86.56	<.0001
score*party	Indep	−0.9426	0.1260	−1.1896	−0.6956	55.95	<.0001
score*party	Repub	0.0000	0.0000	0.0000	0.0000	.	.

```
LR Statistics
     Source            DF        Chi-Square        Pr > ChiSq
     score*party        2          102.85             <.0001
```

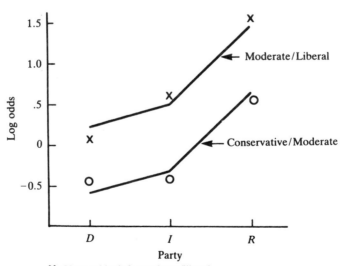

FIGURE 9.3   Observed and predicted logits for adjacent response categories.

the effects of party affiliation on ideology, or a baseline-category logit model to describe linear effects of ideology on party affiliation.

9.5.4 Ordinal Variables in Models for Multiway Tables

Multidimensional tables with ordinal responses can use generalizations of association models. In three dimensions, the rich collection of models includes (1) association models that are more parsimonious than the nominal model (XY, XZ, YZ), and (2) models permitting heterogeneous association that, unlike model (XYZ), are unsaturated.

Models for association that are special cases of (XY, XZ, YZ) replace λ association terms by structured terms that account for ordinality. For instance, when both X and Y are ordinal, alternatives to λ_{ij}^{XY} are a linear-by-linear term $\beta u_i v_j$, a row effects term $\mu_i v_j$, or a column effects term $u_i v_j$; these provide a stochastic ordering of conditional distributions within rows and within columns, or just within rows, or just within columns. With a linear-by-linear term, the model is

$$\log \mu_{ijk} = \lambda + \lambda_i^X + \lambda_j^Y + \lambda_k^Z + \beta u_i v_j + \lambda_{ik}^{XZ} + \lambda_{jk}^{YZ}. \qquad (9.10)$$

The conditional local odds ratios (8.13) then satisfy

$$\log \theta_{ij(k)} = \beta(u_{i+1} - u_i)(v_{j+1} - v_j) \quad \text{for all } k.$$

The association is the same in different partial tables, with *homogeneous linear-by-linear XY association*.

When the association is heterogeneous, structured terms for ordinal variables make effects simpler to interpret than in the saturated model. For instance, the *heterogeneous linear-by-linear XY association model*

$$\log \mu_{ijk} = \lambda + \lambda_i^X + \lambda_j^Y + \lambda_k^Z + \beta_k u_i v_j + \lambda_{ik}^{XZ} + \lambda_{jk}^{YZ} \qquad (9.11)$$

allows the XY association to change across levels of Z. With unit-spaced scores,

$$\log \theta_{ij(k)} = \beta_k \quad \text{for all } i \text{ and } j.$$

It has uniform association within each level of Z, but heterogeneity among levels of Z in the strength of association. Fitting it corresponds to fitting the $L \times L$ model (9.6) separately at each level of Z.

9.5.5 Air Pollution and Breathing Examples

Table 9.7 displays associations among smoking status (S), breathing test results (B), and age (A) for workers in certain industrial plants in Houston,

TABLE 9.7 Cross-Classification of Industrial Workers by Breathing Test Results

Age	Smoking Status	Breathing Test Results		
		Normal	Borderline	Abnormal
< 40	Never smoked	577	27	7
	Former smoker	192	20	3
	Current smoker	682	46	11
40–59	Never smoked	164	4	0
	Former smoker	145	15	7
	Current smoker	245	47	27

Source: From p. 21 of *Public Program Analysis* by R. N. Forthofer and R. G. Lehnen. Copyright © 1981 by Lifetime Learning Publications, Belmont, CA 94002, a division of Wadsworth, Inc. Reprinted by permission of Van Nostrand Reinhold. All rights reserved.

Texas. The loglinear model (SA, SB, BA) fits poorly $(G^2 = 25.9,$ df $= 4)$. Thus, simpler models such as homogeneous linear-by-linear SB association are not plausible $(G^2 = 29.1,$ df $= 7$, using equally spaced scores). The heterogeneous linear-by-linear SB association model fits much better with only one additional parameter $(G^2 = 10.8,$ df $= 6)$. With integer scores for S and B, $\hat{\beta}_1 = 0.115$ for the younger group and $\hat{\beta}_2 = 0.781$ for the older group, with SE $= 0.167$ for the difference. The effect of smoking seems much stronger for the older group, with estimated local odds ratio of $\exp(0.781) = 2.18$ compared to $\exp(0.115) = 1.12$ for the younger group. Here, it may be more natural to use logit models with B as the response variable (Problem 7.11).

When strata are ordered, roughly a linear trend may exist across strata in certain log odds ratios as Table 9.8 illustrates. The data refer to a sample of coal miners, measured on $B =$ breathlessness, $W =$ wheeze, and $A =$ age, where B and W are response variables. One could use a separate logit model to describe effects of age on each response. To study whether the BW association varies by age, we fit model (BW, AB, AW). It has residual $G^2 = 26.7$, with df $= 8$. Table 9.8 reports the standardized Pearson residuals. They show a decreasing tendency as age increases.

This suggests the model

$$\log \mu_{ijk} = (BW, AB, AW) + kI(i = j = 1)\delta, \qquad (9.12)$$

where I is the indicator function. It amends the homogeneous association model by adding δ in the cell for $\mu_{111}, \ldots, 9\delta$ in the cell for μ_{119}. Then, the BW log odds ratio changes linearly in the age category. The model fit has $\hat{\delta} = -0.131$ (SE $= 0.029$). The estimated BW log odds ratio at level k of age is $3.676 - 0.131k$, decreasing from 3.55 to 2.50. The model has residual $G^2 = 6.80$ (df $= 7$). McCullagh and Nelder (1989, Sec. 6.6) showed other analyses.

TABLE 9.8 Coal Miners Classified by Breathlessness, Wheeze, and Age

| | Breathlessness | | | | |
| | Yes | | No | | |
Age	Wheeze Yes	Wheeze No	Wheeze Yes	Wheeze No	Std. Pearson Residual[a]
20–24	9	7	95	1841	0.75
25–29	23	9	105	1654	2.20
30–34	54	19	177	1863	2.10
35–39	121	48	257	2357	1.77
40–44	169	54	273	1778	1.13
45–49	269	88	324	1712	−0.42
50–54	404	117	245	1324	0.81
55–59	406	152	225	967	−3.65
60–64	372	106	132	526	−1.44

[a]Residual refers to yes–yes and no–no cells; reverse sign for yes–no and no–yes cells.

Source: Reprinted with permission from Ashford and Sowden (1970).

9.5.6 Other Ordinal Tests of Conditional Independence

Tests of conditional independence of ordinal classifications can generalize $G^2(I \mid L \times L)$. For instance, one can compare the XY conditional independence model (XZ, YZ) to the homogeneous linear-by-linear XY association model (9.10). It tests $\beta = 0$ in that model, with df = 1. This is an alternative to the ordinal test of conditional independence in Section 7.5.3. Like Mantel's score statistic (7.21), this statistic uses correlation information, since $\Sigma_k(\Sigma_i\Sigma_j u_i v_j n_{ijk})$ is the sufficient statistic for β in model (9.10). In fact, the Mantel statistic provides the score test of H_0: $\beta = 0$ in that model.

Exact, small-sample tests can use likelihood-ratio, score, or Wald statistics for such models. Computations require special algorithms (Agresti et al. 1990; Kim and Agresti 1997).

9.6 ASSOCIATION MODELS, CORRELATION MODELS, AND CORRESPONDENCE ANALYSIS*

The linear-by-linear association $(L \times L)$ model is a special case of the row effects (R) model, which has parameter row scores, and the column effects (C) model, which has parameter column scores. These models are special cases of a more general model with row *and* column parameter scores.

9.6.1 Multiplicative Row and Column Effects Model

Replacing $\{u_i\}$ and $\{v_j\}$ in the $L \times L$ model (9.6) by parameters yields the *row and column effects* (RC) model (Goodman 1979a)

$$\log \mu_{ij} = \lambda + \lambda_i^X + \lambda_j^Y + \beta\mu_i v_j. \tag{9.13}$$

Identifiability requires location and scale constraints on $\{\mu_i\}$ and $\{\nu_j\}$. The residual df $= (I - 2)(J - 2)$. This model is *not* loglinear, because the predictor is a multiplicative (rather than linear) function of parameters μ_i and ν_j. It treats classifications as nominal; the same fit results from a permutation of rows or columns. Parameter interpretation is simplest when at least one variable is ordinal, through the local log odds ratios

$$\log\theta_{ij} = \beta(\mu_{i+1} - \mu_i)(\nu_{j+1} - \nu_j).$$

Although it may seem appealing to use parameters instead of arbitrary scores, the *RC* model presents complications that do not occur with loglinear models. The likelihood may not be concave and may have local maxima. Independence is a special case, but it is awkward to test independence using the *RC* model. Haberman (1981) showed that the null distribution of $G^2(I) - G^2(RC)$ is not chi-squared but rather that of the maximum eigenvalue from a Wishart matrix.

When one set of parameter scores is fixed, the *RC* model simplifies to the *R* or *C* model. Goodman (1979a) suggested an iterative model-fitting algorithm that exploits this. A cycle of the algorithm has two steps. First, for some initial guess of $\{\nu_j\}$, it estimates the row scores as in the *R* model. Then, treating the estimated row scores from the first step as fixed, it estimates the column scores as in the *C* model. Those estimates serve as fixed column scores in the first step of the next cycle, for reestimating the row scores in the *R* model. There is no guarantee of convergence to ML estimates, but this seems to happen when the model fits well. Haberman (1995) provided more sophisticated fitting methods for association models.

Goodman (1985) expressed the association term in the saturated model in a form that generalizes the $\beta\mu_i\nu_j$ term in the *RC* model, namely,

$$\lambda_{ij}^{XY} = \sum_{k=1}^{M} \beta_k \mu_{ik} \nu_{jk} \tag{9.14}$$

where $M = \min(I - 1, J - 1)$. The parameters satisfy constraints such as

$$\sum_i \mu_{ik} \pi_{i+} = \sum_j \nu_{jk} \pi_{+j} = 0 \qquad \text{for all } k,$$

$$\sum_i \mu_{ik}^2 \pi_{i+} = \sum_j \nu_{jk}^2 \pi_{+j} = 1 \qquad \text{for all } k, \tag{9.15}$$

$$\sum_i \mu_{ik} \mu_{ih} \pi_{i+} = \sum_j \nu_{jk} \nu_{jh} \pi_{+j} = 0 \qquad \text{for all } k \neq h.$$

When $\beta_k = 0$ for $k > M^*$, model (9.14) is called the $RC(M^*)$ model. See Becker (1990) for ML model fitting. The *RC* model (9.13) is the case $M^* = 1$.

TABLE 9.9 Cross-Classification of Mental Health Status and Socioeconomic Status

Parents' Socioeconomic Status		Mental Health Status		
	Well	Mild Symptom Formation	Moderate Symptom Formation	Impaired
A (high)	64	94	58	46
B	57	94	54	40
C	57	105	65	60
D	72	141	77	94
E	36	97	54	78
F (low)	21	71	54	71

Source: Reprinted with permission from L. Srole et al. *Mental Health in the Metropolis: The Midtown Manhattan Study*, (New York: NYU Press, 1978), p. 289.

9.6.2 Mental Health Status Example

Table 9.9 describes the relationship between child's mental impairment and parents' socioeconomic status for a sample of residents of Manhattan (Goodman 1979a). The *RC* model fits well ($G^2 = 3.6$, df $= 8$). For scaling (9.15), the ML estimates are $(-1.11, -1.12, -0.37, 0.03, 1.01, 1.82)$ for the row scores, $(-1.68, -0.14, 0.14, 1.41)$ for the column scores, and $\hat{\beta} = 0.17$. Nearly all estimated local log odds ratios are positive, indicating a tendency for mental health to be better at higher levels of parents' SES.

Ordinal loglinear models also fit well. For equal-interval scores, $G^2(L \times L) = 9.9$ (df $= 14$). The statistic $G^2(L \times L | RC) = 6.3$ (df $= 6$) tests that row and column scores in the *RC* model are equal-interval. The parameter scores do not provide a significantly better fit. It is sufficient to use a uniform local odds ratio to describe the table. For unit-spaced scores, $\hat{\beta} = 0.091$ (SE $= 0.015$), so the fitted local odds ratio is $\exp(0.091) = 1.09$. There is strong evidence of positive association, but the degree of association is rather weak, at least locally.

9.6.3 Correlation Models

A *correlation model* for two-way tables has many features in common with the *RC* model (Goodman 1985). In its simplest form, it is

$$\pi_{ij} = \pi_{i+}\pi_{+j}(1 + \lambda\mu_i\nu_j), \tag{9.16}$$

where $\{\mu_i\}$ and $\{\nu_i\}$ are score parameters satisfying

$$\sum \mu_i\pi_{i+} = \sum \nu_j\pi_{+j} = 0 \quad \text{and} \quad \sum \mu_i^2\pi_{i+} = \sum \nu_j^2\pi_{+j} = 1.$$

The parameter λ is the correlation between the scores for joint distribution (9.16).

The correlation model is also called the *canonical correlation model*, because ML estimates of the scores maximize the correlation for (9.16). The general canonical correlation model is

$$\pi_{ij} = \pi_{i+} \pi_{+j} \left(1 + \sum_{k=1}^{M} \lambda_k \mu_{ik} \nu_{jk} \right)$$

where $0 \leq \lambda_M \leq \cdots \leq \lambda_1 \leq 1$ and with constraints such as in (9.15). The parameter λ_k is the correlation between $\{\mu_{ik}, i = 1, \ldots, I\}$ and $\{\nu_{jk}, j = 1, \ldots, J\}$. The $\{\mu_{i1}\}$ and $\{\nu_{j1}\}$ are standardized scores that maximize the correlation λ_1 for the joint distribution; $\{\mu_{i2}\}$ and $\{\nu_{j2}\}$ are standardized scores that maximize the correlation λ_2, subject to $\{\mu_{i1}\}$ and $\{\mu_{i2}\}$ being uncorrelated and $\{\nu_{j1}\}$ and $\{\nu_{j2}\}$ being uncorrelated, and so on.

Unsaturated models result from replacing M by $M^* < \min(I - 1, J - 1)$. Gilula and Haberman (1986) and Goodman (1985) discussed ML fitting. When λ is close to zero in (9.16), Goodman (1981a, 1985, 1986) noted that ML estimates of λ and the score parameters are similar to those of β and the score parameters in the *RC* model. Correlation models can also use fixed scores instead of parameter scores.

Goodman discussed advantages of association models over correlation models. The correlation model is not defined for all possible combinations of score values because of the constraint $0 \leq \pi_{ij} \leq 1$, ML fitted values do not have the same marginal totals as the observed data, and the model is not simply generalizable to multiway tables. Gilula and Haberman (1988) analyzed multiway tables with correlation models by treating explanatory variables as a single variable and response variables as a second variable.

9.6.4   Correspondence Analysis

Correspondence analysis is a graphical way to represent associations in two-way contingency tables. The rows and columns are represented by points on a graph, the positions of which indicate associations. Goodman (1985, 1986) noted that coordinates of the points are reparameterizations of $\{\mu_{ik}\}$ and $\{\nu_{jk}\}$ in the general canonical correlation model. Correspondence analysis uses adjusted scores

$$x_{ik} = \lambda_k \mu_{ik}, \qquad y_{jk} = \lambda_k \nu_{jk}.$$

These are close to zero for dimensions k in which the correlation λ_k is close to zero. A correspondence analysis graph uses the first two dimensions, plotting (x_{i1}, x_{i2}) for each row and (y_{j1}, y_{j2}) for each column.

TABLE 9.10 Scores from Correspondence Analysis Applied to Table 9.9

Column Score	Dimension			Row Score	Dimension		
	1	2	3		1	2	3
1	0.260	0.012	0.023	1	0.181	−0.018	0.028
2	0.030	0.024	−0.019	2	0.185	−0.011	−0.026
3	−0.013	−0.069	−0.002	3	0.059	−0.021	−0.010
4	−0.236	0.019	0.016	4	−0.008	0.042	0.011
				5	−0.164	0.044	−0.009
				6	−0.287	−0.061	0.005

Source: Reprinted with permission from the Institute of Mathematical Statistics, based on Goodman (1985).

Goodman (1985, 1986) used Table 9.9 to illustrate the similarities of correspondence analysis to analyses using correlation models and association models. For the general canonical correlation model, $M = \min(I - 1, J - 1) = 3$. Its estimated squared correlations are (0.0260, 0.0014, and 0.0003). The association is rather weak. Table 9.10 contains estimated row and column scores for the correspondence analysis of these three dimensions. Both sets of scores in the first dimension fall in a monotone increasing pattern, except for a slight discrepancy between the first two row scores. This indicates an overall positive association. The scores for the second and third dimension are close to zero, reflecting the relatively small $\hat{\lambda}_2$ and $\hat{\lambda}_3$.

Figure 9.4 exhibits the results of the correspondence analysis. The horizontal axis has estimates for the first dimension, and the vertical axis has estimates for the second dimension. Six points (circles) represent the six rows, with point i giving $(\hat{x}_{i1}, \hat{x}_{i2})$. Similarly, four points (squares) display the estimates $(\hat{y}_{j1}, \hat{y}_{j2})$. Both sets of points lie close to the horizontal axis, since the first dimension is more important than the second.

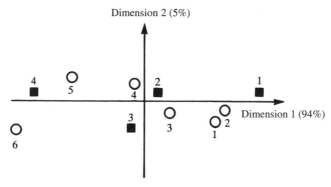

FIGURE 9.4 Graphical display of scores from first two dimensions of correspondence analysis. [Based on Escoufier (1982); reprinted with permission.]

Row points that are close together represent rows with similar conditional distributions across the columns. Close column points represent columns with similar conditional distributions across rows. Row points close to column points represent combinations that are more likely than expected under independence. Figure 9.4 shows a tendency for subjects at the high end of one scale to be at the high end of the other and for subjects at the low end of one to be at the low end of the other.

Correspondence analysis is used mainly as a descriptive tool. Goodman (1986) developed inferential methods for it. For Table 9.9, inferential analysis reveals that the first dimension, accounting for 94% of the total squared correlation, is adequate for describing the association. Goodman argued for choosing the unsaturated model employing only one dimension and having graphics display fitted scores for that dimension alone. Then, correspondence analysis is equivalent to a ML analysis using correlation model (9.16). The estimated scores for that model are $(-1.09, -1.17, -0.37, 0.05, 1.01, 1.80)$ for the rows and $(-1.60, -0.19, 0.09, 1.48)$ for the columns. The model fits well $(G^2 = 2.75, df = 8)$. The quality of fit and the estimated scores are similar to those we saw in Section 9.6.2 for the RC model. More parsimonious correlation models also fit these data well, such as ones using equally spaced scores.

All analyses of Table 9.9 have yielded similar conclusions about the association. They all neglect, however, that mental health is a natural response variable. It may make more sense to use an ordinal logit model.

Like correlation models, a severe limitation of correspondence analysis is nontrivial generalization to multiway tables. Greenacre (1993) showed displays of several pairwise associations in a single plot.

9.6.5 Model Selection and Score Choice for Ordinal Variables

The past three sections showed several ways to use category orderings in model building. With allowance for ordinal effects, the variety of potential models is much greater than standard loglinear models. To choose among models, one approach uses the standard models for guidance. If a standard model fits well, simplify by replacing some parameters with structured terms for ordinal classifications.

Association, correlation, and correspondence analysis models have scores for categories of ordinal variables. Parameter interpretations are simplest for equally spaced scores. With parameter scores, the resulting ML estimates of scores need not be monotone. Constrained versions of the models force monotonicity by maximizing the likelihood subject to order restrictions (e.g., Agresti et al. 1987; Ritov and Gilula 1991). Disadvantages exist, however, of treating scores as parameters. The model becomes less parsimonious, and tests of effects may be less powerful because of a greater df value (recall Section 6.4.3). When one variable alone is a response, cumulative link models

(Sections 7.2 and 7.3) for that response do not require preassigned or parameter scores.

9.7 POISSON REGRESSION FOR RATES

Loglinear models need not refer to contingency tables. In Section 4.3 we introduced Poisson regression for modeling counts. When outcomes occur over time, space, or some other index of size, it is more relevant to model their *rate* of occurrence than their raw number.

9.7.1 Analyzing Rates Using Loglinear Models with Offsets

When a response count n_i has index equal to t_i, the sample rate is n_i/t_i. Its expected value is μ_i/t_i. With an explanatory variable x, a loglinear model for the expected rate has form

$$\log(\mu_i/t_i) = \alpha + \beta x_i. \tag{9.17}$$

This model has equivalent representation

$$\log \mu_i - \log t_i = \alpha + \beta x_i.$$

As noted in Section 8.7.4, the adjustment term, $-\log t_i$, to the log link of the mean is called an *offset*. The fit correspond to using $\log t_i$ as a predictor on the right-hand side and forcing its coefficient to equal 1.0.

For model (9.17), the expected response count satisfies

$$\mu_i = t_i \exp(\alpha + \beta x_i) .$$

The mean is proportional to the index, with proportionality constant depending on the value of x. The identity link is also sometimes useful. The model is then

$$\mu_i/t_i = \alpha + \beta x_i, \quad \text{or} \quad \mu_i = \alpha t_i + \beta x_i t_i.$$

This does not require an offset. It corresponds to an ordinary Poisson GLM using identity link with t_i and $x_i t_i$ as explanatory variables and no intercept. It provides additive, rather than multiplicative, predictor effects. It is less useful with many predictors, as the fitting process may fail because of negative fitted counts at some iteration.

9.7.2 Modeling Death Rates for Heart Valve Operations

Laird and Olivier (1981) analyzed patient survival after heart valve replacement operations. A sample of 109 patients were classified by type of heart

TABLE 9.11 Data on Heart Valve Replacement Operations

Age		Type of Heart Valve	
		Aortic	Mitral
< 55	Deaths	4	1
	Time at risk	1259	2082
	Death rate	0.0032	0.0005
55 +	Deaths	7	9
	Time at risk	1417	1647
	Death rate	0.0049	0.0055

Source: Reprinted with permission, based on data in Laird and Olivier (1981).

valve (aortic, mitral) and by age (< 55, ≥ 55). Follow-up observations occurred until the patient died or the study ended. Operations occurred throughout the study period, and follow-up observations covered lengths of time varying from 3 to 97 months. The response was whether the subject died and the follow-up time. For subjects who died, this is the time after the operation until death; for the others, it is the time until the study ended or the subject withdrew from it.

Table 9.11 lists the numbers of deaths during the follow-up period, by valve type and age. These counts are the first layer of a three-way contingency table that classifies valve type, age, and whether died (yes, no). The subjects not tabulated in Table 9.11 were not observed to die. They are *censored*, since we know only a lower bound for how long they lived after the operation. It is inappropriate to analyze that $2 \times 2 \times 2$ table using binary GLMs for the probability of death, since subjects had differing times at risk; it is not sensible to treat a subject who could be observed for 3 months and a subject who could be observed for 97 months as identical trials with the same probability. To use age and valve type as predictors in a model for frequency of death, the proper baseline is not the number of subjects but rather the total time that subjects were at risk. Thus, we model the *rate* of death.

The *time at risk* for a subject is their follow-up time of observation. For a given age and valve type, the total time at risk is the sum of the times at risk for all subjects in that cell (those who died and those censored). Table 9.11 lists those total times in months. The sample rate, also shown in that table, divides the number of deaths by total time at risk. For instance, 4 deaths in 1259 months of observation occurred for younger subjects with aortic valve replacement, so their sample rate is $4/1259 = 0.0032$.

We now model effects of age and valve type on the rate. Let a be a dummy variable for age, with $a_1 = 0$ for the younger age group and $a_2 = 1$ for the older group. Let v be a dummy variable for valve type, with $v_1 = 0$ for aortic and $v_2 = 1$ for mitral. Let n_{ij} denote the number of deaths for age a_i and valve type v_j, with expected value μ_{ij} for total time at risk t_{ij}. Given t_{ij},

TABLE 9.12 Fit to Table 9.11 for Poisson Regression Models

Age		Log Link		Identity Link	
		Aortic	Mitral	Aortic	Mitral
< 55	Number of deaths	2.28	2.72	3.16	1.19
	Death rate	0.0018	0.0013	0.0025	0.0006
55 +	Number of deaths	8.72	7.28	9.17	7.48
	Death rate	0.0062	0.0044	0.0065	0.0046

the expected rate is μ_{ij}/t_{ij}. The model

$$\log(\mu_{ij}/t_{ij}) = \alpha + \beta_1 a_i + \beta_2 v_j \qquad (9.18)$$

assumes a lack of interaction in the effects.

Model fitting uses standard iterative methods, treating $\{n_{ij}\}$ as independent Poisson variates with means $\{\mu_{ij}\}$. This is done conditional on $\{t_{ij}\}$. Table 9.12 presents the fitted death counts and estimated rates. The estimated effects are

$$\hat{\beta}_1 = 1.221 \quad (\text{SE} = 0.514), \qquad \hat{\beta}_2 = -0.330 \quad (\text{SE} = 0.438).$$

There is evidence of an age effect. Given valve type, the estimated rate for the older age group is $\exp(1.221) = 3.4$ times that for the younger age group. The 95% Wald confidence interval for β_1 of $1.221 \pm 1.96(0.514)$ translates to $(1.2, 9.3)$ for the true multiplicative effect $\exp(\beta_1)$. (The likelihood-ratio confidence interval is $(1.3, 10.4)$.) The study contains much censored data. Of the 109 patients, only 21 died during the study period. Both effect estimates are imprecise. Note, though, that the analysis uses all 109 patients through their contributions to the times at risk.

Goodness-of-fit statistics comparing $\{n_{ij}\}$ to fitted values $\{\hat{\mu}_{ij}\}$ are $G^2 = 3.2$ and $X^2 = 3.1$. The residual df $= 1$, since the four response counts have three parameters. The mild evidence of lack of fit corresponds to evidence of interaction between valve type and age. However, the model without valve-type effects [i.e., $\beta_2 = 0$ in (9.18)] fits nearly as well, with $G^2 = 3.8$ and $X^2 = 3.8$ (df $= 2$). Models omitting age effects fit poorly.

The corresponding model with identity link

$$\mu_{ij} = \alpha t_{ij} + \beta_1 a_i t_{ij} + \beta_2 v_j t_{ij}$$

shows a good fit, with $G^2 = 1.1$ and $X^2 = 1.1$ (df $= 1$). Table 9.12 shows the fit. Substantive conclusions are similar. The estimate $\hat{\beta}_1 = 0.0040$ (SE $= 0.0014$) then represents an estimated difference in death rates between the older and younger age groups for each valve type.

9.7.3 Modeling Survival Times*

A method for modeling survival times relates to the Poisson loglinear model for rates. This method focuses on times until death rather than on numbers of deaths. Let T denote the time to some event, such as death or such as product failure in a reliability study. Let $f(t)$ denote the probability density function (pdf) and $F(t)$ the cdf of T. A connection exists between ML estimation using a Poisson likelihood for numbers of events and a negative exponential likelihood for T (Aitkin and Clayton 1980).

A subject having $T = t$ contributes $f(t)$ to the likelihood. For a subject whose censoring time equals t, we know only that $T > t$. Thus, this subject contributes $P(T > t) = 1 - F(t)$. Using the indicator $w_i = 1$ for death and 0 for censoring for subject i, the survival-time likelihood for n independent observations is

$$\prod_{i=1}^{n} f(t_i)^{w_i} [1 - F(t_i)]^{1-w_i}.$$

The log likelihood equals

$$\sum_i w_i \log[f(t_i)] + \sum_i (1 - w_i) \log[1 - F(t_i)]. \qquad (9.19)$$

Further analysis requires a parametric form for f and a model for the dependence of its parameters on explanatory variables.

Most survival models focus on the *rate* at which death occurs rather than on $E(T)$. The *hazard function*

$$h(t) = \frac{f(t)}{1 - F(t)} = \lim_{\epsilon \downarrow 0} \frac{P[t < T < t + \epsilon \mid T > t]}{\epsilon}$$

represents the instantaneous rate of death for subjects who have survived to time t. A simple density for survival modeling is the negative exponential. The pdf is

$$f(t) = \lambda e^{-\lambda t}, \qquad t > 0.$$

The cdf is $F(t) = 1 - e^{-\lambda t}$ for $t > 0$, and $E(T) = \lambda^{-1}$. The hazard function is

$$h(t) = \lambda, \qquad t > 0,$$

constant for all t.

Now we include explanatory variables $\mathbf{x}$. Suppose that the hazard function for a negative exponential survival distribution is

$$h(t; \mathbf{x}) = \lambda \exp(\boldsymbol{\beta}' \mathbf{x}). \qquad (9.20)$$

That is, the distribution for T has parameter depending on **x** through (9.20). The choice of functional form (9.20) for explanatory variable effects ensures the hazard is nonnegative at all **x**. For instance, loglinear model (9.18) corresponds to a multiplicative model of type (9.20) for the rate itself.

Now, consider the log likelihood (9.19) with $f(t)$ equal to the negative exponential density with parameter $\lambda \exp(\boldsymbol{\beta}'\mathbf{x})$. For subject i, let

$$\mu_i = t_i \lambda \exp(\boldsymbol{\beta}'\mathbf{x}_i).$$

With this substitution, the log likelihood simplifies to

$$\sum_i w_i \log \mu_i - \sum_i \mu_i - \sum_i w_i \log t_i.$$

The first two terms involve $\boldsymbol{\beta}$. This part is identical to the log likelihood for independent Poisson variates $\{w_i\}$ with expected values $\{\mu_i\}$. In this application $\{w_i\}$ are binary rather than Poisson, but that is irrelevant to the process of maximizing with respect to $\boldsymbol{\beta}$. This process is equivalent to maximizing the likelihood for the Poisson loglinear model

$$\log \mu_i - \log t_i = \log \lambda + \boldsymbol{\beta}'\mathbf{x}_i$$

with offset $\log(t_i)$, using observations $\{w_i\}$. When we sum terms in the log likelihood for subjects having a common value of **x**, the observed data are the numbers of deaths $(\sum w_i)$ at each setting of **x**, and the offset is the log of $(\sum t_i)$ at each setting.

The assumption of constant hazard over time is often not sensible. As products wear out, their failure rate increases. A generalization divides the time scale into disjoint time intervals and assumes constant hazard in each, namely,

$$h(t;\mathbf{x}) = \lambda_k \exp(\boldsymbol{\beta}'\mathbf{x})$$

for t in interval k, $k = 1, \ldots$. A separate hazard rate applies to each piece of the time scale. Consider the contingency table for numbers of deaths, in which one dimension is a discrete time scale and other dimensions represent categorical explanatory variables. Holford (1980) and Laird and Olivier (1981) showed that Poisson loglinear models and likelihoods for this table are equivalent to loglinear hazard models and likelihoods that assume piecewise exponential hazards for the survival times.

For short time intervals, the piecewise exponential approach is essentially nonparametric, making no assumption about the dependence of the hazard on time. This suggests the generalization of model (9.20) that replaces λ by an unspecified function $\lambda(t)$, so that

$$h(t;\mathbf{x}) = \lambda(t)\exp(\boldsymbol{\beta}'\mathbf{x}).$$

This is the Cox *proportional hazards* model. Its ratio of hazards

$$h(t;\mathbf{x}_1)/h(t;\mathbf{x}_2) = \exp[\boldsymbol{\beta}'(\mathbf{x}_1 - \mathbf{x}_2)]$$

is the same for all t.

TABLE 9.13 Number of Deaths from Lung Cancer

Follow-up Time Interval (months)	Disease Stage:	Histology[a]								
		I			II			III		
		1	2	3	1	2	3	1	2	3
0–2		9	12	42	5	4	28	1	1	19
		(157	134	212	77	71	130	21	22	101)
2–4		2	7	26	2	3	19	1	1	11
		(139	110	136	68	63	72	17	18	63)
4–6		9	5	12	3	5	10	1	3	7
		(126	96	90	63	58	42	14	14	43)
6–8		10	10	10	2	4	5	1	1	6
		(102	86	64	55	42	21	12	10	32)
8–10		1	4	5	2	2	0	0	0	3
		(88	66	47	50	35	14	10	8	21)
10–12		3	3	4	2	1	3	1	0	3
		(82	59	39	45	32	13	8	8	14)
12 +		1	4	1	2	4	2	0	2	3
		(76	51	29	42	28	7	6	6	10)

[a] Values in parentheses represent total follow-up.

Source: Reprinted with permission from the Biometric Society, based on Holford (1980).

9.7.4 Lung Cancer Survival Example*

Table 9.13 describes survival for 539 males diagnosed with lung cancer. The prognostic factors are histology (H) and stage (S) of disease. For a piecewise exponential hazard approach, the time scale for follow-up (T) was divided into two-month intervals.

Let μ_{ijk} denote the expected number of deaths and t_{ijk} the total time at risk for histology i and state of disease j, in follow-up time interval k. The model

$$\log(\mu_{ijk}/t_{ijk}) = \lambda + \lambda_i^H + \lambda_j^S + \lambda_k^T \tag{9.21}$$

has residual $G^2 = 43.9$ (df = 52). All models assuming no interaction between follow-up time interval and either prognostic factor are proportional hazards models, since they have the same effects of histology and stage of disease for each time interval. Table 9.14 summarizes results of fitting several such models. Although stage of disease is an important prognostic factor, histology did not contribute significant additional information.

For model (9.21), the effects of stage of disease satisfy

$$\hat{\lambda}_2^S - \hat{\lambda}_1^S = 0.470 \quad (\text{SE} = 0.174),$$

$$\hat{\lambda}_3^S - \hat{\lambda}_1^S = 1.324 \quad (\text{SE} = 0.152).$$

**TABLE 9.14   Results for Poisson Regression Models
of Proportional Hazards Form with Table 9.13**

Effects[a]	G^2	df
T	170.7	56
$T + H$	143.1	54
$T + S$	45.8	54
$T + S + H$	43.9	52
$T + S + H + S \times H$	41.5	48

[a]T, time scale for follow-up; H, histology; S, disease stage.

For instance, at a fixed follow-up time for a given histology, the estimated death rate at the third stage of disease is $\exp(1.324) = 3.8$ times that at the first stage. Adding interaction terms between stage and time does not significantly improve the fit (change in $G^2 = 14.9$, change in df $= 12$). The $\{\hat{\lambda}_j^S\}$ are very similar for the simpler model without the histology effects.

9.7.5   Analyzing Weighted Data*

The process of fitting a loglinear model with an offset is also useful in other applications. For expected frequencies $\{\mu_i\}$ and fixed constants $\{t_i\}$, consider a model

$$\log(\mu_i/t_i) = \alpha + \beta_1 x_{i1} + \beta_2 x_{i2} + \cdots .$$

Standard loglinear models have $\{t_i = 1\}$. The general form is useful for the analysis of categorical data with sampling designs more complex than simple random sampling.

Many surveys have sampling designs employing stratification and/or clustering. Case weights inflate or deflate the influence of each observation according to features of that design. Adding the case weights for subjects in a particular cell i provides a total weighted frequency for that cell. The average cell weight z_i is defined to be the total weighted frequency divided by the cell count. Conditional on $\{z_i\}$, loglinear models for the weighted expected frequencies $\{z_i \mu_i = \mu_i/t_i\}$ with $t_i = z_i^{-1}$ express the model as a standard loglinear model for $\{\log \mu_i\}$, with offset $\{\log t_i = -\log z_i\}$. Fitting this model provides appropriate parameter estimates and standard errors (Clogg and Eliason 1987).

9.8   EMPTY CELLS AND SPARSENESS IN MODELING CONTINGENCY TABLES

Contingency tables having small cell counts are said to be *sparse*. We end this chapter by discussing effects of sparse tables on model fitting. Sparse tables occur when the sample size n is small. They also occur when n is large but so is the number of cells. Sparseness is common in tables with many variables. The following discussion refers to a generic contingency table and model, with cell counts $\{n_i\}$ and expected frequencies $\{\mu_i\}$ for n observations in N cells.

9.8.1 Empty Cells: Sampling versus Structural Zeros

Sparse tables usually contain cells with $n_i = 0$. These *empty cells* are of two types: *sampling zeros* and *structural zeros*. In most cases, even though $n_i = 0$, $\mu_i > 0$. It is possible to have observations in the cell, and $n_i > 0$ with sufficiently large n. This empty cell is called a *sampling zero*. The empty cells in Table 9.1 for the student survey are sampling zeros.

An empty cell in which observations are impossible is called a *structural zero*. For such cells $\mu_i = 0$ and necessarily $\hat{\mu}_i = 0$ and $n_i = 0$ regardless of n. For a table that cross classifies cancer patients on their gender, race, and type of cancer, some cancers (e.g., prostate cancer, ovarian cancer) are gender specific. Thus, certain cells have structural zeros. Contingency tables with structural zeros are called *incomplete tables*.

Sampling zeros are part of the data set. A count of 0 is a permissible outcome for a Poisson or multinomial variate. It contributes to the likelihood function and model fitting. A structural zero, on the other hand, is not an observation and is not part of the data. Sampling zeros are much more common than structural zeros, and the remaining discussion refers to them.

9.8.2 Existence of Estimates in Loglinear / Logit Models

Sampling zeros can affect the existence of finite ML estimates of loglinear and logit model parameters. Haberman (1973b, 1974a), generalizing work by Birch (1963) and Fienberg (1970b), studied this. Let $\mathbf{n}$ denote the vector of cell counts and $\boldsymbol{\mu}$ their expected values. Haberman showed results 1 through 5 for Poisson sampling, but by result 6 they apply also to multinomial sampling.

1. The log-likelihood function is a strictly concave function of $\log \boldsymbol{\mu}$.
2. If a ML estimate of $\boldsymbol{\mu}$ exists, it is unique and satisfies the likelihood equations $\mathbf{X'n} = \mathbf{X'\hat{\mu}}$. Conversely, if $\hat{\boldsymbol{\mu}}$ satisfies the model and also the likelihood equations, it is the ML estimate of $\boldsymbol{\mu}$.
3. If all $n_i > 0$, ML estimates of loglinear model parameters exist.
4. Suppose that ML parameter estimates exist for a loglinear model that equates observed and fitted counts in certain marginal tables. Then those marginal tables have uniformly positive counts.
5. If ML estimates exist for a model M, they also exist for any special case of M.

6. For any loglinear model, the ML estimates $\hat{\boldsymbol{\mu}}$ are identical for multinomial and independent Poisson sampling, and those estimates exist in
the same situations.

To illustrate, consider the saturated model. By results 2 and 3, when all
$n_i > 0$, the ML estimate of $\boldsymbol{\mu}$ is **n**. By result 4, parameter estimates do not
exist when any $n_i = 0$. Model parameter estimates are contrasts of $\{\log \hat{\mu}_i\}$,
and since $\hat{\boldsymbol{\mu}} = \mathbf{n}$ for the saturated model, the estimates are finite only when
all $n_i > 0$.

For unsaturated models, by results 3 and 4 ML estimates exist when all
$n_i > 0$ and do not exist when any count is zero in the set of sufficient
marginal tables. Suppose that at least one $n_i = 0$ but the sufficient marginal
counts are all positive. For hierarchical loglinear models, Glonek et al. (1988)
showed that the positivity of the sufficient counts implies the existence of ML
estimates if and only if the model is decomposable (Note 8.2), which includes
the conditional independence models. Models having all pairs of variables
associated, however, are more complex. For model (XY, XZ, YZ), for instance, ML estimates exist when only one $n_i = 0$ but may not exist when at
least two cells are empty. For instance, ML estimates do not exist for Table
9.15, even though all sufficient statistics (the two-way marginal totals) are
positive (Problem 9.47).

Haberman showed that the supremum of the likelihood function is finite.
This motivated him to define *extended ML* estimators of $\boldsymbol{\mu}$. These always
exist but may equal 0 and, falling on the boundary, need not have the same
properties as regular ML estimators [see also Baker et al. (1985)]. A sequence
of estimates satisfying the model that converges to the extended estimate has
log likelihood approaching its supremum. In this extended sense, $\hat{\mu}_i = 0$ is
the ML estimate of μ_i for the saturated model when $n_i = 0$, and one can
have infinite loglinear parameter estimates.

When a sufficient marginal count for a factor equals zero, infinite estimates occur for that term. For instance, when a XY marginal total equals
zero, infinite estimates occur among $\{\hat{\lambda}_{ij}^{XY}\}$ for loglinear models such as
(XY, XZ, YZ), and infinite estimates occur among $\{\hat{\beta}_i^X\}$ for the effect of X
on Y in logit models. Sometimes, however, not even infinite estimates exist.
An example is estimating the log odds ratio when both entries in a row or
column of a 2×2 table equal 0.

**TABLE 9.15   Data for Which ML Estimates Do Not Exist
for Model (XY, XZ, YZ)[a]**

	Z:	1		2	
X	Y:	1	2	1	2
1		0	*	*	*
2		*	*	*	0

[a]Cells containing * may contain any positive numbers.

A value of ∞ (or $-\infty$) for a ML parameter estimate implies that ML fitted values equal 0 in some cells, and some odds ratio estimates equal ∞ or 0. One potential indicator is when the iterative fitting process does not converge, typically because an estimate keeps increasing from cycle to cycle. Most software, however, is fooled after a certain point in the iterative process by the nearly flat likelihood. It reports convergence, but because of the very slight curvature of the log likelihood, the estimated standard errors (based on inverting the information matrix of second partial derivatives) are extremely large and numerically unstable. Slight changes in the data then often cause dramatic changes in the estimates and their standard errors. A danger with sparse data is that one might not realize that a true estimated effect is infinite and, as a consequence, report estimated effects and results of statistical inferences that are invalid and highly unstable.

Many ML analyses are unharmed by empty cells. Even when a parameter estimate is infinite, this is not fatal to data analysis. The likelihood-ratio confidence interval for the true log odds ratio has one endpoint that is finite. For instance, when $n_{11} = 0$ but other $n_{ij} > 0$ in a 2×2 table, $\log \hat{\theta} = -\infty$ and a confidence interval has form $(-\infty, U)$ for some finite upper bound U. When the pattern of empty cells forces certain fitted values for a model to equal 0, this affects the df for testing model fit (Haslett 1990).

9.8.3   Clinical Trials Example

Table 9.16 shows results of a clinical trial conducted at five centers. The purpose was to compare an active drug to placebo for treating fungal infections, with a binary (success, failure) response. For these data, let Y = response, X = treatment ($x_1 = 1$ for active drug and $x_2 = 0$ for placebo), and Z = center.

Centers 1 and 3 had no successes. Thus, the 5×2 marginal table relating response to center, collapsed over treatment, contains zero counts. The last two columns of Table 9.16 show this marginal table. Infinite ML estimates occur for terms in loglinear or logit models containing the YZ association. An example is the logit model

$$\text{logit}[P(Y = 1 | X = i, Z = k)] = \beta x_i + \beta_k^Z.$$

(We omit the intercept, so the $\{\beta_k^Z\}$ need no constraint; then, these refer to center effects rather than contrasts between centers and a baseline center.) The likelihood function increases continually as β_1^Z and β_3^Z decrease toward $-\infty$; that is, as the logit decreases toward $-\infty$, so the fitted probability of success decreases toward the ML estimate of 0 for those centers.

The counts in the 2×2 marginal table relating response to treatment, shown in the bottom panel of Table 9.16, are all positive. The empty cells in Table 9.16 affect the center estimates, but not the treatment estimate, for this logit model. In the limit as the log likelihood increases, the fitted values have a log odds ratio $\hat{\beta} = 1.55$ (SE = 0.70). Most software reports this, but

TABLE 9.16   Clinical Trial Relating Treatment to Response with XY and YZ Marginal Tables[a]

Center	Treatment	Response		YZ Marginal	
		Success	Failure	Success	Failure
1	Active drug	0	5	0	14
	Placebo	0	9		
2	Active drug	1	12	1	22
	Placebo	0	10		
3	Active drug	0	7	0	12
	Placebo	0	5		
4	Active drug	6	3	8	9
	Placebo	2	6		
5	Active drug	5	9	7	21
	Placebo	2	12		
XY	Active drug	12	36		
marginal	Placebo	4	42		

[a]X, Treatment; Y, response; Z, center.

Source: Data courtesy of Diane Connell, Sandoz Pharmaceuticals Corporation.

instead of $\hat{\beta}_1^Z = \hat{\beta}_3^Z = -\infty$ reports large numbers with extremely large standard errors. For instance, PROC GENMOD in SAS reports values of about -26 for $\hat{\beta}_1^Z$ and $\hat{\beta}_3^Z$, with standard errors of about 200,000.

The treatment estimate $\hat{\beta} = 1.55$ also results from deleting centers 1 and 3 from the analysis. When a center contains responses of only one type, it provides no information about this odds ratio. (It does provide information about the size of some other measures, such as the difference of proportions.) In fact, such tables also make no contribution to standard tests of conditional independence, such as the Cochran–Mantel–Haenszel test (Section 6.3.2) and exact test (Section 6.7.5).

An alternative strategy in multicenter analyses combines centers of a similar type. Then, if each resulting partial table has responses with both outcomes, the inferences use all data. For Table 9.16, perhaps centers 1 and 3 are similar to center 2, since the success rate is very low for that center. Combining these three centers and refitting the model to this table and the tables for the other two centers yields $\hat{\beta} = 1.56$ (SE = 0.70). Usually, this strategy produces results similar to deleting the table with no outcomes of a particular type.

9.8.4   Effect of Small Samples on X^2 and G^2

Although empty cells and sparse tables need not affect parameter estimates of interest, they can cause sampling distributions of goodness-of-fit statistics to be far from chi-squared. The true sampling distributions converge to chi-squared as $n \to \infty$, for a fixed number of cells N. The adequacy of the chi-squared approximation depends both on n and N.

Cochran studied the chi-squared approximation for X^2 in several articles. In 1954, he suggested that to test independence with df > 1, a minimum expected value $\mu_i \approx 1$ is permissible as long as no more than about 20% of $\mu_i < 5$. Koehler (1986), Koehler and Larntz (1980), and Larntz (1978) showed that X^2 applies with smaller n and more sparse tables than G^2. The distribution of G^2 is usually poorly approximated by chi-squared when n/N is less than 5. Depending on the sparseness, P-values based on referring G^2 to a chi-squared distribution can be too large or too small. When most μ_i are smaller than 0.5, treating G^2 as chi-squared gives a highly conservative test; when H_0 is true, reported P-values tend to be much larger than true ones. When most μ_i are between 0.5 and 4, G^2 tends to be too liberal; the reported P-value tends to be too small.

The size of n/N that produces adequate approximations for X^2 tends to decrease as N increases (Koehler and Larntz 1980). However, the approximation tends to be poor for sparse tables containing both small and moderately large μ_i (Haberman 1988). It is difficult to give a guideline that covers all cases. For other discussion, see Cressie and Read (1989) and Lawal (1984).

For fixed n and N, the chi-squared approximation is better for tests with smaller df. For instance, in testing conditional independence in $I \times J \times K$ tables, $G^2[(XZ, YZ) | (XY, XZ, YZ)]$ (with df $= (I - 1)(J - 1)$) is closer to chi-squared than $G^2(XZ, YZ)$ [with df $= K(I - 1)(J - 1)$]. The ordinal test of H_0: $\beta = 0$ with the homogeneous linear-by-linear XY association model (9.10) has df $= 1$, and behaves even better.

9.8.5 Model-Based Tests and Sparseness

From (9.3) and (9.4), the model-based statistics $G^2(M_0 | M_1)$ and $X^2(M_0 | M_1)$ depend on the data only through the fitted values, and hence only through minimal sufficient statistics for the more complex model. These statistics have null distributions converging to chi-squared as the expected values of the minimal sufficient statistics grow. For most loglinear models, these sufficient statistics refer to marginal tables. Marginal totals are more nearly normally distributed than are single cell counts. Thus, $G^2(M_0 | M_1)$ and $X^2(M_0 | M_1)$ converge to their limiting chi-squared distribution more quickly than does $G^2(M_0)$ and $X^2(M_0)$, which depend also on individual cell counts.

When $\{\hat{\mu}_i\}$ are small but the sufficient marginal totals for M_1 are mostly in at least the range 5 to 10, the chi-squared approximation is usually adequate for model comparison statistics. Haberman (1977a) provided theoretical justification.

9.8.6 Alternative Asymptotics and Alternative Statistics

When large-sample approximations are inadequate, exact small-sample methods are an alternative. When they are infeasible, it is often possible to approximate exact distributions precisely using Monte Carlo methods (e.g., Booth and Butler 1999; Forster et al. 1996; Kim and Agresti 1997; Mehta et al. 1988).

An alternative approach uses sparse asymptotic approximations that apply when the number of cells N increases as n increases. For this approach, $\{\mu_i\}$ need not increase, as they must do in the usual (fixed N, $n \to \infty$) large-sample theory. For goodness-of-fit testing of a specified multinomial, Koehler and Larntz (1980) showed that a standardized version of G^2 has an approximate normal distribution for very sparse tables. Koehler (1986) presented limiting normal distributions for G^2 for use in testing models having direct ML estimates. McCullagh (1986) reviewed ways of handling sparse tables and presented an alternative approximation for G^2. Zelterman (1987) gave normal approximations for X^2 and proposed an alternative statistic.

9.8.7 Adding Constants to Cells of a Contingency Table

Empty cells and sparse tables can cause problems with existence of estimates for loglinear model parameters, estimation of odds ratios, performance of computational algorithms, and asymptotic approximations of chi-squared statistics. However, they need not be problematic. The likelihood can still be maximized, a point estimate of ∞ for an effect still usually has a finite lower bound for a likelihood-based confidence interval, and one can use small-sample inferential methods rather than asymptotic ones.

One way to obtain finite estimates of all effects and ensure convergence of fitting algorithms is to add a small constant to cell counts. Some algorithms add $\frac{1}{2}$ to each cell, as Goodman (1964b, 1970, 1971a) recommended for saturated models. An example of the beneficial effect of this for a saturated model is bias reduction for estimating an odds ratio in a 2×2 table (Gart 1966; Gart and Zweiful 1967). Adding $\frac{1}{2}$ to each cell before fitting an unsaturated model smooths the data too much, however, causing havoc with sampling distributions. This operation has too conservative an influence on estimated effects and test statistics. The effect is very severe with a large number of cells.

Even for a saturated model, adding $\frac{1}{2}$ to each cell is not a panacea for all purposes. When the ordinary ML estimate of an odds ratio is infinite, the estimate after adding $\frac{1}{2}$ to each cell is finite, as are the endpoints of any confidence interval. However, it is more sensible to use an upper bound of ∞ for the odds ratio, since no sample evidence suggests that the odds ratio falls below any given value.

When in doubt about the effect of sparse data, one should perform a sensitivity analysis. For example, for each possibly influential observation, delete it or move it to another cell to see how results vary with small perturbations to the data. Influence diagnostics for GLMs (Williams 1987) are also useful for this purpose. Often, some associations are not affected by empty cells and give stable results for the various analyses, whereas others that are affected are highly unstable. Use caution in making conclusions about an association if small changes in the data are influential.

Later chapters show ways to smooth data in a less *ad hoc* manner than adding arbitrary constants to cells. These include random effects models (Section 12.3) and Bayesian methods (Section 15.2).

NOTES

Section 9.1: Association Graphs and Collapsibility

9.1. Darroch et al. (1980) defined a class of *graphical models* that contains the family of decomposable models (see Note 8.2). For expositions on graphical models and their relevant *independence graphs*, which show the conditional independence structure, see also Anderson and Böckenholt (2000), Edwards (2000), Edwards and Kreiner (1983), Kreiner (1998), Lauritzen (1996), and Whittaker (1990). Whittaker (1990, Sec. 12.5) summarized connections with various definitions of collapsibility.

9.2 For $I \times J \times 2$ tables, the collapsibility conditions (Section 9.1.2) are necessary as well as sufficient (Simpson 1951; Whittemore 1978). For $I \times J \times K$ tables, Ducharme and Lepage (1986) showed the conditions are necessary and sufficient for the odds ratios to remain the same no matter how the levels of Z are pooled (i.e., no matter how Z is partially collapsed).

Darroch (1962) defined a *perfect* table as one for which for all i, j, k,

$$\sum_i \frac{\pi_{ij+} \pi_{i+k}}{\pi_{i++}} = \pi_{+j+} \pi_{++k}, \qquad \sum_j \frac{\pi_{+jk} \pi_{ij+}}{\pi_{+j+}} = \pi_{i++} \pi_{++k},$$

$$\sum_k \frac{\pi_{i+k} \pi_{+jk}}{\pi_{++k}} = \pi_{i++} \pi_{+j+}.$$

For perfect tables, homogeneous association implies that

$$\left\{ \pi_{ijk} = \pi_{ij+} \pi_{i+k} \pi_{+jk} / \pi_{i++} \pi_{+j+} \pi_{++k} \right\}$$

and conditional odds ratios are identical to marginal odds ratios. Whittemore (1978) used perfect tables to illustrate that for $I \times J \times K$ tables with $K > 2$, conditional and marginal odds ratios can be identical even when no pair of variables is conditionally independent. See also Davis (1986b).

Suppose that the difference of proportions or relative risk, computed for a binary response Y and predictor X, is the same at every level of Z. If Z is independent of X in the marginal XZ table or if Z is conditionally independent of Y given X, the measure has the same value in the marginal XY table (Shapiro 1982). Thus, for factorial designs with the same number of observations at each combination of levels, the difference of proportions and relative risk are collapsible. See also Wermuth (1987).

Section 9.2: Model Selection and Comparison

9.3. Articles on loglinear model selection include Aitkin (1979, 1980), Benedetti and Brown (1978), Brown (1976), Goodman (1970, 1971a), Wermuth (1976), and Whittaker and Aitkin (1978). When a certain model holds, G^2/df has an asymptotic mean of 1. Goodman (1971a) recommended this index for comparing fits. Smaller values represent better fits.

9.4. Kullback et al. (1962) and Lancaster (1951) were among the first to partition chi-squared statistics in multiway tables. Goodman (1970) and Plackett (1962) noted difficulties with their approaches. When observations have distribution in the natural exponential family, Simon (1973) showed $G^2(M_0 \mid M_1) = 2\sum_i \hat{\mu}_{1i} \log(\hat{\mu}_{1i}/\hat{\mu}_{0i})$ whenever models are linear in the natural parameters. See Lang (1996b) for partitionings for more complex models.

Section 9.4: Modeling Ordinal Associations

9.5. Goodman (1979a) stimulated research on loglinear models for ordinal data. His work extended Haberman (1974b), who expressed the λ^{XY} association term with an expansion in orthogonal polynomials. For more general ordinal models for multiway tables, see Agresti (1984), Becker (1989a), Becker and Clogg (1989), and Goodman (1986).

Section 9.6: Association Models, Correlation Models, and Correspondence Analysis

9.6. Early articles on the *RC* model include Goodman (1979a, 1981a, b) and Andersen (1980, pp. 210–216), apparently partly motivated by earlier work of G. Rasch (see Andersen 1995). Anderson and Böckenholt (2000), Becker (1989a, b, 1990), Becker and Clogg (1989), Chuang et al. (1985), and Goodman (1985, 1986, 1996) discussed generalizations for multiway tables. Anderson (1984) discussed a related model. Anderson and Vermunt (2000) showed that *RC* and related association models arise when observed variables are conditionally independent given a latent variable that is conditionally normal, given the observed variables. Their work generalizes results in Lauritzen and Wermuth (1989) and discussion by Whittaker of van der Heijden et al. (1989). See also de Falguerolles et al. (1995). Clogg and Shihadeh (1994) surveyed association models and related correlation models.

9.7. Kendall and Stuart (1979, Chap. 33) surveyed basic canonical correlation methods for contingency tables. See also Williams (1952), who discussed earlier work by R. A. Fisher and others. Karl Pearson often analyzed tables by assuming an underlying bivariate normal distribution (Section 16.1). For estimating that distribution's correlation, see Becker (1989b), Goodman (1981b), Kendall and Stuart (1979, Chaps. 26 and 33), Lancaster (1969, Chap. X), the Pearson (1904) tetrachoric correlation for 2×2 tables, and the Lancaster and Hamdan (1964) polychoric correlation for $I \times J$ tables.

9.8. Correspondence analysis gained popularity in France under the influence of Benzécri (see, e.g., 1973). Goodman (1996) attributed its origins to H. O. Hartley, publishing under his original German name (Hirschfeld, 1935). Greenacre (1993) related it to the singular value decomposition of a matrix. For other discussion, see Escoufier (1982), Friendly (2000, Chap. 5), Goodman (1986, 1996, 2000), Michailidis and de Leeuw (1998), van der Heijden and de Leeuw (1985), and van der Heijden et al. (1989). Gabriel (1971) discussed related work on biplots.

Section 9.7: Poisson Regression for Rates

9.9. Another application using offsets is table standardization (Section 8.7.4). For analyses of rate data, see Breslow and Day (1987, Sec. 4.5), Freeman and Holford (1980), Frome (1983), and Hoem (1987). Articles dealing with grouped survival data, particularly loglinear and logit models for survival probabilities, include Aranda-Ordaz (1983), Larson (1984), Prentice and Gloeckler (1978), Schluchter and Jackson (1989), Stokes et al. (2000, Chap. 17), and Thompson (1977). Aitkin and Clayton (1980) discussed exponential survival models and also presented similar models having hazard functions for Weibull or extreme-value survival distributions. Log likelihood (9.19) actually applies only for *noninformative* censoring mechanisms. It does not make sense if subjects tend to withdraw from the study because of factors related to it, perhaps because of health effects related to one of the treatments.

9.10. Lindsey and Mersch (1992) showed a clever way to use loglinear models to fit exponential family distributions $f(y; \theta)$ of form (4.14) with ϕ known. One breaks the response scale into intervals $\{(y_k - \Delta_k/2, y_k + \Delta_k/2)\}$. Counts in those intervals follow a multinomial with probabilities approximated by $\{f(y_k, \theta)\Delta_k\}$. The log expected count approximations are linear in θ with an offset.

PROBLEMS

Applications

9.1 Use odds ratios in Table 8.3 to illustrate the collapsibility conditions.

    a. For (A, C, M), all conditional odds ratios equal 1.0. Explain why all reported marginal odds ratios equal 1.0.

    b. For (AC, M), explain why **(i)** all conditional odds ratios are the same as the marginal odds ratios, and **(ii)** all $\hat{\mu}_{ac+} = n_{ac+}$.

    c. For (AM, CM), explain why **(i)** the AC conditional odds ratios of 1.0 need not be the same as the AC marginal odds ratio, **(ii)** the AM and CM conditional odds ratios are the same as the marginal odds ratios, and **(iii)** all $\hat{\mu}_{a+m} = n_{a+m}$ and $\hat{\mu}_{+cm} = n_{+cm}$.

    d. For (AC, AM, CM), explain why **(i)** no conditional odds ratios need be the same as the related marginal odds ratios, and **(ii)** the fitted marginal odds ratios must equal the sample marginal odds ratios.

9.2 Table 9.17 summarizes a study with variables age of mother (A), length of gestation (G) in days, infant survival (I), and number of cigarettes smoked per day during the prenatal period (S). Treat G and I as response variables and A and S as explanatory.

    a. Explain why a loglinear model should include the λ^{AS} term.

    b. Fit the models $(AGIS)$, (AGI, AIS, AGS, GIS), (AG, AI, AS, GI, GS, IS), and (AS, G, I). Identify a subset of models nested between two of these that may fit well. Select one such model.

    c. Use **(i)** forward selection, and **(ii)** backward elimination to build a model. Compare the results of the strategies, and interpret the models chosen.

9.3 Refer to Table 2.13. Consider the nested set $\{(DVP), (DP, VP, DV), (VP, DV), (P, DV), (D, V, P)\}$. Partition chi-squared to compare the four pairs, ensuring that the overall type I error probability for the four comparisons does not exceed $\alpha = 0.10$. Which model would you select, using a backward comparison starting with (DVP)? Show that the final

TABLE 9.17 Data for Problem 9.2

Age	Smoking	Gestation	Infant Survival No	Infant Survival Yes
< 30	< 5	≤ 260	50	315
		> 260	24	4012
	5 +	≤ 260	9	40
		> 260	6	459
30 +	< 5	≤ 260	41	147
		> 260	14	1594
	5 +	≤ 260	4	11
		> 260	1	124

Source: N. Wermuth, pp. 279–295 in *Proc. 9th International Biometrics Conference*, Vol. 1 (1976). Reprinted with permission from the Biometric Society.

model selected depends on the choice of nested set, by repeating the analysis with (DP, VP, DV), (DP, DV), (P, DV), (D, V, P).

9.4 Consider the loglinear model selection for Table 6.3.

    a. Why is it not sensible to consider models omitting the λ^{GM} term?

    b. Using forward selection starting with (GM, E, P), show that model (GM, GP, EG, EMP) seems reasonable.

    c. Using backward elimination, show that (GM, GP, EMP) or (GM, GP, EG, EMP) seems reasonable.

    d. The *EMP* interaction seems vital. To describe it, show that the effect of extramarital sex on divorce is greater for subjects who had no premarital sex.

    e. Use residuals to describe the lack of fit of model (GM, EMP).

9.5 For model (AC, AM, CM) with Table 8.3, the standardized Pearson residual in each cell equals ± 0.63. Interpret, and explain why each one has the same absolute value. By contrast, model (AM, CM) has standardized Pearson residual ± 3.70 in each cell where $M = $ yes (e.g., $+3.70$ when $A = C = $ yes) and ± 12.80 in each cell where $M = $ no (e.g., $+12.80$ when $A = C = $ yes). Interpret.

9.6 Refer to Table 8.8. Conduct a residual analysis with the model of no three-factor interaction to describe the nature of the interaction.

9.7 Perform a residual analysis for the independence model with Table 3.2. Explain why it suggests that the linear-by-linear association model may fit better. Fit it, compare to the independence model, and interpret.

9.8 Refer to Problem 9.7.

   a. Using standardized scores, find $\hat{\beta}$. Comment on the strength of association.

   b. Fit a model in which job satisfaction scores are parameters. Interpret the estimated scores, and compare the fit to the $L \times L$ model.

9.9 Refer to Table 9.3.

   a. For the linear-by-linear association model, construct a 95% confidence interval for the odds ratio using the four corner cells. Interpret.

   b. Fit the column effects model. Compare estimated column scores to the equal-interval scores in part (a). Test that the true column scores are equal-interval, given that the model holds. Interpret. Construct a 95% confidence interval for the odds ratio using the four corner cells. Compare to part (a).

9.10 A weak local association may be substantively important for nonlocal categories. Illustrate with the $L \times L$ model for Table 9.9, showing how the estimated odd ratio for the four corner cells compares to the estimated local odds ratio.

9.11 Refer to Table 7.8. Fit the homogeneous linear-by-linear association model, and interpret. Test conditional independence between income (I) and job satisfaction (S), controlling for gender (G), using (**a**) that model, and (**b**) model (IS, IG, SG). Explain why the results are so different.

9.12 Fit the RC model to Table 9.3. Interpret the estimated scores. Does it fit better than the uniform association model?

9.13 Replicate the results in Section 9.6 for the correlation and correspondence models with Table 9.9.

9.14 One hundred leukemia patients were randomly assigned to two treatments. During the study, 10 subjects on treatment A died and 18 subjects on treatment B died. The total time at risk was 170.4 years for treatment A and 147.3 years for treatment B. Test whether the two treatments have the same death rates. Compare the rates with a confidence interval.

9.15 For Table 9.11, fit a model in which death rate depends only on age. Interpret the age effect.

9.16 Consider model (9.18). What is the effect on the model parameter estimates, their standard errors, and the goodness-of-fit statistics when (**a**) the times at risk are doubled, but the numbers of deaths stay the same; (**b**) the times at risk stay the same, but the numbers of deaths double; and (**c**) the times at risk and the numbers of deaths both double.

9.17 Consider Table 9.13. Explain how one could analyze whether the hazard depends on time.

9.18 An article by W. A. Ray et al. (*Amer. J. Epidemiol.* **132**: 873–884, 1992) dealt with motor vehicle accident rates for 16,262 subjects aged 65–84 years, with data on each for up to 4 years. In 17.3 thousand years of observation, the women had 175 accidents in which an injury occurred. In 21.4 thousand years, men had 320 injurious accidents.

   a. Find a 95% confidence interval for the true overall rate of injurious accidents.

   b. Using a model, compare the rates for men and women.

9.19 A table at the text's Web site (*www.stat.ufl.edu/~aa/cda/cda.html*) shows the number of train miles (in millions) and the number of collisions involving British Rail passenger trains between 1970 and 1984. A Poisson model assuming a constant log rate α over the 14-year period has $\hat{\alpha} = -4.177$ (SE = 0.1325) and $X^2 = 14.8$ (df = 13). Interpret.

9.20 Table 9.18 lists total attendance (in thousands) and the total number of arrests in the 1987–1988 season for soccer teams in the Second Division of the British football league. Let Y = number of arrests for a team, and let t = total attendance. Explain why the model $E(Y) = \mu t$

TABLE 9.18 Data for Problem 9.20

Team	Attendance (thousands)	Arrests	Team	Attendance (thousands)	Arrests
Aston Villa	404	308	Shrewsbury	108	68
Bradford City	286	197	Swindon Town	210	67
Leeds United	443	184	Sheffield Utd.	224	60
Bournemouth	169	149	Stoke City	211	57
West Brom	222	132	Barnsley	168	55
Hudderfield	150	126	Millwall	185	44
Middlesbro	321	110	Hull City	158	38
Birmingham	189	101	Manchester City	429	35
Ipswich Town	258	99	Plymouth	226	29
Leicester City	223	81	Reading	150	20
Blackburn	211	79	Oldham	148	19
Crystal Palace	215	78			

Source: The *Independent* (London), Dec. 21, 1988. Thanks to P. M. E. Altham for showing me these data.

might be plausible. Assuming Poisson sampling, fit it and interpret. Plot arrests against attendance, and overlay the prediction equation. Use residuals to identify teams that had arrest counts much different than expected.

TABLE 9.19   Data for Problem 9.21

	Person-Years		Coronary Deaths	
Age	Nonsmokers	Smokers	Nonsmokers	Smokers
35–44	18,793	52,407	2	32
45–54	10,673	43,248	12	104
55–64	5710	28,612	28	206
65–74	2585	12,663	28	186
75–84	1462	5317	31	102

Source: R. Doll and A. B. Hill, *Natl. Cancer Inst. Monogr.* **19**: 205–268 (1966). See also N. R. Breslow in *A Celebration of Statistics*, ed. A. C. Atkinson and S. E. Fienberg, (New York: Springer-Verlag, 1985).

9.21   Table 9.19 is based on a study with British doctors.

    a. For each age, find the sample coronary death rates per 1000 person-years for nonsmokers and smokers. To compare them, take their ratio and describe its dependence on age.

    b. Fit a main-effects model for the log rates having four parameters for age and one for smoking. In discussing lack of fit, show that this model assumes a constant ratio of nonsmokers' to smokers' coronary death rates over age.

    c. From part (a), explain why it is sensible to add a quantitative interaction of age and smoking. For this model, show that the log ratio of coronary death rates changes linearly with age. Assign scores to age, fit the model, and interpret.

9.22   Analyze Table 9.9 using ordinal logit models. Interpret, and discuss advantages/disadvantages compared to loglinear analyses.

9.23   Refer to Problem 8.6. Analyze these data, using methods of this chapter.

Theory and Methods

9.24   In a $2 \times 2 \times K$ table, the true XY conditional odds ratios are identical, but different from the XY marginal odds ratio. Is there three-factor interaction? Is Z conditionally independent of X or Y? Explain.

9.25 Consider loglinear model (WX, XY, YZ). Explain why W and Z are independent given X alone or given Y alone or given both X and Y. When are W and Y conditionally independent? When are X and Z conditionally independent?

9.26 Suppose that loglinear model (XY, XZ) holds.
   a. Find μ_{ij+} and $\log \mu_{ij+}$. Show the loglinear model for the XY marginal table has the same association parameters as $\{\lambda_{ij}^{XY}\}$ in (XY, XZ). Deduce that odds ratios are the same in the XY marginal table as in the partial tables. Using an analogous result for model (XY, YZ), deduce the collapsibility conditions in Section 9.1.2.
   b. Calculate $\log \mu_{ij+}$ for model (XY, XZ, YZ), and explain why marginal associations need not equal conditional associations.

9.27 For a four-way table, is the WX conditional association the same as the WX marginal association for the loglinear model **(a)** (WX, XYZ)? and **(b)** (WX, WZ, XY, YZ)? Why?

9.28 Loglinear model M_0 is a special case of loglinear model M_1.
   a. Explain why the fitted values for the two models are identical in the sufficient marginal distributions for M_0.
   b. Haberman (1974a) showed that when $\{\hat{\mu}_i\}$ satisfy any model that is a special case of M_0, $\sum_i \hat{\mu}_{1i} \log \hat{\mu}_i = \sum_i \hat{\mu}_{0i} \log \hat{\mu}_i$. Thus, $\hat{\boldsymbol{\mu}}_0$ is the orthogonal projection of $\hat{\boldsymbol{\mu}}_1$ onto the linear manifold of $\{\log \boldsymbol{\mu}\}$ satisfying M_0. Using this, show that $G^2(M_0) - G^2(M_1) = 2\sum_i \hat{\mu}_{1i} \log(\hat{\mu}_{1i}/\hat{\mu}_{0i})$.

9.29 Refer to Section 9.2.4. Show that $G^2(M_j | M_{j-1})$ equals G^2 for independence in the 2×2 table comparing columns 1 through $j - 1$ with column j.

9.30 For T categorical variables $X_1, \ldots, X_T$, explain why:
   a. $G^2(X_1, X_2, \ldots, X_T) = G^2(X_1, X_2) + G^2(X_1 X_2, X_3)$
   $+ \cdots + G^2(X_1 X_2 \cdots X_{T-1}, X_T)$.
   b. $G^2(X_1 \cdots X_{T-1}, X_T) = G^2(X_1, X_T) + G^2(X_1 X_T, X_1 X_2)$
   $+ \cdots + G^2(X_1 X_2 \cdots X_{T-1}, X_1 X_2 \cdots X_{T-2} X_T)$.

9.31 For $I \times 2$ contingency tables, explain why the linear-by-linear association model is equivalent to the linear logit model (5.5).

9.32 Consider the $L \times L$ model (9.6) with $\{v_j = j\}$ replaced by $\{v_j = 2j\}$. Explain why $\hat{\beta}$ is halved but $\{\hat{\mu}_{ij}\}$, $\{\hat{\theta}_{ij}\}$, and G^2 are unchanged.

9.33 Lehmann (1966) defined (X, Y) to be *positively likelihood-ratio dependent* if their joint density satisfies $f(x_1, y_1)f(x_2, y_2) \geq f(x_1, y_2)f(x_2, y_1)$ whenever $x_1 < x_2$ and $y_1 < y_2$. Then, the conditional distribution of Y (X) stochastically increases as X (Y) increases (Goodman 1981a).

   a. For the $L \times L$ model, show that the conditional distributions of Y and of X are stochastically ordered. What is its nature if $\beta > 0$?

   b. In row effects model (9.8), if $\mu_i > \mu_h$, show that the conditional distribution of Y is stochastically higher in row i than in row h. Explain why $\mu_1 = \cdots = \mu_I$ is equivalent to the equality of the I conditional distributions within rows.

9.34 Yule (1906) defined a table to be *isotropic* if an ordering of rows and of columns exists such that the local log odds ratios are all nonnegative [see also Goodman (1981a)].

   a. Show that a table is isotropic if it satisfies **(i)** the linear-by-linear association model, **(ii)** the row effects model, and **(iii)** the *RC* model.

   b. Explain why a table that is isotropic for a certain ordering is still isotropic when adjacent rows or columns are combined.

9.35 Consider the log likelihood for the linear-by-linear association model.

   a. Differentiating with respect to β and evaluating at $\beta = 0$ and null estimates of parameters, show that the score function is proportional to

   $$\sum_i \sum_j u_i v_j (p_{ij} - p_{i+} p_{+j}).$$

   b. Use the delta method to show that its null SE is

   $$\left\{ \left[\sum u_i^2 p_{i+} - \left(\sum u_i p_{i+} \right)^2 \right] \left[\sum v_j^2 p_{+j} - \left(\sum v_j p_{+j} \right)^2 \right] \Big/ n \right\}^{1/2}.$$

   c. Construct a score statistic for testing independence. Show that it is essentially the correlation test (3.15). [Hirotsu (1982) discussed a family of score tests for ordered alternatives.]

9.36 Given the parenthetical result in Problem 7.33, show that if cumulative logit model (7.24) holds and $|\beta|$ is small, the linear-by-linear association model should fit well with row scores $\{x_i\}$ and "ridit" column scores $\{v_j = [P(Y \leq j - 1) + P(Y \leq j)]/2\}$, with its β parameter about twice β for model (7.24).

9.37 Consider the row effects model (9.8).

    a. Show that no loss of generality occurs in letting $\lambda_I^X = \lambda_J^Y = \mu_I = 0$.

    b. Show that minimal sufficient statistics are $\{n_{i+}\}$, $\{n_{+j}\}$, and $\{\Sigma_j v_j n_{ij},$ $i = 1, \ldots, I\}$, and derive the likelihood equations.

9.38 Show that the column effects model corresponds to a baseline-category logit model for Y that is linear in scores for X, with slope depending on the paired response categories.

9.39 Refer to the homogeneous linear-by-linear association model (9.10).

    a. Show that the likelihood equations are, for all i, j, and k,

$$\hat{\mu}_{i+k} = n_{i+k}, \quad \hat{\mu}_{+jk} = n_{+jk}, \quad \sum_i \sum_j u_i v_j \hat{\mu}_{ij+} = \sum_i \sum_j u_i v_j n_{ij+}.$$

    b. Show that residual df $= K(I - 1)(J - 1) - 1$.

    c. When $I = J = 2$, explain why it is equivalent to (XY, XZ, YZ).

    d. Show how the last likelihood equation above changes for heterogeneous linear-by-linear XY association (9.11). Explain why, in each stratum, the fitted XY correlation equals the sample correlation.

9.40 When model (XY, XZ, YZ) is inadequate and variables are ordinal, useful models are nested between it and (XYZ). For ordered scores $\{u_i\}$, $\{v_j\}$, and $\{w_k\}$, consider

$$\log \mu_{ijk} = \lambda + \lambda_i^X + \lambda_j^Y + \lambda_k^Z + \lambda_{ij}^{XY} + \lambda_{ik}^{XZ} + \lambda_{jk}^{YZ} + \beta u_i v_j w_k. \quad (9.22)$$

    a. Define $\theta_{ijk} = \theta_{ij(k+1)}/\theta_{ij(k)} = \theta_{i(j+1)k}/\theta_{i(j)k} = \theta_{(i+1)jk}/\theta_{(i)jk}$. For unit-spaced scores, show that $\log \theta_{ijk} = \beta$. Goodman (1979a) called this the *uniform interaction model*.

    b. Show that log odds ratios for any two variables change linearly across levels of the third variable.

    c. Show that the likelihood equations are those for model (XY, XZ, YZ) plus

$$\sum_i \sum_j \sum_k u_i v_j w_k \hat{\mu}_{ijk} = \sum_i \sum_j \sum_k u_i v_j w_k n_{ijk}.$$

    d. Explain why model (9.12) is a special case of model (9.22).

9.41 Construct a model having general XZ and YZ associations, but row effects for the XY association that are **(a)** homogeneous, and **(b)** heterogeneous across levels of Z. Interpret.

9.42 Explain why the *RC* model requires scale constraints for the scores. Show that the residual df $= (I - 2)(J - 2)$. Find and interpret the likelihood equations. Explain why the fit is invariant to category orderings.

9.43 Refer to correlation model (9.16) (Goodman 1985, 1986).
  a. Show that λ is the correlation between the scores.
  b. If this model holds, show that $\sum_i \mu_i(\pi_{ij}/\pi_{+j}) = \lambda \nu_j$ and $\sum_j \nu_j(\pi_{ij}/\pi_{i+}) = \lambda \mu_i$. Interpret.
  c. With λ close to zero, show that $\log(\pi_{ij})$ has form $\gamma_i + \delta_i + \lambda \mu_i \nu_j + o(\lambda)$, where $o(\lambda)/\lambda \to 0$ as $\lambda \to 0$. Thus, when the association is weak, the correlation model is similar to the linear-by-linear association model with $\beta = \lambda$ and scores $\{u_i = \mu_i\}$ and $\{v_j = \nu_j\}$.

9.44 For the general canonical correlation model, show that $\sum \lambda_k^2 = \sum_i \sum_j (\pi_{ij} - \pi_{i+} \pi_{+j})^2 / \pi_{i+} \pi_{+j}$. Thus, the squared correlations partition a dependence measure that is the noncentrality (6.8) of X^2 for the independence model with $n = 1$. [Goodman (1986) stated other partitionings.]

9.45 Refer to model (9.18). Given the times at risk $\{t_{ij}\}$, show that sufficient statistics are $\{n_{i+}\}$ and $\{n_{+j}\}$.

9.46 Refer to Section 9.7.3. Let $T = \sum t_i$ and $W = \sum w_i$. Suppose that survival times have a negative exponential distribution with parameter λ.
  a. Using log likelihood (9.19), show that $\hat{\lambda} = W/T$.
  b. Conditional on T, show that W has a Poisson distribution with mean $T\lambda$. Using the Poisson likelihood, show that $\hat{\lambda} = W/T$.

9.47 Show that ML estimates do not exist for Table 9.15. [*Hint:* Haberman (1973b, 1974a, p. 398): If $\hat{\mu}_{111} = c > 0$, then marginal constraints the model satisfy imply that $\hat{\mu}_{222} = -c$.]

9.48 For a loglinear model, explain heuristically why the ML estimate of a parameter is infinite when its sufficient statistic takes its maximum or minimum possible value, for given values of other sufficient statistics.

CHAPTER 10

Models for Matched Pairs

We next introduce methods for comparing categorical responses for two samples when each observation in one sample pairs with an observation in the other. Such *matched-pairs* data commonly occur in studies with repeated measurement of subjects, such as *longitudinal studies* that observe subjects over time. Because of the matching, the responses in the two samples are statistically dependent. This is the first of four chapters on special methods for handling such dependence.

Table 10.1 illustrates matched-pairs data. For a poll of a random sample of 1600 voting-age British citizens, 944 indicated approval of the Prime Minister's performance in office. Six months later, of these same 1600 people, 880 indicated approval. The two cells with identical row and column response form the main diagonal of the table. These subjects had the same opinion at both surveys. They compose most of the sample, since relatively few people changed opinion. A strong association exists between opinions six months apart, the sample odds ratio being $(794 \times 570)/(150 \times 86) = 35.1$.

For matched pairs with a categorical response, a two-way contingency table with the same row and column categories summarizes the data. The table is *square*. In this chapter we present analyses of square tables. In Section 10.1 we describe methods for comparing proportions with a binary response. In Section 10.2 we discuss logistic regression analyses of such data. For multicategory responses, Section 10.3 covers nominal and ordinal logit

TABLE 10.1 **Rating of Performance of Prime Minister**

First Survey	Second Survey		Total
	Approve	Disapprove	
Approve	794	150	944
Disapprove	86	570	656
Total	880	720	1600

409

models for comparing the response distributions. In Section 10.4 we introduce loglinear models for square tables. In Sections 10.5 and 10.6 we discuss two matched-pairs applications for which models for square tables are useful: analyzing agreement between two observers who rate a common set of subjects, and evaluating preferences of treatments based on their pairwise evaluation.

Section 10.7 extends the models of Sections 10.2 through 10.4 to multiway tables that result from matched sets of observations. In Chapter 11 we extend them further to incorporate explanatory variables.

10.1 COMPARING DEPENDENT PROPORTIONS

For each of n matched pairs, let π_{ab} denote the probability of outcome a for the first observation and outcome b for the second. Let n_{ab} count the number of such pairs, with $p_{ab} = n_{ab}/n$ the sample proportion. We treat $\{n_{ab}\}$ as a sample from a multinomial $(n; \{\pi_{ab}\})$ distribution. Then p_{a+} is the proportion in category a for observation 1, and p_{+a} is the corresponding proportion for observation 2. We compare samples by comparing marginal proportions $\{p_{a+}\}$ with $\{p_{+a}\}$. With matched samples, these proportions are correlated, and methods for independent samples are inappropriate.

In this section we consider binary outcomes. When $\pi_{1+} = \pi_{+1}$, then $\pi_{2+} = \pi_{+2}$ also, and there is *marginal homogeneity*. Since

$$\pi_{1+} - \pi_{+1} = (\pi_{11} + \pi_{12}) - (\pi_{11} + \pi_{21}) = \pi_{12} - \pi_{21},$$

marginal homogeneity in 2×2 tables is equivalent to $\pi_{12} = \pi_{21}$. The table then shows *symmetry* across the main diagonal.

10.1.1 Inference for Dependent Proportions

One comparison of the marginal distributions uses $\delta = \pi_{+1} - \pi_{1+}$. Let

$$d = p_{+1} - p_{1+} = p_{2+} - p_{+2}.$$

From formula (1.3) for multinomial covariances, $\text{cov}(p_{+1}, p_{1+}) = \text{cov}(p_{11} + p_{21}, p_{11} + p_{12})$ simplifies to $(\pi_{11}\pi_{22} - \pi_{12}\pi_{21})/n$. Thus,

$$\text{var}(\sqrt{n}\, d) = \pi_{1+}(1 - \pi_{1+}) + \pi_{+1}(1 - \pi_{+1}) - 2(\pi_{11}\pi_{22} - \pi_{12}\pi_{21}). \tag{10.1}$$

For large samples, d has approximately a normal sampling distribution. A confidence interval for $\delta = \pi_{+1} - \pi_{1+}$ is then

$$d \pm z_{\alpha/2}\,\hat{\sigma}(d),$$

where

$$\hat{\sigma}^2(d) = [p_{1+}(1 - p_{1+}) + p_{+1}(1 - p_{+1}) - 2(p_{11}p_{22} - p_{12}p_{21})]/n$$

$$= [(p_{12} + p_{21}) - (p_{12} - p_{21})^2]/n, \qquad (10.2)$$

with the second formula following after substitution and some algebra. Inverting the score test of H_0: $\delta = \delta_0$ is more complex but provides coverage probabilities closer to the nominal values (Tango 1998), as does adding 1 to each cell before computing d and $\hat{\sigma}(d)$.

The hypothesis of marginal homogeneity is H_0: $\pi_{1+} = \pi_{+1}$ (i.e., $\delta = 0$). The ratio $z = d/\hat{\sigma}(d)$ or its square is a Wald test statistic. Under H_0, an alternative estimated variance is

$$\hat{\sigma}_0^2(d) = \frac{p_{12} + p_{21}}{n} = \frac{n_{12} + n_{21}}{n^2}. \qquad (10.3)$$

The score test statistic $z_0 = d/\hat{\sigma}_0(d)$ simplifies to

$$z_0 = \frac{n_{21} - n_{12}}{(n_{21} + n_{12})^{1/2}}. \qquad (10.4)$$

The square of z_0 is a chi-squared statistic with df = 1. The test using it is called *McNemar's test* (McNemar 1947).

The McNemar statistic depends only on cases classified in *different* categories for the two observations. The $n_{11} + n_{22}$ on the main diagonal are irrelevant to inference about whether π_{1+} and π_{+1} differ. This may seem surprising, but *all* cases contribute to inference about *how much* π_{1+} and π_{+1} differ: for instance, to estimating δ and the standard error.

10.1.2 Prime Minister Approval Rating Example

For Table 10.1, the sample proportions of approval of the prime minister's performance are $p_{1+} = 944/1600 = 0.59$ for the first survey and $p_{+1} = 880/1600 = 0.55$ for the second. Using (10.2), a 95% confidence interval for $\pi_{+1} - \pi_{1+}$ is $(0.55 - 0.59) \pm 1.96(0.0095)$, or $(-0.06, -0.02)$. The approval rating appears to have dropped between 2 and 6%.

For testing marginal homogeneity, the test statistic (10.4) using the null variance is

$$z_0 = \frac{86 - 150}{(86 + 150)^{1/2}} = -4.17.$$

It shows strong evidence of a drop in the approval rating.

10.1.3 Increased Precision with Dependent Samples

The final term of formula (10.1), based on $\text{cov}(p_{+1}, p_{1+})$, reflects the dependence between the marginal proportions. By contrast, for *independent* samples of size n each to estimate binomial probabilities π_1 and π_2, the covariance for the sample proportions is zero, and

$$\text{var}\left[\sqrt{n} \,(\text{difference of sample proportions})\right] = \pi_1(1 - \pi_1) + \pi_2(1 - \pi_2).$$

Dependent samples usually exhibit a positive dependence, with $\log \theta = \log[\pi_{11}\pi_{22}/\pi_{12}\pi_{21}] > 0$; that is, $\pi_{11}\pi_{22} > \pi_{12}\pi_{21}$. From (10.1), positive dependence implies that $\text{var}(d)$ is smaller than when the samples are independent.

A study design using dependent samples can help improve the precision of statistical inferences for within-subject effects. (By contrast, standard errors tend to be larger, per given number of observations, for between-subject group comparisons.) The improvement is substantial when samples are highly correlated. To illustrate, Table 10.1 with dependent samples of size 1600 each has a standard error of 0.0095 for $d = 0.55 - 0.59$. The two observations have strong association, the sample odds ratio being 35.1. *Independent* samples of size 1600 each with $\hat{\pi}_1 - \hat{\pi}_2 = 0.55 - 0.59$ have a standard error of 0.0175 for the difference, nearly twice as large.

10.1.4 Small-Sample Test Comparing Matched Proportions

The null hypothesis of marginal homogeneity for binary matched pairs is, equivalently, $H_0: \pi_{12} = \pi_{21}$ or $\pi_{21}/(\pi_{21} + \pi_{12}) = 0.5$. For small samples, an exact test conditions on $n^* = n_{21} + n_{12}$ (Mosteller 1952). Under H_0, n_{21} has a binomial $(n^*, \frac{1}{2})$ distribution, for which $E(n_{21}) = \frac{1}{2}n^*$. The P-value for the test is a binomial tail probability.

For instance, for Table 10.1, consider $H_a: \pi_{+1} < \pi_{1+}$, or equivalently, $H_a: \pi_{21} < \pi_{12}$. Since $n^* = 86 + 150 = 236$, the reference distribution is $\text{bin}(236, \frac{1}{2})$. The P-value is the probability of at least 150 successes out of 236 trials, which equals 0.00002. The P-value for $H_a: \pi_{+1} \neq \pi_{1+}$ doubles this.

When $n^* > 10$, the reference binomial distribution is approximately normal with mean $\frac{1}{2}n^*$ and variance $n^*(\frac{1}{2})(\frac{1}{2})$. The standardized normal test statistic equals

$$z = \frac{n_{21} - \frac{1}{2}n^*}{\left[n^*(\frac{1}{2})(\frac{1}{2})\right]^{1/2}} = \frac{n_{21} - n_{12}}{(n_{21} + n_{12})^{1/2}}.$$

This is identical to the McNemar statistic (10.4).

10.1.5 Connection between McNemar and Cochran–Mantel–Haenszel Tests

An alternative representation of binary responses for n matched pairs presents the data in n partial tables, one 2×2 table for each pair. It has columns that are the two possible outcomes for each measurement. Row 1 shows the outcome of the first observation, and row 2 shows the outcome of the second.

Table 10.2 shows the four possible partial tables in this representation. For Table 10.1, the full three-way table has 1600 partial tables; 794 look like the one for subject 1 (i.e., "approve" at both surveys), 570 who disapproved at each survey have tables like the one for subject 2, 86 have tables like the one for subject 3, and 150 have tables like the one for subject 4. The 1600 subjects from Table 10.1 provide 3200 observations in a $2 \times 2 \times 1600$ contingency table. Collapsing this table over the 1600 partial tables yields a 2×2 table with first row equal to $(944, 656)$ and second row equal to $(880, 720)$. These are the total number of (approve, disapprove) responses for the two surveys. They form the marginal counts in Table 10.1.

For each subject, suppose that the probability of approval is identical in each survey. Then, conditional independence exists between the opinion outcome and the survey time, controlling for subject. The probability of approval is then also the same for each survey in the marginal table collapsed over the subjects. But this implies that the true probabilities for Table 10.1 satisfy marginal homogeneity. Thus, a test of conditional independence in the $2 \times 2 \times 1600$ table provides a test of marginal homogeneity for Table 10.1.

To test conditional independence in this three-way table, one can use the Cochran–Mantel–Haenszel (CMH) statistic (6.6). The result of that chi-squared statistic is algebraically identical to the squared McNemar's statistic, namely $(n_{21} - n_{12})^2/(n_{12} + n_{21})$ for tables of form (10.1). McNemar's test is a special case of the CMH test applied to the binary responses of n matched pairs displayed in n partial tables. This connection is not helpful for computational purposes, since the McNemar statistic is simple. But it does suggest

TABLE 10.2 Representation of Four Types of Matched Pairs Contributing to Counts in Table 10.1

Subject	Survey	Response	
		Approve	Disapprove
1	First	1	0
	Second	1	0
2	First	0	1
	Second	0	1
3	First	0	1
	Second	1	0
4	First	1	0
	Second	0	1

ways of handling more complex matched data. With several outcome categories or several observations, one can test marginal homogeneity by applying the generalized CMH tests (Section 7.5) using a single stratum for each subject, with each row representing a particular observation (Darroch 1981; Mantel and Byar 1978).

Coming sections refer to the $2 \times 2 \times n$ table representation of matched-pairs data as the *subject-specific* table. They refer to the 2×2 table of form of Table 10.1 as the *population-averaged* table, since its margins provide direct estimates of population marginal proportions.

10.2 CONDITIONAL LOGISTIC REGRESSION FOR BINARY MATCHED PAIRS

In Section 6.7 we introduced *conditional logistic regression* for eliminating nuisance parameters from an analysis. We now study this for binary matched-pairs data. The models refer to subject-specific tables.

10.2.1 Marginal versus Conditional Models for Matched Pairs

The analyses of Section 10.1 occur in the context of models. Let (Y_1, Y_2) denote the pair of observations for a randomly selected subject, where a "1" outcome denotes category 1 (success) and "0" denotes category 2. The difference $\delta = P(Y_2 = 1) - P(Y_1 = 1)$ between marginal probabilities occurs as a parameter in

$$P(Y_t = 1) = \alpha + \delta x_t, \tag{10.5}$$

where $x_1 = 0$ and $x_2 = 1$; then, $P(Y_1 = 1) = \alpha$ and $P(Y_2 = 1) = \alpha + \delta$. Alternatively, the logit link yields

$$\text{logit}[P(Y_t = 1)] = \alpha + \beta x_t. \tag{10.6}$$

The parameter β is a log odds ratio with the marginal distributions.

Models (10.5) and (10.6) are *marginal models*: They focus on the marginal distributions of responses for the two observations. For instance, in terms of the population-averaged table, the ML estimate of β in (10.6) is the log odds ratio of marginal proportions, $\hat{\beta} = \log[p_{+1} p_{2+}/p_{+2} p_{1+}]$. See Problem 10.26 for its asymptotic variance.

By contrast, the subject-specific table having strata like Table 10.2 implicitly allows probabilities to vary by subject. Let (Y_{i1}, Y_{i2}) denote the ith pair of observations, $i = 1, \ldots, n$. A model then has the form

$$\text{link}[P(Y_{it} = 1)] = \alpha_i + \beta x_t. \tag{10.7}$$

This is called a *conditional model*, since the effect β is defined conditional on the subject. Its estimate describes conditional association for the three-way table stratified by subject. The effect is *subject-specific*, since it is defined at the subject level. By contrast, the effects in marginal models (10.5) and (10.6) are *population-averaged*, since they refer to averaging over the entire population rather than to individual subjects.

For the identity link, subject-specific and population-averaged effects are identical. For instance, for the conditional model (10.7) with identity link, $\beta = P(Y_{i2} = 1) - P(Y_{i1} = 1)$ for all i, and averaging this over subjects in the population equates β to the δ parameter in model (10.5). For nonlinear links, however, the effects differ. For model (10.7) with the logit link, for instance,

$$P(Y_{it} = 1) = \exp(\alpha_i + \beta x_t)/[1 + \exp(\alpha_i + \beta x_t)].$$

The average of this for the population does not have the form $\exp(\alpha + \beta x_t)/[1 + \exp(\alpha + \beta x_t)]$ corresponding to the marginal logit model (10.6). We now take a closer look at the conditional model with logit link.

10.2.2 A Logit Model with Subject-Specific Probabilities

Model (10.7) differs from models in earlier chapters by permitting subjects to have their own probability distributions. Cox (1958b, 1970) and Rasch (1961) presented this model with logit link. This model for Y_{it}, observation t for subject i, is

$$\text{logit}[P(Y_{it} = 1)] = \alpha_i + \beta x_t, \tag{10.8}$$

where $x_1 = 0$ and $x_2 = 1$. Although permitting subject-specific distributions, it assumes a common effect β. For subject i,

$$P(Y_{i1} = 1) = \frac{\exp(\alpha_i)}{1 + \exp(\alpha_i)}, \qquad P(Y_{i2} = 1) = \frac{\exp(\alpha_i + \beta)}{1 + \exp(\alpha_i + \beta)}.$$

The parameter β compares the response distributions. For each subject, the odds of success for observation 2 are $\exp(\beta)$ times the odds for observation 1.

Given the parameters, with model (10.8) one normally assumes independence of responses for different subjects and for the two observations on the same subject. However, averaged over all subjects, the responses are nonnegatively associated. Suppose that $|\beta|$ is small compared to $|\alpha_i|$. A subject with a large positive α_i has high $P(Y_{it} = 1)$ for each t and is likely to have a success each time; a subject with a large negative α_i has low $P(Y_{it} = 1)$ for each t and is likely to have a failure each time. The greater the variability in $\{\alpha_i\}$, the greater the overall positive association between responses, successes (failures) for observation 1 tending to occur with successes (failures) for observation 2. This is true for any β. The positive association reflects the shared value of α_i for each observation in a pair. No association occurs only when $\{\alpha_i\}$ are identical. Thus, the model does account for the dependence in matched pairs. Fitting it takes into account nonnegative association through the structure of the model.

For this model, the large number of $\{\alpha_i\}$ causes difficulties with the fitting process and with the properties of ordinary ML estimators (Problem 10.24). The remedy of conditional ML treats them as nuisance parameters and maximizes the likelihood function for a conditional distribution that eliminates them. A note on terminology: We've referred to model (10.8) as a *conditional* model, meaning that its effect β is subject-specific, conditional on the subject. The analyses described below for such models are examples of *conditional* logistic regression; but here the term *conditional* refers to the ML analysis that is performed conditional on sufficient statistics for nuisance parameters, to eliminate those parameters from the likelihood.

10.2.3 Conditional ML Inference for Binary Matched Pairs

For model (10.8), assuming independence of responses for different subjects and for the two observations on the same subject, the joint mass function for $\{(y_{11}, y_{12}), \ldots, (y_{n1}, y_{n2})\}$ is

$$\prod_{i=1}^{n} \left(\frac{\exp(\alpha_i)}{1 + \exp(\alpha_i)} \right)^{y_{i1}} \left(\frac{1}{1 + \exp(\alpha_i)} \right)^{1-y_{i1}}$$

$$\times \left(\frac{\exp(\alpha_i + \beta)}{1 + \exp(\alpha_i + \beta)} \right)^{y_{i2}} \left(\frac{1}{1 + \exp(\alpha_i + \beta)} \right)^{1-y_{i2}}.$$

In terms of the data, this is proportional to

$$\exp\left[\sum_i \alpha_i (y_{i1} + y_{i2}) + \beta\left(\sum_i y_{i2} \right) \right].$$

To eliminate $\{\alpha_i\}$, we condition on their sufficient statistics, the pairwise success totals $\{S_i = y_{i1} + y_{i2}\}$. Given $S_i = 0$, $P(Y_{i1} = Y_{i2} = 0) = 1$, and given $S_i = 2$, $P(Y_{i1} = Y_{i2} = 1) = 1$. The distribution of (Y_{i1}, Y_{i2}) depends on β only when $S_i = 1$; that is, only when outcomes differ for the two responses. Given $y_{i1} + y_{i2} = 1$, the conditional distribution is

$$P(Y_{i1} = y_{i1}, Y_{i2} = y_{i2} \mid S_i = 1)$$

$$= P(Y_{i1} = y_{i1}, Y_{i2} = y_{i2}) / \left[P(Y_{i1} = 1, Y_{i2} = 0) + P(Y_{i1} = 0, Y_{i2} = 1) \right]$$

$$= \frac{\left(\frac{\exp(\alpha_i)}{1 + \exp(\alpha_i)} \right)^{y_{i1}} \left(\frac{1}{1 + \exp(\alpha_i)} \right)^{1-y_{i1}} \left(\frac{\exp(\alpha_i + \beta)}{1 + \exp(\alpha_i + \beta)} \right)^{y_{i2}} \left(\frac{1}{1 + \exp(\alpha_i + \beta)} \right)^{1-y_{i2}}}{\frac{\exp(\alpha_i)}{1 + \exp(\alpha_i)} \frac{1}{1 + \exp(\alpha_i + \beta)} + \frac{1}{1 + \exp(\alpha_i)} \frac{\exp(\alpha_i + \beta)}{1 + \exp(\alpha_i + \beta)}}$$

$$= \exp(\beta) / [1 + \exp(\beta)], \quad y_{i1} = 0, \quad y_{i2} = 1$$

$$= 1 / [1 + \exp(\beta)], \quad y_{i1} = 1, \quad y_{i2} = 0.$$

Again, let $\{n_{ab}\}$ denote the counts for the four possible sequences. For subjects having $S_i = 1$, $\Sigma_i y_{i1} = n_{12}$, the number of subjects having success for observation 1 and failure for observation 2. Similarily, for those subjects, $\Sigma_i y_{i2} = n_{21}$ and $\Sigma_i S_i = n^* = n_{12} + n_{21}$. Since n_{21} is the sum of n^* independent, identical Bernoulli variates, its conditional distribution is binomial with parameter $\exp(\beta)/[1 + \exp(\beta)]$. For testing marginal homogeneity ($\beta = 0$), the parameter equals $\frac{1}{2}$. In summary, the conditional analysis for the logit model implies that pairs in which $y_{i1} = y_{i2}$ are irrelevant to inference about β. When this model is realistic, it provides justification for comparing marginal distributions using only the $n_{12} + n_{21}$ pairings having outcomes in different categories at the two observations.

Conditional on $S_i = 1$, the joint distribution of the matched pairs is

$$\prod_{S_i=1} \left(\frac{1}{1 + \exp(\beta)}\right)^{y_{i1}} \left(\frac{\exp(\beta)}{1 + \exp(\beta)}\right)^{y_{i2}} = \frac{[\exp(\beta)]^{n_{21}}}{[1 + \exp(\beta)]^{n^*}} \quad (10.9)$$

where the product refers to all pairs having $S_i = 1$. Differentiating the log of this conditional likelihood and equating to 0 and solving yields the conditional ML estimator of β in model (10.8). You can check that it and its standard error are

$$\hat{\beta} = \log(n_{21}/n_{12}), \quad SE = \sqrt{1/n_{21} + 1/n_{12}}. \quad (10.10)$$

10.2.4 Random Effects in Binary Matched-Pairs Model

An alternative remedy to handling the huge number of nuisance parameters in logit model (10.8) treats $\{\alpha_i\}$ as *random effects*. This regards $\{\alpha_i\}$ as an unobserved random sample from a probability distribution, usually assumed to be $N(\mu, \sigma^2)$ with unknown μ and σ. It eliminates $\{\alpha_i\}$ by averaging with respect to their distribution, yielding a marginal distribution. The likelihood function then depends on β as well as the $N(\mu, \sigma^2)$ parameters. It has only three parameters and is more manageable. For matched pairs with nonnegative sample log odds ratio, this approach also yields $\hat{\beta} = \log(n_{21}/n_{12})$ (Neuhaus et al. 1994). This model is an example of a *generalized linear mixed model*, containing both random effects and the fixed effect β. Its analysis is presented in Chapter 12.

Model (10.8) implies that the true odds ratio for each of the n subject-specific partial tables equals $\exp(\beta)$. In Section 6.3.5 we presented the Mantel–Haenszel estimate of a common odds ratio for several 2×2 tables. In fact, that estimator applied to subject-specific tables of the form shown in Table 10.2 is algebraically identical to n_{21}/n_{12} for tables of the form shown in Table 10.1. (Recall that partial tables with responses in only one column do not contribute to the CMH test or Mantel–Haenszel estimate.) In summary, the Mantel–Haenszel estimate, the conditional ML estimate, and (with nonnegative log odds ratio) the ML estimate for the random effects version of logit model (10.8) yield $\exp(\hat{\beta}) = n_{21}/n_{12}$.

10.2.5 Logistic Regression for Matched Case–Control Studies

The two observations (y_{i1}, y_{i2}) in a matched pair need not refer to the same subject. For instance, case–control studies that match a single control with each case yield matched-pairs data. For a binary response Y, each case $(Y = 1)$ is matched with a control $(Y = 0)$ according to criteria that could affect the response. Subjects in the matched pairs are measured on the predictor variable(s) of interest, X, and the XY association is analyzed.

Table 10.3 illustrates. A case–control study of acute myocardial infarction (MI) among Navajo Indians matched 144 victims of MI according to age and gender with 144 people free of heart disease. Subjects were asked whether they had ever been diagnosed as having diabetes ($x = 0$, no; $x = 1$, yes). Table 10.3 has the same form as Table 10.1 except that the levels of X rather than the levels of Y form the rows and the columns.

One can display the data for each matched case–control pair using a partial table of the form shown in Table 10.2, but reversing the roles of X and Y. The X values have four possible patterns, shown in Table 10.4. There are 37 partial tables of type a, since for 37 pairs the case had diabetes and the control did not, 16 partial tables of type b, 9 of type c, and 82 of type d.

Now, for subject t in matched pair i, consider the model

$$\text{logit}[P(Y_{it} = 1)] = \alpha_i + \beta x_{it}. \tag{10.11}$$

TABLE 10.3   Previous Diagnoses of Diabetes for Myocardial Infarction (MI) Case–Control Pairs

MI Controls	MI Cases		Total
	Diabetes	No Diabetes	
Diabetes	9	16	25
No diabetes	37	82	119
Total	46	98	144

Source: J. L. Coulehan et al., *Amer. J. Public Health* **76**: 412–414 (1986), reprinted with permission from the American Public Health Association.

TABLE 10.4   Possible Case–Control Pairs for Table 10.3

Diabetes	a		b		c		d	
	Case	Control	Case	Control	Case	Control	Case	Control
Yes	1	0	0	1	1	1	0	0
No	0	1	1	0	0	0	1	1

The probabilities modeled refer to the distribution of Y given X, but the retrospective study provides information only about the distribution of X given Y. One can estimate the odds ratio $\exp(\beta)$, however, since it refers to the XY odds ratio, which relates to both conditional distributions (Sections 2.2.4, 5.1.4). Even though this study reverses the roles of X and Y in terms of which is fixed and which is random, the conditional ML estimate of $\exp(\beta)$ is simply $n_{21}/n_{12} = 37/16 = 2.3$.

10.2.6 Conditional ML for Matched Pairs with Multiple Predictors

When the binary response has p predictors for case–control or subject-specific matched pairs, the model generalizes to

$$\text{logit}\big[P(Y_{it} = 1)\big] = \alpha_i + \beta_1 x_{1it} + \beta_2 x_{2it} + \cdots + \beta_p x_{pit}, \quad (10.12)$$

where x_{hit} denotes the value of predictor h for observation t in pair i, $t = 1, 2$. Typically, one predictor is an explanatory variable of interest, such as diabetes status. The others are covariates being controlled, in addition to those already controlled by virtue of using them to form the matched pairs. The conditional ML approach to estimating $\{\beta_j\}$ conditions on sufficient statistics for α_i to eliminate them from the likelihood.

Let $\mathbf{x}_{it} = (x_{1it}, \ldots, x_{pit})'$ and $\boldsymbol{\beta} = (\beta_1, \ldots, \beta_p)'$. A generalization of the derivation in Section 10.2.3 shows that

$$P(Y_{i1} = 0, Y_{i2} = 1 \mid S_i = 1) = \exp(\mathbf{x}'_{i2}\boldsymbol{\beta})/\big[\exp(\mathbf{x}'_{i1}\boldsymbol{\beta}) + \exp(\mathbf{x}'_{i2}\boldsymbol{\beta})\big],$$

$$P(Y_{i1} = 1, Y_{i2} = 0 \mid S_i = 1) = \exp(\mathbf{x}'_{i1}\boldsymbol{\beta})/\big[\exp(\mathbf{x}'_{i1}\boldsymbol{\beta}) + \exp(\mathbf{x}'_{i2}\boldsymbol{\beta})\big]. \quad (10.13)$$

Dividing numerator and denominator by $\exp(\mathbf{x}'_{i1}\boldsymbol{\beta})$ shows that the first equation has the form of logistic regression with no intercept and with predictor values $\mathbf{x}^*_i = \mathbf{x}_{i2} - \mathbf{x}_{i1}$. In fact, one can obtain conditional ML estimates for model (10.12) by fitting a logistic regression model to those pairs alone, using artificial response $y^* = 1$ when $(y_{i1} = 0, y_{i2} = 1)$, $y^* = 0$ when $(y_{i1} = 1, y_{i2} = 0)$, no intercept, and predictor values $\mathbf{x}^*_i$. This addresses the same likelihood as the conditional likelihood (Breslow et al. 1978; Chamberlain 1980).

To illustrate, for model (10.11) with Table 10.3, let $y^*_i = y_{i2} - y_{i1}$ and $x^*_i = x_{i2} - x_{i1}$. If $t = 1$ refers to the control and $t = 2$ to the case, then $y^*_i = 1$ always. Since $x_{it} = 1$ represents "yes" for diabetes and $x_{it} = 0$ represents "no," $(y^*_i = 1, x^*_i = -1)$ for 16 observations, $(y^*_i = 1, x^*_i = 0)$ for $9 + 82 = 91$ observations, and $y^*_i = 1, x^*_i = +1)$ for 37 observations. The logit model that forces $\hat{\alpha} = 0$ has $\hat{\beta} = 0.84$. With a single binary predictor, the estimate is identical to $\log(n_{21}/n_{12})$.

10.2.7 Marginal Models and Conditional Models: Extensions

For binary matched-pairs data, Section 10.1 presented analyses for a marginal (i.e., population-averaged) model, and this section presented analyses for a conditional (i.e., subject-specific) model. These models generalize to multinomial responses and to matched sets. For instance, Chamberlain (1980) discussed conditional ML for matched pairs on a multinomial response. For binary responses, model (10.12) applies when α_i refers to a set of repeated measurements on subject i. Or, it could refer to a matched set that is a cluster of subjects, such as children from family i or fetuses from litter i.

With extensions of the conditional model to matched-set clusters, the conditional ML approach is restricted to estimating β_j that are within-cluster effects, such as occur in case–control and crossover studies. For these, the explanatory variable varies in t for each i. Conditional ML cannot estimate a between-cluster effect. Statistics providing information about such an effect use subject totals at different levels of the relevant explanatory variable; however, those totals sum the sufficient statistics for $\{\alpha_i\}$, so they are themselves fixed and have degenerate distributions after conditioning on the sufficient statistics. An explanatory variable that is constant in t for each i cancels out of the conditional likelihood. [You can observe this for matched pairs with (10.13) for any j for which $x_{ji1} = x_{ji2}$ all i.] For it, at best one can stratify by its levels and fit a model estimating within-cluster effects separately at each level. An advantage of using the random effects approach instead of conditional ML with the conditional model is that it is not restricted to estimating within-cluster effects.

In the remainder of this chapter we emphasize marginal models for matched pairs with multinomial responses. In the following chapter we deal with marginal model extensions allowing matched sets and explanatory variables. Conditional models using a random effects approach have extra computational complexities. We mention briefly some multinomial conditional models in this chapter, but we defer most discussion to Chapter 12.

10.3 MARGINAL MODELS FOR SQUARE CONTINGENCY TABLES

Matched pairs analyses generalize from binary to $I > 2$ outcome categories. A square $I \times I$ table $\{n_{ab}\}$ shows counts of possible sequences (a, b) of outcomes for (Y_1, Y_2). Let $\pi_{ab} = P(Y_1 = a, Y_2 = b)$. Marginal homogeneity is $P(Y_1 = a) = P(Y_2 = a)$ for $a = 1, \ldots, I$. Marginal models compare $\{P(Y_1 = a)\}$ and $\{P(Y_2 = a)\}$.

10.3.1 Marginal Models for Ordinal Classifications

For ordered categories, marginal model (10.6) for binary matched pairs extends using ordinal logits. With cumulative logits,

$$\text{logit}[P(Y_t \le j)] = \alpha_j + \beta x_t, \quad t = 1, 2, \quad j = 1, \ldots, I - 1, \quad (10.14)$$

where $x_1 = 0$ and $x_2 = 1$. This model has proportional odds structure (Section 7.2.2). The odds of outcome $Y_2 \leq j$ equal $\exp(\beta)$ times the odds of outcome $Y_1 \leq j$. The model implies stochastically ordered marginal distributions, with $\beta > 0$ meaning that Y_1 tends to be higher than Y_2. Marginal homogeneity corresponds to $\beta = 0$.

Model fitting treats (Y_1, Y_2) as dependent. The ML approach maximizes the multinomial likelihood for $\{\pi_{ab}\}$. This is not simple. Since the model refers to marginal probabilities $\{P(Y_1 = a) = \pi_{a+}\}$ and $\{P(Y_2 = b) = \pi_{+b}\}$, one cannot substitute the model formula in the kernel $\Sigma_a \Sigma_b n_{ab} \log \pi_{ab}$ of the log likelihood, which has joint probabilities. We defer discussion of ML model fitting of marginal models to Section 11.2.5. Model (10.14) describes the $2(I - 1)$ marginal probabilities by I parameters, so df $= I - 2$ for testing fit. Alternatively, one can compare margins using summaries such as a difference in means for chosen category scores (Problem 10.38).

10.3.2 Premarital and Extramarital Sex Example

Refer to Table 10.5. For a General Social Survey, subjects gave their opinion about premarital sex (a couple having sex before marriage) and extramarital sex (a married person having sex with someone other than the marriage partner). The response categories are 1 = always wrong, 2 = almost always wrong, 3 = wrong only sometimes, 4 = not wrong at all.

The sample cumulative marginal proportions are $(0.307, 0.389, 0.611)$ for premarital sex and $(0.815, 0.918, 0.987)$ for extramarital sex. This suggests that responses on premarital sex tended to be higher on the ordinal scale than those on extramarital sex. With scores $(1, 2, 3, 4)$, the mean for premarital sex is 2.69, closest to the "wrong only sometimes" score, and the mean response for extramarital sex is 1.28, closest to the "always wrong" score.

The cumulative logit model (10.14) has $\hat{\beta} = 2.51$ (SE $= 0.13$). There is strong evidence that population responses are more positive on premarital than on extramarital sex. The fit of the marginal homogeneity model has $G^2 = 348.1$ (df $= 3$), and the fit of model (10.14) has $G^2 = 35.1$ (df $= 2$). The ordinal model does not fit well, but it fits much better than the marginal homogeneity model. Models to be considered in Section 10.4.7 fit better yet.

TABLE 10.5 Opinions on Premarital Sex and Extramarital Sex

Premarital	Extramarital Sex				
Sex	1	2	3	4	Total
1	144	2	0	0	146
2	33	4	2	0	39
3	84	14	6	1	105
4	126	29	25	5	185
Total	387	49	33	6	475

Source: 1989 General Social Survey, National Opinion Research Center.

10.3.3 Marginal Models for Nominal Classifications

With nominal responses, it is not sensible to assume the same effect for each logit. A baseline-category logit model has form

$$\log[P(Y_t = j)/P(Y_t = I)] = \alpha_j + \beta_j x_t, \qquad t = 1, 2, \quad j = 1, \ldots, I - 1,$$

$$(10.15)$$

where $x_1 = 0$ and $x_2 = 1$. This model has $2(I - 1)$ parameters for the $2(I - 1)$ marginal probabilities. It is saturated.

Marginal homogeneity is the special case $\beta_1 = \cdots = \beta_{I-1} = 0$. To fit it, Lipsitz et al. (1990) and Madanksy (1963) maximized the multinomial likelihood for $\{n_{ab}\}$ subject to these constraints. Iterative methods produce fitted values $\{\hat{\mu}_{ab}\}$. Comparing these to $\{n_{ab}\}$ using G^2 or X^2 tests marginal homogeneity, with df $= I - 1$.

Bhapkar (1966) tested marginal homogeneity by exploiting the asymptotic normality of marginal proportions. Let $d_a = p_{+a} - p_{a+}$, and let $\mathbf{d}' = (d_1, \ldots, d_{I-1})$. It is redundant to include d_I, since $\Sigma d_a = 0$. The sample covariance matrix $\hat{\mathbf{V}}$ of $\sqrt{n}\,\mathbf{d}$ has elements

$$\hat{v}_{ab} = -(p_{ab} + p_{ba}) - (p_{+a} - p_{a+})(p_{+b} - p_{b+}) \qquad \text{for } a \neq b,$$

$$\hat{v}_{aa} = p_{+a} + p_{a+} - 2p_{aa} - (p_{+a} - p_{a+})^2.$$

Now $\sqrt{n}\,[\mathbf{d} - E(\mathbf{d})]$ has an asymptotic multivariate normal distribution with estimated covariance matrix $\hat{\mathbf{V}}$. Under marginal homogeneity, $E(\mathbf{d}) = \mathbf{0}$, and

$$W = n\mathbf{d}' \hat{\mathbf{V}}^{-1} \mathbf{d} \tag{10.16}$$

is asymptotically chi-squared with df $= I - 1$. This is a Wald test for parameters in the analog of model (10.15) using the identity link. Stuart (1955) proposed $W_0 = n\mathbf{d}' \hat{\mathbf{V}}_0^{-1} \mathbf{d}$, which uses the sample *null* covariance matrix $\hat{\mathbf{V}}_0$ and is the score test. This has

$$\hat{v}_{ab0} = -(p_{ab} + p_{ba}) \qquad \text{for } a \neq b,$$

$$\hat{v}_{aa0} = p_{+a} + p_{a+} - 2p_{aa}.$$

Ireland et al. (1969) noted that $W = W_0/(1 - W_0/n)$. For $I = 2$, W_0 is McNemar's statistic, the square of (10.4).

These tests use all $I - 1$ degrees of freedom available for comparisons of I pairs of marginal proportions. With ordered categories, when I is large and the dependence between classifications is strong, ordinal tests (with df $= 1$) can be much more powerful (Agresti 1984, p. 209).

TABLE 10.6   Migration from 1980 to 1985, with Fit of Marginal Homogeneity Model

Residence in 1980	Residence in 1985				
	Northeast	Midwest	South	West	Total
Northeast	11,607	100	366	124	12,197
	(11,607)	(98.1)	(265.7)	(94.0)	(12,064.7)
Midwest	87	13,677	515	302	14,581
	(88.7)	(13,677)	(379.1)	(323.3)	(14,377.1)
South	172	255	17,819	270	18,486
	(276.5)	(350.8)	(17,819)	(287.3)	(18,733.5)
West	63	176	286	10,192	10,717
	(92.5)	(251.3)	(269.8)	(10,192)	(10,805.6)
Total	11,929	14,178	18,986	10,888	55,981
	(12,064.7)	(14,377.1)	(18,733.5)	(10,805.6)	

Source: Data based on Table 12 of U.S. Bureau of the Census, Current Population Reports, Series P-20, No. 420, *Geographical Mobility: 1985* (Washington, DC: U.S. Government Printing Office), 1987.

10.3.4   Migration Example

For a sample of U.S. residents, Table 10.6 compares region of residence in 1985 with 1980. Relatively few people changed region, 95% of the observations falling on the main diagonal. The ML fit of marginal homogeneity, shown in Table 10.6, gives $G^2 = 240.8$ (df = 3). Statistics using differences in sample marginal proportions give similar results. For instance, Bhapkar's statistic (10.16) is $W = 236.5$ (df = 3).

The sample marginal proportions for the four regions were $(0.218, 0.260, 0.330, 0.191)$ in 1980 and $(0.213, 0.253, 0.339, 0.194)$ in 1985. Little change occurred over such a short time period. The large test statistics reflect the huge sample size. To estimate the change for a given region, we apply (10.2) to the collapsed 2×2 table that combines the other regions. A 95% confidence interval for $\pi_{+1} - \pi_{1+}$ is $(0.2131 - 0.2179) \pm 1.96(0.00054)$, or -0.005 ± 0.001. Similarly, a 95% confidence interval for $\pi_{+2} - \pi_{2+}$ is -0.007 ± 0.001, for $\pi_{+3} - \pi_{3+}$ is 0.009 ± 0.001, and for $\pi_{+4} - \pi_{4+}$ is 0.003 ± 0.001. Although strong evidence of change occurs for all four regions, the changes were small.

10.4   SYMMETRY, QUASI-SYMMETRY, AND QUASI-INDEPENDENCE

An alternative analysis of square contingency tables directly models the joint distribution using logit or loglinear models. Some models have marginal homogeneity as a special case.

An $I \times I$ joint distribution $\{\pi_{ab}\}$ satisfies *symmetry* if

$$\pi_{ab} = \pi_{ba} \qquad \text{whenever } a \neq b. \qquad (10.17)$$

Under symmetry, $\pi_{a+} = \Sigma_b \pi_{ab} = \Sigma_b \pi_{ba} = \pi_{+a}$ for all a, so marginal homogeneity occurs. For $I = 2$, symmetry is equivalent to marginal homogeneity, but for $I > 2$, marginal homogeneity can occur without symmetry.

10.4.1 Symmetry as Logit and Loglinear Models

When all $\pi_{ab} > 0$, symmetry is a logit and a loglinear model. In logit form, it is trivially

$$\log(\pi_{ab}/\pi_{ba}) = 0 \qquad \text{for all } a < b.$$

For expected frequencies $\{\mu_{ab} = n\pi_{ab}\}$, it has the loglinear form

$$\log \mu_{ab} = \lambda + \lambda_a + \lambda_b + \lambda_{ab} \qquad (10.18)$$

where all $\lambda_{ab} = \lambda_{ba}$. Both classifications have the same single-factor parameters $\{\lambda_a\}$, so $\log \mu_{ab} = \log \mu_{ba}$. Identifiability requires constraints. A simpler expression is $\log \mu_{ab} = \lambda_{ab}$, with all $\lambda_{ab} = \lambda_{ba}$.

For Poisson or multinomial cell counts $\{n_{ab}\}$, the likelihood equations are

$$\hat{\mu}_{ab} + \hat{\mu}_{ba} = n_{ab} + n_{ba} \quad \text{for all } a < b \quad \text{and} \quad \hat{\mu}_{aa} = n_{aa} \quad \text{for all } a.$$

The main diagonal has perfect fit. The solution that satisfies symmetry is

$$\hat{\mu}_{ab} = \frac{n_{ab} + n_{ba}}{2} \qquad \text{for all } a, b.$$

The logit symmetry model has no parameters for the $\binom{I}{2}$ binomial pairs $\{(n_{ab}, n_{ba})\}$ with $a < b$, so its residual df $= I(I-1)/2$. Equivalently, the loglinear symmetry model $\log \mu_{ab} = \lambda_{ab}$ $(\lambda_{ab} = \lambda_{ba})$ for I^2 Poisson counts $\{n_{ab}\}$ has $\binom{I}{2}$ $\{\lambda_{ab}\}$ with $a < b$ and $I\{\lambda_{aa}\}$, so df $= I^2 - [I + I(I-1)/2] = I(I-1)/2$. For testing symmetry, Bowker (1948) showed that X^2 simplifies to

$$X^2 = \sum_{a<b}\sum \frac{(n_{ab} - n_{ba})^2}{n_{ab} + n_{ba}}.$$

For $I = 2$ this is McNemar's statistic, the square of (10.4). The standardized Pearson residuals equal

$$r_{ab} = (n_{ab} - n_{ba})\big/(n_{ab} + n_{ba})^{1/2}.$$

Only one residual for each pair of categories is nonredundant, since $r_{ab} = -r_{ba}$. They satisfy $\sum_{a<b} r_{ab}^2 = X^2$.

The symmetry model is very simple. Except for a few specialized applications, such as describing intraobserver agreement for pairs of measurements by an observer, it rarely fits well. When the marginal distributions differ substantially, it fits poorly.

10.4.2 Quasi-symmetry

One can accommodate marginal heterogeneity by permitting the main-effect terms in the symmetry model (10.18) to differ. The resulting loglinear model, called *quasi-symmetry*, is

$$\log \mu_{ab} = \lambda + \lambda_a^X + \lambda_b^Y + \lambda_{ab}, \tag{10.19}$$

where $\lambda_{ab} = \lambda_{ba}$ for all $a < b$ (Caussinus 1966). Symmetry is the special case $\lambda_a^X = \lambda_a^Y$ for $a = 1, \ldots, I$, and independence is the special case in which all $\lambda_{ab} = 0$.

The likelihood equations for quasi-symmetry are

$$\hat{\mu}_{a+} = n_{a+}, \qquad a = 1, \ldots, I$$
$$\hat{\mu}_{+b} = n_{+b}, \qquad b = 1, \ldots, I \tag{10.20}$$
$$\hat{\mu}_{ab} + \hat{\mu}_{ba} = n_{ab} + n_{ba} \qquad \text{for } a \leq b.$$

Only one of the first two sets of equations is needed. The other is redundant, given the other two. The residual df $= (I - 1)(I - 2)/2$. From (10.20), $\hat{\mu}_{aa} = n_{aa}$ for $a = 1, \ldots, I$. Otherwise, the likelihood equations do not have a direct solution. They are solved using iterative methods such as Newton–Raphson and IPF (Caussinus 1966).

The quasi-symmetry model has multiplicative form

$$\pi_{ab} = \alpha_a \beta_b \gamma_{ab}, \quad \text{where } \gamma_{ab} = \gamma_{ba} \text{ all } a < b \tag{10.21}$$

and all parameters are positive. The symmetry model is (10.21) with $\alpha_a = \beta_a$ for all a. This equation indicates that a table satisfying quasi-symmetry is the cellwise product of a table satisfying independence with one satisfying symmetry. The association symmetry implies that odds ratios on one side of the main diagonal are identical to corresponding odds ratios on the other side. In fact, the model can be defined by properties such as

$$\frac{\mu_{ab}\,\mu_{II}}{\mu_{aI}\,\mu_{Ib}} = \frac{\mu_{ba}\,\mu_{II}}{\mu_{bI}\,\mu_{Ia}} \qquad \text{for all } a < b \tag{10.22}$$

or $\theta_{ab} = \theta_{ba}$ for local odds ratios. Goodman (1979a) referred to it as the *symmetric association* model.

The meaning of quasi-symmetry is less obvious than symmetry. However, it usually fits much better and has greater scope. One way to interpret its parameters relates to subject-specific logit models. For such models having additivity of subject terms and occasion terms, of which model (10.8) is the simplest case, the joint distribution in the corresponding population-averaged table necessarily satisfies quasi-symmetry (see Darroch 1981; Section 13.2.7 shows this). Consider the generalization of baseline-category logit model (10.15) to a subject-specific model

$$\log[P(Y_{it}=j)/P(Y_{it}=I)] = \alpha_{ij} + \beta_j x_t, \quad t=1,2, \quad j=1,\ldots,I-1.$$

This has the additive form of (10.8) for each j. The model implies, averaging over subjects, that the quasi-symmetry model (10.19) holds for the $I \times I$ population-averaged table with $\{\beta_j = \lambda_j^Y - \lambda_j^X\}$, when one constrains $\lambda_I^X = \lambda_I^Y = 0$. In fact, for the conditional ML analysis that conditions out $\{\alpha_{ij}\}$, the conditional ML estimates of $\{\hat{\beta}_j\}$ relate to the ordinary ML fit of quasi-symmetry by $\{\hat{\beta}_j = \hat{\lambda}_j^Y - \hat{\lambda}_j^X\}$ (Conaway 1989). This provides an interpretation for the main-effect terms in quasi-symmetry.

Related results hold for multiple occasions using a multivariate form (10.33) of quasi-symmetry (e.g., Agresti 1997; Conaway 1989; Darroch 1981; Tjur 1982; see also Section 13.2.7). In addition, quasi-symmetry contains as a special case other useful models. These include the ones in Sections 10.4.3 and 10.6.3.

10.4.3 Quasi-independence

Square tables usually exhibit positive dependence, manifested by larger counts on the main diagonal than the independence model predicts. Conditional on the event that a matched pair falls off the main diagonal, though, the relationship may have a simple structure.

A square contingency table satisfies *quasi-independence* when the variables are independent, given that the row and column outcomes differ. This has the loglinear form

$$\log \mu_{ab} = \lambda + \lambda_a^X + \lambda_b^Y + \delta_a I(a=b), \tag{10.23}$$

where $I(\cdot)$ is the indicator function,

$$I(a=b) = \begin{cases} 1, & a=b \\ 0, & a \neq b. \end{cases}$$

This adds a parameter to the independence model for each cell on the main diagonal. The first three terms in (10.23) specify independence, and $\{\delta_a\}$ permit $\{\mu_{aa}\}$ to depart from this pattern and have arbitrary positive values. When $\delta_a > 0$, μ_{aa} is larger than under independence.

The likelihood equations for quasi-independence are

$$\hat{\mu}_{a+} = n_{a+}, \quad \hat{\mu}_{+a} = n_{+a}, \quad \hat{\mu}_{aa} = n_{aa}, \quad a=1,\ldots,I.$$

A perfect fit occurs on the main diagonal, but independence holds for the remaining cells. The model implies that odds ratios equal 1.0 for all rectangularly formed 2×2 tables in which all cells fall off the main diagonal. One can fit the model using Newton–Raphson or IPF. The model has I more parameters than the independence model, so its residual df = $(I - 1)^2 - I$. It applies to tables with $I \geq 3$.

Quasi-independence is the special case of quasi-symmetry (10.21) in which $\{\gamma_{ab}$ for $a \neq b\}$ are identical. Caussinus (1966, p. 146) showed that they are equivalent when $I = 3$.

10.4.4 Migration Revisited

We now return to Table 10.6 on migration patterns. Not surprisingly, the independence model fits terribly, with $G^2 = 125{,}923$ and $X^2 = 146{,}929$. (The maximum possible value of X^2 is $3n = 167{,}943$; see Problem 3.33.) The symmetry model is also unpromising. For instance, 124 people moved from the northeast to the west, but only 63 people made the reverse move. The deviance for testing symmetry is $G^2 = 243.6$ (df = 6).

Quasi-independence states that for people who moved, residence in 1985 is independent of region in 1980. Table 10.7 contains its fitted values, for which $G^2 = 69.5$ (df = 5). This model fits much better than the independence model, primarily because it forces a perfect fit on the main diagonal, where most observations occur. However, lack of fit is apparent off that diagonal. Many more people moved from the northeast to the south and many fewer moved from the west to the south than quasi-independence predicts.

TABLE 10.7 Fit of Models to Table 10.6

Residence in 1980	Residence in 1985[a]				
	Northeast	Midwest	South	West	Total
Northeast	11,607	100 (126.6)[1] (95.8)[2]	366 (312.9) (370.4)	124 (150.5) (123.8)	12,197
Midwest	87 (117.4) (91.2)	13,677	515 (531.1) (501.7)	302 (255.5) (311.1)	14,581
South	172 (133.2) (167.6)	255 (243.8) (238.3)	17,189	270 (290.0) (261.1)	18,486
West	63 (71.4) (63.2)	176 (130.6) (166.9)	286 (323.0) (294.9)	10,192	10,717
Total	11,929	14,178	18,986	10,888	55,981

[a1] Quasi-independence fit; [2] quasi-symmetry fit; both models giving perfect fit on main diagonal.

The quasi-symmetry model has $G^2 = 3.0$, with df = 3. Table 10.7 displays its fit, which is much better than with quasi-independence. The lack of symmetry in cell probabilities reflects slight marginal heterogeneity. The subject-specific effects can be described using the model's parameter estimates, $\{\hat{\lambda}_1^Y - \hat{\lambda}_1^X = -0.672, \hat{\lambda}_2^Y - \hat{\lambda}_2^X = -0.623, \hat{\lambda}_3^Y - \hat{\lambda}_3^X = 0.122\}$. For instance, for a given subject the estimated odds of living in the south instead of the west in 1985 were $\exp(0.122) = 1.13$ times the odds in 1980. We'll see in Chapter 12 that such subject-specific effects tend to be stronger than those in corresponding marginal models, especially in tables like this with strong association.

A related application with matched samples is the study of occupational mobility. Each observation pairs parent's occupation with child's occupation (Goodman 1979b; Hout et al. 1987).

10.4.5 Marginal Homogeneity and Quasi-symmetry

Marginal homogeneity is not equivalent to a loglinear model. However, quasi-symmetry is a useful model for studying marginal homogeneity. Caussinus (1966) showed that symmetry is equivalent to quasi-symmetry and marginal homogeneity holding simultaneously. We have seen that symmetry implies both quasi-symmetry and marginal homogeneity. Now we give Caussinus's argument for the converse, that the joint occurrence of quasi-symmetry and marginal homogeneity implies symmetry.

From (10.21), if quasi-symmetry holds, $\pi_{ab} = \alpha_a \beta_b \gamma_{ab}$, where $\gamma_{ab} = \gamma_{ba} > 0$ for all $a < b$. Equivalently,

$$\pi_{ab} = \rho_a \delta_{ab},$$

where $\rho_a = \alpha_a / \beta_a$ and $\delta_{ab} = \beta_a \beta_b \gamma_{ab}$ also satisfies $\delta_{ab} = \delta_{ba} > 0$ for all $a < b$. If there is also marginal homogeneity, then

$$\pi_{j+} = \rho_j \sum_b \delta_{jb} = \sum_a \rho_a \delta_{aj} = \pi_{+j},$$

or

$$\rho_j = \left(\sum_a \rho_a \delta_{aj} \right) \Big/ \left(\sum_b \delta_{jb} \right) = \left(\sum_a \rho_a \delta_{aj} \right) \Big/ \left(\sum_b \delta_{bj} \right), \quad j = 1, \ldots, I.$$

Thus, each ρ_j is a weighted average of $\{\rho_a\}$, with weights $\{\delta_{aj} / \sum_b \delta_{bj} > 0, a = 1, \ldots, I\}$. Any set $\{\rho_a\}$ satisfying this must be identical. Otherwise, there would be a ρ_j that is no greater than any ρ_a but smaller than at least one, and hence it could not be a positive weighted average of all of them. But since $\{\rho_a\}$ are identical, $\pi_{ab} = \rho_a \delta_{ab} = \rho_b \delta_{ab} = \rho_b \delta_{ba} = \pi_{ba}$, so symmetry holds. Thus, a table that satisfies both quasi-symmetry and marginal homogeneity also satisfies symmetry. Since the converse holds,

$$\text{quasi-symmetry} + \text{marginal homogeneity} = \text{symmetry}. \quad (10.24)$$

It follows that when quasi-symmetry (QS) holds, marginal homogeneity (MH) is equivalent to symmetry (S), which is $\{\lambda_a^X = \lambda_a^Y, \ a = 1, \ldots, I\}$ in the QS model. Thus, conditional on quasi-symmetry, testing marginal homogeneity is equivalent to testing symmetry. A test of marginal homogeneity compares fit statistics for the symmetry and quasi-symmetry models,

$$G^2(S \mid QS) = G^2(S) - G^2(QS), \quad (10.25)$$

with df $= I - 1$. This is an alternative to approaches using marginal models discussed in Section 10.3.3.

Table 10.6 on migration from 1980 to 1985 has $G^2(S) = 243.6$ and $G^2(QS) = 3.0$. The difference $G^2(S \mid QS) = 240.6$ (df $= 3$) shows extremely strong evidence of marginal heterogeneity. Results are similar to those quoted in Section 10.3.4 for the likelihood-ratio test based on model (10.15), for which $G^2 = 240.8$, or the Wald test, for which $W = 236.5$ (both with df $= 3$).

10.4.6 Ordinal Quasi-symmetry Model

The loglinear models presented so far for square tables treat classifications as nominal. With ordered categories, more parsimonious models are useful. Let $u_1 \leq \cdots \leq u_I$ denote ordered scores for both the row and columns. An *ordinal quasi-symmetry model* is

$$\log \mu_{ab} = \lambda + \lambda_a + \lambda_b + \beta u_b + \lambda_{ab}, \quad (10.26)$$

where $\lambda_{ab} = \lambda_{ba}$ for all $a < b$. It is the special case of the quasi-symmetry model (10.19) in which

$$\lambda_b^Y - \lambda_b^X = \beta u_b$$

has a linear trend. Symmetry is the special case $\beta = 0$.

This model has logit representation,

$$\log(\pi_{ab}/\pi_{ba}) = \beta(u_b - u_a) \qquad \text{for } a \leq b. \quad (10.27)$$

This is the special case of the linear logit model, $\text{logit}(\pi) = \alpha + \beta x$, with $\alpha = 0$, $x = u_b - u_a$ and π equal to the conditional probability of cell (a,b), given response sequence (a, b) or (b, a). The greater the value of $|\beta|$, the greater the difference between π_{ab} and π_{ba} and hence between the marginal distributions.

The likelihood equations for ordinal quasi-symmetry are

$$\sum_a u_a \hat{\mu}_{a+} = \sum_a u_a n_{a+}, \qquad \sum_b u_b \hat{\mu}_{+b} = \sum_b u_b n_{+b},$$

$$\hat{\mu}_{ab} + \hat{\mu}_{ba} = n_{ab} + n_{ba} \qquad \text{for} \quad a < b.$$

The fitted marginal counts need not equal the observed marginal counts. However, dividing the first two equations by n shows that they have the same means.

When $\beta \neq 0$, this model implies stochastically ordered margins. When $\beta > 0$ ($\beta < 0$), responses have a higher mean in the column (row) distribution. Like the ordinal marginal models (Section 10.3.1), this model concentrates the marginal effect on df = 1. A test of marginal homogeneity (H_0: $\beta = 0$) uses

ordinal quasi-symmetry + marginal homogeneity = symmetry.

The likelihood-ratio test statistic compares the deviance for symmetry and ordinal quasi-symmetry.

One can fit this model by fitting (10.27) with logit model software: Identify (n_{ab}, n_{ba}) as binomial with $n_{ab} + n_{ba}$ trials, and fit a logit model with no intercept and predictor $x = u_b - u_a$. One can also fit (10.26) using iterative methods for loglinear models.

10.4.7 Premarital and Extramarital Sex Revisited

For Table 10.5 on attitudes toward premarital and extramarital sex, a cursory glance at the data reveals that the symmetry model is inadequate ($G^2 = 402.2$, df = 6). By comparison, quasi-symmetry fits well ($G^2 = 1.4$, df = 3). The simpler model of ordinal quasi-symmetry also fits well: With scores $\{1, 2, 3, 4\}$, $G^2 = 2.1$ (df = 5).

The ML estimate $\hat{\beta} = -2.86$. From (10.27), the estimated probability that outcome on premarital sex is x categories more positive than the outcome on extramarital sex equals $\exp(2.86x)$ times the reverse probability. For instance, the estimated probability that premarital sex is judged almost always wrong and extramarital sex is always wrong equals $\exp(2.86) = 17.4$ times the estimated probability that premarital sex is always wrong and extramarital sex is almost always wrong.

10.4.8 Other Ordinal Models for Square Tables

For ordered classifications, when symmetry does not hold, often either $\pi_{ab} > \pi_{ba}$ for all $a < b$, or $\pi_{ab} < \pi_{ba}$ for all $a < b$. A generalization of symmetry with this property is the logit model

$$\log(\pi_{ab}/\pi_{ba}) = \tau \quad \text{for } a < b. \tag{10.28}$$

It implies that for all $a < b$,

$$P(Y_{i1} = a, Y_{i2} = b \mid Y_{i1} < Y_{i2}) = P(Y_{i1} = b, Y_{i2} = a \mid Y_{i1} > Y_{i2}).$$

The pattern of probabilities for cells above the main diagonal is a mirror image of the pattern for cells below it. This property is called *conditional symmetry* (McCullagh 1978). Problem 10.35 shows the corresponding loglinear model and its fit. Symmetry is the special case $\tau = 0$.

Another model generalizes quasi-independence. Let $\{u_a\}$ be ordered scores. The model

$$\log \mu_{ab} = \lambda + \lambda_a^X + \lambda_b^Y + \beta u_a u_b + \delta_a I(a = b) \tag{10.29}$$

permits linear-by-linear association [see (9.6)] off the main diagonal. It is a special case of quasi-symmetry, and quasi-independence is the special case $\beta = 0$. For equal-interval scores, it implies uniform local association, given that responses differ. Goodman (1979a) called it *quasi-uniform association*.

For Table 10.5 on opinions about premarital and extramarital sex, the conditional symmetry model has $\hat{\tau} = -4.130$ (SE = 0.451). The estimated probability that extramarital sex is considered more wrong are $\exp(4.13) = 62.2$ times the estimated probability that premarital sex is considered more wrong. The quasi-uniform association model has $\hat{\beta} = 0.632$ (SE = 0.106). Off the main diagonal, the estimated local odds ratio equals $\exp(0.632) = 1.88$.

10.5 MEASURING AGREEMENT BETWEEN OBSERVERS

We now discuss an application, analyzing agreement between two observers, that uses matched-pairs models. We illustrate with Table 10.8. This shows ratings by two pathologists, labeled A and B, who separately classified 118 slides regarding the presence and extent of carcinoma of the uterine cervix. The rating scale has the ordered categories (1) negative, (2) atypical squamous hyperplasia, (3) carcinoma *in situ*, (4) squamous or invasive carcinoma.

10.5.1 Agreement: Departures from Independence

Let π_{ab} denote the probability that observer A classifies a slide in category a and observer B classifies it in category b. Then π_{aa} is the probability that they both choose category a, and $\Sigma_a \pi_{aa}$ is the total probability of agreement. Perfect agreement occurs when $\Sigma_a \pi_{aa} = 1$.

With subjective scales, agreement is less than perfect. Analyses focus on describing strength of agreement and detecting patterns of disagreement.

TABLE 10.8 Diagnoses of Carcinoma

Pathologist A	Pathologist B[a]				Total
	1	2	3	4	
1	22	2	2	0	26
	(8.5)	(−0.5)	(−5.9)	(−1.8)	
2	5	7	14	0	26
	(−0.5)	(3.2)	(−0.5)	(−1.8)	
3	0	2	36	0	38
	(−4.1)	(−1.2)	(5.5)	(−2.3)	
4	0	1	17	10	28
	(−3.3)	(−1.3)	(0.3)	(5.9)	
Total	27	12	69	10	118

[a] Values in parentheses are standardized Pearson residuals for the independence model.
Source: N. S. Holmquist, C. A. McMahon, and O. D. Williams, *Arch. Pathol.* **84**: 334–345 (1967); reprinted with permission from the American Medical Association. See also Landis and Koch (1977).

Agreement and *association* are distinct facets of the joint distribution. Strong agreement requires strong association, but strong association can exist without strong agreement. If observer A consistently rates subjects one category higher than observer B, strength of agreement is poor even though the association is strong.

Evaluations of agreement compare $\{n_{ab}\}$ to the values $\{n_{a+}n_{+b}/n\}$ predicted under independence. That model is a baseline, showing the agreement expected if no association existed between ratings. Normally, it fits poorly if even mild agreement exists, but its cell standardized residuals (Section 3.3.1) show patterns of agreement and disagreement. Ideally, standardized residuals are large positive on the main diagonal and large negative off that diagonal. The sizes are influenced by sample size n, however, larger values tending to occur as n increases.

The independence model fits Table 10.8 poorly ($G^2 = 118.0$, df = 9). That table reports the standardized Pearson residuals in parentheses. The large positive residuals on the main diagonal indicate that agreement for each category is greater than expected by chance, especially for the first category. Off the main diagonal they are primarily negative. Disagreements occurred less than expected under independence, although the evidence of this is weaker for categories closer together. The most common disagreements were observer B choosing category 3 and observer A instead choosing category 2 or 4.

10.5.2 Using Quasi-independence to Analyze Agreement

More complex models add components that relate to agreement beyond that expected under independence. A useful generalization is quasi-independence

TABLE 10.9   Fitted Values for Carcinoma Diagnoses of Table 10.8

Pathologist A	Pathologist B[a]			
	1	2	3	4
1	22	2	2	0
	$(22)^1$	(0.7)	(3.3)	(0.0)
	$(22)^2$	(2.4)	(1.6)	(0.0)
2	5	7	14	0
	(2.4)	(7)	(16.6)	(0.0)
	(4.6)	(7)	(14.4)	(0.0)
3	0		36	0
	(0.8)	(1.2)	(36)	(0.0)
	(0.4)	(1.6)	(36)	(0.0)
4	0	1	17	10
	(1.9)	(3.0)	(13.1)	(10)
	(0.0)	(1.0)	(17.0)	(10)

[a][1] Quasi-independence model; [2] quasi-symmetry model.

(10.23), which adds main-diagonal parameters $\{\delta_a\}$. For Table 10.8, this model has $G^2 = 13.2$ (df = 5). It fits much better than independence, but some lack of fit remains. Table 10.9 shows the fit.

For two subjects, suppose that each observer classifies one in category a and one in category b. The odds that the observers agree rather than disagree on which is in category a and which is in category b equal

$$\tau_{ab} = \frac{\pi_{aa}\pi_{bb}}{\pi_{ab}\pi_{ba}} = \frac{\mu_{aa}\mu_{bb}}{\mu_{ab}\mu_{ba}}. \tag{10.30}$$

As τ_{ab} increases, the observers are more likely to agree for that pair of categories. Under quasi-independence,

$$\tau_{ab} = \exp(\delta_a + \delta_b) .$$

Larger $\{\delta_a\}$ represent stronger agreement. For instance, for Table 10.8, $\hat{\delta}_2 = 0.6$ and $\hat{\delta}_3 = 1.9$, and $\hat{\tau}_{23} = 12.3$. The degree of agreement also seems fairly strong for other pairs of categories.

10.5.3   Quasi-symmetry and Agreement Modeling

For Table 10.8, the quasi-independence model shows some lack of fit. Given that the pathologists disagree, some association remains between ratings. For observer agreement tables, this is common. Quasi-symmetry (10.19) often fits much better, because it permits association. For Table 10.8, it has $G^2 = 1.0$ (df = 2). Table 10.9 displays the fit. It is not unusual for tables to have many empty cells. When $n_{ab} + n_{ba} = 0$ for any pair (such as categories 1 and 4 in Table 10.8), the ML fitted values for quasi-symmetry in those cells must also be zero since one of its likelihood equations is $\hat{\mu}_{ab} + \hat{\mu}_{ba} = n_{ab} + n_{ba}$. One should eliminate those cells from the fitting process to get the proper residual df value.

Under quasi-symmetry, $\hat{\tau}_{ab} = \exp(\hat{\lambda}_{aa} + \hat{\lambda}_{bb} - \hat{\lambda}_{ab} - \hat{\lambda}_{ba})$, where $\hat{\lambda}_{ab} = \hat{\lambda}_{ba}$. For categories 2 and 3 of Table 10.8, for instance, $\hat{\tau}_{23} = 10.7$.

Loglinear models directly address the association component of agreement. The quasi-symmetry model also yields information about similarity of marginal distributions. The simpler symmetry model that forces the margins to be identical fits Table 10.8 poorly ($G^2 = 39.2$, df = 5). The statistic $G^2(S \mid QS) = 39.2 - 1.0 = 38.2$ (df = 3) provides strong evidence of marginal heterogeneity. In Table 10.8, differences in marginal proportions are substantial in each category but the first. The marginal heterogeneity is one reason that the agreement is not stronger.

Models for agreement can take ordering of categories into account. Conditional on observer disagreement, a tendency usually remains for high (low) ratings by one observer to occur with relatively high (low) ratings by the other observer (see Problem 10.41).

10.5.4 Kappa Measure of Agreement

An alternative approach summarizes agreement with a single index. For nominal scales, the most popular measure is *Cohen's kappa* (Cohen 1960). It compares the probability of agreement $\Sigma_a \pi_{aa}$ to that expected if the ratings were independent, $\Sigma_a \pi_{a+} \pi_{+a}$, by

$$\kappa = \frac{\Sigma_a \pi_{aa} - \Sigma_a \pi_{a+} \pi_{+a}}{1 - \Sigma_a \pi_{a+} \pi_{+a}}.$$

The denominator equals the numerator with $\Sigma_a \pi_{aa}$ replaced by its maximum possible value of 1, corresponding to perfect agreement. Kappa equals 0 when the agreement merely equals that expected under independence. It equals 1.0 when perfect agreement occurs. The stronger the agreement, the higher is κ, for given marginal distributions. Negative values occur when agreement is weaker than expected by chance, but this rarely happens.

For multinomial sampling, the sample value $\hat{\kappa}$ has a large-sample normal distribution. Its estimated asymptotic variance (Fleiss et al. 1969) is

$$\hat{\sigma}^2(\hat{\kappa}) = \frac{1}{n} \left\{ \frac{P_o(1 - P_o)}{(1 - P_e)^2} + \frac{2(1 - P_o)[2P_o P_e - \Sigma_a P_{aa}(p_{a+} + p_{+a})]}{(1 - P_e)^3} \right.$$
$$\left. + \frac{(1 - P_o)^2 [\Sigma_a \Sigma_b P_{ab}(p_{b+} + p_{+a})^2 - 4P_e^2]}{(1 - P_e)^4} \right\},$$

where $P_o = \sum_a p_{aa}$ and $P_e = \sum_a p_{a+}p_{+a}$. It is rarely plausible that agreement is no better than expected by chance. Thus, rather than testing $H_0: \kappa = 0$, it is more relevant to estimate strength of agreement by interval estimation of κ.

For Table 10.8, $P_o = 0.636$ and $P_e = 0.281$. Sample kappa equals $(0.636 - 0.281)/(1 - 0.281) = 0.493$. The difference between observed agreement and that expected under independence is about 50% of the maximum possible difference. The estimated standard error is 0.057, so κ apparently falls roughly between 0.4 and 0.6, moderately strong agreement.

10.5.5 Weighted Kappa: Quantifying Disagreement

Kappa treats classifications as nominal. When categories are ordered, the seriousness of a disagreement depends on the difference between the ratings. For nominal classifications also, some disagreements may be considered more severe than others. The measure *weighted kappa* (Spitzer et al. 1967) uses weights $\{w_{ab}\}$ satisfying $0 \le w_{ab} \le 1$, with all $w_{aa} = 1$ and all $w_{ab} = w_{ba}$ to describe closeness of agreement. One possibility is $\{w_{ab} = 1 - |a - b|/(I - 1)\}$, for which agreement is greater for cells nearer the main diagonal. Fleiss and Cohen (1973) suggested $\{w_{ab} = 1 - (a - b)^2/(I - 1)^2\}$. The weighted agreement is $\sum_a \sum_b w_{ab} \pi_{ab}$ and weighted kappa is

$$\kappa_w = \frac{\sum_a \sum_b w_{ab} \pi_{ab} - \sum_a \sum_b w_{ab} \pi_{a+} \pi_{+b}}{1 - \sum_a \sum_b w_{ab} \pi_{a+} \pi_{+b}}.$$

Controversy surrounds the utility of kappa and weighted kappa, partly because their values depend strongly on the marginal distributions. The same diagnostic rating process can yield quite different values, depending on the proportions of cases of the various types (Problem 10.40). In summarizing a contingency table by a single number, the reduction in information can be severe. It is helpful to construct models providing more detailed investigation of the agreement and disagreement structure rather than to depend solely on a summary index.

10.5.6 Extensions to Multiple Observers

With several observers, ordinary loglinear models are not usually relevant. Their description of agreement and association between two observers is conditional on ratings by the others. It is more relevant to study this marginally, without conditioning on the other ratings. Hence, for R observers, modelling simultaneously the pairwise agreement and association structure requires studying the $\binom{R}{2}$ pairs of two-way marginal distributions (Becker and Agresti 1992).

Other approaches have also been used. For instance, generalizations of kappa summarize pairwise agreements or multiple agreements (Fleiss 1981, Sec. 13.2; Landis and Koch 1977). Or, it may make sense to use a mixture model that assumes latent classes of subjects for whom the observers agree and subjects for whom they disagree. Such an analysis is shown in Section 13.1.2.

10.6 BRADLEY–TERRY MODEL FOR PAIRED PREFERENCES

Sometimes, categorical outcomes result from pairwise evaluations. A common example is athletic competitions, when the outcome for a team or player consists of categories (win, lose). Another example is pairwise comparison of product brands, such as two brands of wine of some type. When a wine critic rates I brands of sauvignon blanc, it might be difficult to establish an outright ranking, especially if I is large. However, for any given pair, the critic could probably state a preference after tasting them at the same occasion. An overall ranking of the wines could then be based on the pairwise preferences. We present a model for this in this section.

10.6.1 Bradley–Terry Model

Bradley and Terry (1952) proposed a logit model for paired evaluations. Let Π_{ab} denote the probability that a is preferred to b. Suppose that $\Pi_{ab} + \Pi_{ba} = 1$ for all pairs; that is, a tie cannot occur. The Bradley–Terry model is

$$\log \frac{\Pi_{ab}}{\Pi_{ba}} = \beta_a - \beta_b. \tag{10.31}$$

Alternatively,

$$\Pi_{ab} = \exp(\beta_a) / [\exp(\beta_a) + \exp(\beta_b)].$$

Thus, $\Pi_{ab} = \frac{1}{2}$ when $\beta_a = \beta_b$ and $\Pi_{ab} > \frac{1}{2}$ when $\beta_a > \beta_b$.

Identifiability requires a constraint such as $\beta_I = 0$ or $\Sigma_a \exp(\hat{\beta}_a) = 1$. Since the model describes $\binom{I}{2}$ probabilities ($\{\Pi_{ab}\}$ for $a < b$) by $(I - 1)$ parameters, residual df $= \binom{I}{2} - (I - 1)$.

For $a < b$, let N_{ab} denote the sample number of evaluations, with a preferred n_{ab} times and b preferred $n_{ba} = N_{ab} - n_{ab}$ times. A square contingency table with empty cells on the main diagonal summarizes results. When the N_{ab} comparisons are independent with probability Π_{ab} for each, n_{ab} has a bin(N_{ab}, Π_{ab}) distribution. If evaluations for different pairs are also independent, ordinary methods for logit models apply for fitting the model.

TABLE 10.10 Results of 1987 Season for American League Baseball Teams

Winning Team	Losing Team[a]						
	Milwaukee	Detroit	Toronto	New York	Boston	Cleveland	Baltimore
Milwaukee	—	7 (7.0)	9 (7.4)	7 (7.6)	7 (8.0)	9 (9.2)	11 (10.8)
Detroit	6 (6.0)	—	7 (7.0)	5 (7.1)	11 (7.6)	9 (8.8)	9 (10.5)
Toronto	4 (5.6)	6 (6.0)	—	7 (6.7)	7 (7.1)	8 (8.4)	12 (10.2)
New York	6 (5.4)	8 (5.9)	6 (6.3)	—	6 (7.0)	7 (8.3)	10 (10.1)
Boston	6 (5.0)	2 (5.4)	6 (5.9)	7 (6.0)	—	7 (7.9)	12 (9.8)
Cleveland	4 (3.8)	4 (4.2)	5 (4.6)	6 (4.7)	6 (5.1)	—	6 (8.6)
Baltimore	2 (2.2)	4 (2.5)	1 (2.8)	3 (2.9)	1 (3.2)	7 (4.4)	—

[a] Values in parentheses represent the fit of the Bradley–Terry model.

Source: American League Red Book, 1988 (St. Louis, MO: Sporting News Publishing Co.)

10.6.2 Home Team Advantage in Baseball

Table 10.10 shows results of the 1987 season for the seven baseball teams in the Eastern Division of the American League. For instance, of games between Boston and New York, Boston won 7 and New York won 6. Table 10.10 shows the population of regular-season games. We regard this as a sample estimate of a conceptual distribution representing the long-run performance of teams as constituted in 1987.

We fitted the Bradley–Terry model as a logit model for $\binom{7}{2} = 21$ independent binomial samples, using an appropriate model matrix and no intercept (e.g., for SAS, see Table A.19). The model fits adequately ($G^2 = 15.7$, df = 15). Table 10.10 contains the fitted values $\{\hat{\mu}_{ab}\}$. Table 10.11 displays the sample proportion of games each team won and the model estimates of $\{\hat{\beta}_a\}$ (setting $\hat{\beta}_7 = 0$) and $\{\exp(\hat{\beta}_a)\}$ [setting $\Sigma_a \exp(\hat{\beta}_a) = 1$]. When Boston played New York, the estimated probability that Boston won is

$$\hat{\Pi}_{54} = \exp(\hat{\beta}_5) \big/ \big[\exp(\hat{\beta}_5) + \exp(\hat{\beta}_4)\big] = 0.46.$$

The standard error of each $\hat{\beta}_a$ and of each $\hat{\beta}_a - \hat{\beta}_b$ is about 0.3, so not much evidence exists of a difference among the top five teams.

TABLE 10.11 Results of Fitting Bradley–Terry Models to Baseball Data

Team	Winning Percentage	$\hat{\beta}_i$ (10.31)	$\exp(\hat{\beta}_i)$ (10.31)	$\exp(\hat{\beta}_i)$ (10.32)
Milwaukee	64.1	1.58	0.218	0.220
Detroit	60.2	1.44	0.189	0.190
Toronto	56.4	1.29	0.164	0.164
New York	55.1	1.25	0.158	0.157
Boston	51.3	1.11	0.136	0.137
Cleveland	39.7	0.68	0.089	0.088
Baltimore	23.1	0.00	0.045	0.044

TABLE 10.12   Wins / Losses by Home and Away Team, 1987

Home Team	Away Team						
	Milwaukee	Detroit	Toronto	New York	Boston	Cleveland	Baltimore
Milwaukee	—	4-3	4-2	4-3	6-1	4-2	6-0
Detroit	3-3	—	4-2	4-3	6-0	6-1	4-3
Toronto	2-5	4-3	—	2-4	4-3	4-2	6-0
New York	3-3	5-1	2-5	—	4-3	4-2	6-1
Boston	5-1	2-5	3-3	4-2	—	5-2	6-0
Cleveland	2-5	3-3	3-4	4-3	4-2	—	2-4
Baltimore	2-5	1-5	1-6	2-4	1-6	3-4	—

Source: American League Red Book, 1988 (St. Louis, MO: Sporting News Publishing Co.).

This model does not recognize which team is the home team. Most sports have a home field advantage: A team is more likely to win when it plays at its home city. Table 10.12 contains results for the 1987 season according to the (home team, away team) classification. For instance, when Boston was the home team, it beat New York 4 times and lost 2 times; when New York was the home team, it beat Boston 4 times and lost 3 times. Now for all $a \neq b$, let Π_{ab}^* denote the probability that team a beats team b, when a is the home team. Consider logit model

$$\log \frac{\Pi_{ab}^*}{1 - \Pi_{ab}^*} = \alpha + (\beta_a - \beta_b). \qquad (10.32)$$

When $\alpha > 0$, a home field advantage exists. The home team of two evenly matched teams has probability $\exp(\alpha)/[1 + \exp(\alpha)]$ of winning.

For Table 10.12, model (10.32) describes 42 binomial distributions with 7 parameters. It has $G^2 = 38.6$ (df = 35). Table 10.11 displays $\{\exp(\hat{\beta}_a)\}$, which are similar to those obtained previously. The estimate of the home-field parameter is $\hat{\alpha} = 0.302$. For two evenly matched teams, the home team had estimated probability 0.575 of winning. When Boston played New York, the estimated probability of a Boston win was 0.54 at Boston and 0.39 at New York.

Model (10.32) is a useful generalization of the Bradley–Terry model whenever an *order effect* exists. For instance, in pairwise taste evaluations, the product tasted first may have a slight advantage.

10.6.3   Bradley–Terry Model and Quasi-symmetry

Fienberg and Larntz (1976) showed that the Bradley–Terry model is a logit formulation of the quasi-symmetry model (10.19). For quasi-symmetry, given that an observation is in cell (a, b) or (b, a), the logit of the conditional probability of cell (a, b) equals

$$\log \frac{\mu_{ab}}{\mu_{ba}} = \left(\lambda + \lambda_a^X + \lambda_b^Y + \lambda_{ab}^{XY} \right) - \left(\lambda + \lambda_b^X + \lambda_a^Y + \lambda_{ba}^{XY} \right)$$

$$= \left(\lambda_a^X - \lambda_a^Y \right) - \left(\lambda_b^X - \lambda_b^Y \right) = \beta_a - \beta_b,$$

where $\beta_a = \lambda_a^X - \lambda_a^Y$. Estimates $\{\hat{\lambda}_a^X\}$ and $\{\hat{\lambda}_a^Y\}$ for quasi-symmetry yield $\{\hat{\beta}_a\}$ for the Bradley–Terry model.

10.6.4 Extensions to Ties and Ordinal Evaluations

The Bradley–Terry model extends to ordinal comparisons, such as the evaluation scale (much better, slightly better, the same, slightly worse, much worse) in comparing two products. With cumulative logits and an I-category evaluation scale, let Y_{ab} denote the response for a comparison of a with b. The model is

$$\text{logit} \left[P(Y_{ab} \leq j) \right] = \alpha_j + (\beta_a - \beta_b).$$

Since $P(Y_{ab} \leq j) = P(Y_{ba} > I - j) = 1 - P(Y_{ba} \leq I - j)$, it follows that $\text{logit}[P(Y_{ab} \leq j] = - \text{logit}[P(Y_{ba} \leq I - j)]$. Thus, necessarily, $\alpha_j = -\alpha_{I-j}$.

The most common ordered preference scale is (win, tie, lose). Then, $\alpha_1 = -\alpha_2$.

10.7 MARGINAL AND QUASI-SYMMETRY MODELS FOR MATCHED SETS*

Methods for matched pairs extend to matched sets. Here we present mainly the loglinear modeling approach; in Chapters 11 and 12 we present extensions of the marginal and conditional logit modeling approaches.

10.7.1 Marginal Homogeneity, Complete Symmetry, and Quasi-symmetry

Let $(Y_1, Y_2, \ldots, Y_T)$ denote the T responses in each matched set. With I response categories, a contingency table with I^T cells summarizes the possible outcomes. Let $\mathbf{i} = (i_1, \ldots, i_T)$ denote the cell having $Y_t = i_t$, $t = 1, \ldots, T$. Let $\pi_{\mathbf{i}} = P(Y_t = i_t, t = 1, \ldots, T)$, and let $\mu_{\mathbf{i}} = n\pi_{\mathbf{i}}$. Then

$$P(Y_t = j) = \pi_{+\cdots+j+\cdots+},$$

where the j subscript is in position t, and $\{P(Y_t = j), j = 1, \ldots, I\}$ is the marginal distribution for Y_t.

This T-way table satisfies *marginal homogeneity* if

$$P(Y_1 = j) = P(Y_2 = j) = \cdots = P(Y_T = j) \qquad \text{for } j = 1, \ldots, I.$$

It satisfies *complete symmetry* if

$$\pi_{\mathbf{i}} = \pi_{\mathbf{j}}$$

for any permutation $\mathbf{j} = (j_1, \ldots, j_T)$ of $\mathbf{i} = (i_1, \ldots, i_T)$. Complete symmetry implies marginal homogeneity, but the converse does not hold except when $T = I = 2$.

Complete symmetry is a loglinear model. One representation is

$$\log \mu_{\mathbf{i}} = \lambda_{ab\ldots m},$$

where a is the minimum of $(i_1, \ldots, i_T)$, b is the next smallest, $\ldots$, and m is the maximum. In a three-way table, for instance, $\log \mu_{122} = \log \mu_{212} = \log \mu_{221} = \lambda_{122}$. The number of $\{\lambda_{ab\ldots m}\}$ parameters is the number of ways of selecting T out of I items with replacement, which is $\binom{I + T - 1}{T}$. Thus, residual df $= I^T - \binom{I + T - 1}{T}$ (Haberman 1978, p. 518).

An I^T table satisfies *quasi-symmetry* if

$$\log \mu_{\mathbf{i}} = \lambda_{1i_1} + \lambda_{2i_2} + \cdots + \lambda_{Ti_T} + \lambda_{ab\ldots m} \tag{10.33}$$

where $\lambda_{ab\ldots m}$ is defined as in the complete symmetry model. It has symmetric association and higher-order interaction terms, but permits each single-factor marginal distribution to have its own parameters. Identifiability requires constraints such as $\lambda_{tI} = 0$ for each t. One set of main-effect terms is redundant (Problem 10.31). This model has $(I - 1)(T - 1)$ more parameters than complete symmetry. It is fitted using iterative methods.

For ordinal responses, a simpler model with quantitative main effects uses ordered scores $\{u_a\}$. The *ordinal quasi-symmetry model* is

$$\log \mu_{\mathbf{i}} = \beta_1 u_{i_1} + \beta_2 u_{i_2} + \cdots + \beta_T u_{i_T} + \lambda_{ab\ldots m}$$

where one can set $\beta_T = 0$. Complete symmetry is the special case $\beta_1 = \cdots = \beta_T$.

When quasi-symmetry (10.33) or ordinal quasi-symmetry holds, marginal homogeneity is equivalent to complete symmetry. Marginal heterogeneity occurs if quasi-symmetry (QS) holds but complete symmetry (S) does not. The statistic

$$G^2(S \mid QS) = G^2(S) - G^2(QS)$$

tests marginal homogeneity. Under complete symmetry, it is asymptotically chi-squared with df $= (I - 1)(T - 1)$. The corresponding test for the ordinal quasi-symmetry model has df $= (T - 1)$.

10.7.2 Attitudes toward Legalized Abortion Example

Refer to Table 10.13. Subjects indicated whether they support legalized abortion in three situations: (1) if the family has a very low income and cannot afford any more children, (2) when the woman is not married and does not want to marry the man, and (3) when the woman wants it for any reason. The table also classifies subjects by gender, resulting in a 2^4 table.

Let μ_{ghij} denote the expected frequency for gender g (1 = female; 0 = male) with response sequence (h, i, j) for the three questions. Consider the model

$$\log \mu_{ghij} = \beta g + \lambda_{abc},$$

where the interaction term is λ_{111} when $(h, i, j) = (1, 1, 1)$, λ_{112} when $(h, i, j) = (1, 1, 2)$ or $(1, 2, 1)$ or $(2, 1, 1)$, λ_{122} when $(h, i, j) = (1, 2, 2)$ or $(2, 1, 2)$ or $(2, 2, 1)$, and λ_{222} when $(h, i, j) = (2, 2, 2)$. This model implies the same complete symmetry pattern of probabilities for each gender. Its fit has $G^2 = 39.2$ with df $= 11$.

Adding main-effect terms for the three issues implies the same quasi-symmetric pattern for each gender. It fits much better, having $G^2 = 10.2$ with df $= 9$. Thus, it seems plausible to assume a symmetric association structure. In fact, the loglinear model with only two-factor association terms has fitted log odds ratios of 3.2 for items 1 and 2, 2.6 for items 1 and 3, and 3.3 for items 2 and 3.

One can test marginal homogeneity, given gender, by the likelihood-ratio statistic $39.2 - 10.2 = 29.0$, with df $= 2$. An analysis of the main-effect terms in the quasi-symmetry model shows greater support for legalized abortion when the family has a low income and cannot afford any more children than in the other two instances.

TABLE 10.13 Support for Legalizing Abortion in Three Situations, by Gender

	Sequence of Responses on the Three Items[a]							
Gender	(1, 1, 1)	(1, 1, 2)	(2, 1, 1)	(2, 1, 2)	(1, 2, 1)	(1, 2, 2)	(2, 2, 1)	(2, 2, 2)
Male	342	26	6	21	11	32	19	356
Female	440	25	14	18	14	47	22	457

[a]Items are (1) if the family has a very low income and cannot afford anymore children, (2) when the woman is not married and does not want to marry the man, and (3) when the woman wants it for any reason. 1, yes; 2, no.
Source: Data from 1994 General Social Survey, National Opinion Research Center.

10.7.3 Types of Marginal Symmetry

A general type of symmetry for I^T tables has marginal homogeneity and complete symmetry as special cases. For an I^T table, $P(Y_{t_1} = j_1, \ldots, Y_{t_h} = j_h)$, where h is between 1 and T, is a h-dimensional marginal probability, $h = 1$ giving single-variable marginal probabilities. There is hth-*order marginal symmetry* if for all h-tuples $\mathbf{j} = (j_1, \ldots, j_h)$, this probability is the same for each permutation of $\mathbf{j}$ and for all combinations $\mathbf{t} = (t_1, \ldots, t_h)$ of h of the T responses.

For $h = 1$, first-order marginal symmetry is marginal homogeneity. Second-order marginal symmetry occurs if for all t and u, $P(Y_t = a, Y_u = b)$ is the same and the equality holds for all pairs of outcomes (a, b). In other words, the two-way marginal tables exhibit symmetry, and they are identical. Tth-order marginal symmetry in an I^T table is complete symmetry.

When hth-order symmetry holds, ith-order marginal symmetry holds for any $i < h$. For instance, complete symmetry implies second-order marginal symmetry, which itself implies marginal homogeneity. Although this hierarchy is mathematically attractive, the higher-order symmetries are usually too restrictive to fit well in practice.

10.7.4 Marginal Models: Multiway Tables

In practice, usually the form of the joint distribution is of secondary interest. Research questions pertain instead to the marginal distributions. The marginal models of Section 10.3 for matched pairs extend to matched sets. For instance, with ordinal classifications, a cumulative logit model is

$$\text{logit}[P(Y_t \leq j)] = \alpha_j + \beta_t, \qquad j = 1, \ldots, I - 1, \quad t = 1, \ldots, T. \quad (10.34)$$

In the next chapter we study marginal models in more general contexts, extending the analyses of this chapter to incorporate matched sets and explanatory variables.

NOTES

Section 10.1: Comparing Dependent Proportions

10.1. Miettinen (1969) generalized the McNemar test to case–control sets having several controls per case. The Table 10.2 representation is then useful. Each of n matched sets forms a stratum of a $2 \times 2 \times n$ table with one observation in column 1 (the case) and several observations in column 2 (the controls).

Altham (1971) and Ghosh et al. (2000) presented Bayesian analyses for binary matched pairs. Copas (1973), Gart (1969), Kenward and Jones (1994), and Miettinen (1969) studied generalizations of matched-pairs designs. With some approaches (Ghosh et al. 2000; Liang and Zeger 1988; Suissa and Shuster 1991), inferences about marginal homogeneity also use the main-diagonal observations.

Section 10.4: Symmetry, Quasi-symmetry, and Quasi-independence

10.2. For other discussion of quasi-symmetry, see Darroch (1981) and McCullagh (1982). The term *quasi-independence* originated in Goodman (1968). A more general definition of it is $\pi_{ab} = \alpha_a \beta_b$ for some fixed set of cells. See Caussinus (1966), Fienberg (1970b, 1972), and Goodman (1968). Caussinus used the concept to analyze tables that deleted a certain set of cells from consideration, and Goodman used it in earlier analyses of social mobility. Altham (1975) used it with triangular tables, for which observations occur only above or only below the main diagonal. Stigler (1999, Chap. 19) summarized early uses, including Karl Pearson's handling in 1913 of a triangular array. Booth and Butler (1999) and Smith et al. (1996) discussed exact tests for square-table models.

10.3. The effect β in ordinal quasi-symmetry relates to the occasion effect in a subject-specific adjacent-categories-logit model (Agresti 1993). Conditional symmetry is a special case of *diagonals-parameter symmetry*,

$$\log(\pi_{ab}/\pi_{ba}) = \tau_{b-a}, \quad a < b.$$

See Goodman (1979b, 1985) and Hout et al. (1987).

10.4. In some applications a table is *a priori* symmetric or independent, but one can observe only the pair (i, j) rather than their order, thus leading to an upper-triangular table. See Khamis (1983) for examples and ML fitting of models for such three-way tables that are symmetric within layers.

Section 10.5: Measuring Agreement between Observers

10.5. Kappa and weighted kappa relate to the intraclass correlation, a measure of interrater reliability for interval scales (Fleiss 1981; Fleiss and Cohen 1973; Kraemer 1979). Banerjee et al. (1999) and Fleiss (1981, Chap. 13) reviewed kappa and its generalizations. See Becker and Agresti (1992), Goodman (1979b), Tanner and Young (1985), and Problem 10.41 for examples of modeling agreement with loglinear models. Darroch and McCloud (1986) showed that quasi-symmetry has an important role in agreement modeling.

Section 10.6: Bradley–Terry Model for Paired Preferences

10.6. Zermelo (1929) proposed a model that is equivalent to the Bradley–Terry model. Luce (1959) provided an axiomatic basis for it. Mosteller (1951) and Thurstone (1927) proposed an analogous model with probit link. An interesting interview of Ralph Bradley by M. Hollander (*Stat. Sci.* **16**: 75–100, 2001) discussed food-tasting applications that motivated its development. For extensions, see Bradley (1976). Fienberg and Larntz (1976) and Imrey et al. (1976) related it to quasi-independence. Dittrich et al. (1998) allowed covariates. Matthews and Morris (1995) gave an application with a factorial design, ties, and allowance for dependence among judgments. Böckenholt and Dillon (1997) modeled dependence with ordinal preferences. David (1988) and Imrey (1998) surveyed paired preference methods.

TABLE 10.14   Data for Problem 10.1

Suicide	Let Patient Die	
	Yes	No
Yes	1097	90
No	203	435

Source: 1994 General Social Survey, National Opinion Research Center.

PROBLEMS

Applications

10.1 Table 10.14 shows results when subjects were asked "Do you think a person has the right to end his or her own life if this person has an incurable disease?" and "When a person has a disease that cannot be cured, do you think doctors should be allowed to end the patient's life by some painless means if the patient and his family request it?" The table refers to these variables as "suicide" and "let patient die."

a. Compare the marginal proportions using a confidence interval.

b. Perform McNemar's test, and interpret.

c. Find the conditional ML estimate of β for model (10.8). Interpret.

10.2 Refer to Table 8.16 and Problem 8.1. Treat the data as matched pairs on opinion, stratified by gender. Testing independence for the 2×2 table using entries $(6, 160)$ in row 1 and $(11, 181)$ in row 2 tests equality of β for logit model (10.8) for each gender. Explain why.

10.3 A crossover experiment with 100 subjects compares two drugs for treating migraine headaches. The response scale is success (1) or failure (0). Half the study subjects, randomly selected, used drug A the first time they had a headache and drug B the next time. For them, 6 had outcomes $(1, 1)$ for (A, B), 25 had outcomes $(1, 0)$, 10 had outcomes $(0, 1)$, and 9 had outcomes $(0, 0)$. For the 50 subjects who took the drugs in the reverse order, 10 were $(1, 1)$ for (A, B), 20 were $(1, 0)$, 12 were $(0, 1)$, and 8 were $(0, 0)$.

a. Ignoring treatment order, compare the success probabilities for the two drugs. Interpret.

b. McNemar's test uses only the pairs of outcomes that differ. For this study, Table 10.15 shows such data from both treatment orders. Testing independence for this table tests whether success rates are identical for the treatments (Gart 1969). Explain why. Analyze these data, and interpret.

TABLE 10.15 Data for Problem 10.3

Treatment Order	Treatment That Is Better	
	First	Second
A, then B	25	10
B, then A	12	20

10.4 A case–control study has 8 pairs of subjects. The cases have colon cancer, and the controls are matched with the cases on gender and age. A possible explanatory variable is the extent of red meat in a subject's diet, measured as "1 = high" or "0 = low." The (case, control) observations on this were (1, 1) for 3 pairs, (0, 0) for 1 pair, (1, 0) for 3 pairs, and (0, 1) for 1 pair.

    a. Cross-classify the 8 pairs in terms of diet (1 or 0) for the case against diet (1 or 0) for the control. Call this Table A. Display the $2 \times 2 \times 8$ table with eight partial tables relating diet (1 or 0) to response (case or control) for the 8 pairs. Call this Table B.

    b. Calculate the McNemar z^2 for Table A and the CMH statistic for Table B. Compare.

    c. Show that the Mantel–Haenszel estimate of a common odds ratio for Table B is identical to n_{12}/n_{21} for Table A.

    d. For Table B with pairs deleted in which the case and the control had the same diet, show that the CMH statistic and the Mantel–Haenszel odds ratio estimate do not change.

    e. This sample size is small for large-sample tests. Use the binomial distribution with Table A to find the exact P-value for testing marginal homogeneity against the alternative hypothesis of a higher incidence of colon cancer for the high-red-meat diet.

10.5 Each week *Variety* magazine summarizes reviews of new movies by critics in several cities. Each review is categorized as pro, con, or mixed, according to whether the overall evaluation is positive, negative, or a mixture of the two. Table 10.16 summarizes the ratings from

TABLE 10.16 Data for Problem 10.5

Siskel	Ebert		
	Con	Mixed	Pro
Con	24	8	13
Mixed	8	13	11
Pro	10	9	64

Source: A. Agresti and L. Winner, *CHANCE* **10**: 10–14 (1997), reprinted with permission, copyright 1997 by the American Statistical Association.

April 1995 through September 1996 for Chicago film critics Gene Siskel and Roger Ebert.

a. Fit the symmetry model, quasi-independence model, and quasi-symmetry model. Interpret.

b. Test marginal homogeneity using models, and interpret.

c. Analyze these data using agreement models and/or measures of agreement.

10.6 Refer to Table 10.5. Fit the ordinal quasi-symmetry model using $u_1 = 1$ and $u_4 = 4$ and picking u_2 and u_3 that are unequally spaced but represent sensible choices. Compare results and interpretations to those in Sections 10.3.2 and 10.4.7.

10.7 Refer to all four items in Table 8.19.

a. Fit the complete symmetry and quasi-symmetry models. Test marginal homogeneity. Interpret.

b. Fit the ordinal quasi-symmetry model. Test marginal homogeneity. Interpret the effects.

10.8 Table 10.17 shows subjects' purchase choice of instant decaffeinated coffee at two times.

a. Fit the symmetry model and use residuals to analyze changes.

b. Test marginal homogeneity. Show that the small P-value reflects a decrease in the proportion choosing High Point and an increase in the proportion choosing Sanka, with no evidence of change for the other coffees.

c. Show that quasi-independence has $G^2 = 13.8$ (df = 11). Interpret, and suggest other analyses that might be useful.

TABLE 10.17   Data for Problem 10.8

First Purchase	Second Purchase				
	High Point	Taster's Choice	Sanka	Nescafe	Brim
High Point	93	17	44	7	10
Taster's Choice	9	46	11	0	9
Sanka	17	11	155	9	12
Nescafe	6	4	9	15	2
Brim	10	4	12	2	27

Source: Based on data from R. Grover and V. Srinivasan, *J. Market. Res.* **24**: 139–153 (1987). Reprinted with permission from the American Marketing Association.

TABLE 10.18 Data for Problem 10.9

Father's Status	Son's Status					Total
	1	2	3	4	5	
1	50	45	8	18	8	129
2	28	174	84	154	55	495
3	11	78	110	223	96	518
4	14	150	185	714	447	1510
5	3	42	72	320	411	848
Total	106	489	459	1429	1017	3500

Source: Reprinted with permission from D. V. Glass (ed), *Social Mobility in Britain*, Glencoe, IL: Free Press (1954).

10.9 Table 10.18 relates father's and son's occupational status for a British sample. Analyze these data, using models of (**a**) symmetry, (**b**) quasi-symmetry, (**c**) ordinal quasi-symmetry, (**d**) conditional symmetry, (**e**) marginal homogeneity, (**f**) quasi-independence, and (**g**) quasi-uniform association. Interpret using their fit and lack of fit.

10.10 For Table 10.18, use kappa to describe agreement. Interpret.

10.11 Table 10.19 displays multiple sclerosis diagnoses for two neurologists who classified patients in two sites, Winnipeg and New Orleans. The diagnostic classes are (1) certain; (2) probable; (3) possible; and (4) doubtful, unlikely, or definitely not. For the New Orleans patients, study the agreement using (**a**) the independence model and residuals, (**b**) more complex models, and (**c**) kappa. Interpret each.

TABLE 10.19 Data for Problem 10.11

New Orleans Neurologist	Winnipeg Neurologist							
	Winnipeg Patients				New Orleans Patients			
	1	2	3	4	1	2	3	4
1	38	5	0	1	5	3	0	0
2	33	11	3	0	3	11	4	0
3	10	14	5	6	2	13	3	4
4	3	7	3	10	1	2	4	14

Source: J. R. Landis and G. G. Koch, *Biometrics* **33**: 159–174 (1977). Reprinted with permission from the Biometric Society.

10.12 For Problem 10.11, construct a model that describes agreement between neurologists for the two sites simultaneously.

10.13 Calculate kappa for a 4×4 table having $n_{ii} = 5$ all i, $n_{i, i+1} = 15$, $i = 1, 2, 3$, $n_{41} = 15$, and $n_{ij} = 0$ otherwise. Explain why strong association does not imply strong agreement.

10.14 Refer to Table 10.8. Based on the reported standardized residuals, explain why the linear-by-linear association model (9.6) might fit well. Fit it and describe the association.

10.15 In 1990, a sample of psychology graduate students at the University of Florida made blind, pairwise preference tests of three cola drinks. For 49 comparisons of Coke and Pepsi, Coke was preferred 29 times. For 47 comparisons of Classic Coke and Pepsi, Classic Coke was preferred 19 times. For 50 comparisons of Coke and Classic Coke, Coke was preferred 31 times. Comparisons resulting in ties are not reported.

a. Fit the Bradley–Terry model, analyze the quality of fit, and rank the drinks. Is there sufficient evidence to conclude a preference for one drink?

b. Estimate the probability that Coke is preferred to Pepsi, using the model, and compare to the sample proportion.

10.16 Table 10.20 refers to journal citations among four statistics journals during 1987–1989. The more often articles in a particular journal are cited, the more prestige that journal accrues. For citations involving pair A and B, view it as a victory for A if it is cited by B and a defeat for A if it cites B. Fit the Bradley–Terry model. Interpret the fit, and give a prestige ranking of the journals. For citations involving *Commun. Stat.* and *JRSS-B*, estimate the probability that the *Commun. Stat.* article cites the *JRSS-B* article.

TABLE 10.20 Data for Problem 10.16

Citing Journal	Cited Journal			
	Biometrika	*Commun. Stat.*	*JASA*	*JRSS-B*
Biometrika	714	33	320	284
Commun. Stat.	730	425	813	276
JASA	498	68	1072	325
JRSS-B	221	17	142	188

Source: Stigler (1994). Reprinted with permission from the Institute of Mathematical Statistics.

TABLE 10.21 Data for Problem 10.17

Winner	Loser				
	Seles	Graf	Sabatini	Navratilova	Sanchez
Seles	—	2	1	3	2
Graf	3	—	6	3	7
Sabatini	0	3	—	1	3
Navratilova	3	0	2	—	3
Sanchez	0	1	2	1	—

10.17 Table 10.21 refers to matches for several women tennis players during 1989 and 1990.

   a. Fit the Bradley–Terry model. Interpret, and rank the players.

   b. Estimate the probability of Seles beating Graf. Compare the model estimate to the sample proportion. Construct a 90% confidence interval for the probability.

   c. Which pairs of players are significantly different according to a 80% simultaneous Bonferroni comparison?

10.18 Refer to Problem 3.3 on basketball free-throw shooting. Analyze these data.

10.19 Refer to Table 2.12 and Problem 2.19. Using models, describe the relationship between husband's and wife's sexual fun.

10.20 Refer to Table 8.19. The two-way table relating responses for the environment (as rows) and cities (as columns) has cell counts, by row, $(108, 179, 157 \; / \; 21, 55, 52 \; / \; 5, 6, 24)$. Analyze these data.

Theory and Methods

10.21 Explain the following analogy: McNemar's test is to binary data as the paired difference t test is to normally distributed data.

10.22 For a 2×2 table, derive $\text{cov}(p_{+1}, p_{1+})$, and show that $\text{var}[\sqrt{n}\,(p_{+1} - p_{1+})]$ equals (10.1).

10.23 Refer to the subject-specific model (10.8) for binary matched pairs.

   a. Show that $\exp(\beta)$ is a conditional odds ratio between observation and outcome. Explain the distinction between it and the odds ratio $\exp(\beta)$ for model (10.6).

b. Using the conditional distribution (10.9), show that $\hat{\beta} = \log(n_{21}/n_{12})$.

c. For a random sample of n pairs, explain why

$$E(n_{21}/n) = \frac{1}{n} \sum_{i=1}^{n} \frac{1}{1 + \exp(\alpha_i)} \frac{\exp(\alpha_i + \beta)}{1 + \exp(\alpha_i + \beta)}.$$

Similarily, state $E(n_{12}/n)$. Using their ratio for fixed n and as $n \to \infty$, explain why $n_{21}/n_{12} \xrightarrow{p} \exp(\beta)$. (*Hint:* Apply the law of large numbers due to A. A. Markov for independent but not identically distributed random variables, or use Chebyshev's inequality.)

d. Show that the Mantel–Haenszel estimator (6.7) of a common odds ratio in the $2 \times 2 \times n$ form of the data simplifies to $\exp(\hat{\beta}) = n_{21}/n_{12}$.

e. Use the delta method to show (10.10) for the SE of $\hat{\beta}$.

f. For a table of the form shown in Table 10.2, show that the CMH statistic (6.6) is algebraically identical to the McNemar statistic $(n_{21} - n_{12})^2/(n_{21} + n_{12})$ for tables of Table 10.1 type.

10.24 Refer to Problem 10.23. Unlike the conditional ML estimator of β, the unconditional ML estimator is inconsistent (Andersen 1980, pp. 244–245; first shown by him in 1973). Show this as follows:

a. Assuming independence of responses for different subjects and different observations by the same subject, find the log likelihood. Show that the likelihood equations are $y_{+t} = \Sigma_i P(Y_{it} = 1)$ and $y_{i+} = \Sigma_t P(Y_{it} = 1)$.

b. Substituting $\exp(\alpha_i)/[1 + \exp(\alpha_i)] + \exp(\alpha_i + \beta)/[1 + \exp(\alpha_i + \beta)]$ in the second likelihood equation, show that $\hat{\alpha}_i = -\infty$ for the n_{22} subjects with $y_{i+} = 0$, $\hat{\alpha}_i = \infty$ for the n_{11} subjects with $y_{i+} = 2$, and $\hat{\alpha}_i = -\hat{\beta}/2$ for the $n_{21} + n_{12}$ subjects with $y_{i+} = 1$.

c. By breaking $\Sigma_i P(Y_{it} = 1)$ into components for the sets of subjects having $y_{i+} = 0$, $y_{i+} = 2$, and $y_{i+} = 1$, show that the first likelihood equation is, for $t = 1$, $y_{+1} = n_{22}(0) + n_{11}(1) + (n_{21} + n_{12})\exp(-\hat{\beta}/2)/[1 + \exp(-\hat{\beta}/2)]$. Explain why $y_{+1} = n_{11} + n_{12}$, and solve the first likelihood equation to show that $\hat{\beta} = 2\log(n_{21}/n_{12})$. Hence, as a result of Problem 10.23, $\hat{\beta} \xrightarrow{p} 2\beta$.

10.25 Consider marginal model (10.6) when Y_1 and Y_2 are independent and conditional model (10.8) when $\{\alpha_i\}$ are identical. Explain why they are equivalent.

10.26 Let $\hat{\beta}_M = \log(p_{+1}p_{2+}/p_{+2}p_{1+})$ refer to marginal model (10.6) and $\hat{\beta}_C = \log(n_{21}/n_{12})$ to conditional model (10.8). Using the delta method, show that the asymptotic variance of $\sqrt{n}\,(\hat{\beta}_M - \beta_M)$ is

$$(\pi_{1+}\pi_{2+})^{-1} + (\pi_{+1}\pi_{+2})^{-1} - 2(\pi_{11}\pi_{22} - \pi_{12}\pi_{21})/(\pi_{1+}\pi_{2+}\pi_{+1}\pi_{+2}).$$

Under the independence condition of the previous problem, $\beta_M = \beta_C$. In that case, show that the asymptotic variances satisfy

$$\mathrm{var}\left[\sqrt{n}\,(\hat{\beta}_M)\right] = (\pi_{1+}\pi_{2+})^{-1} + (\pi_{+1}\pi_{+2})^{-1}$$

$$\leq (\pi_{1+}\pi_{+2})^{-1} + (\pi_{+1}\pi_{2+})^{-1}$$

$$= \pi_{12}^{-1} + \pi_{21}^{-1} = \mathrm{var}\left[\sqrt{n}\,(\hat{\beta}_C)\right]$$

10.27 Refer to model (10.12) for a matched-pairs study. For the conditional ML approach, show that the conditional distribution satisfies (10.13) and does not depend on β when $S_i = 0$ or 2. Show what happens to β_j in the conditional distribution for a predictor for which $x_{ji1} = x_{ji2}$ all i.

10.28 Consider model (10.12) for a study with matched sets of T observations rather than matched pairs. Explain how (10.13) generalizes and construct the form of the conditional likelihood.

10.29 Give an example illustrating that when $I > 2$, marginal homogeneity does not imply symmetry.

10.30 Derive the likelihood equations and residual df for (a) symmetry, (b) quasi-symmetry, (c) quasi-independence, and (d) ordinal quasi-symmetry.

10.31 For the quasi-symmetry model (10.19), let $\lambda_a = \lambda_a^X - \lambda_a^Y$. Show that one can express it equivalently as $\log \mu_{ab} = \lambda + \lambda_a + \lambda_{ab}^*$, with $\lambda_{ab}^* = \lambda_{ba}^*$. Hence, one needs only one set of main-effect parameters.

10.32 Show that quasi-symmetry is equivalent (Caussinus 1966) to

$$(\pi_{ab}\pi_{bc}\pi_{ca})/(\pi_{ba}\pi_{cb}\pi_{ac}) = 1 \quad \text{all } a, b, \text{ and } c.$$

10.33 Derive the covariance matrix (10.16) for the difference vector **d**.

10.34 Construct the loglinear model satisfying both marginal homogeneity and statistical independence. Show that $\hat{\pi}_{ab} = (p_{+a} + p_{a+})(p_{+b} + p_{b+})/4$ and residual df $= I(I - 1)$.

10.35 Consider the conditional symmetry (CS) model (10.28).
   a. Show that it has the loglinear representation

$$\log \mu_{ab} = \lambda_{\min(a, b), \max(a, b)} + \tau I(a < b),$$

   where $I(\cdot)$ is an indicator (see also Bishop et al. 1975, pp. 285–286).
   b. Show that the likelihood equations are

$$\hat{\mu}_{ab} + \hat{\mu}_{ba} = n_{ab} + n_{ba} \quad \text{for all } a \le b, \quad \sum\sum_{a<b} \hat{\mu}_{ab} = \sum\sum_{a<b} n_{ab}.$$

   c. Show that $\hat{\tau} = \log[(\sum\sum_{a<b} n_{ab})/(\sum\sum_{a>b} n_{ab})]$, $\hat{\mu}_{aa} = n_{aa}$, $a = 1, \ldots, I$, $\hat{\mu}_{ab} = \exp[\hat{\tau} I(a < b)](n_{ab} + n_{ba})/[\exp(\hat{\tau}) + 1]$ for $a \ne b$.
   d. Show that the estimated asymptotic variance of $\hat{\tau}$ is

$$\left(\sum\sum_{a<b} n_{ab}\right)^{-1} + \left(\sum\sum_{a>b} n_{ab}\right)^{-1}.$$

   e. Show that residual df $= (I + 1)(I - 2)/2$.
   f. Show that conditional symmetry + marginal homogeneity = symmetry. Explain why $G^2(\text{S}|\text{CS})$ tests marginal homogeneity (df = 1). When the model holds $G^2(\text{S}|\text{CS})$ is more powerful asymptotically than $G^2(\text{S}|\text{QS})$. Why?

10.36 Identify loglinear models that correspond to the logit models, for $a < b$, $\log(\pi_{ab}/\pi_{ba}) =$ **(a)** 0, **(b)** τ, **(c)** $\alpha_a - \alpha_b$, and **(d)** $\beta(b - a)$.

10.37 A nonmodel-based ordinal measure of marginal heterogeneity is

$$\hat{\Delta} = \sum\sum_{a<b} p_{a+}p_{+b} - \sum\sum_{a>b} p_{a+}p_{+b}.$$

   Show that $\hat{\Delta}$ estimates $\Delta = P(Y_1 > Y_2) - P(Y_2 > Y_1)$, where Y_1 has distribution $\{\pi_{a+}\}$ and Y_2 is independent from $\{\pi_{+b}\}$. Show that marginal homogeneity implies that $\Delta = 0$. Show that the estimated asymptotic variance of $\hat{\Delta}$ is

$$\left[\sum_a \sum_b \hat{\phi}_{ab}^2 P_{ab} - \left(\sum_a \sum_b \hat{\phi}_{ab} P_{ab}\right)^2\right]\Big/n,$$

where $\hat{\phi}_{ab} = \hat{F}_{b1} + \hat{F}_{b-1,1} - \hat{F}_{a2} - \hat{F}_{a-1,2}$ with $\hat{F}_{a1} = (p_{1+} + \cdots + p_{a+})$ and $\hat{F}_{a2} = (p_{+1} + \cdots + p_{+a})$ (Agresti 1984, pp. 208–209).

10.38 For ordered scores $\{u_a\}$, let $\bar{y}_1 = \sum_a u_a p_{a+}$ and $\bar{y}_2 = \sum_a u_a p_{+a}$. Show that marginal homogeneity implies that $E(\bar{Y}_1) = E(\bar{Y}_2)$ and

$$\left[\sum_a \sum_b (u_a - u_b)^2 P_{ab} - (\bar{y}_1 - \bar{y}_2)^2\right]\Big/n.$$

estimates $\text{var}(\bar{Y}_1 - \bar{Y}_2)$. Construct a test of marginal homogeneity (Bhapkar 1968).

10.39 Consider the multiplicative model for a square table,

$$\pi_{ab} = \begin{cases} \alpha_a \alpha_b (1 - \beta), & a \neq b \\ \alpha_a^2 + \beta \alpha_a (1 - \alpha_a), & a = b. \end{cases}$$

a. Show that the model satisfies (**i**) symmetry, (**ii**) marginal homogeneity, (**iii**) quasi-symmetry, (**iv**) quasi-independence.
b. Show that $\alpha_a = \pi_{a+} = \pi_{+a}$, $a = 1, \ldots, I$.
c. Show that $\beta = $ Cohen's kappa, and interpret $\kappa = 0$ and $\kappa = 1$ for this model.

10.40 A 2×2 table has a true odds ratio of 10. Find the cell probabilities for which (**a**) $\pi_{1+} = \pi_{+1} = 0.5$, (**b**) $\pi_{1+} = \pi_{+1} = 0.3$, and (**c**) $\pi_{1+} = \pi_{+1} = 0.1$. Find the value of kappa for each. (This shows that for a given association, kappa depends strongly on the marginal probabilities; see also Sprott 2000, p. 59.)

10.41 A model for agreement on an ordinal response partitions beyond-chance agreement into that due to a baseline association and a main-diagonal increment (A. Agresti, *Biometrics* **44**: 539–548, 1988). For ordered scores $\{u_a\}$, the model is

$$\log \mu_{ab} = \lambda + \lambda_a^A + \lambda_b^B + \beta u_a u_b + \delta I(a = b). \quad (10.35)$$

a. Show that this is a special case of quasi-symmetry and of quasi-association (10.29).

b. For agreement odds (10.30), show that $\log \tau_{ab} = (u_b - u_a)^2 \beta + 2\delta$. For unit-spaced scores, show the local odds ratios have $\log \theta_{ab} = \beta$ when none of the four cells falls on the main diagonal.

c. Find the likelihood equations and show that $\{\hat{\mu}_{ab}\}$ and $\{n_{ab}\}$ share the same marginal distributions, correlation, and prevalence of exact agreement.

d. For Table 10.8 using $\{u_a = a\}$, show that (10.35) has $G^2 = 4.8$ (df $= 7$), with $\hat{\delta} = 0.842$ (SE $= 0.427$) and $\hat{\beta} = 1.316$ (SE $= 0.420$). Interpret using $\hat{\tau}_{a,\,a+1}$ and $\hat{\theta}_{ab}$ for $|a - b| > 1$.

10.42 Refer to the Bradley–Terry model.

a. Show that $\log(\Pi_{ac}/\Pi_{ca}) = \log(\Pi_{ab}/\Pi_{ba}) + \log(\Pi_{bc}/\Pi_{cb})$.

b. With this model, is it possible that a could be preferred to b (i.e., $\Pi_{ab} > \Pi_{ba}$) and b could be preferred to c, yet c could be preferred to a? Explain.

c. Explain why $\{\beta_a\}$ are not identifiable without a constraint such as $\beta_I = 0$. (*Hint:* Show the model holds when $\{\beta_a^* = \beta_a - c\}$ for any c.)

10.43 Refer to model (10.32).

a. Construct a more general model having home-team parameters $\{\beta_{Hi}\}$ and away-team parameters $\{\beta_{Ai}\}$, such that the probability team i beats team j when i is the home team is $\exp(\beta_{Hi})/[\exp(\beta_{Hi}) + \exp(\beta_{Aj})]$, where $\beta_{AI} = 0$ but β_{Hi} is unrestricted.

b. Interpret the case $\{\beta_{Hi} = \beta_{Ai} + c\}$, when **(i)** $c = 0$, and **(ii)** $c > 0$.

c. Fit the model to Table 10.12. Compare the fit to model (10.32). Compare $\{\hat{\beta}_{Hi}\}$ and $\{\hat{\beta}_{Ai}\}$ to describe how teams play at home and away.

10.44 Find the log likelihood for the Bradley–Terry model. From the kernel, show that (given $\{N_{ab}\}$) the minimal sufficient statistics are $\{n_{a+}\}$. Thus, explain how "victory totals" determine the estimated ranking.

10.45 Explain how to fit the complete symmetry model in T dimensions.

10.46 Prove that if kth-order marginal symmetry holds, jth-order marginal symmetry holds for any $j < k$.

10.47 Suppose that quasi-symmetry holds for an I^T table. When the table is collapsed over a variable, show that the model holds for the I^{T-1} table with the same main effects.

CHAPTER 11

Analyzing Repeated Categorical Response Data

Many studies observe the response variable for each subject repeatedly, at several times or under various conditions. Repeated categorical response data occur commonly in health-related applications, especially in longitudinal studies. For example, a physician might evaluate patients at weekly intervals regarding whether a new drug treatment is successful. In some cases explanatory variables may also vary over time. But the repeated responses need not refer to different times. A dental study might measure whether there is decay for each tooth in a subject's mouth.

Often, the responses refer to matched sets, or *clusters*, of subjects. An example is a (survival, nonsurvival) response for each fetus in a litter, for a sample of pregnant mice exposed to various dosages of a toxin. A multistage sample to study factors affecting obesity in children may regard children from the same family as a cluster. Observations within a cluster tend to be more alike than observations from different clusters. Ordinary analyses that ignore this may be badly inappropriate.

In this chapter we generalize methods of Chapter 10, which referred to matched pairs. In Section 11.1 we compare marginal distributions in *T*-way tables. The remaining sections extend models to include explanatory variables. For instance, many studies compare the repeated measurements for different groups or treatments. In Section 11.2 we use ML methods for fitting marginal models. In Section 11.3 we use *generalized estimating equations* (GEE), a multivariate version of quasi-likelihood that is computationally simpler than ML. Section 11.4 covers technical details about the GEE approach. In the final section we introduce a *transitional* approach that models observations in terms of previous outcomes.

11.1 COMPARING MARGINAL DISTRIBUTIONS: MULTIPLE RESPONSES

Usually, the multivariate dependence among repeated responses is of less interest than their marginal distributions. For instance, in treating a chronic condition (such as a phobia) with some treatment, the primary goal might be to study whether the probability of success increases over the T weeks of a treatment period. The T success probabilities refer to the T first-order marginal distributions. In Sections 10.2.1 and 10.3 we compared marginal distributions for matched pairs ($T = 2$) using models that apply directly to the marginal distributions. In this section we extend this approach to $T > 2$.

11.1.1 Binary Marginal Models and Marginal Homogeneity

Denote T binary responses by $(Y_1, Y_2, \ldots, Y_T)$. The marginal logit model (10.6) for matched pairs extends to

$$\text{logit}[P(Y_t = 1)] = \alpha + \beta_t, \qquad t = 1, \ldots, T, \tag{11.1}$$

with a constraint such as $\beta_T = 0$ or $\alpha = 0$. For a possible sequence of outcomes $\mathbf{i} = (i_1, i_2, \ldots, i_T)$ where each $i_t = 0$ or 1, let

$$\pi_{\mathbf{i}} = P(Y_1 = i_1, Y_2 = i_2, \ldots, Y_T = i_T).$$

Let $\boldsymbol{\pi}$ denote the vector of these probabilities for the possible $\mathbf{i}$. They refer to a 2^T table that cross-classifies the T responses and describes the joint distribution of $(Y_1, \ldots, Y_T)$. The sample cell proportions are the ML estimates of $\boldsymbol{\pi}$, and the sample proportion with $y_t = 1$ is the ML estimate of $P(Y_t = 1)$.

Model (11.1) is saturated, describing T marginal probabilities by T parameters. Marginal homogeneity, for which $P(Y_1 = 1) = \cdots = P(Y_T = 1)$, is the special case $\beta_1 = \cdots = \beta_T$. Even though this case has only one parameter, ML fitting is not simple. The multinomial likelihood refers to the 2^T joint cell probabilities $\boldsymbol{\pi}$ rather than the T marginal probabilities $\{P(Y_t = 1)\}$. Fitting methods are described in Section 11.2.5.

Let $n_{\mathbf{i}}$ denote the sample cell count in cell $\mathbf{i}$. The kernel of the log likelihood $L(\boldsymbol{\pi})$ is $\sum_{\mathbf{i}} n_{\mathbf{i}} \log \pi_{\mathbf{i}}$. Let $L(\mathbf{p})$ denote the log likelihood evaluated at the sample proportions $\{p_{\mathbf{i}} = n_{\mathbf{i}}/n\}$, the ML fit of model (11.1). Let $L(\hat{\boldsymbol{\pi}}^{MH})$ denote the maximized log likelihood assuming marginal homogeneity. The likelihood-ratio test of marginal homogeneity (Lipsitz et al. 1990; Madansky 1963) uses

$$-2[L(\hat{\boldsymbol{\pi}}^{MH}) - L(\mathbf{p})] = 2\sum_{\mathbf{i}} n_{\mathbf{i}} \log(p_{\mathbf{i}}/\hat{\pi}_{\mathbf{i}}^{MH}). \tag{11.2}$$

TABLE 11.1   Responses to Three Drugs in a Crossover Study

	Drug A Favorable		Drug A Unfavorable	
	B Favorable	B Unfavorable	B Favorable	B Unfavorable
C Favorable	6	2	2	6
C Unfavorable	16	4	4	6

Source: Reprinted with permission from the Biometric Society (Grizzle et al. 1969).

The asymptotic null chi-squared distribution has df $= T - 1$, since the general model (11.1) has $T - 1$ more parameters than marginal homogeneity.

11.1.2   Crossover Drug Comparison Example

Table 11.1 comes from a crossover study in which each subject used each of three drugs for treatment of a chronic condition at three times. The response measured the reaction as favorable or unfavorable. The 2^3 table gives the (favorable, unfavorable) classification for reaction to drug A in the first dimension, drug B in the second, and drug C in the third. We assume that the drugs have no carryover effects and that the severity of the condition remained stable for each subject throughout the experiment. These assumptions are reasonable for many chronic conditions, such as migraine headache.

The sample proportion favorable was $(0.61, 0.61, 0.35)$ for drugs (A, B, C). The likelihood-ratio statistic for testing marginal homogeneity is 5.95 (df $= 2$), for a *P*-value of 0.05. For simultaneous confidence intervals comparing pairs of treatments with overall error probability no greater than 0.05, the Bonferroni method uses confidence coefficient $(1 - 0.05/3) = 0.9833$ for each. For instance, from formula (10.1), the estimate $0.261 = 0.609 - 0.348$ of the difference between drugs A and C has an estimated standard error of 0.108. The confidence interval for the true difference is $0.261 \pm 2.39(0.108)$, or $(0.002, 0.520)$. The same interval holds for comparison of drugs B and C. There is some evidence that the proportion of favorable responses is lower for drug C.

The sample size is not large, however, so we view these results with caution. For each pair of drugs, a 2×2 table relates the two responses. An exact binomial test (Section 10.4.1) uses its off-diagonal counts. These yield *P*-values of 1.0 for comparing drugs A and B and 0.036 for comparing A with C and for comparing B with C.

11.1.3   Modeling Margins of a Multicategory Response

The binary marginal model (11.1) extends to multinomial responses. With baseline-category logits for I outcome categories, the saturated model is

$$\log[P(Y_t = j)/P(Y_t = I)] = \beta_{tj}, \qquad t = 1, \ldots, T, \quad j = 1, \ldots, I - 1.$$

$$(11.3)$$

Marginal homogeneity, whereby $P(Y_1 = j) = \cdots = P(Y_T = j)$ for $j = 1, \ldots, I - 1$, is the special case in which

$$\beta_{1j} = \beta_{2j} = \cdots = \beta_{Tj}, \qquad j = 1, \ldots, I - 1.$$

The likelihood-ratio test of marginal homogeneity comparing the two models has form (11.2) and df $= (T - 1)(I - 1)$.

For an ordinal response, an unsaturated model that is more complex than marginal homogeneity focuses on shifts up and down in the T margins. One such model is

$$\text{logit}[P(Y_t \leq j)] = \alpha_j + \beta_t, \qquad t = 1, \ldots, T, \quad j = 1, \ldots, I - 1, \quad (11.4)$$

with constraint such as $\beta_T = 0$. Marginal homogeneity is the special case $\beta_1 = \cdots = \beta_T$. Its test has df $= T - 1$. The $\{\alpha_j\}$ satisfy $\alpha_1 < \ldots < \alpha_{I-1}$ because of the ordering of the cumulative probabilities. These models can be fitted using ML methodology presented in Section 11.2.5.

11.1.4 Wald and Generalized CMH Score Tests of Marginal Homogeneity

In this chapter we focus on modeling the marginal distributions rather than merely testing marginal homogeneity. However, a variety of tests are available besides the likelihood ratio, so we briefly summarize a couple of them.

Let $p_j(t)$ denote the sample proportion in category j for response Y_t, let

$$\bar{p}_j = \sum_t p_j(t)/T, \qquad d_j(t) = p_j(t) - \bar{p}_j,$$

and let $\mathbf{d}$ denote the vector of $\{d_j(t), t = 1, \ldots, T - 1, j = 1, \ldots, I - 1\}$. Let $\hat{\mathbf{V}}$ denote the estimated covariance matrix of $\sqrt{n}\,\mathbf{d}$. Bhapkar (1973) proposed the Wald statistic

$$W = n\mathbf{d}'\hat{\mathbf{V}}^{-1}\mathbf{d}. \qquad (11.5)$$

for the general alternative. This generalizes (10.16) and has a large-sample chi-squared distribution with df $= (I - 1)(T - 1)$.

Other statistics are special cases of the generalized Cochran–Mantel–Haenszel (CMH) statistic (Section 7.5.3). Recall that for the binary case $(I = 2)$ with matched pairs $(T = 2)$, the CMH statistic applies to a three-way table (see, e.g., Table 10.2) in which each stratum shows the two outcomes for a given subject. A generalization of Table 10.2 provides n strata of $T \times I$ tables. The kth stratum gives the T outcomes for subject k. Row t in a stratum has a 1 in the column that is the outcome for observation t, and 0 in all other columns (or 0 in every column if that observation is missing). Probability distributions for the subject-stratified setup naturally relate to subject-specific models such as logit model (10.8), rather than to marginal models. However, conditional independence in this three-way table (given subject) corresponds to an exchangeability among variables in the I^T table that implies marginal homogeneity. A generalized CMH test of conditional independence in the $T \times I \times n$ table also tests marginal homogeneity using a sampling distribution generated under the stronger exchangeability condition (Darroch 1981). For an ordinal response with fixed scores, the generalized CMH statistic for detecting variability among T means is appropriate.

When $I = 2$ and $T = 2$, this CMH approach is equivalent to McNemar's statistic. When $I = 2$ but $T > 2$, the generalized CMH statistic treating the T responses as unordered is identical to a statistic Cochran (1950) proposed. His statistic, called *Cochran's Q*, has df $= T - 1$ (Problem 11.22).

11.2 MARGINAL MODELING: MAXIMUM LIKELIHOOD APPROACH

Analyses above compared marginal distributions, but without accounting for explanatory variables. We now include such predictors. In this section we use ML, but we defer model fitting details to the end of the section.

11.2.1 Longitudinal Mental Depression Example

We use Table 11.2 to illustrate a variety of analyses in this and the next chapter. It refers to a longitudinal study comparing a new drug with a standard drug for treatment of subjects suffering mental depression (Koch et al. 1977). Subjects were classified into two initial diagnosis groups according to whether severity of depression was mild or severe. In each group, subjects were randomly assigned to one of the two drugs. Following 1 week, 2 weeks, and 4 weeks of treatment, each subject's suffering from mental depression was classified as normal or abnormal.

TABLE 11.2 Cross-Classification of Responses on Depression at Three Times by Diagnosis and Treatment

Diagnosis	Treatment	Response at Three Times[a]							
		NNN	NNA	NAN	NAA	ANN	ANA	AAN	AAA
Mild	Standard	16	13	9	3	14	4	15	6
	New drug	31	0	6	0	22	2	9	0
Severe	Standard	2	2	8	9	9	15	27	28
	New drug	7	2	5	2	31	5	32	6

[a]N, normal; A, abnormal.

Source: Reprinted with permission from the Biometric Society (Koch et al. 1977).

Table 11.2 shows four groups, the combinations of categories of the two explanatory variables: treatment type and severity of initial diagnosis. Since the study observed the binary response (depression assessment) at $T = 3$ occasions, Table 11.2 shows a 2^3 table for each group. The three depression assessments form a multivariate response variable with three components, with $Y_t = 1$ for normal and 0 for abnormal. The 12 marginal distributions result from three repeated observations for each of the four groups.

Let s denote the severity of the initial diagnosis, with $s = 1$ for severe and $s = 0$ for mild. Let d denote the drug, with $d = 1$ for new and $d = 0$ for standard. Let t denote the time of measurement. Koch et al. (1977) noted that if the time metric reflects cumulative drug dosage, a logit scale often has a linear effect for the logarithm of time. They used scores (0, 1, 2), the logs to base 2 of the week numbers (1, 2, and 4), for time.

Table 11.3 shows sample proportions of normal responses (i.e., $y_t = 1$) for the 12 marginal distributions. For instance, from Table 11.2, the sample proportion of normal responses after week 1 for subjects with mild initial diagnosis using the standard drug was $(16 + 13 + 9 + 3)/(16 + 13 + 9 + 3 + 14 + 4 + 15 + 6) = 0.51$. The sample proportion of normal responses (1) increased over time for each group; (2) increased at a faster rate for the new drug than the standard, for each fixed initial diagnosis; and (3) was higher for the mild than the severe initial diagnosis, for each treatment at each occasion. In such a study the company that developed the new drug would hope to show that patients have a significantly higher rate of improvement with it.

The marginal logit model

$$\text{logit}[P(Y_t = 1)] = \alpha + \beta_1 s + \beta_2 d + \beta_3 t$$

has the main effects of the explanatory variables (severity of initial diagnosis and drug) and of the variable (time) that specifies the different components of the multivariate response. Its linear time effect β_3 is the same for each group.

The natural sampling assumption is multinomial for the eight cells in the 2^3 cross-classification of the three responses, independently for the four

TABLE 11.3 Sample Marginal Proportions of Normal Response for Depression Data of Table 11.2

		Sample Proportion		
Diagnosis	Treatment	Week 1	Week 2	Week 4
Mild	Standard	0.51	0.59	0.68
	New drug	0.53	0.79	0.97
Severe	Standard	0.21	0.28	0.46
	New drug	0.18	0.50	0.83

groups. However, the model refers to 12 marginal probabilities (for 2 drug treatments $\times$ 2 initial severity diagnoses $\times$ 3 time points) rather than the $4 \times 2^3 = 32$ cell probabilities in the product multinomial likelihood function. The three marginal binomial variates for each group are dependent. ML estimation requires an iterative routine for maximizing the product multinomial likelihood, subject to the constraint that the marginal probabilities satisfy the model. An algorithm for this is given in Section 11.2.5.

A check of model fit compares the 32 cell counts in Table 11.2 to their ML fitted values. Since the model describes 12 marginal logits using four parameters, residual df = 8. The deviance $G^2 = 34.6$. The poor fit is not surprising. The model assumes a common rate of improvement β_3, but the sample shows a higher rate for the new drug.

A more realistic model permits the time effect to differ by drug,

$$\text{logit}[P(Y_t = 1)] = \alpha + \beta_1 s + \beta_2 d + \beta_3 t + \beta_4 dt.$$

Its time effect estimate is $\hat{\beta}_3 = 0.48$ (SE $= 0.12$) for the standard drug ($d = 0$) and $\hat{\beta}_3 + \hat{\beta}_4 = 1.49$ (SE $= 0.14$) for the new one ($d = 1$). For the new drug, the slope is $\hat{\beta}_4 = 1.01$ (SE $= 0.18$) higher than for the standard, giving strong evidence of faster improvement. This model fits much better, with $G^2 = 4.2$ (df $= 7$). The G^2 decrease of $34.6 - 4.2 = 30.4$ compared to the simpler model is the likelihood-ratio test of H_0: $\beta_4 = 0$, a common time effect for each drug.

The severity of initial diagnosis estimate is $\hat{\beta}_1 = -1.29$ (SE $= 0.14$); for each drug–time combination, the estimated odds of a normal response when the initial diagnosis was severe equal $\exp(-1.29) = 0.27$ times the estimated odds when the initial diagnosis was mild. The estimate $\hat{\beta}_2 = -0.06$ (SE $= 0.22$) indicates an insignificant difference between the drugs after 1 week (for which $t = 0$). At time t, the estimated odds of normal response with the new drug are $\exp(-0.06 + 1.01\, t)$ times the estimated odds for the standard drug, for each initial diagnosis level. In summary, severity of initial diagnosis, drug treatment, and time all have substantial effects on the probability of a normal response.

11.2.2 Modeling a Repeated Multinomial Response

Models for marginal distributions of a repeated binary response generalize to multicategory responses. At observation t, the marginal response distribution has $I - 1$ logits. With nominal responses, baseline-category logit models describe the odds of each outcome relative to a baseline. For ordinal responses, one might use cumulative logit models.

For a particular marginal logit, a model has the form

$$\text{logit}_j(t) = \alpha_j + \boldsymbol{\beta}_j' \mathbf{x}_t, \quad j = 1, \ldots, I - 1, \quad t = 1, \ldots .$$

For an ordinal response, perhaps $\text{logit}_j(t) = \text{logit}[P(Y_t \leq j)]$. Then, β_j may simplify to β, in which case the model takes the proportional odds form with the same effects for each logit. Some parameters in β may refer to the variable subscripted by t (e.g., time) that indexes the repeated measurements. One can then compare marginal distributions at particular settings of **x** or evaluate effects of **x** on the response. In either case, checking for interaction is crucial. For instance, are the effects of **x** the same at each t?

11.2.3 Insomnia Example

Table 11.4 shows results of a randomized, double-blind clinical trial comparing an active hypnotic drug with a placebo in patients who have insomnia problems. The response is the patient's reported time (in minutes) to fall asleep after going to bed. Patients responded before and following a two-week treatment period. The two treatments, active and placebo, form a binary explanatory variable. The subjects receiving the two treatments were independent samples.

Table 11.5 displays sample marginal distributions for the four treatment–occasion combinations. From the initial to follow-up occasion, time to falling asleep seems to shift downward for both treatments. The degree of shift seems greater for the active treatment, indicating possible interaction. The response variable is a discrete version of a continuous variable, so by the derivation in Section 7.2.3 a cumulative link model is natural. The proportional odds model

$$\text{logit}[P(Y_t \leq j)] = \alpha_j + \beta_1 t + \beta_2 x + \beta_3 tx \qquad (11.6)$$

permits interaction between t = occasion (0 = initial, 1 = follow-up) and

TABLE 11.4   Time to Falling Asleep, by Treatment and Occasion

Treatment	Initial	Time to Falling Asleep			
		Follow-up			
		< 20	20–30	30–60	> 60
Active	< 20	7	4	1	0
	20–30	11	5	2	2
	30–60	13	23	3	1
	> 60	9	17	13	8
Placebo	< 20	7	4	2	1
	20–30	14	5	1	0
	30–60	6	9	18	2
	> 60	4	11	14	22

Source: From S. F. Francom, C.Chuang-Stein, and J. R. Landis, *Statist. Med.* **8**: 571–582 (1989). Reprinted with permission from John Wiley & Sons Ltd.

TABLE 11.5   Sample Marginal Distributions of Table 11.4

Treatment	Occasion	Response			
		< 20	20–30	30–60	> 60
Active	Initial	0.101	0.168	0.336	0.395
	Follow-up	0.336	0.412	0.160	0.092
Placebo	Initial	0.117	0.167	0.292	0.425
	Follow-up	0.258	0.242	0.292	0.208

x = treatment (0 = placebo, 1 = active), but assumes the same effects for each response cutpoint.

For ML model fitting, $G^2 = 8.0$ (df = 6) for comparing observed to fitted cell counts in modeling the 12 marginal logits using these six parameters. The ML estimates are $\hat{\beta}_1 = 1.074$ (SE = 0.162), $\hat{\beta}_2 = 0.046$ (SE = 0.236), and $\hat{\beta}_3 = 0.662$ (SE = 0.244). This shows evidence of interaction. At the initial observation, the estimated odds that time to falling asleep for the active treatment is below any fixed level equal exp(0.046) = 1.04 times the estimated odds for the placebo treatment; at the follow-up observation, the effect is exp(0.046 + 0.662) = 2.03. In other words, initially the two groups had similar distributions, but at the follow-up those with the active treatment tended to fall asleep more quickly.

For simpler interpretation, it can be helpful to report sample marginal means and their differences. With response scores {10, 25, 45, 75} for time to fall asleep, the initial means were 50.0 for the active group and 50.3 for the placebo. The difference in means between the initial and follow-up responses was 22.2 for the active group and 13.0 for the placebo. The difference between these differences of means equals 9.2, with SE = 3.0, indicating that the change was significantly greater for the active group.

11.2.4   Comparisons That Control for Initial Response

For data such as Table 11.4, suppose that the marginal distributions for initial response are identical for the treatment groups. This is true, apart from sampling error, with random assignment of subjects to the groups. Suppose also that conditional on the initial response, the follow-up response distribution is identical for the treatment groups. Then, the follow-up marginal distributions are also identical.

If the initial marginal distributions are not identical, however, the difference between follow-up and initial marginal distributions may differ between treatment groups, even though their conditional distributions for follow-up response are identical. In such cases, although marginal models can be useful, they may not tell the entire story. It may be more informative to construct models that compare the follow-up responses while controlling for the initial response.

Let Y_2 denote the follow-up response, for treatment x with initial response y_1. In the model

$$\text{logit}[P(Y_2 \leq j)] = \alpha_j + \beta_1 x + \beta_2 y_1, \tag{11.7}$$

β_1 compares the follow-up distributions for the treatments, controlling for initial observation. This is an analog of an analysis-of-covariance model, with ordinal rather than continuous response. This cumulative logit model refers to a univariate response (Y_2) rather than marginal distributions of a multivariate response (Y_1, Y_2). It is an example of a *transitional model*, discussed in the final section of this chapter.

11.2.5 ML Fitting of Marginal Logit Models*

ML fitting of marginal logit models is awkward. For T observations on an I-category response, at each setting of predictors the likelihood refers to I^T multinomial joint probabilities, but the model applies to T sets of marginal multinomial parameters $\{P(Y_t = k), k = 1, \ldots, I\}$. The marginal multinomial variates are not independent.

Let π denote the complete set of multinomial joint probabilities for all settings of predictors. Marginal logit models have the generalized loglinear model form

$$\mathbf{C}\log(\mathbf{A}\pi) = \mathbf{X}\boldsymbol{\beta} \tag{11.8}$$

introduced in Section 8.5.4. In the binary case, the matrix $\mathbf{A}$ applied to π forms the T marginal probabilities $\{P(Y_t = 1)\}$ and their complements at each setting of predictors. The matrix $\mathbf{C}$ applied to the log marginal probabilities forms the T marginal logits for each setting; each row of $\mathbf{C}$ has 1 in the position multiplied by the log numerator probability for a given marginal logit, -1 in the position multiplied by the log denominator probability, and 0 elsewhere.

For instance, for the model of marginal homogeneity in a 2^T table with no covariates, $\boldsymbol{\beta}$ is a single parameter, denoted by α in (11.1). For $T = 2$, π has four elements, and this model is

$$\begin{bmatrix} 1 & -1 & 0 & 0 \\ 0 & 0 & 1 & -1 \end{bmatrix} \log \begin{bmatrix} 1 & 1 & 0 & 0 \\ 0 & 0 & 1 & 1 \\ 1 & 0 & 1 & 0 \\ 0 & 1 & 0 & 1 \end{bmatrix} \begin{bmatrix} \pi_{11} \\ \pi_{12} \\ \pi_{21} \\ \pi_{22} \end{bmatrix} = \begin{bmatrix} 1 \\ 1 \end{bmatrix} \alpha,$$

which sets both $\text{logit}(\pi_{11} + \pi_{12}) = \text{logit}[P(Y_1 = 1)]$ and $\text{logit}(\pi_{11} + \pi_{21}) = \text{logit}[P(Y_2 = 1)]$ equal to α.

The likelihood function $\ell(\pi)$ for a marginal logit model is the product of the multinomial mass functions from the various predictor settings. One approach for ML fitting views the model as a set of constraints and uses methods for maximizing a function subject to constraints. In model (11.8), let U denote a full column rank matrix such that the space spanned by the columns of U is the orthogonal complement of the space spanned by the columns of **X**. Then, $U'X = 0$, and the model has the equivalent constraint form

$$U'C \log(A\pi) = 0.$$

For instance, for marginal homogeneity in a 2×2 table with (11.8) as expressed above, $U' = (1, -1)$. Then U' applied to $C \log(A\pi)$ sets the difference between the row and column marginal logits equal to 0.

This method of maximizing the likelihood incorporates these model constraints as well as identifiability constraints, which constrain the response probabilities at each predictor setting to sum to 1. We express this collection of model constraints $U'C \log(A\pi) = 0$ and identifiability constraints as $f(\pi) = 0$. The method introduces Lagrange multipliers corresponding to these constraints and solves the Lagrangian likelihood equations using a Newton–Raphson algorithm (Aitchison and Silvey 1958; Haber 1985). Let θ be a vector having elements π and the Lagrange multipliers λ. The Lagrangian likelihood equations have form $h(\theta) = 0$, where

$$h(\theta) = h(\pi, \lambda) = \left(f(\pi), \partial \log[\ell(\pi)]/\partial\pi + [\partial f(\pi)/\partial\pi]'\lambda\right)'$$

is a vector with terms involving the contrasts in marginal logits that the model specifies as constraints as well as log-likelihood derivatives.

The Newton–Raphson method then is

$$\theta^{(t+1)} = \theta^{(t)} - \left[\frac{\partial h(\theta^{(t)})}{\partial\theta}\right]^{-1} h(\theta^{(t)}), \quad t = 1, \dots .$$

This can be computationally intensive because the derivative matrix inverted has dimensions larger than the number of elements in π. A refinement (Lang 1996a; Lang and Agresti 1994) uses an asymptotic approximation to a reparameterized derivative matrix that has a much simpler form, requiring inverting only a diagonal matrix and a symmetric positive definite matrix.

This ML marginal fitting method is available in specialized software (Appendix A mentions an S-Plus function). It makes no assumption about the model that describes the joint distribution π. Thus, when the marginal model holds, the ML estimate of β in (11.8) is consistent regardless of the dependence structure for that distribution. Several alternative fitting approaches have been considered. Lang and Agresti (1994) simultaneously fitted a marginal model and an unsaturated loglinear model for π. The complete model can be specified as a special case of (11.8) and fitted using the constraint approach with Lagrange multipliers just described. In standard cases, the marginal and joint model parameters are orthogonal. If the marginal model holds, the ML estimator of the marginal model parameters is consistent even if the model for the joint distribution is incorrect.

Fitzmaurice and Laird (1993) gave a related ML approach. A one-to-one correspondence holds between π and parameters of the saturated loglinear model. They used a further one-to-one correspondence between the main effect and the higher-order parameters of that loglinear model with the marginal probabilities and those same higher-order loglinear parameters. Models were then specified separately for the marginal probabilities and the higher-order (conditional) loglinear parameters. The likelihood is then maximized in terms of the two sets of model parameters. Again, the two sets of parameters are orthogonal, so the ML estimator of marginal model parameters is consistent when the marginal model holds. This *mixed parameter* approach is also available in specialized software (Kastner et al. 1997; see also Appendix A).

Yet another ML approach uses a one-to-one correspondence between π and parameters that describe the marginal distributions, the bivariate distributions, the trivariate distributions, and so on (e.g., Glonek and McCullagh 1995; Molenberghs and Lesaffre 1994). Multivariate logistic models then apply to the component distributions, although some higher-order effects may be assumed to vanish, for simplicity. Glonek (1996) proposed a hybrid of this and the Fitzmaurice and Laird (1993) approach.

11.3 MARGINAL MODELING: GENERALIZED ESTIMATING EQUATIONS (GEE) APPROACH

At each combination of predictor values, ML fitting assumes a multinomial distribution for the I^T cell probabilities for the T observations on an I-category response. As the number of predictors increases, the number of multinomial probabilities increases dramatically. Currently, all the ML approaches described above are not practical when T is large or there are many predictors, especially when some are continuous. Compared to the continuous-response case using the multivariate normal, marginal modeling of multivariate categorical responses is also hindered by the lack of a simple multivariate distribution for describing correlations among the T responses. For instance, with T means and a common variance and correlation, the multivariate normal has only $T + 2$ parameters, compared to the $I^T - 1$ parameters for the multinomial.

An alternative to ML fitting uses a multivariate generalization of quasi-likelihood (Section 4.7). Rather than assuming a particular distribution for Y, the quasi-likelihood method specifies only the first two moments; it links the mean to a linear predictor and also specifies how the variance depends on the mean. The estimates are solutions of estimating equations that are likelihood equations under the further assumption of a distribution in the exponential family with that mean and variance (Wedderburn 1974).

11.3.1 Generalized Estimating Equation Methodology: Basic Ideas

Repeated measurement provides a multivariate response $(Y_1, Y_2, \ldots, Y_T)$, where T sometimes varies by subject. As in the univariate case, the quasi-likelihood method specifies a model for $\mu = E(Y)$ and specifies a variance function $v(\mu)$ describing how var(Y) depends on μ. Now, though, that model applies to the marginal distribution for each Y_t. The method also requires a working guess for the correlation structure among $\{Y_t\}$. The estimates are solutions of quasi-likelihood equations called *generalized estimating equations*. The method is often referred to as the GEE method. Liang and Zeger (1986) proposed it for marginal modeling with GLMs. Their work built on related material in the econometrics literature (e.g., Gourieroux et al. 1984; Hansen 1982; White 1982). We outline concepts here and give more details in Section 11.4.

The GEE approach utilizes an assumed covariance structure for $(Y_1, Y_2, \ldots, Y_T)$, specifying a variance function and a pairwise correlation pattern, without assuming a particular multivariate distribution. The GEE estimates of model parameters are valid even if one misspecifies the covariance structure. Consistency (i.e., estimates converging in probability to the true parameters) depends on the first moment but not the second. Specifically, suppose that the model is correct in the sense that the chosen link function and linear predictor truly describe how $E(Y_t)$ depend on the predictors, $t = 1, \ldots, T$. Then the GEE model parameter estimators are consistent.

In practice, a chosen model is never exactly correct. This result is useful, however, for suggesting that the correlation structure need not adversely affect the quality of estimates for whatever model one uses. Often, no a priori information is available about this structure, and the correlation is regarded as a nuisance. A simple implementation of the GEE method naively treats $\{Y_t\}$ as pairwise independent. Although parameter estimates are usually fine under this naive assumption, standard errors are not. More appropriate standard errors result from an adjustment the GEE method makes using the empirical dependence the data exhibit. The naive standard errors based on the independence assumption are updated using the information the data provide about the actual dependence structure to yield more appropriate (*robust*) standard errors.

As an alternative to estimates that treat $\{Y_t\}$ as pairwise independent, the GEE method can use a working guess about the correlation structure but again empirically adjust the standard error. The *exchangeable* working correlation structure treats corr(Y_t, Y_s) as identical for all s and t. This is more flexible and realistic than the naive independence assumption. Even more realistic is an unstructured working correlation that permits a separate correlation for each pair. When T is large, however, this approach suffers some efficiency loss because of the many additional parameters.

In theory, choosing the working correlation wisely can pay benefits of improved efficiency of estimation. However, Liang and Zeger (1986) noted that estimators based on independence working correlation can have surprisingly good efficiency when the actual correlation is weak to moderate. One can check the sensitivity to the selection by comparing results for different working correlation assumptions. In our experience, when the correlations are modest, all working correlation structures yield similar GEE estimates and standard errors, as the empirical dependence has a large impact on adjusting the naive standard errors. (If they differed substantially, a more careful study of the correlation structure would be necessary.) Unless one expects dramatic differences among the correlations, we recommend the exchangeable working correlation structure. This recognizes the dependence at the cost of only one extra parameter.

The GEE approach is appealing for categorical data because of its computational simplicity compared to ML. Advantages include not requiring a multivariate distribution and the consistency of estimation even with misspecified correlation structure. However, it has limitations. Since the GEE approach does not completely specify the joint distribution, it does not have a likelihood function. Likelihood-based methods are not available for testing fit, comparing models, and conducting inference about parameters. Instead, inference uses Wald statistics constructed with the asymptotic normality of the estimators together with their estimated covariance matrix. However, unless the sample size is quite large, the empirically based standard errors tend to underestimate the true ones (e.g., Firth 1993b). As estimators, those standard errors can also show more variability than parametric estimators (Kauermann and Carroll 2001). Boos (1992) and Rotnitzky and Jewell (1990) proposed analogs of score tests for effects of predictors, using quasi-log-likelihood, that may be more trustworthy than Wald tests. Some statisticians (e.g., Lindsey 1999) are critical of the GEE approach because of the lack of likelihood. Others do not find this problematic, as they regard GEE as an estimation method rather than a model.

11.3.2 Longitudinal Mental Depression Example

For Table 11.2 comparing two treatments for mental depression, ML fitting of a logit model with drug $\times$ time interaction was used in Section 11.2.1. The GEE analysis provides similar results, regardless of the choice of working correlation structure. With the exchangeable structure, the GEE estimated slope (on the logit scale) for the standard drug is $\hat{\beta}_3 = 0.48$ (SE = 0.12). For the new drug the slope increases by $\hat{\beta}_4 = 1.02$ (SE = 0.19). Table 11.6 shows results using the independence working correlations. Estimates are the same to two decimal places. The initial estimates and standard errors there are those that apply if the repeated responses are truly independent. They equal those obtained by using ordinary logistic regression with $3 \times 340 = 1020$ independent observations rather than treating the data as three dependent observations for each of 340 subjects. The empirical standard errors incorporate the sample dependence to adjust the independence-based standard errors.

TABLE 11.6   Output from Using GEE to Fit Logit Model to Table 11.2

Initial Parameter Estimates			GEE Parameter Estimates Empirical Std Error Estimates		
Parameter	Estimate	Std Error	Parameter	Estimate	Std Error
Intercept	−0.0280	0.1639	Intercept	−0.0280	0.1742
diagnose	−1.3139	0.1464	diagnose	−1.3139	0.1460
drug	−0.0596	0.2222	drug	−0.0596	0.2285
time	0.4824	0.1148	time	0.4824	0.1199
drug*time	1.0174	0.1888	drug*time	1.0174	0.1877

	Working Correlation Matrix		
	Col1	Col2	Col3
Row1	1.0000	0.0000	0.0000
Row2	0.0000	1.0000	0.0000
Row3	0.0000	0.0000	1.0000

With exchangeable correlation structure, the estimated common correlation between pairs of the three responses is -0.003. The successive observations apparently have pairwise appearance like independent observations. This is quite unusual for repeated measurement data. For this reason, similar results occur from fitting the model assuming the three observations for a subject actually come from three separate subjects (i.e., assuming 1020 independent observations).

11.3.3   GEE Approach for Multinomial Responses: Insomnia Example

Liang and Zeger (1986) originally specified the GEE methodology for modeling univariate marginal distributions, such as the binomial and Poisson. It extends to marginal modeling of multinomial responses. Lipsitz et al. (1994) outlined a GEE approach for cumulative logit models with repeated ordinal responses. With this approach, for each pair of outcome categories one selects a working correlation matrix for the pairs of repeated observations. Each multinomial response at a fixed observation uses the $(I - 1) \times (I - 1)$ multinomial covariance matrix. Section 11.4.4 has details.

We illustrate for the insomnia data of Table 11.4. In Section 11.2.3 we used ML to fit the marginal model

$$\text{logit}[P(Y_t \leq j)] = \alpha_j + \beta_1 t + \beta_2 x + \beta_3 tx$$

for $Y_t =$ time to fall asleep with treatment x at occasion t. With independence working correlation structure, the GEE estimates are $\hat{\beta}_1 = 1.038$ (SE $= 0.168$), $\hat{\beta}_2 = 0.034$ (SE $= 0.238$), and $\hat{\beta}_3 = 0.708$ (SE $= 0.244$). The estimates are similar to the ML estimates, and the substantive conclusions are the same. Considerable evidence exists that the distribution of time to fall asleep decreased more for the treatment group than for the placebo group.

11.4 QUASI-LIKELIHOOD AND ITS GEE MULTIVARIATE EXTENSION: DETAILS*

A GLM assumes a certain distribution for the response variable. Sometimes it is unclear how to select it. However, often there is a plausible relationship between the mean and variance, such as $v(\mu_i) = \phi\mu_i$ for count data. Then, an alternative to ML estimation is quasi-likelihood estimation (Section 4.7). We next present some details about this method and its GEE extension for marginal modeling of multivariate responses.

We begin with models for a single response and later discuss marginal models for a multivariate response. For subject $i, i = 1, \ldots, n$, let y_i be the outcome on Y with $\mu_i = E(Y_i)$ and variance function $v(\mu_i)$, and let x_{ij} be the value of explanatory variable j. For link function g, the linear predictor is $\eta_i = g(\mu_i) = \sum_j \beta_j x_{ij} = \mathbf{x}_i'\boldsymbol{\beta}$. The quasi-likelihood (QL) parameter estimates $\boldsymbol{\beta}$ are the solutions of quasi-score equations

$$\mathbf{u}(\boldsymbol{\beta}) = \sum_i \left(\frac{\partial\mu_i}{\partial\boldsymbol{\beta}}\right)' v(\mu_i)^{-1}(y_i - \mu_i) = \mathbf{0}, \tag{11.9}$$

where $\mu_i = g^{-1}(\mathbf{x}_i'\boldsymbol{\beta})$. These *estimating equations* are the same as the likelihood equations (4.22) for GLMs when we substitute

$$\frac{\partial\mu_i}{\partial\beta_j} = \frac{\partial\mu_i}{\partial\eta_i}\frac{\partial\eta_i}{\partial\beta_j} = \frac{\partial\mu_i}{\partial\eta_i}x_{ij}.$$

They are not likelihood equations, however, without the extra assumption that $\{y_i\}$ has distribution in the natural exponential family. Under that assumption, $v(\mu_i)$ characterizes the distribution within the natural exponential family (Jørgensen 1987). Another motivation for equations (11.9) is that with $v(\mu_i)$ replaced by known variance v_i, they result from the weighted least squares problem of minimizing $\sum_i(y_i - \mu_i)^2 v_i^{-1}$.

The likelihood equations (4.22) for a GLM depend only on the mean and variance of $\{y_i\}$ and the link function g, which determines $\partial\mu_i/\partial\eta_i$. Thus, Wedderburn (1974) suggested using them as estimating equations for *any* link and variance function, even if they do not correspond to a particular member of the natural exponential family.

11.4.1 Properties of Quasi-likelihood Estimators

In the quasi-likelihood (QL) method, the *quasi-score function* $u_j(\boldsymbol{\beta})$ in (11.9) is called an *unbiased estimating function*; this term refers to any function $h(\mathbf{y}; \boldsymbol{\beta})$ of $\mathbf{y}$ and $\boldsymbol{\beta}$ such that $E[h(\mathbf{Y}; \boldsymbol{\beta})] = 0$ for all $\boldsymbol{\beta}$. The equations (11.9) that determine $\hat{\boldsymbol{\beta}}$ are called *estimating equations*.

The quasi-likelihood method treats the quasi-score function as the derivative of a function called the *quasi-log likelihood*. This function may not be a proper log likelihood function. Nonetheless, McCullagh (1983) showed that QL estimators have properties similar to those of ML estimators. For instance, the QL estimators $\hat{\beta}$ are asymptotically normal with covariance matrix approximated by

$$\mathbf{V} = \left[\sum_i \left(\frac{\partial \mu_i}{\partial \beta} \right)' [v(\mu_i)]^{-1} \left(\frac{\partial \mu_i}{\partial \beta} \right) \right]^{-1}. \tag{11.10}$$

This is equivalent to the formula for the large-sample covariance matrix of the ML estimator in a GLM [which is estimated by (4.28)].

A key result is that the QL estimator $\hat{\beta}$ is consistent for β (i.e., $\hat{\beta} \xrightarrow{p} \beta$) even if the variance function is misspecified, as long as the specification is correct for the link function and linear predictor. That is, assuming that the model form $g(\mu_i) = \sum_j \beta_j x_{ij}$ is correct, the consistency of $\hat{\beta}$ holds even if the true variance function is not $v(\mu_i)$. We now give a heuristic explanation for this.

When truly $\mu_i = g^{-1}(\sum_j \beta_j x_{ij})$, then from (11.9), $E[u_j(\beta)] = 0$ for all j. From (11.9), $\mathbf{u}(\beta)/n$ is a vector of sample means. By a law of large numbers, it converges in probability to its expected value of $\mathbf{0}$. The solution $\hat{\beta}$ of the quasi-score equations is a continuous function of these sample means, so it converges to β, since $\hat{\beta}$ is the value of β for which the sum is exactly equal to $\mathbf{0}$. The consistency also follows from general results for unbiased estimating functions (Liang and Zeger 1995).

11.4.2 Sandwich Covariance Adjustment for Variance Misspecification

If one assumes that $\text{var}(Y_i) = v(\mu_i)$ but the true $\text{var}(Y_i) \neq v(\mu_i)$, then the actual asymptotic covariance matrix of the QL estimator $\hat{\beta}$ is not $\mathbf{V}$ as given in (11.10). Instead, it is (Diggle et al. 2001; White 1982)

$$\mathbf{V} \left[\sum_i \left(\frac{\partial \mu_i}{\partial \beta} \right)' [v(\mu_i)]^{-1} \text{var}(Y_i) [v(\mu_i)]^{-1} \left(\frac{\partial \mu_i}{\partial \beta} \right) \right] \mathbf{V}. \tag{11.11}$$

Even though the variances are scalar, we express the matrices in this form to motivate the GEE multivariate extension discussed below. Matrix (11.11) simplifies to $\mathbf{V}$ if $\text{var}(Y_i) = v(\mu_i)$. In practice, the true variance function is unknown. A consistent estimator of (11.11) is a sample analog, replacing μ_i by $\hat{\mu}_i$ and $\text{var}(Y_i)$ by $(y_i - \hat{\mu}_i)^2$ (Liang and Zeger 1986). The estimated covariance matrix is valid regardless of whether the variance specification $v(\mu_i)$ is correct. This estimated covariance matrix is called a *sandwich estimator*, because the empirical evidence is sandwiched between the model-driven covariance matrices.

In summary, even with incorrect specification of the variance function, one can still consistently estimate β and one can estimate the asymptotic variance of $\hat{\boldsymbol{\beta}}$ by estimating the sandwich adjustment (11.11). However, some efficiency loss occurs when the variance chosen, $v(\mu_i)$, is wildly inaccurate. Also, the number of clusters n may need to be large for the sample version of (11.11) to work well; otherwise, it can be biased downward. Of course, a modeling process never gets anything exactly correct. Just as the variance function chosen only approximates the true one (hopefully, closely), so is the specification for the mean only approximate.

11.4.3 GEE Methodology: Technical Details

Now we consider the generalized estimating equations (GEE) multivariate generalization of QL. For subject i, let $\mathbf{y}_i = (y_{i1}, \ldots, y_{iT_i})'$ and $\boldsymbol{\mu}_i = (\mu_{i1}, \ldots, \mu_{iT_i})'$, where $\mu_{it} = E(Y_{it})$. The number T_i of responses may vary by cluster. Let $\mathbf{x}_{it}$ denote a $p \times 1$ vector of explanatory variable values for y_{it}. The notation allows for cases where explanatory variables also vary for the repeated measurements. The linear predictor of the model is $\eta_{it} = g(\mu_{it}) = \mathbf{x}'_{it}\boldsymbol{\beta}$ for link function g. The model refers to the marginal distribution at each t rather than the joint distribution. Let $\mathbf{X}_i$ be the $T_i \times p$ matrix of predictor values for cluster (or subject) i, for which row t is $\mathbf{x}'_{it}$.

We assume that y_{it} has probability mass function of form

$$f(y_{it}; \theta_{it}, \phi) = \exp\left\{\left[y_{it}\theta_{it} - b(\theta_{it})\right]/\phi + c(y_{it}, \phi)\right\}.$$

When ϕ is known, this is the natural exponential family with natural parameter θ_{it}. From Section 4.4.1,

$$\mu_{it} = E(Y_{it}) = b'(\theta_{it}), \qquad v(\mu_{it}) = \text{var}(Y_{it}) = b''(\theta_{it})\phi.$$

The GEE method also assumes a working correlation matrix $\mathbf{R}(\boldsymbol{\alpha})$ for $\mathbf{Y}_i$, depending on parameters $\boldsymbol{\alpha}$. The exchangeable working correlation has $\text{corr}(Y_{it}, Y_{is}) = \alpha$ for each pair in $\mathbf{Y}_i$. Let $\mathbf{b}_i(\boldsymbol{\theta}) = (b(\theta_{i1}), \ldots, b(\theta_{iT_i}))$, and let $\mathbf{B}_i$ denote a diagonal matrix with main diagonal elements $\mathbf{b}''_i(\boldsymbol{\theta})$. Then the working covariance matrix for $\mathbf{Y}_i$ is

$$\mathbf{V}_i = \mathbf{B}_i^{1/2} \mathbf{R}(\boldsymbol{\alpha}) \mathbf{B}_i^{1/2} \phi. \tag{11.12}$$

Note that $\mathbf{V}_i = \text{cov}(\mathbf{Y}_i)$ if $\mathbf{R}$ is the true correlation matrix for $\mathbf{Y}_i$.

Now let $\boldsymbol{\Delta}_i$ be the diagonal matrix with elements $\partial\theta_{it}/\partial\eta_{it}$ on the main diagonal for $t = 1, \ldots, T_i$. (For the canonical link, this is the identity matrix.) Let $\mathbf{D}_i = \partial\boldsymbol{\mu}_i/\partial\boldsymbol{\beta} = \mathbf{B}_i\boldsymbol{\Delta}_i\mathbf{X}_i$ be a $T_i \times p$ matrix with typical element expressing $\partial\mu_{it}/\partial\beta_j$ in the form $(\partial\mu_{it}/\partial\theta_{it})(\partial\theta_{it}/\partial\eta_{it})(\partial\eta_{it}/\partial\beta_j)$. From (11.9), for univariate GLMs the quasi-likelihood estimating equations have the form

$$\sum_i (\partial\boldsymbol{\mu}_i/\partial\boldsymbol{\beta})' v(\mu_i)^{-1} \left[y_i - \mu_i(\boldsymbol{\beta})\right] = \mathbf{0},$$

where $\mu_i = \mu_i(\boldsymbol{\beta}) = g^{-1}(\mathbf{x}_i'\boldsymbol{\beta})$. The analog of this in the multivariate case is the set of *generalized estimating equations*

$$\sum_{i=1}^{n} \mathbf{D}_i' \mathbf{V}_i^{-1} [\mathbf{y}_i - \boldsymbol{\mu}_i(\boldsymbol{\beta})] = \mathbf{0}.$$

The GEE estimator $\hat{\boldsymbol{\beta}}$ is the solution of these equations.

The naive approach, which sets $\mathbf{R}(\boldsymbol{\alpha}) = \mathbf{I}$, treats pairs of responses as independent. In that case, (11.12) simplifies to $\mathbf{V}_i = \mathbf{B}_i\phi$, and the generalized estimating equations simplify to

$$\sum_i \mathbf{D}_i' \mathbf{V}_i^{-1} [\mathbf{y}_i - \boldsymbol{\mu}_i(\boldsymbol{\beta})] = \sum_i \mathbf{X}_i' \boldsymbol{\Delta}_i \mathbf{B}_i \mathbf{V}_i^{-1} [\mathbf{y}_i - \boldsymbol{\mu}_i(\boldsymbol{\beta})]$$

$$= (1/\phi) \sum_i \mathbf{X}_i' \boldsymbol{\Delta}_i [\mathbf{y}_i - \boldsymbol{\mu}_i(\boldsymbol{\beta})] = \mathbf{0},$$

or $\sum_i \mathbf{X}_i' \boldsymbol{\Delta}_i [\mathbf{y}_i - \boldsymbol{\mu}_i(\boldsymbol{\beta})] = \mathbf{0}$. The solution $\hat{\boldsymbol{\beta}}$ is then the same as the ordinary estimator for a GLM with the chosen link function and variance function, treating $(y_{i1}, \ldots, y_{iT_i})$ as independent observations.

Normally, one selects a working correlation matrix permitting dependence, such as the exchangeable structure. For time-series data, also popular is the autoregressive structure, $\text{corr}(Y_{it}, Y_{is}) = \alpha^{|t-s|}$, which treats observations farther apart in time as more weakly correlated. Liang and Zeger (1986) suggested computing the GEE estimates by iterating between a modified Fisher scoring algorithm for solving the generalized estimating equations for $\boldsymbol{\beta}$ (given current estimates of $\boldsymbol{\alpha}$ and ϕ) and using residuals for moment estimation of $\boldsymbol{\alpha}$ and ϕ (based on the current estimates of $\boldsymbol{\beta}$). They suggested estimates of $\mathbf{R}(\boldsymbol{\alpha})$ for a variety of correlation structures. Alternative algorithms simultaneously solve estimating equations for $\boldsymbol{\beta}$ and for association parameters (e.g., Liang et al. 1992; see also Note 11.8). GEE algorithms need not converge, but often one iteration gives adequate results (Lipsitz et al. 1991).

Liang and Zeger (1986) showed asymptotic normality and consistency as the number of clusters n increases. Under certain regularity conditions,

$$\sqrt{n} \left(\hat{\boldsymbol{\beta}} - \boldsymbol{\beta} \right) \xrightarrow{d} N(\mathbf{0}, \mathbf{V}_G) .$$

Here, generalizing (11.11), $\mathbf{V}_G = \lim_{n \to \infty} \mathbf{V}_{G,n}$ with

$$\mathbf{V}_{G,n} = n \left[\sum_i \mathbf{D}_i' \mathbf{V}_i^{-1} \mathbf{D}_i \right]^{-1} \left[\sum_i \mathbf{D}_i' \mathbf{V}_i^{-1} \text{cov}(\mathbf{Y}_i) \mathbf{V}_i^{-1} \mathbf{D}_i \right] \left[\sum_i \mathbf{D}_i' \mathbf{V}_i^{-1} \mathbf{D}_i \right]^{-1} .$$

The estimated covariance matrix $\hat{\mathbf{V}}_{G,n}/n$ of $\hat{\boldsymbol{\beta}}$ replaces $\boldsymbol{\beta}$ with $\hat{\boldsymbol{\beta}}$, ϕ with $\hat{\phi}$, $\boldsymbol{\alpha}$ with $\hat{\boldsymbol{\alpha}}$, and $\text{cov}(\mathbf{Y}_i)$ by $[\mathbf{y}_i - \boldsymbol{\mu}_i(\hat{\boldsymbol{\beta}})][\mathbf{y}_i - \boldsymbol{\mu}_i(\hat{\boldsymbol{\beta}})]'$. The purpose of the sandwich estimator is to use the data's empirical evidence about covariation to adjust the standard errors in case the true covariance differs substantially from the working guess.

When the working correlation structure is the true one and $\text{cov}(\mathbf{Y}_i) = \mathbf{V}_i$, the asymptotic covariance matrix $\mathbf{V}_{G,n}/n$ simplifies to $(\Sigma_i \mathbf{D}_i' \mathbf{V}_i^{-1} \mathbf{D}_i)^{-1}$. This is the relevant covariance if we put complete faith in our guess about the correlation structure.

With binary data, the correlation may not be the best way to express the within-cluster association. The marginal probabilities constrain the possible correlation values, since the range of possible values for $E(Y_{it}Y_{is}) = P(Y_{it} = 1, Y_{is} = 1)$ depends on $P(Y_{it} = 1)$ and $P(Y_{is} = 1)$. An alternative approach uses the odds ratio, for instance by modeling the log odds ratios for pairs in a cluster as exchangeable. This has the advantage that the association parameters are distinct from the means. See Fitzmaurice et al. (1993) and Lipsitz et al. (1991). Carey et al. (1993) suggested an iterative *alternating logistic regressions* algorithm. It alternates between a GEE step for the regression parameters in the model for the mean and a step for an association model for the log odds ratio. This is useful when the structure of the association is itself a major focus rather than a nuisance.

11.4.4   GEE Approach: Multinomial Responses

We now briefly describe the Lipsitz et al. (1994) GEE approach for marginal modeling with a multinomial response. This is appropriate, for instance, with cumulative logit models. Let $y_{it}(j) = 1$ if observation t in cluster i has outcome j $(j = 1, \ldots, I - 1)$. Let $\mathbf{y}_i$ be the $T_i(I - 1)$ binary indicators for cluster i. Then, one selects a $[T_i(I - 1)] \times [T_i(I - 1)]$ working covariance matrix $\mathbf{V}_i$ for $\mathbf{y}_i$, specifying a pattern for $\text{corr}(Y_{it}(j), Y_{is}(k))$ for each pair of outcome categories (j, k) and each pair (t, s). The $(I - 1) \times (I - 1)$ block of $\mathbf{V}_{it}$ for $(y_{it}(1), \ldots, y_{it}(I - 1))$ is a multinomial covariance matrix with $v_{it}(j) = P(Y_{it}(j) = 1)[1 - P(Y_{it}(j) = 1)]$ on the main diagonal and $-P(Y_{it}(j) = 1)P(Y_{it}(k) = 1)$ off it. The remaining elements of $\mathbf{V}_i$ contain elements $\text{cov}(Y_{it}(j), Y_{is}(k))$. For instance, one possibility is the exchangeable structure, $\text{corr}(Y_{it}(j), Y_{is}(k)) = \rho_{jk}$ for all t and s.

In this approach the generalized estimating equations for $\boldsymbol{\beta}$ again have the form

$$\mathbf{u}(\boldsymbol{\beta}) = \sum_{i=1}^{n} \mathbf{D}_i' \mathbf{V}_i^{-1}(\mathbf{y}_i - \boldsymbol{\mu}_i) = \mathbf{0},$$

where $\boldsymbol{\mu}_i$ is the vector of probabilities associated with $\mathbf{y}_i$, $\mathbf{D}_i' = \partial \boldsymbol{\mu}_i'/\partial \boldsymbol{\beta}$, and the parameters are evaluated at their current estimates. Lipsitz et al. suggested a Fisher scoring algorithm for solving these equations and a method of moments update for estimating $\{\rho_{jk}\}$ at each step of the iteration. An empirically adjusted sandwich covariance matrix of $\hat{\boldsymbol{\beta}}$ is again

$$\left[\sum_{i=1}^{n} \mathbf{D}_i' \mathbf{V}_i^{-1} \mathbf{D}_i\right]^{-1} \left[\sum_{i=1}^{n} \mathbf{D}_i' \mathbf{V}_i^{-1} \text{cov}(\mathbf{Y}_i) \mathbf{V}_i^{-1} \mathbf{D}_i\right] \left[\sum_{i=1}^{n} \mathbf{D}_i' \mathbf{V}_i^{-1} \mathbf{D}_i\right]^{-1}.$$

This is estimated by substituting $\hat{\boldsymbol{\mu}}_i$ from the model fit and replacing cov($\mathbf{Y}_i$) by the empirical covariance matrix of $\mathbf{y}_i$.

11.4.5 Dealing with Missing Data

Unfortunately, studies with repeated measurement often have cases for which at least one response in a cluster is missing. In a longitudinal study, for instance, some subjects may drop out before its conclusion. When data are missing, analyzing the observed data alone as if no data are missing can result in biased estimates.

An advantage of the GEE method is that different clusters can have different numbers of observations. The data input file has a separate line for each observation, and for longitudinal studies, computations use those times for which a subject has an observation. However, bias can arise in GEE estimates unless one can make certain assumptions about why the data are missing.

Let $\mathbf{Y}^{(o)}$ denote the observed responses, $\mathbf{Y}^{(m)}$ the missing responses, and $\mathbf{Y}$ their union. Let M denote a missing data indicator that equals 1 when an observation is missing and 0 otherwise. Little and Rubin (1987) called the data *missing completely at random* if M is statistically independent of $\mathbf{Y}$; that is, the probability that an observation is missing is independent of that observation's value, although it may depend on the explanatory variables. Less restrictively, they called the data *missing at random* if the distribution of $(M \mid \mathbf{Y})$ equals that of $(M \mid \mathbf{Y}^{(o)})$; that is, missingness depends only on $\mathbf{Y}^{(o)}$ and not on the missing values.

When either of these is plausible, with a likelihood-based analysis it is not necessary to model the missingness mechanism. An analysis using only $\mathbf{Y}^{(o)}$ is not systematically biased. The same is true with GEE methods when estimating equations can be weighted by response probabilities (Robins et al. 1995). Otherwise, however, with non-likelihood-based methods such as GEE, the missingness process can be ignored only when data are missing completely at random. Kenward et al. (1994) illustrated the breakdown in GEE estimates when the data are not missing completely at random.

Often, missingness depends on the missing values. For instance, in a longitudinal study measuring pain, perhaps a subject dropped out when the pain got above some threshhold. Then, more complex analyses are needed that model the joint distribution of $\mathbf{Y}$ and M (Little 1998). Let $f(\cdot)$ denote a generic probability mass function, which also depends on explanatory variables $\mathbf{x}$ and parameters. *Selection models* factor the joint distribution of $\mathbf{Y}$

and M as

$$f(\mathbf{y}, M; \mathbf{x}, \boldsymbol{\beta}, \boldsymbol{\psi}) = f(\mathbf{y}; \mathbf{x}, \boldsymbol{\beta}) f(M \mid \mathbf{y}; \mathbf{x}, \boldsymbol{\psi}),$$

where $f(\mathbf{y}; \mathbf{x}, \boldsymbol{\beta})$ is the model in the absence of missing values and $f(M \mid \mathbf{y}; \mathbf{x}, \boldsymbol{\psi})$ is the model for the missing-data mechanism. *Pattern mixture models* use the alternative factorization,

$$f(\mathbf{y}, M; \mathbf{x}, \boldsymbol{\theta}, \boldsymbol{\phi}) = f(\mathbf{y} \mid M, \mathbf{x}, \boldsymbol{\phi}) f(M; \mathbf{x}, \boldsymbol{\theta}),$$

which conditions the distribution of $\mathbf{Y}$ on the missing data pattern. The two specifications are equivalent when M is independent of $\mathbf{Y}$, with $\boldsymbol{\beta} = \boldsymbol{\phi}$ and $\boldsymbol{\psi} = \boldsymbol{\theta}$. For discussion of advantages of each modeling approach and details on ways of modeling missingness, see Little (1998) and references in Note 11.9. See Stokes et al. (2000, p. 524) for an example of building the missingness pattern into a model to check whether it is associated with the response or interacts with effects of explanatory variables.

Analyses in the presence of much missingness should be made with caution. Typically, little is known about the missing data mechanism, and assumptions about it cannot be checked. Since inferences may not be robust, a sensitivity study is necessary to check how results depend on specification of that mechanism. In the absence of a model for the missingness, one should at least compare results of the analysis using all available cases for all clusters to the analysis using only clusters having no missing observations. If results differ substantially, conclusions should be very tentative until the reasons for missingness can be studied.

11.5   MARKOV CHAINS: TRANSITIONAL MODELING

When Y_t denotes the response at time $t, t = 0, 1, 2, \ldots$, the indexed family of random variables $(Y_0, Y_1, Y_2, \ldots)$ is a *stochastic process*. The *state space* of the process is the set of possible values for Y_t. The value Y_0 is the *initial state*. When the state space is categorical and observations occur at a discrete set of times, $\{Y_t\}$ has *discrete state space* and *discrete time*.

11.5.1   Transitional Models

The main focus is usually on the dependence of Y_t on the responses $\{y_0, y_1, \ldots, y_{t-1}\}$ observed previously as well as any explanatory variables. Models of this type are called *transitional models*. Let $f(y_0, \ldots, y_T)$ denote the joint probability mass function of $(Y_0, \ldots, Y_T)$ (ignoring, for now, explanatory variables). Transitional models use the factorization

$$f(y_0, \ldots, y_T) = f(y_0)f(y_1|y_0)f(y_2|y_0, y_1) \cdots f(y_T|y_0, y_1, \ldots, y_{T-1}).$$

Unlike the marginal models in the other sections of this chapter, this modeling is conditional on previous responses.

In this section we introduce discrete-time *Markov chains*, a simple stochastic process having discrete state space. Many transitional models have Markov chain structure for at least part of the model.

11.5.2 First-Order Markov Chains

A *Markov chain* is a stochastic process for which, for all t, the conditional distribution of Y_{t+1}, given $Y_0, \ldots, Y_t$, is identical to the conditional distribution of Y_{t+1} given Y_t alone. That is, given Y_t, Y_{t+1} is conditionally independent of $Y_0, \ldots, Y_{t-1}$. Knowing the present state of a Markov chain, information about past states does not help us predict the future. For Markov chains,

$$f(y_0, \ldots, y_T) = f(y_0)f(y_1|y_0)f(y_2|y_1) \ldots f(y_T|y_{T-1}). \quad (11.13)$$

A stochastic process is a kth-*order Markov chain* if, for all t, the conditional distribution of Y_{t+1}, given $Y_0, \ldots, Y_t$, is identical to the conditional distribution of Y_{t+1}, given $(Y_t, \ldots, Y_{t-k+1})$. Given the states at the previous k times, the future behavior of the chain is independent of past behavior before those k times. Our discussion here focuses mainly on ordinary Markov chains as in (11.13), which are first order ($k = 1$).

Denote the conditional probability $P(Y_t = j | Y_{t-1} = i)$ by $\pi_{j|i}(t)$. The $\{\pi_{j|i}(t)\}$, which satisfy $\sum_j \pi_{j|i}(t) = 1$, are called *transition probabilities*. The $I \times I$ matrix $\{\pi_{j|i}(t), i = 1, \ldots, I, j = 1, \ldots, I\}$ is a *transition probability matrix*. It is called *one-step*, to distinguish it from the matrix of probabilities for k-step transitions from time $t - k$ to time t.

From (11.13), the joint distribution for a Markov chain depends only on one-step transition probabilities and the marginal distribution for the initial state. It also follows that the joint distribution satisfies loglinear model

$$(Y_0Y_1, Y_1Y_2, \ldots, Y_{T-1}Y_T).$$

For a sample of realizations of a stochastic process, a contingency table displays counts of the possible sequences. A test of fit of this loglinear model checks whether the process plausibly satisfies the Markov property.

Statistical inference for Markov chains uses standard methods of categorical data analysis. For example, consider ML estimation of transition probabilities. Let $n_{ij}(t)$ denote the number of transitions from state i at time $t - 1$ to state j at time t. For fixed t, $\{n_{ij}(t)\}$ form the two-way marginal table for dimensions $t - 1$ and t of an I^{T+1} contingency table. For the $n_{i+}(t)$ subjects in category i at time $t - 1$, suppose that $\{n_{ij}(t), j = 1, \ldots, I\}$ have a multinomial distribution with parameters $\{\pi_{j|i}(t)\}$. Let $\{n_{i0}\}$ denote the initial counts. Suppose that they also have a multinomial distribution, with parameters $\{\pi_{i0}\}$. If subjects behave independently, from (11.13) the likelihood function is proportional to

$$\left(\prod_{i=1}^{I} \pi_{i0}^{n_{i0}} \right) \left\{ \prod_{t=1}^{T} \prod_{i=1}^{I} \left[\prod_{j=1}^{I} \pi_{j|i}(t)^{n_{ij}(t)} \right] \right\}. \tag{11.14}$$

The transition probabilities are parameters of IT independent multinomial distributions. From Anderson and Goodman (1957), the ML estimates are

$$\hat{\pi}_{j|i}(t) = n_{ij}(t)/n_{i+}(t) .$$

11.5.3 Respiratory Illness Example

Table 11.7 refers to a longitudinal study at Harvard of effects of air pollution on respiratory illness in children. The children were examined annually at ages 9 through 12 and classified according to the presence or absence of wheeze.

Denote the binary response (wheeze, no wheeze) by Y_t at age t, $t = 9, 10, 11, 12$. The loglinear model $(Y_9Y_{10}, Y_{10}Y_{11}, Y_{11}Y_{12})$ represents a first-order Markov chain. It fits poorly, with $G^2 = 122.9$ (df $= 8$). Given the state at time t, classification at time $t + 1$ depends on states at times previous to time t. The model $(Y_9Y_{10}Y_{11}, Y_{10}Y_{11}Y_{12})$ represents a second-order Markov chain, satisfying conditional independence at ages 9 and 12, given states at ages 10 and 11. This model also fits poorly, with $G^2 = 23.9$ (df $= 4$). The poor fits may partly reflect subject heterogeneity, since these analyses ignore possibly relevant covariates such as parental smoking behavior.

The loglinear model $(Y_9Y_{10}, Y_9Y_{11}, Y_9Y_{12}, Y_{10}Y_{11}, Y_{10}Y_{12}, Y_{11}Y_{12})$ that permits association at each pair of ages fits well, with $G^2 = 1.5$ (df $= 5$). Table

TABLE 11.7 Results of Breath Test at Four Ages[a]

Y_9	Y_{10}	Y_{11}	Y_{12}	Count	Y_9	Y_{10}	Y_{11}	Y_{12}	Count
1	1	1	1	94	2	1	1	1	19
1	1	1	2	30	2	1	1	2	15
1	1	2	1	15	2	1	2	1	10
1	1	2	2	28	2	1	2	2	44
1	2	1	1	14	2	2	1	1	17
1	2	1	2	9	2	2	1	2	42
1	2	2	1	12	2	2	2	1	35
1	2	2	2	63	2	2	2	2	572

[a] 1, wheeze; 2, no wheeze.

Source: Ware et al. (1988).

TABLE 11.8   **Estimated Conditional Log Odds
Ratios for Table 11.7**

Association	Estimate	Simpler Structure
Y_9Y_{10}	1.81	1.75
$Y_{10}Y_{11}$	1.65	1.75
$Y_{11}Y_{12}$	1.85	1.75
Y_9Y_{11}	0.95	1.04
Y_9Y_{12}	1.05	1.04
$Y_{10}Y_{12}$	1.07	1.04

11.8 shows its ML estimates of pairwise conditional log odds ratios. The association seems similar for pairs of ages 1 year apart, and somewhat weaker for pairs of ages more than 1 year apart. The simpler model in which

$$\lambda_{ij}^{Y_9Y_{10}} = \lambda_{ij}^{Y_{10}Y_{11}} = \lambda_{ij}^{Y_{11}Y_{12}} \quad \text{and} \quad \lambda_{ij}^{Y_9Y_{11}} = \lambda_{ij}^{Y_9Y_{12}} = \lambda_{ij}^{Y_{10}Y_{12}}$$

fits well, with $G^2 = 2.3$ (df = 9). The estimated log odds ratios are 1.75 in the first case, and 1.04 in the second.

11.5.4   Transitional Models with Explanatory Variables

Transitional models usually also include explanatory variables **x**. The joint mass function of T sequential responses is then

$$f(y_1, \ldots, y_T; \mathbf{x})$$
$$= f(y_1; \mathbf{x})f(y_2 \mid y_1; \mathbf{x})f(y_3 \mid y_1, y_2; \mathbf{x}) \cdots f(y_T \mid y_1, y_2, \ldots, y_{T-1}; \mathbf{x}) \,.$$

With binary y, for instance, one might specify a logistic regression model for each term in this factorization,

$$f(y_t \mid y_1, \ldots, y_{t-1}; \mathbf{x}_t)$$
$$= \frac{\exp[y_t(\alpha + \beta_1 y_1 + \cdots + \beta_{t-1} y_{t-1} + \boldsymbol{\beta}'\mathbf{x}_t)]}{1 + \exp(\alpha + \beta_1 y_1 + \cdots + \beta_{t-1} y_{t-1} + \boldsymbol{\beta}'\mathbf{x}_t)}, \quad y_t = 0,1.$$

Here, the predictor **x** may take different value for each component. The model treats previous responses as explanatory variables. It is called a *regressive logistic model* (Bonney 1987).

The interpretation and magnitude of $\hat{\boldsymbol{\beta}}$ depends on how many previous observations are in the model. Within-cluster effects may diminish markedly by conditioning on previous responses. This is an important difference from marginal models, for which the interpretation does not depend on the specification of the dependence structure. In the special case of first-order Markov structure, the coefficients of $\{y_1, \ldots, y_{t-2}\}$ equal 0 in the model for y_t (e.g., Azzalini 1994; Bonney 1987). It may help to allow interaction between $\mathbf{x}_t$ and y_{t-1} in their effects on y_t.

For a given subject, the product of the conditional mass functions determines that subject's contribution to the likelihood function. (One usually ignores the contribution of the marginal distribution for the first term.) That is, given the predictor, the model treats repeated transitions by a subject as independent. Thus, one can fit the model with ordinary GLM software, treating each transition as a separate observation (Bonney 1986).

11.5.5 Child's Respiratory Illness and Maternal Smoking

Table 11.9 is also from the Harvard study of air pollution and health. At ages 7 through 10, children were evaluated annually on the presence of respiratory illness. A predictor is maternal smoking at the start of the study, where $s = 1$ for smoking regularly and $s = 0$ otherwise. Let y_t denote the response at age t ($t = 7, 8, 9, 10$). We consider the regressive logistic model

$$\text{logit}[P(Y_t = 1)] = \alpha + \beta_1 s + \beta_2 t + \beta_3 y_{t-1}, \quad t = 8, 9, 10.$$

Each subject contributes three observations to the model fitting. The data set consists of 12 binomials, for the $2 \times 3 \times 2$ combinations of (s, t, y_{t-1}). For instance, for the combination $(0, 8, 0)$, $y_8 = 0$ for $237 + 10 + 15 + 4 =$

TABLE 11.9   Child's Respiratory Illness by Age and Maternal Smoking

			No Maternal Smoking		Maternal Smoking	
			Age 10		Age 10	
Child's Respiratory Illness						
Age 7	Age 8	Age 9	No	Yes	No	Yes
No	No	No	237	10	118	6
		Yes	15	4	8	2
	Yes	No	16	2	11	1
		Yes	7	3	6	4
Yes	No	No	24	3	7	3
		Yes	3	2	3	1
	Yes	No	6	2	4	2
		Yes	5	11	4	7

Source: Data courtesy of James Ware.

266 subjects and $y_8 = 1$ for $16 + 2 + 7 + 3 = 28$ subjects. The ML fit is

$$\text{logit}\left[\hat{P}(Y_t = 1)\right] = -0.293 + 0.296s - 0.243t + 2.211y_{t-1},$$

with SE values $(0.846, 0.156, 0.095, 0.158)$. Not surprisingly, the previous observation has a strong effect. Given that and the child's age, there is slight evidence of a positive effect of maternal smoking: The likelihood-ratio statistic for H_0: $\beta_1 = 0$ is 3.55 (df = 1, $P = 0.06$). The model itself does not show any evidence of lack of fit ($G^2 = 3.1$, df = 8).

NOTES

Section 11.1: Comparing Marginal Distributions: Multiple Responses

11.1. Darroch (1981) surveyed thoroughly the relationships among statistics for testing marginal homogeneity and their connections with generalized CMH analyses. See also Mantel and Byar (1978) and White et al. (1982). Croon et al. (2000) studied a variety of hypotheses for longitudinal data in the context of the generalized loglinear model.

Section 11.2: Marginal Modeling: Maximum Likelihood Approach

11.2. For other work on ML fitting of marginal models, see Bergsma and Rudas (2002), Ekholm et al. (2000), Fitzmaurice et al. (1993), and Lang et al. (1999).

Section 11.3: Marginal Modeling: Generalized Estimating Equations Approach

11.3. Liang et al. (1992) discussed GEE methods for categorical (primarily binary) responses. For multinomial responses, see Heagerty and Zeger (1996), Lipsitz et al. (1994), Miller et al. (1993), and references in Agresti and Natarajan (2001). More general models with ordinal responses allow for dispersion parameters that also depend on covariates (Toledano and Gatsonis 1996).

11.4. LaVange et al. (2001) used GEE methods to adjust for clustered sampling in surveys and clinical trials. Boos (1992) discussed generalized score tests that incorporate empirical variance estimates, illustrating with tests for trend and lack of fit in binary regression.

11.5. Koch et al. (1977) used weighted least squares (WLS) to fit marginal models to Table 11.2. WLS for categorical modeling is described in Section 15.1. It has severe limitations (e.g., covariates must be categorical and marginal tables cannot be sparse) but led naturally to the GEE approach.

Section 11.4: Quasi-likelihood and Its GEE Multivariate Extension: Details

11.6. Firth (1993b) provided a useful overview of quasi-likelihood methods. McCullagh (1983) showed that under correct specification of the mean and the variance function, quasi-likelihood estimators are asymptotically efficient among estimators that are locally linear in $\{y_i\}$. His result generalizes the Gauss–Markov theorem, although in an asymptotic rather than exact manner. See also Heyde (1997) and Liang and Zeger (1995) for discussions of unbiased estimating functions and their connections with asymptotic consistency and efficiency. Godambe showed in 1960 that ML estimators are optimal solutions with an unbiased estimating function. When quasi-likelihood estimators are not ML, Cox (1983) and Firth (1987) suggested that they still retain good efficiency when the departure from the natural exponential family is at most moderate, such as modest overdispersion relative to such a family.

11.7. The generalized estimating equations are likelihood equations, and hence the GEE estimates are also ML, in certain cases. Examples are multivariate normal data or binary data when the working covariance is correct (Fitzmaurice et al. 1993). Results about effects of model misspecification arise in a variety of model-building contexts. For general theory, see Gourieroux et al. (1984), Hansen (1982), Liang and Zeger (1995), and White (1982).

11.8. A GEE2 analysis adds estimating equations for the correlation structure (Prentice and Zhao 1991). This has the potential to increase efficiency. A disadvantage is that, unlike with ordinary GEE, $\hat{\beta}$ is no longer consistent if this part of the model is misspecified. Qu et al. (2000) showed how to increase efficiency by representing the working correlation matrix by a linear combination of basis matrices.

11.9. For surveys of ways to handle missing data, see Little (1998), Little and Rubin (1987, Chap. 9), Schafer (1997), and Verbeke and Molenberghs (2000). See also Baker and Laird (1988), Fay (1986), Fitzmaurice et al. (1994), Forster and Smith (1998), Fuchs (1982), Molenberghs and Goetghebeur (1997), Molenberghs et al. (1997), Park and Brown (1994), and Stokes et al. (2000).

Section 11.5: Markov Chains: Transitional Modeling

11.10. For statistical inference with Markov chains, see Andersen (1980, Sec. 7.7), Anderson and Goodman (1957), Billingsley (1961), Bishop et al. (1975, Chap. 7), and Kalbfleisch and Lawless (1985). See Conaway (1989), Stiratelli et al. (1984), and Ware et al. (1988) for other analyses focusing on the conditional dependence structure.

PROBLEMS

Applications

11.1 Refer to Table 8.3. Viewing the table as matched triplets, construct the marginal distribution for each substance. Find the sample proportions of students who used marijuana, alcohol, and cigarettes. Test the hypothesis of marginal homogeneity. Interpret results.

11.2 Refer to Table 9.1. Fit a marginal model to describe main effects of race, gender, and substance type (marijuana, alcohol, cigarettes) on whether a subject had used that substance. Summarize effects.

11.3 Refer to Problem 11.2. Further study shows evidence of an interaction between gender and substance type. Using GEE with exchangeable working correlation, the model fit for the probability π of using a particular substance is

$$\text{logit}(\hat{\pi}) = -0.57 + 1.93 S_1 + 0.86 S_2 + 0.38 R$$
$$- 0.20 G + 0.37 G \times S_1 + 0.22 G \times S_2,$$

where R, G, S_1, S_2 are dummy variables for race ($1 =$ white), gender ($1 =$ female), and substance type ($S_1 = 1$, $S_2 = 0$ for alcohol; $S_1 = 0$, $S_2 = 1$ for cigarettes; $S_1 = S_2 = 0$ for marijuana). Show that:

a. The estimated odds a nonwhite male has used marijuana are $\exp(-0.57) = 0.57$.

b. Given gender, the estimated odds a white subject used a given substance are 1.46 times the estimated odds for a black subject.

c. Given race, the estimated odds a female has used alcohol are 1.19 times the estimated odds for males; for cigarettes and for marijuana, the estimated odds ratios are 1.02 and 0.82.

d. Given race, the estimated odds a female has used alcohol (cigarettes) are 9.97 (2.94) times the estimated odds she has used marijuana.

e. Given race, the estimated odds a male has used alcohol (cigarettes) are 6.89 (2.36) times the estimated odds he has used marijuana. Interpret the interaction.

11.4 Refer to Table 11.2. Analyze the data using the scores $(1, 2, 4)$ for the week number, using ML or GEE. Interpret estimates and compare substantive results to those in the text with scores $(0, 1, 2)$.

11.5 Analyze Table 11.9 using a marginal logit model with age and maternal smoking as predictors. Compare interpretations to the Markov model of Section 11.5.5.

11.6 Table 11.10 refers to a three-period crossover trial to compare placebo (treatment A) with a low-dose analgesic (treatment B) and high-dose analgesic (treatment C) for relief of primary dysmenorrhea. Subjects in the study were divided randomly into six groups, the possible sequences for administering the treatments. At the end of each period, each subject rated the treatment as giving no relief (0) or some relief (1). Let $y_{i(k)t} = 1$ denote relief for subject i using treatment t ($t = A, B, C$), where subject i is nested in treatment sequence k ($k = 1, \ldots, 6$). Assuming common treatment effects for each sequence, and setting $\beta_A = 0$, obtain and interpret $\{\hat{\beta}_t\}$ (using ML or GEE) for the model

$$\text{logit}\left[P(Y_{i(k)t} = 1) \right] = \alpha_k + \beta_t.$$

How would you order the drugs, taking significance into account?

TABLE 11.10   Data for Problem 11.6

Treatment	Response Pattern for Treatments (A, B, C)							
Sequence	000	001	010	011	100	101	110	111
A B C	0	2	2	9	0	0	1	1
A C B	2	0	0	9	1	0	0	4
B A C	0	1	1	8	1	3	0	1
B C A	0	1	1	8	1	0	0	1
C A B	3	0	0	7	0	1	2	1
C B A	1	5	0	4	0	3	1	0

Source: Jones and Kenward (1987).

11.7   Table 11.11 is from a Kansas State University survey of 262 pig farmers. For the question "What are your primary sources of veterinary information?," the categories were (A) professional consultant, (B) veterinarian, (C) state or local extension service, (D) magazines, and (E) feed companies and reps. Farmers sampled were asked to select all relevant categories. The $2^5 \times 2 \times 4$ table shows the (yes, no) counts for each of these five sources cross-classified with the farmers' education (whether they had at least some college education) and size of farm (number of pigs marketed annually, in thousands).

TABLE 11.11   Data for Problem 11.7

			Response on D														
			A = yes								A = no							
			B = yes				B = no				B = yes				B = no			
			C = yes		C = no		C = yes		C = no		C = yes		C = no		C = yes		C = no	
Educ	Pigs	E	Y	N	Y	N	Y	N	Y	N	Y	N	Y	N	Y	N	Y	N
No	< 1	Y	1	0	0	0	0	0	0	0	2	1	1	2	1	1	5	3
		N	0	0	0	0	0	0	0	1	1	0	0	5	4	7	7	0
	1–2	Y	2	0	0	0	0	0	0	0	4	0	0	4	1	0	0	4
		N	0	0	0	0	0	0	0	0	0	0	0	5	0	3	4	0
	2–5	Y	3	0	0	0	0	0	0	0	3	0	0	1	2	0	1	1
		N	1	0	0	0	0	0	0	3	0	0	0	2	0	1	4	0
	> 5	Y	2	0	0	0	0	0	0	0	1	0	1	0	0	1	0	2
		N	1	0	0	2	1	0	1	6	0	1	1	1	0	0	6	0
Some	< 1	Y	3	0	0	0	0	0	0	0	4	0	1	1	0	0	2	11
		N	0	0	0	0	0	0	0	0	4	0	1	2	4	6	14	0
	1–2	Y	0	0	0	0	0	0	0	0	2	0	0	1	0	0	1	6
		N	0	0	0	0	1	0	0	1	2	1	0	4	2	7	14	0
	2–5	Y	0	0	0	0	0	0	0	0	1	0	0	0	0	1	1	3
		N	1	0	0	0	0	0	0	0	0	0	0	5	0	4	4	0
	> 5	Y	1	0	0	0	0	0	0	0	0	0	1	1	0	0	0	2
		N	1	1	0	0	0	1	0	10	0	0	0	4	1	2	4	0

Source: Data courtesy of Tom Loughin, Kansas State University.

a. Explain why it is not proper to analyze the data by fitting a multinomial model to the counts in the $2 \times 4 \times 5$ contingency table cross-classifying education by size of farm by the source of veterinary information, treating source as the response variable. (This table contains 453 positive responses of sources from the 262 farmers.)

b. For a farmer with education i and size of farm s, let $\pi_j(is)$ denote the probability of responding "yes" on the jth source. Table 11.12 shows output for using GEE with exchangeable working correlation to estimate parameters in the model lacking an education effect,

$$\text{logit}\left[\pi_j(is)\right] = \alpha_j + \beta_j s, \qquad s = 1, 2, 3, 4.$$

Explain how to interpret the working correlation matrix. Explain why the results suggest a strong positive size of farm effect for source A and perhaps a weak negative size effect of similar magnitude for C, D, and E.

c. Constraining $\beta_3 = \beta_4 = \beta_5$, the ML estimate of the common slope is -0.184 (SE $= 0.063$). Explain why it is advantageous to fit the marginal model simultaneously for all sources rather than separately to each. [Agresti and Liu (1999) and Loughin and Scherer (1998) discussed analyses for data of this form.]

TABLE 11.12 Output for Problem 11.7

| | | Working Correlation Matrix | | | |
	Col1	Col2	Col3	Col4	Col5
Row1	1.0000	0.0997	0.0997	0.0997	0.0997
Row2	0.0997	1.0000	0.0997	0.0997	0.0997
Row3	0.0997	0.0997	1.0000	0.0997	0.0997
Row4	0.0997	0.0997	0.0997	1.0000	0.0997
Row5	0.0997	0.0997	0.0997	0.0997	1.0000

Analysis Of GEE Parameter Estimates
Empirical Standard Error Estimates

Parameter		Estimate	Std Error	Z	Pr>|Z|
source	1	-4.4994	0.6457	-6.97	<.0001
source	2	-0.8279	0.2809	-2.95	0.0032
source	3	-0.1526	0.2744	-0.56	0.5780
source	4	0.4875	0.2698	1.81	0.0708
source	5	-0.0808	0.2738	-0.30	0.7680
size*source	1	1.0812	0.1979	5.46	<.0001
size*source	2	0.0792	0.1105	0.72	0.4738
size*source	3	-0.1894	0.1121	-1.69	0.0912
size*source	4	-0.2206	0.1081	-2.04	0.0412
size*source	5	-0.2387	0.1126	-2.12	0.0341

TABLE 11.13 Output for Problem 11.8

```
                    Working Correlation Matrix
                     Col1        Col2        Col3
             Row1    1.0000      0.8173      0.8173
             Row2    0.8173      1.0000      0.8173
             Row3    0.8173      0.8173      1.0000

                Analysis Of GEE Parameter Estimates
                Empirical Standard Error Estimates
Parameter        Estimate        Std Error         Z          Pr>|Z|
Intercept        -0.1253          0.0676         -1.85        0.0637
question 1        0.1493          0.0297          5.02        <.0001
question 2        0.0520          0.0270          1.92        0.0544
question 3        0.0000          0.0000            .            .
female            0.0034          0.0878          0.04        0.9688
```

11.8 Refer to Table 11.13 on attitudes toward legalized abortion. For the response Y_t (1 = support legalization, 0 = oppose) for question t ($t = 1, 2, 3$) and for gender g (1 = female, 0 = male), consider the model logit$[P(Y_t = 1)] = \alpha + \gamma g + \beta_t$, with $\beta_3 = 0$.

    a. A GEE analysis using unstructured working correlation gives correlation estimates 0.826 for questions 1 and 2, 0.797 for 1 and 3, and 0.832 for 2 and 3. What does this suggest about a reasonable working correlation structure?

    b. Table 11.13 shows a GEE analysis with exchangeable working correlation. Interpret effects.

    c. Treating the three responses for each subject as independent observations and performing ordinary logistic regression, $\hat{\beta}_1 = 0.149$ (SE = 0.066), $\hat{\beta}_2 = 0.052$ (SE = 0.066), and $\hat{\gamma} = 0.004$ (SE = 0.054). Give a heuristic explanation of why within-subject standard errors are much larger than with GEE, yet the between-subject standard error is smaller.

11.9 Refer to the air pollution data in Table 11.7. Using ML or GEE, fit marginal logit models that assume (**a**) marginal homogeneity, (**b**) a linear effect of time, and (**c**) no pattern. Interpret and compare.

11.10 Refer to the clinical trials data in Table 12.5, analyzed with random effects models in Section 12.3.4. Use GEE methods to analyze them, treating each center as a correlated cluster.

11.11 Refer to Table 10.5. Using GEE methods with cumulative logits, compare the two marginal distributions. Compare results to those using ML in Section 10.3.2.

11.12 Refer to the 3^4 table on government spending in Table 8.19. Analyze these data with a marginal cumulative logit model. Interpret effects.

11.13 Refer to Table 11.4.

    a. To compare effects while controlling for initial response, fit model (11.7), using scores $\{10, 25, 45, 75\}$ for time to falling asleep. Also fit the interaction model, and describe the lack of fit. (Note that for the first two baseline levels, the active and placebo treatments have similar sample response distributions at the follow-up; at higher baseline levels, the active treatment seems more successful.)

    b. Fit the interaction model

$$\text{logit}\left[P(Y_2 \leq j) \right] = \alpha_j + \beta_1 x + \beta_2 y_1 + \beta_3 x y_1$$

    that constrains effects $\{ \beta_1 x + \beta_2 y_1 + \beta_3 x y_1 \}$ to follow the pattern $(\tau, \tau, \lambda + \sigma, \lambda)$ for the active group and $(\tau, \tau, \sigma, 0)$ for the placebo group. Interpret $\hat{\lambda}$.

11.14 Find a marginal model with another type of logit that fits the insomnia data of Table 11.4 well. Interpret parameter estimates, and compare conclusions to those using cumulative logits.

11.15 Refer to Table 11.9. Combine the data for the two levels of maternal smoking. Does a first-order Markov chain model these data adequately? Find a loglinear model that does fit adequately.

11.16 Analyze Table 11.9 using a transitional model with two previous responses. Does it fit better than the first-order model of Section 11.5.5? Interpret.

11.17 Analyze Table 11.2 using a first-order transitional model. Compare interpretations to those in this chapter using marginal models.

11.18 Table 11.14 is from a longitudinal study of coronary risk factors in schoolchildren (Woolson and Clarke 1984). A sample of children aged 11–13 in 1977 were classified by gender and by relative weight (obese, not obese) in 1977, 1979, and 1981. Analyze these data.

TABLE 11.14 Data for Problem 11.18

Gender	NNN	NNO	NON	NOO	ONN	ONO	OON	OOO
				Responses[a]				
Male	119	7	8	3	13	4	11	16
Female	129	8	7	9	6	2	7	14

[a]NNN indicates not obese in 1977, 1979, and 1981; NNO indicates not obese in 1977 and 1979 but obese in 1981; and so on.

Source: Reproduced with permission from the Royal Statistical Society, London (Woolson and Clarke 1984).

11.19 Refer to the pig farmer survey of Problem 11.7 (Table 11.11). Analyze these data using marginal models with all the variables.

11.20 Refer to the cereal diet and cholesterol study of Problem 7.18 (Table 7.23). Analyze these data with marginal models.

Theory and Methods

11.21 Refer to Problem 11.1. Suppose that we expressed the data with a 3×2 partial table of drug-by-response for each subject, to use a generalized CMH procedure to test marginal homogeneity. Explain why the $911 + 279$ subjects who make the same response for every drug have no effect on the test.

11.22 Let $y_{it} = 1$ or 0 for observation t on subject i, $i = 1, \ldots, n$, $t = 1, \ldots, T$. Let $y_{.t} = \Sigma_i y_{it}/n$, $y_{i.} = \Sigma_t y_{it}/T$, and $y_{..} = \Sigma_i \Sigma_t y_{it}/nT$.

    a. Regard $\{y_{i+}\}$ as fixed. Suppose that each way to allocate the y_{i+} "successes" to y_{i+} of the observations is equally likely. Show that $E(Y_{it}) = y_{i.}$, $\text{var}(Y_{it}) = y_{i.}(1 - y_{i.})$, and $\text{cov}(Y_{it}, Y_{ik}) = -y_{i.}(1 - y_{i.})/(T - 1)$ for $t \neq k$. [*Hint:* The covariance is the same for any pair of cells in the same row, and $\text{var}(\Sigma_t Y_{it}) = 0$ since y_{i+} is fixed.]

    b. Refer to part (a). For large n with independent subjects, explain why $(Y_{.1}, \ldots, Y_{.T})$ is approximately multivariate normal with pairwise correlation $\rho = -1/(T - 1)$. Conclude that Cochran's Q statistic (Cochran 1950)

$$Q = \frac{n^2(T - 1)\Sigma_{t=1}^{T}(y_{.t} - y_{..})^2}{T\Sigma_{i=1}^{n} y_{i.}(1 - y_{i.})}$$

    is approximately chi-squared with df $= (T - 1)$. [One way notes that if $(X_1, \ldots, X_T)$ is multivariate normal with common mean and common variance σ^2 and common correlation ρ for pairs (X_t, X_k), then $\Sigma(X_t - \bar{X})^2/\sigma^2(1 - \rho)$ is chi-squared with df $= (T - 1)$. See Bhapkar and Somes (1977) for slightly weaker conditions for a chi-squared limiting distribution for Q than those in part (a).]

    c. Show that Q is unaffected by deleting cases in which $y_{i1} = \cdots = y_{iT}$.

11.23 Consider the model $\mu_i = \beta$, $i = 1, \ldots, n$, assuming that $v(\mu_i) = \mu_i$. Suppose that actually $\text{var}(Y_i) = \mu_i^2$. Using the univariate version of GEE described in Section 11.4, show that $u(\beta) = \Sigma_i(y_i - \beta)/\beta$ and $\hat{\beta} = \bar{y}$. Show that V in (11.10) equals β/n, the actual asymptotic variance (11.11) simplifies to β^2/n, and its consistent estimate is $\Sigma_i(y_i - \bar{y})^2/n^2$.

11.24 Repeat Problem 11.23 assuming that $v(\mu_i) = \sigma^2$ when actually var(Y_i) = μ_i.

11.25 Consider the model $\mu_i = \beta$, $i = 1, \dots, n$, for independent Poisson observations. For $\hat{\beta} = \bar{y}$, show that the model-based asymptotic variance estimate is $\bar{y}/n$, whereas the robust estimate of the asymptotic variance is $\Sigma_i(y_i - \bar{y})^2/n^2$. Which would you expect to be better **(a)** if the Poisson model holds, and **(b)** if there is severe overdispersion?

11.26 Show that (11.10) is equivalent to the formula for the large-sample covariance of the ML estimator in a GLM, estimated by (4.28).

11.27 **a.** For a univariate response, how is quasi-likelihood (QL) inference different from ML inference? When are they equivalent?

b. Explain the sense in which GEE methodology is a multivariate version of QL.

c. Summarize the advantages and disadvantages of the QL approach.

d. Describe conditions under which GEE parameter estimators are consistent and conditions under which they are not. For conditions in which they are consistent, explain why.

11.28 Formulate a model using adjacent-categories logits or continuation-ratio logits that is analogous to (11.4). Interpret parameters.

11.29 Refer to the analysis of mean time to falling asleep at the end of Section 11.2.3. Explain how to calculate SE for the difference between the difference of means reported there. (Note that one difference uses paired samples and the other uses independent samples.)

11.30 What is wrong with this statement?: "For a first-order Markov chain, Y_t is independent of Y_{t-2}."

11.31 Suppose that loglinear model $(Y_0, Y_1, \dots, Y_T)$ holds. Is this a Markov chain?

11.32 Gamblers A and B have a total of I dollars. They play games of pool repeatedly. Each game they each bet $1, and the winner takes the other's dollar. The outcomes of the games are statistically independent, and A has probability π and B has probability $1 - \pi$ of winning any game. Play stops when one player has all the money. Let Y_t denote A's monetary total after t games.

a. Show that $\{Y_t\}$ is a first-order Markov chain.

b. State the transition probability matrix. (For this *gambler's ruin* problem, 0 and I are *absorbing* states. Eventually, the chain enters one of these and stays. The other states are *transient*.)

11.33 A first-order Markov chain has *stationary* (or *time-homogeneous*) transition probabilities if the one-step transition probability matrices are identical, that is, if for all i and j,

$$\pi_{j|i}(1) = \pi_{j|i}(2) = \cdots = \pi_{j|i}(T) = \pi_{j|i}.$$

Let X, Y, and Z denote the classifications for the $I \times I \times T$ table consisting of $\{n_{ij}(t), i = 1, \ldots, I, j = 1, \ldots, I, t = 1, \ldots, T\}$.

a. Explain why all transition probabilities are stationary if expected frequencies for this table satisfy loglinear model (XY, XZ). [Thus, the likelihood-ratio statistic for testing stationary transition probabilities equals G^2 for testing fit of model (XY, XZ).]

b. Let $n_{ij} = \Sigma_t n_{ij}(t)$. Under the assumption of stationary transition probabilities, show how the likelihood in (11.14) simplifies, and show that the ML estimators are

$$\hat{\pi}_{j|i} = n_{ij}/n_{i+}.$$

c. For a Markov chain with stationary transition probabilities, let y_{ijk} denote the number of transitions from i to j to k over two successive steps. For $\{y_{ijk}\}$, argue that the goodness of fit of loglinear model (Y_1Y_2, Y_2Y_3) tests that the chain is first order against the alternative that it is second order (Anderson and Goodman 1957).

CHAPTER 12

Random Effects: Generalized Linear Mixed Models for Categorical Responses

In Chapter 11 we noted that observations often occur in clusters. For instance, cluster i might consist of repeated measurements on subject i or observations for all subjects in family i. Observations within a cluster tend to be more alike than observations from different clusters. Thus, they are usually positively correlated. Ordinary analyses that ignore the correlation and treat within-cluster observations the same as between-cluster observations produce invalid standard errors.

In Chapter 11 we focused on modeling the *marginal* distributions of clustered responses, treating the joint dependence structure as a nuisance. In this chapter we present an alternative approach using cluster-level terms in the model. These terms take the same value for each observation in a cluster but different values for different clusters. They are unobserved and, when treated as varying randomly among clusters, are called *random effects*. In Section 10.2.4 we introduced this approach in a model for matched pairs. The models have *conditional* interpretations, referred to as *subject-specific* when each cluster is a subject. This contrasts with marginal models, which have *population-averaged* interpretations.

Random effects models for normal responses are well established. By contrast, only recently have random effects been used much in models for categorical data. In this chapter we extend generalized linear models to include random effects. In Section 12.1 we introduce this extension, the *generalized linear mixed model*. In Section 12.2 we discuss an important special case for binary data, the *logistic-normal model*. Several examples are shown in Section 12.3. Section 12.4 covers extensions for multinomial responses, and Section 12.5 covers models with multivariate random effects. In Section 12.6 we discuss model fitting, assuming normality for the random effects. Parts of this chapter are from Agresti et al. (2000).

12.1 RANDOM EFFECTS MODELING OF CLUSTERED CATEGORICAL DATA

Parameters that describe a factor's effects in ordinary linear models are called *fixed effects*. They apply to *all* categories of interest, such as genders, age groupings, or treatments. By contrast, random effects usually apply to a *sample*. For a study using a sample of clinics, for example, the model treats observations from a given clinic as a cluster, and it has a random effect for each clinic.

GLMs extend ordinary regression by allowing nonnormal responses and a link function of the mean. The *generalized linear mixed model* (GLMM) is a further extension that permits random effects as well as fixed effects in the linear predictor.

12.1.1 Generalized Linear Mixed Model

Let y_{it} denote observation t in cluster i, $t = 1, \ldots, T_i$. As in the GEE analyses in Chapter 11, the number of observations may vary by cluster. In a longitudinal study, even if clusters have equal size, many of them may have missing observations. Let $\mathbf{x}_{it}$ denote a column vector of values of explanatory variables, for fixed effect model parameters $\boldsymbol{\beta}$. Let $\mathbf{u}_i$ denote the vector of random effect values for cluster i. This is common to all observations in the cluster. Let $\mathbf{z}_{it}$ denote a column vector of their explanatory variables. Often, the random effect is univariate.

Conditional on $\mathbf{u}_i$, a GLMM resembles an ordinary GLM. Let $\mu_{it} = E(Y_{it} \mid \mathbf{u}_i)$. The linear predictor for a GLMM has the form

$$g(\mu_{it}) = \mathbf{x}'_{it}\boldsymbol{\beta} + \mathbf{z}'_{it}\mathbf{u}_i \qquad (12.1)$$

for link function $g(\cdot)$. The random effect vector $\mathbf{u}_i$ is assumed to have a multivariate normal distribution $N(\mathbf{0}, \boldsymbol{\Sigma})$. The covariance matrix $\boldsymbol{\Sigma}$ depends on unknown *variance components* and possibly also correlation parameters.

Denote $\mathrm{var}(Y_{it} \mid \mathbf{u}_i) = \phi_{it} v(\mu_{it})$, where the variance function $v(\cdot)$ describes how the (conditional) variance depends on the mean. As in Section 4.4, often $\phi_{it} = 1$ or $\phi_{it} = \phi / \omega_{it}$, where ω_{it} is a known weight (e.g., number of trials for a binomial count) and ϕ is an unknown dispersion parameter. Conditional on $\mathbf{u}_i$, the model treats $\{y_{it}\}$ as independent over i and t. As discussed in Section 10.2.2, the variability among $\mathbf{u}_i$ induces a nonnegative association among the responses, for the marginal distribution averaged over the subjects. This is caused by the shared random effect $\mathbf{u}_i$ for each observation in a cluster.

In (12.1), the random effect enters the model on the same scale as the predictor terms. This is convenient but also natural for many applications. For instance, random effects sometimes represent heterogeneity caused by omitting certain explanatory variables. Consider the special case with univariate random effect and $z_{it} = 1$. With u_i replaced by $u_i^* \sigma$ where $\{u_i^*\}$ are $N(0, 1)$, the GLMM has the form

$$g(\mu_{it}) = \mathbf{x}'_{it}\boldsymbol{\beta} + u_i^* \sigma.$$

This has the form of an ordinary GLM with unobserved values $\{u_i^*\}$ of a particular covariate. Thus, random effects models relate to methods of dealing with unmeasured predictors and other forms of missing data. The random effects part of the linear predictor reflects terms that would be in the fixed effects part if those explanatory variables had been included. Random effects also sometimes represent random measurement error in the explanatory variables. If we replace a particular predictor x_{it} by $x_{it}^* + \epsilon_i$, with x_{it}^* the true value and ϵ_i the measurement error, then ϵ_i times the regression parameter can be absorbed in the random effects term. Related to these motivations, random effects also provide a mechanism for explaining overdispersion in basic models not having those effects (Breslow and Clayton 1993).

12.1.2 Logit GLMM for Binary Matched Pairs

We illustrate the GLMM expression (12.1) using a simple case, that of binary matched pairs. The data form two dependent binomial samples (Section 10.1). Cluster i consists of the responses (y_{i1}, y_{i2}) for matched pair i. Observation t in cluster i has $y_{it} = 1$ (a success) or 0 (a failure), $t = 1, 2$.

In Section 10.2.2 we introduced the model (Cox 1958b, Rasch 1961)

$$\text{logit}[P(Y_{it} = 1)] = \alpha_i + \beta x_t \tag{12.2}$$

where $x_1 = 0$ and $x_2 = 1$. For it, β is a cluster-specific log odds ratio. That section treated α_i as a fixed effect and eliminated it using conditional ML. An equivalent representation of (12.2) is

$$\text{logit}[P(Y_{i1} = 1 | u_i)] = \alpha + u_i, \qquad \text{logit}[P(Y_{i2} = 1 | u_i)] = \alpha + \beta + u_i, \tag{12.3}$$

where $u_i = \alpha_i - \alpha$ for some constant α. Now, we treat u_i as a random effect for cluster i, with $\{u_i\}$ independent from a $N(0, \sigma^2)$ distribution with σ unknown. Conditionally on u_i, we assume that y_{i1} and y_{i2} are independent.

Model (12.3) is the special case of (12.1) in which $\mu_{it} = P(Y_{it} = 1 | u_i)$, $g(\cdot)$ is the logit link, $\boldsymbol{\beta}' = (\alpha, \beta)$, $\mathbf{x}'_{i1} = (1, 0)$ and $\mathbf{x}'_{i2} = (1, 1)$ for all i, and $z_{it} = 1$ for all i and t. The univariate random effect adjusts the intercept but does not modify the fixed effect. A GLMM with random effect of this form is called a *random intercept* model. Instead of the usual fixed intercept α, it has a random intercept $\alpha + u_i$.

Let $Y_1 = \Sigma_i y_{i1}$ and $Y_2 = \Sigma_i y_{i2}$. Marginally, Y_1 is binomial with n trials and parameter $E\{\exp(\alpha + U)/[1 + \exp(\alpha + U)]\}$, and Y_2 is binomial with parameter $E\{\exp(\alpha + \beta + U)/[1 + \exp(\alpha + \beta + U)]\}$. The expectations refer to U, a $N(0, \sigma^2)$ random variable. The model implies a nonnegative correlation between Y_1 and Y_2, with greater association resulting from greater heterogeneity (i.e., larger σ). Clusters with a large positive u_i have a relatively large $P(Y_{it} = 1 | u_i)$ for each t, whereas clusters with a large negative u_i have a relatively small $P(Y_{it} = 1 | u_i)$ for each. For this model, Y_1 and Y_2 are independent only if $\sigma = 0$.

A 2×2 population-averaged table with (success, failure) for both the row and column categories summarizes the number of observations for which $(y_{i1}, y_{i2}) = (1, 1)$, $(1, 0)$, $(0, 1)$, or $(0, 0)$. Let $\{n_{ab}\}$ denote these counts. Table 12.1, analyzed first in Section 10.1, is an example. Let $\{\hat{\mu}_{ab}\}$ denote marginal fitted values for model (12.3). We defer discussion of model fitting until Section 12.6. However, model (12.3) is a rare instance in which the fixed effect in a random effects model has a closed-form ML estimate,

$$\hat{\beta} = \log(\hat{\mu}_{21}/\hat{\mu}_{12}).$$

When the sample log odds ratio $\log(n_{11}n_{22}/n_{12}n_{21}) \geq 0$, then $\{\hat{\mu}_{ab} = n_{ab}\}$ and $\hat{\beta} = \log(n_{21}/n_{12})$. This is the same as the conditional ML estimate (Section 10.2.3). Neuhaus et al. (1994) showed that this is true for *any* parametric choice of random effects distribution for which the model (12.3) can generate $\{n_{ab}\}$ as fitted values. Lindsay et al. (1991) showed that this estimate also results with a nonparametric approach discussed in Section 13.2.4. The model implies that the true log odds ratio for this 2×2 table is at least 0. When $\log(n_{11}n_{22}/n_{12}n_{21}) < 0$, however, then $\hat{\sigma} = 0$ and the fitted values $\{\hat{\mu}_{ab} = n_{a+}n_{+b}/n\}$ satisfy independence. Then, $\hat{\beta}$ is identical to the estimate for the marginal model (10.6) by which β is the difference between logits for the two marginal distributions, namely $\hat{\beta} = \log[(n_{2+}n_{+1})/(n_{1+}n_{+2})]$.

12.1.3 Ratings of Prime Minister Revisited

For Table 12.1, the ML fit of model (12.3), treating $\{u_i\}$ as normal, yields $\hat{\beta} = \log(86/150) = -0.556$ (SE = 0.135), with $\hat{\sigma} = 5.16$. This is identical to the conditional ML estimate (10.10), with standard error $[(1/86) + (1/150)]^{1/2}$. For a given subject, the estimated odds of approval at the second

TABLE 12.1 Rating of Performance of Prime Minister

First Survey	Second Survey		Total
	Approve	Disapprove	
Approve	794	150	944
Disapprove	86	570	656
Total	880	720	1600

survey equal $\exp(-0.556) = 0.57$ times those at the first survey. The large $\hat{\sigma}$ reflects the very strong association between the two responses, with sample odds ratio 35.1.

12.1.4 Extension: Rasch Model and Item Response Models

An extension of the logit matched-pairs model (12.3) allows $T > 2$ observations in each cluster. The random intercept model then has form

$$\text{logit}\left[P(Y_{it} = 1 \mid u_i) \right] = u_i + \beta_t, \tag{12.4}$$

where $\{u_i\}$ are independent $N(0, \sigma^2)$. Equivalently, the model can add an intercept α or let $E(u_i) = \alpha$, but then identifiability requires a constraint such as $\beta_T = 0$.

Early applications of this GLMM were in psychometrics. The model describes responses to a battery of T questions on an exam. The probability $P(Y_{it} = 1 \mid u_i)$ that subject i makes the correct response on question t depends on the overall ability of subject i, characterized by u_i, and the easiness of question t, characterized by β_t. Such models are called *item-response models*. The logit form (12.4) is called the *Rasch model* (Rasch 1961). In estimating $\{\beta_t\}$, Rasch treated $\{u_i\}$ as fixed effects and used conditional ML, as outlined in Section 10.2.3 for matched pairs. Later authors used the normal random effects approach for this model and the model with probit link (e.g., Bock and Aitkin 1981).

The $\{\beta_t\}$ in the Rasch model differ from parameters in corresponding marginal models such as (11.1), since the effects are subject specific. The Rasch model refers to a $T \times 2 \times n$ table of observation by outcome by subject, whereas the marginal model refers to the $T \times 2$ observation-by-outcome table of the T marginal distributions, collapsed over subjects. For observations s and t for a given subject i with model (12.4),

$$\beta_s - \beta_t = \text{logit}\left[P(Y_{is} = 1 \mid u_i) \right] - \text{logit}\left[P(Y_{it} = 1 \mid u_i) \right],$$

which is a log odds ratio conditional on the subject. By contrast, the corresponding population-averaged effect in marginal model (11.1) is

$$\beta_s - \beta_t = \text{logit}\left[P(Y_{hs} = 1) \right] - \text{logit}\left[P(Y_{it} = 1) \right],$$

with subject h randomly selected for observation s and subject i randomly selected for observation t (i.e., h and i are *independent* observations).

12.1.5 Random Effects versus Conditional ML Approaches

Suppose that one treated $\{u_i\}$ in model (12.4) as fixed effects instead of random effects. Then, consider ordinary ML estimation of $\{\beta_t\}$ and $\{u_i\}$. As n increases, so does the number of parameters, since each subject has a u_i.

Even though the number of $\{\beta_i\}$ does not increase as n does, the ordinary ML estimators $\{\hat{\beta}_i\}$ are not consistent. This happens in many models when the number of parameters has an order similar to that of the number of subjects. Asymptotic optimality properties of ML estimators, such as consistency, require the number of parameters to be fixed as n increases. For model (12.4), ML estimators of $\{\beta_i\}$ have bias of order $T/(T-1)$ (Andersen 1980, pp. 244–245). For the matched-pairs model (12.2), for instance, $\hat{\beta} \rightarrow 2\beta$ in probability (Problem 10.24).

For this reason, the preferable approach for the fixed effects model is *conditional ML*. One eliminates $\{u_i\}$ by conditioning on their sufficient statistics $\{S_i = \sum_t y_{it}, \ i = 1, \ldots, n\}$. In the item response context, these are the numbers of correct responses for each subject. Conditional on $\{S_i\}$, the distribution of $\{y_{it}\}$ is independent of $\{u_i\}$. Maximizing the resulting likelihood then yields consistent estimators of $\{\beta_i\}$. The analysis generalizes the one in Section 10.2.3 for the subject-specific logistic model (10.8) for matched pairs. See Andersen (1980) for details.

Compared with the random effects approach, the conditional ML approach has certain advantages. One does not need to assume a parametric distribution for $\{u_i\}$. It is difficult to check this assumption in the random effects approach. Conditional ML is also appropriate with retrospective sampling. In that case, bias can occur with a random effects approach because the clusters are not randomly sampled (Neuhaus and Jewell 1990b).

However, the conditional ML approach has severe disadvantages. It is restricted to the canonical link (the logit), for which reduced sufficient statistics exist for $\{u_i\}$. More important, as discussed in Section 10.2.7, it is restricted to inference about within-cluster fixed effects. The conditioning removes the source of variability needed for estimating between-cluster effects in models with explanatory variables such as those considered next. Also, this approach does not provide information about $\{u_i\}$, such as predictions of their values and estimates of their variability or of the probabilities they determine. Finally, in more general models with covariates, conditional ML can be less efficient than the random effects approach for estimating the fixed effects (see Note 12.2).

12.2 BINARY RESPONSES: LOGISTIC-NORMAL MODEL

The item response model (12.4) with random intercept is a special case of an important class of random effects models for binary data called *logistic-normal models*. With univariate random effect, the model form is

$$\text{logit}[P(Y_{it} = 1 \mid u_i)] = \mathbf{x}'_{it}\boldsymbol{\beta} + u_i \tag{12.5}$$

where $\{u_i\}$ are independent $N(0, \sigma^2)$ variates. This is the special case of the GLMM (12.1) in which $g(\cdot)$ is the logit link and the random effects structure simplifies to a random intercept. The logistic-normal model has a long history, dating at least to Cox (1970, Prob. 20 in that text) for the matched-pairs model (12.3) and Pierce and Sands (1975).

More generally, the link function in model (12.5) can be an arbitrary inverse cdf. For such models, Y_{is} and Y_{it} are treated conditionally (given u_i) as independent but are marginally nonnegatively correlated. Let Φ denote the cdf that is the inverse link function. Then, for $s \neq t$,

$$\text{cov}(Y_{is}, Y_{it}) = E\left[\text{cov}(Y_{is}, Y_{it} \mid u_i)\right] + \text{cov}\left[E(Y_{is} \mid u_i), E(Y_{it} \mid u_i)\right]$$

$$= 0 + \text{cov}\left[\Phi(\mathbf{x}'_{is}\boldsymbol{\beta} + u_i), \Phi(\mathbf{x}'_{it}\boldsymbol{\beta} + u_i)\right]. \tag{12.6}$$

The functions in the last covariance term are both monotone increasing in u_i, and hence are nonnegatively correlated. For common predictor value $\mathbf{x}$ at each t, the joint distribution for the model is exchangeable. This is often plausible for clustered data. In longitudinal studies, however, observations closer together in time may tend to be more highly correlated.

Usually, the main focus in using a GLMM is inference about the fixed effects. The random effects part of the model is a mechanism for representing how the positive correlation occurs between observations within a cluster. Parameters pertaining to the random effects may themselves be of interest, however. For instance, the estimate $\hat{\sigma}$ of the standard deviation of a random intercept may be a useful summary of the degree of heterogeneity of a population.

12.2.1 Interpreting Heterogeneity in Logistic-Normal Models

When $\sigma = 0$, the logistic-normal model (12.5) simplifies to the ordinary logistic regression model treating all observations as independent. When $\sigma > 0$, how can we interpret the variability in effects this model implies?

Consider observation y_{it} at setting $\mathbf{x}_{it}$ of predictors and observation y_{hs} at setting $\mathbf{x}_{hs}$. Their log odds ratio is

$$\text{logit}\left[P(Y_{it} = 1 \mid u_i)\right] - \text{logit}\left[P(Y_{hs} = 1 \mid u_h)\right] = (\mathbf{x}_{it} - \mathbf{x}_{hs})'\boldsymbol{\beta} + (u_i - u_h).$$

We cannot observe $(u_i - u_h)$, which has a $N(0, 2\sigma^2)$ distribution. However, $100(1 - \alpha)\%$ of those log odds ratios fall within

$$(\mathbf{x}_{it} - \mathbf{x}_{hs})'\boldsymbol{\beta} \pm z_{\alpha/2}\sqrt{2}\,\sigma. \tag{12.7}$$

When $\sigma = 0$, $(\mathbf{x}_{it} - \mathbf{x}_{hs})'\boldsymbol{\beta}$ is the usual form of log odds ratio for a model without random effects. When $\sigma > 0$, $(\mathbf{x}_{it} - \mathbf{x}_{hs})'\boldsymbol{\beta}$ is the log odds ratio for two observations in the same cluster ($h = i$) or with the same random effect value. Suppose that $\mathbf{x}_{it} = \mathbf{x}_{hs}$ for observations from different clusters. Then, using $z_{0.25} = 0.674$, the middle 50% of the log odds ratios fall within

$\pm 0.674\sqrt{2}\,\sigma = \pm 0.95\sigma$. Hence, the median odds ratio between the observa tion with higher random effect and the observation with lower random effect equals $\exp(0.95\sigma)$. With a single predictor and $x_{it} - x_{hs} = 1$, the median such odds ratio equals $\exp(\beta + 0.95\sigma)$. Larsen et al. (2000) presented related interpretations.

12.2.2 Connections between Conditional Models and Marginal Models

The fixed effects parameters $\boldsymbol{\beta}$ in GLMMs have conditional intepretations, given the random effect. Those fixed effects are of two types. First, consider an explanatory variable that varies in value among observations in a cluster. For instance, in a crossover study comparing T drugs, for each subject the drug taken varies from observation to observation in that subject's cluster of T observations. For such an explanatory variable, its coefficient in the model refers to the effect on the response of a within-cluster (e.g., subject-specific) 1-unit increase of that predictor. The random effect as well as other explana tory variables in the model are constant while that predictor increases by 1. The effect of that explanatory variable is a "within-cluster" or "within-sub ject" one.

Second, consider an explanatory variable with constant value among observations in a cluster. An example is gender when each subject forms a cluster. For such an explanatory variable, its coefficient refers to the effect on the response of a "between-cluster" 1-unit increase of that predictor. An example is a comparison of females and males using a dummy variable and its coefficient. However, this fixed effect in the GLMM applies only when the random effect (as well as other explanatory variables in the model) takes the same value in both groups: for instance, a male and a female with the same value for their random effects.

It is in this sense that random effects models are conditional models, as both within- and between-cluster effects apply conditional on the random effect value. By contrast, effects in marginal models are averaged over all clusters (i.e., population averaged), so those effects do not refer to a compari son at a fixed value of a random effect. In fact, a fundamental difference between the two model types is that when the link function is nonlinear, such as the logit, the population-averaged effects of marginal models often are smaller than the cluster-specific effects of GLMMs.

Specifically, the GLMM (12.1) refers to the conditional mean, $\mu_{it} = E(Y_{it} \mid \mathbf{u}_i)$. By inverting the link function,

$$E(Y_{it} \mid \mathbf{u}_i) = g^{-1}(\mathbf{x}'_{it}\boldsymbol{\beta} + \mathbf{z}'_{it}\mathbf{u}_i).$$

Marginally, averaging over the random effects, the mean is

$$E(Y_{it}) = E[E(Y_{it} \mid \mathbf{u}_i)] = \int g^{-1}(\mathbf{x}'_{it}\boldsymbol{\beta} + \mathbf{z}'_{it}\mathbf{u}_i)f(\mathbf{u}_i; \boldsymbol{\Sigma})\,d\mathbf{u}_i,$$

where $f(\mathbf{u}; \mathbf{\Sigma})$ is the $N(\mathbf{0}, \mathbf{\Sigma})$ density function for the random effects. For the identity link,

$$E(Y_{it}) = \int (\mathbf{x}'_{it}\mathbf{\beta} + \mathbf{z}'_{it}\mathbf{u}_i)f(\mathbf{u}_i; \mathbf{\Sigma})d\mathbf{u}_i = \mathbf{x}'_{it}\mathbf{\beta} .$$

The marginal model has the same model form and effects $\mathbf{\beta}$. This is not true for other links. For instance, for the logistic-normal model (12.5),

$$E(Y_{it}) = E\left[\frac{\exp(\mathbf{x}'_{it}\mathbf{\beta} + u_i)}{1 + \exp(\mathbf{x}'_{it}\mathbf{\beta} + u_i)} \right].$$

This expectation does not have form $\exp(\mathbf{x}'_{it}\mathbf{\beta})/[1 + \exp(\mathbf{x}'_{it}\mathbf{\beta})]$ except when u_i has a degenerate distribution ($\sigma = 0$).

Approximate relationships exist between estimates from the two model types. In the logistic-normal case with effect $\mathbf{\beta}$ and small σ, Zeger et al. (1988) showed that

$$E(Y_{it}) \approx \exp(c\mathbf{x}'_{it}\mathbf{\beta})/[1 + \exp(c\mathbf{x}'_{it}\mathbf{\beta})], \tag{12.8}$$

where $c = [1 + 0.6\sigma^2]^{-1/2}$. Since the effect in the marginal model multiplies that of the conditional model by about c, it is typically smaller in absolute value. The discrepancy increases as σ increases. For $\mathbf{\beta}$ near 0, Neuhaus et al. (1991) showed that the marginal model effect is approximately $\mathbf{\beta}(1 - \rho)$, where $\rho = \text{corr}(Y_{it}, Y_{is})$ at $\mathbf{\beta} = \mathbf{0}$. Again, the discrepancy increases as σ increases, since ρ increases with σ.

For Table 12.1 on ratings of the prime minister, the ML estimate for model (12.3) is $\hat{\beta} = -0.556$, with $\hat{\sigma} = 5.16$ for variability of $\{u_i\}$. Approximation (12.8) suggests that $\hat{\beta} = -0.556$ with $\hat{\sigma} = 5.16$ corresponds to a marginal estimate of about $[1 + 0.6(5.16)^2]^{-1/2}(-0.556) = -0.135$. The actual marginal estimate is the log odds ratio for the sample marginal distributions, equaling

$$\log[(880/720)/(944/656)] = -0.163.$$

In fact, the marginal effect is much smaller than the conditional effect, but this approximation connecting the two estimates works better for smaller $\hat{\sigma}$. At $\beta = 0$, the fit of the model is that of the symmetry model, for which $\hat{\mu}_{12} = \hat{\mu}_{21} = (n_{12} + n_{21})/2$. The correlation for that 2×2 table equals 0.699, from which the conditional estimate of -0.556 suggests a marginal estimate of $-0.556(1 - 0.699) = -0.167$, very close to the actual value of -0.163.

Figure 12.1 illustrates why the marginal effect is smaller than the conditional effect. For a single explanatory variable x, the figure shows subject-specific curves for $P(Y_{it} = 1|u_i)$ for several subjects when considerable heterogeneity exists. This corresponds to a relatively large σ for random effects.

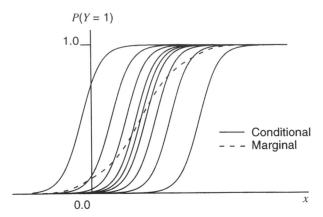

FIGURE 12.1 Logistic random-intercept model, showing the conditional (subject-specific) curves and the marginal (population-averaged) curve averaging over these.

At any fixed value of x, variability occurs in the conditional means, $E(Y_{it} | u_i)$ $= P(Y_{it} = 1 | u_i)$. The average of these is the marginal mean, $E(Y_{it})$. These averages for various x values yield the superimposed curve. It has a shallower slope. In fact, it does not exactly follow the logistic formula. Similar remarks apply to other GLMMs. For the probit link with binary data, however, the conditional probit model with normal random effect does imply a marginal model of probit form (Problem 12.29). With univariate random intercept, the marginal effect equals the conditional effect multiplied by $[1 + \sigma^2]^{-1/2}$ (Zeger et al. 1988). In Section 13.5.1 we explore the conditional–marginal connection for loglinear GLMMs.

12.2.3 Comments about Conditional versus Marginal Models

Random effects models describe conditional (subject-specific) effects, whereas marginal models describe population-averaged effects. Some statisticians prefer one of these types, but most feel that both are useful, depending on the application.

The conditional modeling approach is preferable if one wants to specify a mechanism that could generate positive association among clustered observations, estimate cluster-specific effects, estimate their variability, or model the joint distribution. Latent variable constructions used to motivate model forms (e.g., the tolerance motivation for binary models of Section 6.6.1 and the related threshold motivation in Problem 6.28 and utility motivation in Problem 6.29) usually apply more naturally at the cluster level than at the marginal level. Given a conditional model, one can recover information about marginal distributions. That is, a conditional model implies a marginal model, but a marginal model does not itself imply a conditional model (although see Note 12.10 for an implicit connection).

In many surveys or epidemiological studies, a goal is to compare the relative frequency of occurrence of some outcome for different groups in a population. Then, quantities of primary interest include between-group odds ratios among marginal probabilities for the different groups. That is, effects of interest are between-cluster rather than within-cluster. When marginal effects are the main focus, it is usually simpler and may be preferable to model the margins directly. One can then parameterize the model so that regression parameters have a direct marginal interpretation. Developing a more detailed model of the joint distribution that generates those margins, as a random effects model does, provides greater opportunity for misspecification. For instance, with longitudinal data the assumption that observations are independent, given the random effect, need not be realistic. With the marginal model approach, we showed in Chapter 11 that ML is sometimes possible but that the GEE approach is computationally simpler and more versatile. A drawback of the GEE approach is that it does not explicitly model random effects and therefore does not allow these effects to be estimated. In addition, likelihood-based inferences are not possible because the joint distribution of the responses is not specified.

In Section 12.2.2 it was noted that conditional effects are usually larger than marginal effects, and increase as variance components increase. Usually, though, the significance of an effect (e.g., as measured by the ratio of estimate to standard error) is similar in the two model types. If one effect seems more important than another in a conditional model, the same is usually true with a marginal model. So the choice of the model is usually not crucial to inferential conclusions.

This statement requires a caveat, however, since sizes of effects in marginal models depend on the degree of heterogeneity in conditional models. In comparing effects for two groups or two variables that have quite different variance components, relative sizes of effects will differ for marginal and conditional models. From (12.8), with binary data the attenuation from the conditional to the marginal effect will tend to be greater for the group having the larger variance component. For instance, suppose that two groups, one young in age and the other elderly, both show the same conditional effect in a crossover study comparing two drugs. If the elderly group has more heterogeneity on the response, their marginal effect may be smaller than that for the younger group. The marginal effects differ even though the conditional effects are the same, because of the greater variance component for the elderly. In such cases, the conditional effect (appropriately modeled) may have more relevance.

Finally, with either marginal or conditional models, missing data are a common problem with multivariate responses. Unless data are missing at random, potential bias occurs in ML inference. GEE methods usually require the stronger condition that data are missing completely at random (Section 11.4.5). Thus, modeling missingness or conducting a sensitivity study to discern its potential effects can be an important component of an analysis.

Regardless of the choice of paradigm, it is a challenge for statisticians even to explain to practitioners why marginal and conditional effects differ with a nonlinear link function. Graphics such as Figure 12.1 can help. Neuhaus (1992) and Pendergast et al. (1996) surveyed ways of analyzing clustered binary data, including conditional and marginal models. Agresti and Natarajan (2001) surveyed conditional and marginal modeling of clustered ordinal data.

12.3   EXAMPLES OF RANDOM EFFECTS MODELS FOR BINARY DATA

In the next three sections we present a variety of examples of random effects models. In this section we consider binary responses.

12.3.1   Small-Area Estimation of Binomial Proportions

Small-area estimation refers to estimation of parameters for a large number of geographical areas when each has relatively few observations. For instance, one might want county-specific estimates of characteristics such as the unemployment rate or the proportion of families having health insurance coverage. With a national or statewide survey, some counties may have few observations. Then, sample proportions in the counties may poorly estimate the true countywide proportions. Random effects models that treat each county as a cluster can provide improved estimates. In assuming that the true proportions vary according to some distribution, the fitting process "borrows from the whole"—it uses data from all the counties to estimate the proportion in any given one.

Let π_i denote the true proportion in area i, $i = 1, \ldots, n$. These areas may be all the ones of interest, or only a sample. Let $\{y_i\}$ denote independent $\text{bin}(T_i, \pi_i)$ variates; that is, $y_i = \sum_{t=1}^{T_i} y_{it}$, where $\{y_{it}, t = 1, \ldots, T_i\}$ are independent with $P(Y_{it} = 1) = \pi_i$ and $P(Y_{it} = 0) = 1 - \pi_i$. The sample proportions $\{p_i = y_i/T_i\}$ are ML estimates of $\{\pi_i\}$ for the fixed-effects model

$$\text{logit}(\pi_i) = \alpha + \beta_i, \qquad i = 1, \ldots, n.$$

This model is saturated, having n nonredundant parameters (with a constraint such as $\sum_i \beta_i = 0$) for the n binomial observations.

For small $\{T_i\}$, $\{p_i\}$ have large standard errors. Thus, $\{p_i\}$ may display much more variability than $\{\pi_i\}$, especially when $\{\pi_i\}$ are similar. Then, it is helpful to shrink $\{p_i\}$ toward their overall mean. One can accomplish this with the random effects model

$$\text{logit}\big[P(Y_{it} = 1 \,|\, u_i)\big] = \alpha + u_i, \tag{12.9}$$

where $\{u_i\}$ are independent $N(0, \sigma^2)$ variates. This model is a logit analog of one-way random effects ANOVA. When $\sigma = 0$, all π_i are identical.

For this model,

$$\hat{\pi}_i = \exp(\hat{\alpha} + \hat{u}_i) / \big[1 + \exp(\hat{\alpha} + \hat{u}_i)\big].$$

This estimate differs from the sample proportion p_i. If $\hat{\sigma} = 0$, then all $\hat{u}_i = 0$. Then, the random effects estimate of each π_i is $(\sum_{i=1}^{n} \sum_{t=1}^{T_i} y_{it}) / (\sum_i T_i)$, the overall sample proportion after pooling all n samples. When truly all π_i are equal, this is a much better estimator of that common value than the sample proportion from a single sample.

Generally, the random effects model estimators shrink the separate sample proportions toward the overall sample proportion. The amount of shrinkage decreases as $\hat{\sigma}$ increases. The shrinkage also decreases as the $\{T_i\}$ grow; as each sample has more data, we put more trust in the separate sample proportions. The predicted random effect $\hat{u}_i$ is the estimated mean of the distribution of u_i, given the data (see Section 12.6.7). This prediction depends on all the data, not just data from area i. A benefit is potential reduction in the mean-squared error of the estimates around the true values.

We illustrate model (12.9) with a simulated sample of size 2000 to mimic a poll taken before the 1996 U.S. presidential election. For T_i observations in state i ($i = 1, \ldots, 51$, where $i = 51$ is DC = District of Columbia), y_i is bin(T_i, π_i), where π_i is the actual proportion of votes in state i for Bill Clinton in the 1996 election, conditional on voting for Clinton or the Republican candidate, Bob Dole. Here, T_i is proportional to the state's population size, subject to $\sum_i T_i = 2000$. Table 12.2 shows $\{T_i\}$, $\{\pi_i\}$, and $\{p_i = y_i/T_i\}$.

For the ML fit of model (12.9), $\hat{\alpha} = 0.163$ and $\hat{\sigma} = 0.29$. The predicted random effect values (obtained using PROC NLMIXED in SAS) yield the proportion estimates $\{\hat{\pi}_i\}$, also shown in Table 12.2. Since $\{T_i\}$ are mostly small and since $\hat{\sigma}$ is relatively small, considerable shrinkage of these estimates occurs from the sample proportions toward the overall proportion supporting Clinton, which was 0.548. The $\{\hat{\pi}_i\}$ vary only between 0.468 (for TX = Texas) and 0.696 (for NY = New York), whereas the sample proportions vary between 0.111 (for Idaho) and 1.0 (for DC). Sample proportions based on fewer observations, such as DC, tended to shrink more. Although the estimates incorporating random effects are relatively homogeneous, they tend to be closer than the sample proportions to the true values.

TABLE 12.2   Estimates of Proportion of Vote for Clinton, Conditional on Voting for Clinton or Dole in 1996 U.S. Presidential Election[a]

State	T_i	π_i	p_i	$\hat{\pi}_i$	State	T_i	π_i	p_i	$\hat{\pi}_i$
AK	5	0.394	0.200	0.508	MT	7	0.483	0.429	0.526
AL	32	0.463	0.500	0.524	NC	55	0.475	0.455	0.494
AR	19	0.594	0.526	0.537	ND	5	0.461	0.600	0.546
AZ	34	0.512	0.618	0.573	NE	13	0.395	0.462	0.524
CA	240	0.572	0.538	0.538	NH	9	0.567	0.556	0.543
CO	29	0.492	0.586	0.558	NJ	60	0.600	0.667	0.611
CT	25	0.604	0.720	0.602	NM	13	0.540	0.462	0.524
DC	4	0.903	1.000	0.576	NV	12	0.506	0.500	0.533
DE	5	0.586	0.400	0.527	NY	137	0.660	0.752	0.696
FL	108	0.532	0.602	0.583	OH	84	0.536	0.488	0.507
GA	56	0.494	0.554	0.548	OK	23	0.456	0.478	0.520
HI	9	0.643	0.556	0.543	OR	24	0.547	0.625	0.569
IA	22	0.557	0.500	0.528	PA	90	0.552	0.567	0.558
ID	9	0.391	0.111	0.472	RI	7	0.689	0.571	0.545
IL	89	0.596	0.539	0.540	SC	28	0.469	0.571	0.552
IN	44	0.468	0.432	0.488	SD	6	0.479	0.667	0.555
KS	19	0.400	0.316	0.477	TN	40	0.513	0.500	0.522
KY	29	0.506	0.448	0.506	TX	144	0.473	0.444	0.468
LA	33	0.566	0.667	0.592	UT	15	0.380	0.333	0.490
MA	46	0.686	0.739	0.637	VA	51	0.489	0.412	0.473
MD	38	0.586	0.474	0.511	VT	4	0.633	0.500	0.538
ME	9	0.627	0.778	0.578	WA	42	0.572	0.619	0.578
MI	73	0.573	0.589	0.570	WI	39	0.559	0.487	0.517
MN	35	0.594	0.571	0.554	WV	14	0.584	0.571	0.548
MO	41	0.535	0.561	0.550	WY	4	0.426	0.250	0.518
MS	21	0.472	0.333	0.477					

[a] π_i, True; p_i, sample; $\hat{\pi}_i$, estimate using random effects model.

12.3.2   Modeling Repeated Binary Responses

In Section 12.1.4 we introduced a random effects version of the Rasch model for repeated binary measurement. This model extends to incorporate covariates.

We illustrate using Table 10.13, first analyzed in Section 10.7.2. The subjects indicated whether they supported legalizing abortion in each of three situations. Table 10.13 also classified the subjects by gender. Let y_{it} denote the response for subject i on item t, with $y_{it} = 1$ representing support. Consider the model

$$\text{logit}\left[P(Y_{it} = 1 \mid u_i)\right] = u_i + \beta_t + \gamma x_i, \qquad (12.10)$$

where $x_i = 1$ for females and 0 for males, and where $\{u_i\}$ are independent $N(0, \sigma^2)$. (Equivalently, one could place a constraint on $\{\beta_t\}$ and allow an intercept α.) Here, the gender effect γ is assumed the same for each item, and the $\{\beta_i\}$ refer to the items.

Since model (12.10) implies nonnegative association among responses on the items, one should use items and scales for which this should occur. For opinions about legalized abortion with scale (yes, no), it would not be appropriate for one question to ask "Do you agree that abortion should be legal when a woman is not married?" and another to ask "Do you agree that abortion should be illegal during the last three months of pregnancy?"

Table 12.3 summarizes ML fitting results. The contrasts of $\{\hat{\beta}_i\}$ indicate greater support for legalized abortion with item 1 (when the family has a low income and cannot afford any more children) than with the other two. There is slight evidence of greater support with item 2 (when the woman is not married and does not want to marry the man) than with item 3 (when the woman wants the abortion for any reason). The fixed effects estimates have log odds ratio interpretations. For a given subject of either gender, for instance, the estimated odds of supporting legalized abortion for item 1 equal $\exp(0.83) = 2.3$ times the estimated odds for item 3. Since $\hat{\gamma} = 0.01$, for each item the estimated probability of supporting legalized abortion is similar for females and males with similar random effect values.

For these data, subjects are highly heterogeneous ($\hat{\sigma} = 8.6$). Thus, strong associations exist among responses on the three items. This is reflected by 1595 of the 1850 subjects making the same response on all three items: that is, response patterns $(0, 0, 0)$ and $(1, 1, 1)$. It implies tremendous variability in between-subject odds ratios. From (12.7), for different subjects of a given gender, the middle 50% of odds ratios comparing items 1 and 3 are estimated to vary between about $\exp(0.83 - 0.95 \times 8.6)$ and $\exp(0.83 + 0.95 \times 8.6)$.

For contingency tables, one can obtain cell fitted values. To do this, one must integrate over the estimated random effects distribution to obtain estimated marginal probabilities of any particular sequence of responses. For the ML parameter estimates, the probability of a particular sequence of responses $(y_{i1}, \ldots, y_{iT})$ for a given u_i is the appropriate product of conditional probabilities, $\prod_t P(Y_{it} = y_{it} | u_i)$, since the responses are independent given u_i. Integrating this product probability with respect to u_i for the

TABLE 12.3   Summary of ML Estimates for Random Effects Model (12.10) and ML and GEE Estimates for Corresponding Marginal Model

Effect	Parameter	GLMM ML		Marginal Model ML		Marginal Model GEE	
		Estimate	SE	Estimate	SE	Estimate	SE
Abortion	$\beta_1 - \beta_3$	0.83	0.16	0.148	0.030	0.149	0.030
	$\beta_1 - \beta_2$	0.54	0.16	0.098	0.027	0.097	0.028
	$\beta_2 - \beta_3$	0.29	0.16	0.049	0.027	0.052	0.027
Gender	γ	0.01	0.48	0.005	0.088	0.003	0.088
$\sqrt{\mathrm{var}(u_i)}$	σ	8.6	0.54				

$N(0, \hat{\sigma}^2)$ distribution estimates the marginal probability for a given cell (averaged over subjects). This requires numerical integration methods described in Section 12.6. Multiplying this marginal probability of a given sequence by the sample size for that multinomial gives a fitted value.

Not surprisingly, for these data, the response patterns $(0, 0, 0)$ and $(1, 1, 1)$ also have the largest fitted values for the multinomial for each gender. For instance, for females 440 indicated support under all three circumstances (457 under none of the three), and the fitted value was 436.5 (459.3). Overall chi-squared statistics comparing the 16 observed and fitted counts are $G^2 = 23.2$ and $X^2 = 27.8$ (df = 9). These are not that large considering the very large sample size and the few parameters ($\beta_1, \beta_2, \beta_3, \gamma, \sigma$) used to describe the 14 multinomial cell probabilities ($8 - 1 = 7$ for each gender) in Table 10.13. Here, df = 9 since we are modeling 14 multinomial parameters using five GLMM parameters.

An extended model allows interaction between gender and item. It has different $\{\beta_t\}$ for men and women. However, it does not fit better. The likelihood-ratio statistic = 1.0 (df = 2) for testing that the extra parameters equal 0.

An alternative analysis of these data focuses on the marginal distributions, treating the dependence as a nuisance. A marginal model analog of (12.10) is

$$\text{logit}[P(Y_t = 1)] = \beta_t + \gamma x.$$

For it, Table 12.3 also shows GEE estimates for the exchangeable working correlation structure and ML estimates. The marginal model fits well, with $G^2 = 1.1$; here, df = 2 since the model describes six marginal probabilities (three for each gender) using four parameters. These population-averaged $\{\hat{\beta}_t\}$ are much smaller than the subject-specific $\{\hat{\beta}_t\}$ from the GLMM. This reflects the very large GLMM heterogeneity ($\hat{\sigma} = 8.6$) and the corresponding strong correlations among the three responses. For instance, the GEE analysis estimates a common correlation of 0.82 between pairs of responses. Although the GLMM $\{\hat{\beta}_t\}$ are about five to six times the marginal model $\{\hat{\beta}_t\}$, so are the standard errors. The two approaches provide similar substantive interpretations and conclusions.

12.3.3 Longitudinal Mental Depression Study Revisited

We now revisit Table 11.2 from a longitudinal study to compare a new drug with a standard for treating subjects suffering mental depression. In Section 11.2.1 we analyzed the data using marginal models. The response y_t for measurement t on mental depression equals 1 for normal and 0 for abnormal. For severity of initial diagnosis s (1 = severe, 0 = mild), drug treatment d (1 = new, 0 = standard), and time of measurement t, we used the model

$$\text{logit}[P(Y_t = 1)] = \alpha + \beta_1 s + \beta_2 d + \beta_3 t + \beta_4 \, dt$$

to evaluate the marginal distributions.

TABLE 12.4   Model Parameter Estimates for Marginal and Conditional Logit Models Fitted to Table 11.2

Parameter	ML Marginal Estimate	Std. Error	GEE Marginal Estimate	Std. Error	Random Effects ML Estimate	Std. Error
Diagnosis	−1.29	0.14	−1.31	0.15	−1.32	0.15
Drug	−0.06	0.22	−0.06	0.23	−0.06	0.22
Time	0.48	0.12	0.48	0.12	0.48	0.12
Drug × Time	1.01	0.18	1.02	0.19	1.02	0.19

Now let y_{it} denote observation t for subject i. The model

$$\text{logit}\left[P(Y_{it} = 1 \mid u_i) \right] = \alpha + \beta_1 s + \beta_2 d + \beta_3 t + \beta_4 dt + u_i$$

has subject-specific rather than population-averaged effects. Table 12.4 shows the ML estimates. The time trend estimates are $\hat{\beta}_3 = 0.48$ for the standard drug and $\hat{\beta}_3 + \hat{\beta}_4 = 1.50$ for the new one. These are nearly identical to the ML and GEE estimates for the corresponding marginal model, also shown in the table (these are discussed in Sections 11.2.1 and 11.3.2). The reason is that the repeated observations do not exhibit much correlation, as the GEE analysis observed. Here, this is reflected by $\hat{\sigma} = 0.07$, showing little heterogeneity among subjects.

Based on the model fit, integrating over the $N(0, 0.07^2)$ random effects distribution yields marginal fitted values of the possible response sequences. Comparing these to the sample counts in Table 11.2 indicates a relatively good fit. The model describes the 28 multinomial cell probabilities (seven for the trivariate response at each of the four severity–drug combinations) using six parameters. The usual fit statistics comparing the observed cell counts to their fitted values are $G^2 = 22.0$ and $X^2 = 20.8$ (df $= 28 - 6 = 22$).

The deviance increases by only 0.001 when one assumes that $\sigma = 0$. From results to be discussed in Section 12.6.6, the P-value for comparing models is half what one gets by treating the deviance as chi-squared with df $= 1$, or $P = 0.49$. This simpler model, which gives nearly identical effect estimates and SE values, is adequate. This is also suggested by AIC values (e.g., PROC NLMIXED in SAS reports 1173.9 for the random effects model and 1171.9 for the simpler model with $\sigma = 0$).

12.3.4   Modeling Heterogeneity among Multicenter Clinical Trials

Many applications compare two groups on a response for data stratified on a third variable. With binary outcomes, the data form several 2×2 contingency tables. The main focus relates to studying the association in the 2×2 tables and whether and how it varies among the strata.

The strata are sometimes themselves a sample, such as schools or medical clinics. A random effects approach is then natural. With a random sampling of strata, it enables inferences to extend to the population of strata. The fit of the random effects model provides a simple summary such as an estimated mean and standard deviation of log odds ratios for the population of strata. In each stratum it also provides a predicted log odds ratio that shrinks the sample value toward the mean. This is especially useful when the sample size in a stratum is small and the ordinary sample odds ratio has large standard error. Even when the strata are not a random sample or not even a sample and a random effects approach is not as natural, the model is beneficial for these purposes.

We illustrate using Table 12.5, previously analyzed in Section 6.3, showing the results of a clinical trial at eight centers. The purpose was to compare an active drug and a control, for curing an infection. For a subject in center i using treatment t (1 = active drug; 2 = control), let $y_{it} = 1$ denote success. One possible model is the logistic-normal,

$$\text{logit}[P(Y_{i1} = 1 | u_i)] = \alpha + \beta/2 + u_i$$

$$\text{logit}[P(Y_{i2} = 1 | u_i)] = \alpha - \beta/2 + u_i,$$

(12.11)

TABLE 12.5 Clinical Trial Relating Treatment to Response for Eight Centers

Center	Treatment	Response Success	Response Failure	Sample Odds Ratio	Fitted Odds Ratio
1	Drug	11	25	1.19	2.02
	Control	10	27		
2	Drug	16	4	1.82	2.09
	Control	22	10		
3	Drug	14	5	4.80	2.19
	Control	7	12		
4	Drug	2	14	2.29	2.11
	Control	1	16		
5	Drug	6	11	∞	2.18
	Control	0	12		
6	Drug	1	10	∞	2.12
	Control	0	10		
7	Drug	1	4	2.0	2.11
	Control	1	8		
8	Drug	4	2	0.33	2.06
	Control	6	1		

Source: Beitler and Landis (1985).

where $\{u_i\}$ are independent $N(0, \sigma^2)$ variates. This model assumes that the log odds ratio β between treatment and response is constant over centers. The parameter σ summarizes center heterogeneity in the success probabilities.

A logistic-normal model permitting treatment-by-center interaction is

$$\text{logit}[P(Y_{i1} = 1 \mid u_i, b_i)] = \alpha + (\beta + b_i)/2 + u_i,$$
$$\text{logit}[P(Y_{i2} = 1 \mid u_i, b_i)] = \alpha - (\beta + b_i)/2 + u_i, \tag{12.12}$$

where $\{u_i\}$ are independent $N(0, \sigma_a^2)$, $\{b_i\}$ are independent $N(0, \sigma_b^2)$, and $\{u_i\}$ are independent of $\{b_i\}$. The log odds ratio equals $\beta + b_i$ in center i. These vary among centers according to a $N(\beta, \sigma_b^2)$ distribution. That is, β is the expected center-specific log odds ratio between treatment and response, and σ_b describes variability in those log odds ratios. The model parameters are $(\alpha, \beta, \sigma_a, \sigma_b)$.

In Table 12.5 the sample success rates vary markedly among centers both for the control and drug treatments, but in all except the last center that rate is higher for the drug treatment. In using models with random center and possibly random treatment effects, it is preferable to have more than eight centers. It is difficult to get reliable variance component estimates with so few centers. Keeping this in mind, we use these data to illustrate the models. With a large number of centers it would also be sensible to allow correlation between b_i and u_i, but we shall not attempt that here. The treatment estimates are $\hat{\beta} = 0.739$ (SE $= 0.300$) for the model (12.11) of no interaction and $\hat{\beta} = 0.746$ (SE $= 0.325$) for the model (12.12) permitting interaction. Considerable evidence of a drug effect occurs. With such a small sample, however, it is unclear whether that effect is weak or moderate.

The evidence about association is weaker for the model permitting interaction. The Wald statistics are $(0.739/0.300)^2 = 6.0$ for the no-interaction model and $(0.746/0.325)^2 = 5.3$ for the interaction model. The corresponding likelihood-ratio statistics are 6.3 and 4.6 (df $= 1$). The extra variance component in the interaction model pertains to variability in the log odds ratios. As its estimate $\hat{\sigma}_b$ increases, so does the standard error of the estimated treatment effect $\hat{\beta}$ tend to increase. In this example, $\hat{\sigma}_b = 0.15$ is relatively small and the standard errors of $\hat{\beta}$ are not very different in the two models. When $\hat{\sigma}_b = 0$, the standard errors and the model fits are the same.

To show the effect of larger $\hat{\sigma}_b$ on the standard error of the mean treatment effect estimate $\hat{\beta}$, we alter Table 12.5 slightly. We change three failures to successes for drug in center 3 and three successes to failures for drug in center 8. With these changes, the estimated variability of the treatment effects increases from $\hat{\sigma}_b = 0.15$ to $\hat{\sigma}_b = 1.4$. The ML estimates of the mean treatment effects are then $\hat{\beta} = 0.722$ (SE $= 0.299$) for the no interaction model (12.11) and $\hat{\beta} = 0.767$ (SE $= 0.623$) for the interaction model. The Wald statistics are 5.8 and 1.5. The evidence of a treatment effect is then dramatically weaker for the interaction model (12.12). Not surprisingly, when the treatment effect varies substantially among centers, it is more difficult to estimate the mean of that effect.

For the actual data in Table 12.5, because $\hat{\sigma}_b = 0.15$ for model (12.12) is relatively small, the model shrinks the sample odds ratios considerably. Table 12.5 shows the sample values and the model predicted values. These are based on predicting the random effects (to be explained in Section 12.6), and substituting them and the ML estimates of fixed effects into the model formula to estimate the two response probabilities for each treatment in each center. The sample odds ratios vary from 0.33 to ∞; their random effects model counterparts (computed with PROC NLMIXED in SAS) vary only between 2.0 and 2.2. The smoothed estimates are much less variable and do not have the same ordering as the sample values. For instance, the smoothed estimate of 2.2 for center 3 is greater than the estimate of 2.1 for center 6, even though the sample value is infinite for the latter. This partly reflects the greater shrinkage that occurs when sample sizes are smaller. When $\hat{\sigma}_b = 0$, model (12.12) provides the same fit as model (12.11), and estimated odds ratios are identical in each center.

For related analyses permitting heterogeneity in odds ratios with several 2×2 tables, see Liu and Pierce (1993) and Skene and Wakefield (1990).

12.3.5 Alternative Formulations of Random Effects Models

There are other ways to express the models. For instance, an equivalent expression for interaction model (12.12) is

$$\text{logit}\big[P(Y_{it} = 1 | u_i, b_{it}) \big] = \alpha + \beta x_t + b_{it} + u_i,$$

where x_t is a treatment dummy variable ($x_1 = 1, x_2 = 0$), $\{u_i\}$ are independent $N(0, \sigma_a^2)$, and $\{b_{i1}\}$ and $\{b_{i2}\}$ are independent $N(0, \sigma^2)$. Here, $b_{i1} - b_{i2}$ corresponds to b_i in parameterization (12.12), and $2\sigma^2$ corresponds to σ_b^2.

Formulating a random effects model requires care about implications of the model expression and the random effects correlation structure. Suppose that one expressed the interaction model (12.12) as

$$\text{logit}\big[P(Y_{it} = 1 | u_i, b_i) \big] = \alpha + (\beta + b_i)x_t + u_i, \qquad (12.13)$$

with $\{b_i\}$ from $N(0, \sigma_b^2)$. This is inappropriate, since the model then imposes greater variability for the logit with the first treatment than the second, since $x_2 = 0$ and $\{u_i\}$ and $\{b_i\}$ are uncorrelated. Also, the model should not depend on the definition of the dummy variable x_t. Note, however, that if $z_t = x_t + c$ for some constant c, then model (12.13) is equivalently

$$\text{logit}\big[P(Y_{it} = 1 | u_i, b_i) \big]$$
$$= \alpha + (\beta + b_i)(z_t - c) + u_i = \alpha' + (\beta + b_i)z_t + v_i,$$

where $\alpha' = \alpha - c\beta$ and $v_i = u_i - cb_i$. Thus, (v_i, b_i) are correlated even if (u_i, b_i) are not. In fact, expression (12.13) is sensible only with correlated random effects. It is then equivalent to (12.12) with correlated random effects. See Agresti and Hartzel (2000) for further discussion.

12.3.6 Capture–Recapture Modeling to Predict Population Size

Capture–recapture experiments are a method of using a series of samples to estimate the size of a population. Such methods have traditionally been used to estimate animal abundance in some habitat. At each sampling occasion, animals are captured and marked in some manner. The animals captured for any given sample are freed and all animals are candidates for recapture in a later sample. With T sampling occasions, a 2^T contingency table displays the data, with scale (captured, not captured) at each occasion. The count $n_{22\ldots2}$ is missing for the cell corresponding to noncapture at each occasion. If we knew this cell count, adding it to the others would yield the population size. Models specified for this 2^T table use the $2^T - 1$ observed counts to fit the model. The fit refers to those $2^T - 1$ cells, but extrapolating it yields an estimated count in the unobserved cell. Adding that to the total of the $2^T - 1$ observed counts yields an estimate of population size.

To illustrate, suppose that $T = 2$. We observe n_{11} animals at both occasions, n_{12} at the first but not the second occasion, and n_{21} at the second but not the first. We do not know the number n_{22} not captured either time. If we assumed independence in the 2×2 table, the prediction $\hat{n}_{22}$ would be the value giving an odds ratio of 1.0; but $(n_{11}\hat{n}_{22})/(n_{12}n_{21}) = 1$ implies that $\hat{n}_{22} = n_{12}n_{21}/n_{11}$. This yields a population size prediction (Sekar and Deming 1949) of

$$\hat{N} = n_{11} + n_{12} + n_{21} + n_{12}n_{21}/n_{11}$$

$$= n_{1+}n_{+1}/n_{11} \quad \text{with} \quad \widehat{\text{var}}(\hat{N}) = \frac{n_{1+}n_{+1}n_{12}n_{21}}{n_{11}^3}.$$

The assumption of independence is usually unrealistic, however. With additional sampling occasions, one can try more complex models.

Table 12.6, analyzed by Cormack (1989) and others, refers to a study having $T = 6$ consecutive trapping days for a population of snowshoe hares. The study observed 68 hares. For instance, Table 12.6 indicates that 3 hares were observed on the first day but on none of the other days. For simplicity, models for studies over a brief time period assume that no deaths, births, or immigration into the population occurred during the study period. This is called a *closed population*.

Most methods for capture–recapture treat the probability of capture at a given occasion as identical for each subject (e.g., animal). This is usually

TABLE 12.6 Results of Capture–Recapture of Snowshoe Hares

Capture 6	Capture 5	Capture 4	Capture 3, Capture 2, Capture 1[a]							
			000	001	010	011	100	101	110	111
0	0	0	—	3	6	0	5	1	0	0
			(24.0)	(2.3)	(5.4)	(0.9)	(3.2)	(0.5)	(1.2)	(0.3)
0	0	1	3	2	3	0	0	1	0	0
			(4.8)	(0.8)	(1.8)	(0.5)	(1.1)	(0.3)	(0.6)	(0.3)
0	1	0	4	2	3	1	0	1	0	0
			(3.9)	(0.6)	(1.5)	(0.4)	(0.9)	(0.2)	(0.5)	(0.2)
0	1	1	1	0	0	0	0	0	0	0
			(1.3)	(0.3)	(0.8)	(0.3)	(0.5)	(0.2)	(0.4)	(0.3)
1	0	0	4	1	1	1	2	0	2	0
			(6.8)	(1.1)	(2.6)	(0.6)	(1.5)	(0.4)	(0.9)	(0.4)
1	0	1	4	0	3	0	1	0	2	0
			(2.3)	(0.6)	(1.3)	(0.5)	(0.8)	(0.3)	(0.7)	(0.4)
1	1	0	2	0	1	0	1	0	1	0
			(1.9)	(0.5)	(1.1)	(0.4)	(0.7)	(0.3)	(0.6)	(0.4)
1	1	1	1	1	1	0	0	0	1	2
			(1.0)	(0.4)	(0.9)	(0.5)	(0.5)	(0.3)	(0.7)	(0.7)

[a]Values in parentheses represent the fit of the logistic-normal model.
Source: A. Agresti, *Biometrics* **50**: 494–500 (1994).

unrealistic. One way to allow heterogeneous capture probabilities uses a logit model having subject random effects. For subject i, $i = 1, \ldots, N$ with N unknown, let $y'_i = (y_{i1}, \ldots, y_{iT})$, where $y_{it} = 1$ denotes capture in sample t and $y_{it} = 0$ denotes noncapture. Lacking explanatory variables, one might use the Rasch-type model

$$\text{logit}\left[P(Y_{it} = 1 \mid u_i) \right] = u_i + \beta_t,$$

where $\{u_i\}$ are independent $N(0, \sigma^2)$. The larger the value of β_t, the greater the capture probability at occasion t. The larger is σ, the more heterogeneous are the capture probabilities. When $\sigma = 0$ this logistic-normal model simplifies to mutual independence [i.e., loglinear model (8.6)] for the 2^T table.

As with other random effects models, integrating the random effect from the probability mass function of $(y_i \mid u_i)$ yields the likelihood function (as discussed in Section 12.6). One can consider this likelihood function and the resulting ML estimates of $\{\beta_t\}$ and σ for all possible counts in the unobserved cell. A profile likelihood function views the maximized likelihood as a function of the unobserved cell count. The ML prediction for that unobserved cell count is the value that maximizes this profile likelihood. Lacking specialized software, one can fit the random effects model repeatedly with various counts in the unobserved cell to determine by trial and error the count that maximizes the likelihood function.

ML fitting of this model to Table 12.6 yields a prediction of 24 for the unobserved cell count. Since the study observed 68 hares, the population size estimate is $\hat{N} = 92$. For this fit, $\hat{\sigma} = 1.0$.

Methods for obtaining a confidence interval for N include using the profile likelihood function or a nonparametric bootstrap method. With the profile likelihood approach, the interval for the missing cell count consists of the possible counts for that cell such that the G^2 fit statistic increases by less than $\chi_1^2(\alpha)$ from its value at the ML estimate. Adding the number of subjects observed in the samples to the endpoints of this interval gives the corresponding interval for N. For the snowshoe hares, a 95% profile-likelihood confidence interval for N is (75, 154). It is common for $\hat{N}$ to be nearer the low end of the interval. See Coull and Agresti (1999) for details.

The greater the heterogeneity, as reflected by larger $\hat{\sigma}$, $\hat{N}$ tends to be larger and the confidence interval tends to be wider. Large $\hat{\sigma}$ causes difficulties in estimation, since it results in a relatively flat likelihood surface. This implies imprecise estimates of N. In particular, the upper limit of the profile-likelihood confidence interval for N is essentially infinite when the likelihood function gets sufficiently flat. Also, the ML estimator is then often unstable, with small changes in the data yielding large changes in $\hat{N}$. Difficulties can also arise when probabilities of capture are small. Evidence of this occurs when most subjects captured appear in only one sample. When this happens or when $\hat{\sigma}$ is large, it is unrealistic to expect narrow confidence intervals for N.

Alternative models are discussed in Section 13.1.3. Models that ignore likely heterogeneity can give unrealistically narrow confidence intervals for N. Although traditionally used for animal populations, capture–recapture applications also include estimating population size for human populations, such as estimating population prevalence of injecting drug use and HIV infection. Darroch et al. (1993) considered census population estimation, and Chao et al. (2001) estimated the number of people infected during a hepatitis outbreak (Problem 12.21). An interesting application is estimating the number of files on the World Wide Web relating to some subject by taking samples using several search engines (Fienberg et al. 1999).

12.4 RANDOM EFFECTS MODELS FOR MULTINOMIAL DATA

Random effects models for binary responses extend to multicategory responses. For the multicategory models of Chapter 7, a multinomial observation with I categories is a vector of $I - 1$ indicators, the jth of which is 1 when the observation falls in category j and 0 otherwise. In Section 7.1.5 we defined a multivariate GLM by applying a vector of link functions to this multivariate response. Adding random effects extends this multivariate GLM and the GLMM (12.1) to a multivariate GLMM (Hartzel et al. 2001b; Tutz and Hennevogl 1996). This class includes models for nominal and ordinal responses.

12.4.1 Cumulative Logit Model with Random Intercept

Modeling is simpler with ordinal than nominal responses, since often the same random effect and the same fixed effect can apply to each logit. With cumulative logits, this is the *proportional odds* structure (Section 7.2.2). Denote the possible outcomes for y_{it}, observation t in cluster i, by $1, 2, \ldots, I$. A GLMM for the cumulative logits has the form

$$\text{logit}\left[P(Y_{it} \le j \mid \mathbf{u}_i) \right] = \alpha_j + \mathbf{x}'_{it}\boldsymbol{\beta} + \mathbf{z}'_{it}\mathbf{u}_i, \quad j = 1, \ldots, I - 1. \quad (12.14)$$

Hedeker and Gibbons (1994) discussed model fitting, primarily with $\mathbf{u}_i$ as multivariate normal.

For cumulative logit and probit random intercept models, the same relationship exists between their effects and those in marginal models as presented in Section 12.2.2 for binary-response models. Marginal effects tend to be smaller, increasingly so as σ increases. Also, the same predictor structure as in (12.14) holds with other links for which a common effect for each logit is plausible. For instance, Hartzel et al. (2001a, b) used it with adjacent-categories logits.

12.4.2 Insomnia Study Revisited

Table 11.4 showed results of a clinical trial at two occasions comparing a drug with placebo in treating insomnia patients. In Sections 11.2.3 and 11.3.3 the data were analyzed with marginal models. For y_t = time to fall asleep at occasion t, the marginal model

$$\text{logit}\left[P(Y_t \le j) \right] = \alpha_j + \beta_1 t + \beta_2 x + \beta_3 tx$$

permitted interaction between t = occasion (0 = initial, 1 = follow-up) and x = treatment (1 = active, 0 = placebo). Table 12.7 shows the ML and GEE estimates.

Now, let y_{it} denote the response for subject i at occasion t. Table 12.7 also shows results of fitting the random-intercept model

$$\text{logit}\left[P(Y_{it} \le j \mid u_i) \right] = u_i + \alpha_j + \beta_1 t + \beta_2 x + \beta_3 tx.$$

TABLE 12.7 Fits of Cumulative Logit Models to Table 11.4[a]

Effect	Marginal ML	Marginal GEE	Random Effects (GLMM) ML
Treatment	0.046 (0.236)	0.034 (0.238)	0.058 (0.366)
Occasion	1.074 (0.162)	1.038 (0.168)	1.602 (0.283)
Treatment × occasion	0.662 (0.244)	0.708 (0.244)	1.081 (0.380)

[a]Values in parentheses represent standard errors.

Results are substantively similar to the marginal model, but estimates and standard errors are about 50% larger. This reflects the relatively large heterogeneity ($\hat{\sigma} = 1.90$) and the resultant strong association between the responses at the two occasions.

12.4.3 Cluster Sampling

With surveys that use cluster sampling, standard methods based on simple random sampling (e.g., for a single multinomial sample) require adjustment. Ordinary standard errors are too small. The usual chi-squared test statistics no longer have chi-squared null distributions, but rather, weighted sums of chi-squared. Rao and Thomas (1988) surveyed ways of adjusting standard inferences to take into account complex sampling methods in the analysis and modeling of categorical data.

When the sampling scheme randomly samples clusters, one can account for the clustering using cluster random effects. We illustrate using data from Brier (1980), who reported 96 observations taken from 20 neighborhoods (the clusters) on Y = satisfaction with home and X = satisfaction with neighborhood as a whole. Each variable was measured with the ordinal scale (unsatisfied, satisfied, very satisfied). Brier's analysis adjusted for clustering by reducing the Pearson statistic for testing independence in the 3×3 contingency table relating X and Y from 17.9 to 15.7 (df = 4).

Consider the model for y_{it}, observation t in cluster i,

$$\text{logit}\big[P(Y_{it} \leq j \,|\, u_i)\big] = u_i + \alpha_j + x_{it}\beta, \tag{12.15}$$

with scores (1, 2, 3) for the satisfaction levels of x_{it}. With a $N(0, \sigma^2)$ distribution assumed for u_i, the ML effect estimate is $\hat{\beta} = -1.201$ (SE = 0.407), with $\hat{\sigma} = 0.92$. By contrast, treating the 96 observations as a random sample corresponds to fitting this model with $\sigma = 0$. It has $\hat{\beta} = -1.226$ (SE = 0.370). A slight reduction in significance results from adjusting for clustering.

12.4.4 Baseline-Category Logit Models with Random Effects

For nominal response variables, one can formulate a binary model that pairs each category with a baseline and fit these models simultaneously while allowing separate effects. This requires using a vector of cluster-specific random effects $\mathbf{u}_{ij}$, one for each logit. The general form of the baseline-category logit model with random effects is

$$\log \frac{P(Y_{it} = j)}{P(Y_{it} = I)} = \alpha_j + \mathbf{x}'_{it}\boldsymbol{\beta}_j + \mathbf{z}'_{it}\mathbf{u}_{ij}, \quad j = 1, \ldots, I - 1.$$

The fixed effects $\boldsymbol{\beta}_j$ and the random effects $\mathbf{u}_{ij}$ depend on j, since the baseline category is arbitrary. With nominal responses there is no reason to expect effects to be similar for different j.

Cluster i has a vector $\mathbf{u}'_i = (\mathbf{u}'_{i1}, \ldots, \mathbf{u}'_{i,I-1})$ of random effects. The usual approach treats $\{\mathbf{u}_i\}$ as independent multivariate normal variates. We recommend an unspecified covariance matrix Σ for $\mathbf{u}_i$. For instance, it is sensible to allow different variances for random effects that apply to different logits. With a common variance, that variance would not be the same as that for the implied random effect for a logit for an arbitrary pair of categories, $\log[P(Y_{it} = j)/P(Y_{it} = k)]$. With unspecified covariance the model is structurally the same regardless of the choice of baseline category. See Hartzel et al. (2001b) for an example.

12.5 MULTIVARIATE RANDOM EFFECTS MODELS FOR BINARY DATA

In practice, random effects are often univariate, taking the form of random intercepts. However, we've seen that nominal responses require multivariate random effects and that bivariate random effects are helpful for describing heterogeneity in multicenter clinical trials. In this section we present other examples in which multivariate random effects are natural.

12.5.1 Matched Pairs with a Bivariate Binary Response

Leo Goodman analyzed Table 12.8 in several articles (e.g., Goodman 1974). A sample of schoolboys were interviewed twice, several months apart, and asked about their self-perceived membership in the "leading crowd" and about whether they sometimes needed to go against their principles to belong to that group. Thus, there are two binary response variables, which we refer to as membership and attitude, measured at two interview times for each subject. Table 12.8 labels the categories for attitude as (positive, negative), where "positive" refers to disagreeing with the statement that one must go against his principles.

TABLE 12.8   Membership and Attitude Toward the "Leading Crowd"

(M, A) for First Interview	(M, A) for Second Interview[a]			
	(Yes, Positive)	(Yes, Negative)	(No, Positive)	(No, Negative)
Yes, positive	458	140	110	49
Yes, negative	171	182	56	87
No, positive	184	75	531	281
No, negative	85	97	338	554

[a]M, membership; A, attitude.

Source: J. S. Coleman, *Introduction to Mathematical Sociology* (London: Free Press of Glencoe, 1964), p. 170.

For subject i, let y_{itv} be the response at interview time t on variable v, where $v = M$ for membership and $v = A$ for attitude. The logit model

$$\text{logit}\left[P(Y_{itv} = 1 \mid u_{iv}) \right] = \beta_{tv} + u_{iv} \tag{12.16}$$

is a multivariate form of the Rasch-type model (12.4). It has additive item and subject effects for each variable v. Here, (u_{iM}, u_{iA}) is a bivariate random effect that describes subject heterogeneity for (membership, attitude). We assume that the $\{(u_{iM}, u_{iA})\}$ are independent from a bivariate normal distribution, $N(\mathbf{0}, \Sigma)$, with possibly different variances and nonzero correlation.

The ML fit yields $\hat{\beta}_{2M} - \hat{\beta}_{1M} = 0.379$ (SE $= 0.075$) and $\hat{\beta}_{2A} - \hat{\beta}_{1A} = 0.176$ (SE $= 0.058$). For both variables, the probability of the first outcome category is higher at the second interview. For instance, for a given subject the odds of self-perceived membership in the leading crowd at interview 2 are estimated to be $\exp(0.379) = 1.46$ times the odds at interview 1.

The estimated correlation between the random effects is 0.30. Their estimated standard deviations are $\hat{\sigma}_1 = 3.1$ for $\{u_{iM}\}$ and $\hat{\sigma}_2 = 1.5$ for $\{u_{iA}\}$. Since these are quite different, the relative sizes of membership and attitude effects differ for marginal and conditional models (recall the caveat in Section 12.2.3). The marginal effect is attenuated more for membership. For this conditional model, the ratio of estimated odds ratios is $\exp(0.379)/\exp(0.176) = 1.46/1.19 = 1.22$. For the marginal model, the estimated odds ratios use the marginal distributions of each variable at each time [e.g., this is $(1392/2006)/(1253/2145) = 1.188$ for membership], and the ratio of estimated odds ratios is $1.188/1.133 = 1.05$.

Integrating over the estimated random effects distribution yields fitted values for the 16 possible sequences of responses in Table 12.8. The deviance of $G^2 = 5.5$ (df $= 8$) compares the 16 observed counts to their fitted values. The model, which describes 15 multinomial probabilities with seven parameters, fits well. The model constraining the random effects to be uncorrelated fits poorly ($G^2 = 97.5$, df $= 9$). The model constraining the random effects to be perfectly correlated is equivalent to having a single random effect u_i for each subject. The model is then a Rasch-type model with four items that are the combinations of interviews and variables. That model fits very poorly ($G^2 = 655.5$, df $= 10$). Agresti et al. (2000) gave further details.

12.5.2 Continuation-Ratio Logits for Clustered Ordinal Outcomes: Toxicity Study

For continuation-ratio logit models with ordinal responses, the logits refer to independent binomial variates (Section 7.4.3). Thus, binary logit random effects models apply to clustered ordinal responses using continuation-ratio logits (Ten Have and Uttal 1994). For observation t in cluster i, let $\omega_{ij} = P(Y_{it} = j \mid Y_{it} \geq j, u_{ij})$. (More generally, this probability could also depend on t, but this generality is not needed for the example below.) The continuation-ratio logits are $\{\text{logit}(\omega_{ij}), \ j = 1, \ldots, I - 1\}$.

Let n_{ij} be the number of subjects in cluster i making response j. Let $n_i = \sum_{j=1}^{I} n_{ij}$. For a given cluster in a continuation-ratio logit model, treating $(n_{i1}, \ldots, n_{i,I-1})$ as multinomial is equivalent to treating them as a sequential set of independent binomial variates, where n_{ij} is $\text{bin}(n_i - \sum_{h<j} n_{ih}, \omega_{ij})$, $j = 1, \ldots, I - 1$.

We illustrate with a developmental toxicity study conducted under the U.S. National Toxicology Program. This study examined the developmental effects of ethylene glycol (EG) by administering one of four dosages $(0, 0.75, 1.50, 3.00 \text{ g/kg})$ to pregnant rodents. The four dose groups had $(25, 24, 22, 23)$ pregnant rodents. The clusters are litters of mice. The three possible outcomes (dead/resorption, malformation, normal) for each fetus are ordered, normal being the most desirable result. Table 12.9 shows the data. The continuation-ratio logit is natural here since categories are hierarchically related; an animal must survive before a malformation can take place. The following analyses are from Coull and Agresti (2000).

For litter i in dose group d, let $\text{logit}(\omega_{i(d)1})$ be the continuation-ratio logit for the probability of death and $\text{logit}(\omega_{i(d)2})$ the continuation-ratio logit for the conditional probability of malformation, given survival. [The notation $i(d)$ represents litter i nested within dose d.] Let x_d be the dosage for group d. We account for the litter effect using litter-specific random effects $\mathbf{u}_{i(d)} = (u_{i(d)1}, u_{i(d)2})$ sampled from $N(\mathbf{0}, \Sigma_d)$. This bivariate random effect allows for differing amounts of overdispersion for the probability of death and for the probability of malformation, given survival. A model also permitting different fixed effects for each is

$$\text{logit}(\omega_{i(d)j}) = u_{i(d)j} + \alpha_j + \beta_j x_d. \tag{12.17}$$

TABLE 12.9   Response Counts for 94 Litters of Mice on (Number Dead, Number Malformed, Number Normal)

Dose = 0.00 g/kg	Dose = 0.75 g/kg	Dose = 1.50 g/kg	Dose = 3.00 g/kg
$(1, 0, 7), (0, 0, 14)$	$(0, 3, 7), (1, 3, 11)$	$(0, 8, 2), (0, 6, 5)$	$(0, 4, 3), (1, 9, 1)$
$(0, 0, 13), (0, 0, 10)$	$(0, 2, 9), (0, 0, 12)$	$(0, 5, 7), (0, 11, 2)$	$(0, 4, 8), (1, 11, 0)$
$(0, 1, 15), (1, 0, 14)$	$(0, 1, 11), (0, 3, 10)$	$(1, 6, 3), (0, 7, 6)$	$(0, 7, 3), (0, 9, 1)$
$(1, 0, 10), (0, 0, 12)$	$(0, 0, 15), (0, 0, 11)$	$(0, 0, 1), (0, 3, 8)$	$(0, 3, 1), (0, 7, 0)$
$(0, 0, 11), (0, 0, 8)$	$(2, 0, 8), (0, 1, 10)$	$(0, 8, 3), (0, 2, 12)$	$(0, 1, 3), (0, 12, 0)$
$(1, 0, 6), (0, 0, 15)$	$(0, 0, 10), (0, 1, 13)$	$(0, 1, 12), (0, 10, 5)$	$(2, 12, 0), (0, 11, 3)$
$(0, 0, 12), (0, 0, 12)$	$(0, 1, 9), (0, 0, 14)$	$(0, 5, 6), (0, 1, 11)$	$(0, 5, 6), (0, 4, 8)$
$(0, 0, 13), (0, 0, 10)$	$(1, 1, 11), (0, 1, 9)$	$(0, 3, 10), (0, 0, 13)$	$(0, 5, 7), (2, 3, 9)$
$(0, 0, 10), (1, 0, 11)$	$(0, 1, 10), (0, 0, 15)$	$(0, 6, 1), (0, 2, 6)$	$(0, 9, 1), (0, 0, 9)$
$(0, 0, 12), (0, 0, 13)$	$(0, 0, 15), (0, 3, 10)$	$(0, 1, 2), (0, 0, 7)$	$(0, 5, 4), (0, 2, 5)$
$(1, 0, 14), (0, 0, 13)$	$(0, 2, 5), (0, 1, 11)$	$(0, 4, 6), (0, 0, 12)$	$(1, 3, 9), (0, 2, 5)$
$(0, 0, 13), (1, 0, 14)$	$(0, 1, 6), (1, 1, 8)$		$(0, 1, 11)$
$(0, 0, 14)$			

Source: Study described by C. J. Price, C. A. Kimmel, R. W. Tyl, and M. C. Marr, *Toxicol. Appl. Pharmacol.* **81**: 113–127 (1985).

**TABLE 12.10   Comparisons of Log Likelihoods for Multivariate Random
Effects Models for Developmental Toxicity Study**

Model	Number of Parameters	Change in Parameters	Change in Log Likelihood
Dose-specific Σ_i	16	—	—
Σ_i, Common α, β	14	2	28.4
Common Σ	7	9	7.4
Common Σ, $\rho = 0$	6	10	7.4
Univariate σ^2	5	11	16.7

Table 12.10 reports the change in the maximized log likelihood from fitting
four special cases of this model:

1. Common intercept and slope for the two logits: $\alpha_1 = \alpha_2$ and $\beta_1 = \beta_2$
2. Common covariance matrix for the four doses: $\Sigma_1 = \Sigma_2 = \Sigma_3 = \Sigma_4$
3. Common covariance matrix and uncorrelated random effects
4. Univariate common variance component across dose: $u_{i(d)1} = u_{i(d)2}$ and $\sigma_d = \sigma$

Tests of the first three special cases against the general model (12.17) can
use ordinary likelihood-ratio tests. Little seems to be lost by using the simpler
model having uncorrelated random effects with homogeneous covariance
structure (i.e., the fourth model listed in Table 12.10), as the likelihood-ratio
statistic comparing this to model (12.17) equals $2(7.4) = 14.8$ (df $= 10$). The
model provides a separate univariate logistic-normal model for each conditional binomial outcome, specifying that the proportion of dead pups and the
proportion of malformed pups (given survival) are independent, both within
litter and marginally.

The univariate model in Table 12.10 is the special case of the third model
listed in which the variances are common for the two logits and the random
effects are perfectly correlated. Hence, it reduces to a univariate random
effects model. Comparing the univariate model to a multivariate counterpart
involves testing that correlation parameters fall on the boundary. Ordinary
chi-squared asymptotic theory for likelihood-ratio tests applies only when the
parameter falls in the interior of the parameter space. Tests when a null
model has a correlation of 1 or a variance component of 0 are complex and
beyond our scope here (see Section 12.6.6). However, an informal analysis of
change in log likelihoods suggests that the univariate model is inadequate.

The ML estimated effects for the separate univariate logistic-normal
model for each conditional binomial outcome are $\hat{\beta}_1 = 0.08$ (SE $= 0.21$),
$\hat{\beta}_2 = 1.79$ (SE $= 0.22$). For a given cluster, there is no evidence of a dose
effect on the death rate, but the estimated odds of malformation, given
survival, multiply by $\exp(1.79) = 6.0$ for every additional g/kg of ethylene

glycol. The variance component estimates suggest a stronger litter effect for the malformation outcome given survival ($\hat{\sigma}_2 = 1.6$) than for death ($\hat{\sigma}_1 = 0.5$).

12.5.3 Hierarchical (Multilevel) Modeling

Hierarchical data structures, with units grouped at different levels, are common in education. A statewide study of factors that affect student performance might measure students' scores on a battery of exams but use a model that takes into account the student, the school or school district, and the county. Just as two observations on the same student might tend to be more alike than observations on different students, so might two students in the same school tend to be more alike than students from different schools. Student, school, and county terms might be treated as random effects, with different ones referring to different *levels* of the model. For instance, a model might have students at level 1, schools at level 2, and counties at level 3. GLMMs for data having a hierarchical grouping of this sort are called *multilevel models*. Random effects enter the model at each level of the hierarchy.

We illustrate with a two-level model. Let $\pi_{i(j)t}$ denote the probability that student i in school j passes test t in a battery of tests. A multilevel model with random effects for student and school and fixed effects for explanatory variables has the form

$$\text{logit}[\pi_{i(j)t}] = \mathbf{x}'_{i(j)t}\boldsymbol{\beta} + u_j + v_{i(j)}.$$

Here, the explanatory variables $\mathbf{x}$ might include one that identifies the test in the battery. The random effects u_j for schools and $v_{i(j)}$ for students within schools are independent with different variance components. The level 1 random effects $\{v_{i(j)}\}$ account for variability among students in ability or parents' socioeconomic status or other characteristics not measured by $\mathbf{x}$. When they have a relatively large variance component, there is a strong correlation among the test results for students. The level 2 random effects $\{u_j\}$ account for variability among schools due to possibly unmeasured factors such as per-capita expenditure in the school's budget.

For examples of the use of multivariate random effects in multilevel modeling, see Aitkin et al. (1981), Anderson and Aitkin (1985), Gibbons and Hedeker (1997), Goldstein (1995), Goldstein and Rasbash (1996), and Longford (1993).

12.6 GLMM FITTING, INFERENCE, AND PREDICTION

Model fitting is rather complex for GLMMs. The main difficulty is that the likelihood function does not have a closed form. Numerical methods for approximating it can be computationally intensive for models with multivariate random effects. In this section we outline the basic ideas of ML fitting of GLMMs. Some ML methods are available in software (e.g., PROC NLMIXED in SAS).

12.6.1 Marginal Likelihood and Maximum Likelihood Fitting

The GLMM is a two-stage model. At the first stage, conditional on the random effects, observations are assumed to follow a GLM. That is, observation y_{it} in cluster i has distribution in the exponential family with expected value μ_{it} linked to a linear predictor,

$$g(\mu_{it}) = \mathbf{x}'_{it}\boldsymbol{\beta} + \mathbf{z}'_{it}\mathbf{u}_i.$$

Then, $\mathbf{z}'_{it}\mathbf{u}_i$ is a known offset and observations in a cluster are independent. At the second stage, the random effects $\{\mathbf{u}_i\}$ are assumed independent from a $N(\mathbf{0}, \boldsymbol{\Sigma})$ distribution.

For a discrete variable, denote the vector of all the observations by $\mathbf{y}$ and the vector of all the random effects by $\mathbf{u}$. Let $f(\mathbf{y}|\mathbf{u}; \boldsymbol{\beta})$ denote the conditional mass function of $\mathbf{y}$, given $\mathbf{u}$. Let $f(\mathbf{u}; \boldsymbol{\Sigma})$ denote the normal density function for $\mathbf{u}$. The likelihood function $\ell(\boldsymbol{\beta}, \boldsymbol{\Sigma}; \mathbf{y})$ for a GLMM is the probability mass function $f(\mathbf{y}; \boldsymbol{\beta}, \boldsymbol{\Sigma})$ of $\mathbf{y}$, viewed as a function of $\boldsymbol{\beta}$ and $\boldsymbol{\Sigma}$. This mass function refers to the marginal distribution of $\mathbf{y}$ after integrating out the random effects,

$$\ell(\boldsymbol{\beta}, \boldsymbol{\Sigma}; \mathbf{y}) = f(\mathbf{y}; \boldsymbol{\beta}, \boldsymbol{\Sigma}) = \int f(\mathbf{y}|\mathbf{u}; \boldsymbol{\beta}) f(\mathbf{u}; \boldsymbol{\Sigma}) d\mathbf{u}. \qquad (12.18)$$

It is often called a *marginal likelihood*. For example, the likelihood function $\ell(\boldsymbol{\beta}, \sigma^2; \mathbf{y})$ for the logistic-normal model (12.5) (absorbing α into $\boldsymbol{\beta}$) is

$$\prod_i \left(\int_{-\infty}^{\infty} \prod_t \left[\frac{\exp(\mathbf{x}'_{it}\boldsymbol{\beta} + u_i)}{1 + \exp(\mathbf{x}'_{it}\boldsymbol{\beta} + u_i)} \right]^{y_{it}} \left[\frac{1}{1 + \exp(\mathbf{x}'_{it}\boldsymbol{\beta} + u_i)} \right]^{1-y_{it}} f(u_i; \sigma^2) \, du_i \right).$$

The likelihood function is evaluated numerically and maximized as a function of $\boldsymbol{\beta}$ and $\boldsymbol{\Sigma}$. Many methods have been developed to do this. We next discuss a few of the most popular.

12.6.2 Gauss–Hermite Quadrature Methods

The integral determining the likelihood function has dimension that depends on the random effects structure. When the dimension is small, as in the one-dimensional integral above for the logistic-normal model (12.5), standard numerical integration methods can approximate the likelihood function.

Gauss–Hermite quadrature is a method for approximating the integral of a function $f(\cdot)$ multiplied by another function having the shape of a normal density. The approximation is a finite weighted sum that evaluates the function at certain points. In the univariate normal random effects case, the approximation has the form

$$\int_{-\infty}^{\infty} f(u)\exp(-u^2)\, du \approx \sum_{k=1}^{q} c_k f(s_k) ,$$

with *weights* $\{c_k\}$ and *quadrature points* $\{s_k\}$ that are tabulated. The approximation improves as q, the number of quadrature points, increases.

The approximated likelihood can be maximized with standard algorithms such as Newton–Raphson, yielding ML estimates $\hat{\boldsymbol{\beta}}$ and $\hat{\boldsymbol{\Sigma}}$. Inverting an approximation for the observed information matrix provides standard errors for the ML estimates. For complex models, second partial derivatives for the Hessian may be computed numerically rather than analytically. Adequate approximation usually requires larger q for standard errors than for $\hat{\boldsymbol{\beta}}$. We recommend sequentially increasing q until the changes are negligible in both the estimates and standard errors.

An adaptive version of Gauss–Hermite quadrature (e.g., Liu and Pierce 1994) centers the quadrature points with respect to the mode of the function being integrated and scales them according to the estimated curvature at the mode. This improves efficiency, dramatically reducing the number of quadrature points needed to approximate the integrals effectively. Lesaffre and Spiessens (2001) showed comparisons and warned against using too few points.

12.6.3 Monte Carlo Methods

Multivariate forms of Gauss–Hermite quadrature handle multivariate, correlated random effects. Adequate approximation becomes more difficult, however, when the dimension of the integral exceeds roughly 5. Then, Monte Carlo methods are more feasible computationally than numerical integration. Various Monte Carlo approaches have been studied (e.g., McCulloch 1997), including Monte Carlo in combination with Newton–Raphson, Monte Carlo in combination with the EM algorithm, and simulating the likelihood directly. Here, we briefly describe a Monte Carlo EM (MCEM) algorithm.

The EM algorithm is a popular iterative method of finding ML estimates when data are missing or when filling in some "missing" data simplifies a likelihood (Dempster et al. 1977) [see Laird (1998) for a useful review]. In each cycle an *E*-step takes an expectation over the missing data to approximate the likelihood function and an *M*-step maximizes the likelihood given the working values of the parameter estimates. In GLMMs, one regards the random effects $\mathbf{u}$ as missing data. Then, $h(\mathbf{y}, \mathbf{u}; \boldsymbol{\beta}, \boldsymbol{\Sigma}) = f(\mathbf{y} \mid \mathbf{u}; \boldsymbol{\beta}) f(\mathbf{u}; \boldsymbol{\Sigma})$ specifies the joint distribution of the complete data. The *E*-step in iteration r

of the EM algorithm calculates

$$E\{\log h(\mathbf{y}, \mathbf{u}; \boldsymbol{\beta}, \boldsymbol{\Sigma}) \mid \mathbf{y}; \boldsymbol{\beta}^{(r)}, \boldsymbol{\Sigma}^{(r)}\}.$$

The expectation is with respect to the distribution of $(\mathbf{u} \mid \mathbf{y})$ with parameter values equal to $\boldsymbol{\beta}^{(r)}$ and $\boldsymbol{\Sigma}^{(r)}$, the working estimates for iteration r. The distribution of $(\mathbf{u} \mid \mathbf{y})$ follows from those of $(\mathbf{y} \mid \mathbf{u})$ and $\mathbf{u}$ in the GLMM via Bayes' theorem. The M-step then maximizes the result with respect to $\boldsymbol{\beta}$ and $\boldsymbol{\Sigma}$ to obtain $\boldsymbol{\beta}^{(r+1)}$ and $\boldsymbol{\Sigma}^{(r+1)}$.

The MCEM algorithm approximates the expectation in the E-step using Monte Carlo methods. Possible ways of doing this include using independent simulations from the distribution of $(\mathbf{u} \mid \mathbf{y})$, at the current estimate of parameters, or using Markov chain Monte Carlo (MCMC). For details, including the issue of choosing an appropriate Monte Carlo sample size, see Booth and Hobert (1999), Chan and Kuk (1997), and McCulloch (1994, 1997).

12.6.4 Penalized Quasi-likelihood Approximation

The Gauss–Hermite and Monte Carlo integration methods provide likelihood approximations such that resulting parameter estimates converge to the ML estimates as they are applied more finely (i.e., as the number of quadrature points increases for numerical integration and as the Monte Carlo sample size increases in the MCEM method). This contrasts with other approximate methods that are simpler but need not yield estimates near the ML estimates. These methods maximize an analytical approximation of the likelihood function.

Recall that the likelihood function (12.18) results from integrating out the random effects $\mathbf{u}$ from the joint distribution of $\mathbf{y}$ and $\mathbf{u}$. Using the exponential family representation of each component of that joint distribution, the integrand of (12.18) is an exponential function of $\mathbf{u}$. One approach approximates that function using a second-order Taylor series expansion of its exponent around a point $\tilde{\mathbf{u}}$ at which the first-order term equals 0. [That point $\tilde{\mathbf{u}} \approx E(\mathbf{u} \mid \mathbf{y})$.] The approximating function for the integrand is then exponential with quadratic exponent in $(\mathbf{u} - \tilde{\mathbf{u}})$ and has the form of a constant multiple of a multivariate normal density. Thus, its integral has closed form. This type of integral approximation is called a *Laplace approximation*. The approximation for integral (12.18) is then treated as a likelihood and maximized with respect to $\boldsymbol{\beta}$ and $\boldsymbol{\Sigma}$.

For one such method (Breslow and Clayton 1993), the integral approximation yields a function approximating the log likelihood that has the form

$$q(\boldsymbol{\beta}, \mathbf{y}) - (1/2)\tilde{\mathbf{u}}'\boldsymbol{\Sigma}^{-1}\tilde{\mathbf{u}},$$

where $q(\boldsymbol{\beta}, \mathbf{y})$ resembles a quasi-log-likelihood function for the GLM conditional on $\mathbf{u} = \hat{\mathbf{u}}$. Thus, the approximation results in a penalty for the quasi-log likelihood, with the penalty increasing as elements of $\hat{\mathbf{u}}$ increase in absolute value. This approach is called *penalized quasi-likelihood* (PQL). The calculations for maximizing the penalized quasi-likelihood use methods for linear mixed models with a normal response. This treats a linearization of the logit as a working response and entails iterative solution of sets of likelihood-like equations in $\boldsymbol{\beta}$ and $\mathbf{u}$. PQL methods do not require numerical or Monte Carlo integration and so are simpler than ML methods. They are computationally feasible for large data sets and models with complex random effects structure.

Unfortunately, PQL methods can perform poorly relative to ML (McCulloch 1997). For instance, for the abortion example in Section 12.3.2, the PQL approximations to the ML estimates (obtained using the GLIMMIX macro in SAS) are decent for $\{\beta_i\}$, but the standard errors and the estimate of σ are only about half what they should be (e.g., PQL gives $\hat{\sigma} = 4.3$, compared to the ML estimate of 8.6). When true variance components are large, ordinarily PQL tends to produce variance component estimates with substantial negative bias (Breslow and Lin 1995). The PQL estimators also behave poorly when the response distribution is far from normal (e.g., binary). Adjustments have been developed for some cases to lessen the bias (e.g., Goldstein and Rasbash 1996), but where possible we recommend using ML rather than PQL.

12.6.5 Bayesian Approaches

Another approach to fitting of GLMMs is Bayesian. With it, the distinction between fixed and random effects no longer occurs, as every effect has a probability distribution. Use of a flat prior distribution yields a posterior that is a constant multiple of the likelihood function. Then, Markov chain Monte Carlo (MCMC) methods for approximating intractable posterior distributions can approximate the likelihood function (Zeger and Karim 1991). For instance, an approximation for the mode of the posterior distribution approximates the ML estimate.

A danger is that improper prior distributions have improper posteriors for many models for categorical data (Natarajan and McCulloch 1995). In using MCMC, one may fail to realize that the posterior is improper. It is safer to use a proper but relatively diffuse prior. However, the posterior mode need not be close to the ML estimate, and Markov chains may converge slowly (Natarajan and McCulloch 1998). This is currently an active area of research, not just as a way of approximating ML results but also as an approach preferred over ML by those who adopt the Bayesian paradigm. See, for instance, Daniels and Gatsonis (1999) for multilevel modeling of geographic and temporal trends with clustered longitudinal binary data, which built on earlier hierarchical modeling by Wong and Mason (1985).

12.6.6 Inference for Model Parameters

After fitting the model, inference about fixed effects proceeds in the usual way. For instance, likelihood-ratio tests can compare nested models. Asymptotics for GLMMs apply as the number of clusters increases, rather than as the numbers of observations within the clusters increase. Similarly, resampling methods such as the bootstrap using a large number of clusters should sample clusters rather than individual observations within clusters, to preserve the within-cluster dependence.

Inference about random effects (e.g., their variance components) is more complex. For instance, sometimes one model is a special case of another in which a variance component equals 0. The simpler model then falls on the boundary of the parameter space relative to the more complex model, so ordinary likelihood-based inference does not apply. The asymptotic distribution of the likelihood-ratio statistic is known for the most common situation, testing H_0: $\sigma^2 = 0$ against H_a: $\sigma^2 > 0$ for a model containing a single variance component. The null distribution is an equal mixture of χ_0^2 (i.e., degenerate at 0) and χ_1^2 random variables (Self and Liang 1987). The value of 0 occurs when $\hat{\sigma} = 0$, in which case the maximized likelihoods are identical under H_0 and H_a. When $\hat{\sigma} > 0$ and the observed test statistic equals t, the P-value for this large-sample test is $\frac{1}{2}P(\chi_1^2 > t)$, half the P-value that applies for χ_1^2 asymptotic tests. For testing more than one variance component, the mixture distribution becomes more complex, and it is simpler to use a score test (Lin 1997).

12.6.7 Prediction Using Random Effects

The use of random effects in a model implies heterogeneity of certain effects of interest, such as odds ratios. Estimated effects of interest are often then linear combinations of fixed and random effects. For example, in the clinical trial comparing two treatments with random effects for centers (Section 12.3.4), one can predict the probability of success for each treatment in each center and odds ratios in those centers.

Given the data, the conditional distribution of $(\mathbf{u} \mid \mathbf{y})$ contains the information about the random effects $\mathbf{u}$. A prediction for $\mathbf{u}$ is $E(\mathbf{u} \mid \mathbf{y})$, its *posterior mean* given the data. Calculation of $E(\mathbf{u} \mid \mathbf{y})$ itself requires numerical integration or Monte Carlo approximation. The expectation depends on $\boldsymbol{\beta}$ and $\boldsymbol{\Sigma}$, so in practice one substitutes $\hat{\boldsymbol{\beta}}$ and $\hat{\boldsymbol{\Sigma}}$ in the approximation. The standard error of the predictor of the random effect u_i is the standard deviation of the distribution of $(u_i \mid \mathbf{y})$. When one substitutes $\hat{\boldsymbol{\beta}}$ and $\hat{\boldsymbol{\Sigma}}$ in $E(\mathbf{u} \mid \mathbf{y})$, however, the standard error does not account for the sampling variability in those estimates. Hence, the true standard error tends to be underestimated (Booth and Hobert 1998).

This approach to prediction using posterior means of random effects provides effect estimates that exhibit shrinkage relative to estimates using only data in the specific cluster. In this sense the results are similar to those using an *empirical Bayes* approach (Ten Have and Localio 1999). This adapts an ordinary Bayesian analysis by using the sample data to estimate parameters of the prior distribution. For a vector of mean parameters, this approach yields an estimate of a particular mean that is a weighted average of the sample mean and the overall mean of the sample means. Thus, it shrinks the sample mean toward the overall mean. Shrinkage estimators can be far superior to sample values when the sample size for estimating each parameter is small, when there are many parameters to estimate, or when the true parameter values are roughly equal. The empirical Bayes paradigm has been in use for some time: for instance, for estimating a vector of means or binomial proportions (Efron and Morris 1975).

Although random effects models are natural in many applications, further work is needed. Work continues on the development of methodology for model-fitting and inference with complex GLMMs. In addition, research is needed on model checking and diagnostics. Nonetheless, we believe that GLMMs provide a very useful extension of ordinary GLMs.

NOTES

Section 12.1: Random Effects Modeling of Clustered Categorical Data

12.1. For further discussion of the Rasch model and ways of estimating its parameters, see Andersen (1980, Sec. 6.4) and Fischer and Molenaar (1995). Haberman (1977b) showed ML estimators can achieve consistency when both n and T grow at suitable rates. For multinomial Rasch extensions, see Andersen (1980, pp. 272–284; 1995) and Conaway (1989). Early work on random effects models for a categorical response includes Anderson and Aitkin (1985), Bartholomew (1980), Bock and Aitkin (1981), Chamberlain (1980), Gilmour et al. (1985), Pierce and Sands (1975), and Stiratelli et al. (1984).

12.2. In models with covariates, Neuhaus and Lesperance (1996) noted that conditional ML may lose considerable efficiency compared to the random effects approach when cluster sizes are small and covariates have strong positive within-cluster correlation. As that correlation approaches $+1$, the covariate effect resembles a between-cluster one, which the conditional ML approach cannot estimate. The matched-pairs case referred to in Section 12.1.2 in which the conditional ML estimate equals the random effects estimate has within-cluster covariate correlation $= -1$, as depending on the order of viewing the observations, x_t changes from 0 to 1 or from 1 to 0; then, no efficiency loss occurs.

Section 12.3: Examples of Random Effects Models for Binary Data

12.3. For further discussion of modeling capture–recapture data, see Bishop et al. (1975, Chap. 6), Chao et al. (2001), Cormack (1989), Coull and Agresti (1999), Darroch et al. (1993), Fienberg et al. (1999), and Hook and Regal (1995). Similarities exist between this problem and the related problem of estimating the binomial index n when observing independent $\text{bin}(n, \pi)$ counts with unknown n and π; see Aitkin and Stasinopoulos (1989) and references therein. Relatively flat log likelihoods also occur with other models that permit capture heterogeneity (Burnham and Overton 1978), such as a beta-binomial model.

12.4. King (1997) used random effects models as part of a solution for analyzing aggregated categorical data, the problem of *ecological inference*. Chambers and Steel (2001) discussed early work by Leo Goodman on this problem and proposed a simpler semiparametric approach.

Section 12.4: Random Effects Models for Multinomial Data

12.5. With the complementary log-log link, the likelihood function has closed form with a log gamma random effects distribution (Crouchley 1995, Farewell 1982, Ten Have 1996).

12.6. Chen and Kuo (2001) discussed nominal responses, including discrete choice models (Sec. 7.6) with random effects. See also Brownstone and Train (1999) for discrete choice GLMMs.

Section 12.5: Multivariate Random Effects Models for Binary Data

12.7. Rabe-Hesketh and Skrondal (2001) showed that careful attention must be paid to parameter identification in models with multivariate random effects. Their factor model contains many multivariate random effects models as special cases.

12.8. For longitudinal bivariate binary responses, Ten Have and Morabia (1999) simultaneously modeled bivariate log odds ratios and univariate logits. Multivariate responses sometimes have both continuous and categorical components. For random effects modeling of such data, see Catalano and Ryan (1992) and Gueorguieva and Agresti (2001).

Section 12.6: GLMM Fitting, Inference, and Prediction

12.9. See Fahrmeir and Tutz (2001, Chap. 7) and McCulloch and Searle (2001) for more details on the fitting of GLMMs. Just as the likelihood function for a GLMM is an integral, so do likelihood equations have the form of integral equations (McCulloch and Searle 2001, p. 227). Wolfinger and O'Connell (1993) described a fitting method related to PQL, also motivated by a Laplace approximation.

12.10. A GLMM determines the marginal relationship (averaged over random effects) between the mean response and explanatory variables. Conversely, Heagerty (1999) noted that a marginal model for the mean implicitly determines the form of the fixed portion of the linear predictor in a conditional model. The conditional GLMM (12.1) has linear predictor, $\mathbf{x}'_{it}\boldsymbol{\beta} + \mathbf{z}'_{it}\mathbf{u}_i$. A more general form $\Delta_{it} + \mathbf{z}'_{it}\mathbf{u}_i$ implies a particular marginal model. Here, Δ_{it} is a function of the marginal linear predictor and the random effects distribution. It is implicitly defined by the integral equation that links the marginal and conditional means.

PROBLEMS

Applications

12.1 Refer to the matched-pairs data of Table 10.14 and Problem 10.1.

  a. Fit model (12.3). Interpret $\hat{\beta}$. If your software uses numerical integration, report $\hat{\beta}$, $\hat{\sigma}$, and their standard errors for 5, 10, 25, 100, and 200 quadrature points, and comment on convergence.

    b. Compare $\hat{\beta}$ and its SE for this approach to the conditional ML approach.

12.2 Refer to Table 4.8 on the free-throw shooting of Shaq O'Neal. In game i, suppose that $y_i =$ number made out of n_i attempts is a $\text{bin}(n_i, \pi_i)$ variate and $\{y_i\}$ are independent.

    a. Fit the model, $\text{logit}(\pi_i) = \alpha$. Find and interpret $\hat{\pi}_i$. Does the model appear to fit adequately?

    b. Fit the model, $\text{logit}(\pi_i) = \alpha + u_i$, where $\{u_i\}$ are independent $N(0, \sigma^2)$. Use $\hat{\alpha}$ and $\hat{\sigma}$ to summarize O'Neal's free-throw shooting.

    c. Explain how the model in part (a) is a special case of that in part (b). Is there evidence that the one in part (b) fits better?

12.3 For Table 8.3, let $y_{it} = 1$ when subject i used substance t. Table 12.11 shows output for the logistic-normal model

$$\text{logit}\big[P(Y_{it} = 1 \,|\, u_i) \big] = u_i + \beta_t.$$

Interpret. Illustrate by comparing use of cigarettes and marijuana.

TABLE 12.11 Output for Problem 12.3

Description	Value	Parameter	Estimate	Std Error	t Value
Subjects	2276				
Max Obs Per Subject	3	beta1	4.2227	0.1824	23.15
Parameters	4	beta2	1.6209	0.1207	13.43
Quadrature Points	200	beta3	−0.7751	0.1061	−7.31
Log Likelihood	−3311	sigma	3.5496	0.1627	21.82

12.4 How is the focus different for the model in Problem 12.3 than for the loglinear model (AC, AM, CM) used in Section 8.2.4? If $\hat{\sigma} = 0$, which loglinear model has the same fit as the GLMM?

12.5 For the student survey in Table 9.1, **(a)** analyze using GLMMs, and **(b)** compare results and interpretations to those with marginal models in Problem 11.2.

12.6 Fit model (12.10) to the responses on abortion. If your software uses Gauss–Hermite quadrature, report the approximate number of quadrature points needed for parameter estimates to converge and the number needed for standard error estimates to converge. (This example has large $\hat{\sigma}$ and requires many points.)

12.7 For the crossover study in Table 11.10 (Problem 11.6), fit the model

$$\text{logit}\left[P\left(Y_{i(k)t} = 1 \mid u_{i(k)}\right)\right] = \alpha_k + \beta_t + u_{i(k)}, \qquad (12.19)$$

where $\{u_{i(k)}\}$ are independent $N(0, \sigma^2)$. Interpret $\{\hat{\beta}_t\}$ and $\hat{\sigma}$.

12.8 For Problem 12.7, compare estimates of $\beta_B - \beta_A$ and $\beta_C - \beta_A$ and SE values to those using **(a)** a marginal model (Problem 11.6), and **(b)** conditional logistic regression (Section 10.2), treating subject terms in model (12.19) as fixed effects.

12.9 For Problem 12.7, fit the more general GLMM having treatment effects $\{\beta_{tk}\}$ that vary by sequence. Test whether the fit is better. One could also consider period or carryover effects. Add two period effects to model (12.19) (e.g., the first-period-effect parameter adds to the model when $t = A$ and $k = 1, 2$, $t = B$ and $k = 3, 4$, and $t = C$ and $k = 5, 6$). Check whether the fit improves. Interpret.

12.10 Consider the logistic-normal model (12.10) for the abortion opinion data, under the constraint $\sigma = 0$.
   a. Explain why the fit is the same as an ordinary logit model treating the three responses for each subject as if they were independent responses for three separate subjects.
   b. Explain why the model fit is the same as an ordinary loglinear model (GI_1, GI_2, GI_3) of mutual independence of responses on the three items (I_1, I_2, I_3), given $G = $ gender.
   c. Fit the model. Interpret, and explain why $\{\hat{\beta}_t - \hat{\beta}_u\}$ are quite different from those in Section 12.3.2 allowing $\sigma > 0$.

12.11 For Table 6.7 on admissions decisions for graduate school applicants, let $y_{ig} = 1$ denote a subject in department i of gender g ($1 = $ females, $0 = $ males) being admitted.
   a. For the fixed effects model, $\text{logit}[P(Y_{ig} = 1)] = \alpha + \beta g + \beta_i^D$, $\hat{\beta} = 0.173$ (SE $= 0.112$). Interpret.
   b. The corresponding model (12.12) in which departments are a normal random effect has $\hat{\beta} = 0.163$ (SE $= 0.111$). Interpret.
   c. The model of form (12.12) allowing the gender effect to vary by department has $\hat{\beta} = 0.176$ (SE $= 0.132$), with $\hat{\sigma}_b = 0.20$. Interpret. Explain why the standard error of $\hat{\beta}$ is slightly larger than with the other analyses.
   d. The marginal sample log odds ratio between gender and whether admitted equals -0.07. How could this take different sign from $\hat{\beta}$ in these models?

e. The sample conditional odds ratios between gender and whether admitted vary between 0 and ∞. By contrast, predicted odds ratios for the interaction random effects model do not vary much. Explain why results can be so different.

12.12 For the clinical trial in Table 9.16, let $\pi_{it} = P(Y_{it} = 1 | u_i)$ denote the probability of success for treatment t in center i.

  a. The random intercept model (12.11) has $\hat{\beta} = 1.52$ (SE $= 0.70$) and $\hat{\sigma} = 1.9$. Interpret.

  b. From Section 9.8.3, the fixed effects analog of this model (replacing $\alpha + u_i$ by α_i) has $\hat{\alpha}_1 = \hat{\alpha}_3 = -\infty$, corresponding to $\hat{\pi}_{1t} = \hat{\pi}_{3t} = 0$ for each treatment. By contrast, the random effects model has $\hat{\alpha} + \hat{u}_1 = -3.78$ (using NLMIXED in SAS) and $\hat{\pi}_{11} = 0.047$ and $\hat{\pi}_{12} = 0.011$ in center 1. Explain how this model can have $\hat{\pi}_{it} > 0$ in centers having no successes.

12.13 Refer to the subject-specific model in Section 12.3.3. Verify that the estimated difference in time effect slopes between the new and standard drugs for treating depression are (a) 1.018 (SE $= 0.192$) with the GLMM approach, and (b) 1.156 (SE $= 0.222$) with conditional ML.

12.14 For marginal model (10.14) for Table 10.5 on premarital and extramarital sex, Table 12.12 shows results of fitting a corresponding random intercept model. Interpret $\hat{\beta}$. Compare estimates of and inferences about β to those in Section 10.3.2 for the marginal model.

TABLE 12.12   Output for Problem 12.14

				Std	
		Parameter	Estimate	Error	t Value
Subjects	475	inter1	-1.5422	0.1826	-8.45
Max Obs Per Subject	2	inter2	-0.6682	0.1578	-4.24
Parameters	5	inter3	0.9273	0.1673	5.54
Quadrature Points	100	beta	4.1342	0.3296	12.54
Log Likelihood	-890.1	sigma	2.0757	0.2487	8.35

12.15 A data set from the 1994 General Social Survey on subjects' opinions on four items (the environment, health, law enforcement, education) related to whether they believed government spending on each item should increase, stay the same, or decrease. Subjects were also classified by their gender and race. For subject i, let $G_i = 1$ for females and 0 for males, let $R_{1i} = 1$ for whites and 0 otherwise,

$R_{2i} = 1$ for blacks and 0 otherwise, and $R_{1i} = R_{2i} = 0$ for the other category of race. Let y_{it} denote the response for subject i on spending item t, where outcomes (1, 2, 3) represent (increase, stay the same, decrease).

a. With constraint $\beta_4 = 0$, the random-intercept model

$$\text{logit}\big[P(Y_{it} \leq j | u_i) \big]$$
$$= \alpha_j + \beta_t + \beta_g G_i + \beta_{r1} R_{i1} + \beta_{r2} R_{2i} + u_i, \quad j = 1, 2,$$

has $\hat{\beta}_1 = -0.55$, $\hat{\beta}_2 = -0.60$, $\hat{\beta}_3 = -0.49$, with $\hat{\sigma} = 1.03$. These estimates are greater than five standard errors in absolute value. Interpret.

b. Table 12.13 shows results with a race-by-item interaction. Interpret.

TABLE 12.13 Results for Problem 12.15[a]

Variable	Estimate	SE
Intercept-1	1.065	0.391
Intercept-2	1.919	0.051
Gender	0.409	0.088
Race1-w	−0.055	0.397
Race2-b	0.434	0.452
Item1-envir	−0.357	0.539
Item2-health	−0.319	0.493
Item3-crime	−0.585	0.480
Race1 * Item1	−0.170	0.549
Race1 * Item2	−0.387	0.503
Race1 * Item3	0.197	0.491
Race2 * Item1	−0.452	0.606
Race2 * Item2	0.454	0.598
Race2 * Item3	−0.518	0.560

[a] Coding 0 for item 4 (education) and race 3 (other).

12.16 Refer to Problem 11.12 for Table 8.19 on government spending. Analyze these data using a cumulative logit model with random effects. Interpret. Compare results to those with a marginal model (Problem 11.12).

12.17 For the insomnia example in Section 12.4.2, according to SAS the maximized log likelihood equals -593.0, compared to -621.0 for the simpler model forcing $\sigma = 0$. Compare models, using either a likelihood-ratio test or AIC. What do you conclude?

TABLE 12.14   Results for Problem 12.18

Observer Effect	GEE	Random Effects
A	−0.451 (0.108)	−1.201 (0.300)
B	−0.391 (0.093)	−0.919 (0.299)
C	0.319 (0.118)	0.558 (0.301)
D	0.632 (0.105)	1.545 (0.313)
E	−0.491 (0.098)	−1.379 (0.300)
F	1.252 (0.161)	2.907 (0.344)

12.18   Landis and Koch (1977) showed ratings by seven pathologists who separately classified 118 slides regarding the presence and extent of carcinoma of the uterine cervix, using a five-point ordinal scale. (Table 13.1 is a collapsing of their table that combines the first two categories and the last three categories.) For slide i with rater t, Table 12.14 shows results of fitting model

$$\text{logit}\big[P(Y_{it} \leq j | u_i)\big] = u_i + \alpha_j + \beta_t$$

to the ordinal table (with $\hat{\beta}_G = 0$), assuming that the $\{u_i\}$ are independent $N(0, \sigma^2)$. It also shows GEE estimates, using independence working equations, for the corresponding marginal model. Interpret $\hat{\beta}_F$ for each model. Explain why estimates using the random effects model, for which $\hat{\sigma} = 3.8$, tend to be much larger in absolute value. Discuss the differences in assumptions and interpretations for the two models.

12.19   Refer to Section 12.5.1 on boys' attitudes toward the leading crowd. Table 12.15 shows results for a sample of schoolgirls. Fit model (12.16) and interpret. Summarize the estimated variability and correlation of random effects.

TABLE 12.15   Data for Problem 12.19

(M, A) for First Interview	(M, A) for Second Interview[a]			
	(Yes, Positive)	(Yes, Negative)	(No, Positive)	(No, Negative)
Yes, positive	484	93	107	32
Yes, negative	112	110	30	46
No, positive	129	40	768	321
No, negative	74	75	303	536

[a]M, membership; A, attitude.

Source: J. S. Coleman, *Introduction to Mathematical Sociology* (London: Free Press of Glencoe, 1964), p. 168.

12.20 Generalize model (12.16) to apply simultaneously to Tables 12.8 and 12.15, using a gender main effect but the same membership effect and the same attitude effect for each gender. Fit the model. Use the maximized log likelihood to compare with a more general model having different membership effects and different attitude effects for each gender. Interpret.

12.21 Table 12.16 reports results from a study to estimate the number N of people infected during a 1995 hepatitis A outbreak in Taiwan. The 271 observed cases were reported from records based on a serum test taken by the Institute of Preventive Medicine of Taiwan (P), records reported by the National Quarantine Service (Q), and records based on questionnaires administered by epidemiologists (E). Estimating N is difficult, because many subjects had only one capture.

  a. Find $\hat{N}$ if you observed only (**i**) P and Q, (**ii**) P and E, (**iii**) Q and E.

  b. Find $\hat{N}$ using the model of mutual independence with P, Q, and E.

  c. Find a 95% profile likelihood interval for N using the model in part (b).

  d. The random effects model of Section 12.3.6 has fit shown in Table 12.16, for which $\hat{\sigma} = 2.9$. The log-likelihood is relatively flat, and $\hat{N} = 4551$ with a 95% profile likelihood interval of $(758, \infty)$ (Coull and Agresti 1999). Explain why this model may provide imprecise estimates of N. Since the interval in part (c) is much narrower, is it necessarily more reliable?

TABLE 12.16   Data for Problem 12.21

P Q E	Observed Count	Logistic-Normal ML Fit
0 0 0	—	$(487, \infty)$
0 0 1	63	61.0
0 1 0	55	58.0
0 1 1	18	17.0
1 0 0	69	68.0
1 0 1	17	20.0
1 1 0	21	19.0
1 1 1	28	28.0

Source: Data from Chao et al. (2001).

12.22 Analyze the crossover data of Table 11.1 using a random effects approach. Interpret, and compare results to those in Section 11.1.2.

12.23 The analyses in Section 12.3.2 comparing opinions on some topic extend to ordinal responses. Using an ordinal random effects model, analyze the 4^3 table in Agresti (1993), found also at the book's Web site, *www.stat.ufl.edu/~ aa/cda/cda.html*.

12.24 The analyses in Section 12.3.4 describing heterogeneity in multicenter clinical trials extend to ordinal responses. Using random effects models, analyze the $2 \times 3 \times 8$ table in Hartzel et al. (2001a).

12.25 You are a statistical consultant asked to analyze Table 4 in B. Efron, *Statistical Science* **13**: 95–122 (1998), which shows 2×2 tables from a clinical trial in 41 cities. Analyze, and write a report summarizing your analysis.

12.26 Analyze Table 11.9 with age and maternal smoking as predictors using a (**a**) logistic-normal model, (**b**) marginal model, and (**c**) transitional model. Explain how the interpretation of the maternal smoking effect differs for the three approaches.

Theory and Methods

12.27 Refer to Section 12.3.1. Using supplementary information improves predictions. Let q_i denote the true proportion of votes for Clinton in state i in the 1992 election, conditional on voting for him or Bush. Consider the model

$$\text{logit}[P(Y_{it} = 1 \mid u_i)] = \text{logit}(q_i) + \alpha + u_i,$$

where $\{q_i\}$ are known and $\{u_i\}$ are independent $N(0,\sigma^2)$. When $\hat{\sigma} = 0$, show $\hat{\pi}_i = q_i \exp(\hat{\alpha})/[1 - q_i + q_i \exp(\hat{\alpha})]$. Compared to $\{q_i\}$, explain how $\hat{\pi}_i$ then shifts up or down depending on how the overall Democratic vote compares in the current poll to the previous election (i.e., depending on $\hat{\alpha}$). When also $\hat{\alpha} = 0$, show $\hat{\pi}_i = q_i$.

12.28 For a binary response, consider the random effects model

$$\text{logit}[P(Y_{it} = 1 \mid u_i)] = \alpha + \beta_t + u_i, \qquad t = 1, \ldots, T,$$

where $\{u_i\}$ are independent $N(0, \sigma^2)$, and the marginal model

$$\text{logit}[P(Y_t = 1)] = \alpha + \beta_t^*, \qquad t = 1, \ldots, T.$$

For identifiability, $\beta_T = \beta_T^* = 0$. Explain why all $\beta_t = 0$ implies that all $\beta_t^* = 0$. Is the converse true?

12.29 The GLMM for binary data using probit link function is

$$\Phi^{-1}[P(Y_{it} = 1 \mid \mathbf{u}_i)] = \mathbf{x}'_{it}\boldsymbol{\beta} + \mathbf{z}'_{it}\mathbf{u}_i,$$

where Φ is the $N(0, 1)$ cdf and $\mathbf{u}_i$ has $N(\mathbf{0}, \boldsymbol{\Sigma})$ pdf, $f(\mathbf{u}_i; \boldsymbol{\Sigma})$.
a. Show that the marginal mean is

$$P(Y_t = 1) = \int P(Z - \mathbf{z}'_{it}\mathbf{u}_i \le \mathbf{x}'_{it}\boldsymbol{\beta})f(\mathbf{u}_i; \boldsymbol{\Sigma})d\mathbf{u}_i,$$

where Z is a standard normal variate that is independent of $\mathbf{u}_i$.
b. Since $Z - \mathbf{z}'_{it}\mathbf{u}_i$ has a $N(0, 1 + \mathbf{z}'_{it}\boldsymbol{\Sigma}\mathbf{z}_{it})$ distribution, deduce that

$$\Phi^{-1}[P(Y_t = 1)] = \mathbf{x}'_{it}\boldsymbol{\beta}[1 + \mathbf{z}'_{it}\boldsymbol{\Sigma}\mathbf{z}_{it}]^{-1/2}.$$

Hence, the marginal model is a probit model with attenuated effect. In the univariate random intercept case, show the marginal effect equals that from the GLMM divided by $\sqrt{1 + \sigma^2}$.

12.30 In the Rasch model, $\text{logit}[P(Y_{it} = 1)] = \alpha_i + \beta_t$, α_i is a fixed effect.
a. Assuming independence of responses for different subjects and for different observations on the same subject, show that the log likelihood is

$$\sum_i \sum_t \alpha_i y_{it} + \sum_i \sum_t \beta_t y_{it} - \sum_i \sum_t \log[1 + \exp(\alpha_i + \beta_t)].$$

b. Show that the likelihood equations are $y_{+t} = \Sigma_i P(Y_{it} = 1)$ and $y_{i+} = \Sigma_t P(Y_{it} = 1)$ for all i and t. Explain why conditioning on $\{y_{i+}\}$ yields a distribution that does not depend on $\{\alpha_i\}$.
c. Discuss advantages and disadvantages of, instead, treating α_i as random.

12.31 Consider the matched-pairs random effects model (12.3). For given β_0, let δ_0 be such that $\hat{\mu}_{12} = n_{12} + \delta_0$ and $\hat{\mu}_{21} = n_{21} - \delta_0$ satisfies $\log(\hat{\mu}_{21}/\hat{\mu}_{12}) = \beta_0$. Suppose $\{\hat{\mu}_{ij}\}$ has nonnegative log odds ratio. Explain why:
a. This is the fit of the model assuming $\beta = \beta_0$.
b. The likelihood-ratio statistic for testing $H_0: \beta = \beta_0$ in this model equals

$$2\left(n_{12} \log\frac{n_{12}}{n_{12} + \delta_0} + n_{21} \log\frac{n_{21}}{n_{21} - \delta_0}\right).$$

c. The likelihood-ratio test of $H_0: \beta = 0$ is the test of symmetry.

12.32 Explain why the logistic-normal model is not helpful for capture–recapture experiments with only two captures.

12.33 Refer to the crossover study in Problem 12.7. Kenward and Jones (1991) reported results using the ordinal response scale (none, moderate, complete) for relief. Explain how to formulate an ordinal logit random effects model for these data analogous to model (12.19).

12.34 Formulate a model using adjacent-categories logits that is analogous to model (12.14) for cumulative logits. Interpret parameters.

12.35 For ordinal square $I \times I$ tables of counts $\{n_{ab}\}$, model (12.3) for binary matched-pairs responses (Y_{i1}, Y_{i2}) for subject i extends to

$$\text{logit}[P(Y_{it} \leq j | u_i)] = \alpha_j + \beta x_t + u_i$$

with $\{u_i\}$ independent $N(0, \sigma^2)$ variates and $x_1 = 0$ and $x_2 = 1$.

a. Explain how to interpret β, and compare to the interpretation of β in the corresponding marginal model (10.14).

b. This model implies model (12.3) for each 2×2 collapsing that combines categories 1 through j for one outcome and categories $j + 1$ through I for the other. Use the form of the conditional ML (or random effects ML) estimator for binary matched pairs to explain why

$$\log\left[\left(\sum_{a>j}\sum_{b<j} n_{ab}\right)\Big/\left(\sum_{a<j}\sum_{b>j} n_{ab}\right)\right]$$

is a consistent estimator of β.

c. Treat these $(I - 1)$ collapsed 2×2 tables naively as if they are independent samples. Show that adding the numerators and adding the denominators of the separate estimates of e^β motivates the summary estimator of β,

$$\tilde{\beta} = \log\left\{\left[\sum_{a>b}(a - b)n_{ab}\right]\Big/\left[\sum_{b>a}(b - a)n_{ab}\right]\right\}.$$

Explain why $\tilde{\beta}$ is consistent for β even recognizing the actual dependence.

d. A standard error for $\tilde{\beta}$ that treats the collapsed tables in part (c) as independent is inappropriate. Treating $\{n_{ab}\}$ as a multinomial sample, show that an estimated asymptotic variance of $\tilde{\beta}$ is (Agresti and Lang 1993a)

$$\left\{ \sum_{b>a} (b-a)^2 n_{ab} \Big/ \left[\sum_{b>a} (b-a) n_{ab} \right]^2 \right\}$$

$$+ \left\{ \sum_{a>b} (a-b)^2 n_{ab} \Big/ \left[\sum_{a>b} (a-b) n_{ab} \right]^2 \right\}.$$

12.36 Summarize advantages and disadvantages of using a GLMM approach compared to a marginal model approach. Describe conditions under which parameter estimators are consistent for (**a**) marginal models using GEE, (**b**) marginal models using ML, (**c**) GLMM using PQL, and (**d**) GLMM using ML.

CHAPTER 13

Other Mixture Models for Categorical Data*

In Chapters 10 through 12 we introduced ways of handling correlated observations due to repeated measurement and other forms of clustering. The generalized linear mixed models (GLMMs) of Chapter 12 assume normal random effects. They describe heterogeneity by replacing the linear predictor by a normally distributed mixture of linear predictors. In this chapter we present additional models having connections with GLMMs. Except for one case, these models use nonnormal mixture distributions.

In Section 13.1 we present latent class models. These treat a contingency table as a mixture of unobserved tables at categories of a qualitative latent (unobserved) variable. In Section 13.2 we discuss a related nonparametric approach to fitting GLMMs that uses an unspecified discrete quantitative distribution for the random effects distribution.

In Section 13.3 we model clustered binomial responses using the beta distribution to describe heterogeneity of binomial parameters. The resulting beta-binomial distribution has variance function for which quasi-likelihood methods are also available. In Section 13.4 we model count responses using the gamma distribution to describe heterogeneity of Poisson parameters. The resulting negative binomial regression model corresponds to a Poisson GLMM having a log-gamma distributed random effect. It is an alternative to the GLMM for Poisson responses with normal random effects, a model discussed in Section 13.5.

13.1 LATENT CLASS MODELS

GLMMs create a mixture of linear predictor values using a latent variable, the unobserved random effect vector, having a normal distribution. By contrast, latent class models use a mixture distribution that is qualitative rather than quantitative. The basic model assumes existence of a latent

538

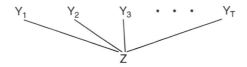

FIGURE 13.1 Association graph for latent class model.

categorical variable such that the observed response variables are conditionally independent, given that variable.

For categorical response variables $(Y_1, Y_2, \ldots, Y_T)$, the latent class model assumes a latent categorical variable Z such that for each possible sequence of response outcomes $(y_1, \ldots, y_T)$ and each category z of Z,

$$P(Y_1 = y_1, \ldots, Y_T = y_T | Z = z) = P(Y_1 = y_1 | Z = z) \cdots P(Y_T = y_T | Z = z).$$

Figure 13.1 shows the association graph for the model. A latent class model summarizes probabilities of classification $P(Z = z)$ in the latent classes as well as conditional probabilities $P(Y_t = y_t | Z = z)$ of outcomes for each Y_t within each latent class. These are the model parameters. More generally, the latent variable $\mathbf{Z}$ can be multivariate. The model is an analog for categorical responses and latent variables of the factor analysis model for multivariate normal responses.

The latent class model is sometimes plausible when the observed variables are several indicators of some concept, such as prejudice, religiosity, or opinion about an issue. An example is Table 10.13, in which subjects gave their opinions about whether abortion should be legal in various situations. Perhaps an underlying latent variable describes one's basic attitude toward legalized abortion, such that given the value of that latent variable, responses on the observed variables are conditionally independent. For instance, the latent variable may be a qualitative variable with three categories: One class for those who always oppose legalized abortion regardless of the situation, one for those who always favor it, and one for those whose response depends on the situation.

The T-dimensional contingency table cross classifying $(Y_1, \ldots, Y_T)$ is observed. The $(T + 1)$-dimensional table that cross-classifies it with the latent variable is an unobserved table. Denote the number of categories of each Y_t by I and the number of latent classes of Z by q. For the observed table, let $\pi_{y_1, \ldots, y_T} = P(Y_1 = y_1, \ldots, Y_T = y_T)$. The model assumes a multinomial distribution over its I^T cells. For a given cell,

$$\pi_{y_1, \ldots, y_T} = \sum_{z=1}^{q} P(Y_1 = y_1, \ldots, Y_T = y_T | Z = z) P(Z = z).$$

The conditional independence factorization for the latent class model states that

$$\pi_{y_1,\ldots,y_T} = \sum_{z=1}^{q} \left[\prod_{t=1}^{T} P(Y_t = y_t | Z = z) \right] P(Z = z). \qquad (13.1)$$

This is a nonlinear model for the I^T multinomial probabilities.

13.1.1 Fitting Latent Class Models

Denote the counts in the observed table by $\{n_{y_1,\ldots,y_T}\}$. Summing over the I^T cells in that table, the kernel of the multinomial log likelihood is

$$\sum n_{y_1,\ldots,y_T} \log \pi_{y_1,\ldots,y_T}. \qquad (13.2)$$

Substituting parameters from (13.1), one can maximize (13.2) with respect to those parameters using Newton–Raphson (Haberman 1979, Chap. 10) or the EM algorithm (Goodman 1974). It is helpful to note that the latent class model states that the loglinear model symbolized by $(Y_1 Z, Y_2 Z, \ldots, Y_T Z)$ holds for the unobserved table. The model makes no assumption about the $\{Y_t Z\}$ associations but assumes that the $\{Y_t\}$ are mutually independent within each category of Z.

The EM algorithm has two steps in each iteration. The E (expectation) step in iteration s calculates pseudo-counts $\{n^{(s)}_{y_1,\ldots,y_T,z}\}$ for the unobserved table using $\{n_{y_1,\ldots,y_T}\}$ and a working conditional distribution for $(Z | Y_1, \ldots, Y_T)$ described shortly. The M (maximization) step treats $\{n^{(s)}_{y_1,\ldots,y_T,z}\}$ as data and applies an algorithm such as iterative reweighted least squares or IPF for fitting the model (i.e., the loglinear model $(Y_1 Z, Y_2 Z, \ldots, Y_T Z)$). The fit $\{\mu^{(s)}_{y_1,\ldots,y_T,z}\}$ of that model in the unobserved table then determines the new working conditional distribution of $(Z | Y_1, \ldots, Y_T)$ to apply to $\{n_{y_1,\ldots,y_T}\}$ for the E-step of the next iteration. This allocates the observed data to pseudo-counts in the unobserved cells in proportion to this fit, using

$$n^{(s+1)}_{y_1,\ldots,y_T,z} = n_{y_1,\ldots,y_T} \frac{\mu^{(s)}_{y_1,\ldots,y_T,z}}{\sum\limits_{k=1}^{q} \mu^{(s)}_{y_1,\ldots,y_T,k}}.$$

These are entries in the unobserved table for iteration $(s + 1)$. They are used as pseudo-data for the M-step of iteration $(s + 1)$.

Eventually, the algorithm converges to fitted values for the unobserved table that provide fitted probabilities that satisfy mutual independence within each latent class, and such that the corresponding fitted probabilities in the observed table (i.e., added over the latent categories) maximize the likelihood (13.2). The fitted probabilities in the unobserved table are an estimated joint distribution for $(Y_1, \ldots, Y_T, Z)$. One can use them to calculate the ML estimates of the latent class model parameters $\{P(Y_t = y_t \mid Z = z)\}$ and $\{P(Z = z)\}$.

The EM algorithm is computationally simple and relatively stable. Each iteration increases the likelihood. However, its convergence can be slow. See Laird (1998) for a review. The log likelihood for a latent class model may have local maxima. Thus, with either the Newton–Raphson or EM algorithm, it is advisable to perform the fitting process a few times with different starting guesses for the parameter values. The EM algorithm tends to be less sensitive to the choice of starting values. Thus, some software begins with the EM algorithm and then switches to the Newton–Raphson algorithm as it approaches the ML estimates to speed the process. As q increases, multiple local maxima are more likely and the danger increases of a lack of identifiability.

Standard errors for model parameter estimates result from inverting the model's estimated information matrix. This is a by-product of the Newton–Raphson algorithm but not the EM algorithm. One way to obtain standard errors with it applies a useful formula of Louis (1982) for the observed information when using the EM algorithm. It equals the expected value of the observed information for the loglinear model for the unobserved table minus the expected value of the information for the conditional distribution of Z given the observed data. Baker (1992) and Lang (1992) gave related results.

Chi-squared statistics comparing observed cell counts to fitted values test the model fit. The residual $df = I^T - qT(I - 1) - q$. This follows since multinomial model (13.1) describes $I^T - 1$ multinomial probabilities using $(I - 1)$ parameters $\{P(Y_t = y_t \mid Z = z), y_t = 1, \ldots, I - 1\}$ at each of qT combinations of z and t, and $q - 1$ parameters $\{P(Z = z)\}$. Often, the nature of the variables suggests a value for q, usually quite small (2 to 4). Otherwise, the usual procedure starts with $q = 2$; if the fit is inadequate, it increases by steps of 1 as long as the fit shows substantive improvement. Specialized software exists for such models (Appendix A).

13.1.2 Latent Class Model for Rater Agreement

Table 13.1 is an expanded data set of the example in Section 10.5. Seven pathologists classified each of 118 slides on the presence or absence of carcinoma in the uterine cervix. For modeling interobserver agreement, the conditional independence assumption of the latent class model is often plausible. With a blind rating scheme, ratings of a given subject or unit by different pathologists are independent. If subjects having true rating in a given category are relatively homogeneous, then ratings by different pathologists may be nearly independent within a given true rating class. Thus, one might posit a latent class model with $q = 2$ classes, one for subjects whose true rating is positive and one for subjects whose true rating is negative. This

TABLE 13.1   Diagnoses of Carcinoma and Fits of Latent Class Models[a]

Pathologist								Fit		
A	B	C	D	E	F	G	Count	$q = 1$	$q = 2$	$q = 3$
0	0	0	0	0	0	0	34	1.1	23.0	33.8
0	0	0	0	1	0	0	2	1.6	6.6	2.0
0	1	0	0	0	0	0	6	2.2	12.7	6.3
0	1	0	0	0	0	1	1	2.8	1.7	1.5
0	1	0	0	1	0	0	4	3.3	3.6	3.0
0	1	0	0	1	0	1	5	4.2	0.5	4.7
1	0	0	0	0	0	0	2	1.4	3.0	2.1
1	0	1	0	1	0	1	1	1.6	0.2	0.2
1	1	0	0	0	0	0	2	2.8	1.7	1.3
1	1	0	0	0	0	1	1	3.5	0.3	1.6
1	1	0	0	1	0	0	2	4.2	0.5	2.9
1	1	0	0	1	0	1	7	5.3	3.7	6.5
1	1	0	0	1	1	1	1	1.4	2.6	1.4
1	1	0	1	0	0	1	1	1.3	0.1	0.1
1	1	0	1	1	0	1	2	2.0	4.3	2.6
1	1	0	1	1	1	1	3	0.5	3.1	2.0
1	1	1	0	1	0	1	13	3.3	11.5	9.6
1	1	1	0	1	1	1	5	0.9	8.4	8.7
1	1	1	1	1	0	1	10	1.2	13.5	13.6
1	1	1	1	1	1	1	16	0.3	9.9	12.3

[a] Fits obtained with Latent Gold (Statistical Innovations, Belmont MA). 1, yes; 0, no.

Source: Based on data in Landis and Koch (1977), not showing empty cells.

model expresses the 2^7 joint distribution of the seven ratings as a mixture of two 2^7 distributions, one for each true rating class.

Table 13.2 shows results of fitting some latent class models (including a mixture model studied in Section 13.2.4). Because the observed table is sparse, the deviance is mainly useful for comparing models. This is an informal comparison, though, since the chi-squared distribution does not apply for comparing deviances of models with different numbers of latent classes. A model with q classes is a special case of a model with $q^* > q$ classes in which $P(Z = z) = 0$ for $z > q$ and hence falls on the boundary of the parameter space. Ordinary chi-squared likelihood-ratio tests require parameters to fall in the interior of the parameter space (i.e., $0 < P(Z = z) < 1$ for $z = 1, \ldots, q^*$).

Table 13.1 also shows the fitted values for latent class models with $q = 1, 2, 3$, for the cells having positive counts. (Each empty cell also has a fitted value, not shown here). The model with $q = 1$ latent class is the model of mutual independence of the seven ratings. Equivalently, it is the loglinear model $(Y_1, Y_2, \ldots, Y_7)$. It fits poorly, as one would expect. With $q = 2$, considerable evidence remains of lack of fit. For instance, the fitted count for

TABLE 13.2   Likelihood-Ratio Statistics for Latent Class Models Fitted to Table 13.1[a]

Number of Latent Classes	Model	Deviance (G^2) Statistic	df
1	Mutual independence	476.8	120
2	Latent class	62.4	112
	Rasch mixture	67.6	118
3	Latent class	15.3	104
	Rasch mixture	27.5	116
4	Latent class	6.4	96
	Rasch mixture (quasi-symmetry)	23.7	114

[a]Models fitted with Latent Gold (Statistical Innovations, Belmont, MA).

a negative rating by each pathologist is 23.0, compared to an observed count of 34. (The small G^2 that Table 13.2 reports for this model does not imply a good fit; in Section 9.8.4 we noted that G^2 tends to be highly conservative when most fitted values are very close to 0.) The model with $q = 3$ seems to fit adequately.

Studying the estimated probability $P(Y_t = 1 \mid Z = z)$ of a carcinoma diagnosis for each pathologist, conditional on a given latent class z, helps illuminate the nature of these classes. Table 13.3 reports these for the three-class model. They suggest that (1) the first latent class refers to cases that all pathologists (except occasionally B) agree show no carcinoma; (2) the third latent class refers to cases in which A, B, E, and G agree show carcinoma and C and D usually agree; and (3) the second latent class refers to cases of strong disagreement, whereby C, D, and F rarely diagnose carcinoma but B, E, and G usually do. The estimated proportions in the three latent classes are $\hat{P}(Z = 1) = 0.37$, $\hat{P}(Z = 2) = 0.18$, and $\hat{P}(Z = 3) = 0.45$. The model estimates that 18% of the cases fall in the problematic class.

TABLE 13.3   Estimated Probabilities of Diagnosing Carcinoma, for Latent Class Model and Rasch Mixture Model with Three Classes[a]

Model	Latent Class	Pathologist A	B	C	D	E	F	G
Latent	1	0.057	0.138	0.000	0.000	0.055	0.000	0.000
Class	2	0.513	1.00	0.000	0.058	0.751	0.000	0.631
	3	1.000	0.981	0.858	0.586	1.000	0.476	1.000
Rasch	1	0.022	0.150	0.001	0.000	0.047	0.000	0.022
Mixture	2	0.611	0.923	0.052	0.015	0.774	0.009	0.611
	3	0.994	0.999	0.853	0.617	0.997	0.483	0.994

[a]Results obtained with Latent Gold (Statistical Innovations, Belmont, MA).

A danger with latent variable models, shared by factor analysis for continuous responses, is the temptation to interpret latent variables too literally. In this example it is tempting to treat latent class 1 (latent class 3) as cases truly without carcinoma (with carcinoma). Thus, it is tempting to treat a rating of no carcinoma (a rating of carcinoma) given that the subject falls in latent class 1 (latent level 3) as necessarily being a correct judgment. One should realize the tentative nature of the latent variable. Be careful not to make the error of reification—treating an abstract construction as if it has actual existence (Gould 1981).

Using model parameter estimates and Bayes' theorem, one can also estimate $P(Z = z \mid Y_t = y_t)$ and $P(Z = z \mid Y_1 = y_1, \ldots, Y_T = y_T)$. If a pathologist makes a "yes" rating, for instance, what is the estimated probability that the subject is in the latent class for which agreement on a positive rating usually occurs? We perform further analysis in Section 13.2.5 after studying a simpler model.

Espeland and Handelman (1989), Uebersax (1993), Uebersax and Grove (1990, 1993), and Yang and Becker (1997) presented various latent variable models for rater agreement and diagnostic accuracy. One could also use methods of Chapters 11 and 12, such as a model with a continuous rather than qualitative latent variable. A logistic-normal random intercept model, for instance, yields subject-specific comparisons of $P(Y_t = 1)$ for various t.

13.1.3 Latent Class Models for Capture–Recapture

We next apply latent class models to capture–recapture modeling for estimating population size. In Section 12.3.6 a logistic-normal GLMM was used for this. With T sampling occasions, a 2^T contingency table displays the data, with scale (captured, not captured) at each occasion. A prediction of the population size equals the prediction for the missing cell count, representing subjects not captured at every occasion, added to the counts in other cells.

With two classes, the latent class model treats the population as a mixture of two types, perhaps determined by genetic or environmental factors. Homogeneity of capture probabilities occurs for subjects within each type, but the type of any given subject is unknown. This model represents a compromise between the mutual independence model, which assumes a single latent class and complete homogeneity, and the logistic-normal GLMM, which assumes a continuous mixture of capture probabilities rather than two classes.

We illustrate with the $T = 6$-capture data set on snowshoe hares in Table 12.6. The model of mutual independence predicts that $\hat{N} = 75$. Its 95% profile-likelihood confidence interval for N is (70, 83). The latent class model with two classes has $\hat{N} = 85$ and a profile-likelihood confidence interval of (74, 106). The latent class model with three classes gives similar results. Since the logistic-normal GLMM in Section 12.3.6 gave the interval (75, 154), these seem too short to be trusted. This simple latent class model may not capture all the existing heterogeneity. It is more plausible to assume a continuous latent variable than a discrete one with a couple of classes. We'll analyze these data further with related models in the next section.

13.2 NONPARAMETRIC RANDOM EFFECTS MODELS

In spite of its popularity and attractive features, the normality assumption for random effects in ordinary GLMMs can rarely be closely checked. For instance, in studying normal GLMMs, Verbeke and Lesaffre (1996) noted that under a normality assumption for random effects, their predicted values often appear normally distributed even when the true values are generated from a highly nonnormal distribution. An obvious concern of this or any parametric assumption for the random effects is possibly harmful effects of misspecification. To check sensitivity to this assumption, one can fit GLMMs using alternative or more general random effects assumptions.

13.2.1 Logit Models with Unspecified Random Effects Distribution

A nonparametric approach (e.g., Aitkin 1999) guards against possibly harmful misspecification effects. This uses an unspecified random effects distribution on a finite set of mass points. The location of the mass points and their probabilities are parameters. The number of mass points can be fixed. When this number is itself unknown, one treats it as fixed in the estimation process but increases it sequentially until the likelihood is maximized. The maximization usually requires relatively few mass points. Even allowing a continuous mixture distribution, the nonparametric estimate of that distribution takes a finite number of points (e.g., Lindsay et al. 1991). In fact, fitting a model having only two mass points often results in fixed effects estimates quite similar to those with the full maximization. This approach is useful primarily when the random effects distribution is not itself of direct interest, since the nonparametric estimate of that distribution tends to be poor even for very large samples.

Model fitting is actually simpler than for models with normal random effects, since the integral that determines the likelihood function simplifies to a finite sum. In Section 13.2.4 we discuss this point with a Rasch-type model. Specialized software can fit nonparametric mixture models (Appendix A). However, this approach also has disadvantages. For instance, with multivariate random effects it cannot provide simple correlation structure as the normal can. Standard inference does not apply for comparing models with different numbers of mass points, since one model is on the boundary of the parameter space compared to the other. Also, the ML estimate of the random effects distribution often places some weight at $\pm\infty$. Although this can be useful with binary data for identifying a subsample for which the estimated response probability equals 1 or equals 0 for all observations in a cluster, it is not then possible to describe heterogeneity with an estimated variance component.

To illustrate this approach, we reanalyze Table 10.13 on attitudes about legalized abortion. In Section 12.3.2 we fitted the logistic-normal model (12.10),

$$\text{logit}[P(Y_{it} = 1 \mid u_i)] = u_i + \beta_t + \gamma x, \tag{13.3}$$

with x = gender and parameters $\{\beta_t\}$ representing three conditions under which abortion might be legal. Treating u_i instead nonparametrically, the likelihood maximizes with a two-point mixture distribution. Estimated abortion item effects are $\hat{\beta}_1 - \hat{\beta}_3 = 0.83$ (SE = 0.16), $\hat{\beta}_2 - \hat{\beta}_3 = 0.30$ (SE = 0.16), and $\hat{\beta}_1 - \hat{\beta}_2 = 0.52$ (SE = 0.16). Results are similar to those that Table 12.3 shows for the normal random effects approach (Section 12.3.2).

13.2.2   Nonparametric Mixing of Logistic Regression

Follman and Lambert (1989) presented an example with a prespecified number of mass points. They analyzed the effect of the dosage of a poison on the probability of death of a protozoan of a particular genus. Table 13.4 shows the data. They assumed two unobserved types of that genus.

Let $\pi_i(x)$ denote the probability of death at log dose level x for genus type i, $i = 1, 2$. Let ρ denote the probability a protozoan belongs to genus type 1. Their model specifies

$$\pi(x) = \rho\pi_1(x) + (1 - \rho)\pi_2(x), \quad \text{where} \quad \text{logit}[\pi_i(x)] = \alpha_i + \beta x,$$

with unknown ρ. The curve for $\pi(x)$ is a weighted average of two curves having the same shapes but different intercepts.

The ordinary logistic regression model is the special case $\rho = 1$. Its fit, $\text{logit}[\hat{\pi}(x)] = -68.4 + 42.1x$ (with SE = 3.8 for $\hat{\beta} = 42.1$), is poor, with deviance $G^2 = 24.7$ (df = 6). The fit of the mixture model is

$$\hat{\pi}(x) = 0.34\hat{\pi}_1(x) + 0.66\hat{\pi}_2(x),$$

with

$$\text{logit}[\hat{\pi}_1(x)] = -196.2 + 124.8x, \quad \text{logit}[\hat{\pi}_2(x)] = -205.7 + 124.8x,$$

TABLE 13.4   Number of Protozoa Exposed to Poison Dose and Number That Died

Poison Dose	Exposed	Dead	Poison Dose	Exposed	Dead
4.7	55	0	5.1	53	22
4.8	49	8	5.2	53	37
4.9	60	18	5.3	51	47
5.0	55	18	5.4	50	50

Source: Follman and Lambert (1989). Reprinted with permission from the *Journal of the American Statistical Association.*

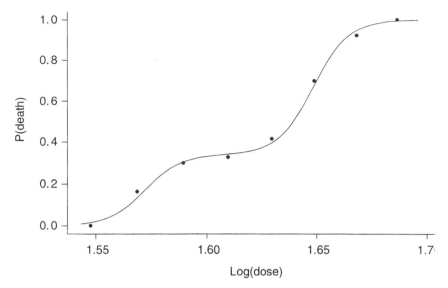

FIGURE 13.2 Fit of binary mixture of logistic regressions to Table 13.4 [model fitted using Latent Gold (Statistical Innovations, Belmont, MA)].

and SE = 25.2 for $\hat{\beta} = 124.8$. Figure 13.2 shows the fit. This is much better, with $G^2 = 3.4$ (df = 4); that is, double the maximized log-likelihood increases by $24.7 - 3.4 = 21.3$ by adding two parameters: an additional intercept and the probability for the mixture. Follman and Lambert noted that with eight dose levels, at most two mixture points are identifiable for this model.

The ordinary GLMM assumes a normal mixture of logistic curves. It gives a deviance reduction of only 1.7 compared to the ordinary logistic model with $\rho = 1$.

13.2.3 Is Misspecification a Serious Problem?

Is it worth the trouble to consider alternatives to the normality assumption for random effects in GLMMs, whether they be parametric or nonparametric? Not much work exists on investigating misspecification effects. For logistic random intercept models, different assumptions for the random effects distribution often provide similar results for estimating the regression effects. Choosing an incorrect random effects distribution does not tend to bias estimators of those effects. The true distribution for the random effects being skewed can result in some bias for the normal intercept estimator (Neuhaus et al. 1992). The choice of random effects distribution also usually has little impact on efficiency of estimation.

When the true random effects distribution is dramatically far from normal, there can be some efficiency loss for the logistic-normal estimator. This can happen when the true distribution is a two-point mixture with large variance component. B. Caffo and I studied this with various models, such as a simple one-way random effects model. In cluster i, let y_{it} be a Bernoulli variate satisfying

$$\text{logit}\left[P(Y_{it} = 1 \mid u_i) \right] = \alpha + u_i, \qquad i = 1, \ldots, n, \quad t = 1, \ldots, T, \quad (13.4)$$

where $\text{var}(u_i) = \sigma^2$. Simulated samples from this model used various n, T, α, and σ, and various true distributions for u_i including normal, uniform, exponential, and binary. Usually, assuming normality does not hurt when the true distribution is nonnormal. Also, using a nonparametric approach when the true distribution is normal does not result in much efficiency loss [Neuhaus and Lesperance (1996) noted this for a related model.] However, when the true distribution is a two-point mixture, the normal approach loses efficiency in estimating $\{\mu_i = P(Y_{it} = 1 \mid u_i)\}$ as σ and T increase. For example, when $n = T = 30$, $\alpha = 0$, and the mixture has probability 0.5 at each point, the expected value of $|\hat{\mu}_i - \mu_i|$ is $(0.06, 0.05)$ for the (normal, nonparametric) approach when $\sigma = 0.5$, $(0.06, 0.02)$ when $\sigma = 1.0$, and $(0.04, 0.01)$ when $\sigma = 2.0$. Differences for estimating α are less dramatic.

The example from Follman and Lambert (1989) discussed in Section 13.2.2, which has a covariate but $T = 1$, illustrates the potential efficiency loss with the logistic-normal GLMM. The two-point mixture model has $\hat{\beta} = 124.8$ with SE $= 25.2$, for which $\hat{\beta}/\text{SE} = 4.9$. The normal mixture model has $\hat{\beta} = 65.5$ with SE $= 19.5$, for which $\hat{\beta}/\text{SE} = 3.4$.

Our study suggested that the random effects distribution has to be rather extremely nonnormal for the normal GLMM to suffer in bias or efficiency. However, Heagerty and Zeger (2000) (see also McCulloch 1997) noted that other types of misspecification can be more crucial. Regarding bias, they argued that sensitivity to the random effects assumption is greater for estimating regression parameters in random effects models than estimating their counterparts in corresponding marginal models. They illustrated this with a model violation by which the variance of the random effects depends on values of covariates. They concluded that between-cluster effects may be more sensitive to correct specification of the random effects distribution than within-cluster effects. This is an advantage of using marginal models for between-cluster effects.

13.2.4 Rasch Mixture Model

From Section 12.1.4, for subject i with item t the Rasch model for a binary response is

$$\text{logit}\left[P(Y_{it} = 1 \mid u_i) \right] = u_i + \beta_t, \quad t = 1, \ldots, T. \qquad (13.5)$$

The GLMM treats $\{u_i\}$ as normal random effects. Lindsay et al. (1991) studied this model when u_i instead can assume only a finite number q of values. Denote the distribution of the latent variable u_i, which is the same for all i, by

$$P(U = a_k) = \rho_k, \quad k = 1, \ldots, q,$$

for unknown $\{a_k\}$ and $\{\rho_k\}$. For identifiability one can either place a constraint on this distribution, such as $\sum_k \rho_k a_k = 0$, or on $\{\beta_t\}$. This model is called a *Rasch mixture model*.

Like other random effects models, the Rasch mixture model is a latent variable model. The random effect u_i is unobserved, and the T responses are assumed conditionally independent at each fixed u_i value. It differs from the ordinary latent class model for binary responses having q latent classes (Section 13.1), since it assumes structure (13.5) for $P(Y_{it} = 1 | u_i)$ whereas latent class model (13.1) assumes no structure for $P(Y_t = y_t | Z = z)$.

This model is simpler to fit than GLMMs with normal random effects because the GLMM's intractable integral that determines the likelihood function is replaced by a finite sum. The marginal probability of a sequence of responses $(y_1, \ldots, y_T)$ is

$$\pi_{y_1, \ldots, y_T} = \sum_{k=1}^{q} \rho_k \left[\prod_{t=1}^{T} \frac{\exp[y_t(a_k + \beta_t)]}{1 + \exp(a_k + \beta_t)} \right].$$

Substituting this in the multinomial log likelihood (13.2), ML estimation of $\{a_k, \rho_k\}$ and $\{\beta_t\}$ can proceed using Newton–Raphson or EM algorithms. As q increases, the maximized likelihood increases and the fit improves. However, Lindsay et al. (1991) showed that with T items, the likelihood no longer changes once $q = (T + 1)/2$. Then, the model gives the same fit to the 2^T observed table as the quasi-symmetry model (10.33). Thus, this simpler latent class model has a symmetric conditional association structure among the observed variables. Arminger et al. (2000) extended the Rasch mixture model to incorporate covariates.

13.2.5 Modeling Rater Agreement

For the ratings of carcinoma by seven pathologists (Table 13.1), Table 13.2 also summarizes the fit of Rasch mixture models. Here, $P(Y_{it} = 1 | u_i)$ in (13.5) denotes the probability of a carcinoma diagnosis for pathologist t evaluating slide i. With $q = 3$ (i.e., u_i can take 3 values), it does not fit significantly more poorly than the latent class model. With $T = 7$ raters, the discrete mixture can take at most $(T + 1)/2 = 4$ points. The model with $q = 4$ is equivalently the quasi-symmetry model. It does not seem to fit better than with $q = 3$.

Pathologist	F	D	C	A	G	E	B
Estimate	−3.70	−3.15	−1.87	1.48	1.48	2.26	3.52
Comparison		———————— ———			————————————— ———		

FIGURE 13.3 Pathologist estimates for Rasch mixture model and results of 90% Bonferroni simultaneous comparison.

Figure 13.3 shows $\{\hat{\beta}_t\}$ for the Rasch mixture model with $q = 3$, setting $\sum_t \hat{\beta}_t = 0$. These describe variation among the pathologists' response distributions at each latent level. For a given latent class, for instance, the estimated odds of a carcinoma diagnosis for pathologist B are $\exp(3.52 - 1.48) = 7.7$ times the estimated odds for pathologist A. Pathologist B tends to make a carcinoma diagnosis most often, and D and F the least. The figure also shows results of a 90% Bonferroni comparison of the 21 pairs of pathologists, based on standard errors of pairwise differences $\hat{\beta}_t - \hat{\beta}_s$.

For pathologist t, conditional on latent level k for a slide,

$$\exp(\hat{a}_k + \hat{\beta}_t)\big/\big[1 + \exp(\hat{a}_k + \hat{\beta}_t)\big]$$

estimates the probability of a carcinoma diagnosis. Table 13.3 reports these, which use $\hat{a}_1 = -5.25$, $\hat{a}_2 = -1.02$, and $\hat{a}_3 = 3.63$. They are similar to the estimates for the ordinary latent class model but a bit smoother, with fewer estimates at the boundary. Again, at latent level 1 pathologists tend not to diagnose carcinoma, at level 2 many disagreements occur, and at level 3 pathologists tend to diagnose carcinoma. The estimated latent class proportions are $\hat{\rho}_1 = 0.37$, $\hat{\rho}_2 = 0.19$, and $\hat{\rho}_3 = 0.43$, with 19% of cases falling in the problematic class.

Model (13.5) implies that the association between each Y_t and U has log odds ratio $(a_k - a_\ell)$ for levels k and ℓ of U. For instance, in the third latent class the estimated odds that a pathologist diagnoses carcinoma are $\exp[3.63 - (-5.25)] > 7000$ times those in the first latent class. In terms of the estimated probabilities in Table 13.3, using pathologist A this is $\exp[(0.994/0.006)/(0.022/0.978)]$. The large $\{\hat{a}_k - \hat{a}_\ell\}$ suggest strong association between each pathologist's rating and the latent variable. This induces strong association between pairs of pathologist ratings. (The model-fitted odds ratios between pairs of raters vary between about 7 and 400.) However, the quite varied $\{\hat{\beta}_t\}$ suggest that substantial marginal heterogeneity exists among the seven ratings. This causes heterogeneity in pairwise levels of agreement.

The mutual independence model is the special case of the Rasch mixture model with $q = 1$; that is, $\rho_1 = 1$. For Table 13.1 the Rasch mixture model with $q = 3$ has only four more parameters than the mutual independence model (i.e., ρ_k and a_k, $k = 1,2$). Yet it fits well and has simple interpretations. See Agresti and Lang (1993b) for further details and a simpler model that sets $a_1 - a_2 = a_2 - a_3$.

13.2.6 Other Models for Capture–Recapture

In Section 13.1.3 latent-class models were used for capture–recapture experiments. Alternatively, one could use the Rasch mixture model. Model (13.5) with two classes gives $\hat{N} = 77$ and a 95% profile-likelihood confidence interval of (71, 87). This seems too short to trust. It is more realistic to allow a continuous distribution for capture probabilities. Model (13.5) treating u_i as normal rather than binary does this, and in Section 12.3.6 we used it for these data.

So, which models might be used other than a parametric random effects model? One possibility is a loglinear model (Cormack 1989). This is a marginal model, applying to probabilities averaged over subjects. Let Y_t denote the binary capture variable for a randomly selected subject at occasion t, with categories (captured, not captured). The simplest model, denoted by $(Y_1, Y_2, \ldots, Y_T)$, assumes that capture events are mutually independent. This is equivalent to the logistic-normal model (13.5) with $\sigma = 0$ and latent class model (13.1) with $q = 1$. A more plausible model allows an association between pairs of capture variables. This is equivalently the loglinear model denoted $(Y_1Y_2, Y_1Y_3, \ldots, Y_{T-1}Y_T)$. Alternatively, a model with Markov structure such as $(Y_1Y_2, Y_2Y_3, \ldots, Y_{T-1}Y_T)$ may be useful. Usually, insufficient data exists to warrant using very complex loglinear models. For any such model, its fit for the $2^T - 1$ observed cells projects to the remaining cell to predict the number unobserved at every occasion.

A connection exists between nonparametric random effects and loglinear approaches. In Section 13.2.7 we show that assuming model (13.5) but using a nonparametric treatment of u_i implies a loglinear model of quasi-symmetric form for the marginal model. The quasi-symmetry model (10.33) itself is not useful for this problem, because *any* count in the missing cell is consistent with it. The model has an interaction parameter pertaining to that cell alone, which results in a likelihood equation equating that cell count to its fitted value. So, information in other cells does not help in the estimation of the expected frequency in that cell. However, special cases of quasi-symmetry are useful (Darroch et al. 1993). An example is the loglinear model with the same association for each pair of occasions. Like the logistic-normal model, this model of *exchangeable association* has only one more parameter than the mutual independence model.

For the snowshoe hare data of Table 12.6, the model with exchangeable two-factor association has $\hat{N} = 90.5$ and a confidence interval of (75, 125). This interval and the one of (71, 87) for the Rasch mixture model with $q = 2$ are substantially narrower than the interval (75, 154) for the logistic-normal model (Section 12.3.6). In capture–recapture experiments, $\hat{N}$ and the confidence interval for N depend strongly on the choice of model. The problem is inherently one of prediction. Estimating N requires extrapolating from the observed numbers of subjects having $1, 2, \ldots, T$ captures to the number of subjects with 0 captures. Standard goodness-of-fit criteria are of limited help. Two models can fit the data well, yet yield quite different estimates for the unobserved count. For instance, for the snowshoe hare data, the loglinear models of mutual independence and of two-factor association both fit the observed cells relatively well ($G^2 = 58.3$, df = 56 for mutual independence and $G^2 = 32.4$, df = 41 for the two-factor model); however, their $\hat{N}$ values are 75 and 105.

Simpler models usually give narrower confidence intervals for N, through the usual benefits of model parsimony. This is not necessarily good. A narrow confidence interval for N is desirable, but not at the expense of severe sacrifice in the actual confidence level. Intervals based on a possibly unrealistic assumption of subject homogeneity may be overly optimistic. Simulations suggest that actual coverage probabilities are often well below nominal levels when even slight model misspecification occurs. Allowance for heterogeneity among subjects results in wider intervals. Severe population heterogeneity makes reaching useful conclusions difficult, as intervals can be very wide (Burnham and Overton 1978, Coull and Agresti 1999).

13.2.7 Nonparametric Mixtures and Quasi-symmetry

A distribution-free approach for u_i with the Rasch form of model (13.5) implies the quasi-symmetry loglinear model marginally (Darroch 1981; Tjur 1982). We now show this result, to which we alluded in Section 10.4.2.

Let $\mathbf{Y}_i$ denote the sequence of T responses for subject i. For possible outcomes $\mathbf{y} = (y_1, \ldots, y_T)$, where each $y_t = 1$ or 0,

$$
P(\mathbf{Y}_i = \mathbf{y} \mid u_i) = \prod_t \left[\frac{\exp(u_i + \beta_t)}{1 + \exp(u_i + \beta_t)} \right]^{y_t} \left[\frac{1}{1 + \exp(u_i + \beta_t)} \right]^{1 - y_t}
$$

$$
= \frac{\exp[u_i(\Sigma_t y_t) + \Sigma_t y_t \beta_t]}{\prod_t [1 + \exp(u_i + \beta_t)]} .
$$

Let F denote the cdf of u_i. The marginal probability of sequence $\mathbf{y}$ for a randomly selected subject is (suppressing the subject label)

$$
\pi_{y_1, \ldots, y_T} = E_U P(\mathbf{Y} = \mathbf{y} \mid U) = \exp\left(\sum_t y_t \beta_t \right) \int \frac{\exp[u(\Sigma_t y_t)]}{\prod_t [1 + \exp(u + \beta_t)]} dF(u) .
$$

This probability contributes to the log likelihood, which is (13.2) for a multinomial distribution over the 2^T cells for possible $\mathbf{y}$. Regardless of the choice for F, the integral is complex. However, it depends on the data only through $\Sigma_t y_t$. A more general model replaces this integral by a separate parameter for each value of $\Sigma_t y_t$. This model has form

$$\log \pi_{y_1, \ldots, y_T} = \sum_t y_t \beta_t + \lambda_{y_1 + \cdots + y_t}. \tag{13.6}$$

The final term represents a separate parameter at each value of $\Sigma_t y_t$.

The implied marginal model (13.6) has interaction term that is invariant to a permutation of the response outcomes **y**, since each such permutation yields the same sum, $\Sigma_t y_t$. Thus, it is the loglinear model of quasi-symmetry (10.33). No matter what form F takes, the marginal model has the same main effect structure, and it has an interaction term that is a special case of the one in (13.6). Thus, one can consistently estimate $\{\beta_t\}$ using the ordinary ML estimates for the loglinear model. In fact, Tjur (1982) showed that these estimates are also the conditional ML estimates, treating $\{u_i\}$ as fixed effects and conditioning on their sufficient statistics. The interaction parameters in model (13.6) result from the dependence in responses among variables, due to heterogeneity in $\{u_i\}$.

We illustrate for the opinions about legalized abortion analyzed in Sections 10.7.2 and 12.3.2 and with a nonparametric random effects approach in Section 13.2.1. For model (13.3), estimated within-subject comparisons $\beta_t - \beta_s$ of items result from fitting a quasi-symmetric loglinear model. Let $\mu_g(y_1, y_2, y_3)$ denote the expected frequency for gender g making response y_t to item t, $t = 1, 2, 3$, where for item t, $y_t = 1$ for approval of legalized abortion and 0 for disapproval. The loglinear model is

$$\log \mu_g(y_1, y_2, y_3) = \beta_1 y_1 + \beta_2 y_2 + \beta_3 y_3 + \gamma g + \lambda_{y_1 + y_2 + y_3}. \tag{13.7}$$

For $y_1 + y_2 + y_3 = k$, λ_k refers to all cells in which subjects voiced approval for k of the three items, $k = 0, 1, 2, 3$. The ML fit, which has $G^2 = 10.2$ with df $= 9$, yields $\hat{\beta}_1 - \hat{\beta}_2 = 0.521$ (SE $= 0.154$), $\hat{\beta}_1 - \hat{\beta}_3 = 0.828$ (SE $= 0.160$), and $\hat{\beta}_2 - \hat{\beta}_3 = 0.307$ (SE $= 0.161$). These are similar to the normal random effects estimates (Table 12.3) and nonparametric random effects estimates in Section 13.2.1. They also are the conditional ML estimates for model (13.3), treating $\{u_i\}$ as fixed. With this approach or conditional ML, however, one cannot estimate between-groups effects, such as the gender effect in model (13.7). [The γ parameter in model (13.7) refers to relative sample sizes of males and females and is not the same as the gender effect in (13.3).]

13.3 BETA-BINOMIAL MODELS

The beta-binomial model is a parametric mixture model that is another alternative to binary GLMMs with normal random effects. As with other mixture models that assume a binomial distribution at a fixed parameter value, the marginal distribution permits more variation than the binomial. Thus, a model using the beta-binomial is a way to handle overdispersion occurring with ordinary binomial models.

13.3.1 Beta-Binomial Distribution

The beta-binomial distribution results from a beta distribution mixture of binomials. Suppose that (a) given π, Y has a binomial distribution, bin(n, π), and (b) π has a beta distribution.

The beta probability density function is

$$f(\pi; \alpha, \beta) = \frac{\Gamma(\alpha + \beta)}{\Gamma(\alpha)\Gamma(\beta)} \pi^{\alpha-1}(1 - \pi)^{\beta-1}, \qquad 0 \le \pi \le 1, \quad (13.8)$$

with parameters $\alpha > 0$ and $\beta > 0$, for the gamma function $\Gamma(\cdot)$. Let

$$\mu = \frac{\alpha}{\alpha + \beta}, \quad \theta = 1/(\alpha + \beta).$$

The beta distribution for π has mean and variance

$$E(\pi) = \mu, \quad \text{var}(\pi) = \mu(1 - \mu)\theta/(1 + \theta).$$

When α and β exceed 1.0, the distribution is unimodal, with skew to the right when $\alpha < \beta$, skew to the left with $\alpha > \beta$, and symmetry when $\alpha = \beta$. It simplifies to the uniform distribution when $\alpha = \beta = 1$.

Marginally, averaging with respect to the beta distribution for π, Y has the *beta-binomial distribution*. Its mass function is

$$p(y; \alpha, \beta) = \binom{n}{y} \frac{B(\alpha + y, n + \beta - y)}{B(\alpha, \beta)}, \qquad y = 0, 1, \ldots, n.$$

In terms of μ and θ, the beta-binomial mass function is

$$p(y; \mu, \theta) = \binom{n}{y} \frac{\left[\prod_{k=0}^{y-1}(\mu + k\theta)\right]\left[\prod_{k=0}^{n-y-1}(1 - \mu + k\theta)\right]}{\prod_{k=0}^{n-1}(1 + k\theta)}. \quad (13.9)$$

It is easier to understand the nature of this distribution from its moments than from its mass function. The first two moments are

$$E(Y) = n\mu, \quad \text{var}(Y) = n\mu(1 - \mu)[1 + (n - 1)\theta/(1 + \theta)].$$

As $\theta \to 0$ in the beta distribution, $\text{var}(\pi) \to 0$ and that distribution converges to a degenerate distribution at μ. Then $\text{var}(Y) \to n\mu(1 - \mu)$ and the beta-binomial distribution converges to the $\text{bin}(n, \mu)$.

13.3.2 Models Using the Beta-Binomial Distribution

Models using the beta-binomial distribution permit μ [and hence $E(Y)$] to depend on explanatory variables. The simplest models let θ be the same unknown constant for all observations. [Prentice (1986) considered extensions where it could also depend on covariates.] Like GLMs, models can use various link functions, but the logit is most common. For observation i with n_i trials, assuming that y_i has a beta-binomial distribution with index n_i and parameters (μ_i, θ), the model links μ_i to predictors by

$$\text{logit}(\mu_i) = \alpha + \beta' x_i.$$

The beta-binomial is not in the natural exponential family, even for known θ. Articles using beta-binomial models have employed a variety of fitting methods (Note 13.4). Crowder (1978) discussed the likelihood behavior for an ANOVA-type model. Hinde and Demétrio (1998) obtained the ML fit by iterating between solving the likelihood equations for the regression parameters β, for fixed θ, and solving the likelihood equation for θ for fixed β. Each part can use Newton–Raphson. McCulloch and Searle (2001, p. 61) showed the asymptotic covariance matrix of $(\hat{\mu}, \hat{\theta})$ and of $(\hat{\alpha}, \hat{\beta})$ for independent observations from a single beta-binomial distribution.

A related but simpler approach for overdispersed binary counts uses quasi-likelihood with similar variance function as the beta-binomial. The quasi-likelihood variance function is

$$v(\mu_i) = n_i \mu_i (1 - \mu_i)\left[1 + (n_i - 1)\rho\right] \tag{13.10}$$

with $|\rho| \le 1$. Although motivated by the beta-binomial model, this variance function results merely from assuming that π_i has a distribution with $\text{var}(\pi_i) = \rho\mu_i(1 - \mu_i)$. It also results from assuming a common correlation ρ between each pair of the n_i individual binary random variables that sum to y_i (Altham 1978). The ordinary binomial variance results when $\rho = 0$. Overdispersion occurs when $\rho > 0$.

For this quasi-likelihood approach, Williams (1982) gave an iterative routine for estimating β and the overdispersion parameter ρ. He let $\hat{\rho}$ be such that the resulting Pearson X^2 that sums the squared Pearson residuals for this variance function equals the residual df for the model. This requires an iterative two-step process of (1) solving the quasi-likelihood equations for β for a given $\hat{\rho}$, and then (2) using the updated $\hat{\beta}$, solving for $\hat{\rho}$ in the equation that equates X^2 (which depends on $\hat{\beta}$ and $\hat{\rho}$) to its df.

An alternative quasi-likelihood approach uses the simpler variance function

$$v(\mu_i) = \phi n_i \mu_i (1 - \mu_i) \tag{13.11}$$

introduced in Section 4.7.3. The ordinary binomial variance has $\phi = 1.0$ and overdispersion has $\phi > 1$. With this approach, $\hat{\beta}$ is the same as its ML estimate for the ordinary binomial model. Commonly, $\hat{\phi} = X^2/\text{df}$, where X^2 is the Pearson fit statistic for the binomial model (Finney 1947). The standard errors for the overdispersion approach multiply those for the binomial model by $\hat{\phi}^{1/2}$.

Liang and McCullagh (1993) showed several examples using these two variance functions. A plot of the standardized residuals for the ordinary binomial model against the indices $\{n_i\}$ can provide insight about which is more appropriate. When the residuals show an increasing trend in their spread as n_i increases, the beta-binomial-type variance function may be more appropriate. This is because when the beta-binomial variance holds, the residuals from an ordinary binomial model have denominator that is progressively too small as n_i increases. The two quasi-likelihood approaches are equivalent when $\{n_i\}$ are identical. Only when the indices vary considerably might results differ much. Because the variance function $v(\mu_i) = \phi n_i \mu_i (1 - \mu_i)$ has a structural problem when $n_i = 1$ (Problem 13.33) and has less direct motivation, we prefer quasi-likelihood with the beta-binomial variance function.

13.3.3 Teratology Overdispersion Example Revisited

Refer back to Table 4.5 on results of a teratology experiment analyzed by Liang and McCullagh (1993) and Moore and Tsiatis (1991). Female rats on iron-deficient diets were assigned to four groups. Group 1 was given only placebo injections. The other groups were given injections of an iron supplement according to various schedules. The rats were made pregnant and then sacrificed after 3 weeks. For each fetus in each rat's litter, the response was whether the fetus was dead. Because of unmeasured covariates, it is natural to permit the probability of death to vary from litter to litter within a particular treatment group.

Let y_i denote the number dead out of the n_i fetuses in litter i. Let π_{it} denote the probability of death for fetus t in litter i. First, suppose that y_i is a bin(n_i, π_{it}) variate, with

$$\text{logit}(\pi_{it}) = \alpha + \beta_2 z_{2i} + \beta_3 z_{3i} + \beta_4 z_{4i},$$

where $z_{gi} = 1$ if litter i is in group g and 0 otherwise. This model treats all litters in a group g as having the same probability of death, $\exp(\alpha + \beta_g)/[1 + \exp(\alpha + \beta_g)]$, where $\beta_1 = 0$. However, it has evidence of overdispersion,

TABLE 13.5 Estimates for Several Logit Models Fitted to Table 4.5

Parameter	Type of Logit Model[a]				
	Binomial ML	QL(1)	QL(2)	GEE	GLMM
Intercept	1.144 (0.129)	1.212 (0.223)	1.144 (0.219)	1.144 (0.276)	1.802 (0.362)
Group 2	−3.322 (0.331)	−3.370 (0.563)	−3.322 (0.560)	−3.322 (0.440)	−4.515 (0.736)
Group 3	−4.476 (0.731)	−4.585 (1.303)	−4.476 (1.238)	−4.476 (0.610)	−5.855 (1.190)
Group 4	−4.130 (0.476)	−4.250 (0.848)	−4.130 (0.806)	−4.130 (0.576)	−5.594 (0.919)
Overdispersion	None	$\hat{\rho} = 0.192$	$\hat{\phi} = 2.86$	$\hat{\rho} = 0.185$	$\hat{\sigma} = 1.53$

[a]Binomial ML assumes no overdispersion, QL(1) is quasi-likelihood with beta-binomial-type variance, QL(2) is quasi-likelihood with inflated binomial variance; QL(2) and GEE (independence working equations) estimates are the same as binomial ML estimates. Values in parentheses are standard errors.

with $X^2 = 154.7$ and $G^2 = 173.5$ (df = 54). Table 13.5 shows ML estimates and standard errors.

Table 13.5 also shows results for the two quasi-likelihood approaches. Estimates and standard errors are qualitatively similar for each. For variance function $v(\mu_i) = \phi n_i \mu_i (1 - \mu_i)$, the estimates equal the binomial ML estimates but standard errors are multiplied by $\hat{\phi}^{1/2} = (X^2/\mathrm{df})^{1/2} = \sqrt{154.7/54} = 1.69$. For the beta-binomial-type variance function, $\hat{\rho} = 0.192$. This fit treats the variance of Y_i as

$$n_i \mu_i (1 - \mu_i)[1 + 0.192(n_i - 1)].$$

This corresponds roughly to a doubling of the variance relative to the binomial with a litter size of 6 and a tripling with $n_i = 11$. Even with these adjustments for overdispersion, Table 13.5 shows that strong evidence remains that the probability of death is substantially lower for each treatment group than the placebo group.

Figure 13.4 plots the standardized Pearson residuals against litter size for the binomial logit model. The apparent increase in their variability as litter size increases suggests that the beta-binomial variance function is plausible. The term ρ in that variance function corresponds to $\theta/(1 + \theta)$ in the variance of the beta-binomial distribution. For that distribution or more generally, $\hat{\rho} = 0.192$ means that the probabilities of death for litters of a particular group have estimated standard deviation $\sqrt{0.192 \mu_i (1 - \mu_i)}$. This equals 0.22 when the mean is 0.5 and 0.13 when the mean is 0.1 or 0.9, which is considerable heterogeneity. More generally, a model could let ρ vary by treatment group or be different for the placebo group than the others. We leave this to the reader.

For comparison, Table 13.5 also shows results with the GEE approach to fitting the logit model, assuming an independence working correlation structure for observations within a litter. The estimates are the same as the ML

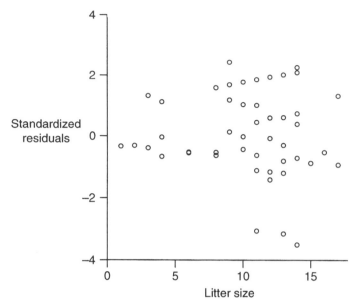

FIGURE 13.4 Standardized Pearson residuals for binomial logit model fitted to Table 4.5.

estimates for the binomial logit model, but the empirical adjustment increases the standard errors. Similar results occur with an exchangeable working correlation structure. For it, the estimated within-litter correlation between the binary responses is 0.185. This is comparable to the value of 0.192 that yields the quasi-likelihood results with beta-binomial variance function. The GEE standard errors are somewhat different from those with the quasi-likelihood approach. It may be that the sample size is insufficient for the GEE sandwich adjustment, which tends to underestimate standard errors unless the number of clusters is quite large. Or, this may simply reflect the different variance function for the GEE approach.

Finally, Table 13.5 also shows results for the GLMM that adds a normal random intercept u_i for litter i to the binomial logit model. Results are also similar in terms of significance of the treatment groups relative to placebo. Estimated effects are larger for this logistic-normal model, since they are subject-specific (i.e., litter-specific) rather than population-averaged.

13.3.4 Conjugate Mixture Models

The beta-binomial model is an example of a *conjugate mixture model*. These are models for which the marginal distribution has closed form. The data have a particular distribution, conditional on a parameter, and then the parameter has its own distribution such that the marginal distribution has closed form.

Similarly, in Bayesian methods the conjugate prior distribution is a distribution that when combined with the likelihood, gives a closed form for the posterior distribution. For instance, for observations from a binomial distribution with beta prior distribution for the binomial parameter, the posterior distribution of that parameter is also beta. Conjugate models were the primary method of conducting Bayesian analysis before the development of computationally intensive methods, such as Markov chain Monte Carlo, for evaluating the integral that determines the posterior distribution.

The beta-binomial conjugate mixture model applies with totals from binary trials. In the next section we study a conjugate mixture model for count data. It uses a gamma distribution to mix the Poisson parameter. A disadvantage of the conjugate mixture approach is the lack of generality and flexibility, requiring a different mixture distribution for each type of problem. In addition, the extra variability need not enter on the same scale as the ordinary predictors, and it can be difficult to have multivariate random effects structure. Lee and Nelder (1996) discussed this approach and considered a variety of hierarchical models of GLMM form in which the random effect need not be normal.

13.4 NEGATIVE BINOMIAL REGRESSION

The negative binomial is a conjugate mixture distribution for count data. It is useful when overdispersion occurs with Poisson GLMs.

13.4.1 Negative Binomial as Gamma Mixture of Poisson Distributions

In Section 4.3.3 we noted that a severe limitation of Poisson models is that the variance of Y must equal the mean. Hence, at a fixed mean the variance cannot decrease as additional predictors enter the model. Count data often show overdispersion, with the variance exceeding the mean. This might happen, for instance, because some relevant explanatory variables are not in the model. A mixture model is a flexible way to account for overdispersion. At a fixed setting of the predictors used, given the mean the distribution of Y is Poisson, but the mean itself varies according to some distribution.

Suppose that (1) given λ, Y has a Poisson distribution with mean λ, and (2) λ has a gamma distribution, $G(k, \mu)$. The gamma probability density function for λ is

$$f(\lambda; k, \mu) = \frac{(k/\mu)^k}{\Gamma(k)} \exp(-k\lambda/\mu) \lambda^{k-1}, \quad \lambda \geq 0. \quad (13.12)$$

This gamma distribution has

$$E(\lambda) = \mu, \quad \text{var}(\lambda) = \mu^2/k.$$

The parameter $k > 0$ describes the shape. The density is skewed to the right, but the degree of skewness decreases as k increases.

Marginally, the gamma mixture of the Poisson distributions yields the negative binomial distribution for Y. Its probability mass function is

$$p(y; k, \mu) = \frac{\Gamma(y + k)}{\Gamma(k)\Gamma(y + 1)} \left(\frac{k}{\mu + k} \right)^k \left(1 - \frac{k}{\mu + k} \right)^y, \qquad y = 0, 1, 2, \dots .$$

$$(13.13)$$

This negative binomial distribution has

$$E(Y) = \mu, \qquad \text{var}(Y) = \mu + \mu^2/k .$$

The index k^{-1} is called the *dispersion parameter*. As $k^{-1} \to 0$, the gamma distribution has var$(\lambda) \to 0$ and it converges to a degenerate distribution at μ; similarly, the negative binomial distribution then has var$(Y) \to \mu$ and it converges to the Poisson distribution with mean μ.

For given k^{-1}, the negative binomial is in the natural exponential family. The natural parameter is log$[\mu/(\mu + k)]$. Usually, though, the dispersion parameter k^{-1} is itself unknown. Estimating it helps to summarize the extent of overdispersion. The greater k^{-1}, the greater the overdispersion compared to the ordinary Poisson GLM. For independent observations, the ML estimate of μ is the sample mean, but ML estimation for k^{-1} requires iterative methods (R. A. Fisher showed this in an appendix of a 1953 *Biometrics* article by C. Bliss). Problem 13.40 shows an alternative gamma parameterization that implies a linear rather than quadratic variance function for the negative binomial.

13.4.2 Negative Binomial Regression Modeling

Negative binomial models for counts permit μ to depend on explanatory variables (Lawless 1987). Such models normally take k^{-1} to be the same for all observations. This corresponds to a constant coefficient of variation in the gamma mixing distribution, $\sqrt{\text{var}(\lambda)}/E(\lambda) = 1/\sqrt{k}$. with the standard deviation increasing as the mean does. Most common is the log link, as in Poisson loglinear models. Sometimes the identity link is adequate. One such case is with a single predictor that is a factor.

For k fixed, a negative binomial model is a GLM. Thus, the likelihood equations for the regression parameters $\boldsymbol{\beta}$ are special cases of those [see (4.22)] for an ordinary GLM with variance function $v(\mu) = \mu + \mu^2/k$. The usual iterative reweighted least squares algorithm applies for ML model fitting. When k is unknown, ML fitting can use a Newton–Raphson routine on all the parameters simultaneously. Or, one can evaluate the profile likelihood for various fixed k (Lawless 1987). Another approach alternates between (1) using iterative reweighted least squares to solve the equations for β, for fixed k, and (2) for fixed $\hat{\beta}$, using Newton–Raphson to estimate k, iterating between them until convergence.

The full log likelihood $L(\beta, k; y)$ for a negative binomial model satisfies

$$\frac{\partial^2 L}{\partial \beta_j \partial k} = \sum_i \frac{y_i - \mu_i}{(k + \mu_i)^2 g'(\mu_i)} x_{it}.$$

Thus, $E(\partial^2 L / \partial \beta_j \partial k) = 0$ for each j. Similarly, the inverse of the expected information matrix has 0 elements connecting k with each β_j. Since this is the asymptotic covariance matrix, $\hat{\beta}$ and $\hat{k}$ are asymptotically independent. It follows that standard errors for $\hat{\beta}$ obtained from part (1) of the iterative scheme above are correct. Cameron and Trivedi (1998, p. 72) showed the asymptotic covariance matrix. They [and Lawless (1987)] considered a moment estimator for k^{-1} and studied robustness properties of estimators. They noted that $\hat{\beta}$ from this model is consistent if the model for the mean is correctly specified, even if the true distribution is not negative binomial.

13.4.3 Frequency of Knowing Homicide Victims Example

Table 13.6 summarizes responses of 1308 subjects to the question: Within the past 12 months, how many people have you known personally that were victims of homicide? The table shows responses by race, for those who identified their race as white or as black. The sample mean for the 159 blacks was 0.522, with a variance of 1.150. The sample mean for the 1149 whites was 0.092, with a variance of 0.155.

A natural first choice for modeling count data is a Poisson GLM, such as a loglinear model with a dummy predictor for race. Let y_{it} denote the response for subject t of race i. For $\mu_{it} = E(Y_{it})$, this model is

$$\log \mu_{it} = \alpha + \beta x_{it},$$

TABLE 13.6 Number of Victims of Murder Known in Past Year, by Race, with Fit of Poisson and Negative Binomial Models

Response	Data		Poisson GLM		Neg. Bin. GLM		Poisson GLMM	
	Black	White	Black	White	Black	White	Black	White
0	119	1070	94.3	1047.7	122.8	1064.9	116.7	1068.3
1	16	60	49.2	96.7	17.9	67.5	24.5	65.3
2	12	14	12.9	4.5	7.8	12.7	8.1	10.1
3	7	4	2.2	0.1	4.1	2.9	3.6	2.8
4	3	0	0.3	0.0	2.4	0.7	1.9	1.1
5	2	0	0.0	0.0	1.4	0.2	1.1	0.5
6	0	1	0.0	0.0	0.9	0.1	0.7	0.3

Source: 1990 General Social Survey, National Opinion Research Center.

with $x_{1t} = 1$ (blacks) and $x_{2t} = 0$ (whites). This model has fit $\log \hat{\mu}_{it} = -2.38 + 1.733x_{it}$. The estimated expected responses are $\exp(-2.38 + 1.733) = 0.522$ for blacks and $\exp(-2.38) = 0.092$ for whites, the sample means. For any link function for this model, the likelihood equations imply that the fitted means equal the sample means. Since $\hat{\beta} = 1.733$ (SE $= 0.147$) is the difference between the log means for blacks and whites, the ratio of sample means is $\exp(1.733) = 5.7 = 0.522/0.092$. However, for each race the sample variance is roughly double the mean. Table 13.6 also shows the fit of this model. The evidence of overdispersion is reflected by the higher observed counts at $y = 0$ and at large y values than the Poisson GLM predicts.

An alternative is the same model form but assuming a negative binomial response. A mixture model does seem plausible. Due to various demographic factors, heterogeneity probably occurs among subjects of a given race in the distribution of Y. For ML fitting, the deviance decreases by 122.2 compared to the ordinary Poisson GLM that is the special case with $k^{-1} = 0$. Table 13.6 also shows this model fit. It is dramatically better at $y = 0$ and 1.

Table 13.7 shows parameter estimates for the negative binomial and Poisson GLMs. For both, $\hat{\beta} = 1.733$ since both models provide fitted means equal to the sample means. However the estimated standard error of $\hat{\beta}$ increases from 0.147 for the Poisson GLM to 0.238 for the negative binomial model. The Wald 95% confidence interval for the ratio of means for blacks and whites goes from $\exp[1.733 \pm 1.96(0.147)] = (4.2, 7.5)$ for the Poisson GLM to $\exp[1.733 \pm 1.96(0.238)] = (3.5, 9.0)$ for the negative binomial. In accounting for the overdispersion, we obtain results that are not as precise as the more naive model suggests.

The negative binomial model has $\hat{k}^{-1} = 4.94$ (SE $= 1.00$). This shows strong evidence that $k^{-1} > 0$, indicating that the negative binomial model is more appropriate than the Poisson GLM. The estimated variance of Y is $\hat{\mu} + \hat{\mu}^2/\hat{k} = \hat{\mu} + 4.94\hat{\mu}^2$, which is 0.13 for whites and 1.87 for blacks, much closer to the sample values than the Poisson model provides.

Table 13.7 also shows results for negative binomial and Poisson models using the identity link. The fits $\hat{\mu}_{it} = 0.092 + 0.430x_{it}$ reproduce the sample means. Now $\hat{\beta}$ refers to the difference in means rather than their log ratio. The estimated difference $\hat{\beta} = 0.430$ has SE $= 0.058$ for the Poisson model and SE $= 0.109$ for the negative binomial. Results are more imprecise but

TABLE 13.7   Parameter Estimates for Models Fitted to Homicide Data

| Term | Models with Log Link | | | Models with Identity Link | |
	Neg. Binom. GLM	Poisson GLM	Poisson GLMM	Neg. Binom. GLM	Poisson GLM
α	-2.38	-2.38	-3.69	0.092	0.092
β	1.733	1.733	1.897	0.430	0.430
SE($\hat{\beta}$)	0.238	0.147	0.246	0.109	0.058

more realistic with the negative binomial model. For this link also the estimated dispersion parameter is $\hat{k}^{-1} = 4.94$.

13.5 POISSON REGRESSION WITH RANDOM EFFECTS

The GLMMs introduced in Chapter 12 referred to categorical responses. GLMMs are also useful for other types of discrete responses, such as counts. This section illustrates with Poisson regression modeling of count data.

We've seen that a flexible way to account for overdispersion is with a mixture model. In Section 13.4 we mixed the Poisson using the gamma distribution, yielding the negative binomial marginally. Breslow (1984) and Hinde (1982) suggested the GLMM structure (12.1) with the log link and normal random intercept. The model for the mean for observation t in cluster i is

$$\log[E(Y_{it}|u_i)] = \mathbf{x}'_{it}\boldsymbol{\beta} + u_i, \tag{13.14}$$

where $\{u_i\}$ are independent $N(0, \sigma^2)$. Conditional on u_i, y_{it} has a Poisson distribution. Marginally, the distribution has variance greater than the mean whenever $\sigma > 0$.

Applications of Poisson GLMMs include the analysis of maps of cancer rates in epidemiology (Breslow and Clayton 1993) and modeling variability in bacteria counts (Aitchison and Ho 1989). Although links other than the log are possible, the identity link (and any other link having range only the positive real line) has a structural problem. With a normal random effect with $\sigma > 0$, a positive probability exists that the linear predictor is negative, but the Poisson mean must be nonnegative.

The negative binomial model (for fixed k) is a GLMM with nonnormal random effect. With the log link, it results from a loglinear model of form (13.14) with random intercept, where $\exp(u_i)$ has a gamma distribution with mean 1 and variance k^{-1}. With identity link, negative binomial models usually work better than Poisson GLMMs. Regardless of the gamma mixture distribution, the resulting marginal mean is nonnegative for the negative binomial.

13.5.1 Marginal Model Implied by Poisson GLMM

The Poisson GLMM (13.14) implies a relatively simple marginal model, averaging out the random effect. The mean of the marginal distribution is

$$E(Y_{it}) = E[E(Y_{it}|u_i)] = E[e^{\mathbf{x}'_{it}\boldsymbol{\beta} + u_i}] = e^{\mathbf{x}'_{it}\boldsymbol{\beta} + \sigma^2/2}.$$

Here $E[\exp(u_i)] = \exp(\sigma^2/2)$ because a $N(0, \sigma^2)$ variate u_i has moment generating function $E[\exp(tu_i)] = \exp(t^2\sigma^2/2)$. So, for the Poisson GLMM the log of the mean conditionally equals $\mathbf{x}'_{it}\boldsymbol{\beta} + u_i$ and marginally equals $\mathbf{x}'_{it}\boldsymbol{\beta} + \sigma^2/2$. A loglinear model still applies. The marginal effects of the explanatory variables are the same as the cluster-specific effects. Thus, the *ratio* of means at two different settings of $\mathbf{x}'_{it}$ is the same conditionally and marginally. However, marginally the intercept is offset. (Note that Jensen's inequality applies, since the link is not linear.)

The variance of the marginal distribution is

$$\text{var}(Y_{it}) = E\big[\text{var}(Y_{it}\,|\,u_i)\big] + \text{var}\big[E(Y_{it}\,|\,u_i)\big] = E\big[e^{\mathbf{x}'_{it}\boldsymbol{\beta}+u_i}\big] + e^{2\mathbf{x}'_{it}\boldsymbol{\beta}}\,\text{var}(e^{u_i})$$

$$= e^{\mathbf{x}'_{it}\boldsymbol{\beta}+\sigma^2/2} + e^{2\mathbf{x}'_{it}\boldsymbol{\beta}}\big(e^{2\sigma^2} - e^{\sigma^2}\big) = E(Y_{it}) + \big[E(Y_{it})\big]^2\big(e^{\sigma^2} - 1\big).$$

Here, $\text{var}(e^{u_i}) = E(e^{2u_i}) - [E(e^{u_i})]^2 = e^{2\sigma^2} - e^{\sigma^2}$ by evaluating the moment generating function at $t = 2$ and $t = 1$. As in the negative binomial model, the marginal variance is a quadratic function of the marginal mean. It exceeds the marginal mean when $\sigma > 0$. The ordinary Poisson model results when $\sigma = 0$. When $\sigma > 0$ the marginal distribution is not Poisson, and the extent to which the variance exceeds the mean increases as σ increases.

As in binary GLMMs, Y_{it} and Y_{is} are independent given u_i but are marginally nonnegatively correlated. For $t \neq s$,

$$\text{cov}(Y_{it}, Y_{is}) = E\big[\text{cov}(Y_{it}, Y_{is}\,|\,u_i)\big] + \text{cov}\big[E(Y_{it}\,|\,u_i), E(Y_{is}\,|\,u_i)\big]$$

$$= 0 + \text{cov}\big[\exp(\mathbf{x}'_{it}\boldsymbol{\beta} + u_i), \exp(\mathbf{x}'_{is}\boldsymbol{\beta} + u_i)\big]. \qquad (13.15)$$

The functions in the last covariance term are both monotone increasing functions of u_i, and hence are nonnegatively correlated (Problem 13.44).

13.5.2 Frequency of Knowing Homicide Victims Example

We now return to Table 13.6 on responses, classified by race, of the number of victims of homicide within the past 12 months that subjects knew personally. Models permitting subject heterogeneity are sensible. For the response y_{it} for subject t of race i, the Poisson GLMM is

$$\log\big[E(Y_{it}\,|\,u_{it})\big] = \alpha + \beta x_{it} + u_{it},$$

where $\{u_{it}\}$ are independent $N(0, \sigma^2)$. The log means vary according to a $N(\alpha, \sigma^2)$ distribution for whites and a $N(\alpha + \beta, \sigma^2)$ distribution for blacks. Given u_{it}, y_{it} has a Poisson distribution.

Table 13.6 also shows this model fit, and Table 13.7 shows estimates. The random effects have $\hat{\sigma} = 1.63$ (SE $= 0.15$). The deviance decreases by 116.6 compared to the Poisson GLM, indicating a better fit by allowing heterogeneity. For subjects at the means of the random effects distributions ($u_{it} = 0$) the estimated expected responses are $\exp(-3.69 + 1.90) = 0.167$ for blacks and

$\exp(-3.69) = 0.025$ for whites. The fitted marginal mean is $\exp(\hat{\alpha} + \hat{\beta}x_{it} + \hat{\sigma}^2/2)$, or 0.63 for blacks and 0.09 for whites. The fitted marginal variances are 0.21 for blacks and 5.78 for whites. These are somewhat larger than the sample means and variances, perhaps because the fitted distribution has nonnegligible mass above the largest observed response of 6.

13.5.3 Negative Binomial Models versus Poisson GLMMs

The Poisson GLMM with normal random effects has the advantage, relative to the negative binomial GLM, of easily permitting multivariate random effects and multilevel models. However, the negative binomial has properties that can make interpretation simpler. We've seen that the identity link is valid for it, which is useful for simple examples such as the preceding one with a factor predictor. With any link and a factor predictor, its ML fitted means equal the sample means. This is not the case for the Poisson GLMM.

Besides the Poisson GLMM and the negative binomial model, an alternative way of accounting for overdispersion with count data is quasi-likelihood with variance function

$$v(\mu_i) = \phi\mu_i,$$

for some constant ϕ. This is often adequate for exploratory analyses.

NOTES

Section 13.1: Latent Class Models

13.1. Aitkin et al. (1981), Bartholomew and Knott (1999), Clogg (1995), Clogg and Goodman (1984), Goodman (1974), Haberman (1979, Chap. 10), Hagenaars (1998), Heinen (1996), and Lazarsfeld and Henry (1968) discussed fitting and intrepetation of latent class and related latent variable models.

13.2. Rudas et al. (1994) proposed a clever mixture method for summarizing goodness of fit. For a model M for a contingency table with true probabilities π, they used the mixture $\pi = (1 - \rho)\pi_1 + \rho\pi_2$, with π_1 the model-based probabilities and π_2 unconstrained. Their index of lack of fit is the smallest such ρ possible for which this holds. It is the fraction of the population that cannot be described by the model. This recognizes that any given model does not truly hold but is useful if ρ is close to 0. The mixture contrasts with the latent class model in which both π_1 and π_2 correspond to independence.

Section 13.2: Nonparametric Random Effects Models

13.3. For connections between Rasch-type models and quasi-symmetry models, see Agresti (1993), Conaway (1989), Darroch (1981), Darroch et al. (1993), Hatzinger (1989), and Kelderman (1984). For the matched-pairs random effects model (12.16), a nonparametric or conditional ML treatment of (u_{i1}, u_{i2}) implies a multivariate quasi-symmetry model (Agresti 1997). Model (12.16) with correlated normal random effects is a continuous analog to discrete latent class models that Goodman (1974) proposed, based on two associated binary latent variables.

Section 13.3: Beta-Binomial Models

13.4. Skellam (1948) introduced the beta-binomial distribution and discussed parameter estimation. For modeling using this distribution or related quasi-likelihood approaches, see Brooks et al. (1997), Crowder (1978), Hinde (1996), Lee and Nelder (1996), Liang and Hanfelt (1994), Liang and McCullagh (1993), Lindsey and Altham (1998), Moore (1986a), Moore and Tsiatis (1991), Nelder and Pregibon (1987), Prentice (1986), Rosner (1984, 1989) [with critique by Neuhaus and Jewell (1990a)], Slaton et al. (2000), and Williams (1975, 1982). For beta-binomial type variance, Ryan (1995) and Williams (1988) showed advantages of the quasi-likelihood approach over ML. Often, it helps to permit the quasi-likelihood scale parameter ρ (or the related parameter θ in the beta-binomial) to vary among groups.

The beta-binomial generalizes to a Dirichlet-multinomial. Conditional on the probabilities, the distribution is multinomial. The probabilities themselves have a Dirichlet distribution, which is a generalization of the beta defined on vectors of probabilities that sum to 1. See Mosimann (1962) and Paul et al. (1989).

13.5. Kupper et al. (1986) and Ryan (1992) discussed modeling overdispersion caused by litter effects in developmental toxicity studies. See Follman and Lambert (1989), Kupper and Haseman (1978), and Lefkopoulou et al. (1989) for related material.

Section 13.4: Negative Binomial Regression

13.6. Greenwood and Yule (1920) derived the negative binomial as a gamma mixture of Poissons. Johnson et al. (1992) summarized its properties. Biggeri (1998), Cameron and Trivedi (1998), Hinde and Demétrio (1998), and Lawless (1987) discussed modeling using it.

PROBLEMS

Applications

13.1 For the 2^3 table of opinions about legalized abortion (Table 10.13) collapsed over gender, fit a latent class model with two classes. Show that it is saturated. For each latent class, report the estimated probability of supporting legalized abortion in each of the three situations. Give a tentative interpretation for the classes.

13.2 Analyze Table 8.3 using a latent class model with $q = 2$.

    a. For a subject in the first latent class, estimate the probability of having used (**i**) marijuana, (**ii**) alcohol, (**iii**) cigarettes, (**iv**) all three, and (**v**) none of them.

    b. Estimate the probability a subject is in the first latent class, given they have used (**i**) marijuana, (**ii**) alcohol, (**iii**) cigarettes, (**iv**) all three, and (**v**) none of them.

13.3 Analyze Table 8.19 on government spending using latent class models.

13.4 For capture–recapture experiments, Coull and Agresti (1999) used a loglinear model with exchangeable association and no higher-order terms. Explain why the model expected frequencies satisfy

$$\log \mu(y_1, \ldots, y_T) = \lambda + \beta_1 y_1 + \cdots + \beta_T y_T$$

$$+ \beta(y_1 y_2 + y_1 y_3 + \cdots + y_{T-1} y_T) .$$

Show that the fit of this model to Table 12.6 yields $\hat{N} = 90.5$ and a 95% profile-likelihood confidence interval for N of $(75, 125)$.

13.5 Use or write software to replicate the analyses of the opinions about abortion data in Section 13.2 using **(a)** nonparametric random effects fitting of logit model (13.3), and **(b)** the quasi-symmetry model.

13.6 A data set on pregnancy rates among girls under 18 years of age in 13 north central Florida counties has information on a 3-year total for each county i on n_i = number of births and y_i = number of those for which mother had age under 18 (see J. Booth, in *Statistical Modelling: Lecture Notes in Statistics*, *104*, Springer, 43–52, 1995).

   a. A beta-binomial model states that given $\{\pi_i\}$, $\{Y_i\}$ are independent $\{\text{bin}(n_i, \pi_i)\}$ variates, and $\{\pi_i\}$ are independent from a beta(α, β) distribution. The ML estimated parameters are $\hat{\alpha} = 9.9$ and $\hat{\beta} = 240.8$ (thanks to J. Booth for this analysis). Use the mean and variance to describe the estimated beta distribution and the estimated marginal distribution of Y_i (as a function of n_i).

   b. Quasi-likelihood using variance function (13.10) for the model $\text{logit}(\mu_i) = \alpha$ has $\hat{\alpha} = -3.18$ and $\hat{\rho} = 0.005$. Describe the estimated mean and variance of Y_i.

   c. Quasi-likelihood using variance (13.11) for the model $\text{logit}(\mu_i) = \alpha$ has $\hat{\alpha} = -3.35$ and $\hat{\phi} = 8.3$. Describe the estimated mean and variance of Y_i.

   d. The logistic-normal GLMM, $\text{logit}(\pi_i) = \alpha + u_i$, yields $\hat{\alpha} = -3.24$ and $\hat{\sigma} = 0.33$. Describe the estimated mean of Y_i [Recall (12.8)].

13.7 In Problem 12.2 about Shaq O'Neal's free-throw shooting, the simple binomial model, $\pi_i = \alpha$, has lack of fit. Fit the beta-binomial model, or use the quasi-likelihood approach with that variance structure. Use the fit to summarize his free-throw shooting, by giving an estimated mean and standard deviation for π_i.

13.8 For the toxicity study of Table 12.9, collapsing to a binary response, consider linear logit models for the probability a fetus is normal.

  a. Does the ordinary binomial model show evidence of overdispersion?

  b. Fit the linear logit model using the quasi-likelihood approach with inflated binomial variance. How do the standard errors change?

  c. Fit the linear logit model using quasi-likelihood with beta-binomial variance. Interpret and compare with previous results.

  d. Fit the linear logit model using a GEE approach with exchangeable working correlation among fetuses in the same litter. Interpret and compare with previous results, including comparing the estimated GEE correlation with the estimate $\hat{\rho}$ from part (c).

  e. Fit the linear logit GLMM after adding a litter-specific normal random effect. Interpret and compare with previous results.

13.9 Extend the various analyses of the teratology data (Table 4.5) in Section 13.3.3 as follows:

  a. Include a predictor for litter size (as well as group). Interpret, and compare results to those without this predictor.

  b. Fit a model with beta-binomial variance (13.10) in which ρ varies by treatment group. Use results to motivate a model that allows overdispersion only in the placebo group. Interpret and compare results to those with common ρ for each group.

13.10 Table 13.8 reports the results of a study of fish hatching under three environments. Eggs from seven clutches were randomly assigned to three treatments, and the response was whether an egg hatched by day 10. The three treatments were (1) carbon dioxide and oxygen removed, (2) carbon dioxide only removed, and (3) neither removed.

TABLE 13.8   Data for Problem 13.10

	Treatment 1		Treatment 2		Treatment 3	
Clutch	Number Hatched	Total	Number Hatched	Total	Number Hatched	Total
1	0	6	3	6	0	6
2	0	13	0	13	0	13
3	0	10	8	10	6	9
4	0	16	10	16	9	16
5	0	32	25	28	23	30
6	0	7	7	7	5	7
7	0	21	10	20	4	20

Source: Data courtesy of Becca Hale, Zoology Department, University of Florida.

a. Let π_{it} denote the probability of hatching for an egg from clutch i in treatment t. Assuming independent binomial observations, fit the model

$$\text{logit}(\pi_{it}) = \beta_1 \, z_1 + \beta_2 \, z_2 + \beta_3 \, z_3,$$

where $z_t = 1$ for treatment t and 0 otherwise. What does your software report for $\hat{\beta}_1$, and what should it be? (*Hint:* Note that treatment 1 has no successes.)

b. Analyze these data using an approach that allows overdispersion. Interpret. Indicate whether evidence of overdispersion occurs for treatments 2 and 3.

13.11 For the train accidents in Problem 9.19, a negative binomial model assuming constant log rate over the 14-year period has estimate -4.177 (SE = 0.153) and estimated dispersion parameter 0.012. Interpret.

13.12 One question in the 1990 General Social Survey asked subjects how many times they had sexual intercourse in the preceding month. Table 13.9 shows responses, classified by gender.

a. The sample means were 5.9 for males and 4.3 for females; the sample variances were 54.8 and 34.4. The mode for each gender was 0. Does an ordinary Poisson GLM seem appropriate? Explain.

b. The Poisson GLM with log link and a dummy variable for gender (1 = males, 0 = females) has gender estimate 0.308 (SE = 0.038). Explain why this implies a ratio of 1.36 for the fitted means. (This is also the ratio of sample means, since this model has fitted means equal to sample means.) Show that the Wald 95% confidence interval for the ratio of means for males and females is (1.26, 1.47).

TABLE 13.9   Data for Problem 13.12

Response	Male	Female	Response	Male	Female	Response	Male	Female
0	65	128	9	2	2	20	7	6
1	11	17	10	24	13	22	0	1
2	13	23	12	6	10	23	0	1
3	14	16	13	3	3	24	1	0
4	26	19	14	0	1	25	1	3
5	13	17	15	3	10	27	0	1
6	15	17	16	3	1	30	3	1
7	7	3	17	0	1	50	1	0
8	21	15	18	0	1	60	1	0

*Source:*1990 General Social Survey, National Opinion Research Center.

    c. For the negative binomial model, the log likelihood increases by 248.7 (deviance decreases by 497.3). The estimated difference between the log means is also 0.308, but now SE = 0.127. Show that the 95% confidence interval for the ratio of means is (1.06, 1.75). Compare to the Poisson GLM, and interpret.

    d. The mode for the Poisson distribution is the integer part of the mean, rather than 0. Argue that a possibly more realistic mixture model assumes for gender i a proportion ρ_i that has a Poisson distribution with mean 0 and a proportion $1 - \rho_i$ that has distribution that is a gamma mixture of Poissons. Explain why the corresponding marginal distribution for each gender is a mixture of a degenerate distribution at 0 and a negative binomial distribution.

13.13 Refer to Problem 13.12. Fit the Poisson and negative binomial GLMs using identity link. Show that the estimated differences in means between males and females are identical for the two GLMs but the SE values are very different. Explain why. Use the more appropriate one to form a confidence interval for the true difference in means.

13.14 For the counts of horseshoe-crab satellites in Table 4.3, Table 13.10 shows the results of ML fitting of the negative binomial model using width as the predictor, with the identity link.

    a. State and interpret the prediction equation.

    b. Show that at a predicted $\hat{\mu}$, the estimated variance is roughly $\hat{\mu} + \hat{\mu}^2$.

    c. The corresponding Poisson GLM has fit $\hat{\mu} = -11.53 + 0.55x$ (SE = 0.06). Compare 95% confidence intervals for the slopes for the two models. Interpret, and indicate whether overdispersion seems to exist relative to the Poisson GLM.

TABLE 13.10 Results for Problem 13.14

Parameter	Estimate	Standard Error	Wald 95% Confidence Limits		Chi-Square
Intercept	−11.1471	2.8275	−16.6890	−5.6052	15.54
width	0.5308	0.1132	0.3089	0.7528	21.97
Dispersion	0.9843	0.1822	0.6847	1.4149	

13.15 Refer to Problem 13.14.

    a. Fit a negative binomial model with log link. Interpret. Plot the counts against width and indicate which link seems more appropriate.

    b. Fit a Poisson GLMM with log link, using width predictor. Interpret.

c. Compare results for the various models, including those in Section 4.3.2 for a Poisson GLM. Indicate your preferred model. Justify.

13.16 Refer to Problems 13.14 and 13.15. Using width and qualitative color as predictors, fit a (**a**) negative binomial GLM, and (**b**) Poisson GLMM, checking for interaction and interpreting the final model.

13.17 Refer to Table 13.6. For those with race classified as "other," the sample counts for $(0, 1, 2, 3, 4, 5, 6)$ homicides were $(55, 5, 1, 0, 1, 0, 0)$. Fit an appropriate model simultaneously to these data and those for white and black race categories. Interpret by making pairwise comparisons of the three pairs of means.

13.18 Use a quasi-likelihood approach to analyze Table 13.6 on counts of murder victims.

13.19 Conduct the analyses of Problem 4.6 on defects in the fabrication of computer chips, but use a negative binomial GLM. Compare results to those for the Poisson GLM. Indicate why results are similar.

13.20 With data at the book's Web site (*www.stat.ufl.edu/∼aa/cda/cda.html*), use methods of this chapter to analyze how the countywide vote for the Reform Party candidate Pat Buchanan in the 2000 presidential election related to the vote for Reform Party candidate Ross Perot in the 1996 presidential election. Note that Palm Beach County is an enormous outlier (apparently mainly reflecting votes intended for Al Gore but cast for Buchanan because of a confusing ballot). Model with and without that observation and compare results.

13.21 Conduct a latent class analysis of the data in Espeland and Handelman (1989).

13.22 Refer to the teratology study in Liang and Hanfelt (1994). Analyze these data using at least two different approaches for overdispersed binary data. Compare results and interpret.

13.23 Refer to Problem 13.14. Using an appropriate subset of width, weight, color, and spine condition as predictors, find and interpret a reasonable model for predicting the number of satellites.

Theory and Methods

13.24 Derive residual df for a latent class model with q latent classes. When $I = 2$, for $q \geq 2$ show one needs $T \geq 4$ for the model to be unsaturated. Then, find the maximum value for q when $T = 4, 5$. For an I^2 table, show one needs $q < I^2/(2I - 1)$.

13.25 Express the log likelihood for latent class model (13.1) in terms of the model parameters. Derive likelihood equations (Goodman 1974, Haberman 1979).

13.26 Let Π denote an $I \times J$ matrix of cell probabilities for the joint distribution of X and Y. Suppose that there exist $I \times 1$ column vectors π_{1k} and $J \times 1$ column vectors π_{2k} of probabilities, $k = 1, \ldots, q$, and a set of probabilities $\{\rho_k\}$ such that

$$\Pi = \sum_{k=1}^{q} \rho_k \pi_{1k} \pi_{2k}'.$$

Explain why this implies that there is a latent variable Z such that X and Y are conditionally independent, given Z.

13.27 In Section 13.2.2, under the null that the ordinary logistic regression model holds, explain why it is inappropriate to treat the difference between the deviances for that model and the mixture of two logistic regressions as a chi-squared statistic.

13.28 Refer to Problem 12.7. Let $\mu_k(a, b, c)$ denote the expected frequency of outcomes (a, b, c) for treatments (A, B, C) under treatment sequence k, where outcome 1 = relief and 0 = nonrelief. With a non-parametric random effects approach, show that one can estimate treatment effects in model (12.19) by fitting the quasi-symmetry model

$$\log \mu_k(a, b, c) = a\beta_A + b\beta_B + c\beta_C + \lambda_k(a, b, c),$$

where $\lambda_k(a, b, c) = \lambda_k(a, c, b) = \lambda_k(b, a, c) = \lambda_k(b, c, a) = \lambda_k(c, a, b) = \lambda_k(c, b, a)$. Fit the model, and show that $\hat{\beta}_B - \hat{\beta}_A = 1.64$ (SE = 0.34), $\hat{\beta}_C - \hat{\beta}_A = 2.23$ (SE = 0.39), $\hat{\beta}_C - \hat{\beta}_B = 0.59$ (SE = 0.39). Interpret. Compare results with Problem 12.7 for model (12.19).

13.29 Show that the beta-binomial distribution (13.9) simplifies to the binomial when $\theta = 0$.

13.30 Express the numerator of the beta density in terms of μ and θ. Using this, show that it is (**a**) unimodal when $\theta < \min(\mu, 1 - \mu)$, and (**b**) the uniform density when $\mu = \theta = \frac{1}{2}$.

13.31 Suppose that $\pi_i = P(Y_{it} = 1) = 1 - P(Y_{it} = 0)$, for $t = 1, \ldots, n_i$, and $\text{corr}(Y_{it}, Y_{is}) = \rho$ for $t \neq s$. Show that $\text{var}(Y_{it}) = \pi_i(1 - \pi_i)$,

$\text{cov}(Y_{it}, Y_{is}) = \rho \pi_i (1 - \pi_i)$, and

$$\text{var}\left(\sum_t Y_{it} \right) = n_i \pi_i (1 - \pi_i)[1 + \rho(n_i - 1)].$$

13.32 When $n = 1$, show that the beta-binomial distribution is no different from the binomial (i.e., Bernoulli). Explain why overdispersion cannot occur when $n = 1$.

13.33 When y_i is the sum of n_i binary responses each having mean μ_i, refer to the quasi-likelihood approach with $v(\mu_i) = \phi n_i \mu_i (1 - \mu_i)$. Explain why this variance function has a structural problem, with only $\phi = 1$ making sense when $n_i = 1$.

13.34 Liang and Hanfelt (1994) described a teratology study comparing control and treatment groups in which the ML estimate of the treatment effect in a beta-binomial model differs by a factor of 2 depending on whether one assumes the same overdispersion parameter for each group. By contrast, with variance function (13.11), the quasi-likelihood estimate of the treatment effect is the same whether one assumes the same or different ϕ for the two groups. Explain why, and discuss whether this is an advantage or disadvantage of that method.

13.35 Consider the logistic-normal model, $\text{logit}(\pi_i) = \alpha + x_i' \beta + u_i$. For small σ, show that it corresponds approximately to a mixture model for which the mixture distribution has $\text{var}(\pi_i) = [\mu_i(1 - \mu_i)]^2 \sigma^2$. (*Hint:* See Problem 6.33.)

13.36 Altham (1978) introduced the discrete distribution

$$f(y; \pi, \psi) = c(\pi, \psi) \binom{n}{y} \pi^y (1 - \pi)^{n-y} \exp[\psi y (n - y)],$$

$$y = 0,1,\ldots,n,$$

where $c(\pi, \psi)$ is a normalizing constant. Show that this is in the exponential family. Show that the binomial occurs when $\psi = 0$. [Altham noted that overdispersion occurs when $\psi < 0$. Corcoran et al. (2001) and Lindsey and Altham (1998) used this as the basis of an alternative model to the beta-binomial.]

13.37 When $y_1, \ldots, y_N$ are independent from the negative binomial distribution (13.13) with k fixed, show that $\hat{\mu} = \bar{y}$.

13.38 Using $E(Y) = E[E(Y|X)]$ and $\text{var}(Y) = E[\text{var}(Y|X)] + \text{var}[E(Y|X)]$, derive the mean and variance of the **(a)** beta-binomial distribution,
and **(b)** negative binomial distribution.

13.39 Suppose that given u, Y is Poisson with $E(Y|u) = u\mu$, where μ may depend on predictors. Suppose that u is a positive random variable with $E(u) = 1$ and $\text{var}(u) = \tau$. Show that $E(Y) = \mu$ and $\text{var}(Y) = \mu + \tau\mu^2$. Explain how negative binomial GLMs and Poisson GLMMs with log link can follow as special cases.

13.40 An alternative negative binomial parameterization results from the gamma density formula,

$$f(\lambda; k, \mu) = \frac{(k)^{k\mu}}{\Gamma(k\mu)} \exp(-k\lambda)\lambda^{k\mu-1}, \qquad \lambda \geq 0,$$

for which $E(\lambda) = \mu$, $\text{var}(\lambda) = \mu/k$. Show that this gamma mixture of Poissons yields a negative binomial with

$$E(Y) = \mu, \quad \text{var}(Y) = \mu(1 + k)/k.$$

For what limiting value of k does this reduce to the Poisson? [See Nelder and Lee (1996) for ML model fitting. Cameron and Trivedi (1998, p. 75) pointed out that, unlike with quadratic variance, consistency does not occur for parameter estimators when the model for the mean holds but the true distribution is not negative binomial.]

13.41 The negative binomial distribution is unimodal with a mode at the integer part of $\mu(k - 1)/k$ (Johnson et al. 1992, pp. 208–209). Show that the mode is 0 when $\mu \leq 1$, and that when $\mu > 1$ the mode is still 0 if $k < \mu/(\mu - 1)$. (This gives greater scope than the Poisson, since its mode equals the integer part of the mean.)

13.42 Consider the loglinear random effects model

$$\log[E(Y_{it}|\mathbf{u}_i)] = \mathbf{x}'_{it}\boldsymbol{\beta} + \mathbf{z}'_{it}\mathbf{u}_i,$$

where $\{\mathbf{u}_i\}$ are independent $N(\mathbf{0}, \boldsymbol{\Sigma})$. Show that this implies the marginal loglinear model

$$\log[E(Y_{it})] - \frac{1}{2}\mathbf{z}'_{it}\boldsymbol{\Sigma}\mathbf{z}_{it} = \mathbf{x}'_{it}\boldsymbol{\beta},$$

with the same fixed effects but with offset term. For the random-intercept case, indicate the role of σ on the size of the offset. Explain what happens when $\sigma = 0$.

13.43 In Section 13.5.1 and Problem 13.42 we saw that for Poisson GLMMs, the marginal effects are the same as the cluster-specific effects. This does not imply that ML estimates of effects are the same for a Poisson GLMM and a Poisson GLM. Explain why. (*Hint*: For the GLMM, is the marginal distribution Poisson?)

13.44 For the Poisson GLMM (13.14), use the normal mgf to show that for $t \neq s$,

$$\text{cov}(Y_{it}, Y_{is}) = \exp\left[(\mathbf{x}'_{it} + \mathbf{x}'_{is})\boldsymbol{\beta}\right]\left[\exp(\sigma^2)(\exp(\sigma^2) - 1)\right]$$

Hence, find $\text{corr}(Y_{it}, Y_{is})$.

13.45 Consider a Poisson GLMM using the identity link. Relate the marginal mean and variance to the conditional mean and variance. Explain the structural problem that this model has.

CHAPTER 14

Asymptotic Theory for Parametric Models

This chapter has a more theoretical flavor than others. It presents asymptotic theory for parametric models for categorical data, with emphasis on multinomial models for contingency tables. In Section 14.1 we review and extend the delta method. This is used to derive large-sample normal distributions for many statistics. In Section 14.2 we apply the delta method to ML estimation of parameters in models for contingency tables, later illustrated in Section 14.4 for logit and loglinear models. In Section 14.3 we derive asymptotic distributions of cell residuals and the X^2 and G^2 goodness-of-fit statistics.

The results in this chapter have a long history. Pearson (1900) derived the asymptotic chi-squared distribution of X^2 for testing a specified multinomial distribution. Fisher (1922, 1924) showed the adjustment in degrees of freedom when multinomial probabilities are functions of unknown parameters. Cramér (1946, pp. 424–434) formally proved this result, under the assumption that ML estimators of the parameters are consistent. Rao (1957) proved consistency of the ML estimators under general conditions. He also gave the asymptotic distribution of the ML estimators, although the primary emphasis of his articles was on proving consistency. Birch (1964a) proved these results under weaker conditions. Andersen (1980), Bishop et al. (1975), Cox (1984), Haberman (1974a), and Watson (1959) provided other proofs or considered related cases.

As in Cramér's and Rao's proofs, our derivation regards the ML estimator as a point in the parameter space where the derivative of the log likelihood function is zero. Birch regarded it as a point at which the likelihood takes value arbitrarily near its supremum. Although his approach is more powerful, the proofs are more complex. We avoid a formal "theorem–proof" style of exposition. Instead, we show that powerful results follow from simple mathematical ideas, such as Taylor series expansions.

14.1   DELTA METHOD

Suppose that a statistic used as an estimator of a parameter has a large-sample normal distribution. Then, in this section we show that many functions of that statistic are also asymptotically normal.

14.1.1 O, o Rates of Convergence

Big O and little o notation is useful for describing limiting behavior of sequences. For real numbers $\{z_n\}$, the *little o* notation $o(z_n)$ represents a term that has *smaller* order than z_n as $n \to \infty$, in the sense that $o(z_n)/z_n \to 0$ as $n \to \infty$. For instance, $\sqrt{n}$ is $o(n)$ as $n \to \infty$, since $\sqrt{n}/n \to 0$ as $n \to \infty$. A sequence that is $o(1)$ satisfies $o(1)/1 = o(1) \to 0$; for instance, $n^{-1/2}$ is $o(1)$ as $n \to \infty$.

The *big O* notation $O(z_n)$ represents terms that have the *same* order of magnitude as z_n, in the sense that $|O(z_n)/z_n|$ is bounded as $n \to \infty$. For instance, $(3/n) + (8/n^2)$ is $O(n^{-1})$ as $n \to \infty$; dividing it by n^{-1} gives a ratio that takes value close to 3 as n increases.

Similar notation applies to sequences of random variables. This notation uses a subscript p to indicate that the sequence has probabilistic rather than deterministic behavior. The symbol $o_p(z_n)$ denotes a random variable of *smaller* order than z_n for large n, in the sense that $o_p(z_n)/z_n$ *converges in probability* to 0; that is, for any fixed $\epsilon > 0$, $P(|o_p(z_n)/z_n| \le \epsilon) \to 1$ as $n \to \infty$. The notation $O_p(z_n)$ represents a random variable such that for every $\epsilon > 0$, there is a constant K and an integer n_0 such that $P[|O_p(z_n)/z_n| < K] > 1 - \epsilon$ for all $n > n_0$.

To illustrate, let $\bar{Y}_n$ denote the sample mean of n independent observations $Y_1, \ldots, Y_n$ from a distribution having $E(Y_i) = \mu$. Then $(\bar{Y}_n - \mu) = o_p(1)$, since $(\bar{Y}_n - \mu)/1$ converges in probability to zero as $n \to \infty$ by the law of large numbers. By Tchebychev's inequality, the difference between a random variable and its expected value has the same order of magnitude as the standard deviation of that random variable. Since $\bar{Y}_n - \mu$ has standard deviation $\sigma/\sqrt{n}$, $(\bar{Y}_n - \mu) = O_p(n^{-1/2})$.

A random variable that is $O_p(n^{-1/2})$ is also $o_p(1)$. An example is $(\bar{Y}_n - \mu)$. Multiplication affects the order in the way one expects intuitively (Problem 14.1). For instance, $\sqrt{n}(\bar{Y}_n - \mu) = n^{1/2}O_p(n^{-1/2}) = O_p(n^{1/2}n^{-1/2}) = O_p(1)$. If the difference between two random variables is $o_p(1)$ as $n \to \infty$, Slutzky's theorem states that those random variables have the same limiting distribution.

14.1.2   Delta Method for Function of Random Variable

Let T_n denote a statistic, the subscript expressing its dependence on the sample size n. For large samples, suppose that T_n is approximately normally distributed about θ, with approximate standard error $\sigma/\sqrt{n}$. More precisely, as $n \rightarrow \infty$, suppose that the cdf of $\sqrt{n}\,(T_n - \theta)$ converges to a $N(0, \sigma^2)$ cdf. This limiting behavior is an example of *convergence in distribution*, denoted

$$\sqrt{n}\,(T_n - \theta) \xrightarrow{d} N(0, \sigma^2). \tag{14.1}$$

For a function g, we now derive the limiting distribution of $g(T_n)$. Suppose that g is at least twice differentiable at θ. We use the Taylor series expansion for $g(t)$ in a neighborhood of θ. For some θ^* between t and θ,

$$g(t) = g(\theta) + (t - \theta)g'(\theta) + (t - \theta)^2 g''(\theta^*)/2$$
$$= g(\theta) + (t - \theta)g'(\theta) + O\big(|t - \theta|^2\big).$$

Substituting the random variable T_n for t, we have

$$\sqrt{n}\,[g(T_n) - g(\theta)] = \sqrt{n}\,(T_n - \theta)g'(\theta) + \sqrt{n}\,O\big(|T_n - \theta|^2\big)$$
$$= \sqrt{n}\,(T_n - \theta)g'(\theta) + O_p(n^{-1/2}) \tag{14.2}$$

since

$$\sqrt{n}\,O\big(|T_n - \theta|^2\big) = \sqrt{n}\,O\big[O_p(n^{-1})\big] = O_p(n^{-1/2}).$$

Since the $O_p(n^{-1/2})$ term is asymptotically negligible, $\sqrt{n}\,[g(T_n) - g(\theta)]$ has the same limiting distribution as $\sqrt{n}\,(T_n - \theta)g'(\theta)$; that is, $g(T_n) - g(\theta)$ behaves like the constant multiple $g'(\theta)$ of $(T_n - \theta)$. Now, $(T_n - \theta)$ is approximately normal with variance σ^2/n. Thus, $g(T_n) - g(\theta)$ is approximately normal with variance $\sigma^2[g'(\theta)]^2/n$. More precisely,

$$\sqrt{n}\,[g(T_n) - g(\theta)] \xrightarrow{d} N\big(0, \sigma^2[g'(\theta)]^2\big). \tag{14.3}$$

Figure 3.1 illustrated this result, and in Section 3.1.6 it was applied to the sample logit.

Result (14.3) is called the *delta method* for obtaining asymptotic distributions. Since $\sigma^2 = \sigma^2(\theta)$ and $g'(\theta)$ usually depends on θ, the asymptotic variance is unknown. Let $\sigma^2(T_n)$ and $g'(T_n)$ denote these terms evaluated at the sample estimator T_n of θ. When $g'(\cdot)$ and $\sigma = \sigma(\cdot)$ are continuous at θ, $\sigma(T_n)g'(T_n)$ is a consistent estimator of $\sigma(\theta)g'(\theta)$. Thus, confidence intervals and tests use the result that $\sqrt{n}\,[g(T_n) - g(\theta)]/\sigma(T_n)\,|g'(T_n)|$ is asymptotically standard normal. For instance,

$$g(T_n) \pm z_{\alpha/2}\,\sigma(T_n)|g'(T_n)|/\sqrt{n}$$

is a large-sample $100(1 - \alpha)\%$ confidence interval for $g(\theta)$.

When $g'(\theta) = 0$, (14.3) is uninformative because the limiting variance equals 0. In that case, $\sqrt{n}\,[g(T_n) - g(\theta)] = o_p(1)$, and higher-order terms in the Taylor series expansion yield the asymptotic distribution (see Note 14.1).

14.1.3 Delta Method for Function of Random Vector

The delta method generalizes to functions of random *vectors*. Suppose that $\mathbf{T}_n = (T_{n1}, \ldots, T_{nN})'$ is asymptotically multivariate normal with mean $\boldsymbol{\theta} = (\theta_1, \ldots, \theta_N)'$ and covariance matrix $\boldsymbol{\Sigma}/n$. Suppose that $g(t_1, \ldots, t_N)$ has a nonzero differential $\boldsymbol{\phi} = (\phi_1, \ldots, \phi_N)'$ at $\boldsymbol{\theta}$, where

$$\phi_i = \frac{\partial g}{\partial t_i}\bigg|_{\mathbf{t} = \boldsymbol{\theta}}.$$

Then,

$$\sqrt{n}\,[g(\mathbf{T}_n) - g(\boldsymbol{\theta})] \xrightarrow{d} N(0, \boldsymbol{\phi}'\boldsymbol{\Sigma}\boldsymbol{\phi}). \tag{14.4}$$

For large n, $g(\mathbf{T}_n)$ has distribution similar to the normal with mean $g(\boldsymbol{\theta})$ and variance $\boldsymbol{\phi}'\boldsymbol{\Sigma}\boldsymbol{\phi}/n$.

The proof of (14.4) follows from the expansion

$$g(\mathbf{T}_n) - g(\boldsymbol{\theta}) = (\mathbf{T}_n - \boldsymbol{\theta})'\boldsymbol{\phi} + o(\|\mathbf{T}_n - \boldsymbol{\theta}\|),$$

where $\|\mathbf{z}\| = (\Sigma z_i^2)^{1/2}$ denotes the length of vector $\mathbf{z}$. For large n, $g(\mathbf{T}_n) - g(\boldsymbol{\theta})$ behaves like a linear function of the approximately normal random vector $(\mathbf{T}_n - \boldsymbol{\theta})$. Thus, it itself is approximately normal.

14.1.4 Asymptotic Normality of Functions of Multinomial Counts

The delta method for random vectors implies asymptotic normality of many functions of cell counts in contingency tables. Suppose that cell counts $(n_1, \ldots, n_N)$ have a multinomial distribution with cell probabilities $\boldsymbol{\pi} = (\pi_1, \ldots, \pi_N)'$. Let $n = n_1 + \cdots + n_N$, and let $\mathbf{p} = (p_1, \ldots, p_N)'$ denote the sample proportions, where $p_i = n_i/n$.

Denote observation i of the n cross-classified in the contingency table by $\mathbf{Y}_i = (Y_{i1}, \ldots, Y_{iN})$, where $Y_{ij} = 1$ if it falls in cell j, and $Y_{ij} = 0$ otherwise, $i = 1, \ldots, n$. For instance, $\mathbf{Y}_6 = (0, 0, 1, 0, 0, \ldots, 0)$ means that observation 6 is in the third cell of the table. Now, since each observation falls in just one cell, $\Sigma_j Y_{ij} = 1$ and $Y_{ij} Y_{ik} = 0$ when $j \neq k$. Also, $p_j = \Sigma_i Y_{ij}/n$, and

$$E(Y_{ij}) = P(Y_{ij} = 1) = \pi_j = E(Y_{ij}^2), \qquad E(Y_{ij}Y_{ik}) = 0 \quad \text{if } j \neq k.$$

It follows that

$$E(\mathbf{Y}_i) = \boldsymbol{\pi} \quad \text{and} \quad \text{cov}(\mathbf{Y}_i) = \boldsymbol{\Sigma}, \quad i = 1, \ldots, n,$$

where $\Sigma = (\sigma_{jk})$ with

$$\sigma_{jj} = \mathrm{var}(Y_{ij}) = E(Y_{ij}^2) - [E(Y_{ij})]^2 = \pi_j(1 - \pi_j),$$

$$\sigma_{jk} = \mathrm{cov}(Y_{ij}, Y_{ik}) = E(Y_{ij}Y_{ik}) - E(Y_{ij})E(Y_{ik}) = -\pi_j\pi_k \quad \text{for } j \neq k.$$

The matrix Σ has form

$$\Sigma = \mathbf{diag}(\boldsymbol{\pi}) - \boldsymbol{\pi}\boldsymbol{\pi}'$$

where $\mathbf{diag}(\boldsymbol{\pi})$ is the diagonal matrix with the elements of $\boldsymbol{\pi}$ on the main diagonal.

Since $\mathbf{p}$ is a sample mean of n independent observations, namely

$$\mathbf{p} = \frac{\sum_{i=1}^n \mathbf{Y}_i}{n},$$

$$\mathrm{cov}(\mathbf{p}) = [\mathbf{diag}(\boldsymbol{\pi}) - \boldsymbol{\pi}\boldsymbol{\pi}']/n. \tag{14.5}$$

This covariance matrix is singular, because of the linear dependence $\sum p_i = 1$. The multivariate central limit theorem (Rao 1973, p. 128) implies

$$\sqrt{n}\,(\mathbf{p} - \boldsymbol{\pi}) \xrightarrow{d} N[\mathbf{0}, \mathbf{diag}(\boldsymbol{\pi}) - \boldsymbol{\pi}\boldsymbol{\pi}']. \tag{14.6}$$

By the delta method, functions of $\mathbf{p}$ having nonzero differential at $\boldsymbol{\pi}$ are also asymptotically normal. Let $g(t_1, \ldots, t_N)$ be a differentiable function, and let

$$\phi_i = \partial g/\partial \pi_i, \qquad i = 1, \ldots, N,$$

denote $\partial g/\partial t_i$ evaluated at $\mathbf{t} = \boldsymbol{\pi}$. By the delta method (14.4),

$$\sqrt{n}\,[g(\mathbf{p}) - g(\boldsymbol{\pi})] \xrightarrow{d} N(0, \boldsymbol{\phi}'[\mathbf{diag}(\boldsymbol{\pi}) - \boldsymbol{\pi}\boldsymbol{\pi}']\boldsymbol{\phi}). \tag{14.7}$$

The asymptotic variance equals

$$\boldsymbol{\phi}'\mathbf{diag}(\boldsymbol{\pi})\boldsymbol{\phi} - (\boldsymbol{\phi}'\boldsymbol{\pi})^2 = \sum \pi_i \phi_i^2 - \left(\sum \pi_i \phi_i\right)^2.$$

In Section 3.1.7 we used this formula to derive the large-sample variance of the sample log odds ratio.

14.1.5 Delta Method for Vector Function of Random Vector

The delta method generalizes further to a *vector* of functions of an asymptotically normal random vector. Let $\mathbf{g}(\mathbf{t}) = (g_1(\mathbf{t}), \ldots, g_q(\mathbf{t}))'$ and let $(\partial \mathbf{g}/\partial \boldsymbol{\theta})$ denote the $q \times N$ Jacobian matrix for which the entry in row i and column j

is $\partial g_i(\mathbf{t})/\partial t_j$ evaluated at $\mathbf{t} = \boldsymbol{\theta}$. Then,

$$\sqrt{n}\left[\mathbf{g}(\mathbf{T}_n) - \mathbf{g}(\boldsymbol{\theta})\right] \xrightarrow{d} N\left[\mathbf{0}, (\partial \mathbf{g}/\partial \boldsymbol{\theta})\Sigma(\partial \mathbf{g}/\partial \boldsymbol{\theta})'\right]. \qquad (14.8)$$

The rank of the limiting normal distribution equals the rank of the asymptotic covariance matrix.

Expression (14.8) is useful for finding large-sample joint distributions. For instance, from (14.6), (14.7), and (14.8), the asymptotic distribution of several functions of multinomial proportions has covariance matrix of the form

$$\text{asymp. cov}\left\{\sqrt{n}\left[\mathbf{g}(\mathbf{p}) - \mathbf{g}(\boldsymbol{\pi})\right]\right\} = \boldsymbol{\Phi}\left[\text{diag}(\boldsymbol{\pi}) - \boldsymbol{\pi}\boldsymbol{\pi}'\right]\boldsymbol{\Phi}',$$

where $\boldsymbol{\Phi}$ is the Jacobian $(\partial \mathbf{g}/\partial \boldsymbol{\pi})$.

14.1.6 Joint Asymptotic Normality of Log Odds Ratios

We illustrate formula (14.8) by finding the joint asymptotic distribution of a set of log odds ratios in a contingency table. We use the log scale because convergence to normality is more rapid for it.

Let $\mathbf{g}(\boldsymbol{\pi}) = \log(\boldsymbol{\pi})$ denote the vector of natural logs of cell probabilities, for which

$$\partial \mathbf{g}/\partial \boldsymbol{\pi} = \text{diag}(\boldsymbol{\pi})^{-1}.$$

The covariance of the asymptotic distribution of $\sqrt{n}\left[\log(\mathbf{p}) - \log(\boldsymbol{\pi})\right]$ is

$$\text{diag}(\boldsymbol{\pi})^{-1}\left[\text{diag}(\boldsymbol{\pi}) - \boldsymbol{\pi}\boldsymbol{\pi}'\right]\text{diag}(\boldsymbol{\pi})^{-1} = \text{diag}(\boldsymbol{\pi})^{-1} - \mathbf{1}\mathbf{1}'$$

where $\mathbf{1}$ is an $N \times 1$ vector of 1 elements.

For a $q \times N$ matrix of constants $\mathbf{C}$, it follows that

$$\sqrt{n}\,\mathbf{C}\left[\log(\mathbf{p}) - \log(\boldsymbol{\pi})\right] \xrightarrow{d} N\left[\mathbf{0}, \mathbf{C}\,\text{diag}(\boldsymbol{\pi})^{-1}\mathbf{C}' - \mathbf{C}\mathbf{1}\mathbf{1}'\mathbf{C}'\right]. \qquad (14.9)$$

Now, suppose $\mathbf{C}\log(\mathbf{p})$ is a set of sample log odds ratios. Then, each row of $\mathbf{C}$ contains zeros except for two $+1$ elements and two -1 elements in the positions multiplied by the relevant elements of $\log(\mathbf{p})$ to form the given log odds ratio. The second term in the covariance matrix in (14.9) is then zero. If a particular odds ratio uses the cells numbered h, i, j, and k, the variance of the asymptotic distribution is

$$\text{asymp. var}\left[\sqrt{n}\,(\text{sample log odds ratio})\right] = \pi_h^{-1} + \pi_i^{-1} + \pi_j^{-1} + \pi_k^{-1}.$$

When two log odds ratios have no cells in common, their asymptotic covariance in the limiting normal distribution equals zero.

14.2 ASYMPTOTIC DISTRIBUTIONS OF ESTIMATORS OF MODEL PARAMETERS AND CELL PROBABILITIES

We now derive basic results of large-sample model-based inference for contingency tables. The delta method is the key tool. The derivations apply to a single multinomial distribution. They extend directly to products of multinomials, when the parameter space stays fixed as the sample size increases.

The observations are counts $\mathbf{n} = (n_1, \ldots, n_N)'$ in N cells of a contingency table. The asymptotics regard N as fixed and let $n = \Sigma n_i \to \infty$. We assume that $\mathbf{n} = n\mathbf{p}$ has a multinomial distribution with probabilities $\boldsymbol{\pi} = (\pi_1, \ldots, \pi_N)'$. The model is

$$\boldsymbol{\pi} = \boldsymbol{\pi}(\boldsymbol{\theta}),$$

where $\boldsymbol{\pi}(\boldsymbol{\theta})$ denotes a function that relates $\boldsymbol{\pi}$ to a smaller number of parameters $\boldsymbol{\theta} = (\theta_1, \ldots, \theta_q)'$.

As $\boldsymbol{\theta}$ ranges over its parameter space, $\boldsymbol{\pi}(\boldsymbol{\theta})$ ranges over a subset of the space of $\boldsymbol{\pi}$ for N probabilities. Adding components to $\boldsymbol{\theta}$, the model becomes more complex and the space of $\boldsymbol{\pi}$ that satisfy the model is larger. We use $\boldsymbol{\theta}$ and $\boldsymbol{\pi}$ to denote generic parameter and probability values, and $\boldsymbol{\theta}_0 = (\theta_{10}, \ldots, \theta_{q0})'$ and $\boldsymbol{\pi}_0 = (\pi_{10}, \ldots, \pi_{N0})' = \boldsymbol{\pi}(\boldsymbol{\theta}_0)$ to denote true values for a particular application. When the model does not hold, no $\boldsymbol{\theta}_0$ exists for which $\boldsymbol{\pi}(\boldsymbol{\theta}_0) = \boldsymbol{\pi}_0$; that is, $\boldsymbol{\pi}_0$ falls outside the subset of $\boldsymbol{\pi}$ values that is the range of $\boldsymbol{\pi}(\boldsymbol{\theta})$ for the space of possible $\boldsymbol{\theta}$. We consider this case in Section 14.3.5.

We first derive the asymptotic distribution of the ML estimator $\hat{\boldsymbol{\theta}}$ of $\boldsymbol{\theta}$. We use that to derive the asymptotic distribution of the model-based ML estimator $\hat{\boldsymbol{\pi}} = \boldsymbol{\pi}(\hat{\boldsymbol{\theta}})$ of $\boldsymbol{\pi}$. The approach follows Rao (1973, Sec. 5e) and Bishop et al. (1975, Secs. 14.7 and 14.8). The assumed regularity conditions are:

1. $\boldsymbol{\theta}_0$ is not on the boundary of the parameter space.
2. All $\pi_{i0} > 0$.
3. $\boldsymbol{\pi}(\boldsymbol{\theta})$ has continuous first-order partial derivatives in a neighborhood of $\boldsymbol{\theta}_0$.
4. The Jacobian matrix $(\partial \boldsymbol{\pi} / \partial \boldsymbol{\theta})$ has full rank q at $\boldsymbol{\theta}_0$.

These conditions ensure that $\boldsymbol{\pi}(\boldsymbol{\theta})$ is locally smooth and one-to-one at $\boldsymbol{\theta}_0$ and Taylor series expansions exist in neighborhoods around $\boldsymbol{\theta}_0$ and $\boldsymbol{\pi}_0$. When the Jacobian does not have full rank, often it does with reformulation of the model using fewer parameters.

14.2.1 Distribution of Model Parameter Estimator

The key to deriving the asymptotic distribution of $\hat{\boldsymbol{\theta}}$ is to express $\hat{\boldsymbol{\theta}}$ as a linearized function of $\mathbf{p}$. Then the delta method applies, using the asymptotic normality of $\mathbf{p}$. The linearization has two steps, first relating $\mathbf{p}$ to $\hat{\boldsymbol{\pi}}$, and then $\hat{\boldsymbol{\pi}}$ to $\hat{\boldsymbol{\theta}}$.

The kernel of the multinomial log likelihood is

$$L(\boldsymbol{\theta}) = \log \prod_{i=1}^{N} \pi_i(\boldsymbol{\theta})^{n_i} = n \sum_{i=1}^{N} p_i \log \pi_i(\boldsymbol{\theta}).$$

The likelihood equations are

$$\frac{\partial L(\boldsymbol{\theta})}{\partial \theta_j} = n \sum_i \frac{p_i}{\pi_i(\boldsymbol{\theta})} \frac{\partial \pi_i(\boldsymbol{\theta})}{\partial \theta_j} = 0, \quad j = 1, \ldots, q. \qquad (14.10)$$

These depend on the functional form $\boldsymbol{\pi}(\boldsymbol{\theta})$ used in the model. Note that

$$\sum_i \frac{\partial \pi_i(\boldsymbol{\theta})}{\partial \theta_j} = \frac{\partial}{\partial \theta_j} \left[\sum_i \pi_i(\boldsymbol{\theta}) \right] = \frac{\partial}{\partial \theta_j} (1) = 0. \qquad (14.11)$$

Let $\partial \pi_i / \partial \hat{\theta}_j$ represent $\partial \pi_i(\boldsymbol{\theta}) / \partial \theta_j$ evaluated at $\hat{\boldsymbol{\theta}}$. Subtracting a common term from both sides of the jth likelihood equation (14.10),

$$\sum_i \frac{n(p_i - \pi_{i0})}{\hat{\pi}_i} \frac{\partial \pi_i}{\partial \hat{\theta}_j} = \sum_i \frac{n(\hat{\pi}_i - \pi_{i0})}{\hat{\pi}_i} \frac{\partial \pi_i}{\partial \hat{\theta}_j}, \qquad (14.12)$$

since the first sum on the right-hand side equals zero from (14.11).

Next we express $\hat{\boldsymbol{\pi}}$ in terms of $\hat{\boldsymbol{\theta}}$ using

$$\hat{\pi}_i - \pi_{i0} = \sum_k (\hat{\theta}_k - \theta_{k0}) \frac{\partial \pi_i}{\partial \bar{\theta}_k}$$

where $\partial \pi_i / \partial \bar{\theta}_k$ represents $\partial \pi_i / \partial \theta_k$ evaluated at some point $\bar{\boldsymbol{\theta}}$ falling between $\hat{\boldsymbol{\theta}}$ and $\boldsymbol{\theta}_0$. Substitution of this into the right-hand side of (14.12) and division of both sides by $\sqrt{n}$ yields, for each j,

$$\sum_i \frac{\sqrt{n}(p_i - \pi_{i0})}{\hat{\pi}_i} \frac{\partial \pi_i}{\partial \hat{\theta}_j} = \sum_k \sqrt{n}(\hat{\theta}_k - \theta_{k0}) \left(\sum_i \frac{1}{\hat{\pi}_i} \frac{\partial \pi_i}{\partial \hat{\theta}_j} \frac{\partial \pi_i}{\partial \bar{\theta}_k} \right). \qquad (14.13)$$

Some notation lets us express more simply the dependence of $\hat{\boldsymbol{\theta}}$ on $\mathbf{p}$. Let $\mathbf{A}$ denote the $N \times q$ matrix having elements

$$a_{ij} = \pi_{i0}^{-1/2} \frac{\partial \pi_i(\boldsymbol{\theta})}{\partial \theta_{j0}}.$$

The matrix expression for $\mathbf{A}$ is

$$\mathbf{A} = \mathbf{diag}(\boldsymbol{\pi}_0)^{-1/2}(\partial\boldsymbol{\pi}/\partial\boldsymbol{\theta}_0), \tag{14.14}$$

where $(\partial\boldsymbol{\pi}/\partial\boldsymbol{\theta}_0)$ denotes the Jacobian $(\partial\boldsymbol{\pi}/\partial\boldsymbol{\theta})$ evaluated at $\boldsymbol{\theta}_0$. As $\hat{\boldsymbol{\theta}}$ converges to $\boldsymbol{\theta}_0$, the term in brackets on the right-hand side of (14.13) converges to the element in row j and column k of $\mathbf{A}'\mathbf{A}$. As $\hat{\boldsymbol{\theta}} \to \boldsymbol{\theta}_0$, the set of equations (14.13) has the form

$$\mathbf{A}'\mathbf{diag}(\boldsymbol{\pi}_0)^{-1/2}\sqrt{n}\,(\mathbf{p} - \boldsymbol{\pi}_0) = (\mathbf{A}'\mathbf{A})\sqrt{n}\left(\hat{\boldsymbol{\theta}} - \boldsymbol{\theta}_0\right) + o_p(1).$$

Since the Jacobian has full rank at $\boldsymbol{\theta}_0$, $\mathbf{A}'\mathbf{A}$ is nonsingular. Thus,

$$\sqrt{n}\left(\hat{\boldsymbol{\theta}} - \boldsymbol{\theta}_0\right) = (\mathbf{A}'\mathbf{A})^{-1}\mathbf{A}'\mathbf{diag}(\boldsymbol{\pi}_0)^{-1/2}\sqrt{n}\,(\mathbf{p} - \boldsymbol{\pi}_0) + o_p(1). \tag{14.15}$$

Now, the asymptotic distribution of $\mathbf{p}$ determines that of $\hat{\boldsymbol{\theta}}$. From (14.6), $\sqrt{n}\,(\mathbf{p} - \boldsymbol{\pi}_0)$ is asymptotically normal, with covariance matrix $[\mathbf{diag}(\boldsymbol{\pi}_0) - \boldsymbol{\pi}_0\boldsymbol{\pi}_0']$. By the delta method, $\sqrt{n}\,(\hat{\boldsymbol{\theta}} - \boldsymbol{\theta}_0)$ is also asymptotically normal, with asymptotic covariance matrix

$$(\mathbf{A}'\mathbf{A})^{-1}\mathbf{A}'\mathbf{diag}(\boldsymbol{\pi}_0)^{-1/2} \times [\mathbf{diag}(\boldsymbol{\pi}_0) - \boldsymbol{\pi}_0\boldsymbol{\pi}_0'] \times \mathbf{diag}(\boldsymbol{\pi}_0)^{-1/2}\mathbf{A}(\mathbf{A}'\mathbf{A})^{-1}.$$

Using (14.11) and (14.14), the term subtracted in this expression disappears because

$$\boldsymbol{\pi}_0'\mathbf{diag}(\boldsymbol{\pi}_0)^{-1/2}\mathbf{A} = \boldsymbol{\pi}_0'\mathbf{diag}(\boldsymbol{\pi}_0)^{-1/2}\mathbf{diag}(\boldsymbol{\pi}_0)^{-1/2}(\partial\boldsymbol{\pi}/\partial\boldsymbol{\theta}_0)$$

$$= \mathbf{1}'(\partial\boldsymbol{\pi}/\partial\boldsymbol{\theta}_0) = \left(\sum_i \partial\pi_i/\partial\boldsymbol{\theta}_0\right)' = \mathbf{0}'.$$

Thus, this asymptotic covariance expression for $\sqrt{n}\,(\hat{\boldsymbol{\theta}} - \boldsymbol{\theta}_0)$ simplifies to $(\mathbf{A}'\mathbf{A})^{-1}$.

In summary, this argument establishes the general result

$$\sqrt{n}\left(\hat{\boldsymbol{\theta}} - \boldsymbol{\theta}_0\right) \xrightarrow{d} N\left[\mathbf{0}, (\mathbf{A}'\mathbf{A})^{-1}\right]. \tag{14.16}$$

The asymptotic covariance matrix of $\hat{\boldsymbol{\theta}}$ depends on $(\partial\boldsymbol{\pi}/\partial\boldsymbol{\theta}_0)$ and hence on the function for modeling $\boldsymbol{\pi}$ in terms of $\boldsymbol{\theta}$. Let $\hat{\mathbf{A}}$ denote $\mathbf{A}$ evaluated at the ML estimate $\hat{\boldsymbol{\theta}}$. The estimated covariance matrix is

$$\widehat{\text{cov}}(\hat{\boldsymbol{\theta}}) = (\hat{\mathbf{A}}'\hat{\mathbf{A}})^{-1}/n.$$

The asymptotic normality and covariance of $\hat{\boldsymbol{\theta}}$ follows more simply from general results for ML estimators. However, those results require stronger regularity conditions (Rao 1973, p. 364) than the ones assumed here. Suppose that observations are independent from $f(\mathbf{y}; \boldsymbol{\theta})$, some probability mass function. The ML estimator $\hat{\boldsymbol{\theta}}$ is efficient, in the sense that

$$\sqrt{n}\,(\hat{\boldsymbol{\theta}} - \boldsymbol{\theta}) \overset{d}{\to} N(\mathbf{0}, \mathcal{J}^{-1}),$$

where $\mathcal{J}$ is the information matrix for a single observation. The (j, k) element of $\mathcal{J}$ is

$$-E\left(\frac{\partial^2 \log f(\mathbf{y}, \boldsymbol{\theta})}{\partial \theta_j \partial \theta_k}\right) = E\left[\frac{\partial \log f(\mathbf{y}, \boldsymbol{\theta})}{\partial \theta_j} \cdot \frac{\partial \log f(\mathbf{y}, \boldsymbol{\theta})}{\partial \theta_k}\right].$$

When f is the probability of a single observation having multinomial probabilities $\{\pi_1(\boldsymbol{\theta}), \dots, \pi_N(\boldsymbol{\theta})\}$, this element of $\mathcal{J}$ equals

$$\sum_{i=1}^{N} \frac{\partial \log(\pi_i(\boldsymbol{\theta}))}{\partial \theta_j} \frac{\partial \log(\pi_i(\boldsymbol{\theta}))}{\partial \theta_k} \pi_i(\boldsymbol{\theta}) = \sum_{i=1}^{N} \frac{\partial \pi_i(\boldsymbol{\theta})}{\partial \theta_j} \frac{\partial \pi_i(\boldsymbol{\theta})}{\partial \theta_k} \frac{1}{\pi_i(\boldsymbol{\theta})}.$$

This is the (j, k) element of $\mathbf{A'A}$. Thus the asymptotic covariance is $\mathcal{J}^{-1} = (\mathbf{A'A})^{-1}$.

For results of this section to apply, a ML estimator of $\boldsymbol{\theta}$ must exist and be a solution of the likelihood equations. This requires the following *strong identifiability* condition: For every $\epsilon > 0$, there exists a $\delta > 0$ such that if $\|\boldsymbol{\theta} - \boldsymbol{\theta}_0\| > \epsilon$, then $\|\boldsymbol{\pi}(\boldsymbol{\theta}) - \boldsymbol{\pi}_0\| > \delta$. This condition implies a weaker one that two $\boldsymbol{\theta}$ values cannot have the same $\boldsymbol{\pi}$ value. When strong identifiability and the other regularity conditions hold, the probability an ML estimator is a root of the likelihood equations converges to 1 as $n \to \infty$. That estimator has the asymptotic properties given above of a solution of the likelihood equations. For proofs, see Birch (1964a) and Rao (1973, pp. 360–362).

14.2.2 Asymptotic Distribution of Cell Probability Estimators

The asymptotic distribution of the model-based estimator $\hat{\boldsymbol{\pi}}$ follows from the Taylor-series expansion

$$\hat{\boldsymbol{\pi}} = \boldsymbol{\pi}(\hat{\boldsymbol{\theta}}) = \boldsymbol{\pi}(\boldsymbol{\theta}_0) + \frac{\partial \boldsymbol{\pi}}{\partial \boldsymbol{\theta}_0}(\hat{\boldsymbol{\theta}} - \boldsymbol{\theta}_0) + o_p(n^{-1/2}). \qquad (14.17)$$

The size of the remainder term follows from $(\hat{\boldsymbol{\theta}} - \boldsymbol{\theta}_0) = O_p(n^{-1/2})$. Now $\boldsymbol{\pi}(\boldsymbol{\theta}_0) = \boldsymbol{\pi}_0$, and $\sqrt{n}\,(\hat{\boldsymbol{\theta}} - \boldsymbol{\theta}_0)$ is asymptotically normal with asymptotic covariance $(\mathbf{A'A})^{-1}$. By the delta method,

$$\sqrt{n}\,(\hat{\boldsymbol{\pi}} - \boldsymbol{\pi}_0) \xrightarrow{d} N\left[\mathbf{0}, \frac{\partial\boldsymbol{\pi}}{\partial\boldsymbol{\theta}_0}(\mathbf{A'A})^{-1}\frac{\partial\boldsymbol{\pi}}{\partial\boldsymbol{\theta}_0}{}'\right]. \qquad (14.18)$$

When the model holds with $\boldsymbol{\theta}$ having $q < N - 1$ elements, $\hat{\boldsymbol{\pi}} = \boldsymbol{\pi}(\hat{\boldsymbol{\theta}})$ is more efficient than the sample proportion $\mathbf{p}$ for estimating $\boldsymbol{\pi}$. More generally, for estimating a smooth function $g(\boldsymbol{\pi})$ of $\boldsymbol{\pi}$, $g(\hat{\boldsymbol{\pi}})$ has smaller asymptotic variance than $g(\mathbf{p})$. We next derive this result, discussed in Section 6.4.5. The derivation deletes the Nth component from $\mathbf{p}$ and $\hat{\boldsymbol{\pi}}$, so their covariance matrices are positive definite (Problem 14.16). The Nth proportion is linearly dependent on the first $N - 1$ since they sum to 1. Let $\boldsymbol{\Sigma} = \mathbf{diag}(\boldsymbol{\pi}) - \boldsymbol{\pi}\boldsymbol{\pi}'$ denote the $(N - 1) \times (N - 1)$ covariance matrix of $\sqrt{n}\,\mathbf{p}$. The inverse of $\boldsymbol{\Sigma}$ is

$$\boldsymbol{\Sigma}^{-1} = \mathbf{diag}(\boldsymbol{\pi})^{-1} + \mathbf{11'}\pi_N, \qquad (14.19)$$

which can be verified by evaluating $\boldsymbol{\Sigma}\boldsymbol{\Sigma}^{-1}$ and showing that it equals the identity matrix.

Let $(\partial g/\partial\boldsymbol{\pi}_0) = (\partial g/\partial\pi_1,\ldots,\partial g/\partial\pi_{N-1})'$, evaluated at $\boldsymbol{\pi} = \boldsymbol{\pi}_0$. By the delta method,

$$\text{asymp. var}\left[\sqrt{n}\,g(\mathbf{p})\right] = \left(\frac{\partial g}{\partial\boldsymbol{\pi}_0}\right)'\left[\text{cov}(\sqrt{n}\,\mathbf{p})\right]\frac{\partial g}{\partial\boldsymbol{\pi}_0} = \left(\frac{\partial g}{\partial\boldsymbol{\pi}_0}\right)'\boldsymbol{\Sigma}\frac{\partial g}{\partial\boldsymbol{\pi}_0}$$

and

$$\text{Asymp. var}\left[\sqrt{n}\,g(\hat{\boldsymbol{\pi}})\right] = \left(\frac{\partial g}{\partial\boldsymbol{\pi}_0}\right)'\left[\text{Asymp. cov}(\sqrt{n}\,\hat{\boldsymbol{\pi}})\right]\frac{\partial g}{\partial\boldsymbol{\pi}_0}$$

$$= \left(\frac{\partial g}{\partial\boldsymbol{\pi}_0}\right)'\frac{\partial\boldsymbol{\pi}}{\partial\boldsymbol{\theta}_0}\left[\text{Asymp. cov}(\sqrt{n}\,\hat{\boldsymbol{\theta}})\right]\left(\frac{\partial\boldsymbol{\pi}}{\partial\boldsymbol{\theta}_0}\right)'\frac{\partial g}{\partial\boldsymbol{\pi}_0}.$$

Using (14.11) and (14.19) yields

$$\text{Asymp. cov}(\sqrt{n}\,\hat{\boldsymbol{\theta}}) = (\mathbf{A'A})^{-1} = \left[(\partial\boldsymbol{\pi}/\partial\boldsymbol{\theta}_0)'\mathbf{diag}(\boldsymbol{\pi}_0)^{-1}(\partial\boldsymbol{\pi}/\partial\boldsymbol{\theta}_0)\right]^{-1}$$

$$= \left[(\partial\boldsymbol{\pi}/\partial\boldsymbol{\theta}_0)'\boldsymbol{\Sigma}^{-1}(\partial\boldsymbol{\pi}/\partial\boldsymbol{\theta}_0)\right]^{-1}.$$

Since $\boldsymbol{\Sigma}$ is positive definite and $(\partial\boldsymbol{\pi}/\partial\boldsymbol{\theta}_0)$ has rank q, $\boldsymbol{\Sigma}^{-1}$ and $[(\partial\boldsymbol{\pi}/\partial\boldsymbol{\theta}_0)'\boldsymbol{\Sigma}^{-1}(\partial\boldsymbol{\pi}/\partial\boldsymbol{\theta}_0)]^{-1}$ are also positive definite.

To show that asymp. $\text{var}[\sqrt{n}\,g(\mathbf{p})] \geq$ asymp. $\text{var}[\sqrt{n}\,g(\hat{\boldsymbol{\pi}})]$, we show that

$$\left(\frac{\partial g}{\partial \boldsymbol{\pi}_0}\right)'\left\{\boldsymbol{\Sigma} - \frac{\partial \boldsymbol{\pi}}{\partial \boldsymbol{\theta}_0}\left[\left(\frac{\partial \boldsymbol{\pi}}{\partial \boldsymbol{\theta}_0}\right)'\boldsymbol{\Sigma}^{-1}\frac{\partial \boldsymbol{\pi}}{\partial \boldsymbol{\theta}_0}\right]^{-1}\left(\frac{\partial \boldsymbol{\pi}}{\partial \boldsymbol{\theta}_0}\right)'\right\}\frac{\partial g}{\partial \boldsymbol{\pi}_0} \geq 0.$$

But this quadratic form is identical to

$$(\mathbf{Y} - \mathbf{B}\boldsymbol{\zeta})'\boldsymbol{\Sigma}^{-1}(\mathbf{Y} - \mathbf{B}\boldsymbol{\zeta})$$

where $\mathbf{Y} = \boldsymbol{\Sigma}(\partial g/\partial \boldsymbol{\pi}_0)$, $\mathbf{B} = (\partial \boldsymbol{\pi}/\partial \boldsymbol{\theta}_0)$, and $\boldsymbol{\zeta} = (\mathbf{B}'\boldsymbol{\Sigma}^{-1}\mathbf{B})^{-1}\mathbf{B}'\boldsymbol{\Sigma}^{-1}\mathbf{Y}$. The result then follows from the positive definiteness of $\boldsymbol{\Sigma}^{-1}$.

This proof is based on one given by Altham (1984). Her proof uses standard properties of ML estimators. It applies whenever regularity conditions hold that guarantee those properties. The proof applies not only to categorical data but to any situation in which a model describes the dependence of a set of parameters $\boldsymbol{\pi}$ on some smaller set $\boldsymbol{\theta}$.

14.3 ASYMPTOTIC DISTRIBUTIONS OF RESIDUALS AND GOODNESS-OF-FIT STATISTICS

We next study the distribution of Pearson X^2 and likelihood-ratio G^2 goodness-of-fit statistics for the multinomial model $\boldsymbol{\pi} = \boldsymbol{\pi}(\boldsymbol{\theta})$. We first derive the asymptotic joint distribution of the sample proportions $\mathbf{p}$ and model-based estimator $\hat{\boldsymbol{\pi}}$. This distribution determines large-sample distributions of statistics that depend on both $\mathbf{p}$ and $\hat{\boldsymbol{\pi}}$. For instance, it determines the asymptotic joint distribution of the Pearson residuals, which compare $\mathbf{p}$ with $\hat{\boldsymbol{\pi}}$. Deriving the large-sample chi-squared distribution for X^2, which is the sum of squared Pearson residuals, is then straightforward. We also show that X^2 and G^2 are asymptotically equivalent, when the model holds. Our presentation borrows from Bishop et al. (1975, Chap 14), Cox (1984), Cramér (1946, pp. 432–433), and Rao (1973, Sect. 6b).

14.3.1 Joint Asymptotic Normality of p and $\hat{\boldsymbol{\pi}}$

We first express the joint dependence of $\mathbf{p}$ and $\hat{\boldsymbol{\pi}}$ on $\mathbf{p}$, in order to show the joint asymptotic normality of $\mathbf{p}$ and $\hat{\boldsymbol{\pi}}$. Let

$$\mathbf{D} = \text{diag}(\boldsymbol{\pi}_0)^{1/2}\mathbf{A}(\mathbf{A}'\mathbf{A})^{-1}\mathbf{A}'\text{diag}(\boldsymbol{\pi}_0)^{-1/2}.$$

From (14.15) and (14.17),

$$\hat{\boldsymbol{\pi}} - \boldsymbol{\pi}_0 = \frac{\partial \boldsymbol{\pi}}{\partial \boldsymbol{\theta}_0}(\hat{\boldsymbol{\theta}} - \boldsymbol{\theta}_0) + \mathbf{o}_p(n^{-1/2})$$

$$= \mathbf{D}(\mathbf{p} - \boldsymbol{\pi}_0) + \mathbf{o}_p(n^{-1/2}).$$

Therefore,

$$\sqrt{n}\begin{pmatrix} \mathbf{p} - \boldsymbol{\pi}_0 \\ \hat{\boldsymbol{\pi}} - \boldsymbol{\pi}_0 \end{pmatrix} = \begin{pmatrix} \mathbf{I} \\ \mathbf{D} \end{pmatrix}\sqrt{n}\,(\mathbf{p} - \boldsymbol{\pi}_0) + \mathbf{o}_p(1)\,,$$

where $\mathbf{I}$ is a $N \times N$ identity matrix. By the delta method,

$$\sqrt{n}\begin{pmatrix} \mathbf{p} - \boldsymbol{\pi}_0 \\ \hat{\boldsymbol{\pi}} - \boldsymbol{\pi}_0 \end{pmatrix} \xrightarrow{d} N(\mathbf{0}, \boldsymbol{\Sigma}^*) \tag{14.20}$$

where

$$\boldsymbol{\Sigma}^* = \begin{pmatrix} \mathbf{diag}(\boldsymbol{\pi}_0) - \boldsymbol{\pi}_0\boldsymbol{\pi}_0' & [\mathbf{diag}(\boldsymbol{\pi}_0) - \boldsymbol{\pi}_0\boldsymbol{\pi}_0']\mathbf{D}' \\ \mathbf{D}[\mathbf{diag}(\boldsymbol{\pi}_0) - \boldsymbol{\pi}_0\boldsymbol{\pi}_0'] & \mathbf{D}[\mathbf{diag}(\boldsymbol{\pi}_0) - \boldsymbol{\pi}_0\boldsymbol{\pi}_0']\mathbf{D}' \end{pmatrix}. \tag{14.21}$$

The two matrix blocks on the main diagonal of $\boldsymbol{\Sigma}^*$ are $\mathrm{cov}(\sqrt{n}\,\mathbf{p})$ and asymp. $\mathrm{cov}(\sqrt{n}\,\hat{\boldsymbol{\pi}})$, derived previously. The new information here is that asymp. $\mathrm{cov}(\sqrt{n}\,\mathbf{p}, \sqrt{n}\,\hat{\boldsymbol{\pi}}) = [\mathbf{diag}(\boldsymbol{\pi}_0) - \boldsymbol{\pi}_0\boldsymbol{\pi}_0']\mathbf{D}'$.

14.3.2 Asymptotic Distribution of Pearson and Standardized Residuals

For cell counts $\{n_i\}$ the Pearson statistic is $X^2 = \Sigma e_i^2$, where

$$e_i = \frac{n_i - \hat{\mu}_i}{\hat{\mu}_i^{1/2}} = \frac{\sqrt{n}\,(p_i - \hat{\pi}_i)}{\hat{\pi}_i^{1/2}}.$$

We next derive the asymptotic distribution of $\mathbf{e} = (e_1, \ldots, e_N)'$, which is a diagnostic measure of lack of fit. For Poisson models it is the Pearson residual. Dividing it by its standard error gives the standardized residual. The distribution of $\mathbf{e}$ is also helpful in deriving the distribution of X^2.

The residuals $\mathbf{e}$ are functions of $\mathbf{p}$ and $\hat{\boldsymbol{\pi}}$, which are jointly asymptotically normal from (14.20). To use the delta method, we calculate

$$\partial e_i/\partial p_i = \sqrt{n}\,\hat{\pi}_i^{-1/2}, \qquad \partial e_i/\partial\hat{\pi}_i = -\sqrt{n}\,(p_i + \hat{\pi}_i)/2\hat{\pi}_i^{-3/2}$$

$$\partial e_i/\partial p_j = \partial e_i/\partial\hat{\pi}_j = 0 \qquad \text{for } i \neq j.$$

That is,

$$\frac{\partial\mathbf{e}}{\partial\mathbf{p}} = \sqrt{n}\,\mathbf{diag}(\hat{\boldsymbol{\pi}})^{-1/2} \qquad \text{and}$$

$$\frac{\partial\mathbf{e}}{\partial\hat{\boldsymbol{\pi}}} = -(\tfrac{1}{2})\sqrt{n}\,[\mathbf{diag}(\mathbf{p}) + \mathbf{diag}(\hat{\boldsymbol{\pi}})]\mathbf{diag}(\hat{\boldsymbol{\pi}})^{-3/2}. \tag{14.22}$$

Evaluated at $\mathbf{p} = \boldsymbol{\pi}_0$ and $\hat{\boldsymbol{\pi}} = \boldsymbol{\pi}_0$, these matrices equal $\sqrt{n}\,\mathbf{diag}(\boldsymbol{\pi}_0)^{-1/2}$ and $-\sqrt{n}\,\mathbf{diag}(\boldsymbol{\pi}_0)^{-1/2}$. Using (14.21), (14.22), and $\mathbf{A}'\boldsymbol{\pi}_0^{1/2} = \mathbf{0}$ [which follows from (14.11)], the delta method implies that

$$\mathbf{e} \xrightarrow{d} N\left(\mathbf{0}, \mathbf{I} - \boldsymbol{\pi}_0^{1/2} \boldsymbol{\pi}_0^{1/2\prime} - \mathbf{A}(\mathbf{A'A})^{-1}\mathbf{A'}\right). \qquad (14.23)$$

The limiting distribution has form $N(\mathbf{0}, \mathbf{I} - \mathbf{Hat})$, where **Hat** is the *hat matrix* (Section 4.5.5). Although asymptotically normal, **e** behaves less variably than standard normal random variables. The standardized Pearson residual (Haberman 1973a) divides **e** by its estimated standard error. This statistic, which *is* asymptotically standard normal, equals

$$r_i = \frac{e_i}{\left[1 - \hat{\pi}_i - \Sigma_j \Sigma_k (1/\hat{\pi}_i)\left(\partial \pi_i / \partial \hat{\theta}_j\right)\left(\partial \pi_i / \partial \hat{\theta}_k\right) \hat{v}^{jk}\right]^{1/2}}, \qquad (14.24)$$

where $\hat{v}^{jk}$ denotes the element in row j and column k of $(\hat{\mathbf{A}}'\hat{\mathbf{A}})^{-1}$. The denominator of r_i is $\sqrt{1 - \hat{h}_i}$, where the leverage $\hat{h}_i$ for observation i estimates the ith diagonal element of the hat matrix. This simplifies to (3.13) for testing independence in two-way tables.

14.3.3 Asymptotic Distribution of Pearson Statistic

The proof that the Pearson X^2 statistic has an asymptotic chi-squared distribution uses the following relationship between normal and chi-squared distributions (Rao 1973, p. 188):

> Let **X** be multivariate normal with mean $\boldsymbol{\nu}$ and covariance matrix **B**. A necessary and sufficient condition for $(\mathbf{X} - \boldsymbol{\nu})'\mathbf{C}(\mathbf{X} - \boldsymbol{\nu})$ to have a chi-squared distribution is $\mathbf{BCBCB} = \mathbf{BCB}$. The degrees of freedom equal the rank of **CB**.

When **B** is nonsingular, the condition simplifies to $\mathbf{CBC} = \mathbf{C}$.

The Pearson statistic relates to **e** by $X^2 = \mathbf{e'e}$, so we apply this result by identifying **X** with **e**, $\boldsymbol{\nu} = \mathbf{0}$, $\mathbf{C} = \mathbf{I}$, and $\mathbf{B} = \mathbf{I} - \boldsymbol{\pi}_0^{1/2} \boldsymbol{\pi}_0^{1/2\prime} - \mathbf{A}(\mathbf{A'A})^{-1}\mathbf{A'}\}$. Since $\mathbf{C} = \mathbf{I}$, the condition for $(\mathbf{X} - \boldsymbol{\nu}')\mathbf{C}(\mathbf{X} - \boldsymbol{\nu}) = \{\mathbf{e'e}\} = X^2$ to have a chi-squared distribution simplifies to $\mathbf{BBB} = \mathbf{BB}$. A direct computation using $\mathbf{A}'\boldsymbol{\pi}_0^{1/2} = \mathbf{0}$ shows that **B** is idempotent, so the condition holds. Since **e** is asymptotically multivariate normal, X^2 is asymptotically chi-squared.

For symmetric idempotent matrices, the rank equals the trace. The trace of **I** is N; the trace of $\boldsymbol{\pi}_0^{1/2} \boldsymbol{\pi}_0^{1/2\prime}$ equals the trace of $\boldsymbol{\pi}_0^{1/2\prime}\boldsymbol{\pi}_0^{1/2} = \Sigma \pi_{i0} = 1$, which is 1; the trace of $\mathbf{A}(\mathbf{A'A})^{-1}\mathbf{A'}$ equals the trace of $(\mathbf{A'A})^{-1}(\mathbf{A'A}) =$ identity matrix of size $q \times q$, which is q. Thus, the rank of $\mathbf{B} = \mathbf{CB}$ is $N - q - 1$, and the asymptotic chi-squared distribution has df $= N - q - 1$.

This result, due to Fisher (1922), is remarkably simple. When the sample size is large, the distribution of X^2 does not depend on $\boldsymbol{\pi}_0$ or the model form. It depends only on the difference between the dimension of $\boldsymbol{\pi}$, which is $N - 1$, and the dimension of $\boldsymbol{\theta}$. With $q = 0$ parameters, X^2 is Pearson's

(1900) statistic (1.15) for testing that multinomial probabilities equal certain specified values, and df $= N - 1$ as Pearson claimed. Watson (1959) showed that the same result holds for the asymptotic conditional distribution, given a sufficient statistic for nuisance parameters.

14.3.4 Asymptotic Distribution of Likelihood-Ratio Statistic

When the model holds, the likelihood-ratio statistic G^2 is asymptotically equivalent to X^2 as $n \to \infty$. To show this, we express

$$G^2 = 2 \sum_i n_i \log \frac{n_i}{\hat{\mu}_i} = 2n \sum_i p_i \log\left(1 + \frac{p_i - \hat{\pi}_i}{\hat{\pi}_i}\right)$$

and apply the expansion

$$\log(1 + x) = x - x^2/2 + x^3/3 - \cdots \quad \text{for } |x| < 1.$$

We identify x with $(p_i - \hat{\pi}_i)/\hat{\pi}_i$, which converges in probability to 0 when the model holds. For large n,

$$G^2 = 2n \sum_i [\hat{\pi}_i + (p_i - \hat{\pi}_i)] \left[\frac{p_i - \hat{\pi}_i}{\hat{\pi}_i} - \left(\frac{1}{2}\right)\frac{(p_i - \hat{\pi}_i)^2}{\hat{\pi}_i^2} + \cdots\right]$$

$$= 2n \sum_i \left[(p_i - \hat{\pi}_i) - \left(\frac{1}{2}\right)\frac{(p_i - \hat{\pi}_i)^2}{\hat{\pi}_i} + \frac{(p_i - \hat{\pi}_i)^2}{\hat{\pi}_i} + O_p(p_i - \hat{\pi}_i)^3\right]$$

$$= n \sum_i \frac{(p_i - \hat{\pi}_i)^2}{\hat{\pi}_i} + 2nO_p(n^{-3/2}) = X^2 + O_p(n^{-1/2}) = X^2 + o_p(1),$$

since $\sum(p_i - \hat{\pi}_i) = 0$ and $(p_i - \hat{\pi}_i) = (p_i - \pi_i) - (\hat{\pi}_i - \pi_i)$, both of which are $O_p(n^{-1/2})$. Thus, when the model holds, the difference between X^2 and G^2 converges in probability to 0. As a consequence, G^2, like X^2, has an asymptotic chi-squared distribution with df $= N - q - 1$.

The parameter value that maximizes the likelihood is the one that minimizes G^2. To show this, we let

$$G^2(\boldsymbol{\pi}; \mathbf{p}) = 2n \sum p_i \log(p_i/\pi_i).$$

The kernel of the multinomial log likelihood is

$$L(\boldsymbol{\theta}) = n \sum p_i \log \pi_i(\boldsymbol{\theta})$$

$$= -n \sum p_i \log \frac{p_i}{\pi_i(\boldsymbol{\theta})} + n \sum p_i \log p_i$$

$$= -\left(\frac{1}{2}\right) G^2(\boldsymbol{\pi}(\boldsymbol{\theta}); \mathbf{p}) + n \sum p_i \log p_i.$$

The second term in the last expression does not depend on $\boldsymbol{\theta}$, so maximizing $L(\boldsymbol{\theta})$ is equivalent to minimizing G^2 with respect to $\boldsymbol{\theta}$.

A fundamental result for G^2 concerns comparisons of nested models. Suppose that model M_0 is a special case of model M_1. Let q_0 and q_1 denote the numbers of parameters in the two models. Let $\{\hat{\pi}_{0i}\}$ and $\{\hat{\pi}_{1i}\}$ denote ML estimators of cell probabilities for the two models. Then

$$G^2(M_0) - G^2(M_1) = 2n \sum p_i \log(\hat{\pi}_{1i}/\hat{\pi}_{0i})$$

has the form of $-2(\log$ likelihood ratio$)$ for testing that M_0 holds against the alternative that M_1 holds. Theory for likelihood-ratio tests suggests that when the simpler model holds, the asymptotic distribution of $G^2(M_0) - G^2(M_1)$ is chi-squared with $q_1 - q_2$ degrees of freedom. For details, see Bishop et al. (1975, pp. 525–526), Haberman (1974a, p. 108), and Rao (1973, pp. 418–419). The statistic $X^2(M_0|M_1)$ defined in (9.4) is a quadratic approximation for the G^2 difference. Haberman (1977a) noted that these tests can perform well even for large, sparse tables, as long as $q_1 - q_0$ is small compared to the sample size and no expected frequency has larger order of magnitude than the others.

14.3.5 Asymptotic Noncentral Distributions

Results in this chapter assume that a certain parametric model holds. In practice, any unsaturated model almost surely does not hold perfectly, so one might question the scope of these results. This is not problematic if we regard models merely as convenient approximations for reality. For instance, the ML estimator $\hat{\boldsymbol{\theta}}$ converges to a value $\boldsymbol{\theta}_0$ that describes the best fit of the chosen model to reality. In this sense, inferences for $\boldsymbol{\theta}$ give us information about a useful approximation for reality. Similarly, model-based inferences about cell probabilities are inconsistent for the true probabilities when the model does not hold; nevertheless, those inferences are consistent for describing a useful smoothing of reality.

For goodness-of-fit statistics, a relevant distinction exists between limiting behavior when the model holds and when it does not hold. When the model holds, we've seen X^2 and G^2 have a limiting chi-squared distribution, and the difference between them disappears as n increases. When the model does not hold, X^2 and G^2 tend to grow unboundedly as n increases, and $|X^2 - G^2|$ need not go to zero. One method for obtaining proper limiting distributions considers a sequence of situations π_n for which the lack of fit diminishes as n increases. Specifically, the model is $\pi = \mathbf{f}(\boldsymbol{\theta})$, but in reality

$$\pi_n = \mathbf{f}(\boldsymbol{\theta}) + \boldsymbol{\delta}/\sqrt{n}. \tag{14.25}$$

The best fit of the model to the population has ith probability equal to $f_i(\boldsymbol{\theta})$, but the true value differs from that by $\delta_i/\sqrt{n}$.

For this representation, Mitra (1958) showed that the Pearson X^2 has a limiting noncentral chi-squared distribution, with df $= N - q - 1$ and noncentrality parameter

$$\lambda = n \sum_{i=1}^{n} \frac{\left[\pi_{ni} - f_i(\boldsymbol{\theta})\right]^2}{f_i(\boldsymbol{\theta})}.$$

This has the form of X^2, with the sample values p_i and $\hat{\pi}_i$ replaced by population values π_{ni} and $f_i(\boldsymbol{\theta})$. Similarly, the noncentrality of the likelihood-ratio statistic has the form of G^2, with the same substitution. Haberman (1974a, pp. 109–112) showed that under certain conditions G^2 and X^2 have the same limiting distribution; that is, their noncentrality values converge to a common value as $n \to \infty$.

Representation (14.25) means that for large n, the noncentral chi-squared approximation is valid when the model is just barely incorrect. In practice, it is often reasonable to adopt (14.25) for fixed, finite n to approximate the distribution of X^2, even though (14.25) would not be plausible as we obtain more data. The alternative representation

$$\boldsymbol{\pi} = \mathbf{f}(\boldsymbol{\theta}) + \boldsymbol{\delta} \tag{14.26}$$

in which $\boldsymbol{\pi}$ differs from $\mathbf{f}(\boldsymbol{\theta})$ by a *fixed* amount as $n \to \infty$ may seem more natural. In fact, this is more appropriate than (14.25) for proving the test to be consistent (i.e., for convergence to 1 of the probability of rejecting the hypothesis that the model holds). For (14.26), however, the noncentrality parameter λ grows unboundedly as $n \to \infty$, and a proper limiting distribution does not result for X^2 and G^2.

When the model holds, $\boldsymbol{\delta} = \mathbf{0}$ in either representation (14.25) or (14.26). That is, $\mathbf{f}(\boldsymbol{\theta}) = \boldsymbol{\pi}(\boldsymbol{\theta})$, $\lambda = 0$, and the results in Sections 14.3.3 and 14.3.4 apply.

14.4 ASYMPTOTIC DISTRIBUTIONS FOR LOGIT / LOGLINEAR MODELS

For loglinear models, formulas in Section 8.6 for the asymptotic covariance matrices of $\hat{\boldsymbol{\theta}}$ and $\hat{\boldsymbol{\pi}}$ are special cases of ones derived in Section 14.2. We present these for the multinomial form of the models, which relates directly to that section. Then we discuss the connection to Poisson loglinear models.

To constrain probabilities to sum to 1, we express loglinear models for multinomial sampling as

$$\boldsymbol{\pi} = \exp(\mathbf{X}\boldsymbol{\theta}) / \left[\mathbf{1}'\exp(\mathbf{X}\boldsymbol{\theta})\right] \tag{14.27}$$

where $\mathbf{X}$ is a model matrix and $\mathbf{1}' = (1, \ldots, 1)$. Letting $\mathbf{x}_i$ denote row i of $\mathbf{X}$,

$$\pi_i = \pi_i(\boldsymbol{\theta}) = \frac{\exp(\mathbf{x}_i\boldsymbol{\theta})}{\sum_k \exp(\mathbf{x}_k\boldsymbol{\theta})}.$$

14.4.1 Asymptotic Covariance Matrices

A model affects covariance matrices through the Jacobian. Since

$$\frac{\partial \pi_i}{\partial \theta_j} = \frac{\left[\sum_k \exp(\mathbf{x}_k \boldsymbol{\theta})\right]\left[\exp(\mathbf{x}_i \boldsymbol{\theta})\right] x_{ij} - \left[\exp(\mathbf{x}_i \boldsymbol{\theta})\right]\left[\sum_k x_{kj}\exp(\mathbf{x}_k \boldsymbol{\theta})\right]}{\left[\sum_k \exp(\mathbf{x}_k \boldsymbol{\theta})\right]^2}$$

$$= \pi_i x_{ij} - \pi_i \sum_k x_{kj}\pi_k ,$$

the matrix of these elements has the form

$$\partial \boldsymbol{\pi}/\partial \boldsymbol{\theta} = [\text{diag}(\boldsymbol{\pi}) - \boldsymbol{\pi}\boldsymbol{\pi}']\mathbf{X}.$$

Using this with (14.14) and (14.16), the information matrix at $\boldsymbol{\theta}_0$ is

$$\mathbf{A}'\mathbf{A} = (\partial \boldsymbol{\pi}/\partial \boldsymbol{\theta}_0)'\text{diag}(\boldsymbol{\pi}_0)^{-1}(\partial \boldsymbol{\pi}/\partial \boldsymbol{\theta}_0)$$

$$= \mathbf{X}'[\text{diag}(\boldsymbol{\pi}_0) - \boldsymbol{\pi}_0\boldsymbol{\pi}_0']'\text{diag}(\boldsymbol{\pi}_0)^{-1}[\text{diag}(\boldsymbol{\pi}_0) - \boldsymbol{\pi}_0\boldsymbol{\pi}_0']\mathbf{X}$$

$$= \mathbf{X}'[\text{diag}(\boldsymbol{\pi}_0) - \boldsymbol{\pi}_0\boldsymbol{\pi}_0']\mathbf{X}.$$

Thus, for multinomial loglinear models, $\hat{\boldsymbol{\theta}}$ is asymptotically normally distributed with estimated covariance matrix

$$\widehat{\text{cov}}(\hat{\boldsymbol{\theta}}) = \{\mathbf{X}'[\text{diag}(\hat{\boldsymbol{\pi}}) - \hat{\boldsymbol{\pi}}\hat{\boldsymbol{\pi}}']\mathbf{X}\}^{-1}/n. \tag{14.28}$$

Similarly, from (14.23) the estimated asymptotic covariance matrix of $\hat{\boldsymbol{\pi}}$ is

$$\widehat{\text{cov}}(\hat{\boldsymbol{\pi}}) = [\text{diag}(\hat{\boldsymbol{\pi}}) - \hat{\boldsymbol{\pi}}\hat{\boldsymbol{\pi}}']\mathbf{X}\{\mathbf{X}'[\text{diag}(\hat{\boldsymbol{\pi}}) - \hat{\boldsymbol{\pi}}\hat{\boldsymbol{\pi}}']\mathbf{X}\}^{-1}$$

$$\times \mathbf{X}'[\text{diag}(\hat{\boldsymbol{\pi}}) - \hat{\boldsymbol{\pi}}\hat{\boldsymbol{\pi}}']/n.$$

From (14.23), the Pearson residuals $\mathbf{e}$ are asymptotically normal with

$$\text{asymp. cov}(\mathbf{e}) = \mathbf{I} - \boldsymbol{\pi}_0^{1/2}(\boldsymbol{\pi}_0^{1/2})' - \mathbf{A}(\mathbf{A}'\mathbf{A})^{-1}\mathbf{A}'$$

$$= \mathbf{I} - \boldsymbol{\pi}_0^{1/2}(\boldsymbol{\pi}_0^{1/2})' - \text{diag}(\boldsymbol{\pi}_0)^{-1/2}[\text{diag}(\boldsymbol{\pi}_0) - \boldsymbol{\pi}_0\boldsymbol{\pi}_0']\mathbf{X}$$

$$\times \{\mathbf{X}'[\text{diag}(\boldsymbol{\pi}_0) - \boldsymbol{\pi}_0\boldsymbol{\pi}_0']\mathbf{X}\}^{-1}\mathbf{X}'$$

$$\times [\text{diag}(\boldsymbol{\pi}_0) - \boldsymbol{\pi}_0\boldsymbol{\pi}_0']\text{diag}(\boldsymbol{\pi}_0)^{-1/2}.$$

14.4.2 Connection with Poisson Loglinear Models

This book expressed loglinear models in terms of Poisson expected cell frequencies $\boldsymbol{\mu} = (\mu_1, \dots, \mu_N)'$, using formulas of the form

$$\log \boldsymbol{\mu} = \mathbf{X}_a \boldsymbol{\theta}_a. \tag{14.29}$$

The model matrix $\mathbf{X}_a$ and parameter vector $\boldsymbol{\theta}_a$ in this formula are slightly different from $\mathbf{X}$ and $\boldsymbol{\theta}$ in multinomial model (14.27). The Poisson expression (14.29) does not have constraints on $\boldsymbol{\mu}$. For multinomial model (14.27), $\sum_i \mu_i = n$ is fixed, and $\boldsymbol{\pi} = \boldsymbol{\mu}/n$ satisfies

$$\log \boldsymbol{\mu} = \log n\,\boldsymbol{\pi} = \mathbf{X}\boldsymbol{\theta} + \big[\log n - \log(\mathbf{1}'\exp(\mathbf{X}\boldsymbol{\theta}))\big]\mathbf{1}$$
$$= \mathbf{X}\boldsymbol{\theta} + \mathbf{1}\mu$$

where $\mu = \log n - \log(\mathbf{1}'\exp(\mathbf{X}\boldsymbol{\theta}))]$. In other words, multinomial model (14.27) implies Poisson model (14.29) with

$$\mathbf{X}_a = [\mathbf{1}:\mathbf{X}] \quad \text{and} \quad \boldsymbol{\theta}_a = (\mu, \boldsymbol{\theta}')'.$$

The columns of $\mathbf{X}$ in the multinomial representation must be linearly independent of $\mathbf{1}$; that is, the parameter μ, which relates to the total sample size, does not appear in $\boldsymbol{\theta}$. The dimension of $\boldsymbol{\theta}$ is 1 less than the number of parameters reported in this text for Poisson loglinear models. For instance, for the saturated model, $\boldsymbol{\theta}$ has $N - 1$ elements for the multinomial representation, reflecting the sole constraint on $\boldsymbol{\pi}$ of $\sum \pi_i = 1$.

NOTES

Section 14.1: Delta Method

14.1. For detailed discussion of large-sample theory including the delta method, see Bishop et al. (1975, Chap. 14) and Sen and Singer (1993).

14.2. In applying the delta method to a function g of an asymptotically normal random vector $\mathbf{T}_n$, suppose that the first-order, $\ldots, (a - 1)$st-order differentials of the function are zero at $\boldsymbol{\theta}$, but the ath-order differential is nonzero. A generalization of the delta method implies that $n^{a/2}[g(\mathbf{T}_n) - g(\boldsymbol{\theta})]$ has limiting distribution involving products of order a of components of a normal random vector. When $a = 2$, the limiting distribution is a quadratic form in a multivariate normal vector, which often relates to a chi-squared distribution; in the univariate case, it is $\sigma^2(g''(\theta))/2$ times a χ_1^2 variable (Casella and Berger 2001, p. 244).

Resampling methods such as the jackknife and the bootstrap are alternative tools for estimating standard errors and obtaining confidence intervals. They can be helpful when use of the delta method is questionable—for instance, for small samples, highly sparse data, or complex sampling designs. For details, see Davison and Hinkley (1997), Fay (1985), Parr and Tolley (1982), and Simonoff (1986).

Section 14.3: Asymptotic Distributions of Residuals and Goodness-of-Fit Statistics

14.3. If Y is Poisson with $E(Y) = \mu$, then for large μ the delta method implies $Y^{1/2}$ is approximately normal with standard deviation $\frac{1}{2}$. This motivates an alternative goodness-of-fit statistic, the *Freeman–Tukey statistic*, $\text{FT} = 4\Sigma(\sqrt{y}_i - \sqrt{\hat{\mu}_i})^2$. When the model holds, $\text{FT} - X^2$ is also $o_p(1)$ as $n \to \infty$. See Bishop et al. (1975, p. 514) for details.

Results of this chapter do not apply when the number of cells N grows as $n \to \infty$, or when different expected frequencies grow at different rates. Haberman (1988) showed the consistency of X^2 breaks down with non-standard asymptotics.

14.4. Drost et al. (1989) showed noncentral approximations using other sequences of alternatives than the local and fixed ones (14.25) and (14.26).

PROBLEMS

14.1 Explain why:

    a. If $c > 0$, $n^{-c} = o(1)$ as $n \to \infty$.

    b. If $c \neq 0$, cz_n has the same order as z_n; that is, $o(cz_n)$ is equivalent to $o(z_n)$ and $O(cz_n)$ is equivalent to $O(z_n)$.

    c. $o(y_n)o(z_n) = o(y_n z_n)$, $O(y_n)O(z_n) = O(y_n z_n)$, $o(y_n)O(z_n) = o(y_n z_n)$.

14.2 If X^2 has an asymptotic chi-squared distribution with fixed df as $n \to \infty$, then explain why $X^2/n = o_p(1)$.

14.3 **a.** Use Tchebychev's inequality to show that if $E(X_n) = \mu_n$ and $\text{var}(X_n) = \sigma_n^2 < \infty$, then $(X_n - \mu_n) = O_p(\sigma_n)$.

    b. Suppose that $Y_1, \ldots, Y_n$ are independent with $E(Y_i) = \mu$ and $\text{var}(Y_i) = \sigma^2$ for $i = 1, \ldots, n$. Let $\bar{Y}_n = (\Sigma_i Y_i)/n$. Apply part (a) to show that $\bar{Y}_n - \mu = O_p(n^{-1/2})$.

14.4 Let Y be a Poisson random variable with mean μ.

    a. For a constant $c > 0$, show that

$$E[\log(Y + c)] = \log \mu + \left(c - \tfrac{1}{2}\right)/\mu + O(\mu^{-2})$$

    (*Hint:* Note that $\log(Y + c) = \log \mu + \log[1 + (Y + c - \mu)/\mu]$.)

    b. Cell counts in a 2×2 table are independent Poisson random variables. Use part (a) to argue that to reduce bias in estimating the log odds ratio, a sensible estimator is the sample log odds ratio after adding $\tfrac{1}{2}$ to each cell.

14.5 Let p denote the sample proportion for n independent Bernoulli trials. Find the asymptotic distribution of the estimator $[p(1 - p)]^{1/2}$ of the standard deviation. What happens when $\pi = 0.5$?

14.6 Suppose that T_n has a Poisson distribution with mean $\lambda = n\mu$, for fixed $\mu > 0$. For large n, show that the distribution of $\log T_n$ is approximately normal with mean $\log(\lambda)$ and variance λ^{-1}. [*Hint:* By the central limit theorem, T_n/n is approximately $N(\mu, \mu/n)$ for large n.]

14.7 **a.** Refer to Problem 14.6. If T_n is Poisson, show $\sqrt{T_n}$ has asymptotic variance $\frac{1}{4}$.

   b. For a binomial sample with n trials and sample proportion p, show the asymptotic variance of $\sin^{-1}(\sqrt{p})$ is $1/4n$. [This transformation and the one in part (a) are *variance stabilizing*, producing variates with asymptotic variances that are the same for all values of the parameter. Traditionally, these transformations were employed to make ordinary least squares applicable to count data. See Cochran 1940 for discussion and ML analyses.]

14.8 For a multinomial $(n, \{\pi_i\})$ distribution, show the correlation between p_i and p_j is $-[\pi_i\pi_j/(1 - \pi_i)(1 - \pi_j)]^{1/2}$. What does this equal when $\pi_i = 1 - \pi_j$ and $\pi_k = 0$ for $k \neq i, j$?

14.9 An animal population has N species, with population proportion π_i of species i. *Simpson's index of ecological diversity* (Simpson 1949) is $I(\boldsymbol{\pi}) = 1 - \Sigma\pi_i^2$. [Rao (1982) surveyed diversity measures.]

   a. Two animals are randomly chosen from the population, with replacement. Show $I(\boldsymbol{\pi})$ is the probability they are different species.

   b. For proportions $\mathbf{p}$ for a random sample, show that the estimated asymptotic standard error of $I(\mathbf{p})$ is

$$2\left\{\left[\sum_i p_i^3 - \left(\sum_i p_i^2\right)^2\right]\Big/n\right\}^{1/2}.$$

14.10 Let $\{Y_i\}$ be independent Poisson random variables. Show by the delta method that the estimated asymptotic variance of $\Sigma a_i\log(Y_i)$ is $\Sigma a_i^2/y_i$. [This formula applies to ML estimators of parameters for the saturated loglinear model, which are contrasts of $\{\log(y_i)\}$. Formula (14.9) yields the asymptotic covariance structure of such estimators; see Lee (1977).]

14.11 Assuming two independent binomial samples, derive the asymptotic standard error of the log relative risk (Section 3.1.4).

14.12 Refer to Problem 3.27. The sample size may need to be quite large for the sampling distribution of $\hat{\gamma}$ to be approximately normal, especially if $|\gamma|$ is large. The Fisher-type transform $\hat{\xi} = \frac{1}{2}\log[(1 + \hat{\gamma})/(1 - \hat{\gamma})]$ (Agresti 1984, pp. 166–167, 177; O'Gorman and Woolson 1988) converges more quickly to normality.

a. Show that the asymptotic variance of $\hat{\xi}$ equals the asymptotic variance of $\hat{\gamma}$ multiplied by $(1 - \gamma^2)^{-2}$.

b. Explain how to construct a confidence interval for ξ and use it to obtain one for γ.

c. Show that $\hat{\xi} = \frac{1}{2} \log(C/D)$. For 2×2 tables, show that this is half the log odds ratio.

14.13 Let $\phi^2(\mathbf{T}) = \Sigma_i (T_i - \pi_{i0})^2 / \pi_{i0}$. Then $\phi^2(\mathbf{p}) = X^2/n$, where X^2 is the Pearson statistic (1.15) for testing $H_0: \pi_i = \pi_{i0}$, $i = 1, \ldots, N$, and $n\phi^2(\boldsymbol{\pi})$ is the noncentrality for that test when $\boldsymbol{\pi}$ is the true value. Under H_0, why does the delta method not yield an asymptotic normal distribution for $\phi^2(\mathbf{p})$? (See Note 14.2.)

14.14 In an $I \times J$ contingency table, let θ_{ij} denote local odds ratio (2.10), and let $\hat{\theta}_{ij}$ denote its sample value.

a. Show that asymp. $\text{cov}(\sqrt{n} \log \hat{\theta}_{ij}, \sqrt{n} \log \hat{\theta}_{i+1, j}) = -[\pi_{i+1, j}^{-1} + \pi_{i+1, j+1}^{-1}]$.

b. Show that asymp. $\text{cov}(\sqrt{n} \log \hat{\theta}_{ij}, \sqrt{n} \log \hat{\theta}_{i+1, j+1}) = \pi_{i+1, j+1}^{-1}$.

c. When $\hat{\theta}_{ij}$ and $\hat{\theta}_{hk}$ use mutually exclusive sets of cells, show that asymp. $\text{cov}(\sqrt{n} \log \hat{\theta}_{ij}, \sqrt{n} \log \hat{\theta}_{hk}) = 0$.

d. State the asymptotic distribution of $\log \hat{\theta}_{ij}$.

14.15 For loglinear model (XY, XZ, YZ), ML estimates of $\{\mu_{ijk}\}$ and hence the X^2 and G^2 statistics are not direct. Alternative approaches may yield direct analyses. For $2 \times 2 \times 2$ tables, find a statistic for testing the hypothesis of no three-factor interaction, using the delta method with the asymptotic normality of $\log \hat{\theta}_{111}$, where

$$\hat{\theta}_{111} = \frac{p_{111} \, p_{221} / p_{121} \, p_{211}}{p_{112} \, p_{222} / p_{122} \, p_{212}}.$$

14.16 Refer to Section 14.2.2, with $\boldsymbol{\Sigma} = \text{diag}(\boldsymbol{\pi}) - \boldsymbol{\pi}\boldsymbol{\pi}'$ the covariance matrix of $\sqrt{n}(p_1, \ldots, p_{N-1})'$. Let

$$Z = \begin{cases} c_i & \text{with probability } \pi_i, \quad i = 1, \ldots, N - 1 \\ 0 & \text{with probability } \pi_N \end{cases}$$

and let $\mathbf{c} = (c_1, \ldots, c_{N-1})'$.

a. Show that $E(Z) = \mathbf{c}'\boldsymbol{\pi}$, $E(Z^2) = \mathbf{c}'\text{diag}(\boldsymbol{\pi})\mathbf{c}$, and $\text{var}(Z) = \mathbf{c}'\boldsymbol{\Sigma}\mathbf{c}$.

b. Suppose that at least one $c_i \neq 0$, and all $\pi_i > 0$. Show $\text{var}(Z) > 0$, and deduce that $\boldsymbol{\Sigma}$ is positive definite.

   c. If $\boldsymbol{\pi} = (\pi_1, \ldots, \pi_N)'$, so $\boldsymbol{\Sigma}$ is $N \times N$, prove that $\boldsymbol{\Sigma}$ is not positive definite.

14.17 Consider the model for a 2×2 table, $\pi_{11} = \theta^2$, $\pi_{12} = \pi_{21} = \theta(1 - \theta)$, $\pi_{22} = (1 - \theta)^2$, where θ is unknown (Problems 3.31 and 10.34).

   a. Find the matrix $\mathbf{A}$ in (14.14) for this model.

   b. Use $\mathbf{A}$ to obtain the asymptotic variance of $\hat{\theta}$. (As a check, it is simple to find it directly using the inverse of $-E\partial^2 L / \partial\theta^2$, where L is the log likelihood.) For which θ value is the variance maximized? What is the distribution of $\hat{\theta}$ if $\theta = 0$ or $\theta = 1$?

   c. Find the asymptotic covariance matrix of $\sqrt{n}\,\hat{\boldsymbol{\pi}}$.

   d. Find df for testing fit using X^2.

14.18 Refer to the model for the calf data in Section 1.5.6. Obtain the asymptotic variance of $\hat{\pi}$.

14.19 Justify the use of *estimated* asymptotic covariance matrices. For instance, for large samples, why is $\hat{\mathbf{A}}'\hat{\mathbf{A}}$ close to $\mathbf{A}'\mathbf{A}$?

14.20 Cell counts $\{Y_i\}$ are independent Poisson random variables, with $\mu_i = E(Y_i)$. Consider the Poisson loglinear model

$$\log \boldsymbol{\mu} = \mathbf{X}_a \boldsymbol{\theta}_a, \quad \text{where } \boldsymbol{\mu} = (\mu_1, \ldots, \mu_N).$$

Using arguments similar to those in Section 14.2, show that the large-sample covariance matrix of $\hat{\boldsymbol{\theta}}_a$ can be estimated by $[\mathbf{X}_a' \text{diag}(\hat{\boldsymbol{\mu}})\mathbf{X}_a]^{-1}$, where $\hat{\boldsymbol{\mu}}$ is the ML estimator of $\boldsymbol{\mu}$.

14.21 For a given set of parameter constraints, show that weak identifiability conditions hold for the independence loglinear model for a two-way table; that is, when two values for $\boldsymbol{\theta}$ give the same $\boldsymbol{\pi}$, those parameter vectors must be identical.

14.22 Use the delta method, with derivatives (14.22), to derive the asymptotic covariance matrix in (14.23) for residuals. Show that this matrix is idempotent.

14.23 In some situations, X^2 and G^2 take very similar values. Explain the joint influence on this event of (**a**) whether the model holds, (**b**) whether the sample size n is large, and (**c**) whether the number of cells N is large.

14.24 Show $\mathbf{X}$ and $\boldsymbol{\theta}$ in multinomial representation (14.27) for the independence model for an $I \times J$ table. By contrast, show $\mathbf{X}_a$ for the corresponding Poisson loglinear model (14.29).

14.25 Using (14.18) and (14.28), derive the asymptotic $\widehat{\text{cov}}(\hat{\boldsymbol{\pi}})$ for a multinomial loglinear model.

14.26 Consider the ML estimator $\hat{\pi}_{ij} = p_{i+}p_{+j}$ of π_{ij} for the independence model, when that model does not hold. Show that $E(p_{i+}p_{+j}) = \pi_{i+}\pi_{+j}(n-1)/n + \pi_{ij}/n$. To what does $\hat{\pi}_{ij}$ converge as n increases?

14.27 Let ζ denote a generic measure of association. For K independent multinomial samples of sizes $\{n_k\}$, suppose that $\sqrt{n_k}(\hat{\zeta}_k - \zeta_k) \overset{d}{\to} N(0, \sigma_k^2)$ as $n_k \to \infty$. A summary measure is

$$\bar{\zeta} = \frac{\Sigma_k(n_k/\hat{\sigma}_k^2)\hat{\zeta}_k}{\Sigma_k(n_k/\hat{\sigma}_k^2)}.$$

a. Show that $\Sigma_k z_k^2 = V + [\bar{\zeta}^2/\hat{\sigma}^2(\bar{\zeta})]$, where

$$V = \sum_k \frac{n_k(\hat{\zeta}_k - \bar{\zeta})^2}{\hat{\sigma}_k^2}, \quad z_k = \frac{n_k^{1/2}\hat{\zeta}_k}{\hat{\sigma}_k}, \quad \hat{\sigma}^2(\bar{\zeta}) = \left(\sum_k \frac{n_k}{\hat{\sigma}_k^2}\right)^{-1}.$$

b. Suppose that $n \to \infty$ with $n_k/n \to \rho_k > 0$, $k = 1, \ldots, K$. State the asymptotic chi-squared distribution for each component in the partitioning in part (a). Indicate the hypothesis that each tests.

CHAPTER 15

Alternative Estimation Theory for Parametric Models

In this book we have used the maximum likelihood (ML) approach to inference. This is by far the most common approach for categorical data analysis. Other paradigms have been used, however. In this chapter we discuss some of them. These methods have similar asymptotic properties as maximum likelihood, so the large-sample theory of Chapter 14 applies also to them.

In Section 15.1 we discuss weighted least squares for fitting models for categorical data. This and related quasi-likelihood methods introduced in Sections 4.7 and 11.4 are sometimes simpler to apply than ML.

The Bayesian paradigm is increasingly popular as computations become easier to implement. A full discussion of modern developments with this approach is beyond our scope, but in Section 15.2 we present Bayesian methods of estimating cell probabilities in a contingency table. Four other methods of estimation for categorical data are described in the final section.

15.1 WEIGHTED LEAST SQUARES FOR CATEGORICAL DATA

Weighted least squares (WLS) is an extension of ordinary least squares that permits responses to be correlated and to have nonconstant variance. Familiarity with the WLS method is useful because:

1. WLS computations have a standard form that is simple to apply for a wide variety of models.

2. Algorithms for calculating ML estimates often consist of iterative use of WLS. An example is the Fisher scoring method for generalized linear models (Section 4.6.3).

3. When the model holds, WLS and ML estimators are asymptotically equivalent, both falling in the class of best asymptotically normal (BAN)

estimators. For large samples, the estimators are approximately normally distributed around the parameter value, and the ratio of their variances converges to 1.

Grizzle, Starmer, and Koch (1969) popularized WLS for categorical data analyses. In honor of them, WLS for such analyses is often called the *GSK method*. This section summarizes the ingredients of this approach.

15.1.1 Notation and Preliminaries for WLS Approach

For a response variable Y with J categories, consider multinomial samples of sizes $n_1, \ldots, n_I$ at I levels of an explanatory variable or combinations of levels of several explanatory variables. Let $\boldsymbol{\pi} = (\boldsymbol{\pi}_1', \ldots, \boldsymbol{\pi}_I')'$, where

$$\boldsymbol{\pi}_i = (\pi_{1|i}, \pi_{2|i}, \ldots, \pi_{J|i})' \quad \text{with} \quad \sum_j \pi_{j|i} = 1$$

denotes the conditional distribution of Y at level i. Let $\mathbf{p}$ denote corresponding sample proportions, with $\mathbf{V}$ their $IJ \times IJ$ covariance matrix. When the I samples are independent,

$$\mathbf{V} = \begin{bmatrix} \mathbf{V}_1 & & & \mathbf{0} \\ & \mathbf{V}_2 & & \\ & & \ddots & \\ \mathbf{0} & & & \mathbf{V}_I \end{bmatrix}$$

From Section 14.1.4, the covariance matrix of $\sqrt{n_i}\,\mathbf{p}_i$ is

$$n_i \mathbf{V}_i = \begin{bmatrix} \pi_{1|i}(1 - \pi_{1|i}) & -\pi_{1|i}\pi_{2|i} & \cdots & -\pi_{1|i}\pi_{J|i} \\ -\pi_{2|i}\pi_{1|i} & \pi_{2|i}(1 - \pi_{2|i}) & \cdots & -\pi_{2|i}\pi_{J|i} \\ \vdots & \vdots & & \vdots \\ -\pi_{J|i}\pi_{1|i} & -\pi_{J|i}\pi_{2|i} & \cdots & \pi_{J|i}(1 - \pi_{J|i}) \end{bmatrix}.$$

Each set of proportions has $(J - 1)$ linearly independent elements.

Let $\mathbf{F}$ be a vector of $u \le I(J - 1)$ response functions

$$\mathbf{F}(\boldsymbol{\pi}) = [F_1(\boldsymbol{\pi}), \ldots, F_u(\boldsymbol{\pi})]'.$$

The WLS approach applies to linear models for $\mathbf{F}$ of form

$$\mathbf{F}(\boldsymbol{\pi}) = \mathbf{X}\boldsymbol{\beta}, \tag{15.1}$$

where $\boldsymbol{\beta}$ is a $q \times 1$ vector of parameters and $\mathbf{X}$ is a $u \times q$ model matrix of known constants having rank q. From Section 8.5.4, loglinear and logit response functions are special cases of $\mathbf{F}(\boldsymbol{\pi}) = \mathbf{C}\log(\mathbf{A}\boldsymbol{\pi})$ for certain matrices $\mathbf{C}$ and $\mathbf{A}$.

Let $\mathbf{F}(\mathbf{p})$ denote the sample response functions. We assume that $\mathbf{F}$ has continuous second-order partial derivatives in an open region containing $\boldsymbol{\pi}$. This assumption enables the delta method to determine the large-sample normal distribution for $\mathbf{F}(\mathbf{p})$. The asymptotic covariance matrix of $\mathbf{F}(\mathbf{p})$ depends on the $u \times IJ$ matrix

$$\mathbf{Q} = \frac{\partial F_k(\boldsymbol{\pi})}{\partial \pi_{j|i}}$$

for $k = 1, \ldots, u$ and all IJ combinations (i, j). Linear response models have response functions of form $\mathbf{F}(\boldsymbol{\pi}) = \mathbf{A}\boldsymbol{\pi}$ for a matrix of known constants $\mathbf{A}$, in which case $\mathbf{Q} = \mathbf{A}$. For the generalized loglinear model $\mathbf{F}(\boldsymbol{\pi}) = \mathbf{C}\log(\mathbf{A}\boldsymbol{\pi})$ (recall Sections 8.5.4 and 11.2.5), $\mathbf{Q} = \mathbf{C}[\text{diag}(\mathbf{A}\boldsymbol{\pi})]^{-1}\mathbf{A}$. [See Magnus and Neudecker 1988 for matrix differential calculus.] By the multivariate delta method (Section 14.1.5), the asymptotic covariance matrix of $\mathbf{F}(\mathbf{p})$ is

$$\mathbf{V}_F = \mathbf{Q}\mathbf{V}\mathbf{Q}'.$$

Let $\hat{\mathbf{V}}_F$ denote the sample version of $\mathbf{V}_F$, substituting sample proportions in $\mathbf{Q}$ and $\mathbf{V}$. For subsequent formulas, this matrix must be nonsingular.

15.1.2   Inference Using the WLS Approach to Model Fitting

For the general model (15.1), the WLS estimate of $\boldsymbol{\beta}$ is

$$\mathbf{b} = \left(\mathbf{X}'\hat{\mathbf{V}}_F^{-1}\mathbf{X}\right)^{-1}\mathbf{X}'\hat{\mathbf{V}}_F^{-1}\mathbf{F}(\mathbf{p}).$$

This is the $\boldsymbol{\beta}$ value that minimizes the quadratic form

$$[\mathbf{F}(\mathbf{p}) - \mathbf{X}\boldsymbol{\beta}]'\hat{\mathbf{V}}_F^{-1}[\mathbf{F}(\mathbf{p}) - \mathbf{X}\boldsymbol{\beta}].$$

The ordinary least squares estimate, for uncorrelated responses with constant variance, results when $\hat{\mathbf{V}}_F$ is a constant multiple of the identity matrix.

The WLS estimator has an asymptotic multivariate normal distribution, with estimated covariance matrix

$$\widehat{\text{cov}}(\mathbf{b}) = \left(\mathbf{X}'\hat{\mathbf{V}}_F^{-1}\mathbf{X}\right)^{-1}.$$

The normal distribution improves as the sample size increases and $\mathbf{F}(\mathbf{p})$ is more nearly normally distributed.

The estimate **b** yields predicted values $\hat{\mathbf{F}} = \mathbf{Xb}$ for the response functions. Since they satisfy the model, these predicted values are smoother than the sample response functions $\mathbf{F(p)}$. When the model holds, $\hat{\mathbf{F}}$ is asymptotically better than $\mathbf{F(p)}$ as an estimator of $\mathbf{F}(\boldsymbol{\pi})$ (Section 14.2.2). The estimated covariance matrix of the predicted values is

$$\hat{\mathbf{V}}_{\hat{F}} = \mathbf{X}\left(\mathbf{X}'\hat{\mathbf{V}}_F^{-1}\mathbf{X}\right)^{-1}\mathbf{X}'.$$

The test of model goodness of fit uses the residual term

$$W = [\mathbf{F(p)} - \mathbf{Xb}]'\hat{\mathbf{V}}_F^{-1}[\mathbf{F(p)} - \mathbf{Xb}] = \mathbf{F(p)}'\hat{\mathbf{V}}_F^{-1}\mathbf{F(p)} - \mathbf{b}'\left(\mathbf{X}'\hat{\mathbf{V}}_F^{-1}\mathbf{X}\right)\mathbf{b},$$

which compares the sample response functions with their model predicted values. Under H_0: $\mathbf{F}(\boldsymbol{\pi}) - \mathbf{X}\boldsymbol{\beta} = \mathbf{0}$ that the model holds, W is asymptotically chi-squared with df $= u - q$, the difference between the number of response functions and the number of model parameters.

One can more closely check the model fit by studying the residuals, $\mathbf{F(p)} - \hat{\mathbf{F}}$. They are orthogonal to the fit $\hat{\mathbf{F}}$, so

$$\text{cov}[\mathbf{F(p)}] = \text{cov}\left\{\left[\mathbf{F(p)} - \hat{\mathbf{F}}\right] + \hat{\mathbf{F}}\right\} = \text{cov}\left[\mathbf{F(p)} - \hat{\mathbf{F}}\right] + \text{cov}(\hat{\mathbf{F}}) .$$

Thus, the estimated covariance matrix of the residuals equals

$$\text{cov}[\mathbf{F(p)}] - \text{cov}(\hat{\mathbf{F}}) = \hat{\mathbf{V}}_F - \hat{\mathbf{V}}_{\hat{F}} = \hat{\mathbf{V}}_F - \mathbf{X}\left(\mathbf{X}'\hat{\mathbf{V}}_F^{-1}\mathbf{X}\right)^{-1}\mathbf{X}'.$$

Dividing the residuals by their standard errors yields standardized residuals having large-sample standard normal distributions.

Hypotheses about contrasts and other effects of explanatory variables have form H_0: $\mathbf{C}\boldsymbol{\beta} = \mathbf{0}$, where $\mathbf{C}$ is a known $c \times q$ matrix with $c \le q$, having rank c. The estimator $\mathbf{Cb}$ of $\mathbf{C}\boldsymbol{\beta}$ is asymptotically normal with mean $\mathbf{0}$ under H_0 and with covariance matrix estimated by $\mathbf{C}(\mathbf{X}'\hat{\mathbf{V}}_F^{-1}\mathbf{X})^{-1}\mathbf{C}'$. The Wald statistic

$$W_C = \mathbf{b}'\mathbf{C}'\left[\mathbf{C}\left(\mathbf{X}'\hat{\mathbf{V}}_F^{-1}\mathbf{X}\right)\mathbf{C}'\right]^{-1}\mathbf{Cb} \tag{15.2}$$

has an approximate chi-squared null distribution with df $= c$. This statistic also equals the difference between residual chi-squared statistics for the reduced model implied by H_0 and the full model. For the special case H_0: $\beta_i = 0$, $W_C = b_i^2/\text{var}(b_i)$ has df $= 1$.

15.1.3 Scope of WLS versus ML Estimation

The WLS approach requires estimating the multinomial covariance matrix of sample responses at each setting of the explanatory variables. It is inapplicable when explanatory variables are continuous, since there may be only one observation at each such setting. WLS also becomes less appropriate as the number of explanatory variables increases, since few observations may occur at each of the many combinations of settings. By contrast, in principle, continuous explanatory variables or many explanatory settings are not problematic to ML.

When a certain model holds, with large cell expected frequencies ML and WLS give similar results. Both estimators are in the class of best asymptotically normal estimators. However, practical considerations often favor ML estimation. For example, zero cell counts often adversely affect the WLS approach. The sample response functions may then be ill-defined or have a singular estimated covariance matrix.

WLS shares with quasi-likelihood the feature that inferential results depend only on specifying a model for the mean responses and specifying a variance function and covariance structure (here, based on the multinomial). It does not use the likelihood function for the complete distribution. Thus, inference uses Wald methods.

Historically, an advantage of the WLS approach was computational simplicity. This is not relevant now that software is available for ML analyses and for extensions of WLS (e.g., quasi-likelihood methods such as GEE) that do not have some of its disadvantages. Thus, WLS is now used much less frequently than it was about 25 years ago. Nonetheless, it has close connections with more sophisticated methods. Some algorithms for calculating ML estimates iteratively use WLS. Also, Miller et al. (1993) showed that under certain conditions the solution of the first iteration in the GEE fitting process gives the WLS estimate. This equivalence uses initial estimates based directly on sample values and assumes a saturated association structure that allows a separate correlation parameter for each pair of response categories and each pair of observations in a cluster. In this sense, GEE is an iterated form of WLS. Moreover, in this case, the covariance matrix for the estimates is the same in both approaches.

15.2   BAYESIAN INFERENCE FOR CATEGORICAL DATA

Methodology using the Bayesian paradigm has advanced tremendously in the past decade. New computational methods make it easier to evaluate posterior distributions for model parameters. Nonetheless, Bayesian inference is not as fully developed or commonly used for categorical data analysis as in many other areas of statistics. For multiway contingency table analysis, partly this is because of the plethora of parameters for multinomial models, often necessitating substantial prior specification. Bayesian theory and methods are beyond the scope of this book. We present only relatively elementary problems in which the Bayesian approach applies quite naturally and is sometimes more appealing than ML. We then briefly summarize more complex developments.

The first applications of Bayesian methods to contingency tables involved smoothing cell counts to improve estimation of cell probabilities (e.g., Good 1965). The sample proportions are ordinary ML estimators for the saturated model. When data are sparse, these can have undesirable features. Large sparse tables often contain many sampling zeros, for which 0.0 is unappealing as a probability estimate. In addition, Stein's results for estimating multivariate normal means suggest that lower total mean-squared error occurs with Bayes estimators that shrink the sample proportions toward some average value (Efron and Morris 1975).

In considering Bayesian estimators, we cannot hope to find one that is uniformly better than ML. For instance, suppose that a true cell probability $\pi_i = 0$. Then the sample proportion $p_i = 0$ with probability 1, and the sample proportion is better than any other estimator. Because parameter values exist for which the sample proportion is optimal, no other estimator is uniformly better over the entire parameter space. Here the criterion of comparison is the expected value of a *loss function* that measures distance between the estimator and the parameter, such as squared error. In decision-theoretic terms the sample proportion is an *admissible* estimator, for standard loss functions (Johnson 1971). In this sense, the sample mean for the multinomial or multivariate binomial differs from the sample mean for the multivariate normal, which is inadmissible (dominated by Bayes estimators) when the dimension of the mean vector is at least three (Ferguson 1967, p. 170). Meeden et al. (1998) gave related results for decomposable loglinear models.

Another approach for estimating cell probabilities fits an unsaturated model. Often, though, there is no particular model expected to describe the table well. For $I \times J$ cross-classifications of nominal variables, for instance, the independence model rarely fits well. When unsaturated models approximate the true relationship poorly, model-based estimators also have undesirable properties. Although they smooth the data, the smoothing is too severe for large samples. The model-based estimators are inconsistent, converging to values that may be far from the true cell probabilities as n increases.

A Bayesian approach to estimating cell probabilities compromises between sample proportions and model-based estimators. A model still provides part of the smoothing mechanism, with the Bayes estimators shrinking the sample proportions toward a set of proportions satisfying the model.

15.2.1 Bayesian Estimation of Binomial Parameter

We illustrate basic ideas with Bayesian inference for a binomial parameter. Let y denote a bin(n, π) variate. Since π falls between 0 and 1, a natural prior density for π is the beta [(13.8) in Section 13.3.1] for some choice of $\alpha > 0$ and $\beta > 0$. This satisfies $E(\pi) = \alpha/(\alpha + \beta)$.

In Bayesian inference the posterior density of a parameter, given the data, is proportional to the product of the prior density with the likelihood function. Here, the beta prior depends on π through $\pi^{\alpha-1}(1 - \pi)^{\beta-1}$, and the binomial likelihood has kernel depending on π through $\pi^y(1 - \pi)^{n-y}$. Thus, the posterior density $h(\pi|y)$ of π is proportional to

$$h(\pi|y) \propto \left[\pi^y(1 - \pi)^{n-y}\right]\left[\pi^{\alpha-1}(1 - \pi)^{\beta-1}\right] = \pi^{y+\alpha-1}(1 - \pi)^{n-y+\beta-1},$$

for $0 \le \pi \le 1$. The beta is the conjugate prior distribution. The posterior density is also beta, with parameters $\alpha^* = y + \alpha$ and $\beta^* = n - y + \beta$.

The mean of the posterior distribution is a Bayesian estimator of a parameter. This is optimal when a squared-error loss function $(T - \pi)^2$ describes the consequence of estimating π by an estimator T (Ferguson 1967, p. 46). The mean of the beta posterior distribution for π is

$$E(\pi|y) = \alpha^*/(\alpha^* + \beta^*) = (y + \alpha)/(n + \alpha + \beta)$$

$$= w(y/n) + (1 - w)[\alpha/(\alpha + \beta)],$$

where $w = n/(n + \alpha + \beta)$. This is a weighted average of the sample proportion $p = y/n$ and the mean of the prior distribution. For fixed (α, β), the weight given the sample increases as n increases. The standard deviation of the posterior distribution describes the accuracy of this estimator. This equals the square root of

$$\text{var}(\pi|y) = \alpha^*\beta^*/(\alpha^* + \beta^*)^2(\alpha^* + \beta^* + 1).$$

For large n the standard deviation is roughly $\sqrt{p(1 - p)/n}$, the ordinary standard error for the ML estimator $\hat{\pi} = p$.

The Bayes estimator requires selecting parameters (α, β) for the prior distribution. Complete ignorance about π might suggest a uniform prior distribution. This is the beta distribution with $\alpha = \beta = 1$. The posterior distribution then has the same shape as the binomial likelihood function. The Bayes estimator is then

$$E(\pi|y) = (y + 1)/(n + 2).$$

This shrinks the sample proportion slightly toward $\frac{1}{2}$.

Alternatively, a popular prior with Bayesians is the *Jeffreys prior*. This is proportional to the square root of the determinant of the Fisher information matrix for the parameters of interest, for a single observation. With a single parameter θ, this is $[E(\partial^2 \log f(y|\theta)/\partial\theta^2)]^{1/2}$. In the binomial case with $\theta = \pi$ and $n = 1$, this equals $[\pi(1 - \pi)]^{-1/2}$ and the prior is beta with $\alpha = \beta = .5$. Brown et al. (2001) showed that the posterior generated by this prior yields a confidence interval for π with good performance. It approximates the Clopper–Pearson interval with the mid-P adjustment (Sections

1.4.4 and 1.4.5). For a test of H_0: $\pi \geq \frac{1}{2}$ against H_a: $\pi < \frac{1}{2}$, a Bayesian *P*-value is the posterior probability that $\pi \geq \frac{1}{2}$. Routledge (1994) showed that with the Jeffreys prior, this posterior probability approximately equals the one-sided mid-*P*-value for the ordinary binomial test.

15.2.2 Dirichlet Prior and Posterior for Multinomial Parameters

These ideas generalize from the binomial to the multinomial (Good 1965). Suppose that cell counts $(n_1, \ldots, n_N)$ have a multinomial distribution with $n = \Sigma n_i$ and parameters $\boldsymbol{\pi} = (\pi_1, \ldots, \pi_N)'$. The multinomial likelihood is proportional to

$$\prod_{i=1}^{N} \pi_i^{n_i}.$$

For a prior distribution over potential $\boldsymbol{\pi}$ values, the multivariate generalization of the beta is the *Dirichlet density*

$$g(\boldsymbol{\pi}) = \frac{\Gamma(\Sigma \beta_i)}{\Pi_i \Gamma(\beta_i)} \prod_{i=1}^{N} \pi_i^{\beta_i - 1} \quad \text{for } 0 \leq \pi_i \leq 1 \text{ all } i, \qquad \sum_i \pi_i = 1,$$

where $\{\beta_i > 0\}$. For it, $E(\pi_i) = \beta_i / (\Sigma_j \beta_j)$.

The posterior density is also Dirichlet, with parameters $\{n_i + \beta_i\}$. The Bayes estimator of π_i is

$$E(\pi_i \mid n_1, \ldots, n_N) = (n_i + \beta_i) \Big/ \Big(n + \sum_j \beta_j\Big). \qquad (15.3)$$

Let $K = \Sigma \beta_j$ and $\gamma_i = E(\pi_i) = \beta_i / K$. The $\{\gamma_i\}$ are prior guesses for the cell probabilities. Bayes estimator (15.3) equals the weighted average

$$[n/(n + K)] p_i + [K/(n + K)] \gamma_i. \qquad (15.4)$$

From (15.3) the Bayes estimator is a sample proportion when the prior information corresponds to $\Sigma_j \beta_j$ trials with β_i outcomes of type i, $i = 1, \ldots, N$. This interpretation may provide guidance for choosing $\{\beta_i\}$. The Jeffreys prior sets all $\beta_i = 0.5$. Good referred to K as a *flattening constant*, since with identical $\{\beta_i\}$ (15.4) shrinks each sample proportion toward the uniform value $\gamma_i = 1/N$. Greater flattening occurs as K increases, for fixed n. Hierarchical models treat $\{\beta_i\}$ as unknown and specify a second-stage prior for them (e.g., Albert and Gupta 1982).

Bayes estimators combine good characteristics of sample proportions and model-based estimators. Like sample proportions and unlike model-based estimators, they are consistent even when the model does not hold. Unless the model holds, the weight given the sample proportion increases to 1.0 as the sample size increases. Like model-based estimators and unlike sample proportions, the Bayes estimators smooth the data. The resulting estimates, although slightly biased, usually have smaller total mean-squared error than the sample proportions.

15.2.3 Development of Bayesian Methods for Categorical Data

We now summarize the development of Bayesian methods for categorical data since Good's (1965) work on smoothing multinomial proportions. Leonard and Hsu (1994) provided a more detailed review. We begin with methods for two-way contingency tables.

For 2×2 tables, Altham (1969) gave a Bayesian analysis comparing parameters for two independent binomial samples. She tested $H_0 : \pi_1 \leq \pi_2$ against $\pi_1 > \pi_2$ using independent beta(α_i, β_i) priors for π_1 and π_2. Altham showed that the P-value that is the posterior probability that $\pi_1 \leq \pi_2$ can equal the one-sided P-value for Fisher's exact test. This happens when one uses improper prior distributions $(\alpha_1, \beta_1) = (1, 0)$ and $(\alpha_2, \beta_2) = (0, 1)$. These represent prior belief favoring the null hypothesis, in effect penalizing against concluding that $\pi_1 > \pi_2$. That is, Fisher's exact test corresponds to a conservative prior distribution.

If $\alpha_i = \beta_i = \gamma$, $i = 1, 2$, with $0 \leq \gamma \leq 1$, Altham showed that the Bayesian P-value is smaller than the Fisher P-value. The difference between the two is no greater than the null probability of the observed data. Use of Jeffreys priors with $\alpha_i = \beta_i = 0.5$ provides a type of continuity correction to Fisher's exact test in much the way the mid-P-value does for the frequentist approach. Howard (1998) showed that with these priors the posterior probability that $\pi_1 \leq \pi_2$ approximates the one-sided P-value for the large-sample z test using pooled variance (i.e., the signed square root of the Pearson statistic; see Problem 3.30) for testing $H_0 : \pi_1 = \pi_2$ against $H_a : \pi_1 > \pi_2$. Howard also discussed other priors for 2×2 tables, including ones that treat π_1 and π_2 as dependent.

Altham (1971) showed Bayesian analyses for binomial proportions from matched-pairs data. For a simple model in which the probability of success is the same for each subject at a given occasion, she again showed that the classical exact P-value (Section 10.1.4, using the binomial distribution) is a Bayesian P-value for a prior distribution favoring H_0. For a model similar to (10.8) in which the probability varies by subject but the occasion effect is constant, she showed that the Bayesian evidence against the null is weaker as the number of pairs giving the same response at both occasions increases, for fixed values of the numbers of pairs giving different responses at the two occasions. This differs from the conditional ML result, which does not depend on such pairs (Section 10.2.3). Ghosh et al. (2000) showed related results.

The Bayesian approaches presented so far focused directly on cell probabilities by using a prior distribution for them. Lindley (1964) did this with $I \times J$ contingency tables. He considered the posterior distribution of contrasts of log probabilities, such as the log odds ratio. An alternative approach (Laird 1978; Leonard 1975) focused on parameters of the saturated loglinear model, using normal priors. This is not a conjugate prior, but normal distributions can approximate the posterior. Using independent normal $N(0, \sigma^2)$ distributions for the association parameters is a way of inducing shrinkage toward the independence model (Laird 1978). A hierarchical approach puts second-stage priors on the parameters of the prior distribution (Leonard 1975).

Historically, a barrier for the Bayesian approach has been the difficulty of calculating the posterior distribution when the prior is not conjugate. This is less problematic with modern ways of approximating posterior distributions by simulating samples from them. These include the importance sampling generalization of Monte Carlo simulation (Zellner and Rossi 1984) and Markov chain Monte Carlo methods such as Gibbs sampling (Gelfand and Smith 1990). Zellner and Rossi used Bayesian methods for logistic regression and Gelfand and Smith considered a class of multinomial models with Dirichlet prior. Zeger and Karim (1991) fitted generalized linear mixed models (GLMMs) essentially using a Bayesian framework with priors for fixed and random effects.

The focus on distributions for random effects in GLMMs in articles such as Zeger and Karim (1991) led to the treatment of parameters in GLMs as random variables with a fully Bayesian approach. Dey et al. (2000) edited a collection of articles that provided Bayesian analyses for GLMs. For instance, in that volume Gelfand and Ghosh surveyed the subject, Albert and Ghosh reviewed item response modeling, Chib modeled correlated binary data, and Chen and Dey modeled correlated ordinal data.

Bayesian methods are used increasingly in applications. For instance, Skene and Wakefield (1990) modeled multicenter binary response studies with a logit model that allows the treatment–response log odds ratio to vary among centers. This gives a Bayesian alternative to the GLMM analysis presented in Section 12.3.4. Daniels and Gatsonis (1999) used multi-level GLMs to analyze geographic and temporal trends with clustered longitudinal binary data. This built on hierarchical modeling ideas introduced by Wong and Mason (1985). An article by Landrum and Normand in Dey et al. (2000) gave a case study using Bayesian ordinal probit and logit models. Chaloner and Larntz (1989) used a Bayesian approach to determining optimal design for experiments using logistic regression. J. Albert has suggested Bayesian models for a variety of categorical data analyses. For instance, Albert (1997) modeled associations in two-way tables and Albert and Chib (1993) studied binary regression modeling, focusing on the probit case with extensions to ordered multinomial responses.

15.2.4 Data-Dependent Choice of Prior Distribution

With Bayesian analyses, careful prior specification is necessary. The use of an improper prior, such as the uniform prior over the entire or positive real line, sometimes results in improper posteriors. One may not realize this from the output of software for Bayesian fitting. In addition, with simulation methods it may not be obvious when convergence has occurred. Be suspicious if results are dramatically different from ordinary ML frequentist results.

Some dislike the subjectivity of the Bayesian approach inherent in selecting a prior distribution. Instead of choosing particular parameters for a prior distribution, it is increasingly popular to use a hierarchical approach in which those parameters themselves have a second-stage prior distribution. Alternatively, the empirical Bayes approach lets the data suggest parameter values for use in the prior distribution (e.g., Efron and Morris 1975). This approach uses the prior that maximizes the marginal probability of the observed data, integrating out with respect to the prior. Laird (1978) did this for the loglinear model, estimating σ^2 in normal priors for association parameters by finding the value that maximizes an approximation for the marginal distribution of the cell counts, evaluated at the observed data. A disadvantage of empirical Bayes compared to the hierarchical approach is that it does not take into account the source of variability due to substituting estimates for prior parameters.

Fienberg and Holland (1973) proposed analyses for contingency tables with data-dependent priors. For a particular choice of Dirichlet means $\{\gamma_i\}$ for the Bayes estimator (15.4), they showed that the minimum total mean-squared error occurs when

$$K = \left(1 - \sum \pi_i^2\right) \Big/ \left[\sum (\gamma_i - \pi_i)^2\right]. \tag{15.5}$$

The optimal $K = K(\gamma, \pi)$ depends on π, so they used the estimate $K(\gamma, \mathbf{p})$ of K in which the sample proportion $\mathbf{p}$ replaces π. As $\mathbf{p}$ falls closer to the prior guess γ, $K(\gamma, \mathbf{p})$ increases and the prior guess receives more weight in the posterior estimate. They selected the prior pattern $\{\gamma_i\}$ for the cell probabilities based on the fit of a simple model. For two-way tables, they used the independence fit $\{\gamma_{ij} = p_{i+}p_{+j}\}$. The Bayes estimator then shrinks sample proportions toward that fit.

As in other inference, Bayesian modeling should normally account for any ordering in the response categories. For instance, in the method just mentioned for smoothing contingency tables, one could shrink toward an ordinal model.

15.3   OTHER METHODS OF ESTIMATION

In this final section we describe some alternative estimation methods for categorical data. Consider estimation of π or θ, assuming a model $\pi = \pi(\theta)$. Let $\tilde{\theta}$ denote a generic estimator of θ, for which $\tilde{\pi} = \pi(\tilde{\theta})$ estimates π. The ML estimator $\hat{\theta}$ maximizes the likelihood. It also minimizes the deviance statistic G^2 comparing observed and fitted proportions (Section 14.3.4).

15.3.2   Minimum Chi-Squared Estimators

Other estimators minimize other measures of distance between $\pi(\theta)$ and p. The value $\tilde{\theta}$ that minimizes the Pearson statistic

$$X^2[\pi(\theta), p] = n \sum \frac{[p_i - \pi_i(\theta)]^2}{\pi_i(\theta)}$$

is called the *minimum chi-squared* estimate. It is simpler to calculate the estimate that minimizes the modified statistic

$$X^2_{\text{mod}}[\pi(\theta), p] = n \sum \frac{[p_i - \pi_i(\theta)]^2}{p_i} \tag{15.6}$$

that replaces the denominator by the sample proportion. This is called the *minimum modified chi-squared* estimate. It is the solution for θ to the equations

$$\sum_i \frac{\pi_i(\theta)}{p_i}\left(\frac{\partial \pi_i(\theta)}{\partial \theta_j}\right) = 0, \qquad j = 1, \ldots, q.$$

Neyman (1949) introduced minimum modified chi-squared estimators. He showed that they and minimum chi-squared estimators are best asymptotically normal (BAN) estimators. When the model holds, they are asymptotically (as $n \to \infty$) equivalent to ML estimators. Under the model, different estimation methods (ML, WLS, minimum chi-squared, etc.) yield nearly identical estimates of parameters when n is large. This happens partly because the estimators are consistent, converging in probability to θ as n increases. When the model does not hold, estimates for different methods can be quite different, even when n is large. The estimators converge to values for which the model gives the best approximation to reality, and this approximation is different when best is defined in terms of minimizing G^2 rather than minimizing X^2 or some other measure.

For any n, minimum modified chi-squared estimates are sometimes identical to WLS estimates. The connection refers to an alternative way of specifying a model, using a set of *constraint equations* for π,

$$\{g_j(\pi_1, \ldots, \pi_N) = 0\}.$$

For instance, for an $I \times J$ table, the $(I - 1)(J - 1)$ constraint equations

$$\log \pi_{ij} - \log \pi_{i,j+1} - \log \pi_{i+1,j} + \log \pi_{i+1,j+1} = 0$$

specify the model of independence. The number of constraint equations equals the residual df for the model.

Neyman (1949) noted that minimum modified chi-squared estimates result from minimizing

$$\sum_{i=1}^{N} \frac{(p_i - \pi_i)^2}{p_i} + \sum_{j=1}^{N-q} \lambda_j g_j(\pi_1, \ldots, \pi_N)$$

with respect to π, where the $\{\lambda_j\}$ are Lagrange multipliers. When the constraint equations are linear in π, the resulting estimating equations are linear. Then Bhapkar (1966) showed that these estimators are identical to WLS estimators. The statistic (15.6) then equals the WLS residual statistic (Section 15.1.2) for testing model fit.

Usually, however, constraint equations are nonlinear in π, such as for the independence model. The WLS estimator is then the minimum modified chi-squared estimator based on a linearized version of the constraints,

$$g_j(\mathbf{p}) + (\pi - \mathbf{p})' \partial g_j(\pi) / \partial \pi = 0,$$

with differential vector evaluated at $\mathbf{p}$.

Berkson (1944, 1955, 1980) was a strong advocate of minimum chi-squared methods. For logistic regression, his *minimum logit chi-squared* estimators minimized a weighted sum of squares between sample logits and linear predictions. Mantel (1985) criticized such methods, noting that their consistency requires group sizes to grow large, whereas ML (or conditional ML, when there are many nuisance parameters) is consistent however information goes to the limit (see also Problem 15.14).

15.3.2 Minimum Discrimination Information

Kullback (1959) formulated estimation by *minimum discrimination information* (MDI). The discrimination information for two probability vectors π and γ is

$$I(\pi; \gamma) = \sum_{i=1}^{N} \pi_i \log(\pi_i / \gamma_i). \tag{15.7}$$

This directed measure of distance between π and γ is nonnegative, equaling 0 only when $\pi = \gamma$. Gokhale and Kullback (1978) studied MDI estimates that minimize $I(\pi; \gamma)$, subject to model constraints, using $\gamma = \mathbf{p}$ for some problems and γ with $\gamma_1 = \gamma_2 = \cdots = \gamma_N = 1/N$ for others. Good (1963) conducted related work in the area of *maximum entropy*.

In some cases with $\{\gamma_i = 1/N\}$, the MDI estimator is identical to the ML estimator (Simon 1973). With $\gamma = \mathbf{p}$ it is not ML, but it has similar asymptotic properties, being best asymptotically normal (BAN). Then Gokhale and Kullback recommended testing goodness of fit using twice the minimized value of $I(\pi; \mathbf{p})$. This statistic reverses the roles of $\mathbf{p}$ and π relative to G^2, much as X_{mod}^2 in (15.6) reverses their roles relative to X^2. Both statistics fall in the class of power divergence statistics (Cressie and Read 1984; see also Problem 3.34) and have similar asymptotic properties. More generally, one could choose any member of the power divergence statistics and define estimates to be the values minimizing it. Under regularity conditions, they are all BAN.

15.3.3 Kernel Smoothing

Kernel estimation is a smoothing method that estimates a probability density or mass function without assuming a parametric distribution. Let $\mathbf{K}$ denote a matrix containing nonnegative elements and having column sums equal to 1. Kernel estimates of cell probabilities in a contingency table have form

$$\tilde{\pi} = \mathbf{Kp}. \tag{15.8}$$

For unordered multinomials with N categories, Aitchison and Aitken (1976) used

$$k_{ij} = \lambda, \quad i = j$$

$$= (1 - \lambda)/(N - 1), \quad i \neq j$$

for $(1/N) \leq \lambda \leq 1$. The resulting kernel estimator of π has form

$$(1 - \alpha)\mathbf{p} + \alpha(1/N), \tag{15.9}$$

where $\alpha = N(1 - \lambda)/(N - 1)$. This estimator shrinks the sample proportion toward $(1/N, \ldots, 1/N)$. As λ decreases from 1 to $1/N$, the smoothing parameter α increases from 0 to 1. Brown and Rundell (1985) proved that when no $\pi_i = 1$, $\lambda < 1$ exists such that the total mean squared error is smaller for this kernel estimator than for the sample proportions. Results for other shrinkage estimators applied to multivariate means suggest that the improvement for the kernel estimator can be large when n is small and the true cell probabilities are roughly equal.

Brown and Rundell generalized kernel smoothing for multiway contingency tables that may contain both nominal and ordinal variables. For a

T-way table, let $\mathbf{L}_k$ be a stochastic matrix (i.e., row and column sums equal to 1) with elements

$$\ell_{k,ij} = \begin{cases} \lambda_k, & i = j \\ d_k(i,j)(1 - \lambda_k), & i \neq j, \end{cases}$$

$k = 1, \ldots, T$. They let $\mathbf{K}$ in (15.8) be the Kronecker product

$$\mathbf{K} = \mathbf{L}_1 \otimes \cdots \otimes \mathbf{L}_T.$$

When variable k is ordinal, shrinkage alone is not enough, and it helps to borrow information from nearby cells. Then $d_k(i, j)$ is chosen to be smaller for greater distances between categories i and j. If variable k is nominal, the natural choice is $d_k(i, j) = 1/(I_k - 1)$, where I_k is the number of categories for variable k. For fixed $\{\lambda_k\}$, collapsing the smoothed table gives the same result as smoothing the corresponding collapsing of the original table. With $\{\lambda_k = \lambda, k = 1, \ldots, T\}$, Brown and Rundell described ways of finding λ to minimize an unbiased estimate of the total mean squared error.

Dong and Simonoff (1995) and Simonoff (1986) described other approaches for ordered categories. Most such kernels yield probability estimates of the form

$$\tilde{\pi}_i = (1 - \alpha)p_i + \alpha \times \text{smoother}_i,$$

where the smoothing is designed to work well when true probabilities in nearby cells are similar.

15.3.4 Penalized Likelihood

Good and Gaskins (1971) introduced the *penalized likelihood* method for density estimation. For log likelihood $L(\boldsymbol{\pi})$, the estimator maximizes

$$L^*(\boldsymbol{\pi}) = L(\boldsymbol{\pi}) - \alpha(\boldsymbol{\pi})$$

where $\alpha(\cdot)$ is a function that provides a roughness penalty. That is, $\alpha(\boldsymbol{\pi})$ decreases as elements of $\boldsymbol{\pi}$ are smoother, in some sense. The penalized likelihood estimator has a Bayesian interpretation. With prior density proportional to $\exp[-\alpha(\boldsymbol{\pi})]$, the posterior density is proportional to the penalized likelihood function. Hence, the mode of the posterior distribution equals the penalized likelihood estimator.

Simonoff (1983) applied penalized likelihood to estimating cell probabilities $\boldsymbol{\pi}$. Like Bayesian and kernel methods, it provides estimates that are smoother than the sample proportions. For a single multinomial with ordered categories, Simonoff (1983) used penalty function $\alpha(\boldsymbol{\pi}) = \lambda \sum_{i=1}^{N-1}(\log \pi_i - \log \pi_{i+1})^2$, which encourages adjacent category estimates to be similar. For two-way contingency tables, Simonoff suggested using $\alpha(\boldsymbol{\pi}) = \lambda\Sigma_i\Sigma_j(\log \theta_{ij})^2$ with the local odds ratios. This provides shrinkage toward the independence estimator. One chooses the smoothing parameter λ to minimize an approximation for the mean-squared error of the estimator.

In evaluating smoothing methods such as kernel smoothing and penalized likelihood, it is useful to distinguish between large-sample asymptotics with a fixed number of cells N and sparse-data asymptotics for which N grows with n (recall Section 6.3.4). For the former, these smoothing methods and Bayesian inference behave asymptotically like ordinary ML (i.e., the sample proportions). They have the same rate of convergence to true probabilities. These methods then improve over ML primarily for small samples, where the benefit of "borrowing from the whole" occurs. For sparse-data asymptotics, however, smoothing is particularly beneficial. As the dimensions of a table increase, the number of cells grows exponentially and the "curse of dimensionality" occurs. Accurate estimation becomes more difficult, with estimators converging more slowly to true values. The table then has an increasing proportion of empty cells. Smoothing can be better than ML even asymptotically. For such results, see Fienberg and Holland (1973) for the Dirichlet-based Bayes multinomial estimator and Simonoff (1983) for penalized likelihood with the multinomial. Simonoff showed that consistency can occur with the latter estimator in the sense that $\sup_i|\hat{\pi}_i/\pi_i - 1| \xrightarrow{p} 0$ as n and N grow and the probabilities themselves approach 0.

For surveys of smoothing methods, see Fahrmeir and Tutz (2001, Chap. 5), Lloyd (1999, Chap. 5), and Simonoff (1996, Chap. 6; 1998). As Simonoff noted, all smoothing methods attempt to balance the low bias of undersmoothing with the low variability of oversmoothing. The methods require input from the user about the degree of smoothness, whether it be determined by a prior distribution or some type of smoothing parameter.

In summary, many methods exist for smoothing categorical data. Besides those discussed in this section, there are traditional model-building methods. Some of these, such as generalized additive models (Section 4.8), are also specifically directed toward smoothing. A particular type of smoothing method may seem most natural for a given application. An advantage of the Bayesian approach is that its entire formulation seems less ad hoc than some others.

NOTES

Section 15.1: Weighted Least Squares for Categorical Data

15.1. Applications of WLS include fitting mean response models (Grizzle et al. 1969) and models for marginal distributions (Koch et al. 1977). For general discussion, see Bhapkar and Koch (1968), Imrey et al. (1981), and Koch et al. (1985).

Section 15.2: Bayesian Inference for Categorical Data

15.2. Other literature on Bayesian analyses of categorical responses includes Fienberg et al. (1999), Forster and Smith (1998), Good (1976), Knuiman and Speed (1988), Spiegelhalter and Smith (1982), and Walley (1996).

Section 15.3: Other Methods of Estimation

15.3. For further discussion of minimum chi-squared methods, see Bhapkar (1966), Koch et al. (1985), Neyman (1949), and Rao (1963).

15.4. For the use of minimum discrimination information, see Gokhale and Kullback (1978), Ireland and Kullback (1968a, b), Ireland et al. (1969), and Ku et al. (1971).

15.5. Hall and Titterington (1987) studied rates of convergence for multinomial kernel estimators. They defined one that achieves the optimal rate. Ordinary kernel estimators tend to be biased toward zero at the boundary of a table. Dong and Simonoff (1994) dealt with improving kernel estimates on the boundary of large sparse tables. Kernel methods are also useful for discrete regression modeling. For binary response data, Copas (1983) used one to display in a nonparametric manner the dependence of $P(Y = 1)$ on x.

PROBLEMS

Applications

15.1 Consider the mean response model fitted in Section 7.4.6. Show how to use WLS for this analysis. Identify the number of multinomial samples I, the number of response categories J, the response functions $\mathbf{F}$, the model matrix $\mathbf{X}$, the parameter vector $\boldsymbol{\beta}$, and the estimated covariance matrix $\hat{\mathbf{V}}_F$.

15.2 Use WLS to conduct the longitudinal analysis of depression in Section 11.2.1. Using software (e.g., SAS: PROC CATMOD), obtain WLS estimates and standard errors and compare to the ML results.

15.3 Refer to Problem 15.2. Using these data, describe the differences between **(a)** WLS and ML, and **(b)** WLS and GEE methods for marginal models with multivariate categorical response data.

15.4 Using data from Section 1.4.3, obtain a Bayesian estimate of the proportion of vegetarians. Explain how you chose the prior distribution. Compare results to those with ML.

15.5 Refer to Table 9.8. Consider the model that simultaneously assumes (9.12) as well as linear logit relationships for the marginal effects of age on breathlessness and on wheeze.

a. Specify $\mathbf{C}$, $\mathbf{A}$, and $\mathbf{X}$ for which this model has form $\mathbf{C}\log\mathbf{A}\boldsymbol{\pi} = \mathbf{X}\boldsymbol{\beta}$.

b. Using software, fit the model and interpret estimates.

Theory and Methods

15.6 Consider marginal homogeneity for an $I \times I$ table.

a. Letting $\mathbf{F}(\boldsymbol{\pi}) = \mathbf{A}\boldsymbol{\pi}$, explain how (i) $\mathbf{F}(\boldsymbol{\pi}) = \mathbf{0}$, where $\mathbf{A}$ has $I - 1$ rows, and (ii) $\mathbf{F}(\boldsymbol{\pi}) = \mathbf{X}\boldsymbol{\beta}$, where $\mathbf{A}$ has $2(I - 1)$ rows and $\boldsymbol{\beta}$ has $I - 1$ elements. In part (ii), show $\mathbf{A}$, $\boldsymbol{\pi}$, $\mathbf{X}$, $\boldsymbol{\beta}$ when $I = 3$.

b. Explain how to use WLS to test marginal homogeneity. [This is Bhapkar's test (10.16).]

15.7 For WLS with $\mathbf{F}(\boldsymbol{\pi}) = \mathbf{C}[\log(\mathbf{A}\boldsymbol{\pi})]$, show that $\mathbf{Q} = \mathbf{C}[\mathrm{diag}(\mathbf{A}\boldsymbol{\pi})]^{-1}\mathbf{A}$.

15.8 With WLS, show that $[\mathbf{F}(\mathbf{p}) - \mathbf{X}\boldsymbol{\beta}]'\hat{\mathbf{V}}_F^{-1}[\mathbf{F}(\mathbf{p}) - \mathbf{X}\boldsymbol{\beta}]$ is minimized by $\boldsymbol{\beta} = (\mathbf{X}'\hat{\mathbf{V}}_F^{-1}\mathbf{X})^{-1}\mathbf{X}'\hat{\mathbf{V}}_F^{-1}\mathbf{F}(\mathbf{p})$.

15.9 The response functions $\mathbf{F}(\mathbf{p})$ have asymptotic covariance matrix $\mathbf{V}_F$. Derive the asymptotic covariance matrix of the WLS model parameter estimator $\mathbf{b}$ and the predicted values $\hat{\mathbf{F}} = \mathbf{X}\mathbf{b}$.

15.10 Consider the Bayes estimator of a binomial parameter π using a beta prior distribution.

a. Does any beta prior distribution produce a Bayes estimator that coincides with the ML estimator?

b. Show that the ML estimator is a limit of Bayes estimators, for a certain sequence of beta prior parameter values.

c. Find an improper prior density (one for which its integral is not finite) such that the Bayes estimator coincides with the ML estimator. (In this sense, the ML estimator is a *generalized Bayes estimator*.)

d. For Bayesian inference using loss function $w(\theta)(T - \theta)^2$, the Bayes estimator of θ is the posterior expected value of $\theta w(\theta)$ divided by the posterior expected value of $w(\theta)$ (Ferguson 1967, p. 47). With loss function $(T - \pi)^2/[\pi(1 - \pi)]$, show the ML estimator of π is a Bayes estimator for the uniform prior distribution.

e. The risk function is the expected loss, treated as a function of π. For the loss function in part (d), show the risk function is constant. (Bayes' estimators with constant risk are *minimax*; their maximum risk is no greater than the maximum risk for any other estimator.)

f. Show that the Jeffreys prior for π equals the beta density with $\alpha = \beta = .5$.

15.11 For the Dirichlet prior for multinomial probabilities, show the posterior expected value of π_i is (15.3). Derive the expression for this Bayes estimator as a weighted average of p_i and $E(\pi_i)$.

15.12 For Bayes estimator (15.4), show that the total mean squared error is

$$[K/(n + K)]^2 \left[\sum (\pi_i - \gamma_i)^2 \right] + [n/(n + K)]^2 \left(1 - \sum \pi_i^2\right).$$

Show that (15.5) is the value of K that minimizes this.

15.13 Refer to Problem 15.6. For marginal homogeneity, explain why the minimum modified chi-squared estimates are identical to WLS estimates.

15.14 Let y_i be a $\mathrm{bin}(n_i, \pi_i)$ variate for group i, $i = 1, \ldots, N$, with $\{y_i\}$ independent. Consider the model that $\pi_1 = \cdots = \pi_N$. Denote that common value by π.
  a. Show that the ML estimator of π is $p = (\sum_i y_i)/(\sum_i n_i)$.
  b. The minimum chi-squared estimator $\tilde{\pi}$ is the value of π minimizing

$$\sum_{i=1}^{N} \frac{[(y_i/n_i) - \pi]^2}{\pi} + \sum_{i=1}^{N} \frac{[(y_i/n_i) - \pi]^2}{1 - \pi}.$$

  The second term results from comparing $(1 - y_i/n_i)$ to $(1 - \pi)$, the proportions in the second category. If $n_1 = \cdots = n_N = 1$, show that $\tilde{\pi}$ minimizes $Np(1 - \pi)/\pi + N(1 - p)\pi/(1 - \pi)$. Hence show

$$\tilde{\pi} = p^{1/2}/\left[p^{1/2} + (1 - p)^{1/2}\right].$$

  Note the bias toward $\frac{1}{2}$ in this estimator.
  c. Argue that as $N \to \infty$ with all $n_i = 1$, the ML estimator is consistent but the minimum chi-squared estimator is not (Mantel 1985).

15.15 Refer to Problem 15.14. For $N = 2$ groups with n_1 and n_2 independent observations, find the minimum modified chi-squared estimator of π. Compare it to the ML estimator.

15.16 Show that the kernel estimator (15.9) is the same as the Bayes estimator (15.3) for the Dirichlet prior with $\{\beta_i = \alpha n/(1 - \alpha)N\}$. Using this result, suggest a way of letting the data determine the value of α in the kernel estimator.

CHAPTER 16

Historical Tour of Categorical Data Analysis*

This book concludes with an informal historical overview of the evolution of methods for categorical data analysis (CDA). We have seen that categorical scales are pervasive in the social sciences and the biomedical sciences. Not surprisingly, the development of GLMs for categorical responses was fostered by statisticians having ties to the social sciences or to the biomedical sciences.

Only in the last quarter of the twentieth century did these models receive the attention given early in the century to models for continuous data. Regression models for continuous variables evolved out of Francis Galton's breakthroughs in the 1880s. The strong influence of R. A. Fisher, G. Udny Yule, and other statisticians on experimentation in agriculture and biological sciences ensured widespread adoption of regression and ANOVA modeling by the mid-twentieth century. On the other hand, despite influential articles around 1900 by Karl Pearson and Yule on association between categorical variables, models for categorical responses received scant attention until the 1960s.

The beginnings of CDA were often shrouded in controversy. Key figures in the development of statistical science made groundbreaking contributions, but these statisticians were often in heated disagreement with one another.

16.1 PEARSON–YULE ASSOCIATION CONTROVERSY

Much of the early development of methods for CDA took place in England, and it is fitting that we begin our historical tour in London at the beginning of the twentieth century. The year 1900 is an apt starting point, since in that year Karl Pearson introduced his chi-squared statistic (X^2) and G. Udny Yule presented the odds ratio and related measures of association. Before then most work focused on descriptive aspects for relatively simple measures. For instance, Goodman and Kruskal (1959) noted that the Belgian social statistician Adolphe Quetelet used the relative risk in 1849.

By 1900, Karl Pearson (1857–1936) was already well known in the statistics community. He was head of a statistical laboratory at University College in London. His work the previous decade included developing a family of skewed probability distributions (called *Pearson curves*), obtaining the product-moment estimate of the correlation coefficient and finding its standard error, and extending Galton's work on linear regression. In fact, Pearson was a true renaissance man, writing on a wide variety of topics that included art, religion, philosophy, law, socialism, women's rights, physics, genetics, eugenics, and evolution. Pearson's motivation for developing the chi-squared test included testing whether outcomes on a roulette wheel in Monte Carlo varied randomly, checking the fit to various data sets of normal distributions and Pearson curves, and testing statistical independence in two-way contingency tables.

Much of the literature on CDA early in the twentieth century consisted of vocal debates about appropriate ways to summarize association. Pearson's approach assumed that continuous bivariate distributions underlie two-way contingency tables (Pearson 1904, 1913). He argued in favor of approximating a measure, such as the correlation, for the underlying continuum. In 1904, Pearson introduced the term *contingency* as a "measure of the total deviation of the classification from independent probability," and he introduced measures to describe its extent. The *tetrachoric correlation* is a ML estimate of the correlation for a bivariate normal distribution assumed to underlie counts in 2×2 tables. It is the correlation value ρ in the bivariate normal density that would produce cell probabilities equal to the sample cell proportions when that density is collapsed to a 2×2 table having the same marginal proportions as the observed table. The *mean-square contingency* and the *contingency coefficient* are normalizations of X^2 to the $(0, 1)$ scale. Pearson's contingency coefficient (Problem 3.33) for $I \times J$ tables standardized X^2 to approximate an underlying correlation.

George Udny Yule (1871–1951), a British contemporary of Pearson's, took a different approach. Having completed pioneering work developing multiple regression models and multiple and partial correlation coefficients, Yule turned his attention between 1900 and 1912 to association in contingency tables. He believed that many categorical variables, such as (vaccinated, unvaccinated) and (died, survived), are inherently discrete. Yule defined indices directly using cell counts without assuming an underlying continuum. He popularized the odds ratio θ [which Goodman (2000) noted may first have been proposed by a Hungarian statistician, J. Kőrösy] and a transformation of it to the $[-1, +1]$ scale, $Q = (\theta - 1)/(\theta + 1)$, now called *Yule's Q* (Problem 2.36). Discussing one of Pearson's measures that assumes underlying normality, Yule argued (1912, p. 612) that "at best the normal coefficient can only be said to give us in cases like these a hypothetical correlation between supposititious variables. The introduction of needless and unverifiable hypotheses does not appear to me a desirable proceeding in scientific work." Yule (1903) also showed the potential discrepancy between marginal and conditional associations in contingency tables, later studied by E. H. Simpson (1951) and now called *Simpson's paradox.*

In the first quarter of the twentieth century, Karl Pearson was the rarely challenged leader of statistical science in Britain. Pearson's strong personality did not take kindly to criticism, and he reacted negatively to Yule's ideas. He argued that Yule's own coefficients were unsuitable. For instance, Pearson claimed that their values were unstable, since different collapsings of $I \times J$ tables to 2×2 tables could produce quite different values of the measures. Pearson and D. Heron (1913) filled more than 150 pages of *Biometrika*, a journal he co-founded and edited, with a scathing reply to Yule's criticism. In a passage critical also of Yule's well-received book *An Introduction to the Theory of Statistics*, they stated "If Mr. Yule's views are accepted, irreparable damage will be done to the growth of modern statistical theory. ... [Yule's Q] has never been and never will be used in any work done under his [Pearson's] supervision. ... We regret having to draw attention to the manner in which Mr. Yule has gone astray at every stage in his treatment of association, but criticism of his methods has been thrust on us not only by Mr. Yule's recent attack, but also by the unthinking praise which has been bestowed on a text-book which at many points can only lead statistical students hopelessly astray." Pearson and Heron attacked Yule's "half-baked notions" and "specious reasoning" and argued that Yule would have to withdraw his ideas "if he wishes to maintain any reputation as a statistician."

In retrospect, Pearson and Yule both had valid points. Some classifications, such as most nominal variables, have no apparent underlying continuous distribution. On the other hand, many applications relate naturally to an underlying continuum, and that fact can motivate models and inference (e.g., Section 7.2.3). Goodman (1981a, b) noted that the ordinal models presented in Sections 9.4.1 and 9.6.1 provide a sort of reconciliation between Yule and Pearson, since Yule's odds ratio characterizes models that fit well when underlying distributions are approximately normal.

Half a century after the Pearson—Yule controversy, Leo Goodman and William Kruskal surveyed the development of association measures for contingency tables and made many contributions of their own. Their 1979 book reprinted four influential articles of theirs from the *Journal of the American Statistical Association* on this topic. Initial development of many measures occurred in the nineteenth century. Their 1959 article contains the following quote from M. H. Doolittle in 1887, which illustrates the lack of precision in early attempts to quantify the meaning of *association* even in 2×2 tables: "Having given the number of instances respectively in which things are both thus and so, in which they are thus but not so, in which they are so but not thus, and in which they are neither thus nor so, it is required to eliminate the general quantitative relativity inhering in the mere thingness of the things, and to determine the special quantitative relativity subsisting between the thusness and the soness of the things." Goodman (2000) added to the historical survey and proposed a new measure.

16.2   R. A. FISHER'S CONTRIBUTIONS

Pearson's disagreements with Yule were minor compared to his later ones with Ronald A. Fisher (1890–1962). Using a geometric representation, Fisher (1922) introduced *degrees of freedom* to characterize the family of chi-squared distributions. Fisher claimed that for tests of independence in $I \times J$ tables, X^2 has df $= (I - 1)(J - 1)$. By contrast, Pearson (1900, 1904) had argued that for any application of X^2, the index that Fisher later identified as df equals the number of cells minus 1, or $IJ - 1$ for two-way tables. Fisher pointed out, however, that estimating hypothesized cell probabilities using estimated row and column probabilities resulted in an additional $(I - 1) + (J - 1)$ constraints on the fitted values, thus affecting the distribution of X^2.

Not surprisingly, Pearson (1922) reacted critically to Fisher's suggestion that his df formula was incorrect. He stated: "I hold that such a view [Fisher's] is entirely erroneous, and that the writer has done no service to the science of statistics by giving it broad-cast circulation in the pages of the *Journal of the Royal Statistical Society*. . . . I trust my critic will pardon me for comparing him with Don Quixote tilting at the windmill; he must either destroy himself, or the whole theory of probable errors, for they are invariably based on using sample values for those of the sampled population unknown to us." Pearson claimed that using row and column sample proportions to estimate unknown probabilities had negligible effect on large-sample distributions, although he had realized (Pearson 1917) that df must be adjusted when the cell counts have linear constraints. Fisher was unable to get his rebuttal published by the Royal Statistical Society, and he ultimately resigned his membership.

Statisticians soon realized that Fisher was correct, but he maintained much bitterness over this and other dealings with Pearson. In the preface to a later volume of his collected works, he remarked that his 1922 article "had to find its way to publication past critics who, in the first place, could not believe that Pearson's work stood in need of correction, and who, if this had to be admitted, were sure that they themselves had corrected it." Writing about Pearson: he stated: "If peevish intolerance of free opinion in others is a sign of senility, it is one which he had developed at an early age." In Fisher (1926), he was able to dig the knife a bit deeper into the Pearson family using 11,688 2×2 tables randomly generated assuming independence by Karl Pearson's son, E. S. Pearson. Fisher showed that the sample mean of X^2 for these tables was 1.00001, much closer to the 1.0 predicted by his formula for $E(X^2)$ of df $= (I - 1)(J - 1) = 1$ than Pearson's $IJ - 1 = 3$. His daughter,

Joan Fisher Box (1978), discussed this and other conflicts between Fisher and Pearson. Hald (1998, pp. 652–663), Plackett (1983), and Stigler (1999, Chap. 19) summarized the chi-squared controversy.

Fisher's preeminent reputation among statisticians today accrues mainly from his theoretical work (introducing concepts such as sufficiency, information, and optimal properties of ML estimators) and his methodological contributions to the design of experiments and the analysis of variance. Although not so well known for work in CDA, he made other interesting contributions. Moreover, he made good use of the methods in his applied work. For instance, Fisher was also a famed geneticist. In one article, he used Pearson's goodness-of-fit test to check Mendel's theories of natural inheritance and showed that the fit was *too* good (Section 1.5.3).

Fisher realized the limitations of large-sample methods for laboratory work, and he was at the forefront of advocating specialized small-sample methods. Writing about large-sample methods in the preface to the first edition of his classic text *Statistical Methods for Research Workers*, he stated: "[T]he traditional machinery of statistical processes is wholly unsuited to the needs of practical research. Not only does it take a cannon to shoot a sparrow, but it misses the sparrow! The elaborate mechanism built on the theory of infinitely large samples is not accurate enough for simple laboratory data. Only by systematically tackling small sample problems on their merits does it seem possible to apply accurate tests to practical data." Fisher was among the first to promote the work by W. S. Gosset (pseudonym "Student") on the t distribution. The fifth edition of *Statistical Methods for Research Workers* (1934) introduced Fisher's exact test for 2×2 contingency tables. In his 1935 book *The Design of Experiments*, Fisher described the tea-tasting experiment (Section 3.5.2) motivated by his experience at an afternoon tea break while employed at Rothamsted Experiment Station.

The mid-1930s finally saw some model building for categorical responses. Chester Bliss (1934, 1935), following up a 1933 report on quantal response methods by J. H. Gaddum, popularized the probit model for applications in toxicology with a binary response. Bliss introduced the term *probit* but used the inverse normal cdf with mean 5 (rather than 0, in order to avoid negative values) and standard deviation 1. In the appendix of Bliss (1935), Fisher (1935b) outlined an algorithm for finding ML estimates of model parameters. That algorithm was a Newton–Raphson type of method using expected information, today commonly called *Fisher scoring* (Section 4.6.2). Stigler (1986, p. 246) and Finney (1971) attributed the first use of inverse normal cdf transformations of proportions to the German physicist Gustav Fechner in his 1860 book *Elemente der Psychophysik*. See Finney (1971) and McCulloch (2000) for other history of the probit method.

The definition for homogeneous association (no interaction) in contingency tables originated in an article by the British statistician Maurice Bartlett (1935) about $2 \times 2 \times 2$ tables. Bartlett showed how to find ML estimates of cell probabilities satisfying the property of equality of odds ratios between two variables at each level of the third. He attributed the idea to Fisher.

In 1940, Fisher developed canonical correlation methods for contingency tables. He showed how to assign scores to rows and columns of a contingency table to maximize the correlation. His work relates to the later development, particularly in France, of *correspondence analysis* methods (e.g., Benzécri 1973).

R. A. Fisher has had the greatest influence on the practice of modern statistical science. The biography by his daughter (Box 1978) gives a fascinating account of his impressive contributions to statistics and genetics. Fienberg (1980) summarized his contributions to CDA.

16.3 LOGISTIC REGRESSION

Bartlett (1937) used $\log[y/(1 - y)]$ in regression and ANOVA to transform observations y that are continuous proportions (Problem 6.33). In a book of statistical tables published in 1938, R. A. Fisher and Frank Yates suggested it as a possible transformation of a binomial parameter for analyzing binary data. In 1944, the physician and statistician Joseph Berkson introduced the term *logit* for this transformation. Berkson showed that the model using the logit fitted similarly to the probit model, and his subsequent work did much to popularize logistic regression. In 1951, Jerome Cornfield, another statistician with strong medical ties, used the odds ratio to approximate relative risks in case–control studies. Dyke and Patterson (1952) apparently first used the logit in models with qualitative predictors.

Sir David R. Cox introduced many statisticians to logistic regression, through his 1958 article and 1970 book, *The Analysis of Binary Data*. About the same time, an article by the Danish statistician and mathematician Georg Rasch sparked an enormous literature on item response models. The most important of these is the logit model with subject and item parameters, now called the *Rasch model* (Section 12.1.4). This work was highly influential in the psychometric community of northern Europe (especially in Denmark, the Netherlands, and Germany) and spurred many generalizations in the educational testing community in the United States.

The extension of logistic regression to multicategory responses received occasional attention before 1970 (e.g., Mantel 1966) but substantial work after about that date. For nominal responses, early work was mainly in the econometrics literature. See Bock (1970), McFadden (1974), Nerlove and Press (1973), and Theil (1969, 1970). In 2000, Daniel McFadden won the Nobel Prize in Economics for his work in the 1970s and 1980s on the discrete-choice model (Section 7.6). For cumulative logit models for ordinal responses, see Bock and Jones (1968), Simon (1974), Snell (1964), Walker and Duncan (1967), and Williams and Grizzle (1972). The cumulative probit case, based on an underlying normal response, has a longer history; see, for instance, Aitchison and Silvey (1957) and Bock and Jones (1968, Chap. 8). Cumulative logit and probit models received much more attention following publication of McCullagh (1980), which provided a Fisher scoring algorithm for ML fitting of all cumulative link models.

The next major advances with logistic regression dealt with its application to case–control studies (e.g., Breslow 1996; Mantel 1973; Prentice 1976a; Prentice and Pyke 1979; see also Section 5.1.4) and the conditional ML approach to model fitting for those studies and others with numerous nuisance parameters (Breslow et al. 1978, with related work in Breslow 1976, 1982; Breslow and Day 1980; Breslow and Powers 1978; Cox 1970; Farewell 1979; Prentice 1976a; Prentice and Breslow 1978; Zelen 1971; see also Sections 6.7 and 10.2). The conditional approach was later exploited in small-sample exact inference (Hirji et al. 1987; Mehta and Patel 1995; see also Section 6.7).

Nathan Mantel, whose name appears in the preceding two paragraphs, made a variety of interesting contributions to CDA. Although best known for the 1959 Mantel–Haenszel test and related odds ratio estimator, he also discussed trend tests (1963), multinomial logit and loglinear modeling (1966), logistic regression for case–control data (1973), the number of contingency tables having fixed margins (Gail and Mantel 1977), the analysis of square contingency tables (Mantel and Byar 1978), and problems with minimum chi-squared and Wald tests (1985, 1987a).

More recently, attention has focused on fitting logistic models to corre-lated responses for clustered data. One strand of this is marginal modeling of longitudinal data (Diggle et al. 2002; Liang and Zeger 1986; Liang et al. 1992). Much of this literature focuses on quasi-likelihood methods such as generalized estimating equations (GEE). Another strand is generalized linear mixed models (e.g., Breslow and Clayton 1993).

Perhaps the most far-reaching contribution of the past half century has been the introduction by British statisticians John Nelder and R. W. M. Wedderburn in 1972 of the concept of *generalized linear models*. This unifies the logistic and probit regression models for binomial data with loglinear models for Poisson data and with long-established regression and ANOVA models for normal-response data. Interestingly, the algorithm they used to fit GLMs is Fisher scoring, which R. A. Fisher introduced in 1935 for ML fitting of probit models. McCulloch (2000) reviewed the journey from probit models to GLMs and their further generalizations such as quasi-likelihood.

16.4   MULTIWAY CONTINGENCY TABLES AND LOGLINEAR MODELS

The quarter century following the end of World War II saw the development of a theoretical underpinning for models for contingency tables. H. Cramér

(1946) derived general expressions for large-sample distributions of parameter estimators. C. R. Rao (1957, 1963) conducted related work.

In 1949, the Berkeley-based statistician Jerzy Neyman, who had already performed fundamental work on hypothesis testing and interval estimation methods with E. S. Pearson, introduced the family of *best asymptotically normal* (BAN) estimators. These have the same optimal large-sample properties as ML estimators. The BAN family includes estimators obtained by minimizing chi-squared-type measures comparing observed proportions to proportions predicted by the model (Section 15.3.1). This type of estimator itself includes some *weighted least squares* (WLS) estimators. The simplicity of their computation, compared to ML estimators, was an important consideration before the advent of modern computing. Neyman's (1949) only mention of Fisher was the suggestion that Fisher did not realize that estimators other than ML could be BAN, stating that "the results ... contradict the assertion of R. A. Fisher, not a very clear one, that 'the maximum likelihood equation may indeed be derived from the conditions that it shall be linear in frequencies, and efficient for all values of θ'." Fisher, of course, returned the compliment: for instance, writing (1956) about proposals for an unconditional test for 2×2 tables, "the Principles of Neyman and Pearson's 'Theory of Testing Hypotheses' are liable to mislead those who follow them into much wasted effort."

In the early 1950s, William Cochran published work dealing with a variety of important topics in CDA. Scottish-born, Cochran spent most of his career at American universities: Iowa State, North Carolina State, Johns Hopkins, and Harvard. He (1940) modeled Poisson and binomial responses with variance-stabilizing transformations. He (1943) recognized and discussed ways of dealing with overdispersion. He (1950) introduced a generalization (Cochran's Q) of McNemar's test for comparing proportions in several matched samples. His classic 1954 article is a mixture of new methodology and advice for applied statisticians. It gave sample-size guidelines for chi-squared approximations to work well for the X^2 statistic. It also stressed the importance of directing inferences toward narrow (e.g., single-degree-of-freedom) alternatives and partitioning chi-squared statistics into components. One instance of this was Cochran's proposed test of conditional independence in several 2×2 tables, which was closely related to the Mantel and Haenszel (1959) test (Section 6.3.2). Another was a test for a linear trend in proportions across quantitatively defined rows of an $I \times 2$ table (Section 5.3.5). See also Cochran (1955). Fienberg (1984) reviewed Cochran's contributions to CDA.

Bartlett's work on interaction structure in $2 \times 2 \times 2$ contingency tables had relatively little impact for 20 years. Indeed, in presenting methods for partitioning X^2 in $2 \times 2 \times 2$ tables, Lancaster (1951) noted that "Doubtless little use will ever be made of more than a three-dimensional classification." However, in the mid-1950s and early 1960s, Bartlett's work was extended in many ways to multiway tables. See, for instance, Darroch (1962), Good

(1963), Goodman (1964b), Plackett (1962), Roy and Kastenbaum (1956), and Roy and Mitra (1956). These articles as well as influential articles by Martin W. Birch (1963, 1964a, b, 1965) were the genesis of research work on loglinear models between about 1965 and 1975. Birch's work was part of a never-submitted Ph.D. thesis at the University of Glasgow. He showed how to obtain ML estimates of cell probabilities in three-way tables, under various conditions. He showed the equivalence of those ML estimates for Poisson and multinomial sampling. He (and Watson 1959) extended theoretical results of Cramér and Rao on large-sample distributions for contingency table models. Mantel (1966) discussed early results and made the loglinear model formula explicit. A survey article by the French statistician Henri Caussinus (1966), based partly on his Ph.D. thesis, provides a good glimpse of the state-of-the-art of CDA just before this decade of advances. There, Caussinus introduced the quasi-symmetry model for square tables.

Much of the work in the next decades on loglinear and related logit modeling took place at three American universities: the University of Chicago, Harvard University, and the University of North Carolina. At Chicago, Leo Goodman wrote a series of groundbreaking articles, dealing with such topics as partitionings of chi-squared, models for square tables (e.g., quasi-independence), stepwise logit and loglinear model-building procedures, deriving asymptotic variances of ML estimates of loglinear parameters, latent class models, association models, correlation models, and correspondence analysis. For surveys of his early work, see Goodman (1968, an R. A. Fisher memorial lecture, 1970). For later work, see Goodman (1985, 1996, 2000). Goodman also wrote a stream of articles for social science journals that had a substantial impact on popularizing loglinear and logit methods for applications (e.g., Goodman 1969b).

Over the past 50 years, Goodman has been the most prolific contributor to the advancement of CDA methodology. The field owes tremendous gratitude to his steady and impressive body of work. In addition, some of Goodman's students at Chicago also made fundamental contributions. In 1970, Shelby Haberman completed a Ph.D. dissertation (the basis of his 1974a monograph) making substantial theoretical contributions to loglinear modeling. Among topics he considered were residual analyses, existence of ML estimates, loglinear models for ordinal variables, and theoretical results for models (such as the Rasch model) for which the number of parameters grows with the sample size. Clifford Clogg followed in Goodman's steps by having influence in the social sciences and in statistics with his work on association models, demography, models for rates, the census, and various other topics.

Simultaneously with Goodman's work, related research on ML methods for loglinear-logit models occurred at Harvard by students of Frederick Mosteller (such as Stephen Fienberg) and William Cochran. Much of this research was inspired by problems arising in analyzing large, multivariate data sets in the National Halothane Study (Bishop and Mosteller 1969; see also p. 345 of an interview with Lincoln Moses in *Statist. Sci.* **14**, 1999). That

Karl Pearson

G. Udny Yule

Ronald A. Fisher

Leo Goodman

FIGURE 16.1   Four leading figures in the development of categorical data analysis.

study investigated whether halothane was more likely than other anesthetics to cause death due to liver damage. A presidential address by Mosteller (1968) to the American Statistical Association described early uses of loglinear models for smoothing multidimensional discrete data sets. Fienberg and his own students advanced this work further. A landmark book in 1975 by him with Yvonne Bishop and Paul Holland, *Discrete Multivariate Analysis*, was largely responsible for introducing loglinear models to the general statistical community and remains an excellent reference.

Research at North Carolina by Gary Koch and several students and co-workers has been highly influential in the biomedical sciences. Their research developed WLS methods for categorical data models (Section 15.1). The 1969 article by Koch with J. Grizzle and F. Starmer popularized this approach. Koch and colleagues extended it in later articles to an impressive variety of problems, including problems for which ML methods are awkward to use, such as the analysis of repeated categorical measurement data (Koch et al. 1977). In 1966, Vasant Bhapkar showed that the WLS estimator is often identical to Neyman's minimum modified chi-squared estimator.

The early literature on loglinear models treated all classifications as nominal. Haberman (1974b) and Simon (1974) showed how to exploit ordinality of classifications in loglinear models. This work was extended in several articles by Leo Goodman (1979a, 1981a, b, 1983, 1985, 1986). The extensions included association models, which replace ordered scores in loglinear models by parameters (Section 9.5). Goodman (1985, 1986, 1996) also discussed related correlation models and provided a model-based perspective for the closely related correspondence analysis methods.

Certain loglinear models with conditional independence structure provide graphical models for contingency tables. These relate to the association graphs used in Section 9.1. Darroch et al. (1980) was the genesis of much of this work.

16.5 RECENT (AND FUTURE?) DEVELOPMENTS

The most active area of new research in CDA in the past decade has been the modeling of clustered data, such as occur in longitudinal studies and other forms of repeated measurement. A variety of ways now exist of modeling while accounting for the correlation among responses in the same cluster.

As discussed in Chapters 11 and 12, ML estimation is difficult for such models. For complex forms of generalized linear mixed models, for instance, it is a challenge to estimate well regression parameters and variance components. Integrating out the random effect to obtain the likelihood function requires an approximation such as numerical integration. Not surprisingly, various Monte Carlo approaches are applied increasingly here. A promising approach is a Monte Carlo EM algorithm that uses a Monte Carlo approximation for the E step (Booth and Hobert 1999). The Monte Carlo error can be assessed at each iteration, and one can accurately reproduce the ML estimates with sufficiently many iterations.

The modeling of clustered correlated data is likely to be an active area of research in coming years. The class of generalized linear mixed models is certain to see substantial work and further generalization. One extension is generalized *additive* mixed models. Time-series models for categorical responses have so far received relatively little attention. For all such models with correlated responses, model diagnostics are of vital importance and need development. For longitudinal data, missing data are a common problem. This area currently has much activity.

Another important recent advance is the development of efficient algorithms for exact small-sample methods. With such methods, one can guarantee that the size of a test is no greater than some prespecified level and that the coverage probability for a confidence interval is at least the nominal level. The "exactness" refers only to inference being based on probability distributions that do not depend on unknown parameters. There is no unique way to do this, and certain methods can be highly conservative because of discreteness. Most literature deals with the conditional approach, which eliminates nuisance parameters by conditioning on their sufficient statistics. Hence, the basic idea builds on Fisher's exact test. Conditional methods are versatile, applying to exponential family linear models that use the canonical link function, such as loglinear models for Poisson responses and logit models for binomial responses. Many of the computational advances with the exact conditional approach occurred in a series of articles by Cyrus Mehta, Nitin Patel, and colleagues at Harvard (e.g., Mehta and Patel 1983), using the network algorithm. See surveys by Agresti (1992), Mehta (1994), Mehta and Patel (1995), and the *StatXact* and *LogXact* manuals (Cytel Software, Cambridge, MA, founded by Mehta and Patel).

Although the development of "exact" methods has seen considerable progress, certain analyses are still infeasible and likely to be so for some time because of the exponential increase in computing time as the table size or sample size increases. There are an ever-increasing variety of methods for accurate approximation of exact methods. These include simple Monte Carlo (e.g., Agresti et al. 1979), Monte Carlo with importance sampling (e.g., Booth and Butler 1999; Mehta et al. 1988), Markov chain Monte Carlo (MCMC; Forster et al. 1996), saddlepoint approximations (Pierce and Peters 1992, Strawderman and Wells 1998), and related work on an *approximate* conditioning approach (Pierce and Peters 1999) in which discreteness is not so problematic.

Finally, the development of Bayesian approaches to CDA is an increasingly active area. The multiplicity of parameters complicates Bayesian modeling. For early use of Bayesian estimation of probabilities, see Good (1965) and Lindley (1964). Good's (1965) article apparently evolved from his work during World War II with Alan Turing at Bletchley Park, England, on breaking Nazi codes. The development of the Bayesian approach for CDA is discussed in Section 15.2.3.

Predicting the future is always dangerous. However, it is likely that much future research will focus on computationally intensive methods such as generalized linear mixed models. Another hot topic, largely outside the realm of traditional modeling, is the development of algorithmic methods for huge data sets with large numbers of variables. Such methods, often referred to as *data mining*, deal with the handling of complex data structures, with a premium on predictive power at the sacrifice of simplicity and interpretability of structure. Important areas of application include genetics, such as the analysis of discrete DNA sequences in the form of very high-dimensional contingency tables, and business applications such as credit scoring and tree-structured methods for predicting future behavior of customers.

Sources for the historical tour in this chapter include Stigler (1986), *Studies in the History of Probability and Statistics*, edited by E. S. Pearson and M. G. Kendall (London: Griffin, 1970), and personal conversations over the years with several statisticians, including Erling Andersen, R. L. Anderson, Henri Caussinus, William Cochran, Sir David Cox, John Darroch, Leo Goodman, Gary Koch, Frederick Mosteller, John Nelder, C. R. Rao, Stephen Stigler, Geoffrey Watson, and Marvin Zelen. To readers who have made it this far, I congratulate your perseverance! To develop a more complete understanding of the historical development of CDA, you may want to study the following chronological list of 25 sources. These convey a sense of how methodology has evolved. Alternatively, look at some early books on this topic, such as A. E. Maxwell's *Analysing Qualitative Data* (New York: Methuen, 1961), R. L. Plackett's *The Analysis of Categorical Data* (London: Griffin, 1974), and the Bishop, Fienberg, and Holland *Discrete Multivariate Analysis* (Cambridge, MA: MIT Press 1975).

Pearson (1900)	Caussinus (1966)
Yule (1912)	Goodman (1968)
Fisher (1922)	Mosteller (1968)
Bartlett (1935)	Grizzle et al. (1969)
Berkson (1944)	Goodman (1970)
Neyman (1949)	Haberman (1974a)
Cochran (1954)	Nelder and Wedderburn (1972)
Goodman and Kruskal (1954)	McFadden (1974)
Roy and Mitra (1956)	Goodman (1979a)
Cox (1958a)	McCullagh (1980)
Mantel and Haenszel (1959)	Liang and Zeger (1986)
Birch (1963)	Breslow and Clayton (1993)
Birch (1964b)	

Using Computer Software to Analyze Categorical Data

In this appendix we discuss statistical software for categorical data analysis, with emphasis on SAS. We begin by mentioning major software that can perform the analyses discussed in this book. Then we illustrate, by chapter, SAS code for the analyses. Information about other packages (such as S-Plus, R, SPSS, and Stata), as well as updated information about SAS, is at the Web site (*www.stat.ufl.edu/~aa/cda/cda.html.*) Section A.2 on SAS also lists other software for analyses not currently available in SAS.

A.1 SOFTWARE FOR CATEGORICAL DATA ANALYSIS

A.1.1 SAS

SAS is general-purpose software for a wide variety of statistical analyses. The main procedures (PROCs) for categorical data analyses are FREQ, GEN-MOD, LOGISTIC, NLMIXED, and CATMOD.

PROC FREQ computes measures of association and their estimated standard errors. It also performs generalized Cochran–Mantel–Haenszel tests of conditional independence, and exact tests of independence in $I \times J$ tables.

PROC GENMOD fits generalized linear models. It fits cumulative link models for ordinal responses. It can perform GEE analyses for marginal models. One can form one's own variance function and allow scale parameters, making it suitable for quasi-likelihood analyses.

PROC LOGISTIC gives ML fitting of binary response models, cumulative link models for ordinal responses, and baseline-category logit models for nominal responses. It incorporates model selection procedures, regression diagnostic options, and exact conditional inference. PROC PROBIT also conducts ML fitting of binary and cumulative link models as well as quantal response models that permit a strictly positive probability as the linear predictor decreases to $-\infty$.

PROC CATMOD fits baseline-category logit models. It is also useful for WLS fitting of a wide variety of models for categorical data.

PROC NLMIXED fits generalized linear mixed models (GLMMs). It approximates the likelihood using adaptive Gauss–Hermite quadrature.

Other programs run on SAS that are not specifically supported by the SAS Institute. For further details about SAS for categorical data analyses, see the very helpful guide by Stokes et al. (2000). Also useful are SAS publications on logistic regression (Allison 1999) and graphics (Friendly 2000).

A.1.2 Other Software Packages

Most major statistical software has procedures for categorical data analyses. For instance, see SPSS (*SPSS Regression Models 10.0* by M. J. Norusis, SPSS Inc., 1999), Stata (*A Handbook of Statistical Analyses Using Stata*, 2nd ed., by S. Rabe-Hesketh and B. Everitt, CRC Press, Boca Raton, FL, 2000), S-Plus (*Modern Applied Statistics with S-Plus*, 3rd ed., by W. N. Venables and B. D. Ripley, Springer-Verlag, New York, 1999), and the related free package, R, and GLIM (Aitkin et al. 1989). Most major software now follows the lead of GLIM and includes a generalized linear models routine. Examples are PROC GENMOD in SAS and the glm function in R and S-Plus.

For certain analyses, specialized software is better than the major packages. A good example is StatXact (Cytel Software, Cambridge, Massachusetts), which provides exact analysis for categorical data methods and some nonparametric methods. Among its procedures are small-sample confidence intervals for differences and ratios of proportions and for odds ratios, and Fisher's exact test and its generalizations for $I \times J$ tables. It can also conduct exact tests of conditional independence and of equality of odds ratios in $2 \times 2 \times K$ tables, and exact confidence intervals for the common odds ratio in several 2×2 tables. StatXact uses Monte Carlo methods to approximate exact P-values and confidence intervals when a data set is too large for exact inference to be computationally feasible. Its companion, LogXact, performs exact conditional logistic regression.

Other examples of specialized software are SUDAAN for GEE-type analyses that handle clustering in survey data (Research Triangle Institute, Research Triangle Park, North Carolina), Latent GOLD for latent class modeling (Statistical Innovations, Belmont, Massachusetts), MLn (Institute of Education, London) and HLM (Scientific Software, Chicago) for multilevel models, and PASS for power analyses (NCSS Statistical Software, Kaysville, Utah). S-Plus and R functions are also available from individuals or from published work for particular analyses. For instance, *Statistical Models in S* by J. M. Chambers and T. J. Hastie (Wadsworth, Belmont, California, 1993, p. 227) showed the use of S-Plus in quasi-likelihood analyses using the quasi and make.family functions.

TABLE A.1 SAS Code for Chi-Squared, Measures of Association, and Residuals for Education–Religion Data in Table 3.2

```
data table;
    input degree religion $ count @@;
datalines;
1 fund 178     1 mod 138     1 lib 108
2 fund 570     2 mod 648     2 lib 442
3 fund 138     3 mod 252     3 lib 252
    ;
proc freq order = data; weight count;
  tables degree*religion/ chisq expected measures cmh1;
proc genmod order = data; class degree religion;
  model count = degree religion/ dist = poi link = log residuals;
```

A.2 EXAMPLES OF SAS CODE BY CHAPTER

The examples below show SAS code (Version 8.1). We focus on basic model fitting rather than the great variety of options. The material is organized by chapter of presentation. For convenience, data for examples are entered in the form of the contingency table displayed in the text. In practice, one would usually enter data at the subject level. These tables and the full data sets are available at *www.stat.ufl.edu/~aa/cda/cda.html*.

Chapters 1–3: Introduction, Two-Way Contingency Tables

Table A.1 uses SAS to analyze Table 3.2. The @@ symbol indicates that each line of data contains more than one observation. Input of a variable as characters rather than numbers requires an accompanying $ label in the INPUT statement. PROC FREQ forms the table with the TABLES statement, ordering row and column categories alphanumerically. To use instead the order in which the categories appear in the data set (e.g., to treat the variable properly in an ordinal analysis), use the ORDER = DATA option in the PROC statement. The WEIGHT statement is needed when one enters the cell counts instead of subject-level data. PROC FREQ can conduct chi-squared tests of independence (CHISQ option), show its estimated expected frequencies (EXPECTED), provide a wide assortment of measures of association and their standard errors (MEASURES), and provide ordinal statistic (3.15) with a "nonzero correlation" test (CMH1). One can also perform chi-squared tests using PROC GENMOD (using loglinear models discussed in the Chapters 8–9 section of this appendix), as shown. Its RESIDUALS option provides cell residuals. The output labeled "StReschi" is the standardized Pearson residual (3.13).

Table A.2 analyzes Table 3.8. With PROC FREQ, for 2×2 tables the MEASURES option in the TABLES statement provides confidence intervals

TABLE A.2 SAS Code for Fisher's Exact Test and Confidence Intervals for Odds Ratio for Tea-Tasting Data in Table 3.8

```
data fisher;
input poured guess count @@;
datalines;
1 1 3    1 2 1    2 1 1    2 2 3
;
proc freq;   weight count;
  tables poured*guess / measures riskdiff;
  exact fisher or / alpha = .05;
proc logistic descending; freq count;
  model guess = poured / clodds = pl;
```

for the odds ratio (labeled "case-control" on output) and the relative risk, and the RISKDIFF option provides intervals for the proportions and their difference. For tables having small cell counts, the EXACT statement can provide various exact analyses. These include Fisher's exact test and its generalization for $I \times J$ tables, treating variables as nominal, with keyword FISHER. The OR keyword gives the odds ratio and its large-sample confidence interval (3.2) and the small-sample interval based on (3.20). Other EXACT statement keywords include binomial tests for 1×2 tables (keyword BINOMIAL), exact trend tests for $I \times 2$ tables (TREND), and exact chi-squared tests (CHISQ) and exact correlation tests for $I \times J$ tables (MHCHI). One can use Monte Carlo simulation (option MC) to estimate exact P-values when the exact calculation is too time consuming. Table A.2 also uses PROC LOGISTIC to get a profile-likelihood confidence interval for the odds ratio (CLODDS = PL). LOGISTIC uses FREQ to serve the same purpose as PROC FREQ uses WEIGHT.

Other

StatXact provides small-sample confidence intervals for a binomial parameter, the difference of proportions, relative risk, and odds ratio. Blaker (2000) gave S-Plus functions that provide his confidence interval for a binomial parameter.

Chapter 4: Models for Binary Response Variables

PROC GENMOD fits GLMs. It specifies the response distribution in the DIST option ("poi" for Poisson, "bin" for binomial, "mult" for multinomial, "negbin" for negative binomial) and specifies the link in the LINK option. Table A.3 illustrates for Table 4.2. For binomial models with grouped data, the response in the model statements takes the form of the number of "successes" divided by the number of cases.

TABLE A.3   SAS Code for Binary GLMs for Snoring Data in Table 4.2

```
data glm;
input snoring disease total @@;
datalines;
0 24 1379    2 35 638    4 21 213    5 30 254
;
proc genmod; model disease / total = snoring / dist = bin link = identity;
proc genmod; model disease / total = snoring / dist = bin link = logit;
proc genmod; model disease / total = snoring / dist = bin link = probit;
```

TABLE A.4   SAS Code for Poisson and Negative Binomial GLMs for Horseshoe Crab Data in Table 4.3

```
data crab;
input color spine width satell weight;
datalines;
3   3   28.3  8   3.05
4   3   22.5  0   1.55
...
3   2   24.5  0   2.00
;
proc genmod;
  model satell = width / dist = poi link = log;
proc genmod;
  model satell = width / dist = poi link = identity;
proc genmod;
  model satell = width / dist = negbin link = identity;
```

Table A.4 uses GENMOD for count modeling of Table 4.3. Each observation refers to a single crab. Using width as the predictor, the first two models use Poisson regression. The third model uses the identity link assuming a negative binomial distribution.

Table A.5 uses GENMOD for the overdispersed data of Table 4.5. A CLASS statement requests dummy variables for the groups. With no intercept in the model (option NOINT) for the identity link, the estimated parameters are the four group probabilities. The ESTIMATE statement provides an estimate, confidence interval, and test for a contrast of model parameters, in this case the difference in probabilities for the first and second groups. The second analysis uses the Pearson statistic to scale standard errors to adjust for overdispersion. PROC LOGISTIC can also provide overdispersion modeling of binary responses; see Table A.27 in the Chapter 13 part of this appendix.

PROC GAM (starting in Version 8.2) fits generalized additive models.

TABLE A.5 SAS Code for Overdispersion Modeling of Teratology Data in Table 4.5

```
data moore;
  input litter group n y @@;
datalines;
 1 1 10 1    2 1 11 4    3 1 12 9    4 1 4 4    5 1 10 10
 ...
55 4 14 1    56 4  8 0    58 4 17 0
 ;
proc genmod;  class group;
  model y/n = group / dist = bin link = identity noint;
estimate 'pi1- pi2 ' group 1 -1 0 0;
proc genmod;  class group;
  model y/n = group / dist - bin link - identity noint scale = pearson;
```

Chapters 5 and 6: Logistic Regression

One can fit logistic regression models using either software for GLMs or specialized software for logistic regression. PROC GENMOD uses Newton-Raphson, whereas PROC LOGISTIC uses Fisher scoring. Both yield ML estimates, but SE values use observed information in GENMOD and expected information in LOGISTIC. These are the same for the logit link.

Table A.6 applies GENMOD and LOGISTIC to Table 5.2, when "y" out of "n" crabs had satellites at a given width level. In GENMOD, the LRCI option provides profile likelihood confidence intervals. The ALPHA = option can specify an error probability other than the default of 0.05. The TYPE3 option provides likelihood-ratio tests for each parameter. (In the Chapter 8–9 section we discuss the second GENMOD analysis.)

TABLE A.6 SAS Code for Modeling Grouped Crab Data in Table 5.2

```
data crab;
input width y n satell;  logcases = log(n);
datalines;
22.69  5 14 14
 ...
30.41 14 14 72
 ;
proc genmod;
  model y/n = width / dist = bin link = logit lrci alpha = .01 type3;
proc logistic;
  model y/n = width / influence stb;
  output out = predict p = pi_hat lower = LCL upper = UCL;
proc print data = predict;
proc genmod;
  model satell = width / dist = poi link = log offset = logcases residuals;
```

TABLE A.7 SAS Code for Logit Modeling of AIDS Data in Table 5.5

```
data aids;
input race $ azt $ y n @@;
datalines;
  White Yes 14 107    White No 32 113    Black Yes 11 63    Black No 12 55
;
proc genmod; class race azt;
  model y/n = azt race / dist = bin type3 lrci residuals obstats;
proc logistic; class race azt / param = reference;
  model y/n = azt race / aggregate scale = none clparm = both clodds = both;
  output out = predict p = pi_hat lower = lower upper = upper;
proc print data = predict;
proc logistic; class race azt (ref = first) / param = ref;
  model y/n = azt / aggregate = (azt race) scale = none;
```

With PROC LOGISTIC, logistic regression is the default for binary data. LOGISTIC has a built-in check of whether logistic regression ML estimates exist. It can detect a complete separation of data points with 0 and 1 outcomes. LOGISTIC can also apply other links, such as the probit. Its INFLUENCE option provides Pearson and deviance residuals and diagnostic measures (Pregibon 1981). The STB option provides standardized estimates by multiplying by $s_{x_j}\sqrt{3}/\pi$ (Section 5.4.7 and Note 5.9). Following the model statement, Table A.6 requests predicted probabilities and lower and upper 95% confidence limits for the probabilities.

Table A.7 uses GENMOD and LOGISTIC to fit a logit model with qualitative predictors to Table 5.5. In GENMOD, the OBSTATS option provides various "observation statistics," including predicted values and their confidence limits. The RESIDUALS option requests residuals such as the Pearson and standardized Pearson residuals (labeled "Reschi" and "StReschi"). A CLASS statement requests dummy variables for the factor. By default, in GENMOD the parameter estimate for the last level of each factor equals 0. In LOGISTIC, estimates sum to zero. That is, dummies take the effect coding $(1, -1)$ of 1 when in the category and -1 when not, for which parameters sum to 0. In the CLASS statement in LOGISTIC, the option PARAM = REF requests $(1, 0)$ dummy variables with the last category as the reference level. Also putting REF = FIRST next to a variable name requests its first category as the reference level. The CLPARM = BOTH and CLODDS = BOTH options provide Wald and profile likelihood confidence intervals for parameters and odds ratio effects of explanatory variables. With AGGREGATE SCALE = NONE in the model statement, LOGISTIC reports Pearson and deviance tests of fit; it forms groups by aggregating data into the possible combinations of explanatory variable values, without overdispersion adjustments. Adding variables in parentheses after AGGREGATE (as in the second use of LOGISTIC in Table A.7) specifies the predictors used for forming the table on which to test fit, even when some predictors may have no effect in the model.

TABLE A.8 SAS Code for Logistic Regression Models with Horseshoe Crab Data in Table 4.3

```
data crab;
input color spine width satell weight;
if satell>0 then y = 1; if satell = 0 then y = 0;
if color = 4 then light = 0; if color<4 then light = 1;
datalines;
2 3 28.3 8 3.05
...
2 2 24.5 0 2.00
;
proc genmod descending; class color;
  model y = width color / dist = bin link = logit lrci type3 obstats;
  contrast 'a-d' color 1 0 0 -1;
proc genmod descending;
  model y = width color / dist = bin link = logit;
proc genmod descending;
  model y = width light / dist = bin link = logit;
proc genmod descending; class color spine;
  model y = width weight color spine / dist = bin link = logit type3;
proc logistic descending; class color spine / param = ref;
  model y = width weight color spine / selection = backward lackfit
      outroc = classif1;
proc plot data = classif1; plot _sensit_*_1mspec_ ;
```

Table A.8 shows logistic regression analyses for Table 4.3. The models refer to a constructed binary variable Y that equals 1 when a horseshoe crab has satellites and 0 otherwise. With binary data entry, GENMOD and LOGISTIC order the levels alphanumerically, forming the logit with $(1, 0)$ responses as $\log[P(Y = 0)/P(Y = 1)]$. Invoking the procedure with DESCENDING following the PROC name reverses the order. The first two GENMOD statements use both color and width as predictors; color is qualitative in the first model (by the CLASS statement) and quantitative in the second. A CONTRAST statement tests contrasts of parameters, such as whether parameters for two levels of a factor are identical. The statement shown contrasts the first and fourth color levels. The third GENMOD statement uses a dummy variable for color, indicating whether a crab is light or dark (color = 4). The fourth GENMOD statement fits the main effects model using all the predictors from Table 4.3. LOGISTIC has options for stepwise selection of variables, as the final model statement shows. The LACKFIT option yields the Hosmer–Lemeshow statistic. Using the OUTROC option, LOGISTIC can output a data set for plotting a ROC curve.

Table A.9 analyzes Table 6.9. The CMH option in PROC FREQ specifies the CMH statistic, the Mantel–Haenszel estimate of a common odds ratio and its confidence interval, and the Breslow–Day statistic. FREQ uses the

TABLE A.9   SAS Code for CMH Analysis of Clinical Trial Data in Table 6.9

```
data crab;
input center $ treat response count @@ ;
datalines;
a  1  1 11    a  1  2 25    a  2  1 10    a  2  2 27
...
h  1  1  4    h  1  2  2    h  2  1  6    h  2  2  1
;
proc freq; weight count;
  tables center*treat*response/ cmh chisq;
```

two rightmost variables in the TABLES statement as the rows and columns for each partial table; the CHISQ option yields chi-square tests of independence for each partial table. For $I \times 2$ tables the TREND keyword in the TABLES statement provides the Cochran–Armitage trend test.

Exact conditional logistic regression is available in PROC LOGISTIC with the EXACT statement. It provides ordinary and mid-P-values as well as confidence limits for each model parameter and the corresponding odds ratio with the ESTIMATE = BOTH option. One can also conduct the exact conditional version of the Cochran–Armitage test using the TREND option in the EXACT statement with PROC FREQ. Version 9 of SAS will include asymptotic conditional logistic regression, using a STRATA statement to indicate the stratification parameters to be conditioned out. One can also use PROC PHREG to do this (Stokes et al. 2000).

Models with probit and complementary log-log (CLOGLOG) links are available with PROC GENMOD, PROC LOGISTIC, or PROC PROBIT. O'Brien (1986) gave a SAS macro for computing powers using the noncentral chi-squared distribution.

Other

LogXact provides exact conditional logistic regression and StatXact provides exact inference about the odds ratio in $2 \times 2 \times K$ tables. PASS (NCSS Statistical Software) provides power analyses.

Chapter 7: Multinomial Response Models

PROC LOGISTIC fits baseline-category logit models (as of Version 8.2) using the LINK = GLOGIT option. The final response category is the default baseline for the logits. Exact inference is also available using the conditional distribution to eliminate nuisance parameters. PROC CATMOD also fits baseline-category logit models, as Table A.10 shows. CATMOD codes estimates for a factor so that they sum to zero. The PRED = PROB and PRED = FREQ options provide predicted probabilities and fitted values and their standard errors. The POPULATION statement provides the

TABLE A.10   SAS Code for Baseline-Category Logit Models with Alligator Data in Table 7.1

```
data gator;
input lake gender size food count @@;
datalines;
1 1 1 1 7  1 1 1 2 1  1 1 1 3 0  1 1 1 4 0  1 1 1 5 5
...
4 2 2 1 8  4 2 2 2 1  4 2 2 3 0  4 2 2 4 0  4 2 2 5 1
;
proc logistic; freq count; class lake size / param = ref;
  model food(ref = '1') = lake size / link = glogit
      aggregate scale = none;
proc catmod; weight count;
  population lake size gender;
  model food = lake size / pred = freq pred = prob;
```

variables that define the predictor settings. For instance, with "gender" in that statement, the model with lake and size effects is fitted to the full table also classified by gender.

PROC GENMOD can fit the proportional odds version of cumulative logit models using the DIST = MULTINOMIAL and LINK = CLOGIT options. Table A.11 fits it to Table 7.5. When the number of response categories exceeds 2, by default PROC LOGISTIC fits this model. It also gives a score test of the proportional odds assumption of identical effect parameters for each cutpoint. Both procedures use the $\alpha_j + \beta x$ form of the model. Cox (1995) used PROC NLIN for the more general model (7.8) having a scale parameter.

Both GENMOD and LOGISTIC can use other links in cumulative link models. GENMOD uses LINK = CPROBIT for the cumulative probit model and LINK = CCLL for the cumulative complementary log-log model. Table A.11 uses LINK = PROBIT in LOGISTIC to fit a cumulative probit model.

TABLE A.11   SAS Code for Cumulative Logit and Probit Models with Mental Impairment Data in Table 7.5

```
data impair;
input mental ses life;
datalines;
1 1 1
...
4 0 9
;
proc genmod ;
  model mental = life ses / dist = multinomial link = clogit lrci type3;
proc logistic;
  model mental = life ses / link = probit;
```

TABLE A.12   SAS Code for Adjacent-Categories Logit and Mean Response Models and CMH Analysis of Job Satisfaction Data in Table 7.8

```
data jobsat;
input gender income satisf count @@;
count2 = count + .01;
datalines;
1 1 1 1  1 1 2 3  1 1 3 11  1 1 4 2
...
0 4 1 0  0 4 2 1  0 4 3  9  0 4 4 6
;
proc catmod order = data; * ML analysis of adj-cat logit (ACL) model;
    weight count;
    population gender income;
    model satisf =
        (1 0 0   3 3,   0 1 0 2 2,   0 0 1 1 1,
         1 0 0   6 3,   0 1 0 4 2,   0 0 1 2 1,
         1 0 0   9 3,   0 1 0 6 2,   0 0 1 3 1,
         1 0 0 12 3,   0 1 0 8 2,   0 0 1 4 1,
         1 0 0   3 0,   0 1 0 2 0,   0 0 1 1 0,
         1 0 0   6 0,   0 1 0 4 0,   0 0 1 2 0,
         1 0 0   9 0,   0 1 0 6 0,   0 0 1 3 0,
         1 0 0 12 0,   0 1 0 8 0,   0 0 1 4 0)
            /ml pred = freq;
proc catmod order = data; weight count2; * WLS analysis of ACL model;
  response alogits; population gender income; direct gender income;
  model satisf = _response_ gender income;
proc catmod; weight count; * mean response model;
  population gender income; response mean; direct gender income;
  model satisf = gender income / covb;
proc freq; weight count;
  tables gender*income*satisf / cmh scores = table;
```

One can fit adjacent-categories logit models in CATMOD by fitting equivalent baseline-category logit models. Table A.12 uses it for Table 7.8, where each line of code in the model statement specifies the predictor values (for the three intercepts, income, and gender) for the three logits. The income and gender predictor values are multiplied by 3 for the first logit, 2 for the second, and 1 for the third, to make effects comparable in the two models. PROC CATMOD has options (CLOGITS and ALOGITS) for fitting cumulative logit and adjacent-categories logit models to ordinal responses; however, those options provide weighted least squares (WLS) rather than ML fits. A constant must be added to empty cells for WLS to run. CATMOD treats zero counts as structural zeros, so they must be replaced by small constants when they are actually sampling zeros. The DIRECT statements identify predictors treated as quantitative. The second analysis in Table A.12 uses the ALOGITS option. CATMOD can also fit mean response models using WLS, as the third analysis in Table A.12 shows.

With the CMH option, PROC FREQ provides the generalized CMH tests of conditional independence. The statistic for the "general association" alternative treats X and Y as nominal [statistic (7.20)], the statistic for the "row mean scores differ" alternative treats X as nominal and Y as ordinal, and the statistic for the "nonzero correlation" alternative treats X and Y as ordinal [statistic (7.21)]. Table A.12 analyzes Table 7.8, using scores $(1, 2, 3, 4)$ for each variable.

PROC MDC fits multinomial discrete choice models, with logit and probit links. One can also use PROC PHREG, which is designed for the Cox proportional hazards model for survival analysis, because the partial likelihood for that analysis has the same form as the likelihood for the multinomial model (Allison 1999, Chap. 7; Chen and Kuo 2001).

Other

LogXact provides exact conditional analyses for baseline-category logit models. Joseph Lang (*jblang@stat.uiowa.edu*) has an R function that can fit mean response models by ML.

Chapters 8 and 9: Loglinear Models

Table A.13 uses GENMOD to fit model (AC, AM, CM) to Table 8.3. Table A.14 uses GENMOD for table raking of Table 8.15. Table A.15 uses GENMOD to fit the linear-by-linear association model (9.6) and the row effects model (9.8) to Table 9.3 (with column scores $1, 2, 4, 5$). The defined

TABLE A.13   SAS Code for Fitting Loglinear Models to Drug Survey Data in Table 8.3

```
data drugs;
input a c m count @@;
datalines;
1 1 1 911    1 1 2 538    1 2 1 44    1 2 2 456
2 1 1   3    2 1 2  43    2 2 1  2    2 2 2 279
;
proc genmod;  class a c m;
  model count = a c m a*m a*c c*m / dist = poi link = log lrci type3 obstats;
```

TABLE A.14   SAS Code for Raking Table 8.15

```
data rake;
input school atti count @@;
log_c = log(count); pseudo = 100 / 3;
data lines;
1 1 209    1 2 101    1 3 237
  ...
;
proc genmod; class school atti;
  model pseudo = school atti / dist = poi link = log offset = log_c obstats;
```

TABLE A.15 SAS Code for Fitting Association Models to GSS Data in Table 9.3

```
data sex;
input premar birth u v count @@; assoc = u*v ;
datalines;
1 1 1 1 38    1 2 1 2 60    1 3 1 4 68    1 4 1 5 81
...
;
proc genmod; class premar birth;
  model count = premar birth assoc / dist = poi link = log;
proc genmod; class premar birth;
  model count = premar birth premar*v / dist = poi link = log;
```

variable "assoc" represents the cross-product of row and column scores, which has β parameter as coefficient in model (9.6). Table A.6 uses GENMOD to fit the Poisson regression model with log link for the grouped data of Table 5.2. It models the total number of satellites at each width level (variable "satell"), using the log of the number of cases as offset.

Correspondence analysis is available with PROC CORRESP.

Other

Prof. Joseph Lang (*jblang@stat.uiowa.edu*) has R and S-Plus functions for ML fitting of the generalized loglinear model (8.18). Becker (1990) gave a FORTRAN program that fits the $RC(M)$ model.

Chapter 10: Models for Matched Pairs

Table A.16 analyzes Table 10.1. For square tables, the AGREE option in PROC FREQ provides the McNemar chi-squared statistic for binary matched pairs, the X^2 test of fit of the symmetry model (also called *Bowker's test*),

TABLE A.16 SAS Code for McNemar's Test and Comparing Proportions for Matched Samples in Table 10.1

```
data matched;
input first second count @@;
datalines;
1 1 794    1 2 150    2 1 86    2 2 570
;
proc freq; weight count;
  tables first*second / agree; exact mcnem;
proc catmod; weight count;
  response marginals;
  model first*second = (1   0 ,
                        1   1 ;
```

TABLE A.17   SAS Code for Testing Marginal Homogeneity with Migration Data in Table 10.6

```
data migrate;
input then $ now $ count m11 m12 m13 m21 m22 m23 m31 m32 m33 m44 m1 m2 m3;
datalines;
  ne   ne 11607  1  0  0  0  0  0  0  0  0  0  0  0  0
  ne   mw   100  0  1  0  0  0  0  0  0  0  0  0  0  0
  ne    s   366  0  0  1  0  0  0  0  0  0  0  0  0  0
  ne    w   124 -1 -1 -1  0  0  0  0  0  0  0  1  0  0
  mw   ne    87  0  0  0  1  0  0  0  0  0  0  0  0  0
  mw   mw 13677  0  0  0  0  1  0  0  0  0  0  0  0  0
  mw    s   515  0  0  0  0  0  1  0  0  0  0  0  0  0
  mw    w   302  0  0  0 -1 -1 -1  0  0  0  0  0  1  0
   s   ne   172  0  0  0  0  0  0  1  0  0  0  0  0  0
   s   mw   225  0  0  0  0  0  0  0  1  0  0  0  0  0
   s    s 17819  0  0  0  0  0  0  0  0  1  0  0  0  0
   s    w   270  0  0  0  0  0  0 -1 -1 -1  0  0  0  1
   w   ne    63 -1  0  0 -1  0  0 -1  0  0  0  1  0  0
   w   mw   176  0 -1  0  0 -1  0  0 -1  0  0  0  1  0
   w    s   286  0  0 -1  0  0 -1  0  0 -1  0  0  0  1
   w    w 10192  0  0  0  0  0  0  0  0  0  1  0  0  0
  ;
proc genmod;
  model count = m11 m12 m13 m21 m22 m23 m31 m32 m33 m44 m1 m2 m3
      / dist = poi  link = identity;
proc catmod;  weight count;  response marginals;
  model then*now = _response_ / freq;
  repeated time 2;
```

and Cohen's kappa and weighted kappa with SE values. The MCNEM keyword in the EXACT statement provides a small-sample binomial version of McNemar's test. PROC CATMOD can provide the confidence interval for the difference of proportions. The code forms a model for the marginal proportions in the first row and the first column, specifying a model matrix in the model statement that has an intercept parameter (the first column) that applies to both proportions and a slope parameter that applies only to the second; hence the second parameter is the difference between the second and first marginal proportions.

PROC LOGISTIC can conduct conditional logistic regression.

Table A.17 shows ways of testing marginal homogeneity for Table 10.6. The GENMOD code shows the Lipsitz et al. (1990) approach, expressing the I^2 expected frequencies in terms of parameters for the $(I - 1)^2$ cells in the first $I - 1$ rows and $I - 1$ columns, the cell in the last row and last column, and $I - 1$ marginal totals (which are the same for rows and columns). Here, m11 denotes expected frequency μ_{11}, m1 denotes $\mu_{1+} = \mu_{+1}$, and so on. This parameterization uses formulas such as $\mu_{14} = \mu_{1+} - \mu_{11} - \mu_{12} - \mu_{13}$ for terms in the last column or last row. CATMOD provides the Bhapkar test (10.16) of marginal homogeneity, as shown.

TABLE A.18   SAS Code Showing Square-Table Analysis of Table 10.5

```
data sex;
input premar extramar symm qi count @@;
unif = premar*extramar;
datalines;
1 1 1 1 144    1 2 2 5  2    1 3 3 5  0    1 4  4 5 0
2 1 2 5  33    2 2 5 2  4    2 3 6 5  2    2 4  7 5 0
3 1 3 5  84    3 2 6 5 14    3 3 8 3  6    3 4  9 5 1
4 1 4 5 126    4 2 7 5 29    4 3 9 5 25    4 4 10 4 5
;
proc genmod; class symm;
  model  count = symm / dist = poi  link = log; * symmetry;
proc genmod; class extramar premar symm;
  model  count = symm extramar premar / dist = poi  link = log; *QS;
proc genmod; class symm;
  model  count = symm extramar premar / dist = poi  link = log; * ordinal QS;
proc genmod; class extramar premar qi;
  model  count = extramar premar qi / dist = poi  link = log; * quasi indep;
proc genmod; class extramar premar;
  model  count = extramar premar unif / dist = poi  link = log;
data sex2;
input score below above @@;   trials = below + above;
datalines;
1 33 2    1 14 2    1 25 1    2 84 0    2 29 0    3 126 0
;
proc genmod data = sex2;
  model above / trials = score / dist = bin link = logit noint;
 proc genmod data = sex2;
  model  above / trials = / dist = bin  link = logit noint;
proc genmod data = sex2;
  model  above / trials = / dist = bin  link = logit;
```

Table A.18 shows various square-table analyses of Table 10.5. The "symm" factor indexes the pairs of cells that have the same association terms in the symmetry and quasi-symmetry models. For instance, "symm" takes the same value for cells $(1, 2)$ and $(2, 1)$. Including this term as a factor in a model invokes a parameter λ_{ij} satisfying $\lambda_{ij} = \lambda_{ji}$. The first model fits this factor alone, providing the symmetry model. The second model looks like the third except that it identifies "premar" and "extramar" as class variables (for quasi-symmetry), whereas the third model statement does not (for ordinal quasi-symmetry). The fourth model fits quasi-independence. The "qi" factor invokes the δ_i parameters. It takes a separate level for each cell on the main diagonal and a common value for all other cells. The fifth model fits the quasi-uniform association model (10.29).

The bottom of Table A.18 fits square-table models as logit models. The pairs of cell counts (n_{ij}, n_{ji}), labeled as "above" and "below" with reference to the main diagonal, are six sets of binomial counts. The variable defined as "score" is the distance $(u_j - u_i) = j - i$. The first two cases are symmetry

TABLE A.19   SAS Code for Fitting Bradley–Terry Model to Table 10.10

```
data baseball;
input wins games milw detr toro newy bost clev balt;
datalines;
7 13 1 -1  0  0  0  0  0
  ...
6 13 0  0  0  0  0  1 -1
;
proc genmod;
  model wins / games = milw detr toro newy bost clev balt /
  dist = bin link = logit noint covb;
```

and ordinal quasi-symmetry. Neither model contains an intercept (NOINT), and the ordinal model uses "score" as the predictor. The third model allows an intercept and is the conditional symmetry model (10.28).

Table A.19 uses GENMOD for logit fitting of the Bradley–Terry model to Table 10.10 by forming an artificial explanatory variable for each team. For a given observation, the variable for team i is 1 if it wins, -1 if it loses, and 0 if it is not one of the teams for that match. Each observation lists the number of wins ("wins") for the team with variate-level equal to 1 out of the number of games ("games") against the team with variate-level equal to -1. The model has these artificial variates, one of which is redundant, as explanatory variables with no intercept term. The COVB option provides the estimated covariance matrix of parameter estimators.

Chapter 11: Analyzing Repeated Categorical Response Data

Table A.20 uses GENMOD for the likelihood-ratio test of marginal homogeneity for Table 11.1, where for instance *m11p* denotes μ_{11+}. The marginal homogeneity model expresses the eight cell expected frequencies in terms of

TABLE A.20   SAS Code for Testing Marginal Homogeneity with Crossover Study of Table 11.1

```
data crossover;
input a b c count m111 m11p m1p1 mp11 m1pp m222 @@;
datalines;
1 1 1 6  1  0  0  0  0  0   1 1 2 16 -1  1  0  0  0  0
1 2 1 2 -1  0  1  0  0  0   1 2 2  4  1 -1 -1  0  1  0
2 1 1 2 -1  0  0  1  0  0   2 1 2  4  1 -1  0 -1  1  0
2 2 1 6  1  0 -1 -1  1  0   2 2 2  6  0  0  0  0  0  1
;
proc genmod;
  model count = m111 m11p m1p1 mp11 m1pp m222 / dist = poi link = identity;
proc catmod; weight count; response marginals;
  model a*b*c = _response_ /freq;
  repeated drug 3;
```

TABLE A.21   SAS Code for Marginal Modeling of Depression Data in Table 11.2

```
data depress;
input  case  diagnose  drug  time  outcome  @@;  * outcome = 1 is normal;
datalines;
  1  0  0  0 1   1  0  0  1 1   1  0  0  2 1
...
340  1  1  0  0 340  1  1  1  0 340  1  1  2  0
;
proc genmod descending;  class case;
  model; outcome = diagnose drug time drug*time / dist = bin  link = logit type3;
  repeated subject = case / type = exch corrw;
proc nlmixed  qpoints = 200;
  parms alpha = -.03 beta1 = -1.3 beta2 = -.06 beta3 = .48 beta4 = 1.02 sigma = .066;
  eta = alpha + beta1*diagnose  + beta2*drug + beta3*time + beta4*drug*time  + u;
  p = exp(eta) / (1 + exp(eta));
  model outcome ~ binary(p);
  random u ~ normal(0, sigma*sigma)  subject = case;
```

TABLE A.22   SAS Code for GEE and Random Intercept Cumulative Logit Analysis of Insomnia Data in Table 11.4

```
data francom;
  input  case  treat  time  outcome  @@;
datalines;
  1  1  0  1   1  1  1  1
...
239  0  0  4 239  0  1  4
;
proc genmod; class case;
  model outcome = treat time treat*time / dist = multinomial
      link = clogit;
  repeated subject = case / type = indep corrw;
proc nlmixed  qpoints = 40;
  bounds  i2>0;  bounds  i3>0;
  eta1 = i1 + treat*beta1 + time*beta2 + treat*time*beta3 + u;
  eta2 = i1 + i2 + treat*beta1 + time*beta2 + treat*time*beta3 + u;
  eta3 = i1 + i2 + i3 + treat*beta1 + time*beta2 + treat*time*beta3 + u;
  p1 = exp(eta) / (1 + exp(eta1));
  p2 = exp(eta2) / (1 + exp(eta2)) - exp(eta1) / (1 + exp(eta1));
  p3 = exp(eta3) / (1 + exp(eta3)) - exp(eta2) / (1 + exp(eta2));
  p4 = 1 - exp(eta3) / (1 + exp(eta3));
  ll = y1*log(p1) + y2*log(p2) + y3*log(p3) + y4*log(p4);
  model y1 ~ general(ll);
  estimate 'interc2 ' i1 + i2; * this is alpha_2 in model, and
      i1 is alpha_1;
  estimate 'interc3 ' i1 + i2 + i3; * this is alpha_3 in model;
  random u ~ normal(0, sigma*sigma) subject = case;
```

μ_{111}, μ_{11+}, μ_{1+1}, μ_{+11}, μ_{1++}, and μ_{222} (since $\mu_{+1+} = \mu_{++1} = \mu_{1++}$). Note, for instance, that $\mu_{112} = \mu_{11+} - \mu_{111}$ and $\mu_{122} = \mu_{111} + \mu_{1++} - \mu_{11+} - \mu_{1+1}$. CATMOD provides the generalized Bhapkar test (11.5) of marginal homogeneity.

Table A.21 uses GENMOD to analyze Table 11.2 using GEE. Possible working correlation structures are TYPE = EXCH for exchangeable, TYPE = AR for autoregressive, TYPE = INDEP for independence, and TYPE = UNSTR for unstructured. Output shows estimates and standard errors under the naive working correlation and based on the sandwich matrix incorporating the empirical dependence. Alternatively, the working association structure in the binary case can use the log odds ratio (e.g., using LOGOR = EXCH for exchangeability). The type 3 option in GEE provides score tests about effects. See Stokes et al. (2000, Sec. 15.11) for the use of GEE with missing data.

Table A.22 uses GENMOD to implement GEE for a cumulative logit model for Table 11.4. For multinomial responses, independence is currently the only working correlation structure.

Other

Joseph Lang (*jblang@stat.uiowa.edu*) has R and S-Plus functions for ML fitting of marginal models through the generalized loglinear model (11.8), using the constraint approach with Lagrange multipliers. The program MAREG (Kastner et al. 1997) provides GEE fitting and ML fitting of marginal models with the Fitzmaurice and Laird (1993) approach, allowing multicategory responses. See *www.stat.uni-muenchen.de/~andreas/mareg/ winmareg.html*.

Chapter 12: Random Effects: Generalized Linear Mixed Models

PROC NLMIXED extends GLMs to GLMMs by including random effects. Table A.23 analyzes the matched pairs model (12.3). Table A.24 analyzes the election data in Table 12.2.

TABLE A.23   SAS Code for Fitting Model (12.3) for Matched Pairs to Table 12.1

```
data matched;
input case occasion response count @@;
datalines;
2  0  1 794    1  1  1 794    2  0  1 150    2  1  0 150
3  0  0  86    3  1  1  86    4  0  0 570    4  1  0 570
;
proc nlmixed;
  eta = alpha + beta*occasion + u;  p = exp(eta) / (1 + exp(eta));
  model  response ~ binary(p);
  random u ~ normal(0, sigma*sigma) subject = case;
  replicate count;
```

TABLE A.24   SAS Code for GLMM Analysis of Election Data in Table 12.2

```
data vote;
input y n;
case = _n_;
datalines;
 1    5
16   32
...
 1    4
;
proc nlmixed;
  eta = alpha + u;   p = exp(eta) / (1 + exp(eta));
  model y ~ binomial(n,p);
  random u ~ normal (0, sigma*sigma) subject = case;
  predict p out = new;
proc print data = new;
```

TABLE A.25   SAS Code for GLMM Modeling of Opinions in Table 10.13

```
data new;
input sex poor single any count;
datalines;
1  1  1  1 342
...
2  0  0  0 457
;
data new;  set new;
  sex = sex - 1;   case = _n_;
  q1 = 1;  q2 = 0;  resp = poor;  output;
  q1 = 0,  q2 = 1;  resp = single;  output;
  q1 = 0;  q2 = 0;  resp = any;  output;
drop poor single any;
proc nlmixed  qpoints = 50;
  parms alpha = 0 beta1 = .8 beta2 = .3 gamma = 0 sigma = 8.6;
  eta = alpha + beta1*q1 + beta2*q2 + gamma*sex + u;
  p = exp(eta) / (1 + exp(eta));
  model resp ~ binary(p);
  random u ~ normal(0, sigma*sigma) subject = case;
  replicate count;
```

TABLE A.26 SAS Code for GLMM for Leading Crowd Data in Table 12.8

```
data crowd;
input mem1  att1  mem2  att2  count;
datalines;
  1  1  1  1 458
 ...
  0  0  0  0 554
;
data new; set crowd;
  case = _n_;
  x1m = 1;   x1a = 0;   x2m = 0;   x2a = 0;   var = 1;   resp = mem1;   output;
  x1m = 0;   x1a = 1;   x2m = 0;   x2a = 0;   var = 0;   resp = att1;   output;
  x1m = 0;   x1a = 0;   x2m - 1;   x2a = 0;   var = 1;   resp = mem2;   output;
  x1m = 0;   x1a = 0;   x2m = 0;   x2a = 1;   var = 0;   resp = att2;   output;
  drop mem1  att1  mem2  att2;
proc  nlmixed  data = new;
  eta = beta1m*x1m + beta1a*x1a + beta2m*x2m + beta2a*x2a + um*var +
    ua*(1-var);
  p = exp(eta) / (1 + exp(eta));
  model resp ~ binary(p);
  random  um  ua ~ normal([0,0],[s1*s1, cov12, s2*s2]) subject = case;
  replicate count;
  estimate 'mem change' beta2m-beta1m; estimate 'att change'
    beta2a-beta1a;
```

Table A.25 fits model (12.10) to Table 10.13. This shows how to set initial values and set the number of quadrature points for Gauss–Hermite quadrature (e.g., QPOINTS =). One could let SAS fit without initial values but then take that fit as initial values in further runs, increasing QPOINTS until estimates and standard errors converge to the necessary precision.

Table A.21 uses NLMIXED for Table 11.2. Table A.22 uses NLMIXED for ordinal modeling of Table 11.4, defining a general multinomial log likelihood. Table A.26 shows a correlated bivariate random effect analysis of Table 12.8. Agresti et al. (2000) showed NLMIXED examples for clustered data, Agresti and Hartzel (2000) showed code for multicenter trials such as Table 12.5, and Hartzel et al. (2001a) showed code for multicenter trials with an ordinal response. The Web site for the journal *Statistical Modelling* shows NLMIXED code for an adjacent-categories logit model and a nominal model at the data archive for Hartzel et al. (2001b). Chen and Kuo (2001) discussed fitting multinomial logit models, including discrete-choice models, with random effects.

Other

MLn (Institute of Education, London) and HLM (Scientific Software, Chicago) fit multilevel models. MIXOR is a FORTRAN program for ML

TABLE A.27   SAS Code for Overdispersion Analysis of Table 4.5

```
data moore;
input litter group n y @@;
  z2 = 0;  z3 = 0;  z4 = 0;
  if group = 2 then z2 = 1;  if group = 3 then z3 = 1;  if group = 4
     then z4 = 1;
datalines;
 1  1 10  1    2  1 11  4    3  1 12  9    4  1  4  4
...
55  4 14  1   56  4  8  0   57  4  6  0   58  4 17  0
;
proc logistic;
  model y / n = z2 z3 z4 / scale = williams;
proc logistic;
  model y / n = z2 z3 z4 / scale = pearson;
proc nlmixed  qpoints = 200;
  eta = alpha + beta2*z2 + beta3*z3 + beta4*z4 + u;
  p = exp(eta) / (1 + exp(eta));
  model y ~ binomial(n,p);
  random u ~ normal(0, sigma*sigma) subject = litter;
```

TABLE A.28   SAS Code for Fitting Models to Murder Data in Table 13.6

```
data new;
input white black other response;
datalines;
1070  119  55   0
  60   16   5   1
...
   1    0    0   6
;
data new; set new; count = white; race = 0; output;
  count = black; race = 1; output; drop white black other;
data new2; set new; do i = 1 to count; output; end; drop i;
proc genmod data = new2;
  model response = race / dist = negbin link = log;
proc genmod data = new2;
  model response = race / dist = poi  link = log scale = pearson;
data new; set new; case = _n_;
proc nlmixed data = new  qpoints = 400;
  parms alpha = -3.7 beta = 1.90 sigma = 1.6;
  eta = alpha + beta*race + u; mu = exp(eta);
  model response ~ poisson(mu);
  random u ~ normal(0, sigma*sigma) subject = case;
  replicate count;
```

fitting of binary and ordinal random effects models available from Don Hedeker (*www.uic.edu/~hedeker/mix.html*).

Chapter 13: Other Mixture Models for Categorical Data

PROC LOGISTIC provides two overdispersion approaches for binary data. The SCALE = WILLIAMS option uses variance function of the beta-binomial form (13.10), and SCALE = PEARSON uses the scaled binomial variance (13.11). Table A.27 illustrates for Table 4.5. That table also uses NLMIXED for adding litter random intercepts.

For Table 13.6, Table A.28 uses GENMOD to fit a negative binomial model and a quasi-likelihood model with scaled Poisson variance using the Pearson statistic, and NLMIXED to fit a Poisson GLMM. PROC NLMIXED can also fit negative binomial models.

Other

Latent GOLD (developed by J. Vermunt and J. Magidson for Statistical Innovations, Belmont, Massachusetts) can fit a wide variety of mixture models, including latent class models, nonparametric mixtures of logistic regression, and some Rasch mixture models.

APPENDIX B

Chi-Squared Distribution Values

df	Right-Tailed Probability						
	0.250	0.100	0.050	0.025	0.010	0.005	0.001
1	1.32	2.71	3.84	5.02	6.63	7.88	10.83
2	2.77	4.61	5.99	7.38	9.21	10.60	13.82
3	4.11	6.25	7.81	9.35	11.34	12.84	16.27
4	5.39	7.78	9.49	11.14	13.28	14.86	18.47
5	6.63	9.24	11.07	12.83	15.09	16.75	20.52
6	7.84	10.64	12.59	14.45	16.81	18.55	22.46
7	9.04	12.02	14.07	16.01	18.48	20.28	24.32
8	10.22	13.36	15.51	17.53	20.09	21.96	26.12
9	11.39	14.68	16.92	19.02	21.67	23.59	27.88
10	12.55	15.99	18.31	20.48	23.21	25.19	29.59
11	13.70	17.28	19.68	21.92	24.72	26.76	31.26
12	14.85	18.55	21.03	23.34	26.22	28.30	32.91
13	15.98	19.81	22.36	24.74	27.69	29.82	34.53
14	17.12	21.06	23.68	26.12	29.14	31.32	36.12
15	18.25	22.31	25.00	27.49	30.58	32.80	37.70
16	19.37	23.54	26.30	28.85	32.00	34.27	39.25
17	20.49	24.77	27.59	30.19	33.41	35.72	40.79
18	21.60	25.99	28.87	31.53	34.81	37.16	42.31
19	22.72	27.20	30.14	32.85	36.19	38.58	43.82
20	23.83	28.41	31.41	34.17	37.57	40.00	45.32
25	29.34	34.38	37.65	40.65	44.31	46.93	52.62
30	34.80	40.26	43.77	46.98	50.89	53.67	59.70
40	45.62	51.80	55.76	59.34	63.69	66.77	73.40
50	56.33	63.17	67.50	71.42	76.15	79.49	86.66
60	66.98	74.40	79.08	83.30	88.38	91.95	99.61
70	77.58	85.53	90.53	95.02	100.4	104.2	112.3
80	88.13	96.58	101.8	106.6	112.3	116.3	124.8
90	98.65	107.6	113.1	118.1	124.1	128.3	137.2
100	109.1	118.5	124.3	129.6	135.8	140.2	149.5

Source: Calculated using *StaTable*, Cytel Software, Cambridge, MA.

References

Adelbasit, K. M., and R. L. Plackett. 1983. Experimental design for binary data. *J. Amer. Statist. Assoc.* **78**: 90–98.

Agresti, A. 1984. *Analysis of Ordinal Categorical Data.* New York: Wiley.

Agresti, A. 1992. A survey of exact inference for contingency tables. *Statist. Sci.* **7**: 131–153.

Agresti, A. 1993. Computing conditional maximum likelihood estimates for generalized Rasch models using simple loglinear models with diagonal parameters. *Scand. J. Statist.* **20**: 63–71.

Agresti, A. 1997. A model for repeated measurements of a multivariate binary response. *J. Amer. Statist. Assoc.* **92**: 315–321.

Agresti, A. 1999. On logit confidence intervals for the odds ratio with small samples. *Biometrics* **55**: 597–602.

Agresti, A. 2001. Exact inference for categorical data: Recent advances and continuing controversies. *Statist. Medic.* **20**: 2709–2722.

Agresti, A., and B. Caffo. 2000. Simple and effective confidence intervals for proportions and difference of proportions result from adding two successes and two failures. *Amer. Statist.* **54**: 280–288.

Agresti, A., and B. A. Coull. 1998. Approximate is better than exact for interval estimation of binomial parameters. *Amer. Statist.* **52**: 119–126.

Agresti, A., and J. Hartzel. 2000. Strategies for comparing treatments on a binary response with multi-centre data. *Statist. Medic.* **19**(8): 1115–1139.

Agresti, A., and J. Lang. 1993a. A proportional odds model with subject-specific effects for repeated ordered categorical responses. *Biometrika* **80**: 527–534.

Agresti, A., and J. Lang. 1993b. Quasi-symmetric latent class models, with application to rater agreement. *Biometrics* **49**: 131–139.

Agresti, A., and I. Liu. 1999. Modeling a categorical variable allowing arbitrarily many category choices. *Biometrics* **55**: 936–943.

Agresti, A., and Y. Min. 2001. On small-sample confidence intervals for parameters in discrete distributions. *Biometrics* **57**: 963–971.

Agresti, A., and R. Natarajan. 2001. Modeling clustered ordered categorical data: A survey. *Internal. Statist. Rev.* **69**: 345–371.

Agresti, A., D. Wackerly, and J. Boyett. 1979. Exact conditional tests for cross-classifications: Approximation of attained significance levels. *Psychometrika* **44**: 75–84.

Agresti, A., C. Chuang, and A. Kezouh. 1987. Order-restricted score parameters in association models for contingency tables. *J. Amer. Statist. Assoc.* **82**: 619–623.

Agresti, A., C. R. Mehta, and N. R. Patel. 1990. Exact inference for contingency tables with ordered categories. *J. Amer. Statist. Assoc.* **85**: 453–458.

Agresti, A., J. Booth, J. Hobert, and B. Caffo. 2000. Random-effects modeling of categorical response data. *Sociol. Methodol.* **30**: 27–81.

Aitchison, J., and C. G. G. Aitken. 1976. Multivariate binary discrimination by the kernel method. *Biometrika* **63**: 413–420.

Aitchison, J., and C. H. Cho. 1989. The multivariate Poisson-log normal distribution. *Biometrika* **76**: 643–653.

Aitchison, J., and S. M. Shen. 1980. Logistic-normal distributions: Some properties and uses. *Biometrika* **67**: 261–272.

Aitchison, J., and S. D. Silvey. 1957. The generalization of probit analysis to the case of multiple responses. *Biometrika* **44**: 131–140.

Aitchison, J., and S. D. Silvey. 1958. Maximum likelihood estimation of parameters subject to restraints. *Ann. Math. Statist.* **29**: 813–828.

Aitkin, M. 1979. A simultaneous test procedure for contingency table models. *Appl. Statist.* **28**: 233–242.

Aitkin, M. 1980. A note on the selection of log-linear models. *Biometrics* **36**: 173–178.

Aitkin, M. 1999. A general maximum likelihood analysis of variance components in generalized linear models. *Biometrics* **55**: 117–128.

Aitkin, M., and D. Clayton. 1980. The fitting of exponential, Weibull, and extreme value distributions to complex censored survival data using GLIM. *Appl. Statist.* **29**: 156–163.

Aitkin, M., and M. Stasinopoulos. 1989. Likelihood analysis of a binomial sample size problem. Pp. 399–411 in *Contributions to Probability and Statistics*: *Essays in Honor of Ingram Olkin*, ed. L. J. Gleser, M. D. Perlman, S. J. Press, and A. R. Sampson. New York: Springer-Verlag.

Aitkin, M., D. Anderson, and J. Hinde. 1981. Statistical modelling of data on teaching styles. *J. Roy. Statist. Soc. Ser. A* **144**: 419–461.

Aitkin, M., D. Anderson, B. Francis, and J. Hinde. 1989. *Statistical Modeling in GLIM*. Oxford: Clarendon Press.

Albert, J. H. 1997. Bayesian testing and estimation of association in a two-way contingency table. *J. Amer. Statist. Assoc.* **92**: 685–693.

Albert, A., and J. A. Anderson. 1984. On the existence of maximum likelihood estimates in logistic models. *Biometrika* **71**: 1–10.

Albert, J. H., and S. Chib. 1993. Bayesian analysis of binary and polychotomous response data. *J. Amer. Statist. Assoc.* **88**: 669–679.

Albert, J. H., and A. K. Gupta. 1982. Mixtures of Dirichlet distributions and estimation in contingency tables. *Ann. Statist.* **10**: 1261–1268.

Allison, P. D. 1999. *Logistic Regression Using the SAS System*. Cary, NC: SAS Institute.

Altham, P. M. E. 1969. Exact Bayesian analysis of a 2×2 contingency table and Fisher's "exact" significance test. *J. Roy. Statist. Soc. Ser B* **31**: 261–269.

Altham, P. M. E. 1970. The measurement of association of rows and columns for an $r \times s$ contingency table. *J. Roy. Statist. Soc. Ser B* **32**: 63–73.

Altham, P. M. E. 1971. The analysis of matched proportions. *Biometrika* **58**: 561–576.

Altham, P. M. E. 1975. Quasi-independent triangular contingency tables. *Biometrics* **31**: 233–238.

Altham, P. M. E. 1978. Two generalizations of the binomial distribution. *Appl. Statist.* **27**: 162–167.

Altham, P. M. E. 1984. Improving the precision of estimation by fitting a model. *J. Roy. Statist. Soc. Ser B* **46**: 118–119.

Amemiya, T. 1981. Qualitative response models: A survey. *J. Econom. Literature* **19**: 1483–1536.

Andersen, E. B. 1970. Asymptotic properties of conditional maximum-likelihood estimators. *J. Roy. Statist. Soc. Ser B* **32**: 283–301.

Andersen, E. B. 1980. *Discrete Statistical Models with Social Science Applications*. Amsterdam: North-Holland.

Andersen, E. B. 1995. Polytomous Rasch models and their estimation. Pp. 272–291 in *Rasch Models: Foundations, Recent Developments, and Applications*, eds. G. Fischer and I. Molenaar. New York: Springer-Verlag.

Anderson, J. A. 1972. Separate sample logistic discrimination. *Biometrika* **59**: 19–35.

Anderson, J. A. 1975. Quadratic logistic discrimination. *Biometrika* **62**: 149–154.

Anderson, J. A. 1984. Regression and ordered categorical variables. *J. Roy. Statist. Soc. Ser B* **46**: 1–30.

Anderson, D. A., and M. Aitkin. 1985. Variance component models with binary response: Interviewer variability. *J. Roy. Statist. Soc. Ser B* **47**: 203–210.

Anderson, C. J., and U. Böckenholt. 2000. Graphical regression models for polytomous variables. *Psychometrika* **65**: 497–509.

Anderson, T. W., and L. A. Goodman. 1957. Statistical inference about Markov chains. *Ann. Math. Statist.* **28**: 89–110.

Anderson, J. A., and P. R. Philips. 1981. Regression, discrimination, and measurement models for ordered categorical variables. *Appl. Statist.* **30**: 22–31.

Anderson, C. J., and J. K. Vermunt. 2000. Log-multiplicative models as latent variable models for nominal and/or ordinal data. *Sociol. Methodol.* **30**: 81–121.

Aranda-Ordaz, F. J. 1981. On two families of transformations to additivity for binary response data. *Biometrics* **68**: 357–363.

Aranda-Ordaz, F. J. 1983. An extension of the proportional hazards model for grouped data. *Biometrics* **39**: 109–117.

Arminger, G., C. C. Clogg, and T. Cheng. 2000. Regression analysis of multivariate binary response variables using Rasch-type models and finite mixture methods. *Sociol. Methodol.* **30**: 1–26.

Armitage, P. 1955. Tests for linear trends in proportions and frequencies. *Biometrics* **11**: 375–386.

Ashford, J. R., and R. D. Sowden. 1970. Multivariate probit analysis. *Biometrics* **26**: 535–546.

Asmussen, S., and D. Edwards. 1983. Collapsibility and response variables in contingency tables. *Biometrika* **70**: 567–578.

Azzalini, A. 1994. Logistic regression for autocorrelated data with application to repeated measures. *Biometrika* **81**: 767–775.

Baglivo, J., D. Olivier, and M. Pagano. 1992. Methods for exact goodness-of-fit tests. *J. Amer. Statist. Assoc.* **87**: 464–469.

Baker, S. G. 1992. A simple method for computing the observed information matrix when using the EM algorithm with categorical data. *J. Comput. Graph. Statist.* **1**: 63–76.

Baker, S. G., and N. M. Laird. 1988. Regression analysis for categorical variables with outcome subject to nonignorable nonresponse. *J. Amer. Statist. Assoc.* **83**: 62–69.

Baker, R. J., M. R. B. Clarke, and P. W. Lane. 1985. Zero entries in contingency tables. *Comput. Statist. Data Anal.* **3**: 33–45.

Banerjee, C., M. Capozzoli, L. McSweeney, and D. Sinha. 1999. Beyond kappa: A review of interrater agreement measures. *Canad. J. Statist.* **27**: 3–23.

Baptista, J., and M. C. Pike. 1977. Algorithm AS115: Exact two-sided confidence limits for the odds ratio in a 2 × 2 table. *Appl. Statist.* **26**: 214–220.

Barnard, G. A. 1945. A new test for 2 × 2 tables. *Nature* **156**: 177.

Barnard, G. A. 1947. Significance tests for 2 × 2 tables. *Biometrika* **34**: 123–138.

Barnard, G. A. 1949. Statistical inference. *J. Roy. Statist. Soc. Ser B* **11**: 115–139.

Barnard, G. A. 1979. In contradiction to J. Berkson's dispraise: Conditional tests can be more efficient. *J. Statist. Plann. Inference* **3**: 181–188.

Barndorff-Nielsen, O. E., and B. Jörgensen. 1991. Some parametric models on the simplex. *J. Multivariate Anal.* **39**: 106–116.

Bartholomew, D. J. 1980. Factor analysis for categorical data. *J. Roy. Statist. Soc. Ser B* **42**: 293–321.

Bartholomew, D. J., and M. Knott. 1999. *Latent Variable Models and Factor Analysis*, 2nd ed. London: Edward Arnold.

Bartlett, M. S. 1935. Contingency table interactions. *J. Roy. Statist. Soc. Suppl.* **2**: 248–252.

Bartlett, M. S. 1937. Some examples of statistical methods of research in agriculture and applied biology. *J. Roy. Statist. Soc. Suppl.* **4**: 137–183.

Becker, M. 1989a. Models for the analysis of association in multivariate contingency tables. *J. Amer. Statist. Assoc.* **84**: 1014–1019.

Becker, M. 1989b. On the bivariate normal distribution and association models for ordinal categorical data. *Statist. Probab. Lett.* **8**: 435–440.

Becker, M. 1990. Maximum likelihood estimation of the RC(M) association model. *Appl. Statist.* **39**: 152–167.

Becker, M., and A. Agresti. 1992. Log-linear modelling of pairwise interobserver agreement on a categorical scale. *Statist. Medic.* **11**: 101–114.

Becker, M., and C. C. Clogg. 1989. Analysis of sets of two-way contingency tables using association models. *J. Amer. Statist. Assoc.* **84**: 142–151.

Bedrick, E. J. 1983. Chi-squared tests for cross-classified tables of survey data. *Biometrika* **70**: 591–595.

Bedrick, E. J. 1987. A family of confidence intervals for the ratio of two binomial proportions. *Biometrics* **43**: 993–998.

Begg, C. B., and R. Gray. 1984. Calculation of polytomous logistic regression parameters using individualized regressions. *Biometrika* **71**: 11–18.

Beitler, P. J., and J. R. Landis. 1985. A mixed-effects model for categorical data. *Biometrics* **41**: 991–1000.

Benedetti, J. K., and M. B. Brown. 1978. Strategies for the selection of loglinear models. *Biometrics* **34**: 680–686.

Benichou, J. 1998. Attributable risk. Pp. 216–229 in *Encyclopedia of Biostatistics*. Chichester, UK: Wiley.

Benzécri, J.-P. 1973. *L'Analyse des Données*, Vol. 1, *La Taxonomie*; Vol. 2, *L'Analyse des Correspondances*. Paris: Dunod.

Berger, R., and D. D. Boos. 1994. *p*-Values maximized over a confidence set for the nuisance parameter. *J. Amer. Statist. Assoc.* **89**: 1012–1016.

Bergsma, W. P., and T. Rudas. 2002. Marginal models for categorical data. *Ann. Statist.* **30**: 140–159.

Berkson, J. 1938. Some difficulties of interpretation encountered in the application of the chi-square test. *J. Amer. Statist. Assoc.* **33**: 526–536.

Berkson, J. 1944. Application of the logistic function to bio-assay. *J. Amer. Statist. Assoc.* **39**: 357–365.

Berkson, J. 1951. Why I prefer logits to probits. *Biometrics* **7**: 327–339.

Berkson, J. 1953. A statistically precise and relatively simple method of estimating the bioassay with quantal response, based on the logistic function. *J. Amer. Statist. Assoc.* **48**: 565–599.

Berkson, J. 1955. Maximum likelihood and minimum logit χ^2 estimation of the logistic function. *J. Amer. Statist. Assoc.* **50**: 130–162.

Berkson, J. 1978. In dispraise of the exact test. *J. Statist. Plann. Inference* **2**: 27–42.

Berkson, J. 1980. Minimum chi-square, not maximum likelihood! *Ann. Statist.* **8**: 457–487.

Berry, G., and P. Armitage. 1995. Mid-*P* confidence intervals: A brief review. *The Statistician* **44**: 417–423.

Bhapkar, V. P. 1966. A note on the equivalence of two test criteria for hypotheses in categorical data. *J. Amer. Statist. Assoc.* **61**: 228–235.

Bhapkar, V. P. 1968. On the analysis of contingency tables with a quantitative response. *Biometrics* **24**: 329–338.

Bhapkar, V. P. 1973. On the comparison of proportions in matched samples. *Sankhya Ser A* **35**: 341–356.

Bhapkar, V. P. 1989. Conditioning on ancillary statistics and loss of information in the presence of nuisance parameters. *J. Statist. Plann. Inference.* **21**: 139–160.

Bhapkar, V. P., and G. G. Koch. 1968. On the hypothesis of "no interaction" in multidimensional contingency tables. *Biometrics* **24**: 567–594.

Bhapkar, V. P., and G. W. Somes. 1977. Distribution of *Q* when testing equality of matched proportions. *J. Amer. Statist. Assoc.* **72**: 658–661.

Biggeri, A. 1998. Negative binomial distribution. Pp. 2962–2967 in *Encyclopedia of Biostatistics*. Chichester, UK: Wiley.

Billingsley, P. 1961. Statistical methods in Markov chains. *Ann. Math. Statist.* **32**: 12–40.

Birch, M. W. 1963. Maximum likelihood in three-way contingency tables. *J. Roy. Statist. Soc. Ser. B* **25**: 220–233.

Birch, M. W. 1964a. A new proof of the Pearson–Fisher theorem. *Ann. Math. Statist.* **35**: 817–824.

Birch, M. W. 1964b. The detection of partial association I: The 2 × 2 case. *J. Roy. Statist. Soc. Ser. B* **26**: 313–324.

Birch, M. W. 1965. The detection of partial association II: The general case. *J. Roy. Statist. Soc. Ser B* **27**: 111–124.

Bishop, Y. M. M. 1971. Effects of collapsing multidimensional contingency tables. *Biometrics* **27**: 545–562.

Bishop, Y. M. M., and F. Mosteller. 1969. Smoothed contingency table analysis. Chap. IV-3 in *The National Halothane Study*. Washington, DC: U.S. Government Printing Office.

Bishop, Y. M. M., S. E. Fienberg, and P. W. Holland. 1975. *Discrete Multivariate Analysis*. Cambridge, MA: MIT Press.

Blaker, H. 2000. Confidence curves and improved exact confidence intervals for discrete distributions. *Canad. J. Statist.* **28**: 783–798.

Bliss, C. I. 1934. The method of probits. *Science* **79**: 38–39.

Bliss, C. I. 1935. The calculation of the dosage–mortality curve. *Ann. Appl. Biol.* **22**: 134–167.

Blyth, C. R. 1972. On Simpson's paradox and the sure-thing principle. *J. Amer. Statist. Assoc.* **67**: 364–366.

Blyth, C. R., and H. A. Still. 1983 Binomial confidence intervals. *J. Amer. Statist. Assoc.* **78**: 108–116.

Bock, R. D. 1970. Estimating multinomial response relations. Pp. 453–479 in *Contributions to Statistics and Probability*, ed. R. C. Bose. Chapel Hill, NC: University of North Carolina Press.

Bock, R. D., and M. Aitkin. 1981. Marginal maximum likelihood estimation of item parameters: Application of an EM algorithm. *Psychometrika* **46**: 443–459.

Bock, R. D., and L. V. Jones. 1968. *The Measurement and Prediction of Judgement and Choice*. San Francisco: Holden-Day.

Böckenholt, U., and W. Dillon. 1997. Modelling within-subject dependencies in ordinal paired comparison data. *Psychometrika* **62**: 411–434.

Bonney, G. E. 1987. Logistic regression for dependent binary observations. *Biometrics* **43**: 951–973.

Boos, D. D. 1992. On generalized score tests. *Amer. Statist.* **46**: 327–333.

Booth, J., and R. Butler. 1999. An importance sampling algorithm for exact conditional tests in log-linear models. *Biometrika* **86**: 321–332.

Booth, J. G., and J. P. Hobert. 1998. Standard errors of prediction in generalized linear mixed models. *J. Amer. Statist. Assoc.* **93**: 262–272.

Booth, J. G., and J. P. Hobert. 1999. Maximizing generalized linear mixed model likelihoods with an automated Monte Carlo EM algorithm. *J. Roy. Statist. Soc. Ser. B* **61**: 265–285.

Bowker, A. H. 1948. A test for symmetry in contingency tables. *J. Amer. Statist. Assoc.* **43**: 572–574.

Box, J. F. 1978. *R. A. Fisher: The Life of a Scientist*. New York: Wiley

Bradley, R. A. 1976. Science, statistics, and paired comparisons. *Biometrics* **32**: 213–240.

Bradley, R. A., and M. E. Terry. 1952. Rank analysis of incomplete block designs I. The method of paired comparisons. *Biometrika* **39**: 324–345.

Breslow, N. 1976. Regression analysis of the log odds ratio: A method for retrospective studies. *Biometrics* **32**: 409–416.

Breslow, N. 1981. Odds ratio estimators when the data are sparse. *Biometrika* **68**: 73–84.

Breslow, N. 1982. Covariance adjustment of relative-risk estimates in matched studies. *Biometrics* **38**: 661–672.

Breslow, N. 1984. Extra-Poisson variation in log-linear models. *Appl. Statist.* **33**: 38–44.

Breslow, N. 1996. Statistics in epidemiology: The case–control study. *J. Amer. Statist. Assoc.* **91**: 14–28.

Breslow, N., and D. G. Clayton. 1993. Approximate inference in generalized linear mixed models. *J. Amer. Statist. Assoc.* **88**: 9–25.

Breslow, N., and N. E. Day. 1980, 1987. *Statistical Methods in Cancer Research*, Vol. I, *The Analysis of Case–Control Studies*; Vol. II. *The Design and Analysis of Cohort Studies*. Lyon: IARC.

Breslow, N., and X. Lin. 1995. Bias correction in generalised linear mixed models with a single component of dispersion. *Biometrika* **82**: 81–91.

Breslow, N., and W. Powers. 1978. Are there two logistic regressions for retrospective studies? *Biometrics* **34**: 100–105.

Breslow, N., N. Day, K. Halvorsen, R. Prentice, and C. Sabai. 1978. Estimation of multiple relative risk functions in matched case–control studies. *Amer. J. Epidemiol.* **108**: 299–307.

Brier, S. S. 1980. Analysis of contingency tables under cluster sampling. *Biometrika* **67**: 591–596.

Brooks, S. P., B. J. T. Morgan, M. S. Ridout, and S. E. Pack. 1997. Finite mixture models for proportions. *Biometrics* **53**: 1097–1115.

Bross, I. D. J. 1958. How to use ridit analysis. *Biometrics* **14**: 18–38.

Brown, M. B. 1976. Screening effects in multidimensional contingency tables. *Appl. Statist.* **25**: 37–46.

Brown, M. B., and J. K. Benedetti. 1977. Sampling behavior of tests for correlation in two-way contingency tables. *J. Amer. Statist. Assoc.* **72**: 309–315.

Brown, P. J., and P. W. K. Rundell. 1985. Kernel estimates for categorical data. *Technometrics* **27**: 293–299.

Brown, L. D., T. T. Cai, and A. Das Gupta. 2001. Interval estimation for a binomial proportion. *Statist. Sci.* **16**: 101–133.

Brownstone, D., and K. F. Train. 1999. Forecasting new product penetration with flexible substitution patterns. *J. Econometrics* **89**: 109–129.

Bull, S. B., and A. Donner. 1987. The efficiency of multinomial logistic regression compared with multiple group discriminant analysis. *J. Amer. Statist. Assoc.* **82**: 1118–1122.

Burnham, K. P., and D. R. Anderson. 1998. *Model Selection and Inference: A Practical Information-Theoretic Approach.* New York: Springer-Verlag.

Burnham, K. P. and W. S. Overton. 1978. Estimation of the size of a closed population when capture probabilities vary among animals. *Biometrika* **65**: 625–633.

Burridge, J. 1981. A note on maximum likelihood estimation for regression models using grouped data. *J. Roy. Statist. Soc. Ser. B* **43**: 41–45.

Cameron, A. C., and P. K. Trivedi. 1998. *Regression Analysis of Count Data.* Cambridge, U.K.: Cambridge University Press.

Carey, V., S. L. Zeger, and P. Diggle. 1993. Modelling multivariate binary data with alternating logistic regressions. *Biometrika* **80**: 517–526.

Carroll, R. J., S. Wang, and C. Y. Wang. 1995. Prospective analysis of logistic case–control pairs. *J. Amer. Statist. Assoc.* **90**: 157–169.

Casella, G., and R. Berger. 2001. *Statistical Inference*, 2nd ed. Pacific Grove, CA: Wadsworth.

Catalano, P. J., and L. M. Ryan. 1992. Bivariate latent variable models for clustered discrete and continuous outcomes. *J. Amer. Statist. Assoc.* **87**: 651–658.

Caussinus, H. 1966. Contribution à l'analyse statistique des tableaux de corrélation. *Ann. Fac. Sci. Univ. Toulouse* **29**: 77–182.

Chaloner, K., and K. Larntz. 1989. Optimal Bayesian design applied to logistic regression experiments. *J. Statist. Plann. Inference* **21**: 191–208.

Chamberlain, G. 1980. Analysis of covariance with qualitative data. *Rev. Econ. Stud.* **47**: 225–238.

Chambers, E. A., and D. R. Cox. 1967. Discrimination between alternative binary response models. *Biometrika* **54**: 573–578.

Chambers, R. L., and D. G. Steel. 2001. Simple methods for ecological inference in 2×2 tables. *J. Roy. Statist. Soc. Ser. A* **164**: 175–192.

Chan, I. 1998. Exact tests of equivalence and efficacy with non-zero lower bound for comparative studies. *Statist. Medic.* **17**: 1403–1413.

Chan, J. S. K., and A. Y. C. Kuk. 1997. Maximum likelihood estimation for probit-linear mixed models with correlated random effects. *Biometrics* **53**: 86–97.

Chao, A., P. K. Tsay, S.-H. Lin, W.-Y. Shau, and D.-Y. Chao. 2001. The applications of capture–recapture models to epidemiological data. *Statist. Medic.* **20**: 3123–3157.

Chapman, D. G., and R. C. Meng. 1966. The power of chi-square tests for contingency tables. *J. Amer. Statist. Assoc.* **61**: 965–975.

Chen, Z. and L. Kuo. 2001. A note on the estimation of the multinomial logit model with random effects. *Amer. Statist.* **55**: 89–95.

Christensen, R. 1997. *Log–Linear Models and Logistic Regression.* New York: Springer-Verlag.

Chuang, C., D. Gheva, and C. Odoroff. 1985. Methods for diagnosing multiplicative-interaction models for two-way contingency tables. *Commun. Statist. Ser. A* **14**: 2057–2080.

Clogg, C. C. 1995. Latent class models. Pp. 311–359 in *Handbook of Statistical Modeling for the Social and Behavioral Sciences*, ed. G. Arminger and C. C. Clogg. New York: Plenum Press.

Clogg, C. C., and S. R. Eliason. 1987. Some common problems in log-linear analysis. *Sociol. Methods Res.* **15**: 4–44.

Clogg, C. C., and L. A. Goodman. 1984. Latent structure analysis of a set of multidimensional contingency tables. *J. Amer. Statist. Assoc.* **79**: 762–771.

Clogg, C. C., and E. S. Shihadeh. 1994. *Statistical Models for Ordinal Variables*. Thousand Oaks, CA: Sage Publications.

Clopper, C. J., and E. S. Pearson. 1934. The use of confidence or fiducial limits illustrated in the case of the binomial. *Biometrika* **26**: 404–413.

Cochran, W. G. 1940. The analysis of variance when experimental errors follow the Poisson or binomial laws. *Ann. Math. Statist.* **11**: 335–347.

Cochran, W. G. 1943. Analysis of variance for percentages based on unequal numbers. *J. Amer. Statist. Assoc.* **38**: 287–301.

Cochran, W. G. 1950. The comparison of percentages in matched samples. *Biometrika* **37**: 256–266.

Cochran, W. G. 1952. The χ^2 test of goodness-of-fit. *Ann. Math. Statist.* **23**: 315–345.

Cochran, W. G. 1954. Some methods of strengthening the common χ^2 tests. *Biometrics* **10**: 417–451.

Cochran, W. G. 1955. A test of a linear function of the deviations between observed and expected numbers. *J. Amer. Statist. Assoc.* **50**: 377–397.

Coe, P. R., and A. C. Tamhane. 1993. Small sample confidence intervals for the difference, ratio and odds ratio of two success probabilities. *Commun. Statist. Ser. B* **22**: 925–938.

Cohen, J. 1960. A coefficient of agreement for nominal scales. *Educ. Psychol. Meas.* **20**: 37–46.

Cohen, J. 1968. Weighted kappa: Nominal scale agreement with provision for scaled disagreement or partial credit. *Psychol. Bull.* **70**: 213–220.

Cohen, A., and H. B. Sackrowitz. 1991. Tests for independence in contingency tables with ordered alternatives. *J. Multivariate Anal.* **36**: 56–67.

Cohen, A., and H. B. Sackrowitz. 1992. An evaluation of some tests of trend in contingency tables. *J. Amer. Statist. Assoc.* **87**: 470–475.

Collett, D. 1991. *Modelling Binary Data*. London: Chapman & Hall.

Conaway, M. R. 1989. Analysis of repeated categorical measurements with conditional likelihood methods. *J. Amer. Statist. Assoc.* **84**: 53–62.

Cook, R. D., and S. Weisberg. 1999. *Applied Regression Including Computing and Graphics*. New York: Wiley.

Copas, J. B. 1973. Randomization models for the matched and unmatched 2×2 tables. *Biometrika* **60**: 467–476.

Copas, J. B. 1983. Plotting p against x. *Appl. Statist.* **32**: 25–31.

Copas, J. B. 1988. Binary regression models for contaminated data. *J. Roy. Statist. Soc. Ser B* **50**: 225–265.

Corcoran, C., L. Ryan, P. Senchaudhuri, C. Mehta, N. Patel, and G. Molenberghs. 2001. An exact trend test for correlated binary data. *Biometrics* **57**: 941–948.

Cormack, R. M. 1989. Log-linear models for capture–recapture. *Biometrics* **45**: 395–413.

Cornfield, J. 1951. A method of estimating comparative rates from clinical data: Applications to cancer of the lung, breast and cervix. *J. Natl. Cancer Inst.* **11**: 1269–1275.

Cornfield, J. 1956. A statistical problem arising from retrospective studies. In *Proc. 3rd Berkeley Symposium on Mathematics, Statistics and Probability*, ed. J. Neyman, **4**: 135–148.

Cornfield, J. 1962. Joint dependence of risk of coronary heart disease on serum cholesterol and systolic blood pressure: A discriminant function analysis. *Fed. Proc.* **21**, *Suppl.* **11**: 58–61.

Coull, B. A., and A. Agresti. 1999. The use of mixed logit models to reflect heterogeneity in capture–recapture studies. *Biometrics* **55**: 294–301.

Coull, B. A., and A. Agresti. 2000. Random effects modeling of multiple binomial responses using the multivariate binomial logit-normal distribution. *Biometrics* **56**: 73–80.

Cox, C. 1984. An elementary introduction to maximum likelihood estimation for multinomial models: Birch's theorem and the delta method. *Amer. Statist.* **38**: 283–287.

Cox, C. 1995. Location-scale cumulative odds models for ordinal data: A generalized non-linear model approach. *Statist. Medic.* **14**: 1191–1203.

Cox, C. 1996. Nonlinear quasi-likelihood models: Applications to continuous proportions. *Comput. Statist. Data Anal.* **21**: 449–461.

Cox, D. R. 1958a. The regression analysis of binary sequences. *J. Roy. Statist. Soc. Ser. B* **20**: 215–242.

Cox, D. R. 1958b. Two further applications of a model for binary regression. *Biometrika* **45**: 562–565.

Cox, D. R. 1970. *The Analysis of Binary Data* (2nd ed. 1989, by D. R. Cox and E. J. Snell). London: Chapman & Hall.

Cox, D. R. 1972. The analysis of multivariate binary data. *Appl. Statist.* **21**: 113–120.

Cox, D. R. 1983. Some remarks on overdispersion. *Biometrika* **70**: 269–274.

Cox, D. R., and D. V. Hinkley. 1974. *Theoretical Statistics*. London: Chapman & Hall.

Cramér, H. 1946. *Mathematical Methods of Statistics*. Princeton, NJ: Princeton University Press.

Cressie, N., and T. R. C. Read. 1984. Multinomial goodness-of-fit tests. *J. Roy. Statist. Soc. Ser. B* **46**: 440–464.

Cressie, N., and T. R. C. Read. 1989. Pearson X^2 and the loglikelihood ratio statistic G^2: A comparative review. *Internat. Statist. Rev.* **57**: 19–43.

Croon, M., W. Bergsma, and J. Hagenaars. 2000. Analyzing change in categorical variables by generalized log-linear models. *Sociol. Methods Res.* **29**: 195–229.

Crouchley, R. 1995. A random-effects model for ordered categorical data. *J. Amer. Statist. Assoc.* **90**: 489–498.

Crowder, M. J. 1978. Beta-binomial ANOVA for proportions. *Appl. Statist.* **27**: 34–37.

D'Agostino, R. B., Jr. 1998. Propensity score methods for bias reduction in the comparison of a treatment to a non-randomized control group. *Statist. Medic.* **17**: 2265–2281.

Daniels, M. J., and C. Gatsonis. 1999. Hierarchical generalized linear models in the analysis of variations in health care utilization. *J. Amer. Statist. Assoc.* **94**: 29–42.

Dardanoni, V., and A. Forcina. 1998. A unified approach to likelihood inference on stochastic orderings in a nonparametric context. *J. Amer. Statist. Assoc.* **93**: 1112–1123.

Darroch, J. N. 1962. Interactions in multi-factor contingency tables. *J. Roy. Statist. Soc. Ser. B* **24**: 251–263.

Darroch, J. N. 1981. The Mantel–Haenszel test and tests of marginal symmetry; Fixed-effects and mixed models for a categorical response. *Internat. Statist. Rev.* **49**: 285–307.

Darroch, J. N., and P. I. McCloud. 1986. Category distinguishability and observer agreement. *Austral. J. Statist.* **28**: 371–388.

Darroch, J. N., and D. Ratcliff. 1972. Generalized iterative scaling for log-linear models. *Ann. Math. Statist.* **43**: 1470–1480.

Darroch, J. N., S. L. Lauritzen, and T. P. Speed. 1980. Markov fields and log-linear interaction models for contingency tables. *Ann. Statist.* **8**: 522–539.

Darroch, J. N., S. E. Fienberg, G. F. V. Glonek, and B. W. Junker. 1993. A three-sample multiple-recapture approach to census population estimation with heterogeneous catchability. *J. Amer. Statist. Assoc.* **88**: 1137–1148.

Das Gupta, S., and M. D. Perlman. 1974. Power of the noncentral F-test: Effect of additional variates on Hotelling's T^2-test. *J. Amer. Statist. Assoc.* **69**: 174–180.

David, H. A. 1988. *The Method of Paired Comparisons*, 2nd ed. Oxford: Oxford University Press.

Davis, L. J. 1986a. Exact tests for 2 by 2 contingency tables. *Amer. Statist.* **40**: 139–141.

Davis, L. J. 1986b. Relationship between strictly collapsible and perfect tables. *Statist. Probab. Lett.* **4**: 119–122.

Davis, L. J. 1989. Intersection union tests for strictly collapsibility in three-dimensional contingency tables. *Ann. Statist.* **17**: 1693–1708.

Davison, A. C., and D. V. Hinkley. 1997. *Bootstrap Methods and Their Application*. Cambridge, U.K. Cambridge University Press.

Dawson, R. B., Jr. 1954. A simplified expression for the variance of the χ^2-function on a contingency table. *Biometrika* **41**: 280.

Day, N. E., and D. P. Byar. 1979. Testing hypotheses in case–control studies: Equivalence of Mantel–Haenszel statistics and logit score tests. *Biometrics* **35**: 623–630.

de Falguerolles, A., S. Jmel, and J. Whittaker. 1995. Correspondence analysis and association models constrained by a conditional independence graph. *Psychometrika* **60**: 161–180.

Deming, W. E. 1964. *Statistical Adjustment of Data* (reprint of 1943 Wiley text). New York: Dover.

Deming, W. E., and F. F. Stephan. 1940. On a least squares adjustment of a sampled frequency table when the expected marginal totals are known. *Ann. Math. Statist.* **11**: 427–444.

Dempster, A. P., N. M. Laird, and D. B. Rubin. 1977. Maximum likelihood from incomplete data via the EM algorithm. *J. Roy. Statist. Soc. Ser. B* **39**: 1–38.

Dey, D. K., S. K. Ghosh, and B. K. Mallick (editors). 2000. *Generalized Linear Models: A Bayesian Perspective*. New York: Marcel Dekker.

Diaconis, P., and B. Efron. 1985. Testing for independence in a two-way table: New interpretations of the chi-square statistic. *Ann. Statist.* **13**: 845–874.

Diaconis, P., and B. Sturmfels. 1998. Algebraic algorithms for sampling from conditional distributions. *Ann. Statist.* **26**: 363–397.

Diggle, P. J., P. Heagerty, K.-Y. Liang, and S. L. Zeger. 2002. *Analysis of Longitudinal Data*, 2nd ed. Oxford: Clarendon Press.

Dittrich, R., R. Hatzinger, and W. Katzenbeisser. 1998. Modeling the effect of subject-specific covariates in paired comparison studies with an application to university rankings. *Appl. Statist.* **47**: 511–525.

Dobson, A. J. 2001. *An Introduction to Generalized Linear Models*, 2nd ed. London: Chapman & Hall.

Dong, J. 1998. Simpson's paradox. Pp. 4108–4110 in *Encyclopedia of Biostatistics*, Vol. 5. Chichester, UK: Wiley.

Dong, J., and J. S. Simonoff. 1994. The construction and properties of boundary kernels for smoothing sparse multinomials. *J. Computat. Graph. Statist.* **3**: 57–66.

Dong, J., and J. S. Simonoff. 1995. A geometric combination estimator for d-dimensional ordinal sparse contingency tables. *Ann. Statist.* **23**: 1143–1159.

Donner, A., and W. W. Hauck. 1986. The large-sample efficiency of the Mantel–Haenszel estimator in the fixed-strata case. *Biometrics* **42**: 537–545.

Doolittle, M. H. 1888. Association ratios. *Bull. Philos. Soc. Washington* **10**: 83–87, 94–96.

Drost, F. C., W. C. M. Kallenberg, D. S. Moore, and J. Oosterhoff. 1989. Power approximations to multinomial tests of fit. *J. Amer. Statist. Assoc.* **84**: 130–141.

Ducharme, G. R., and Y. Lepage. 1986. Testing collapsibility in contingency tables. *J. Roy. Statist. Soc. Ser B* **48**: 197–205.

Dupont, W. D. 1986. Sensitivity of Fisher's exact test to minor perturbations in 2×2 contingency tables. *Statist. Medic.* **5**: 629–635.

Dyke, G. V., and H. D. Patterson. 1952. Analysis of factorial arrangements when the data are proportions. *Biometrics* **8**: 1–12.

Edwardes, M. D. deB. 1997. Univariate random cut-points theory for the analysis of ordered categorical data. *J. Amer. Statist. Assoc.* **92**: 1114–1123.

Edwards, A. W. F. 1963. The measure of association in a 2 × 2 table. *J. Roy. Statist. Soc. Ser A* **126**: 109–114.

Edwards, D. 2000. *Introduction to Graphical Modelling*, 2nd ed. New York: Springer-Verlag.

Edwards, D., and S. Kreiner. 1983. The analysis of contingency tables by graphical models. *Biometrika* **70**: 553–565.

Efron, B. 1975. The efficiency of logistic regression compared to normal discriminant analysis. *J. Amer. Statist. Assoc.* **70**: 892–898.

Efron, B. 1978. Regression and ANOVA with zero–one data: Measures of residual variation. *J. Amer. Statist. Soc.* **73**: 113–121.

Efron, B., and D. V. Hinkley. 1978. Assessing the accuracy of the maximum likelihood estimator: Observed versus expected Fisher information. *Biometrika* **65**: 457–482.

Efron, B., and C. Morris. 1975. Data analysis using Stein's estimator and its generalizations. *J. Amer. Statist. Assoc.* **70**: 311–319.

Ekholm, A., J. W. McDonald, and P. W. F. Smith. 2000. Association models for a multivariate binary response. *Biometrics* **56**: 712–718.

Escoufier, Y. 1982. L'analyse des tableaux de contingence simples et multiples. In *Proc. International Meeting on the Analysis of Multidimensional Contingency Tables* (Rome, 1981), ed. R. Coppi. *Metron* **40**: 53–77.

Espeland, M. A., and S. L. Handelman. 1989. Using latent class models to characterize and assess relative error in discrete measurements. *Biometrics* **45**: 587–599.

Fahrmeir, L., and G. Tutz. 2001. *Multivariate Statistical Modelling based on Generalized Linear Models*, 2nd ed. New York: Springer-Verlag.

Farewell, V. T. 1979. Some results on the estimation of logistic models based on retrospective data. *Biometrika* **66**: 27–32.

Farewell, V. T. 1982. A note on regression analysis of ordinal data with variability of classification. *Biometrika* **69**: 533–538.

Fay, R. 1985. A jackknifed chi-squared test for complex samples. *J. Amer. Statist. Assoc.* **80**: 148–157.

Fay, R. 1986. Causal models for patterns of nonresponse. *J. Amer. Statist. Assoc.* **81**: 354–365.

Ferguson, T. S. 1967. *Mathematical Statistics: A Decision Theoretic Approach*. New York: Academic Press.

Fienberg, S. E. 1970a. An iterative procedure for estimation in contingency tables. *Ann. Math. Statist.* **41**: 907–917.

Fienberg, S. E. 1970b. Quasi-independence and maximum likelihood estimation in incomplete contingency tables. *J. Amer. Statist. Soc.* **65**: 1610–1616.

Fienberg, S. E. 1972. The analysis of incomplete multi-way contingency tables. *Biometrics* **28**: 177–202.

Fienberg, S. E. 1980. Fisher's contributions to the analysis of categorical data. Pp. 75–84 in *R. A. Fisher: An Appreciation*, ed. S. E. Fienberg and D. V. Hinkley. Berlin: Springer-Verlag.

Fienberg, S. E. 1984. The contributions of William Cochran to categorical data analysis. Pp. 103–118 in *W. G. Cochran's Impact on Statistics*, ed. P. S. R. S. Rao and J. Sedransk. New York: Wiley.

Fienberg, S. E., and P. W. Holland. 1973. Simultaneous estimation of multinomial cell probabilities. *J. Amer. Statist. Assoc.* **68**: 683–690.

Fienberg, S. E., and K. Larntz. 1976. Loglinear representation for paired and multiple comparison models. *Biometrika* **63**: 245–254.

Fienberg, S. E., M. A. Johnson, and B. J. Junker. 1999. Classical multilevel and Bayesian approaches to population size estimation using multiple lists. *J. Roy. Statist. Soc. Ser. A* **162**: 383–405.

Finney, D. J. 1947. The estimation from individual records of the relationship between dose and quantal response. *Biometrika* **34**: 320–334.

Finney, D. J. 1971. *Probit Analysis*, 3rd ed. Cambridge: Cambridge University Press.

Firth, D. 1987. On the efficiency of quasi-likelihood estimation. *Biometrika* **74**: 233–245.

Firth, D. 1989. Marginal homogeneity and the superposition of Latin squares. *Biometrika* **76**: 179–182.

Firth, D. 1991. Generalized linear models. Pp. 55–82 in *Statistical Theory and Modelling. In Honour of Sir David Cox, FRS*, D. V. Hinkley, N. Reid, and E. J. Snell, eds. London: Chapman & Hall.

Firth, D. 1993a. Bias reduction of maximum likelihood estimates. *Biometrika* **80**: 27–38.

Firth, D. 1993b. Recent developments in quasi-likelihood methods. *Proc. ISI 49th Session*, pp. 341–358.

Firth, D., and J. Kuha. 2000. On the index of dissimilarity for lack of fit in log linear models. Unpublished manuscript.

Fischer, G. H., and I. W. Molenaar. 1995.*Rasch Models: Foundations, Recent Developments, and Applications*. New York: Springer-Verlag.

Fisher, R. A. 1922. On the interpretation of chi-square from contingency tables, and the calculation of *P. J. Roy. Statist. Soc.* **85**: 87–94.

Fisher, R. A. 1924. The conditions under which chi-square measures the discrepancy between observation and hypothesis. *J. Roy. Statist. Soc.* **87**: 442–450.

Fisher, R. A. 1926. Bayes' theorem and the fourfold table. *Eugenics Rev.* **18**: 32–33.

Fisher, R. A. 1934, 1970. *Statistical Methods for Research Workers* (originally published 1925, 14th ed., 1970.) Edinburgh: Oliver & Boyd.

Fisher, R. A. 1935a. *The Design of Experiments* (8th ed., 1966). Edinburgh: Oliver & Boyd.

Fisher, R. A. 1935b. Appendix to article by C. Bliss. *Ann. Appl. Biol.* **22**: 164–165.

Fisher, R. A. 1935c. The logic of inductive inference. *J. Roy. Statist. Soc.* **98**: 39–82.

Fisher, R. A. 1945. A new test for 2 × 2 tables (Letter to the Editor). *Nature* **156**: 388.

Fisher, R. A. 1956. *Statistical Methods for Scientific Inference*. Edinburgh: Oliver & Boyd.

Fisher, R. A., and F. Yates. 1938. *Statistical Tables*. Edinburgh: Oliver and Boyd.

Fitzmaurice, G. M., and N. M. Laird. 1993. A likelihood-based method for analysing longitudinal binary responses. *Biometrika* **80**: 141–151.

Fitzmaurice, G. M., N. M. Laird, and S. Lipsitz. 1994. Analysing incomplete longitudinal binary responses: A likelihood-based approach. *Biometrics* **50**: 601–612.

Fitzmaurice, G. M., N. M. Laird, and A. G. Rotnitzky. 1993. Regression models for discrete longitudinal responses. *Statist. Sci.* **8**: 284–299.

Fitzpatrick, S., and A. Scott. 1987. Quick simultaneous confidence intervals for multinomial proportions. *J. Amer. Statist. Assoc.* **82**: 875–878.

Fleiss, J. L. 1981. *Statistical Methods for Rates and Proportions*, 2nd ed. New York: Wiley.

Fleiss, J. L., and J. Cohen. 1973. The equivalence of weighted kappa and the intraclass correlation coefficient as measures of reliability. *Educ. Psychol. Meas.* **33**: 613–619.

Fleiss, J. L., J. Cohen, and B. S. Everitt. 1969. Large-sample standard errors of kappa and weighted kappa. *Psychol. Bull.* **72**: 323–327.

Follman, D. A., and D. Lambert. 1989. Generalizing logistic regression by nonparametric mixing. *J. Amer. Statist. Assoc.* **84**: 295–300.

Forster, J. J., and P. W. F. Smith. 1998. Model-based inference for categorical survey data subject to non-ignorable non-response. *J. Roy. Statist. Soc. Ser B* **60**: 57–70.

Forster, J. J., J. W. McDonald, and P. W. F. Smith. 1996. Monte Carlo exact conditional tests for log-linear and logistic models. *J. Roy. Statist. Soc. Ser B* **58**: 445–453.

Fowlkes, E. B. 1987. Some diagnostics for binary logistic regression via smoothing. *Biometrika* **74**: 503–515.

Fowlkes, E. B., A. E. Freeny, and J. Landwehr. 1988. Evaluating logistic models for large contingency tables. *J. Amer. Statist. Assoc.* **83**: 611–622.

Freedman, D., R. Pisani, and R. Purves. 1978. *Statistics*. New York: W. W. Norton.

Freeman, G. H., and J. H. Halton. 1951. Note on an exact treatment of contingency, goodness-of-fit and other problems of significance. *Biometrika* **38**: 141–149.

Freeman, D. H., Jr. and T. R. Holford. 1980. Summary rates. *Biometrics* **36**: 195–205.

Freeman, M. F., and J. W. Tukey. 1950. Transformations related to the angular and the square root. *Ann. Math. Statist.* **21**: 607–611.

Freidlin, B., and J. L. Gastwirth. 1999. Unconditional versions of several tests commonly used in the analysis of contingency tables. *Biometrics* **55**: 264–267.

Friendly, M. 2000. *Visualizing Categorical Data*. Cary, NC: SAS Institute.

Frome, E. L. 1983. The analysis of rates using Poisson regression models. *Biometrics* **39**: 665–674.

Fuchs, C. 1982. Maximum likelihood estimation and model selection in contingency tables with missing data. *J. Amer. Statist. Assoc.* **77**: 270–278.

Gabriel, K. R. 1966. Simultaneous test procedures for multiple comparisons on categorical data. *J. Amer. Statist. Assoc.* **61**: 1081–1096.

Gabriel, K. R. 1971. The biplot graphic display of matrices with applications to principal component analysis. *Biometrika* **58**: 453–467.

Gail, M. H., and J. J. Gart. 1973. The determination of sample sizes for use with the exact conditional test in 2 × 2 comparative trials. *Biometrics* **29**: 441–448.

Gail, M., and N. Mantel. 1977. Counting the number of $r \times c$ contingency tables with fixed margins. *J. Amer. Statist. Assoc.* **72**: 859–862.

Gart, J. J. 1966. Alternative analyses of contingency tables. *J. Roy. Statist. Soc. Ser B* **28**: 164–179.

Gart, J. J. 1969. An exact test for comparing matched proportions in crossover designs. *Biometrika* **56**: 75–80.

Gart, J. J. 1970. Point and interval estimation of the common odds ratio in the combination of 2 × 2 tables with fixed margins. *Biometrika* **57**: 471–475.

Gart, J. J. 1971. The comparison of proportions: A review of significance tests, confidence intervals and adjustments for stratification. *Rev. Internat. Statist. Rev.* **39**: 148–169.

Gart, J. J., and J. Nam. 1988. Approximate interval estimation of the ratio of binomial parameters: A review and corrections for skewness. *Biometrics* **44**: 323–338.

Gart, J. J., and J. R. Zweifel. 1967. On the bias of various estimators of the logit and its variance with applications to quantal bioassay. *Biometrika* **54**: 181–187.

Gelfand, A. E., and A. F. Smith. 1990. Sampling-based approaches to calculating marginal densities. *J. Amer. Statist. Assoc.* **85**: 398–409.

Genter, F. C., and V. T. Farewell. 1985. Goodness-of-link testing in ordinal regression models. *Canad. J. Statist.* **13**: 37–44.

Ghosh, B. K. 1979. A comparison of some approximate confidence intervals for the binomial parameter. *J. Amer. Statist. Assoc.* **74**: 894–900.

Ghosh, M., M. Chen, A. Ghosh, and A. Agresti. 2000. Hierarchical Bayesian analysis of binary matched pairs data. *Statist. Sin.* **10**: 647–657.

Gibbons, R. D., and D. Hedeker. 1997. Random-effects probit and logistic regression models for three-level data. *Biometrics* **53**: 1527–1537.

Gill, J. 2000. *Generalized Linear Models: A Unified Approach*. Thousand Oaks, CA: Sage Publications.

Gilmour, A. R., R. D. Anderson, and A. L. Rae. 1985. The analysis of binomial data by a generalized linear mixed model. *Biometrika* **72**: 593–599.

Gilula, Z., and S. Haberman. 1986. Canonical analysis of contingency tables by maximum likelihood. *J. Amer. Statist. Assoc.* **81**: 780–788.

Gilula, Z., and S. Haberman. 1988. The analysis of multivariate contingency tables by restricted canonical and restricted association models. *J. Amer. Statist. Assoc.* **83**: 760–771.

Gilula, Z., and S. Haberman. 1998. Chi-square, partition of. Pp. 622–627 in *Encyclopedia of Biostatistics*. Chichester, UK: Wiley.

Gleser, L. J., and D. S. Moore. 1985. The effect of positive dependence on chi-squared tests for categorical data. *J. Roy. Statist. Soc. Ser B* **47**: 459–465.

Glonek, G. 1996. A class of regression models for multivariate categorical responses. *Biometrika* **83**: 15–28.

Glonek, G. F. V., and P. McCullagh. 1995. Multivariate logistic models. *J. Roy. Statist. Soc. Ser. B* **57**: 533–546.

Glonek, G., J. N. Darroch, and T. P. Speed. 1988. On the existence of maximum likelihood estimators for hierarchical loglinear models. *Scand. J. Statist.* **15**: 187–193.

Gokhale, D. V., and S. Kullback. 1978. *The Information in Contingency Tables*. New York: Marcel Dekker.

Goldstein, H. 1995. *Multilevel Statistical Models*, 2nd ed. London: Edward Arnold.

Goldstein, H., and J. Rasbash. 1996. Improved approximations for multilevel models with binary responses. *J. Roy. Statist. Soc. Ser A* **159**: 505–513.

Good, I. J. 1963. Maximum entropy for hypothesis formulation, especially for multi-dimensional contingency tables. *Ann. Math. Statist.* **34**: 911–934.

Good, I. J. 1965. *The Estimation of Probabilities: An Essay on Modern Bayesian Methods*. Cambridge, MA: MIT Press.

Good, I. J. 1976. On the application of symmetric Dirichlet distributions and their mixtures to contingency tables. *Ann. Statist.* **4**: 1159–1189.

Good, I. J., and R. A. Gaskins. 1971. Nonparametric roughness penalties for probability densities. *Biometrika* **58**: 255–277.

Good, I. J., and Y. Mittal. 1987. The amalgamation and geometry of two-by-two contingency tables. *Ann. Statist.* **15**: 694–711.

Good, I. J., T. N. Gover, and G. J. Mitchell. 1970. Exact distributions for χ^2 and for the likelihood-ratio statistic for the equiprobable multinomial distribution. *J. Amer. Statist. Assoc.* **65**: 267–283.

Goodman, L. A. 1964a. Simultaneous confidence intervals for cross-product ratios in contingency tables. *J. Roy. Statist. Soc. Ser B* **26**: 86–102.

Goodman, L. A. 1964b. Interactions in multi-dimensional contingency tables. *Ann. Math. Statist.* **35**: 632–646.

Goodman, L. A. 1965. On simultaneous confidence intervals for multinomial proportions. *Technometrics* **7**: 247–254.

Goodman, L. A. 1968. The analysis of cross-classified data: Independence, quasi-independence, and interactions in contingency tables with or without missing entries. *J. Amer. Statist. Assoc.* **63**: 1091–1131.

Goodman, L. A. 1969a. On partitioning chi-square and detecting partial association in three-way contingency tables. *J. Roy. Statist. Soc. Ser B* **31**: 486–498.

Goodman, L. A. 1969b. How to ransack social mobility tables and other kinds of cross-classification tables. *Amer. J. Sociol.* **75**: 1–40.

Goodman, L. A. 1970. The multivariate analysis of qualitative data: Interaction among multiple classifications. *J. Amer. Statist. Assoc.* **65**: 226–256.

Goodman, L. A. 1971a. The analysis of multidimensional contingency tables: Stepwise procedures and direct estimation methods for building models for multiple classifications. *Technometrics* **13**: 33–61.

Goodman, L. A. 1971b. The partitioning of chi-square, the analysis of marginal contingency tables, and the estimation of expected frequencies in multidimensional contingency tables. *J. Amer. Statist. Assoc.* **66**: 339–344.

Goodman, L. A. 1973. The analysis of multidimensional contingency tables with some variables are posterior to others: A modified path analysis approach. *Biometrika* **60**: 179–192.

Goodman, L. A. 1974. Exploratory latent structure analysis using both identifiable and unidentifiable models. *Biometrika* **61**: 215–231.

Goodman, L. A. 1979a. Simple models for the analysis of association in cross-classifications having ordered categories. *J. Amer. Statist. Assoc.* **74**: 537–552.

Goodman, L. A. 1979b. Multiplicative models for square contingency tables with ordered categories. *Biometrika* **66**: 413–418.

Goodman, L. A. 1981a. Association models and canonical correlation in the analysis of cross-classifications having ordered categories. *J. Amer. Statist. Assoc.* **76**: 320–334.

Goodman, L. A. 1981b. Association models and the bivariate normal for contingency tables with ordered categories. *Biometrika* **68**: 347–355.

Goodman, L. A. 1983. The analysis of dependence in cross-classification having ordered categories, using log-linear models for frequencies and log-linear models for odds. *Biometrics* **39**: 149–160.

Goodman, L. A. 1985. The analysis of cross-classified data having ordered and/or unordered categories: Association models, correlation models, and asymmetry models for contingency tables with or without missing entries. *Ann. Statist.* **13**: 10–69.

Goodman, L. A. 1986. Some useful extensions of the usual correspondence analysis approach and the usual log-linear models approach in the analysis of contingency tables. *Internat. Statist. Rev.* **54**: 243–309.

Goodman, L. A. 1996. A single general method for the analysis of cross-classified data: Reconciliation and synthesis of some methods of Pearson, Yule, and Fisher, and also some methods of correspondence analysis and association analysis. *J. Amer. Statist. Assoc.* **91**: 408–427.

Goodman, L. A. 2000. The analysis of cross-classified data: Notes on a century of progress in contingency table analysis, and some comments on its prehistory and its future. Pp. 189–231 in *Statistics for the 21st Century*, ed. C. R. Rao and G. J. Székely. New York: Marcel Dekker.

Goodman, L. A., and W. H. Kruskal. 1979. *Measures of Association for Cross Classifications*. New York: Springer-Verlag (contains articles appearing in *J. Amer. Statist. Assoc.* in 1954, 1959, 1963, 1972).

Gould, S. J. 1981. *The Mismeasure of Man*. New York: W. W. Norton.

Gourieroux, C., A. Monfort, and A. Trognon. 1984. Pseudo maximum likelihood methods: Theory. *Econometrica* **52**: 681–700.

Graubard, B. I., and E. L. Korn. 1987. Choice of column scores for testing independence in ordered $2 \times K$ contingency tables. *Biometrics* **43**: 471–476.

Green, P. J. 1984. Iteratively weighted least squares for maximum likelihood estimation and some robust and resistant alternatives. *J. Roy. Statist. Soc. Ser B* **46**: 149–192.

Greenacre, M. J. 1993. *Correspondence Analysis in Practice*. New York: Academic Press.

Greenland, S. 1991. On the logical justification of conditional tests for two-by-two contingency tables. *Amer. Statist.* **45**: 248–251.

Greenland, S., and J. M. Robins. 1985. Estimation of a common effect parameter from sparse follow-up data. *Biometrics* **41**: 55–68.

Greenwood, M., and G. U. Yule. 1920. An inquiry into the nature of frequency distributions representative of multiple happenings with particular reference to the occurrence of multiple attacks of disease or of repeated accidents. *J. Roy. Statist. Soc. Ser A* **83**: 255–279.

Greenwood, P. E., and M. S. Nikulin. 1996. *A Guide to Chi-Squared Testing.* New York: Wiley.

Grizzle, J. E., C. F. Starmer, and G. G. Koch. 1969. Analysis of categorical data by linear models. *Biometrics* **25**: 489–504.

Gross, S. T. 1981. On asymptotic power and efficiency of tests of independence in contingency tables with ordered classifications. *J. Amer. Statist. Assoc.* **76**: 935–941.

Gueorguieva, R., and A. Agresti. 2001. A correlated probit model for joint modeling of clustered binary and continuous responses. *J. Amer. Statist. Assoc.* **96**: 1102–1112.

Haber, M. 1980. A comparison of some continuity corrections for the chi-squared test on 2×2 tables. *J. Amer. Statist. Assoc.* **75**: 510–515.

Haber, M. 1982. The continuity correction and statistical testing. *Internat. Statist. Rev.* **50**: 135–144.

Haber, M. 1985. Maximum likelihood methods for linear and log-linear models in categorical data. *Comput. Statist. Data Anal.* **3**: 1–10.

Haber, M. 1986. An exact unconditional test for the 2×2 comparative trial. *Psychol. Bull.* **99**: 129–132.

Haber, M. 1989. Do the marginal totals of a 2×2 contingency table contain information regarding the table proportions? *Commun. Statist. Ser A* **18**: 147–156.

Haberman, S. J. 1973a. The analysis of residuals in cross-classification tables. *Biometrics* **29**: 205–220.

Haberman, S. J. 1973b. Log-linear models for frequency data: Sufficient statistics and likelihood equations. *Ann. Statist.* **1**: 617–632.

Haberman, S. J. 1974a. *The Analysis of Frequency Data.* Chicago: University of Chicago Press.

Haberman, S. J. 1974b. Log-linear models for frequency tables with ordered classifications. *Biometrics* **36**: 589–600.

Haberman, S. J. 1977a. Log-linear models and frequency tables with small expected cell counts. *Ann. Statist.* **5**: 1148–1169.

Haberman, S. J. 1977b. Maximum likelihood estimation in exponential response models. *Ann. Statist.* **5**: 815–841.

Haberman, S. J. 1978, 1979. *Analysis of Qualitative Data*, Vols. 1 and 2. New York: Academic Press.

Haberman, S. J. 1981. Tests for independence in two-way contingency tables based on canonical correlation and on linear-by-linear interaction. *Ann. Statist.* **9**: 1178–1186.

Haberman, S. J. 1982. The analysis of dispersion of multinomial responses. *J. Amer. Statist. Assoc.* **77**: 568–580.

Haberman, S. J. 1988. A warning on the use of chi-squared statistics with frequency tables with small expected cell counts. *J. Amer. Statist. Assoc.* **83**: 555–560.

Haberman, S. J. 1995. Computation of maximum likelihood estimates in association models. *J. Amer. Statist. Assoc.* **90**: 1438–1446.

Hagenaars, J. A. 1998. Categorical causal modeling: Latent class analysis and directed log-linear models with latent variables. *Sociol. Methods Res.* **26**: 436–486.

Hald, A. 1998. *A History of Mathematical Statistics from 1750 to 1930.* New York: Wiley.

Haldane, J. B. S. 1940. The mean and variance of χ^2, when used as a test of homogeneity, when expectations are small. *Biometrika* **31**: 346–355.

Haldane, J. B. S. 1956. The estimation and significance of the logarithm of a ratio of frequencies. *Ann. Human Genet.* **20**: 309–311.

Hall, P., and D. M. Titterington. 1987. On smoothing sparse multinomial data. *Austral. J. Statist.* **29**: 19–37.

Hamada, M., and C. F. J. Wu. 1990. A critical look at accumulation analysis and related methods. *Technometrics* **32**: 119–130.

Hansen, L. P. 1982. Large sample properties of generalized-method of moments estimators. *Econometrica* **50**: 1029–1054.

Harkness, W. L., and L. Katz. 1964. Comparison of the power functions for the test of independence in 2 × 2 contingency tables. *Ann. Math. Statist.* **35**: 1115–1127.

Harrell F. E., R. M. Califf, D. B. Pryor, K. L. Lee, and R. A. Rosati. 1982. Evaluating the yield of medical tests. *J. Amer. Medic. Assoc.* **247**: 2543–2546.

Hartzel, J., I.-M. Liu, and A. Agresti. 2001a. Describing heterogeneous effects in stratified ordinal contingency tables, with application to multi-center clinical trials. *Computat. Statist. Data Anal.* **35**: 429–449.

Hartzel, J., A. Agresti, and B. Caffo. 2001b. Multinomial logit random effects models. *Statistical Modelling* **1**: 81–102.

Haslett, S. 1990. Degrees of freedom and parameter estimability in hierarchical models for sparse complete contingency tables. *Computat. Statist. Data Anal.* **9**: 179–195.

Hastie, T., and R. Tibshirani. 1987. Non-parametric logistic and proportional odds regression. *Appl. Statist.* **36**: 260–276.

Hastie, T., and R. Tibshirani. 1990. *Generalized Additive Models*. London: Chapman & Hall.

Hatzinger, R. 1989. The Rasch model, some extensions and their relation to the class of generalized linear models. *Statistical Modelling: Lecture Notes in Statistics*, Vol. 57. Berlin: Springer-Verlag.

Hauck, W. W. 1979. The large sample variance of the Mantel–Haenszel estimator of a common odds ratio. *Biometrics* **35**: 817–819.

Hauck, W. W. 1983. A note on confidence bands for the logistic response curve. *Amer. Statist.* **37**: 158–160.

Hauck, W. W., and A. Donner. 1977. Wald's test as applied to hypotheses in logit analysis. *J. Amer. Statist. Assoc.* **72**: 851–853.

Heagerty, P. J. 1999. Marginally specified logistic-normal models for longitudinal binary data. *Biometrics* **55**: 688–698.

Heagerty, P. J., and S. L. Zeger. 1996. Marginal regression models for clustered ordinal measurements. *J. Amer. Statist. Assoc.* **91**: 1024–1036.

Heagerty, P. J., and S. L. Zeger. 2000. Marginalized multilevel models and likelihood inference. *Statist. Sci.* **15**: 1–19.

Hedeker, D., and R. D. Gibbons. 1994. A random-effects ordinal regression model for multilevel analysis. *Biometrics* **50**: 933–944.

Heinen, T. 1996. *Latent Class and Discrete Latent Trait Models*. Thousand Oaks, CA: Sage Publications.

Heyde, C. C. 1997. *Quasi-likelihood and Its Application*. New York: Springer-Verlag.

Hinde, J. 1982. Compound Poisson regression models. Pp. 109–121 in *GLIM 82: Proc. International Conference on Generalised Linear Models*, ed. R. Gilchrist. New York: Springer-Verlag.

Hinde, J., and C. G. B. Demétrio. 1998. Overdispersion: Models and estimation. *Comput. Statist. Data Anal.* **27**: 151–170.

Hirji, K. F. 1991. A comparison of exact, mid-P, and score tests for matched case-control studies. *Biometrics* **47**: 487–496.

Hirji, K. F., C. R. Mehta, and N. R. Patel. 1987. Computing distributions for exact logistic regression. *J. Amer. Statist. Assoc.* **82**: 1110–1117.

Hirotsu, C. 1982. Use of cumulative efficient scores for testing ordered alternatives in discrete models. *Biometrika* **69**: 567–577.

Hirschfeld, H. O. 1935. A connection between correlation and contingency. *Cambridge Philos. Soc. Proc.* (*Math. Proc.*) **31**: 520–524.

Hodges, J. L., Jr. 1958. Fitting the logistic by maximum likelihood. *Biometrics* **14**: 453–461.

Hoem, J. M. 1987. Statistical analysis of a multiplicative model and its application to the standardization of vital rates: A review. *Internat. Statist. Rev.* **5**: 119–152.

Holford, T. R. 1980. The analysis of rates and of survivorship using log-linear models. *Biometrics* **36**: 299–305.

Holt, D., A. J. Scott, and P. D. Ewings. 1980. Chi-squared tests with survey data. *J. Roy. Statist. Soc. Ser. A* **143**: 303–320.

Hook, E. B., and R. R. Regal. 1995. Capture–recapture methods in epidemiology: Methods and limitations. *Epidemiol. Rev.* **17**: 243–264.

Hosmer, D. W., and S. Lemeshow. 1980. A goodness-of-fit test for multiple logistic regression model. *Commun. Statist. Ser A* **9**: 1043–1069.

Hosmer, D. W., and S. Lemeshow. 2000. *Applied Logistic Regression*, 2nd ed. New York: Wiley.

Hosmer, D. W., T. Hosmer, S. le Cessie, and S. Lemeshow. 1997. A comparison of goodness-of-fit tests for the logistic regression model. *Statist. Medic.* **16**: 965–980.

Hout, M., O. D. Duncan, and M. E. Sobel. 1987. Association and heterogeneity: Structural models of similarities and differences. *Sociol. Methodol.* **17**: 145–184.

Howard, J. V. 1998. The 2×2 table: A discussion from a Bayesian viewpoint. *Statist. Sci.* **13**: 351–367.

Hsieh, F. Y. 1989. Sample size tables for logistic regression. *Statist. Medic.* **8**: 795–802.

Hsieh, F. Y., D. A. Bloch, and M. D. Larsen. 1998. A simple method of sample size calculation for linear and logistic regression. *Statist. Medic.* **17**: 1623–1634.

Hwang, J. T. G., and M. T. Wells. 2002. Optimality results for mid P-values. To appear.

Hwang, J. T. G., and M.-C. Yang. 2001. An optimality theory for mid P-values in 2×2 contingency tables. *Statist. Sin.* **11**: 807–826.

Imrey, P. B. 1998. Bradley–Terry model. Pp. 437–443 in *Encyclopedia of Biostatistics*. Chichester, UK: Wiley.

Imrey, P. B., W. D. Johnson, and G. G. Koch. 1976. An incomplete contingency table approach to paired-comparison experiments. *J. Amer. Statist. Assoc.* **71**: 614–623.

Imrey, P. B., G. G. Koch, and M. E. Stokes. 1981. Categorical data analysis: Some reflections on the log linear model and logistic regression. I: Historical and methodological overview. *Internat. Statist. Rev.* **49**: 265–283.

Ireland, C. T., and S. Kullback. 1968a. Minimum discrimination information estimation. *Biometrics* **24**: 707–713.

Ireland, C. T., and S. Kullback. 1968b. Contingency tables with given marginals. *Biometrika* **55**: 179–188.

Ireland, C. T., H. H. Ku, and S. Kullback. 1969. Symmetry and marginal homogeneity of an $r \times r$ contingency table. *J. Amer. Statist. Assoc.* **64**: 1323–1341.

Irwin, J. O. 1935. Tests of significance for differences between percentages based on small numbers. *Metron* **12**: 83–94.

Jennison, C., and B. W. Turnbull. 2000. *Group Sequential Methods with Applications to Clinical Trials*. London: Chapman & Hall.

Johnson, B. M. 1971. On the admissible estimators for certain fixed sample binomial problems. *Ann. Math. Statist.* **42**: 1579–1587.

Johnson, W. 1985. Influence measures for logistic regression: Another point of view. *Biometrika* **72**: 59–65.

Johnson, N. L., S. Kotz, and A. W. Kemp. 1992. *Univariate Discrete Distributions*, 2nd ed. New York: Wiley.

Jones, B., and M. G. Kenward. 1987. Modelling binary data from a three-period cross-over trial. *Statist. Medic.* **6**: 555–564.

Jones, M. P., T. W. O'Gorman, J. H. Lemke, and R. F. Woolson. 1989. A Monte Carlo investigation of homogeneity tests of the odds ratio under various sample size considerations. *Biometrics* **45**: 171–181.

Jørgensen, B. 1983. Maximum likelihood estimation and large-sample inference for generalized linear and nonlinear regression models. *Biometrika* **70**: 19–28.

Jørgensen, B. 1987. Exponential dispersion models. *J. Roy. Statist. Soc. Ser. B* **49**: 127–162.

Kalbfleisch, J. D., and J. F. Lawless. 1985. The analysis of panel data under a Markov assumption. *J. Amer. Statist. Assoc.* **80**: 863–871.

Kastner, C., A. Fieger, and C. Heumann. 1997. MAREG and WinMAREG: A tool for marginal regression models. *Comput. Statist. Data Anal.* **24**: 237–241.

Kauermann, G., and R. J. Carroll, 2001. A note on the efficiency of sandwich covariance matrix estimation. *J. Amer. Statist. Assoc.* **96**: 1387–1397.

Kauermann, G., and G. Tutz. 2001. Testing generalized linear and semiparametric models against smooth alternatives. *J. Roy. Statist. Soc. Ser. B* **63**: 147–166.

Kelderman, H. 1984. Loglinear Rasch model tests. *Psychometrika* **49**: 223–245.

Kempthorne, O. 1979. In dispraise of the exact test: Reactions. *J. Statist. Plann. Inference* **3**: 199–213.

Kendall, M. G. 1945. The treatment of ties in rank problems. *Biometrika* **33**: 239–251.

Kendall, M., and A. Stuart. 1979. *The Advanced Theory of Statistics*, Vol. 2; *Inference and Relationship*, 4th ed. New York: Macmillan.

Kenward, M. G., and B. Jones. 1991. The analysis of categorical data from cross-over trials using a latent variable model. *Statist. Medic.* **10**: 1607–1619.

Kenward, M. G., and B. Jones. 1994. The analysis of binary and categorical data from crossover trials. *Statist. Methods Medic. Res.* **3**: 325–344.

Kenward, M. G., E. Lesaffre, and G. Molenberghs. 1994. An application of maximum likelihood and estimating equations to the analysis of ordinal data from a longitudinal study with cases missing at random. *Biometrics* **50**: 945–953.

Khamis, H. J. 1983. Log-linear model analysis of the semi-symmetric intraclass contingency table. *Commun. Statist. Ser. A* **12**: 2723–2752.

Kim, D., and A. Agresti. 1995. Improved exact inference about conditional association in three-way contingency tables. *J. Amer. Statist. Assoc.* **90**: 632–639.

Kim, D., and A. Agresti. 1997. Nearly exact tests of conditional independence and marginal homogeneity for sparse contingency tables. *Comput. Statist. Data Anal.* **24**: 89–104.

King, G. 1997. *A Solution to the Ecological Inference Problem*. Princeton, NJ: Princeton University Press.

Knuiman, M. W., and T. P. Speed. 1988. Incorporating prior information into the analysis of contingency tables. *Biometrics* **44**: 1061–1071.

Koch, G. G., and V. P. Bhapkar. 1982. Chi-square tests. Pp. 442–457 in *Encyclopedia of Statistical Sciences*, Vol. 1. New York: Wiley.

Koch, G. G., J. R. Landis, J. L. Freeman, D. H. Freeman, and R. G. Lehnen. 1977. A general methodology for the analysis of experiments with repeated measurement of categorical data. *Biometrics* **33**: 133–158.

Koch, G. G., I. A. Amara, G. W. Davis, and D. B. Gillings. 1982. A review of some statistical methods for covariance analysis of categorical data. *Biometrics* **38**: 563–595.

Koch, G. G., P. B. Imrey, J. M. Singer, S. S. Atkinson, and M. E. Stokes. 1985. *Lecture Notes for Analysis of Categorical Data*. Montreal: Les Presses de L'Université de Montréal.

Koehler, K. 1986. Goodness-of-fit tests for log-linear models in sparse contingency tables. *J. Amer. Statist. Assoc.* **81**: 483–493.

Koehler, K. 1998. Chi-square tests. Pp. 608–622 in *Encyclopedia of Biostatistics*. Chichester, UK: Wiley.

Koehler, K., and K. Larntz. 1980. An empirical investigation of goodness-of-fit statistics for sparse multinomials. *J. Amer. Statist. Assoc.* **75**: 336–344.

Koehler, K., and J. Wilson. 1986. Chi-square tests for comparing vectors of proportions for several cluster samples. *Commun. Statist. Ser. A* **15**: 2977–2990.

Koopman, P. A. R. 1984. Confidence limits for the ratio of two binomial proportions. *Biometrics* **40**: 513–517.

Kraemer, H. C. 1979. Ramifications of a population model for κ as a coefficient of reliability. *Psychometrika* **44**: 461–472.

Kreiner, S. 1987. Analysis of multidimensional contingency tables by exact conditional tests: Techniques and strategies. *Scand. J. Statist.* **14**: 97–112.

Kreiner, S. 1998. Interaction models. Pp. 2063–2068 in *Encyclopedia of Biostatistics*. Chichester, UK: Wiley.

Kruskal, W. H. 1958. Ordinal measures of association. *J. Amer. Statist. Assoc.* **53**: 814–861.

Ku, H. H., R. N. Varner, and S. Kullback. 1971. Analysis of multidimensional contingency tables. *J. Amer. Statist. Assoc.* **66**: 55–64.

Kuha, J., and C. Skinner. 1997. Categorical data analysis and misclassification. Pp. 633–670 in *Survey Measurement and Process Quality*, ed. L. Lyberg et al. New York: Wiley.

Kuha, J., C. Skinner, and J. Palmgren. 1998. Misclassification error. Pp. 2615–2621 in *Encyclopedia of Biostatistics*. Chichester, UK: Wiley.

Kullback, S. 1959. *Information Theory and Statistics*. New York: Wiley.

Kullback, S., M. Kupperman, and H. H. Ku. 1962. Tests for contingency tables and Markov chains. *Technometrics* **4**: 573–608.

Kupper, L. L., and J. K. Haseman. 1978. The use of a correlated binomial model for the analysis of certain toxicological experiments. *Biometrics* **34**: 69–76.

Kupper, L. L., C. Portier, M. D. Hogan, and E. Yamamoto. 1986. The impact of litter effects on dose–response modeling in teratology. *Biometrics* **42**: 85–98.

Läärä, E., and J. N. S. Matthews. 1985. The equivalence of two models for ordinal data. *Biometrika* **72**: 206–207.

Lachin, J. M. 1977. Sample-size determinations for $r \times c$ comparative trials. *Biometrics* **33**: 315–324.

Laird, N. M. 1978. Empirical Bayes methods for two-way contingency tables. *Biometrika* **65**: 581–590.

Laird, N. M. 1998. EM algorithm. Pp. 1300–1313 in *Encyclopedia of Biostatistics*. Chichester, UK: Wiley.

Laird, N. M., and D. Olivier. 1981. Covariance analysis of censored survival data using log-linear analysis techniques. *J. Amer. Statist. Assoc.* **76**: 231–240.

Lancaster, H. O. 1949. The derivation and partition of χ^2 in certain discrete distributions. *Biometrika* **36**: 117–129.

Lancaster, H. O. 1951. Complex contingency tables treated by partition of χ^2. *J. Roy. Statist. Soc. Ser. B* **13**: 242–249.

Lancaster, H. O. 1961. Significance tests in discrete distributions. *J. Amer. Statist. Assoc.* **56**: 223–234.

Lancaster, H. O. 1969. *The Chi-Squared Distribution*. New York: Wiley.

Lancaster, H. O., and M. A. Hamdan. 1964. Estimation of the correlation coefficient in contingency tables with possible nonmetrical characters. *Psychometrika* **29**: 383–391.

Landis, J. R., and G. G. Koch. 1977. An application of hierarchical kappa-type statistics in the assessment of majority agreement among multiple observers. *Biometrics* **33**: 363–374.

Landis, J. R., E. R. Heyman, and G. G. Koch. 1978. Average partial association in three-way contingency tables: A review and discussion of alternative tests. *Internat. Statist. Rev.* **46**: 237–254.

Landis, J. R., T. J. Sharp, S. J. Kuritz, and G. G. Koch. 1998. Mantel-Haenszel methods. Pp. 2378–2691 in *Encyclopedia of Biostatistics*. Chichester, UK: Wiley.

Landwehr, J. M., D. Pregibon, and A. C. Shoemaker. 1984. Graphical methods for assessing logistic regression models. *J. Amer. Statist. Assoc.* **79**: 61–71.

Lang, J. B. 1992. Obtaining the observed information matrix for the Poisson log linear model with incomplete data. *Biometrika* **79**: 405–407.

Lang, J. B. 1996a. Maximum likelihood methods for a generalized class of log-linear models. *Ann. Statist.* **24**: 726–752.

Lang, J. B. 1996b. On the partitioning of goodness-of-fit statistics for multivariate categorical response models. *J. Amer. Statist. Assoc.* **91**: 1017–1023.

Lang, J. B. 1996c. On the comparison of multinomial and Poisson log-linear models. *J. Roy. Statist. Soc. Ser. B* **58**: 253–266.

Lang, J. B., and A. Agresti. 1994. Simultaneously modeling joint and marginal distributions of multivariate categorical responses. *J. Amer. Statist. Assoc.* **89**: 625–632.

Lang, J. B., J. W. McDonald, and P. W. F. Smith. 1999. Association-marginal modeling of multivariate categorical responses: A maximum likelihood approach. *J. Amer. Statist. Assoc.* **94**: 1161–1171.

Laplace, P. S. 1812. *Théorie Analytique des Probabilités*. Paris: Courcier.

Larntz, K. 1978. Small-sample comparison of exact levels for chi-squared goodness-of-fit statistics. *J. Amer. Statist. Assoc.* **73**: 253–263.

Larsen, K., J. H. Petersen, E. Budtz-Jórgensen, and L. Endahl. 2000. Interpreting parameters in the logistic regression model with random effects. *Biometrics* **56**: 909–914.

Larson, M. G. 1984. Covariate analysis of competing-risks data with log-linear models. *Biometrics* **40**: 459–469.

Lauritzen, S. L. 1996. *Graphical Models*. New York: Oxford University Press.

Lauritzen, S. L., and N. Wermuth. 1989. Graphical models for associations between variables, some of which are qualitative and some quantitative. *Ann. Statist.* **17**: 31–57.

LaVange, L. M., G. G. Koch, and T. A. Schwartz. 2001. Applying sample survey methods to clinical trials data. *Statist. Medic.* **20**: 2609–2623.

Lawal, H. B. 1984. Comparisons of the X^2, Y^2, Freeman–Tukey and Williams improved G^2 test statistics in small samples of one-way multinomials. *Biometrika* **71**: 415–418.

Lawless, J. F. 1987. Negative binomial and mixed Poisson regression. *Canad. J. Statist.* **15**: 209–225.

Lazarsfeld, P. F., and N. W. Henry. 1968. *Latent Structure Analysis*. Boston: Houghton Mifflin.

Lee, S. K. 1977. On the asymptotic variances of û terms in loglinear models of multidimensional contingency tables. *J. Amer. Statist. Assoc.* **72**: 412–419.

Lee, Y., and J. A. Nelder. 1996. Hierarchical generalized linear models. *J. Roy. Statist. Soc. Ser B* **58**: 619–678.

Lefkopoulou, M., D. Moore, and L. Ryan. 1989. The analysis of multiple correlated binary outcomes: Application to rodent teratology experiments. *J. Amer. Statist. Assoc.* **84**: 810–815.

Lehmann, E. L. 1966. Some concepts of dependence. *Ann. Math. Statist.* **37**: 1137–1153.

Lehmann, E. L. 1986. *Testing Statistical Hypotheses*, 2nd ed. New York: Wiley.

Leonard, T. 1975. Bayesian estimation methods for two-way contingency tables. *J. Roy. Statist. Soc. Ser. B* **37**: 23–37.

Leonard, T. and J. S. J. Hsu. 1994. The Bayesian analysis of categorical data: A selective review. Pp. 283–310 in *Aspects of Uncertainty: A Tribute to D. V. Lindley.* P. R. Freeman and A. F. M. Smith, eds. New York: Wiley.

Lesaffre, E., and A. Albert. 1989. Multiple-group logistic regression diagnostics. *Appl. Statist.* **38**: 425–440.

Lesaffre, E., and G. Molenberghs. 1991. Multivariate probit analysis: A neglected procedure in medical statistics. *Statist. Medic.* **10**: 1391–1403.

Lesaffre, E., and B. Spiessens. 2001. On the effect of quadrature points in a logistic random-effects model: An example. *Appl. Statist.* **50**: 325–335.

Lewis, T., I. W. Saunders, and M. Westcott. 1984. The moments of the Pearson chi-squared statistic and the minimum expected value in two-way tables. *Biometrika* **71**: 515–522.

Liang, K. Y. 1984. The asymptotic efficiency of conditional likelihood methods. *Biometrika* **71**: 305–313.

Liang, K. Y., and J. Hanfelt. 1994. On the use of the quasi-likelihood method in teratological experiments. *Biometrics* **50**: 872–880.

Liang, K. Y., and P. McCullagh. 1993. Case studies in binary dispersion. *Biometrics* **49**: 623–630.

Liang, K. Y., and S. G. Self. 1985. Tests for homogeneity of odds ratios when the data are sparse. *Biometrika* **72**: 353–358.

Liang, K. Y., and S. L. Zeger. 1986. Longitudinal data analysis using generalized linear models. *Biometrika* **73**: 13–22.

Liang, K. Y., and S. L. Zeger. 1988. On the use of concordant pairs in matched case–control studies. *Biometrics* **44**: 1145–1156.

Liang, K. Y, and S. L. Zeger. 1995. Inference based on estimating functions in the presence of nuisance parameters. *Statist. Sci.* **10** 158–173.

Liang, K. Y., S. L. Zeger, and B. Qaqish. 1992. Multivariate regression analyses for categorical data. *J. Roy. Statist. Soc.Ser. B* **54**: 3–24.

Lin, X. 1997. Variance component testing in generalized linear models with random effects. *Biometrika* **84**: 309–326.

Lindley, D. V. 1964. The Bayesian analysis of contingency tables. *Ann. Math. Statist.* **35**: 1622–1643.

Lindsay, B., C. Clogg, and J. Grego. 1991. Semi-parametric estimation in the Rasch model and related exponential response models, including a simple latent class model for item analysis. *J. Amer. Statist. Assoc.* **86**: 96–107.

Lindsey, J. K. 1999. *Models for Repeated Measurements*, 2nd ed. Oxford: Oxford University Press.

Lindsey, J. K., and P. M. E. Altham. 1998. Analysis of the human sex ratio by using overdispersion models. *Appl. Statist.* **47**: 149–157.

Lindsey, J. K., and G. Mersch. 1992. Fitting and comparing probability distributions with log linear models. *Comput. Statist. Data Anal.* **13**: 373–384.

Lipsitz, S. 1992. Methods for estimating the parameters of a linear model for ordered categorical data. *Biometrics* **48**: 271–281.

Lipsitz, S. R., and G. Fitzmaurice. 1996. The score test for independence in $R \times C$ contingency tables with missing data. *Biometrics* **52**: 751–762.

Lipsitz, S., N. Laird, and D. Harrington. 1990. Finding the design matrix for the marginal homogeneity model. *Biometrika* **77**: 353–358.

Lipsitz, S., N. Laird, and D. Harrington. 1991. Generalized estimating equations for correlated binary data: Using the odds ratio as a measure of association. *Biometrika* **78**: 153–160.

Lipsitz, S. R., K. Kim, and L. Zhao. 1994. Analysis of repeated categorical data using generalized estimating equations. *Statist. Medic.* **13**: 1149–1163.

Little, R. J. 1989. Testing the equality of two independent binomial proportions. *Amer. Statist.* **43**: 283–288.

Little, R. J. 1998. Missing data. Pp. 2622–2635 in *Encyclopedia of Biostatistics*. Chichester, UK: Wiley.

Little, R. J., and D. B. Rubin. 1987. *Statistical Analysis with Missing Data*. New York: Wiley.

Little, R. J. A., and M.-M. Wu. 1991. Models for contingency tables with known margins when target and sampled populations differ. *J. Amer. Statist. Assoc.* **86**: 87–95.

Liu, Q., and D. A. Pierce. 1993. Heterogeneity in Mantel–Haenszel-type models. *Biometrika* **80**: 543–556.

Liu, Q., and D. A. Pierce. 1994. A note on Gauss–Hermite quadrature. *Biometrika* **81**: 624–629.

Lloyd, C. J. 1988a. Some issues arising from the analysis of 2×2 contingency tables. *Austral. J. Statist.* **30**: 35–46.

Lloyd, C. J. 1988b. Doubling the one-sided P-value in testing independence in 2×2 tables against a two-sided alternative. *Statist. Medic.* **7**: 1297–1306.

Lloyd, C. J. 1999. *Statistical Analysis of Categorical Data*. New York: Wiley.

Longford, N. T. 1993. *Random Coefficient Models*. New York: Oxford University Press.

Loughin, T. M., and P. N. Scherer. 1998. Testing for association in contingency tables with multiple column responses. *Biometrics* **54**: 630–637.

Louis, T. A. 1982. Finding the observed information matrix when using the EM algorithm. *J. Roy. Statist. Soc. Ser. B* **44**: 226–233.

Luce, R. D. 1959. *Individual Choice Behavior*. New York: Wiley.

Madansky, A. 1963. Tests of homogeneity for correlated samples. *J. Amer. Statist. Assoc.* **58**: 97–119.

Maddala, G. S. 1983. *Limited-Dependent and Qualitative Variables in Econometrics*. Cambridge: Cambridge University Press.

Magnus, J. R., and H. Neudecker. 1988. *Matrix Differential Calculus with Applications in Statistics and Econometrics*. New York: Wiley.

Mantel, N. 1963. Chi-square tests with one degree of freedom: Extensions of the Mantel–Haenszel procedure. *J. Amer. Statist. Assoc.* **58**: 690–700.

Mantel, N. 1966. Models for complex contingency tables and polychotomous dosage response curves. *Biometrics* **22**: 83–95.

Mantel, N. 1973. Synthetic retrospective studies and related topics. *Biometrics* **29**: 479–486.

Mantel, N. 1985. Maximum likelihood vs. minimum chi-square. *Biometrics* **41**: 777–781.

Mantel, N. 1987a. Understanding Wald's test for exponential families. *Amer. Statist.* **41**: 147–148.

Mantel, N. 1987b. Exact tests for 2×2 contingency tables (Letter). *Amer. Statist.* **41**: 159.

Mantel, N., and D. P. Byar. 1978. Marginal homogeneity, symmetry and independence. *Commun. Statist. Ser. A* **7**: 953–976.

Mantel, N., and W. Haenszel. 1959. Statistical aspects of the analysis of data from retrospective studies of disease. *J. Natl. Cancer Inst.* **22**: 719–748.

Martín Andrés, A., and Silva Mato, A. 1994. Choosing the optimal unconditional test for comparing two independent proportions. *Comput. Statist. Data Anal.* **17**: 555–574.

Matthews, J. N. S., and K. P. Morris. 1995. An application of Bradley–Terry-type models to the measurement of pain. *Appl. Statist.* **44**: 243–255.

McCullagh, P. 1978. A class of parametric models for the analysis of square contingency tables with ordered categories. *Biometrika* **65**: 413–418.

McCullagh, P. 1980. Regression models for ordinal data. *J. Roy. Statist. Soc. Ser. B* **42**: 109–142.

McCullagh, P. 1982. Some applications of quasisymmetry. *Biometrika* **69**: 303–308.

McCullagh, P. 1983. Quasi-likelihood functions. *Ann. Statist.* **11**: 59–67.

McCullagh, P. 1986. The conditional distribution of goodness-of-fit statistics for discrete data. *J. Amer. Statist. Assoc.* **81**: 104–107.

McCullagh, P., and J. A. Nelder. 1983; 2nd ed., 1989. *Generalized Linear Models*. London: Chapman & Hall.

McCulloch, C. E. 1994. Maximum likelihood variance components estimation for binary data. *J. Amer. Statist. Assoc.* **89**: 330–335.

McCulloch, C. E. 1997. Maximum likelihood algorithms for generalized linear mixed models. *J. Amer. Statist. Assoc.* **92**: 162–170.

McCulloch, C. E. 2000. Generalized linear models. *J. Amer. Statist. Assoc.* **95**: 1320–1324.

McCulloch, C. E., and S. Searle. 2001. *Generalized, Linear, and Mixed Models*. New York: Wiley.

McFadden, D. 1974. Conditional logit analysis of qualitative choice behavior. Pp. 105–142 in *Frontiers in Econometrics*, ed. P. Zarembka. New York: Academic Press.

McFadden, D. 1982. Qualitative response models. Pp. 1–37 in *Advances in Econometrics*, ed. W. Hildebrand. Cambridge: Cambridge University Press.

McNemar, Q. 1947. Note on the sampling error of the difference between correlated proportions or percentages. *Psychometrika* **12**: 153–157.

Mee, R. W. 1984. Confidence bounds for the difference between two probabilities (letter). *Biometrics* **40**: 1175–1176.

Meeden, G., C. Geyer, J. Lang, and E. Funo. 1998. The admissibility of the maximum likelihood estimator for decomposable log-linear interaction models for contingency tables. *Commun. Statist. Ser. A* **27**: 473–493.

Mehta, C. R. 1994. The exact analysis of contingency tables in medical research. *Statist. Methods Medic. Res.* **3**: 135–156.

Mehta, C. R., and N. R. Patel. 1983. A network algorithm for performing Fisher's exact test in $r \times c$ contingency tables. *J. Amer. Statist. Assoc.* **78**: 427–434.

Mehta, C. R., and N. R. Patel. 1995. Exact logistic regression: Theory and examples. *Statist. Medic.* **14**: 2143–2160.

Mehta, C. R., and S. J. Walsh. 1992. Comparison of exact, mid-*P*, and Mantel–Haenszel confidence intervals for the common odds ratio across several 2×2 contingency tables. *Amer. Statist.* **46**: 146–150.

Mehta, C. R., N. R. Patel, and R. Gray. 1985. Computing an exact confidence interval for the common odds ratio in several 2 by 2 contingency tables. *J. Amer. Statist. Assoc.* **80**: 969–973.

Mehta, C. R., N. R. Patel, and P. Senchaudhuri. 1988. Importance sampling for estimating exact probabilities in permutational inference. *J. Amer. Statist. Assoc.* **83**: 999–1005.

Mehta, C. R., N. R. Patel, and P. Senchaudhuri. 2000. Efficient Monte Carlo methods for conditional logistic regression. *J. Amer. Statist. Assoc.* **95**: 99–108.

Michailidis, G., and J. de Leeuw. 1998. The Gifi system of descriptive multivariate analysis. *Statist. Sci.* **13**: 307–336.

Miettinen, O. S. 1969. Individual matching with multiple controls in the case of all-or-none responses. *Biometrics* **25**: 339–355.

Miettinen, O. S., and M. Nurminen. 1985. Comparative analysis of two rates. *Statist. Medic.* **4**: 213–226.

Miller, M. E., C. S. Davis, and J. R. Landis. 1993. The analysis of longitudinal polytomous data: Generalized estimating equations and connections with weighted least squares. *Biometrics* **49**: 1033–1044.

Minkin, S. 1987. On optimal design for binary data. *J. Amer. Statist. Assoc.* **82**: 1098–1103.

Mirkin, B. 2001. Eleven ways to look at the chi-squared coefficient for contingency tables. *Amer. Statist.* **55**: 111–120.

Mitra, S. K. 1958. On the limiting power function of the frequency chi-square test. *Ann. Statist.* **29**: 1221–1233.

Molenberghs, G., and E. Goetghebeur. 1997. Simple fitting algorithms for incomplete categorical data. *J. Roy. Statist. Soc. Ser. B* **59**: 401–414.

Molenberghs, G., and E. Lesaffre. 1994. Marginal modeling of correlated ordinal data using a multivariate Plackett distribution. *J. Amer. Statist. Assoc.* **89**: 633–644.

Molenberghs, G., M. G. Kenward, and E. Lesaffre. 1997. The analysis of longitudinal ordinal data with nonrandom drop-out. *Biometrika* **84**: 33–44.

Moore, D. F. 1986a. Asymptotic properties of moment estimates for overdispersed counts and proportions. *Biometrika* **35**: 583–588.

Moore, D. S. 1986b. Tests of chi-squared type. Pp. 63–95 in *Goodness-of-Fit Techniques*, ed. R. D'Agostino and M. A. Stephens. New York: Marcel Dekker.

Moore, D. F., and A. Tsiatis. 1991. Robust estimation of the variance in moment methods for extra-binomial and extra-Poisson variation. *Biometrics* **47**: 383–401.

Morgan, B. J. T. 1992. *Analysis of Quantal Response Data*. London: Chapman & Hall.

Morgan, W. M., and B. A. Blumenstein. 1991. Exact conditional tests for hierarchical models in multidimensional contingency tables. *Appl. Statist.* **40**: 435–442.

Mosimann, J. E. 1962. On the compound multinomial distribution, the multivariate β-distribution and correlations among proportions. *Biometrika* **49**: 65–82.

Mosteller, F. 1951. Remarks on the method of paired comparisons I: The least-squares solution assuming equal standard deviations and equal correlations. *Psychometrika* **16**: 3–9.

Mosteller, F. 1952. Some statistical problems in measuring the subjective response to drugs. *Biometrics* **8**: 220–226.

Mosteller, F. 1968. Association and estimation in contingency tables. *J. Amer. Statist. Assoc.* **63**: 1–28.

Nair, V. N. 1987. Chi-squared-type tests for ordered alternatives in contingency tables. *J. Amer. Statist. Assoc.* **82**: 283–291.

Natarajan, R., and C. McCulloch. 1995. A note on the existence of the posterior distribution for a class of mixed models for binomial responses. *Biometrika* **82**: 639–643.

Natarajan, R., and C. McCulloch. 1998. Gibbs sampling with diffuse proper priors: A valid approach to data-driven inference? *J. Comput. Graph. Statist.* **7**: 267–277.

Nelder, J., and D. Pregibon. 1987. An extended quasi-likelihood function. *Biometrika* **74**: 221–232.

Nelder, J., and R. W. M. Wedderburn. 1972. Generalized linear models. *J. Roy. Statist. Soc. Ser. A* **135**: 370–384.

Nerlove, M., and S. J. Press. 1973. Univariate and multivariate log-linear and logistic models. Technical Report R-1306-EDA/NIH, Rand Corporation, Santa Monica, CA.

Neuhaus, J. M. 1992. Statistical methods for longitudinal and clustered designs with binary responses. *Statist. Methods Medic. Res.* **1**: 249–273.

Neuhaus, J. M., and N. P. Jewell. 1990a. Some comments on Rosner's multiple logistic model for clustered data. *Biometrics* **46**: 523–534.

Neuhaus, J. M., and N. P. Jewell. 1990b. The effect of retrospective sampling on binary regression models for clustered data. *Biometrics* **46**: 977–990.

Neuhaus, J. M., and M. L. Lesperance. 1996. Estimation efficiency in a binary mixed-effects model setting. *Biometrika* **83**: 441–446.

Neuhaus, J. M., J. D. Kalbfleisch, and W. W. Hauck. 1991. A comparison of cluster-specific and population-averaged approaches for analyzing correlated binary data. *Internat. Statist. Rev.* **59**: 25–35.

Neuhaus, J. M., W. W. Hauck, and J. D. Kalbfleisch. 1992. The effects of mixture distribution misspecification when fitting mixed-effects logistic models. *Biometrika* **79**: 755–762.

Neuhaus, J. M., J. D. Kalbfleisch, and W. W. Hauck. 1994. Conditions for consistent estimation in mixed-effects models for binary matched-pairs data. *Canad. J. Statist.* **22**: 139–148.

Newcombe, R. 1998a. Two-sided confidence intervals for the single proportion: Comparison of seven methods. *Statist. Medic.* **17**: 857–872.

Newcombe, R. 1998b. Interval estimation for the difference between independent proportions: Comparison of eleven methods. *Statist. Medic.* **17**: 873–890.

Newcombe, R. 2001. Logit confidence intervals and the inverse sinh transformation. *Amer. Statist.* **55**: 200–202.

Neyman, J. 1935. On the problem of confidence limits. *Ann. Math. Statist.* **6**: 111–116.

Neyman, J. 1949. Contributions to the theory of the χ^2 test. Pp. 239–273 in *Proc. First Berkeley Symposium on Mathematical Statistics and Probability*, ed. J. Neyman. Berkeley, CA: University of California Press.

Nurminen, M. 1986. Confidence intervals for the ratio and difference of two binomial proportions. *Biometrics* **42**: 675–676.

O'Brien, P. C. 1988. Comparing two samples: Extensions of the t, rank-sum, and log-rank tests. *J. Amer. Statist. Assoc.* **83**: 52–61.

O'Brien, R. G. 1986. Using the SAS system to perform power analyses for log-linear models. Pp. 778–784 in *Proc. 11th Annual SAS Users Group Conference*. Cary, NC: SAS Institute.

Ochi, Y., and R. Prentice. 1984. Likelihood inference in a correlated probit regression model. *Biometrika* **71**: 531–543.

O'Gorman, T. W., and R. F. Woolson. 1988. Analysis of ordered categorical data using the SAS system. Pp. 957–963 in *Proc. 13th Annual SAS Users Group Conference*. Cary, NC: SAS Institute.

Paik, M. 1985. A graphic representation of a three-way contingency table: Simpson's paradox and correlation. *Amer. Statist.* **39**: 53–54.

Palmgren, J. 1981. The Fisher information matrix for log-linear models arguing conditionally in the observed explanatory variables. *Biometrika* **68**: 563–566.

Palmgren, J., and A. Ekholm. 1987. Exponential family non-linear models for categorical data with errors of observation. *Appl. Stochastic Models Data Anal.* **3**: 111–124.

Park, T., and M. B. Brown. 1994. Models for categorical data with nonignorable nonresponse. *J. Amer. Statist. Assoc.* **89**: 44–52.

Parr, W. C., and H. D. Tolley. 1982. Jackknifing in categorical data analysis. *Austral. J. Statist.* **24**: 67–79.

Parzen, E. 1997. Concrete statistics. Pp. 309–332 in *Statistics of Quality*. New York: Marcel Dekker.

Patefield, W. M. 1982. Exact tests for trends in ordered contingency tables. *Appl. Statist. Ser B* **31**: 32–43.

Patnaik, P. B. 1949. The non-central χ^2 and F-distributions and their applications. *Biometrika* **36**: 202–232.

Paul, S. R., K. Y. Liang, and S. G. Self. 1989. On testing departure from the binomial and multinomial assumptions. *Biometrics* **45**: 231–236.

Pearson, E. S. 1947. The choice of a statistical test illustrated on the interpretation of data classified in 2 × 2 tables. *Biometrika* **34**: 139–167.

Pearson, K. 1900. On a criterion that a given system of deviations from the probable in the case of a correlated system of variables is such that it can be reasonably supposed to have arisen from random sampling. *Philos. Mag. Ser.* 5 **50**: 157–175. (Reprinted in *Karl Pearson's Early Statistical Papers*, ed. E. S. Pearson. Cambridge: Cambridge University Press, 1948.)

Pearson, K. 1904. Mathematical contributions to the theory of evolution XIII: On the theory of contingency and its relation to association and normal correlation. *Draper's Co. Research Memoirs, Biometric Series*, no. 1. (Reprinted in *Karl Pearson's Early Papers*, ed. E. S. Pearson, Cambridge: Cambridge University Press, 1948.)

Pearson, K. 1913. On the probable error of a correlation coefficient as found from a fourfold table. *Biometrika* **9**: 22–27.

Pearson, K. 1917. On the general theory of multiple contingency with special reference to partial contingency. *Biometrika* **11**: 145–158.

Pearson, K. 1922. On the χ^2 test of goodness of fit. *Biometrika* **14**: 186–191.

Pearson, K., and D. Heron. 1913. On theories of association. *Biometrika* **9**: 159–315.

Peduzzi, P., J. Concato, E. Kemper, T. R. Holford, and A. R. Feinstein. 1996. A simulation study of the number of events per variable in logistic regression analysis. *J. Clin. Epidemiol.* **49**: 1373–1379.

Pendergast, J. F., S. J. Gange, M. A. Newton, M. J. Lindstrom, M. Palta, and M. R. Fisher. 1996. A survey of methods for analyzing clustered binary response data. *Internat. Statist. Rev.* **64**: 89–118.

Pepe, M. S. 2000. Receiver operating characteristic methodology. *J. Amer. Statist. Assoc.* **95**: 308–311.

Peterson, B., and F. E. Harrell, Jr. 1990. Partial proportional odds models for ordinal response variables. *Appl. Statist.* **39**: 205–217.

Pierce, D. A., and D. Peters. 1992. Practical use of higher order asymptotics for multiparameter exponential families. *J. Roy. Statist. Soc. Ser. B* **54**: 701–725.

Pierce, D. A., and D. Peters. 1999. Improving on exact tests by approximate conditioning. *Biometrika* **86**: 265–277.

Pierce, D. A., and B. R. Sands. 1975. Extra-Bernoulli variation in regression of binary data. Technical Report 46, Statistics Deptartment, Oregon State University, Cornwallis, OR.

Pierce, D. A., and D. W. Schafer. 1986. Residuals in generalized linear models. *J. Amer. Statist. Assoc.* **81**: 977–983.

Plackett, R. L. 1962. A note on interactions in contingency tables. *J. Roy. Statist. Soc. Ser. B* **24**: 162–166.

Plackett, R. L. 1964. The continuity correction in 2 × 2 tables. *Biometrika* **51**: 327–337.

Plackett, R. L. 1983. Karl Pearson and the chi-squared test. *Internat. Statist. Rev.* **51**: 59–72.

Podgor, M. J., J. L. Gastwirth, and C. R. Mehta. 1996. Efficiency robust tests of independence in contingency tables with ordered classifications. *Statist. Medic.* **15**: 2095–2105.

Poisson, S.-D. 1837. *Recherches sur la probabilité des jugements en matieère criminelle et en matière civile, précédées des règles générales du calcul des probabilités*. Paris: Bachelier.

Pratt, J. W. 1981. Concavity of the log likelihood. *J. Amer. Statist. Assoc.* **76**: 103–106.

Pregibon, D. 1980. Goodness of link tests for generalized linear models. *Appl. Statist.* **29**: 15–24.

Pregibon, D. 1981. Logistic regression diagnostics. *Ann. Statist.* **9**: 705–724.

Pregibon, D. 1982. Score tests in GLIM with application. Pp. 87–97 in *Lecture Notes in Statistics*, 14: *GLIM 82, Proc. International Conference on Generalised Linear Models*, ed. R. Gilchrist. New York: Springer-Verlag.

Prentice, R. 1976a. Use of the logistic model in retrospective studies. *Biometrics* **32**: 599–606.

Prentice, R. 1976b. Generalization of the probit and logit methods for dose response curves. *Biometrics* **32**: 761–768.

Prentice, R. 1986. Binary regression using an extended beta-binomial distribution, with discussion of correlation induced by covariate measurement errors. *J. Amer. Statist. Assoc.* **81**: 321–327.

Prentice, R., and N. Breslow. 1978. Retrospective studies and failure time models. *Biometrika* **65**: 153–158.

Prentice, R., and L. A. Gloeckler. 1978. Regression analysis of grouped survival data with application to breast cancer data. *Biometrics* **34**: 57–67.

Prentice, R., and R. Pyke. 1979. Logistic disease incidence models and case-control studies. *Biometrika* **66**: 403–412.

Prentice, R., and L. P. Zhao. 1991. Estimating equations for parameters in means and covariances of multivariate discrete and continuous responses. *Biometrics* **47**: 825–839.

Press, S. J., and S. Wilson. 1978. Choosing between logistic regression and discriminant analysis. *J. Amer. Statist. Assoc.* **73**: 699–705.

Qu, A., B. G. Lindsay, and B. Li. 2000. Improving generalised estimating equations using quadratic inference functions. *Biometrika* **87**: 823–836.

Quine, M. P., and E. Seneta. 1987. Bortkiewicz's data and the law of small numbers. *Internat. Statist. Rev.* **5**: 173–181.

Rabe-Hesketh, S., and A. Skrondal. 2001. Parameterisation of multivariate random effects models for categorical data. *Biometrics* **57**:–.

Raftery, A. E. 1986. Choosing models for cross-classification. *Amer. Sociol. Rev.* **51**: 145–146.

Rao, C. R. 1957. Maximum likelihood estimation for the multinomial distribution. *Sankhya* **18**: 139–148.

Rao, C. R. 1963. Criteria of estimation in large samples. *Sankhya* **25**: 189–206.

Rao, C. R. 1973. *Linear Statistical Inference and Its Applications*, 2nd ed. New York: Wiley.

Rao, C. R. 1982. Diversity: Its measurement, decomposition, apportionment, and analysis. *Sankhya Ser. A* **44**: 1–22.

Rao, J. N. K., and A. J. Scott. 1987. On simple adjustments to chi-square tests with sample survey data. *Ann. Statist.* **15**: 385–397.

Rao, J. N. K., and D. R. Thomas. 1988. The analysis of cross-classified categorical data from complex sample surveys. *Sociol. Methodol.* **18**: 213–270.

Rasch, G. 1961. On general laws and the meaning of measurement in psychology. Pp. 321–333 in *Proc. 4th Berkeley Symposium on Mathematics, Statistics, and Probability*, Vol. 4, ed. J. Neyman. Berkeley, CA: University of California Press.

Rayner, J. C. W., and D. J. Best. 2001. *A Contingency Table Approach to Nonparametric Testing*. London: Chapman & Hall.

Read, T. R. C., and N. A. C. Cressie. 1988. *Goodness-of-Fit Statistics for Discrete Multivariate Data*. New York: Springer-Verlag.

Rice, W. R. 1988. A new probability model for determining exact *P*-values for 2×2 contingency tables when comparing binomial proportions. *Biometrics* **44**: 1–22.

Ritov, Y., and Z. Gilula. 1991. The order-restricted RC model for ordered contingency tables: Estimation and testing for fit. *Ann. Statist.* **19**: 2090–2101.

Robins, J., N. Breslow, and S. Greenland. 1986. Estimators of the Mantel–Haenszel variance consistent in both sparse data and large-strata limiting models. *Biometrics* **42**: 311–323.

Robins, J., A. Rotnitzky, and L. P. Zhao. 1995. Analysis of semiparametric regression models for repeated outcomes in the presence of missing data. *J. Amer. Statist. Assoc.* **90**: 106–121.

Röhmel, J., and U. Mansmann. 1999. Unconditional non-asymptotic one-sided tests for independent binomial proportions when the interest lies in showing non-inferiority and/or superiority. *Biometrical J.* **41**: 149–170.

Rosenbaum, P. R., and D. R. Rubin. 1983. The central role of the propensity score in observational studies for causal effects. *Biometrika* **70**: 41–55.

Rosner, B. 1984. Multivariate methods in ophthalmology with application to other paired-data situations. *Biometrics* **40**: 1025–1035.

Rosner, B. 1989. Multivariate methods for clustered binary data with more than one level of nesting. *J. Amer. Statist. Assoc.* **84**: 373–380.

Rotnitzky, A., and N. P. Jewell. 1990. Hypothesis testing of regression parameters in semiparametric generalized linear models for cluster correlated data. *Biometrika* **77**: 485–497.

Routledge, R. D. 1992. Resolving the conflict over Fisher's exact test. *Canad. J. Statist.* **20**: 201–209.

Routledge, R. D. 1994. Practicing safe statistics with the mid-P^*. *Canad. J. Statist.* **22**: 103–110.

Roy, S. N., and M. A. Kastenbaum. 1956. On the hypothesis of no "interaction" in a multiway contingency table. *Ann. Math. Statist.* **27**: 749–757.

Roy, S. N., and S. K. Mitra. 1956. An introduction to some nonparametric generalizations of analysis of variance and multivariate analysis. *Biometrika* **43**: 361–376.

Rudas, T., C. C. Clogg, and B. G. Lindsay. 1994. A new index of fit based on mixture methods for the analysis of contingency tables. *J. Roy. Statist. Soc.* **56**: 623–639.

Ryan, L. 1992. Quantitative risk assessment for developmental toxicity. *Biometrics* **48**: 163–174.

Ryan, L. 1995. Comment on article by Liang and Zeger. *Statist. Sci.* **10**: 189–193.

Samuels, M. L. 1993. Simpson's paradox and related phenomena. *J. Amer. Statist. Assoc.* **88**: 81–88.

Santner, T. J., and M. K. Snell. 1980. Small-sample confidence intervals for p_1-p_2 and p_1/p_2 in 2×2 contingency tables. *J. Amer. Statist. Assoc.* **75**: 386–394.

Santner, T. J., and S. Yamagami. 1993. Invariant small sample confidence intervals for the difference of two success probabilities. *Commun. Statist. Ser. B* **22**: 33–59.

Schafer, J. L. 1997. *Analysis of Incomplete Multivariate Data.* London: Chapman & Hall.

Schluchter, M. D., and K. L. Jackson. 1989. Log-linear analysis of censored survival data with partially observed covariates. *J. Amer. Statist. Assoc.* **84**: 42–52.

Scott, A., and C. Wild. 2001. Case–control studies with complex sampling. *Appl. Statist.* **50**: 389–401.

Seeber, G. 1998. Poisson regression. Pp. 3404–3412 in *Encyclopedia of Biostatistics.* Chichester, UK: Wiley.

Sekar, C. C., and W. E. Deming. 1949. On a method of estimating birth and death rates and the extent of registration. *J. Amer. Statist. Assoc.* **44**: 101–115.

Self, S. G., and K.-Y. Liang. 1987. Asymptotic properties of maximum likelihood estimators and likelihood-ratio tests under nonstandard conditions. *J. Amer. Statist. Assoc.* **82**: 605–610.

Sen, P. K., and J. M. Singer. 1993. *Large Sample Methods in Statistics: An Introduction with Applications.* London: Chapman & Hall.

Shapiro, S. H. 1982. Collapsing contingency tables: A geometric approach. *Amer. Statist.* **36**: 43–46.

Shuster, J., and D. Downing. 1976. Two-way contingency tables for complex sampling schemes. *Biometrika* **63**: 271–276.

Silvapulle, M. J. 1981. On the existence of maximum likelihood estimators for the binomial response models. *J. Roy. Statist. Soc. Ser. B* **43**: 310–313.

Simon, G. 1973. Additivity of information in exponential family probability laws.)*J. Amer. Statist. Assoc.* **68**: 478–482.

Simon, G. 1974. Alternative analyses for the singly-ordered contingency table. *J. Amer. Statist. Assoc.* **69**: 971–976.

Simon, G. 1978. Efficacies of measures of association for ordinal contingency tables. *J. Amer. Statist. Assoc.* **73**: 545–551.

Simonoff, J. 1983. A penalty function approach to smoothing large sparse contingency tables. *Ann. Statist.* **11**: 208–218.

Simonoff, J. 1986. Jackknifing and bootstrapping goodness-of-fit statistics in sparse multinomials. *J. Amer. Statist. Assoc.* **81**: 1005–1111.

Simonoff, J. S. 1996. *Smoothing Methods in Statistics*. New York: Springer-Verlag.

Simonoff, J. S. 1998. Three sides of smoothing: Categorical data smoothing, nonparametric regression, and density estimation. *Internat. Statist. Rev.* **66**: 137–156.

Simpson, E. H. 1949. The measurement of diversity. *Nature* **163**: 699.

Simpson, E. H. 1951. The interpretation of interaction in contingency tables. *J. Roy. Statist. Soc. Ser. B* **13**: 238–241.

Skellam, J. G. 1948. A probability distribution derived from the binomial distribution by regarding the probability of success as variable between the sets of trials. *J. Roy. Statist. Soc. Ser. B* **10**: 257–261.

Skene, A. M., and J. C. Wakefield. 1990. Hierarchical models for multicentre binary response studies. *Statist. Medic.* **9**: 919–929.

Slaton, T. L., W. W. Piegorsch, and S. D. Durham. 2000. Estimation and testing with overdispersed proportions using the beta-logistic regression model of Heckman and Willis. *Biometrics* **56**: 125–133.

Small, K. A. 1987. A discrete choice model for ordered alternatives. *Econometrica* **55**: 409–424.

Smith, K. W. 1976. Table standardization and table shrinking: Aids in the traditional analysis of contingency tables. *Social Forces* **54**: 669–693.

Smith, P. W. F., J. J. Forster, and J. W. McDonald. 1996. Monte Carlo exact tests for square contingency tables. *J. Roy. Statist. Soc. Ser. A* **159**: 309–321.

Snell, E. J. 1964. A scaling procedure for ordered categorical data. *Biometrics* **20**: 592–607.

Somers, R. H. 1962. A new asymmetric measure of association for ordinal variables. *Amer. Sociol. Rev.* **27**: 799–811.

Speed, T. 1998. Iterative proportional fitting. Pp. 2116–2119 in *Encyclopedia of Biostatistics*. Chichester, UK: Wiley.

Spiegelhalter, D. J., and A. F. M. Smith. 1982. Bayes factors for linear and log-linear models with vague prior information. *J. Roy. Statist. Soc. Ser. B* **44**: 377–387.

Spitzer, R. L., J. Cohen, J. L. Fleiss, and J. Endicott. 1967. Quantification of agreement in psychiatric diagnosis. *Arch. Gen. Psychiatry* **17**: 83–87.

Sprott, D. A. 2000. *Statistical Inference in Science*. New York: Springer-Verlag.

Stern, S. 1997. Simulation-based estimation. *J. Econ. Literature* **35**: 2006–2039.

Sterne, T. E. 1954. Some remarks on confidence or fiducial limits. *Biometrika* **41**: 275–278.

Stevens, S. S. 1951. Mathematics, measurement, and psychophysics. Pp. 1–49 in *Handbook of Experimental Psychology*, ed. S. S. Stevens. New York: Wiley.

Stevens, W. L. 1950. Fiducial limits of the parameter of a discontinuous distribution. *Biometrika* **37**: 117–129.

Stigler, S. 1986. *The History of Statistics: The Measurement of Uncertainty before* 1900. Cambridge, MA: Harvard University Press.

Stigler, S. 1994. Citation patterns in the journals of statistics and probability. *Statist. Sci.* **9**: 94–108.

Stigler, S. 1999. *Statistics on the Table*. Cambridge, MA: Harvard University Press.

Stiratelli, R., N. Laird, and J. H. Ware. 1984. Random-effects models for serial observations with binary response. *Biometrics* **40**: 1025–1035.

Stokes, M. E., C. S. Davis, and G. G. Koch. 2000. *Categorical Data Analysis Using the SAS System*, 2nd ed. Cary, NC: SAS Institute.

Strawderman, R. L., and M. T. Wells. 1998. Approximately exact inference for the common odds ratio in several 2×2 tables. *J. Amer. Statist. Assoc.* **93**: 1294–1307.

Stuart, A. 1955. A test for homogeneity of the marginal distributions in a two-way classification. *Biometrika* **42**: 412–416.

Stukel, T. A. 1988. Generalized logistic models. *J. Amer. Statist. Assoc.* **83**: 426–431.

Suissa, S., and J. J. Shuster. 1984. Are uniformly most powerful unbiased tests really best? *Amer. Statist.* **38**: 204–206.

Suissa, S., and J. J. Shuster. 1985. Exact unconditional samples sizes for the 2 by 2 binomial trial. *J. Roy. Statist. Soc. Ser. A* **148**: 317–327.

Suissa, S., and J. J. Shuster. 1991. The 2×2 matched-pairs trial: Exact unconditional design and analysis. *Biometrics* **47**: 361–372.

Sundberg, R. 1975. Some results about decomposable (or Markov-type) models for multidimensional contingency tables: Distribution of marginals and partitioning of tests. *Scand. J. Statist.* **2**: 71–79.

Tango, T. 1998. Equivalence test and confidence interval for the difference in proportions for the paired-sample design. *Statist. Medic.* **17**: 891–908.

Tanner, M. A., and M. A. Young. 1985. Modelling agreement among raters. *J. Amer. Statist. Assoc.* **80**: 175–180.

Tarone, R. E. 1985. On heterogeneity tests based on efficient scores. *Biometrika* **72**: 91–95.

Tarone, R. E., and J. J. Gart. 1980. On the robustness of combined tests for trends in proportions. *J. Amer. Statist. Assoc.* **75**: 110–116.

Tarone, R. E., J. J. Gart, and W. W. Hauck. 1983. On the asymptotic relative efficiency of certain noniterative estimators of a common relative risk or odds ratio. *Biometrika* **70**: 519–522.

Tavaré, S., and P. M. E. Altham. 1983. Serial dependence of observations leading to contingency tables, and corrections to chi-squared statistics. *Biometrika* **70**: 139–144.

Ten Have, T. R. 1996. A mixed effects model for multivariate ordinal response data including correlated discrete failure times with ordinal responses. *Biometrics* **52**: 473–491.

Ten Have, T. R., and A. R. Localio. 1999. Empirical Bayes estimation of random effects parameters in mixed effects logistic regression models. *Biometrics* **55**: 1022–1029.

Ten Have, T. R., and A. Morabia. 1999. Mixed effects models with bivariate and univariate association parameters for longitudinal bivariate binary response data. *Biometrics* **55**: 85–93.

Ten Have, T. R., and D. H. Uttal. 1994. Subject-specific and population-averaged continuation ratio logit models for multiple discrete time survival profiles. *Appl. Statist.* **43**: 371–384.

Theil, H. 1969. A multinomial extension of the linear logit model. *Internat. Econ. Rev.* **10**: 251–259.

Theil, H. 1970. On the estimation of relationships involving qualitative variables. *Amer. J. Sociol.* **76**: 103–154.

Thompson, R., and R. J. Baker. 1981. Composite link functions in generalized linear models. *Appl. Statist.* **30**: 125–131.

Thompson, W. A. 1977. On the treatment of grouped observations in life studies. *Biometrics* **33**: 463–470.

Thurstone, L. L. 1927. The method of paired comparisons for social values. *J. Abnormal Social Psych.* **21**: 384–400.

Tjur, T. 1982. A connection between Rasch's item analysis model and a multiplicative Poisson model. *Scand. J. Statist.* **9**: 23–30.

Tocher, K. D. 1950. Extension of the Neyman–Pearson theory of tests to discontinuous variates. *Biometrika* **37**: 130–144.

Toledano, A., and C. Gatsonis. 1996. Ordinal regression methodology for ROC curves derived from correlated data. *Statist. Medic.* **15**: 1807–1826.

Train, K. 1986. *Qualitative Choice Analysis: Theory, Econometrics, and an Application.* Cambridge, MA: MIT Press.

Tsiatis, A. A. 1980. A note on the goodness-of-fit test for the logistic regression model. *Biometrika* **67**: 250–251.

Tutz, G. 1989. Compound regression models for ordered categorical data. *Biometrical J.* **31**: 259–272.

Tutz, G. 1991. Sequential models in categorical regression. *Comput. Statist. Data Anal.* **11**: 275–295.

Tutz, G., and W. Hennevogl. 1996. Random effects in ordinal regression models. *Comput. Statist. Data Anal.* **22**: 537–557.

Uebersax, J. S. 1993. Statistical modeling of expert ratings on medical treatment appropriateness. *J. Amer. Statist. Assoc.* **88**: 421–427.

Uebersax, J. S., and W. M. Grove. 1990. Latent class analysis of diagnostic agreement. *Statist. Medic.* **9**: 559–572.

Uebersax, J. S., and W. M. Grove. 1993. A latent trait finite mixture model for the analysis of rating agreement. *Biometrics* **49**: 823–835.

van der Heijden, P. G. M., and J. de Leeuw. 1985. Correspondence analysis: A complement to log-linear analysis. *Psychometrika* **50**: 429–447.

van der Heijden, P. G. M., A. de Falguerolles, and J. de Leeuw. 1989. A combined approach to contingency table analysis using correspondence analysis and log-linear analysis. *Appl. Statist.* **38**: 249–292.

Verbeke, G., and E. Lesaffre. 1996. A linear mixed-effects model with heterogeneity in the random-effects population. *J. Amer. Statist. Assoc.* **91**: 217–221.

Verbeke, G., and G. Molenberghs. 2000. *Linear Mixed Models for Longitudinal Data.* New York: Springer-Verlag.

Wald, A. 1943. Tests of statistical hypotheses concerning several parameters when the number of observations is large. *Trans. Amer. Math. Soc.* **54**: 426–482.

Walker, S. H., and D. B. Duncan. 1967. Estimation of the probability of an event as a function of several independent variables. *Biometrika* **54**: 167–179.

Walley, P. 1996. Inferences from multinomial data: Learning about a bag of marbles. *J. Roy. Statist. Soc. Ser. B* **58**: 3–34.

Wardrop, R. L. 1995. Simpson's paradox and the hot hand in basketball. *Amer. Statist.* **49**: 24–28.

Ware, J. H., S. Lipsitz, and F. E. Speizer. 1988. Issues in the analysis of repeated categorical outcomes. *Statist. Medic.* **7**: 95–107.

Watson, G. S. 1956. Missing and "mixed up" frequencies in contingency tables. *Biometrics* **12**: 47–50.

Watson, G. S. 1959. Some recent results in chi-square goodness-of-fit tests. *Biometrics* **15**: 440–468.

Wedderburn, R. W. M. 1974. Quasi-likelihood functions, generalized linear models, and the Gauss–Newton method. *Biometrika* **61**: 439–447.

Wedderburn, R. W. M. 1976. On the existence and uniqueness of the maximum likelihood estimates for certain generalized linear models. *Biometrika* **63**: 27–32.

Wermuth, N. 1976. Model search among multiplicative models. *Biometrics* **32**: 253–263.

Wermuth, N. 1987. Parametric collapsibility and the lack of moderating effects in contingency tables with a dichotomous response variable. *J. Roy. Statist. Soc. Ser. B* **49**: 353–364.

Westfall, P. H., and R. D. Wolfinger. 1997. Multiple tests with discrete distributions. *Amer. Statist.* **51**: 3–8.

Westfall, P. H., and S. S. Young. 1993. *Resampling-Based Multiple Testing: Examples and Methods for p-Value Adjustment*. New York: Wiley.

White, H. 1982. Maximum likelihood estimation of misspecified models. *Econometrica* **50**: 1–26.

White, A. A., J. R. Landis, and M. M. Cooper. 1982. A note on the equivalence of several marginal homogeneity test criteria for categorical data. *Internat. Statist. Rev.* **50**: 27–34.

Whitehead, J. 1993. Sample size calculations for ordered categorical data. *Statist. Medic.* **12**: 2257–2271.

Whittaker, J. 1990. *Graphical Models in Applied Multivariate Statistics*. New York: Wiley.

Whittaker, J., and M. Aitkin. 1978. A flexible strategy for fitting complex log-linear models. *Biometrics* **34**: 487–495.

Whittemore, A. S. 1978. Collapsibility of multidimensional tables. *J. Roy. Statist. Soc. Ser. B* **40**: 328–340.

Whittemore, A. S. 1981. Sample size for logistic regression with small response probability. *J. Amer. Statist. Assoc.* **76**: 27–32.

Wilks, S. S. 1935. The likelihood test of independence in contingency tables. *Ann. Math. Statist.* **6**: 190–196.

Wilks, S. S. 1938. The large-sample distribution of the likelihood ratio for testing composite hypotheses. *Ann. Math. Statist.* **9**: 60–62.

Williams, D. A. 1975. The analysis of binary responses from toxicological experiments involving reproduction and teratogenicity. *Biometrics* **31**: 949–952.

Williams, D. A. 1982. Extra-binomial variation in logistic linear models. *Appl. Statist.* **31**: 144–148.

Williams, D. A. 1987. Generalized linear model diagnostics using the deviance and single-case deletions. *Appl. Statist.* **36**: 181–191.

Williams, D. A. 1988. Comments on "The impact of litter effects on dose–response modeling in teratology." *Biometrics* **44**: 305–308.

Williams, E. J. 1952. Use of scores for the analysis of association in contingency tables. *Biometrika* **39**: 274–289.

Williams, O. D., and J. E. Grizzle. 1972. Analysis for contingency tables having ordered response categories. *J. Amer. Statist. Assoc.* **67**: 55–63.

Wilson, E. B. 1927. Probable inference, the law of succession, and statistical inference. *J. Amer. Statist. Assoc.* **22**: 209–212.

Wolfinger, R., and M. O'Connell. 1993. Generalized linear mixed models: A pseudo-likelihood approach. *J. Statist. Comput. Simul.* **48**: 233–243.

Wong, G. Y., and W. M. Mason. 1985. The hierarchical logistic regression model for multilevel analysis. *J. Amer. Statist. Assoc.* **80**: 513–524.

Woolf, B. 1955. On estimating the relation between blood group and disease. *Ann. Human Genet. (London)* **19**: 251–253.

Woolson, R. F., and W. R. Clarke. 1984. Analysis of categorical incomplete longitudinal data. *J. Roy. Statist. Soc. Ser. A* **147**: 87–99.

Wu, C. F. J. 1985. Efficient sequential designs with binary data. *J. Amer. Statist. Soc.* **80**: 974–984.

Yang, I., and M. P. Becker. 1997. Latent variable modeling of diagnostic accuracy. *Biometrics* **53**: 948–958.

Yates, F. 1934. Contingency tables involving small numbers and the χ^2 test. *J. Roy. Statist. Soc. Suppl.* **1**: 217–235.

Yates, F. 1948. The analysis of contingency tables with grouping based on quantitative characters. *Biometrika* **35**: 176–181.

Yates, F. 1984. Tests of significance for 2×2 contingency tables. *J. Roy. Statist. Soc. Ser. A* **147**: 426–463.

Yee, T. W., and C. J. Wild. 1996. Vector generalized additive models. *J. Roy. Statist. Soc. Ser. B* **58**: 481–493.

Yerushalmy, J. 1947. Statistical problems in assessing methods of medical diagnosis, with special reference to x-ray techniques. *Public Health Rep.* **62**: 1432–1449.

Yule, G. U. 1900. On the association of attributes in statistics. *Philos. Trans. Roy. Soc. London Ser. A* **194**: 257–319.

Yule, G. U. 1903. Notes on the theory of association of attributes in statistics. *Biometrika* **2**: 121–134.

Yule, G. U. 1906. On a property which holds good for all groupings of a normal distribution of frequency for two variables, with application to the study of contingency tables for the inheritance of unmeasured qualities. *Proc. Roy. Soc. Ser A* **77**: 324–336.

Yule, G. U. 1912. On the methods of measuring association between two attributes. *J. Roy. Statist. Soc.* **75**: 579–642.

Zeger, S. L., and M. R. Karim. 1991. Generalized linear models with random effects: A Gibbs sampling approach. *J. Amer. Statist. Assoc.* **86**: 79–86

Zeger, S. L., K.-Y. Liang, and P. S. Albert. 1988. Models for longitudinal data: A generalized estimating equation approach. *Biometrics* **44**: 1049–1060.

Zelen, M. 1971. The analysis of several 2×2 contingency tables. *Biometrika* **58**: 129–137.

Zelen, M. 1991. Multinomial response models. *Comput. Statist. Data Anal.* **12**: 249–254.

Zellner, A., and P. E. Rossi. 1984. Bayesian analysis of dichotomous quantal response models. *J. Economet.* **25**: 365–393.

Zelterman. D. 1987. Goodness-of-fit tests for large sparse multinomial distributions. *J. Amer. Statist. Soc.* **82**: 624–629.

Zermelo, E. 1929. Die Berechnung der Turnier-Ergebnisse als ein Maximumproblem der Wahrscheinlichkeitsrechnung. *Math. Z.* **29**: 436–460.

Zhang, H., J. Crowley, H. Sox, and R. Olshen. 1998. Tree-structured statistical methods. Pp. 4561–4573 in *Encyclopedia of Biostatistics*. Chichester, UK: Wiley.

Zheng, B., and A. Agresti. 2000. Summarizing the predictive power of a generalized linear model. *Statist. Medic.* **19**: 1771–1781.

Zhu, Y., and N. Reid. 1994. Information, ancillarity, and sufficiency in the presence of nuisance parameters. *Canad. J. Statist.* **22**: 111–123.

Examples Index

Author Index

Subject Index

WILEY SERIES IN PROBABILITY AND STATISTICS

ESTABLISHED BY WALTER A. SHEWHART AND SAMUEL S. WILKS

The *Wiley Series in Probability and Statistics* is well established and authoritative. It covers many topics of current research interest in both pure and applied statistics and probability theory. Written by leading statisticians and institutions, the titles span both state-of-the-art developments in the field and classical methods.

Reflecting the wide range of current research in statistics, the series encompasses applied, methodological and theoretical statistics, ranging from applications and new techniques made possible by advances in computerized practice to rigorous treatment of theoretical approaches.

This series provides essential and invaluable reading for all statisticians, whether in academia, industry, government, or research.

ABRAHAM and LEDOLTER · Statistical Methods for Forecasting
AGRESTI · Analysis of Ordinal Categorical Data
AGRESTI · An Introduction to Categorical Data Analysis
AGRESTI · Categorical Data Analysis, *Second Edition*
ANDĚL · Mathematics of Chance
ANDERSON · An Introduction to Multivariate Statistical Analysis, *Second Edition*
*ANDERSON · The Statistical Analysis of Time Series
ANDERSON, AUQUIER, HAUCK, OAKES, VANDAELE, and WEISBERG ·
    Statistical Methods for Comparative Studies
ANDERSON and LOYNES · The Teaching of Practical Statistics
ARMITAGE and DAVID (editors) · Advances in Biometry
ARNOLD, BALAKRISHNAN, and NAGARAJA · Records
*ARTHANARI and DODGE · Mathematical Programming in Statistics
*BAILEY · The Elements of Stochastic Processes with Applications to the Natural
    Sciences
BALAKRISHNAN and KOUTRAS · Runs and Scans with Applications
BARNETT · Comparative Statistical Inference, *Third Edition*
BARNETT and LEWIS · Outliers in Statistical Data, *Third Edition*
BARTOSZYNSKI and NIEWIADOMSKA-BUGAJ · Probability and Statistical Inference
BASILEVSKY · Statistical Factor Analysis and Related Methods: Theory and
    Applications
BASU and RIGDON · Statistical Methods for the Reliability of Repairable Systems
BATES and WATTS · Nonlinear Regression Analysis and Its Applications
BECHHOFER, SANTNER, and GOLDSMAN · Design and Analysis of Experiments for
    Statistical Selection, Screening, and Multiple Comparisons
BELSLEY · Conditioning Diagnostics: Collinearity and Weak Data in Regression
BELSLEY, KUH, and WELSCH · Regression Diagnostics: Identifying Influential
    Data and Sources of Collinearity
BENDAT and PIERSOL · Random Data: Analysis and Measurement Procedures,
    Third Edition

*Now available in a lower priced paperback edition in the Wiley Classics Library.

BERRY, CHALONER, and GEWEKE · Bayesian Analysis in Statistics and
Econometrics: Essays in Honor of Arnold Zellner
BERNARDO and SMITH · Bayesian Theory
BHAT and MILLER · Elements of Applied Stochastic Processes, *Third Edition*
BHATTACHARYA and JOHNSON · Statistical Concepts and Methods
BHATTACHARYA and WAYMIRE · Stochastic Processes with Applications
BILLINGSLEY · Convergence of Probability Measures, *Second Edition*
BILLINGSLEY · Probability and Measure, *Third Edition*
BIRKES and DODGE · Alternative Methods of Regression
BLISCHKE AND MURTHY · Reliability: Modeling, Prediction, and Optimization
BLOOMFIELD · Fourier Analysis of Time Series: An Introduction, *Second Edition*
BOLLEN · Structural Equations with Latent Variables
BOROVKOV · Ergodicity and Stability of Stochastic Processes
BOULEAU · Numerical Methods for Stochastic Processes
BOX · Bayesian Inference in Statistical Analysis
BOX · R. A. Fisher, the Life of a Scientist
BOX and DRAPER · Empirical Model-Building and Response Surfaces
*BOX and DRAPER · Evolutionary Operation: A Statistical Method for Process
Improvement
BOX, HUNTER, and HUNTER · Statistics for Experimenters: An Introduction to
Design, Data Analysis, and Model Building
BOX and LUCEÑO · Statistical Control by Monitoring and Feedback Adjustment
BRANDIMARTE · Numerical Methods in Finance: A MATLAB-Based Introduction
BROWN and HOLLANDER · Statistics: A Biomedical Introduction
BRUNNER, DOMHOF, and LANGER · Nonparametric Analysis of Longitudinal Data in
Factorial Experiments
BUCKLEW · Large Deviation Techniques in Decision, Simulation, and Estimation
CAIROLI and DALANG · Sequential Stochastic Optimization
CHAN · Time Series: Applications to Finance
CHATTERJEE and HADI · Sensitivity Analysis in Linear Regression
CHATTERJEE and PRICE · Regression Analysis by Example, *Third Edition*
CHERNICK · Bootstrap Methods: A Practitioner's Guide
CHILÈS and DELFINER · Geostatistics: Modeling Spatial Uncertainty
CHOW and LIU · Design and Analysis of Clinical Trials: Concepts and Methodologies
CLARKE and DISNEY · Probability and Random Processes: A First Course with
Applications, *Second Edition*
*COCHRAN and COX · Experimental Designs, *Second Edition*
CONGDON · Bayesian Statistical Modelling
CONOVER · Practical Nonparametric Statistics, *Second Edition*
COOK · Regression Graphics
COOK and WEISBERG · Applied Regression Including Computing and Graphics
COOK and WEISBERG · An Introduction to Regression Graphics
CORNELL · Experiments with Mixtures, Designs, Models, and the Analysis of Mixture
Data, *Third Edition*
COVER and THOMAS · Elements of Information Theory
COX · A Handbook of Introductory Statistical Methods
*COX · Planning of Experiments
CRESSIE · Statistics for Spatial Data, *Revised Edition*
CSÖRGŐ and HORVÁTH · Limit Theorems in Change Point Analysis
DANIEL · Applications of Statistics to Industrial Experimentation
DANIEL · Biostatistics: A Foundation for Analysis in the Health Sciences, *Sixth Edition*
*DANIEL · Fitting Equations to Data: Computer Analysis of Multifactor Data,
Second Edition

*Now available in a lower priced paperback edition in the Wiley Classics Library.

*Now available in a lower priced paperback edition in the Wiley Classics Library.

*Now available in a lower priced paperback edition in the Wiley Classics Library.

*Now available in a lower priced paperback edition in the Wiley Classics Library.

MYERS and MONTGOMERY · Response Surface Methodology: Process and Product Optimization Using Designed Experiments, *Second Edition*

MYERS, MONTGOMERY, and VINING · Generalized Linear Models. With Applications in Engineering and the Sciences

NELSON · Accelerated Testing, Statistical Models, Test Plans, and Data Analyses

NELSON · Applied Life Data Analysis

NEWMAN · Biostatistical Methods in Epidemiology

OCHI · Applied Probability and Stochastic Processes in Engineering and Physical Sciences

OKABE, BOOTS, SUGIHARA, and CHIU · Spatial Tesselations: Concepts and Applications of Voronoi Diagrams, *Second Edition*

OLIVER and SMITH · Influence Diagrams, Belief Nets and Decision Analysis

PANKRATZ · Forecasting with Dynamic Regression Models

PANKRATZ · Forecasting with Univariate Box-Jenkins Models: Concepts and Cases

*PARZEN · Modern Probability Theory and Its Applications

PEÑA, TIAO, and TSAY · A Course in Time Series Analysis

PIANTADOSI · Clinical Trials: A Methodologic Perspective

PORT · Theoretical Probability for Applications

POURAHMADI · Foundations of Time Series Analysis and Prediction Theory

PRESS · Bayesian Statistics: Principles, Models, and Applications

PRESS and TANUR · The Subjectivity of Scientists and the Bayesian Approach

PUKELSHEIM · Optimal Experimental Design

PURI, VILAPLANA, and WERTZ · New Perspectives in Theoretical and Applied Statistics

PUTERMAN · Markov Decision Processes: Discrete Stochastic Dynamic Programming

*RAO · Linear Statistical Inference and Its Applications, *Second Edition*

RENCHER · Linear Models in Statistics

RENCHER · Methods of Multivariate Analysis, *Second Edition*

RENCHER · Multivariate Statistical Inference with Applications

RIPLEY · Spatial Statistics

RIPLEY · Stochastic Simulation

ROBINSON · Practical Strategies for Experimenting

ROHATGI and SALEH · An Introduction to Probability and Statistics, *Second Edition*

ROLSKI, SCHMIDLI, SCHMIDT, and TEUGELS · Stochastic Processes for Insurance and Finance

ROSENBERGER and LACHIN · Randomization in Clinical Trials: Theory and Practice

ROSS · Introduction to Probability and Statistics for Engineers and Scientists

ROUSSEEUW and LEROY · Robust Regression and Outlier Detection

RUBIN · Multiple Imputation for Nonresponse in Surveys

RUBINSTEIN · Simulation and the Monte Carlo Method

RUBINSTEIN and MELAMED · Modern Simulation and Modeling

RYAN · Modern Regression Methods

RYAN · Statistical Methods for Quality Improvement, *Second Edition*

SALTELLI, CHAN, and SCOTT (editors) · Sensitivity Analysis

*SCHEFFE · The Analysis of Variance

SCHIMEK · Smoothing and Regression: Approaches, Computation, and Application

SCHOTT · Matrix Analysis for Statistics

SCHUSS · Theory and Applications of Stochastic Differential Equations

SCOTT · Multivariate Density Estimation: Theory, Practice, and Visualization

*SEARLE · Linear Models

SEARLE · Linear Models for Unbalanced Data

SEARLE · Matrix Algebra Useful for Statistics

SEARLE, CASELLA, and McCULLOCH · Variance Components

SEARLE and WILLETT · Matrix Algebra for Applied Economics

SEBER · Linear Regression Analysis

*Now available in a lower priced paperback edition in the Wiley Classics Library.

SEBER · Multivariate Observations

SEBER and WILD · Nonlinear Regression

SENNOTT · Stochastic Dynamic Programming and the Control of Queueing Systems

*SERFLING · Approximation Theorems of Mathematical Statistics

SHAFER and VOVK · Probability and Finance: It's Only a Game!

SMALL and McLEISH · Hilbert Space Methods in Probability and Statistical Inference

SRIVASTAVA · Methods of Multivariate Statistics

STAPLETON · Linear Statistical Models

STAUDTE and SHEATHER · Robust Estimation and Testing

STOYAN, KENDALL, and MECKE · Stochastic Geometry and Its Applications, *Second Edition*

STOYAN and STOYAN · Fractals, Random Shapes and Point Fields: Methods of Geometrical Statistics

STYAN · The Collected Papers of T. W. Anderson: 1943–1985

SUTTON, ABRAMS, JONES, SHELDON, and SONG · Methods for Meta-Analysis in Medical Research

TANAKA · Time Series Analysis: Nonstationary and Noninvertible Distribution Theory

THOMPSON · Empirical Model Building

THOMPSON · Sampling, *Second Edition*

THOMPSON · Simulation: A Modeler's Approach

THOMPSON and SEBER · Adaptive Sampling

TIAO, BISGAARD, HILL, PEÑA, and STIGLER (editors) · Box on Quality and Discovery: with Design, Control, and Robustness

TIERNEY · LISP-STAT: An Object-Oriented Environment for Statistical Computing and Dynamic Graphics

TSAY · Analysis of Financial Time Series

UPTON and FINGLETON · Spatial Data Analysis by Example, Volume II: Categorical and Directional Data

VAN BELLE · Statistical Rules of Thumb

VIDAKOVIC · Statistical Modeling by Wavelets

WEISBERG · Applied Linear Regression, *Second Edition*

WELSH · Aspects of Statistical Inference

WESTFALL and YOUNG · Resampling-Based Multiple Testing: Examples and Methods for p-Value Adjustment

WHITTAKER · Graphical Models in Applied Multivariate Statistics

WINKER · Optimization Heuristics in Economics: Applications of Threshold Accepting

WONNACOTT and WONNACOTT · Econometrics, *Second Edition*

WOODING · Planning Pharmaceutical Clinical Trials: Basic Statistical Principles

WOOLSON and CLARKE · Statistical Methods for the Analysis of Biomedical Data, *Second Edition*

WU and HAMADA · Experiments: Planning, Analysis, and Parameter Design Optimization

YANG · The Construction Theory of Denumerable Markov Processes

*ZELLNER · An Introduction to Bayesian Inference in Econometrics

ZHOU, OBUCHOWSKI, and McCLISH · Statistical Methods in Diagnostic Medicine